Quantitative Corporate Finance

John B. Guerard Jr. • Anureet Saxena
Mustafa N. Gültekin

Quantitative Corporate Finance

Third Edition

 Springer

John B. Guerard Jr.
McKinley Capital Management, LLC
Anchorage, AK, USA

Anureet Saxena
McKinley Capital Mgmt, LLC
Stamford, CT, USA

Mustafa N. Gültekin
Kenan-Flagler Business School
University of North Carolina Chapel Hill
Chapel Hill, NC, USA

ISBN 978-3-030-87268-7   ISBN 978-3-030-87269-4   (eBook)
https://doi.org/10.1007/978-3-030-87269-4

This Springer imprint is published by the registered company Springer Nature Switzerland AG
The registered company address is: Gewerbestrasse 11, 6330 Cham, Switzerland

Endorsements

Quantitative Corporate Finance is impressive in its treatments of corporate finance and risk management. It uses the concept of the Markowitz risk and return, the basis of financial theory, to explain problems in corporate management, investments, and capital markets. Topics include ratio analysis, financial statement analysis, capital budgeting, investments, mergers and acquisitions, econometric forecasting models, and Socially Responsible Investing. I highly recommend this book to everyone interested in applied financial analysis and theory of financial decision making.

Harry Markowitz
Recipient of the 1990 Nobel Memorial Prize in Economic Sciences

Quantitative Corporate Finance is impressive in both its scope and depth. It uses the concept of risk and return, the basis of financial theory, to explain problems in corporate management and puzzles in capital markets. Topics covered range from ratio analysis to sophisticated econometric forecasting models. The use of modern theories of financial economics such as multi-index portfolio theory and option pricing theory to explain optimum corporate management decision-making is impressive. I recommend this book to everyone interested in the practice and theory of financial decision-making.

Martin J. Gruber
Professor Emeritus of Finance
Scholar in Residence
Leonard N. Stern School of Business of New York University

This book is a must have for practitioners and students of corporate finance. Nowhere will you find the cutting edge of research laid out as clearly and as expertly as here. Building on the masterly exposition of the second edition, the third edition takes you into a post-Covid world, where trailblazing methods show just what is possible in empirical analysis – your horizons will be expanded. If you want to get inside the brain of the most influential quantitative financial economists, this book is a must read.

Jennifer L. Castle, Magdalen College, University of Oxford

An eminently wise and readable guide to current thinking about how financial markets work and appropriate financial decisions for corporations and investors.

Ed Tower
Professor Emeritus of Economics at Duke University

John Guerard's *Quantitative Corporate Finance* provides a modern overview of corporate finance in a quantitative approach. Through his masterful writing, he seamlessly connects the theory to the application and illustrates all the concepts and techniques in a very understandable way. The comprehensiveness of this book is exceptional. It will most likely become a standard reference in this subject.

Tim Leung
Computational Finance & Risk Management (CFRM) Director
Boeing Endowed Professor of Applied Mathematics
University of Washington, Seattle

The third edition of *Quantitative Corporate Finance*, by John Guerard, Anureet Saxena and Mustafa Gultekin, is a book of many talents. In over 20 chapters the readers go through an escalating series of terms, concepts, methods and techniques that are used in modern corporate finance – building from foundational material up to and including the most recent advances in corporate management and quantitative analysis. I believe that the sequencing of the material, when read in order, can be an eye-opener for the beginner or a tasteful refresher (and reminder) to the seasoned professional for the value of having an all-rounded, general overview of the field. The book can serve equally well as a text for many different courses in corporate finance, as a graduate textbook, as a reference work for the professional and also as a reference work for the data analyst who knows more statistics and quant methods than finance. In the latter case, it can serve as a significant booster on the quality, implications and interpretation of quantitative work in the real world, as needed. This is a rare gem, highly contemporary but with a classic flair of an old-school book which requires a certain devotion to go over without skipping pages – highly recommended!

Dimitrios D. Thomakos
Professor of Applied Econometrics
National and Kapodistrian University of Athens

The first edition of *Quantitative Corporate Finance* was designed to prepare students with advanced quantitative skills for professional careers in investment and portfolio management. Because many students with graduate level quantitative skills have limited backgrounds in finance, the first edition provided extensive coverage of basic financial topics, such as financial statement analysis; financing current business operations, capital markets, and debt/equity financing; investment decision-making; and mergers and acquisitions, before describing how advanced statistical techniques can be used to identify mean variance efficient portfolios of risky assets such as stocks and bonds. The thorough discussion of key basic financial topics in this book provides a solid foundation for quantitatively oriented students. Guerard has successfully used the first edition of his text in teaching professional management skills to students in quantitative corporate finance programs, including programs at Georgia Tech and Rutgers.

When the first edition of *Quantitative Corporate Finance* was published in 2007, one of the authors, Eli Schwartz, aged 89, died. Guerard has now revised the text with two new authors, Anureet Saxena and Mustafa Gultekin, both of whom have extensive experience in using statistical techniques to estimate the expected return and risk on alternative portfolios of risky assets. Guerard and Saxena have employed these advanced techniques in collaboration with Nobel prize-winning economist Harry Markowitz and others to improve portfolio management decisions at McKinley Capital Management, LLC., and Gultekin has employed similar techniques at other investment management companies. The extensive real-world experience of Guerard, Saxena, and Gultekin in portfolio management and research and their ability to explain how their research can be used to improve portfolio performance is a key strength of the third edition of *Quantitative Corporate Finance*. Another strength of the updated edition arises from the authors' extensive new research, conducted since the time of the publication of the first and second editions, on the economic factors affecting security returns and risk. Their research identifies the basic economic factors that affect the expected returns and risk of portfolios, estimates the statistical relationship between these basic economic factors and expected returns and risk, and allows the analyst to create an efficient frontier of security portfolios that maximize the expected portfolio return for a given level of risk and minimize expected risk for a given level of expected return.

In addition to describing the results of their own statistical model of risk and return, the authors compare their statistical results to the results of other statistical models of risk and return described in the economic literature. They demonstrate that their model provides superior estimates of portfolio risk and return than other models proposed in the literature. This finding is significant because estimates of risk and return are the key inputs in the portfolio selection decision. Better estimates of risk and return produce better choices of risky securities.

A book that presents basic financial skills and advanced statistical techniques that have been used to improve financial performance in real world investment management would not be useful if it were not also readable by target audiences. The updated edition of *Quantitative Corporate Finance* satisfies this requirement.

James H. Vander Weide, PhD

Why a Third Edition of *Quantitative Corporate Finance*?

The third edition of our book became essential because of the COVID-19 pandemic of 2020, and the appropriate regression and time series applications reported in recent peer-reviewed publications during the 2020–2021 period that enhanced Chaps. 12, 13, 14, 15, 18, and 22. These publications were co-authored with Harry Markowitz, Ganlin Xu, and Shijie Deng, members of the McKinley Capital Management (MCM) Scientific Advisory Board, and with long-time friends John Blin, Andrew Mark, Chris Geczy, Bijan Beheshti, and Dimitrios Thomakos. We continue to stress applied investment research in our research projects. I want to thank my co-authors for their hard work, patience (not too much for any of them), and belief that statistically significant and consistent implementation separate quantitative researchers from the merchants of malarkey so common among sell-side analysts on Wall Street. I want to specially thank Professor James Vander Weide, my first finance professor at Duke, and Professor Jennifer Castle, from Oxford, for their help in clarifying presentations reported in Chaps. 12, 13, 14, and 15. Any errors remaining are purely mine.

I want to thank Richard Brealey, Ed Elton, Martin Gruber, Rong Chen, and Doug Martin for our research interactions. Perhaps in the 2024–2029 time period, these interactions may become peer-reviewed applied investment publications.

At the conclusion of the second edition, we closed with thoughts on the business and political worlds being at odds on corporate stock buybacks, which we discuss in Chaps. 8 and 15. Buybacks, a component, of corporate exports, enhanced domestic and global portfolios during the 1997–2015 time period. Whether buybacks have retarded corporate investment is a great question, one that requires a far greater effort and longer time period of analysis than afforded by our revised edition contract. Thus far, political inertia has not (yet) hurt effective corporate financial decision-making.

I used the second edition to supplement decision science models and financial ethics courses in the winter and spring terms of 2021 in the computational financial and risk management program, CFRM, University of Washington. I thank Navid, Bahadoran, Chris Conner, Syed Aun Haider, Borrego Higareda, and Josie Le, in particular, for leading the class in pointing out typos in the second edition. I thank the CFRM program director, Professor Tim Leung, for his support. Professor Doug Martin, founder of the CFRM program, has been a great research resource since we started corresponding on robust time series modeling about 1985. Eli Schwartz and Phoebus Dhrymes both told me before they passed that the problems of corporate finance and financial econometrics problems have been present since 1955, only research methodologies and their corresponding answers have changed.

I thank Robert A. Gillam, my long-time chief investment officer (CIO) and now chief executive officer (CEO) of McKinley Capital Management, for his support. I thank Julie, my wife of almost 40 years, and children Richard, Katherine, and Stephanie, for their love and support.

Scientific Advisory Board, McKinley   John Guerard
Capital Management, LLC, Anchorage,
AK, USA

Why a Second Edition of *Quantitative Corporate Finance?*

Eli Schwartz and I were very pleased with our *Quantitative Corporate Finance* text, published by Kluwer Academic Publishers in 2007. Eli was an economist and elder statesman, a sage, of 86. Eli had written a great book, *Corporate Finance,* published by St. Martin's in 1962. We combined my *Handbook of Financial Modeling* (co-authored with H.T. Vaught, for Probus, in 1989) and updated all materials for the Kluwer text. Eli was fond of telling me that when undergraduate economics students at MIT asked Robert Solow, the famed Nobel Prize laurate about what they needed to know to get a job on Wall Street, Professor Solow handed them a copy of Eli's book and told them to read and know the text. Eli earned his Ph.D. from Brown University and taught at Lehigh for 38 years. Eli was a visiting professor at the University of Pennsylvania, New York University, the London School of Economics, and Tel Aviv University. Eli published over 50 articles, including papers in the *American Economic Review, the Journal of Finance* (five), *Journal of Business, and the National Tax Journal.*

In 2010, Lehigh honored Eli with a festschrift for his 85th birthday. The festschrift included works by two Nobel laurates, Harry Markowitz and Robert Solow; Murray Widenbaum, President Reagan's Chairman of the Council of Economic Advisors, and John Hilley, several Lehigh colleagues, and one practitioner, me.

https://www2.lehigh.edu/news/schwartz-honored-with-festschrift

Nick Philipson, our publisher, attended the Lehigh celebration and an academic version of the festschrift was published by Springer, who had purchased Kluwer, in 2010. Nick made a great call to publish the book.

Eli Schwartz, 89, of Allentown, passed away in his home on August 31, 2010. Born in New York City, he was the husband of Renee (Kartiganer) Schwartz and they celebrated 62 years of marriage on August 29.

Eli was brilliant, a great conversationalist, and a very wise man. He is greatly missed.

There are other reasons for a second edition. First, the papers of the Schwartz festschrift are integrated into the revised edition. Second, our book was 99% complete when I became Director of Quantitative Research at McKinley Capital

Management, MCM, in Anchorage, Alaska. The final ten pages of Chap. 15 took me 18 months to complete. We successfully used the multi-factor APT risk model featured in Chap. 15 at McKinley for portfolio construction at McKinley for over 10 years (validated in the forthcoming *Wilmott Magazine* piece). Moreover, I taught in-house classes at McKinley Capital, and two-day, 15-hour seminars at Rutgers University, where I prepared the first edition of this book for my MBA classes, and the Georgia Institute of Technology Quantitative and Computational Finance program (QCF) programs, where I recruited my MCM Quants. The Tech QCF Program is a top-ten Financial Engineering program. I grew up near Atlanta, earned my MSIM (Masters in Industrial Management) degree from Tech, and I am completely biased in hiring exceptionally bright and hard-working Tech QCF Quants. You will see references to Sundaram Chettiappan, Eli Krauklis, Manish Kumar, and Elaine Wang, all Tech QCF graduates with whom I published MCM Horse Race studies, to validate APT and Axioma risk models. These references are in Chap. 15. The extensions to Chaps. 14 and 15 reflect multi-factor risk modeling at McKinley Capital, 2006 -2019.

Third, Harry Markowitz and I are presenting an extension to our financial anomalies, stock selection, and equity modeling work of 1993 referenced in Chaps. 14 and 15 in October 2019, at the Q-Group meeting. The models we showcased in the first edition continue to be statistically significant. I will bet that there is a 65-75% chance that these stock selection models will continue to be statistically significant, passing the Markowitz-Xu (1994) Data Mining Corrections test, during the 2020-2029 time period, and will be reported in the third edition of the text. Fourth, a new edition allows me to use a more current text in my Tech and Rutgers one and two-day classes. We have had 56,140 PDF downloads of our text, as reported by the Springer website, putting our book in the top 20%. My Springer book to honor Harry Markowitz, with 94,086 downloads, is my personal best. Finally, and most importantly, several major texts and associated papers have been published that compel our second edition. These texts are:

Castle, J. and Shepard, N. (2009). *The Methodology and Practice of Econometrics.* Oxford, UK: Oxford University Press.

Castle, J., Doornik, J. A., and Hendry, D.F. (2013). 'Model selection in equations with many 'small' effects,. *Oxford Bulletin of Economics and Statistics,* Vol. 75, pp. 6-22.

Castle, J., Clements, M. P., and Hendry, D.F. (2015). 'Robust approaches to forecasting', *International Journal of Forecasting*, Vol. 31, pp. 99-112.

Clements, M. P. and D. F. Hendry. (1998). *Forecasting Economic Time Series.* Cambridge, UK: Cambridge University Press.

Connor, G., Goldberg, L. & Korajczyk, R.A. (2010). *Portfolio Risk Analysis.* Princeton: Princeton University Press.

Elton, E.J., Gruber, M.J. Brown, S.J. & Goetzman, W.N. (2007). *Modern Portfolio Theory and Investment Analysis.* John Wiley & Sons, Inc., Seventh Edition.

Dhrymes, P.J. (2017) *Introductory Econometrics.* New York, Springer, Revised Edition.

Hendry, D.F. and Doornik, J.A. (2014), *Empirical Model Discovery and Theory Evaluation,* Cambridge, MA: MIT Press.

Levy, H. (2012). *The Capital Asset Pricing Model in the 21st Century.* Cambridge: Cambridge University Press.

Lo, A.W., Mamaysky, H. & Wang, J. (2000). Foundations of technical analysis: Computational algorithms, statistical inference, and empirical implementation. *Journal of Finance* 60 (2000), 1705-1764.

Lo, A. (2017). *Adaptive Markets.* Princeton: Princeton University Press.

Markowitz, H. M. (2013). *Risk-Return Analysis.* New York: McGraw-Hill. The first volume of a four-volume set on the risk-return trade-off.

Liu, L.M. (2006). Time Series *Forecasting and Time Series Analysis.* Chicago: Scientific Computing Associates Corp.

Maronna, R.A., R. D. Martin, V.J. Yohai, and M. Salibian-Barrera. 2019. *Robust Statistics: Theory and Methods (with R).* New York: Wiley.

Nikolopoulos, K.I. and Thomakos, D.D. (2019). *Forecasting with the Theta Method,* New York: Wiley.

Tsay, R.S. (2010). *Analysis of Financial Time Series.* New York, NY: Wiley, Third Edition.

Tsay, R.S. and Chen, R. (2019). *Nonlinear Time Series Analysis.* New York, NY: Wiley.

In the years post our first edition, several good friends and co-authors passed. Eli Schwartz and Phoebus Dhrymes were over 80. Phoebus Dhrymes was a great Springer author of four major econometric texts. Eli and Phoebus were magnificent, highly independent thinkers, who left very high standards or us to try to reach. Bob Gillam, the founder of McKinley Capital, passed away last year at far too young at age, 72, and is missed. Bob taught me and others one of the greatest lessons an individual can teach young Americans: what it means to be an entrepreneur and keep the lights on in industry! Rob Gillam, our CEO / CIO, has big shoes to fill and big innovative ideas for McKinley.

As with the first edition, this text has been written, 95% of my own time, on Saturday and Sunday afternoons. To Julie, my wife of 38 years, I say thank you. Richard, Katherine, and Stephanie, my children in the first edition, are now real-world working millennials who give me great pride and hope for the future. To my co-authors of papers used in the second edition, Dimitrios Thomakos, Harry Markowitz, Ganlin Xu, Anureet Saxena, Shijie Deng, Andrew Mark, Rob Gillam, Sundaram Chettiappan, Eli Krauklis, Manish Kumar, and Elaine Wang, I say thank you. To my professors of Finance and Economics, Jim Vander Weide, Steve Maier, Bernell Stone, Henry Latane (deceased), and Jan Mossin (deceased), in Finance, and Ed Tower, Larry Moss, Rick Ashley, and Thomas Havrilesky (deceased) in Economics, I say thank you. Any errors remaining in this text are not their fault!

We gratefully acknowledge the typing and editorial assistance of Allison Capps. Please enjoy the second edition. Eli is not with us, but I bet he would be very pleased!

Anchorage, AK
Labor Day, 9/2/2019

As I send this book to press in January 2020, I am very pleased to acknowledge my two co-authors who joined me on the revised text, Anureet Saxena and Mustafa Gultekin. Anureet is brilliant, a Carnegie Mellon Ph.D. in Operations Research, with publications in *Mathematical Programming*, the *Journal of Portfolio Management*, *Journal of Investing*, and *Frontiers in Applied Mathematical and Statistics*. Mustafa, a Ph.D. in Finance, New York University, has published in the *Journal of Finance, Journal of Financial Economics,* the *Journal of Financial and Quantitative Analysis, Management Science*, and *Research in Finance*. Allison Capps helped us with editorial assistance and manuscript preparation. Thank you very much.

The goal of this book is the same goal as the first edition. We seek to introduce the reader to the world of Harry Markowitz, Bill Sharpe, Marty Gruber, and Jack Treynor. Quantitative financial economics should be fun and properly educated (young) bright Quants can make a very good living. My Tech QCF graduates, three years post-graduation, should be earning in the top 10% of the income distributions before they are 30; in the top 5% within 5-10 years. Have you any Grey Poupon? But of course. My goal is that my Quants will be the victors of income-inequalities. As they say downhome, "to the victors go the spoils". Why eat yellow mustard on a hot dog when you can have Grey Poupon on a steak sandwich or Reuben? The fact that the models of Chaps. 8, 13, 14, 15, 21, and 22 of the first edition have been statistically significant, post-publication, was expected but nevertheless celebrated. I have similar expectations for these chapters for 2020-2029. Factor and model statistical significance and ensuring proper portfolio implementation are the only meaningful professional life measures for Quants.

As we are making revisions in our text, March 28, 2020, the business and political worlds at odds on corporate stock buybacks, which we discuss in Chaps. 8 and 15. Buybacks, a component, of corporate exports, enhanced domestic and global portfolios during the 1997 – 2015 time periods. Whether buybacks have retarded corporate investment is a great question, one that requires a far greater effort and longer time period of analysis that afforded by our revised edition contract. Perhaps we will answer this question in a third edition.

Director of Quantitative Research,   John Guerard
McKinley Capital Management, LLC,
Anchorage, AK,

Contents

Chapter 1
Introduction: Capital Formation, Risk, and the Corporation

The corporation is the major institution for private capital formation in our economy. The corporate firm acquires funds from many different sources to purchase or hire economic resources, which are then used to produce marketable goods and services. Investors in the corporation expect to be rewarded for the use of their funds; they also take losses if the investment does not succeed. The study of corporation finance deals with the legal arrangement of the corporation (i.e., its structure as an economic institution), the instruments and institutions through which capital can be raised, the management of the flow of funds through the individual firm, and the methods of dividing the risks and returns among the various contributors of funds. The goal of corporate management is to maximize stockholder wealth. A major societal function of the firm is to accumulate capital, provide productive employment, and distribute wealth. The firm distributes wealth by compensating labor, paying interest on loans, purchasing goods and services, and accumulating capital by making investments in real productive facilities. The goal of corporate finance is to maximize the firm's stock price. We will discuss management—stockholder relations. We will discuss, in considerable detail, socially responsible investing and the firm. Companies can "do good while doing good." However, we firmly believe that management creates its corporate investment, dividend, research and development, acquisitions, and debt policies to maximize the stock price, which maximizes the market value of the firm.

1.1 Financial Mathematics and Theory

The financial market is basically free of the frictions of imperfect competition. Securities of a given class and grade are largely homogenous, and the traders and investors do not have strong label or brand preferences. Because the markets are large and have a long history, there is a large mass of data that can be evaluated. This means that financial theory and derived mathematical models may be better evaluated, more appropriate, and better applied than elsewhere in the continuum of

J. B. Guerard Jr. et al., *Quantitative Corporate Finance*,
https://doi.org/10.1007/978-3-030-87269-4_1

business and economic studies. We make extensive use of data in this text, using the data found on the MSN (Money), FactSet (for Institutional Managers), Wharton Research Data Services (WRDS) Compustat, Global Compustat, IBES, and Center for Research in Security Prices (CRSP) databases. These databases are well-known and respected sources of financial data. The investor or financial researcher can find 4 years on MSN, 10–25 years on FactSet, 20 years of income statement and balance sheet data on an industrial Compustat database, and data for the 1950–2020 period with the WRDS Compustat data facilitates. The most obvious source of information regarding the firm is its annual report and its letter from the CEO.[1] IBM's letter highlights its hybrid cloud and artificial intelligence (AI). Boeing's 2020 Annual Report contains 3 years (2018–2020) of financial data.[2] We specifically selected Boeing, Dominion Resources, IBM, as firms to study with respect to their respective financial statements, ratios, valuation, and the cost of capital. These three firms are large, respected firms in their industries, and are familiar to many readers of this text. Guerard and Schwartz (2007) featured IBM, D, and DuPont (DD) in the first edition. So very much has changed with IBM since 2007, with its Red Hat acquisition and emphasis on AI. Boeing has changed dramatically since the second edition went to press in February 2020. Dominion Energy, a utility, in its 2020 Annual report, committed itself to achieving sustainable, reliable, affordable, and safe energy and to achieving net zero carbon dioxide and methane emissions from its power generation and gas infrastructure operations by 2050. An investor must read the annual reports of individual stocks owned. The reader is introduced to regression and time series analysis to facilitate quantitative analysis, such as estimating betas, or measures of market risk, forecasting earnings per share, predicting stock rankings, analyzing the predictive power of the leading economic indicators.

---

[1] In its 2020 annual report, on page 4, Boeing issued the following statement on its 737 plane.

"Boeing will never forget the 346 victims of the Lion Air Flight 610 and Ethiopian Airlines Flight 302 accidents. Their memories underpin our commitments to our core values of safety, quality and integrity. In November 2020, the U.S. Federal Aviation Administration (FAA) lifted the order grounding the 737–8 and 737–9. The FAA validated that once new software was loaded and other defined steps were completed, the airplanes would be safe and ready to fly. The announcement followed a comprehensive, robust and transparent certification process over 20 months. Our Boeing teams continue to work closely with global regulators and customers, and our employees are deeply committed to the safe operation of the worldwide 737 fleet. As we focus on supporting our customers in safely returning their fleets to service, we are pleased with the confidence our customers have placed in us and the airplane."

[2] John Guerard required a term project from his students at undergraduate and MBA Rutgers students where they compared annual report data to the WRDS Compustat database for their firm of study. The students calculated the ratios of Chap. 5 and performed a valuation analysis, as we report in Chap. 8. The students submitted the firm's latest annual report, its data, and its letter from the CEO helped put the firm's mission in perspective for the students, with their project. Three years of annual report data, versus 58 years of WRDS Compustat data, and one wonders if firms are too short-term oriented.

1.2   Growth and Survival of the Firm

Finance focuses on the flow of values through the firm. Corporate finance is concerned with how the firm produces goods and services, generates cash flow, and generates returns for its investors. It explores the effects that different levels of the flow of values over time will have upon the complex legal and accounting entities making up the firm. It is interested in the relations between the legal owners and the various classes of creditors, and it explores the circumstances under which the claims of the original owners can grow and survive and be augmented, or, in contrast, those circumstances under which the legal claims of the owners must be forfeited to the claims of the creditors.

1.3   Risk and Uncertainty Inherent in Finance

The financial management has the tasks of minimizing the total cost of financial funds to the firm, providing adequate resources for expansion at a cost low enough to make it profitable at a risk low enough to maximize the firm's chance of survival and its stock price. Sometimes the problem of providing adequate funds at minimum financial costs can be separated from the problem of survival or risk and be dealt with simply as a problem in costing or economics. Where this is possible, however, it is usually because some earlier decision set the bounds of the problem. An earlier decision assumed the risks, and historically assumed risks can no more be discarded than historically assumed sunk costs. Whatever the initial approach, the heart of financial theory is the problem of risk and risk-bearing.[3] The flow of values generated by an initial commitment of productive resources cannot be exactly predicted. Risk is always two sided. Asset prices may be better or lower than the original value of the assets invested. If the long-run flow of values is larger than estimated, the value of the owners' original investment is augmented; if, on the other hand, the flow of values falls below the original estimate, the owners will have to accept a lower valuation or a forfeiture of their claims. Should the stream of income fall low enough, even the creditors, depending on their legal position and status as risk-bearers, might have to accept some of the losses. One of the goals of this text is to acquaint the reader with the tools necessary to quantify risk and return.

The financial student must always grapple with the problem of uncertainty. The financial results of any single productive attempt are never sure. There is risk not only when a new enterprise is launched but also for established firms as long as uncertainty exists continually and concurrently with the productive cycle.

---

[3]Risk is used here in its broader sense to include the problems of uncertainty and not in its more precise definition, where it includes only those phenomena whose probabilities are reasonably known and which therefore can be insured against.

1.4   Types of Business Risk

All other things equal, the element of risk intensifies the longer the production cycle. Risk varies directly with the length of time required from the first application of resources to the final appearance of the product. One type of time risk is of natural or physical origin. It includes fire, theft, flood, drought, and machinery breakdown. These risks can sometimes be insured against if the probabilities of their recurrence are known, if they are independent events, and if society is sufficiently well organized to have set up the insurance mechanism. But the full losses entailed by these events are often uninsurable, for though the value of the physical assets may be recovered, lost time in production or sales may not be. Furthermore, a company may suffer a loss through disasters to other firms it relies on as suppliers or customers. Insurance companies may not write regular policies against certain hazards (such as flood) or the insurance costs may be prohibitive.

Another type of time risk involves unexpected political or economic changes. A recession may occur, government policies may change, a rival product appears, or tastes may change before the process of recovering the investment is completed. This type of risk is double edged; there is always the possibility that conditions may change in a favorable direction and lead to a greater return than anticipated.

There is a greater chance of the economy veering over a longer time span. Any productive organization utilizes many factors to which its commitment varies in length of time, and the risks to the firm depend partly upon the composition of these productive resources. Some resources, such as labor, may perhaps be withdrawn quickly; fixed plant or machinery, however, may have to be used for a long period before their value can be recovered. The complete production cycle for different resources (assets) varies, and for the firm, the composition of the resources it uses entails different degrees of risk.

Another risk is that the original estimate of demand or costs may be wrong. The possibility of error in either the forecast or in the original estimate of the economy as it currently exists can be minimized by competent research before production is launched. But error can only be minimized, never eliminated. Moreover, research is not costless; the outlay for research and its value in the reduction of hazards must be weighed against all the other possible outlays and their probable returns. The operational model involves comparing the marginal cost of additional research to the marginal value of the additional risk allayed. The optimum point is reached when the marginal cost of additional research equals the marginal value of the additional risk averted.

The costs of investigation are the reason many beginning small businesses make perfunctory estimates or forecasts. The small businessman cannot spend let us say, $20,000, to investigate the potential of an enterprise for which he has only $20,000 to invest. The initial estimate on returns and costs behind the launching of many small enterprises is likely to exist largely in the head of its founder.[4]

Another class of risks is called exploratory or technical risks. The exploration for mineral deposits is a risky enterprise; one can never be certain that the mineral will be found—or found in amounts sufficient to justify commercial exploitation. Similarly, a new product or a new method of production always involves uncertainty, for no one can be sure exactly how the innovation will behave when put into commercial production. No matter what precautions are taken, the characteristics of a product in the test tube or pilot plant often change unaccountably when the item is produced commercially. Neither the mine operators nor the innovators can be sure of the economic feasibility of their activity until production is launched.[5]

1.5   Financial Risk

We pay particular attention to financial risk in this text. Stockholder returns have been subject to larger variability than bond returns during the 1926–2018 period. One must discern between systematic risk, the risk of the market that cannot be diversified away, and total risk, as measured by the variance of stockholder returns. We estimate the stock beta, or measure of systematic risk, and illustrate how securities can be combined into portfolios to minimize portfolio risk to stockholders. Management calculates the firm's cost of issuing equity, and its cost of capital. The cost of capital is the minimum acceptable rate of return for projects to be acceptable to enhance stockholder wealth. Management must decide whether the expansion of the firm is financed by using reinvesting corporate profits or by issuing debt or new shares of stock. The debt or equity decision should be made such that the firm's earnings per share (eps) is maximized.

1.6   Division of Risk, Income, and Control

Although risks and uncertainties are inherent in every business enterprise, different classes of investors in the firm do not shoulder the same degree of risk. The investor earns returns from bonds and stocks and incurs risk in the form of uncertainty of those returns. An investor in bonds earns returns from interest paid on the debt and

---

[4]Even where resources may be more abundant, the costs of testing or obtaining statistics should be weighed against the degree of certainty likely to be obtained. This concept as first elaborated by Abraham Wald, e.g., "Basic Ideas of a General Theory of Statistical Decision Rules," Proceedings of the International Congress of Mathematicians, Vol. I, 1950, has led to the development of sequential analysis and sampling techniques.

[5]This differs in degree from the error or risk in forecasting demand. In this case, it is the characteristics of the production function that are uncertain.

possible bond price appreciation. The stock investor earns a return dependent upon the dividends paid on the shares of equity and the stock price appreciation. The interest paid on bonds is determined by the coupon rate of the bond, whereas dividends are paid from the earnings of the corporation. Firms may experience losses for several years and not pay dividends. Moreover, the firm's stock price may decline, as we saw from 2000 to 2003, such that total returns may be negative. Traditionally, the total return on stocks over time has exceeded the return on bonds, but with much greater variability of returns.

Modern finance has created different formal legal (or accounting) relationships that distinguish different classes of investors contributing to the funds of the business organization; the intensity of risk which each investor class bears varies considerably. The legal classifications on the liability and capital side of the balance sheet represent a division of risk. For example, the element of risk is less for current creditors. They advance funds for only a few weeks or months. The possibility of deterioration in the credit position of the borrower is limited because of the short-term maturity of the loan. A trade creditor or banker making a short-term advance of credit funds usually looks to the flow of cash from current operations or from the liquidation of receivables or inventory to provide the repayment to the firm. Because of his minimal risk position, the short-term creditor generally offers funds at a low cost.

Moreover, because the risks are relatively short in duration, the supplier of current funds generally does not seek a voice in shaping company policies. Since the advance is temporary, the short-term creditor can do little, in any case, to enhance the probabilities of repayment at maturity. (Occasionally, however, if a firm is slow in meeting its current obligations, the trade creditors may combine in an informal committee to set terms of extended payment and to establish an operating framework for the firm until it is once more current.)

Usually, a distinction through superiority of claim and restrictive covenants is made between the long-term debt (bonds, funded debt, etc.) and the current liabilities. These distinctions are not inherent in the original legal accounting position of the long-term creditor. At the start, a long-term creditor would have no more control over the policies of the firm than the current creditors, and in case of actual insolvency, the longer-term creditors do not initially receive a position superior to short-term creditors. In a failure, all general creditors share the option to collect from the general assets equally. Therefore, since uncertainty usually increases as the time period increases, the long-term creditors, committed to the enterprise for a considerable period, would be at an essential disadvantage. In the case of default, long-term creditors would have only an equal claim shared with the current creditors to the assets. The time risk could be so great that a mere priority of claim to the general assets against the owners, but not against the short-term creditors, would not be sufficient to induce creditors to lend at long-term without a wide differential in interest charges. If the long-term creditors were to invest funds on a basis of equality with the short-term lender, they might require so high a rate of interest that the earnings left to the owners would not be sufficient to induce them to take the remaining risk. This dilemma is solved by granting the long-term lender certain distinct advantages. In the original debt contract, the long-term creditors may be promised that certain financial policies will be followed; in this sense, the long-term creditors are given a limited amount of control over the financial policy of the firm. The long-term creditors may be granted a prior claim over other creditors to specific assets plus an additional undifferentiated claim against the general assets if the value of the pledged asset should prove inadequate. These concessions, coupled with a prior but limited claim to earnings, lower the risk to the long-term lender to the point where he is willing to lend at a usable price. The bondholders have a prior claim on assets relative to stockholders in failure.

The contractual right to a prior claim on earnings is the most important part of the guarantee to the long-term creditors. They are to be paid the contractual interest even though the total flow of earnings does not justify the original cost of the total assets (productive factors) dedicated to the enterprise. The lien against assets is a secondary guarantee or option which comes into play only if the cash flow is insufficient to cover the contracted interest rate and the repayment of principal. The option to acquire the assets is nowhere a full guarantee against loss. The assets of a falling firm would bring full value only if they could be transferred to another enterprise or use where their expected earnings will be higher than that in the present firm. This is not a likely occurrence; nevertheless, the right to seize essential assets gives the long-term creditors a strong bargaining position in the event of a reorganization of the existing failed firm. In most instances, a lien or option against the assets provides a reinforcement of the claim against earnings (or cash flow) and not a guarantee against, only some hope of a limitation to losses.

Since no change in the formal accounting relationships can change the totality of risk or uncertainty, the privileged position of some financial contributors, creditors or preferred shareholders, can only heighten the risk of the other investors or shareholders. In financial practice, the most unsheltered position belongs to the residual owners (the shareholders), for they contract to use outside funds, and they negotiate the terms under which the risks of the various creditors or preferred shareholders are reduced.[6] The owners' investment serves as a buffer to protect the contributions of the various creditors. There is no pro-rata sharing of losses; if the business experiences an operational loss, the owners bear it. Should losses keep recurring, however, and the firm fails, the owners' investment may be wiped out, and bankruptcy may result. Only then, with the owners' equity completely dissipated, would the creditors suffer financial losses. Logically, then, the bearers of the greatest risks, the owners, are also given the ultimate control of the enterprise. The brunt of the loss resulting from an erroneous decision must be borne first by the initial risk bearers, the owners or common stockholders; they cannot create losses for the other

---

[6] Although risk, income, and control can be divided in almost any manner in the various stages of single proprietorship, general partnership, and limited partnerships, combined with various short-run creditors, mortgage holders, and term loans, the corporation formalizes and illustrates these relationships better than the other business organizations. For this reason most discussion of financial structure uses examples from the corporate form. The flexibility of the corporation as a financing device is practically unlimited.

contributors without bringing a fair measure of ruin to themselves. The owners of a firm can be expected to behave rationally and, in protecting their position, shield the creditors from inordinate hazards.[7] In compensation for taking the most exposed risk position, the owners receive the best return if the operations are profitable.

Because it is accounting that classifies the amounts in the various legal and formal risk categories, the student of finance must have some acquaintance with operational accounting. The study of corporate finance draws upon both economics and accounting. Accounting provides the data needed to depict the relations of the capital and liability side of the balance sheet (the financial structure of the firm); it also provides the data necessary to derive the cash flow—the gradual conversion of assets to cash and back again.

1.7 Profitability and Risk

Risk and uncertainty pervade the field of finance. Assuming stockholder wealth maximization as a first approximation of the producer's goal, and developing marginal analysis as a tool, economics can provide useful insights into the organization of society and business decision-making. However, the typical financial decision is made under varying degrees of uncertainty, and the problem is to change these uncertainties, as far as possible, into objective probabilities, and arrive at the proper charge or premium for risk.[8] Uncertainty is moved toward more stable probability only as more and more individual but similar events are averaged together. This means that financial decision-making tends to deal in averages and probabilities.

In practice, financial decisions are basically choices between potential return and risk. Consider the structure of the firm's asset holdings. The greater the liquidity of the assets, the less is the risk of loss. The larger the percentage of assets the firm borrows, the greater is the risk of loss. Greater use of borrowed funds may raise the net return to the owners, but only at the cost of increasing the uncertainty of the firm's survival. The problem of the composition of asset holdings is characteristically posed as liquidity vs. earnings. The problem of choosing sources of funds is a question of maximizing earnings per share and the stock price. The choice between a conservative financial structure and an aggressive financial structure influences returns to the common shareholders.

---

[7] The owners' position of control and greatest risk might be compared to a captain of a ship. In case disaster forces abandoning the vessel, the captain is traditionally the last person overboard. This tradition probably evolved as a result of the captain's control over operations, for the vulnerability of his position (the greatest risk-taker) ensures that he will direct the ship in the most responsible manner unless, of course, he is irrationally bent on self-destruction.

[8] Economic theorists allow for risk as a cost. But the problem is more often recognized than discussed.

Investors hold stocks as a major part of their wealth. We pay particular attention to the investment theories of Harry Markowitz and William F. ("Bill") Sharpe who shared the Nobel Prize in Economic Sciences in 1991. We develop the Markowitz concept of the efficient frontier, where expected return is maximized for a given level of risk, or risk is minimized for a given level of return. Extensive use is made of the Sharpe-Lintner-Mossin Capital Asset Pricing Model (CAPM) and its implications for stock valuation and portfolio construction and management.

The most rational choice is not necessarily in favor of minimum risk. An extremely stable enterprise, which shows a minimum rate of return, may not be as valuable as one, which earned at a high rate for some years even if eventually its assets should dwindle or be dissipated through losses. The most stable firm would hold no assets but cash and it would have a capital structure of all common stock. It would also show no profit.

1.8   Areas Covered in This Book

We conceive of the corporation as the major institution for capital formation in our economy. It is also a method for formalizing the division of risk-bearing, control, and income among the various contributors of funds.

The book develops the legal and organizational framework of the corporation from a functional point of view. The discussion of the major accounting statements provides an understanding of the source of financial data. A description and analysis of the capital markets and the instruments of long-term financing follow. Throughout we are interested in setting up the interplay of risk, control, and the cost of funds.

Although policy considerations (both internal and external) are introduced early, their discussion dominates the last half of the book. The authors explore the factors governing the choice between return and risk in the various areas of financial decision-making and discuss the area of choice in the deployment investment in working assets and the retention of funds for emergencies or improved opportunities. The text deals with the analysis of expansion programs, analyzing their effect on cash flow, in the determination of whether the resultant risk is justified by the projected return. It should explore the models depicting the optimum capital structure, i.e., the composition of borrowed and equity funds, which would maximize the value of the ownership shares. Lastly, the authors discuss the circumstances under which a firm would be disbanded, liquidated, or reorganized. We discuss the financial decisions of firms, dividend and stock buyback strategies, liquidity strategies, leverage policies, merger and acquisition strategies from the perspective of stock price maximization. Robichek and Myers (1965) dealt with financial optimality in their outstanding monograph. Mao (1969); Weston and Copeland (1986); Van Horne (2002); Brealey and Myers (2003); Brealey et al. (2006); and Guerard and Schwartz (2007) continue to address financial decision-making and optimality. The Guerard and Schwartz (2007) developed from integrating Schwartz (1962) and Guerard and Vaught (1989) and updating relevant financial data and analysis. The

Brealey, Myers, and Allen and Guerard and Schwartz are integrated with Compustat databases. We continue to use real-world business data in this text, particularly in financial statement and ratio analysis, dividend, stock buyback, corporate export analysis, domestic and global multi-factor portfolio construction and management, merger and acquisition analysis, and socially responsible investing policies. As we look back on Schwartz, Guerard, and Vaught, and Guerard and Schwartz, we are very pleased to note how relevant the chapter topics were, and continue to be. Guerard, Markowitz, and Xu (2014) address financial decision-making optimality. The authors believe that capital and labor products are paid their respective marginal products. Capitalism effectively allocates scarce resources in our economy. The authors firmly agree with Wright (1951), Schwartz (1962), Chandler (1977), and Meltzer (2012) that capitalism produces optimality in economic and financial decision-making. The authors have used the previous editions of this book in two-day and three-day professional education courses in MBA and MS in Computational Finance programs. This book is primarily concerned introducing the reader to optimal decision-making techniques in quantitative finance. We are under no illusions that our book is a typical MBA text. The authors used the first edition of Van Horne (2002) and Weston and Copeland (1986) in our graduate courses. In the authors' minds, Brealey, Myers, and Allen (2006) in whatever edition is the most outstanding MBA corporate finance text. The Mansfield et al. (2002) text is the best managerial economics text, and Vander Weide and Maier (1985), Ackoff (1999), and Meltzer (2012) are specialty books that MBA students should read. The authors have one further bias, we believe that firm management should engage with financial planning models such developed in Pogue and Bussard (1972), Carleton and McInnes (1982), and Brealey, Myers, and Allen (2006) to produce develop more efficient corporations to maximize stockholder wealth!

The text deals with the corporation as a social and economic institution. It covers the effects of the corporation operations on the economy: capital formation, economic stability, and growth. It deals with the problem of the evolving relations between management, stockholders, and society. Lastly, the text carries a discussion of the different views of the effects of the large corporation in a competitive economy.

References

Ackoff, R. L. (1999). *Re-creating the corporation*. Oxford University Press.

Brealey, R. A., & Myers, S. C. (2003). *Principles of corporate finance* (7th ed.). McGraw-Hill/Irwin, Chapter 1.

Brealey, R. A., Myers, S. C., & Allen, F. (2006). *Principles of corporate finance* (8th ed.). McGraw-Hill/Irwin, Chapter 1.

Buchanan, N. S. (1940). *The economics of corporate enterprise*. Holt, Rinehart and Winston, Chapter 2.

Carleton, W. T., & McInnes, J. M. (1982). Theory, models, and implementation in financial management. *Management Science, 28*, 957–978.

Chandler, A. D., Jr. (1977). *The visible hand: The managerial revolution in American business*. The Belknap Press of Harvard University Press. Chapter 13.

Guerard, J. B., Jr., Markowitz, H. M., & Xu, G. (2014). The role of effective corporate decisions in the creation of efficient portfolios. *IBM Journal of Research and Development, 58*(6), 1–11.

Guerard, J. B., Jr., & Schwartz, E. (2007). *Quantitative corporate finance*. Springer.

Guerard, J. B., Jr., & Vaught, H. T. (1989). *The handbook of financial modeling*. Probus, Chapters 1, 2, 4.

Jensen, M. C., & Meckling, W. H. (1976). Theory of the firm: Managerial behavior, agency costs, and ownership structure. *Journal of Financial Economics, 3*, 305–360.

Mansfield, E. W., Allen, B., Doherty, N. A., & Weigelt, K. (2002). *Managerial Economics* (5th ed.). W.W. Norton and Company.

Mao, J. C. T. (1969). *Quantitative analysis of financial decisions*. The Macmillan Company. Chapters 13 and 14.

Meltzer, A. H. (2012). *Why capitalism?* Oxford University Press.

Pogue, G. A., & Bussard, R. N. (1972). A linear programming model for short-term financial planning under uncertainty. *Sloan Management Review, 13*, 69–99.

Robichek, A. A., & Myers, S. C. (1965). *Optimal financing decisions*. Prentice-Hall.

Schwartz, E. (1962). *Corporate finance*. St. Martin's Press. Chapter 13.

Van Horne, J. C. (2002). *Financial management & policy* (12th ed.). Prentice Hall.

Vander Weide, J. H., & Maier, S. F. (1985). *Managing corporate liquidity: An introduction to working capital management*. John Wiley & Sons. Chapter 10.

Weston, J. F., & Copeland, T. E. (1986). *Managerial finance* (8th ed.). The Dryden Press, Chapters 8, 12.

Wright, D. C. (1951). *Capitalism*. Mc-Graw Hill Book Company.

Chapter 2
The Corporation and Other Forms of Business Organization

The corporate structure was the dominant form of business organization, as reported in Schwartz (1962). In Guerard and Schwartz (2007), the authors reported that in 2000 there were somewhat more than 25 million nonfarm business firms in the United States. About 5.045 million of these were corporations of all classes; the other 2.058 million were partnerships, and 17.805 million were nonfarm proprietorships. We also reported that 66 percent of the gross national product originated in the business sector flows through the corporate sector in 2000.[1]

In 2019, using the latest online US Census Bureau, there were 26.485 million establishments in the United States, employing 132.979 million employees, with total employment payrolls of $7428.6 billion.[2] Firms employing 1–4 persons total 3.1 million and employed 6.8 million employees. Firms with more than 10,000 employees totaled 1129 and employed 37.4 million employees. The American economy is dominated in employment and payrolls, by corporations. In this book, we concentrate on corporations because they employ the most people, raise the most capital, and generate the most profits. American business is the study of corporations. The overriding advantage of the corporation, the limited liability of the shareholder, has made it the dominant form of business enterprise in the world in economic terms. The corporate form has the ability to amass capital from many sources and put it to productive use.

Much of this book is concerned with the complex relationships and the nexus of institutions through which firms (corporations) can raise funds to finance growth and operations. The reader is reminded that management seeks to maximize stockholder wealth in its financial decisions. This chapter is mainly concerned with the institutional and legal arrangements that have enabled the corporation to be more effective than other forms of business organization in performing this function on a large

---

[1] U.S. Department of Commerce, U.S. Census Bureau, *Statistical Abstract of the United States*, 2003, p. 495

[2] https://www.census.gov/quickfacts/fact/dashboard/US/RHI625219

J. B. Guerard Jr. et al., *Quantitative Corporate Finance*,
https://doi.org/10.1007/978-3-030-87269-4_2

scale. Accordingly, in the next section of this chapter, we examine the major forms of business firms. We will examine the legal formality of the organization, the personal liability of the owners, the responsibility for management, and the tax status of the major business forms.

2.1   The Sole or Single Proprietorship

The greatest numbers of firms are single proprietorships. The organization of the single proprietorship involves little legal formality. The owner and the business firm he owns are legally one. No special legal permission is required by the state to set up a sole proprietorship.[3] The proprietor has legal title to the assets of his business; she or he personally also assumes all debts. If in the course of operations, the assets of the business fail to satisfy all of the business liabilities, a proprietor's personal wealth or holdings may be used to help cover the claims of business creditors. Moreover, conversely, the net business assets are subject to the unfulfilled claims of personal creditors. This constitutes the basic rule of "unlimited liability for all debts whether personal or business." This rule actually strengthens the relative credit position of the single proprietor because the proprietor's personal wealth acts as a sort of a second guarantee for the safety of the business debts; it also is the major drawback of this business form, because a failing venture may cost an individual not only the funds directly risked in the business but also the rest of the moneys, assets, or wealth he may have reserved for personal use.[4]

The single proprietor is responsible for running and managing his own business, but he may hire managers and agents as he sees fit. The single proprietor is taxed at the applicable individual income tax rate for all the net income arising from his business. After making proper arrangements with his creditors, the owner may dissolve his business any time he desires.

The single proprietorship has the advantages of simplicity, flexibility, and direct responsibility in management. It is limited, however, in its sources of ownership capital (i.e., risk capital) and in its ability to attract specialized managerial talent.

---

[3] Some kinds of enterprise may require a license. A license, however, is necessary only in certain activities – usually occupations where special skills or sanitary or hygienic considerations are involved – and not to the form of business organization.

[4] In case of bankruptcy, the rule of "marshaling the assets" is followed. Business assets are used first to satisfy business debt and personal assets satisfy personal debt first. Thus, if the ratio of debt to assets differs in the two categories, the rate of settlement for the different classes of creditors are not the same.

2.2 The Partnership

A partnership is an agreement by two or more individuals to own and run a business jointly. The agreement can be oral, but in most cases, it is in writing, to prevent possible subsequent disputes. In some instances, it has even been created (constructed) by the courts where individuals have so acted as to lead others to believe that a partnership existed. The usual clauses in a written agreement are fairly well standardized.

Forming a partnership does not require any specific permission from the state. If the partners wish, however, they are generally entitled to file a copy of the agreement at the courthouse, which thus serves as a reliable neutral depository.

Each partner may bind the others to contracts incurred in the normal operations of the business.[5] Thus the partnership has often been defined as a contract of mutual agency. The legal profession repeatedly advises that one should have confidence in the reliability and judgment of the other party before entering a partnership.

The ordinary partner (strictly defined as a general partner unless the general partner is an LLC, a limited liability corporation) has unlimited liability for the partnership's business debts.[6] Thus if the partnership fails and the firm's assets fail to cover its liabilities, the creditors may seek to recoup their losses from the partners' private assets. Moreover, it is not the duty of the creditors to apportion losses among the partners. They may seek compensation for their claims where they can find it, regardless of any loss-sharing agreement among the partners. If one partner's personal assets are greater, or simply more available, he may well suffer dispropor- tionate losses. A partner who loses more than his agreed share may have a counter- claim against the other partners,[7] but, of course, collection under these circumstances may be delayed considerably.

It is generally held that a rich man should be careful about entering a risky venture in partnership with a poorer individual. On the other hand, although their unlimited liability entails some financial dangers to the partners, it gives the partnership a stronger credit base than would a corporation of the same size.

No matter what formula has been set up for sharing returns out of the company, each partner will be taxed at the appropriate individual income tax rate for all the income realized or imputed to him out of the operations of the firm. Thus, if a partner receives a salary for his services to the partnership, interest on loans to the

---

[5]For example, one partner may, without the express consent of the others, sell part of the normal inventory to a third party. The contract would hold, even if afterward the other partners should disapprove of the price or any other term of the sale. But a contract for the sale of major equipment might not be considered binding without the consent of the other partners if such a sale appeared to be outside the normal operations of the business.

[6]There is a variant of the partnership known as the limited partnership where one or more partners may limit their liability to their investment in the firm. There still must be at least one general partner.

[7]Unless stipulated otherwise, losses are shared on the same basis as profits. If there is no express agreement, profits and losses are shared equally by the partners.

partnership, and a share of the remaining pro forma profits,[8] he will pay tax on all of these items. He must pay tax on his total share of partnership profits, whether he draws them out for personal use or reinvests them in the business. The partnership as a firm files a Form 1065 Information return and distributes K-1 s to the partners. Its report is made purely for informational purposes so that 100 percent of the partnership income can be imputed to the various partners.

However, provisions of the federal income tax laws allow a partnership that satisfies certain conditions to elect to be taxed as if it were a corporation. In this case, retained earnings may (for the time) bear a lower rate of tax than the applicable personal tax bracket of the partners.

Because the partnership rests on a foundation of mutual trust, its dissolution is made very easy. If no specific provision is written into the agreement, any partner can call for dissolution, usually with a required notice of, say, 30 days. Then a procedure called an "accounting" takes place; the assets are liquidated and each partner is paid his ownership share (equity) according to the agreement. Since the assets upon liquidation are unlikely to bring in their "going concern" value, dissolution usually entails an economic loss. Often the partnership agreement sets a definite liquidation amount to be paid to a partner who requests dissolution so that the whole going-concern need not be liquidated.

Since the partnership is a contract of close personal relationship, a partner's interest in the firm cannot be easily transferred or sold to a third party although the partner's interest can be assigned on death. A new party cannot enter the partnership without the consent of all the partners. If with the consent of the other partners, one partner sells out his interest in favor of someone else, the old partnership is actually legally dissolved, and a new firm is formed.

Again, since the partnership is constructed on the basis of personal contracts, the death of any partner dissolves the relationship. The heir of the deceased partner does not automatically become a member of the firm. If the surviving partners wish, they may let him in, actually forming a new partnership. Otherwise, the heir is entitled to dissolution and accounting. It is quite usual, however, for the payment to the heir to be pre-set in the partnership agreement. The difficulties that may occur because of the death of a partner give a partnership an insurable interest in the life of its principals. The insurance benefits may help in settling the partnership accounts.

The main advantages of the partnership form of business firm are flexibility in operation and ease of organization. More managerial talent can be assembled than is possible under the single proprietorship. The partnership can bring together more capital than is likely in the single proprietorship, and it may have a better credit standing relative to that of a small corporation because of the unlimited liability of the partners.

---

[8]Legally any payment of interest or salaries to the partners is not an expense but merely a way of distributing the partnership profits. However, these are generally subtracted on the accounting statements as an operating expense in order to obtain a pro forma profit figure.

On the other hand, the partnership may engender tensions between its principals. It may be unwieldy to operate if there are too many general partners since each partner is a general agent of the firm. This factor, together with the unlimited personal liability of the partners, tends to limit the total capital that can be raised. Because a partner may risk his personal fortune on the activities of the partnership, he generally wants a hand in the partnership affairs. To the drawbacks of the partnership, the difficulties of selling a partnership interest and the possible losses from a forced dissolution caused by either the death or withdrawal of one of the partners must be added.

Nevertheless, the partnership has proved serviceable where the total capital needs of the firm are not extremely large. The partnership is very common in smaller companies in the marketing field, in small manufacturing firms, in the professions, and in agriculture. The partnership form is very suitable where the personal skills and reliability of the partners is more important than capital in engendering income for the firm. (Thus doctors, lawyers, accountants, etc. have often been quite successful practicing as partners.) In the past, in those areas of financial activity where the bulk of the assets are carried in a liquid or highly marketable form and the problems of dissolution are not difficult (such as investment banking or security brokerage), the partnership was quite common.

2.3   The Limited Partnership

Given the versatility and imagination, with which persons direct business affairs and the ingenuity of the legal profession, it follows that there are many forms of business organization. Some of them are not very common today and are mainly of historical interest. The limited partnership, however, is an interesting and fairly common variant of the partnership form.

A limited partnership contains one or more general partners; one or more partners who are limited in their liability for business debts to the amount of their investment in the firm. The general partner or partners manage the firm and are subject to unlimited personal liability for the firm's debt. The limited partner, of course, receives some agreed-upon share of the profits. Sometimes called the "silent" or "sleeping partner," he cannot take an overt part in running the business, for if he does and the creditors come to believe he is an active partner, he may lose his limited status in the eyes of the law.

If a general partner wishes to retire from active management and become a silent partner, all interested creditors must be notified. Thereafter the limited partner cannot present himself in any way that could lead an "innocent third party" to believe he is a general partner.

The limited partnership is not widely applicable. The silent partner must rely heavily on the acumen and integrity of the general partners. He should, of course, keep himself informed of the progress of the business. If he is not satisfied with the performance of the enterprise, that is, the treatment he receives or the results achieved with his capital, his sole legal course of action would be to ask for dissolution and to withdraw his capital. This may involve a loss for all concerned.[9]

2.4   The Corporation, Its Basic Characteristics

The corporation is a complex organization and takes many shapes. For the moment, we shall present a bare outline of the legal characteristics and organizational structure of the corporation.

The corporation is defined as a fictitious person created by the state. It can engage in certain defined activities, and in pursuing its purposes, it can obtain title to or dispose of property, enter into contracts, and engage agents to work and act for it. Of course, the corporation is not a human being and can act only through humans hired to work for it, yet the legal fiction of the "corporate person" is nevertheless a highly ingenious device. It enables the corporation (i.e., the properly state-sanctioned organization) to take responsibility and liability for legitimate activities which otherwise would be the final liability of the individuals in the organization. Unless something illegal has occurred, people dealing with the corporation must satisfy their claims against the corporation and may not look to those behind the company for ultimate settlement.

The limited liability of the owners is the single most important characteristic of the corporation. Limited liability simply means that in case of failure the owner or stockholder may lose what he has ventured in the firm, but even if the company cannot pay its creditors in full, the shareholder's personal assets cannot be endangered.[10] (The feature of limited liability is denoted in Great Britain by the abbreviation LTD (limited), placed after the company's name.) In France, it is denoted by the abbreviation S.A., (Societé Anomie), the company of nameless ones. (The concept, of course, is that the owners cannot be named in a legal suit.) Because it possesses limited liability, the modem business corporation provides a method for many people to risk their funds in a distant organization, perhaps with the chance of gain if the business goes well, but with a definite limit of loss if the worst should happen. We will briefly trace the evolution of the modern corporation in Chap. 18. It is hard to imagine how the large corporations with their funds amassed from many individuals could have developed without this provision.

Another important aspect of the corporate form is that its ownership shares are transferable. Unlike the partner, the owner of a share in a corporation has the inherent right to sell or give away his holdings without the consent of the other

---

[9]At that, the defensive weapon of calling for dissolution may be better than anything available to the minority stockholder in a small corporation.

[10]Unless, as sometimes happens in a closely held corporation, the major shareholder personally endorses the corporation's note in order to obtain additional credit, or he has failed to pay in the full par value of the stock. This last possibility is not too likely in practice.

shareholders,[11] and the new holder acquires the same rights and privileges as the original owners. The characteristic of transferability is especially important to the development of modern free enterprise economies since it provides a high degree of capital mobility. Indeed, the ready transferability of corporate shares makes possible the whole intricate structure of capital markets.

The length of life of the corporate firm is independent of its stockholders. If a shareholder dies, his heirs take title to his shares, and the other surviving stockholders have to accept the new owners; there is no legal problem of "succession." The corporation virtually has "perpetual life." Although most corporation charters granted by the states run from twenty to forty years and only a few are perpetual, most charters may be renewed with ease.

A corporation is a stable form of business organization because it can be dissolved only by a majority vote of the stockholders. In contrast, the partnership is dissolved on the death of a partner or on the request of any partner if he does not approve of the way the partnership is run or if he is displeased with his present or potential partners. In exchange for stability, the shareholder sacrifices the greater flexibility and the choice of associates inherent in the partnership. If the shareholder does not care for the other stockholders or if he thinks he can use his capital more productively elsewhere, his only remedy is to sell his shares for what he can get. The corporation shareholder cannot obtain a dissolution of the firm merely by asking.

The single proprietor or the partners take direct responsibility for managing their firm. The owners of the corporation (i.e., the stockholders) have only an indirect role in the actual management of the company. The stockholders have the right to vote to elect a board of directors, who then appoint corporation officers, a president, secretary, treasurer, and various vice presidents, to run the business and make the day-to-day decisions for the firm. In a closely held company, where one man or a small closely connected group holds a majority of the shares, the major stockholders, the board of directors, and the management (i.e., the major officers) may all be the same people. In general, a widely held corporation is one where no single stockholder or close group of stockholders holds anywhere near a majority of the shares. Nevertheless, the management (though it usually holds only a small minority of the total shares) is the dominant voice in the corporation's affairs. In theory, the board of directors has strong powers. In actuality, the directors are nominated by the management and upon election reappoint the corporate officers.[12] The stockholders almost invariably vote for the management slate; usually, it is the only group

---

[11]Rights such as these can usually be voluntarily abridged. Thus a family-held corporation may require that a stockholder offer his shares to the corporation or other existing stockholders (at some fixed price) before he may sell them to someone else.

[12]In exercising independent choice in the appointment of officers, most boards resemble the electoral college since they are already pledged to a given set of candidates. In the rare instances of a contest for control, the management runs one slate of directors and the dissidents nominate an opposing slate.

running. The voting rights of the stockholders have become in practice a remote, ultimate power that might be used if the firm is badly mismanaged.[13]

The prevalence of management power over that of the shareholders has led to the development of agency-principal theory, which we discuss in Chap. 22. Agency theory deals with the methods, the types of compensation, and other encouragement that the shareholders-owners (the principals) might devise to tie the operations of the management (the agents) to the basic interests of the shareholders.

Although the lack of influence the average stockholder has on the actual direction of the firm is often considered a drawback of the corporate form, it is not entirely a disadvantage. Most investors are not interested in the ordinary business decisions of the firm. They usually have other concerns, and they would consider it burdensome to devote any considerable time to the operation of the company. The large corporation develops its own management, which, after time achieves certain autonomy, may develop a professional outlook, policies, traditions, and very likely a considerable devotion to the affairs and success of the company. Although the remuneration of management may be considerable, and even if at times it may take unfair advantage of its power, in the majority of corporations, the economic interests of the shareholders are well served by the corporate structure.

Regarding federal income taxation, the status of the corporation is quite unique. Since the corporation is a legal personality in its own right, it is taxed independently on its profits. The stockholders are further taxed at the individual tax rates for the profits they receive in the form of dividends or at the capital gains rate when reinvested earnings raise the value of shares and these shares are sold at a profit. Because profits have already been taxed once, corporations are described as being subject to "double taxation." In certain circumstances, however, the corporation form may actually offer some tax advantages. For example, if a firm whose owners are in a high-income tax bracket is in a profitable growth stage, it may retain a considerable proportion of its profits for reinvestment in the business. These may later be taxed at the lower capital gains rate. In any case, the other favorable aspects of the corporate form may decisively outweigh any extra tax burden.[14]

2.4.1 Chartering the Corporation

Since the corporation is a "creature" of the state, its formation requires a more complicated legal procedure than other forms of business. The legal "soul" of the corporate "personality" is the charter granted to it by the state.

---

[13]This is an old story. The gap between the theoretical right of the stockholders to run the firm through the election of the board of directors and their "general helplessness in practice" was labeled "the separation between ownership and control" by Berle and Means in The Modern Corporation and Private Property, Macmillan, 1933.

[14]Under the present tax laws, some closely held corporations which satisfy the requirements may elect to be taxed as partnerships. It depends on the individual situation whether this is advantageous.

The history of the corporation charter and of the relationship between the corporation and the state is a long one. The notion of some semiautonomous body functioning under a charter was not unknown in Roman law. Charters bestowing a grant of power by the sovereign to a subsidiary body existing under his control were used in medieval times to define the relationship between the crown and monasteries, certain guilds, universities, towns, etc. In England, the king later granted charters to various trading companies (actually independent merchants who formed a market) to engage in certain trades, deal in specific commodities, pre-empt certain geographical markets, or settle colonies. Lloyds of London, an association of insurance firms, is a survivor of these early-chartered companies. Some companies were financed with transferable stock and developed a definite resemblance to the modern corporation. The early charters almost always contained some type of monopoly privilege.

In time, British companies wishing to engage in regular commercial activities began to request charters from Parliament, who had begun to exercise this power on behalf of the Crown by about 1688. Each of these requests had to be passed upon separately. Action on these charters not only became a burden upon other legislative activities, but suspicion arose that favoritism existed in the granting of charters. In 1845, a general enabling act set up an administrative office to pass upon corporate charters. Any company fulfilling the standard requirements received a charter; the enabling act, moreover, redefined the concept of limited liability.

In the United States, developments ran parallel to those in England. The right to grant charters rested with the state legislatures. At first, each request for a corporate charter had to be passed on separately.[15] The politicking, favoritism, and sheer bribery that went into securing a charter became an overt scandal. In 1825 Connecticut passed the first inclusive general enabling act. In time all the states passed similar laws; they provided a procedure whereby the promoters of a business corporation who submitted their application in the proper form to a designated state official were automatically granted a charter.

Just how the grant of the charter defines the relation between the state and the corporation is still a matter of theoretical legal discussion. According to the "fiction theory" of the corporation that comes from the old common law, the corporate personality is a legal fiction of the state subordinate in every way to its creator. The "contract" theory, which has its roots in Roman law, holds that there was a vague but nevertheless, existing antecedent body to the final corporation with which the state has made a binding contract by granting a charter. This theory would seem to limit severely the powers of the state to intervene in any way in the corporate structure. Yet in practice, the two theories have moved closer together. The corporation charter is almost inviolate and cannot be abrogated unless the parties behind the corporation have used it for fraudulent or criminal purposes. Under extreme

---

[15] Government corporations, e.g., municipalities, authorities, etc., still usually require individual action by the legislature to get a charter. The various federal corporations such as the Federal Reserve Banks and the Federal Deposit Insurance Corporation were chartered by separate acts of Congress.

provocation, the courts may pierce the "corporate veil" and hold the people maneuvering the corporation personally responsible for illegal actions, as we are currently seeing with many of the individuals associated with WorldCom, Enron, and Tyco.

A firm that plans to operate as a corporation can charter in any state; there is no requirement that its charter is in the state where its business is located. The procedure of obtaining a charter and the power it gives the corporation varies from state to state. By making chartering a matter, of course, the general enabling acts removed the competition of different groups within a state to secure a charter for their enterprise and possibly block the charter application of some rival. However, competition soon developed between the states to write easier enabling acts to induce corporations to charter with them. In their desire to obtain fees and annual franchise taxes (and bring business to local lawyers), some states reduced the difficulties in securing a charter, lowered fees, and, most importantly, widened the corporation privileges and powers allowed in their charters.[16] The striving by individual states for corporation chartering fees (which realistically have never amounted to much) by making incorporation easier and permitting wider corporate powers has been called "competitive laxity." One may wonder whether the United States would not have developed a simpler, more uniform and possibly more responsible corporation managerial and administrative structure if the right to grant corporate charters had been reserved to the federal government in the Constitution.

Among the factors, a firm may consider before deciding upon the state in which it charters are differences in taxes such as chartering fees and the annual franchise taxes. The federal corporation profit tax does not differ no matter where the corporation is chartered; neither do the local property taxes, which are based on where the fixed assets of the company are located, nor the state corporation income taxes, which are set on where the firm does business. Chartering fees and annual franchise taxes are a minor fraction of the taxes a corporation is subject to and hardly the major influence in the choice of state in which to charter.[17] More likely the "liberality" of the charter provisions influences the choice of domicile (i.e., the state in which the charter is obtained), but even this was more crucial in the past than it is today. Most newly formed corporations obtain their charter in the state in which they intend to locate the major part of their business.

A single proprietor or partnership whose principals have legal residence in the United States have the constitutional right to operate in any state. However, a corporation, although a legal personality is not a citizen and has no inherent constitutional right to locate in states other than the state in which it is domiciled. We briefly trace the development of incorporation of holding companies, and the creation of the modern corporation in Chap. 18. In legal terminology, a corporation

---

[16]Many national corporations are chartered in Delaware, not so much because the fees are low but because Delaware was one of the first states to allow extensive holdings of the shares of other corporations. Thus Delaware became an ideal state to charter controlling and holding companies.

[17]Taxes may influence the location of the business, but that is a different matter from where it is chartered.

operating in the state in which it is chartered is called a domestic corporation, one operating in a state other than that in which it is chartered is a foreign corporation. A corporation of another nation is an alien corporation. Actually, a high degree of reciprocity prevails, and the states have not placed severe discriminatory restrictions on foreign corporations. A foreign corporation is generally required to pay an annual registration fee; the amount is usually equivalent to the annual franchise taxes paid by domestic corporations.

2.4.2 Administrative Organization

The overriding law applying to the corporation's government in any state is the state constitution, statutes passed on corporation matters, and court decisions that establish precedents. When a corporation's charter or by-laws conflict with existing state laws, the charter or the by-laws must give way. Therefore, it is necessary to know that state law (or external law) in order to interpret properly the charter and the by-laws (or internal law) of the corporation.

The charter (or certificate of incorporation) is the corporation's right to legal existence, and it also states the general organization, authorized capitalization, the business, and purposes of the company. Ordinarily, the charter does not make detailed provisions for running the affairs of the corporation. At the first meeting of the stockholders, after the charter is obtained, a set of by-laws is adopted, covering such topics as the election and remuneration of a board of directors, the establishment of the corporate offices and the definition of their powers and duties, the issuance and transfer of stock, methods of voting, and other matters pertaining to the management and administration of the company.

Although all corporations go through the formality of passing by-laws and electing directors, the practical importance of this procedure is determined by the structure of the ownership. By far the vast majority of corporations are closely held, closed, or family-held companies where a single individual, family, or closely related group holds all the shares. The closely held corporations are generally small, the principals are in close contact, and there is a minimum of formality in the procedures of administration. Much of the discussion that follows pertains to the widely held corporation.

After the board of directors has been elected (pursuant to the by-laws), it appoints the corporate officers. These might consist of a president, secretary, treasurer, controller, and several vice presidents. The president is the chief executive officer (CEO) of the firm; he and the other officers direct, manage, and take responsibility for the regular operations of the business. In some firms, the chairman of the board of directors has a powerful position as the overall strategist of the company; nevertheless, the president is usually in charge of the major internal affairs of the corporation. In some companies, basic administrative power may rest in an executive committee, consisting perhaps of the major officers and selected members of the board. Often the major officers are also on the board of directors. As we have already indicated, in the going corporate concern, power to control the corporate affairs usually shifts from the stockholders through the board of directors finally to rest largely with the management.[18] The board is generally composed of the management's nominees, and the management's choice is voted down only if a dissenting group obtains enough stockholder votes to install new management (and this seldom occurs).

A board of directors composed mostly of persons holding executive positions in the company is known as an inside board. A board of directors with some outside members is known as a mixed board.[19] The IBM Board of Directors was a mixed board, as reported in Guerard and Schwartz (2007). In December 2020, IBM substantially reduced its board of directors from 24 to 12 members. Arvind Krishna was named Chairman and Chief Executive Officer, replacing Virginia Rometty. He was a driving force behind IBM's $34 billion acquisition of Red Hat, which closed in July 2019. Arvind has an undergraduate degree from the Indian Institute of Technology Kanpur (IITK) and a PhD from the University of Illinois at Urbana-Champaign. Dr. Krishna is the coauthor of 15 patents, has been the editor of IEEE and ACM journals, and has published extensively in technical conferences and journals.[20] The IBM Board continued to include Alex Gorsky, CEO of Johnson & Johnson; David Farr, CEO of Emerson Electric; and Martha Pollack, President of Cornell University, and several retired CEOs, including Michael Eskew (UPS), Andrew Liveris (Dow Chemical), Joseph Swedish (Anthem), Peter Voser (Royal Dutch Shell), and Frank Waddell (Northern Trust). Michelle Howard, Retired Admiral, US Navy, was named to the Board.[21]

The American Management Association has been in favor of outside directors; they feel such people may bring constructive, fresh, and critical insights to the board and provide a useful review of the company and managerial policies.[22] The inside board is defended on grounds of efficiency; it is felt that the members of the board are able to act expeditiously if they are intimately acquainted with the problems they must deal with.

---

[18] Of course, all sorts of possibilities have to be allowed for. In some companies a dominant stockholder may be the real ultimate power, although he may elect not to be an officer or perhaps not even hold a seat on the board.

[19] As reported in Guerard and Schwartz (2007), in its March 10, 2003, 10-K filing for the year ended December 31, 2002, IBM listed its Board of Directors. The Chairman of the Board, President, and Chief Executive Officer, is Samuel Palmisano. The executive officers of IBM are elected by the Board of Directors and serve until the next election (annually). Who is the Board of Directors of IBM? In 2002, the IBM Board had 15 members and included Kenneth Chenault, the Chairman and CEO of American Express Company; Nannerl Keohane, the President of Duke University; Sidney Taurel, the Chairman and CEO of Eli Lilly and Company; John Thompson, the Vice Chairman of the Board at IBM; Charles Vest, President of MIT; and Lockewijk C. VonWachem, Chairman of the Supervisory Board of the Royal Dutch Petroleum Company.

[20] https://www.livemint.com/companies/people/meet-iit-kanpur-graduate-arvind-krishna-who-will-be-new-ceo-of-ibm-11580434283523.html

[21] https://www.ibm.com/investor/governance/board-of-directors.html

[22] On the other hand, outside directors are likely to have many interests, perhaps as officers of other corporations or serving on the boards of many other companies. They may be able to give no more than perfunctory attention to the matters at hand.

Contrary to popular impression, generally, directors as such are not extremely highly compensated, although many firms pay directors annual fees of $50,000, plus additional fees for committee meetings. There is a question when the directors are excessively complimented, $350,000 plus annually, e.g., Enron, whether they will exercise any discretion on the corporate activities. Being on the board of a company is often considered an honorific position. (In most corporations a director does not initially have to be a stockholder of the firm.) Many prominent directors hold positions on a large number of boards. Under these circumstances, it is no wonder that the directors are often willing to accept the management's account of its stewardship and generally approve the programs that management proposes.

Just what are the responsibilities of the directors? The matter is not yet clearly decided in law. On a formal basis, the directors appoint the company officers, approve executive compensation, approve dividend declarations and stock splits, and pass on major expansion plans or changes in company policies. In practice, the directors seldom initiate any policies but abide by management's suggestion. Moreover, matters of special interest are usually put to a special vote of the stockholders. The question is the degree of attention and care the directors owe to the affairs of the corporation. In the past, directors have often been cavalier in looking after the interests of the public stockholders. For some time, however, there has been a move in practice and in law regarding the directors as being in a quasi-fiduciary position. Holding a position of quasi-trust, the directors should exercise due care in protecting the equity of the stockholders. He should protect the stockholder against fraud and obvious malfeasance and make sure that he is fully informed on the corporation's condition and policies. A necessity for full disclosure holds with special force when the director himself engages in a transaction with the corporation; it is important that the director not take advantage of his position to underpay the corporation or overpay himself or other interests he might represent.

Because they can be subject to stockholder suit, it can at times be difficult to recruit directors. To overcome this difficulty most corporations purchase liability insurance for their directors.

2.5   Major Rights of the Shareholders

The main value in owning stock in a publicly held corporation is the dividends and/or capital gains that may be reasonably expected over time. The average shareholder depends on the innate fairness, reliability, skill, and goodwill of the management to give him a reasonable distribution of the gains accruing to a successful enterprise. Rights inhering to the stockholders protect them, to some degree, against arbitrary or capricious management. In practice, these rights give only negative protection; they may be a distant threat to some managements which might otherwise forget their obligations entirely.

The stockholders have the right to vote for the board of directors. Voting is done by share; each share equals one vote. An owner of 100 shares of IBM stock votes his or her opinion, although, with an average of 1720.4 million shares outstanding in 2003, and 885.6 million shares outstanding in 2019, the owner might not feel terribly important. The stockholders may vote by attending the annual meeting, or they may authorize someone else to vote their shares for them. Such authorization is called a proxy. The majority of proxies go to the management representatives. An innovation proposed some years back by students of corporation organization, and actually adopted by a few companies, is cumulative voting for the directors. Cumulative voting is a system of proportional representation. It allows a sufficiently large minority of the shareholders to concentrate their votes on one or more directors in order to achieve representation on the board. Under a straight voting system, where each position on the board is voted for separately, it is clearly possible for a bare majority to place every one of their candidates on the board, thus excluding any minority views entirely.

The proponents of cumulative voting argue that it is in the firm's interest to have a different point of view represented and that the minority directors will tend to check excesses of the majority. Opponents of cumulative voting point out that a dissident minority could get representation on the board merely to harass the normal operations of the company.

As has been pointed out, the shareholders almost always vote to retain the existing management. (This may be a fairly wise procedure; managerial talent is not necessarily in oversupply.) On occasion, however, a rebellion of stockholders (often dominated and financed by major shareholders who would like to be the new management) has succeeded in taking over a company. The stockholders generally have to be well aroused by extraordinarily poor management for this to occur. Nevertheless, although such rebellions are rare, the lesson is not entirely lost on other managements.

The shareholders are usually called upon to vote on any major change in the corporate structure. These changes may entail modifications or amendments of the by-laws, an increase in authorized shares, permission to float a convertible issue, a major expansion, a merger, a bonus or profit-sharing plan, a pension program, or stock options for the company executives. Generally, the shareholders approve the management's proposals.

The shareholder has the right to examine the company's books. This gives him some protection against possible fraud or manipulation of the accounts. It also provides him with the list of shareholders he needs to attempt a proxy drive to change the management. On the other hand, the right to examine the records can be put to unethical uses: to obtain a list of customers for a rival firm, to obtain a list of stockholders to solicit for other enterprises, to look for trade secrets, or otherwise to harry the management. But the shareholder that wishes to examine the company's records at a reasonable time and place is allowed to do so.[23]

An important right of the stockholder is the right to institute a stockholder suit against the management on behalf of himself and the other stockholders. The right of the stockholders to sue on behalf of the company may be a powerful deterrent against errant management. Many times such suits are mere nuisances, started in hope the management will make some sort of settlement rather than fight through the courts; often, however, the results of the suits have been startling, uncovering fraud, manipulation, conflicts of interest, or pure stubborn mismanagement. If the stockholder wins his suit, he collects only court costs for himself; the damages are paid by the parties liable (the company executives and directors, as the case may be) to the corporate treasury.

In the vast majority of states, corporation stockholders enjoy the preemptive right or the right of privileged subscription. Existing shareholders are given the privilege of purchasing—in proportion to their holdings—the shares of a new issue of stock before these are offered on the general market. The preemptive right is said to give the stockholder the opportunity to preserve his proportionate interest in the corporation.[24] By exercising his or her rights, the existing shareholder can maintain the same proportionate claim to the firm's assets as he already has or maintains the same proportionate voting strength.[25] Although often described in political-legal terminology, the major function of the preemptive right is economic. It provides the shareholder with a compensating device if new stock is being floated at a price below the present going market price. (New shares are usually issued below the market price because it makes their flotation easier.) If the shareholder wishes, he can sell his rights, or he can exercise his rights and buy the new shares at the "offering" price. In either case, he loses no financial advantage to the newcomer shareholders.

2.6   The Advantages of the Corporate Form

When set against the possible machinations of the management, the protective rights of the shareholders hardly seems very strong. And it is true that innumerable cases can be cited where the management has taken advantage of the apathy, helplessness, or lack of knowledge of the stockholders to enrich themselves. On the other hand, such things have "also happened in government, politics, labor unions, social

---

[23] Stories have arisen that some company's books are kept in cobwebby, cold vaults to discourage nosey stockholders. These stories are not wholly accurate.

[24] Both the preemptive right and general voting rights are commonly denied to preferred shareholders by charter provision. In a few states, it is possible to deprive the common stockholders of the preemptive right in the charter. The preemptive right on a particular issue can be waived by a vote of the shareholders.

[25] Rights are not given on the flotation of a security other than common stock unless the new issue is a convertible one (i.e., it can be changed into common stock at the option of the holder).

groups, and even religious organizations, and it is no more inherent in one [the corporation] than it is in another. Whatever may be the possibilities of unfair or unethical practices, it is probably true that the overwhelming majority of corporate managements are faithful to the trust they hold."[26]

In spite of its weaknesses, the corporate form gives many advantages to its stockholders (owners). If this were not so, it would be hard to account for the growth and economic dominance of the corporation over other forms of private business organization. Essentially, the corporate form permits the pooling of large amounts of capital from many sources (institutional and private) because it does not obligate the individual investor to be active in the affairs of the business nor subject him to worry over personal liability for business debts.

At most periods in our history, the publicly held corporation has been able to raise capital with relative ease because over time investment in a diversified portfolio of corporation shares has shown a comparatively good rate of return. As a corollary, in our economy characterized by the growth of industries requiring heavy capital investment, large-scale operations, and specialized professional managements, the corporation's ability to amass funds and to develop responsible administrative staff has made it the dominant business form.

2.7   The Corporation in 2020: Trying to Maintain Economic Growth in the COVID-19 Pandemic with the Payroll Protection Plan

Corporate functions were greatly affected by the COVID-19 pandemic in the United States starting in March 2020 when the unemployment rate rose from a 50-year low of 3.50 percent in February 2020 to 14.70 percent in April 2020. As reported in the Economic Report of the President, President Trump signed the CARES Act, Coronavirus Aid, Relief, and Economic Security (CARES) Act to address the economic fallout from the pandemic on March 27, 2020.[27] The CARES Act provided $2.2 trillion in relief to households and businesses affected by COVID-19. The CARES Act provided an additional $347 billion in targeted relief to Americans who lost their jobs. Federal Pandemic Unemployment Compensation (FPUC) offered every beneficiary of unemployment insurance an additional $600 a week in unemployment

---

[26] Chelcie C. Bosland, *Corporate Finance and Regulation*, Ronald Press, 1949, p.69.

[27] *Economic Report of the President,* January 2021, p. 77. On March 27, 2020, President Trump signed the Coronavirus Aid, Relief, and Economic Security (CARES) Act into law, providing $2.2 trillion in relief for households and businesses. A family with two children and an income below $150,000 received an Economic Impact Payment of $3400, almost twice as much as the maximum $1800 stimulus checks provided during the Great Recession. And unlike the Great Recession stimulus payments, full Economic Impact Payments were available to the lowest-income households with no tax liability. The CARES Act also provided unprecedented relief to those workers who lost their jobs.

benefits from March 29, 2020, through July 31, 2020.[28] Under FPUC, the worker would receive. an additional $600, for a total of $800 per week. Pandemic Emergency Unemployment Compensation provided an additional 13 weeks of UI benefits for workers who exhaust their regular State benefits, for a total of 39 weeks of coverage in most States (in addition to potential coverage by Extended Benefits). Pandemic Unemployment Assistance granted UI benefits to workers not eligible for regular State unemployment insurance benefits, such as self-employed workers, gig workers, business owners, independent contractors not participating in the UI elective coverage program, and workers with insufficient work history to normally receive unemployment benefits. This assistance for unemployed workers was complemented by the Paycheck Protection Program (PPP), which helped ensure that businesses could keep their workers on payroll and avoid the need to draw unemployment assistance.

References

Bosland, C. C. (1949). *Corporate finance and regulation.* Ronald Press. Chapters. 2, 3.

Brealey, R. A., & Myers, S. C. (2003). *Principles of corporate finance* (7th ed.). McGraw-Hill/ Irwin. Chapter 1.

Dewing, A. S. (1953). *The financial policy of corporations* (5th ed.). Ronald Press. Vol. 1, Chapters 1, 2, 3, 4.

Economic Report of the President, January 2021.

Gerstenberg, C. W. (1959). *Financial organization and management of business* (4th rev ed.). Prentice-Hall. Chapters 1, 2, 3, 4.

Guerard, J. B., Jr., & Schwartz, E. (2007). *Quantitative corporate finance.* Springer.

Guerard, J. B., Jr., & Vaught, H. T. (1989). *The handbook of financial modeling.* Probus, Chapters 1, 2, 4.

Guthman, H. G., & Dougall, H. E. (1955). *Corporate financial policy* (3rd ed.). Prentice-Hall. Chapters. 2, 3, 4.

Samuelson, P. A., & Nordhaus, W. D. (2002). *Economics* (16th ed.) Chapter 6.

Schwartz, E. (1962). *Corporate finance.* St. Martin's Press. Chapter 13.

Van Horne, J. C. (2002). *Financial management & policy* (12th ed.). Prentice-Hall, Inc. Chapter 1.

---

[28] *Economic Report of the President,* January 2021, Chapter 2

Chapter 3
The Corporation Balance Sheet

This chapter on the balance sheet and the following one on the income statement are designed to serve two modest purposes: to acquaint the student with accounting and financial terminology and concepts used throughout the book and to explain the two important accounting statements on an uncomplicated level so that the student can appreciate and use some of the information presented by the accountants. It is not the purpose of these chapters to review all the procedures of accounting.

3.1 The Balance Sheet

The balance sheet is the financial picture of the firm at a precise point in time. The left side of the balance sheet lists, categorizes, and sums the assets, the items of value the firm owns at a given point of time. The right side lists the debts and other payment obligations that the firm owes; these are the liabilities. The difference between the liabilities and the assets is the net worth, equity, or ownership capital. The liabilities plus the net worth of the firm must equal the sum of the firm's assets. The balance sheet then presents this equation:

The sum of the assets = The sum of liabilities plus stockholder equity

Although the balance sheet or position statement is a useful quantitative picture of a firm's financial position, it is not an exact reflection of the firm's economic worth. The balance sheet is constructed on the basis of formal rules and does not necessarily represent the market value of the firm as either a growing concern or liquidated (sold off) entirely. Moreover, the balance sheet represents the financial position' exactly at 12:00 midnight on the balance sheet date. The assets and liabilities shown are those accountants have ascertained to exist at that point in time. Since a business is always changing (buying a little here, selling something there, borrowing, lending, paying, and getting repaid), the balance sheet never represents more than an approximate

J. B. Guerard Jr. et al., *Quantitative Corporate Finance*,
https://doi.org/10.1007/978-3-030-87269-4_3

financial truth, and even at the moment it is finally put together. The fact, however, that the balance sheet cannot tell us all we should like to know is not an indictment of accounting. The accountant's prime functions are to keep legal claims straight, present his data as consistently as possible, and stay as close as possible to objectively determined costs. The financial statements, a compromise between conflicting functions and demands made on the accountant, are the data very often used in making business decisions.

Several points are provided by the balance sheet equation, which holds that total assets = total liabilities + stockholder equity (or total debits = total credits). Therefore, if the firm increases its total assets (buys more goods, acquires new equipment, increases the money owed it by extending more sales credit), it follows that either the liability or stockholder equity (ownership accounts) must also increase to balance the rise in assets. The firm may increase the amount it owes its suppliers, borrow from the banks, and float a mortgage bond issue, or it may increase its net worth by floating additional common stock or retaining additional earnings in the business. The issuance of debt increases the firm's interest expense, in which it must pay to remain in business, but debt issuance does not possibly dilute the ownership and voting interests of management, inherent in an equity issuance. The problem of whether or not to acquire additional assets and the related question of choosing the best source out of which to finance the additional assets the balance sheet reports how the firm is financed, a central area of financial decision making, as the reader will see in Chap. 11. That is, management should finance the expansion of assets by pursuing the debt or equity strategy that maximizes the stock price of the firm.

3.2 Assets

The assets which the nonfinancial firm may acquire or own are usually broken down into three major categories: current assets, fixed assets, and other assets. The current assets and the fixed assets are usually much larger than the other assets.

Current Assets The current assets consist of cash, items that in the normal course of business will be turned into cash within a short time (the accountants use the rough measure of one year), and prepaid items that will be used up in the operation of the business within the year. The three largest accounts making up the current assets are usually in cash, receivables, and inventories.

Cash Cash is the sum of the cash on hand, held out to make changes and provide petty cash funds, and the deposits in the bank. IOU's, advances to employees, etc. are not counted in the cash account.

Receivables The receivables are amounts due to the firm from customers who have bought on credit. They are often segregated into accounts receivable and notes receivable. An account receivable is the usual way credit is given in American business practice. It simply means that the buyer of the goods is charged for his purchases on the books of the seller—no other "formal" legal evidence is prepared. This type of credit is often called "book" or "ledger" credit.

If a note receivable is used, the purchaser of the goods has signed a promissory note in favor of the seller. A note, except in certain lines of business where they are customary, is generally required only of customers with weaker credit ratings or those who are already overdue on their accounts.

An account called "reserve for bad debts" or "allowance for doubtful accounts" is generally subtracted from the receivables. This is called a valuation reserve; it is an attempt to estimate the amount of receivables that may turn out to be uncollectible. The receivables minus this reserve, the net receivables, is counted as an asset on the balance sheet.

The receivables do not normally contain advances or loans that are not part of the normal business or sales operations of the firm. Loans to officers or employees or advances to subsidiaries are generally included in the other assets. Also, except for financial firms—banks and finance companies—such items as accrued interest receivable are usually not included with the other receivables.

Inventories Inventories are items making up the finished stock in trade of the business and the raw materials which in a manufacturing firm will in due course be turned into the finished products. In a mercantile or distributing firm, the inventory consists basically of "finished goods," that is, items which the company does not have to process further. In manufacturing companies, the inventory divides into three categories, raw materials, work in process, and finished goods. If we consider the current assets from the "flow of funds" aspect, that is, how close they are to being turned into cash, cash will be listed first, of course, next receivables (representing generally sales made but not yet collected), and then inventory. Obviously, for the going concern, finished goods are more current or liquid than work in process, and work in process more so than raw materials. The relative composition of the inventory can become a matter of importance—sometimes unfortunately overlooked in analyzing the current credit position of a manufacturing firm. An inventory of 286 or 386 personal computers (PCs) are of little value to a computer firm.

A problem in presenting inventory values on the balance sheet is to keep separate the amount properly ascribed to supplies. Supplies are not part of the normal stock in trade nor are they processed directly into finished goods. Stationery and stenographic supplies, coal used for heat or power, wrapping material used in a store, are obviously supplies. It is difficult, however, to classify packaging items that are a distinctive part of the final product in a manufacturing concern, or the coal pile of a steel plant. In general, an item that is an integral part of the final product is part of the raw material inventory, whereas items used in corollary functions are supplies. Supplies are usually placed with the miscellaneous current assets; like the prepaid expenses, they represent expenditures made currently which save outlays in the future.

Valuation of the inventory is an additional problem with which the accountants must wrestle. The usual rule of valuation is "cost or market, whichever is lower." This rule gives a conservative value to the inventory. The inventory is marked down in value if prices have declined since the items were purchased, but if prices have risen items are valued at their cost to the firm.[1]

Firms may now make a choice between the rule of "first-in, first-out" (FIFO) and "last-in, first-out" (LIFO) as methods of inventory valuation. Under FIFO (which most firms still use), it is considered (whether physically true or not) that sales have been made of the older items and that the items most recently manufactured or purchased compose the inventory. Conversely, if LIFO is used to value the inventory, then the new items coming in are considered to enter the cost of goods sold, and the cost of the older stock sets the value of the inventory. Under the first method, FIFO, the value given the inventory on the balance sheet is meaningful, but the cost-of-goods-sold figure used on the income statement may not truly represent current economic costs if price levels have been changing rapidly. Under FIFO the accounting figure for cost-of-goods-sold tends to lag behind price level changes, so that reported accounting profits are large on an upturn and decrease rapidly (or turn into reported losses) on a downturn in prices.[2] On the other hand, LIFO reduces the lag in accounting for cost-of-goods-sold when price level changes, thus modifying the swing of reported accounting profits during the trade cycle. The LIFO method of inventory valuation, however, tends to develop an inventory figure on the balance sheet that may not be at all representative of any current cost or price levels. The asset value of the inventory may become more and more fictitious or meaningless as time passes. In fact, more large companies surveyed annually by the American Institute of CPAs (AICPA) had switched to LIFO by 1981 [Horngren, 1984]. In 1994, the corresponding AICPA survey had 351 large firms using LIFO, with 186 firms using LIFO for 50% or more of its inventories.[3] The larger the firm's inventory balance, the greater is the potential tax savings for LIFO companies. Moreover, in defense of FIFO, any distortion it produces on the profit and loss statement is not very great for firms that turn over their inventory rapidly, i.e., for firms whose stock is replaced rapidly in relation to their sales. In fact, one of the earlier studies of market efficiency by Sunder (1973) found no significant excess returns for firms switching from FIFO to LIFO in inflationary times (from the date of the change), whereas firms switching from LIFO to FIFO experienced a drop in excess returns over 8 percent.

Miscellaneous Current Assets Other current assets besides cash, receivables, and inventories are accruals, prepaid expenses, and temporary investments. Accrued items are amounts which the firm has earned over the accounting period but which are not yet collectible or legally due. For example, a firm may have earned interest on

---

[1]Essentially unrealized profits are ignored but unrealized losses affect the accounting results.

[2]White et al. (1998), pp.274–275

[3]White et al. (1998), pp.289–290

a note receivable given to it in the past even though the note is not yet due. The proportionate amount of interest earned on the note from the time it was issued to the date of the balance sheet is called accrued interest and under modern accounting procedures is brought on to the books as an asset.

Prepaid expenses are amounts the company has paid in advance for services still to be rendered. The company may have paid part of its rent in advance or paid in on an advertising campaign yet to get underway. Until the service is rendered, the prepayment is properly considered an asset (i.e., something of value due to the firm). When the service is rendered, the proportionate share of the prepaid item is charged off as an expense.

Temporary investments are holdings of highly marketable and liquid securities representing the investment of temporary excess cash balances. If these are to be classified as a current asset, the firm must intend eventually to use these funds in current operations. If, however, the securities are to be sold to finance the purchase of fixed assets or to cover some long-term obligation, such holdings are more correctly grouped with the other assets or miscellaneous assets.

Fixed Assets   The fixed assets and the current assets are the two important asset classes. The fixed assets are items from which the funds invested are recovered over a relatively longer period than those invested in the current assets. Long-term receivables are included in fixed assets. The fixed assets are also called capital assets, capital equipment, or the fixed plant and equipment of the firm. The annual balance sheet usually lists a Property, Plant & Equipment-Net figure, composed of Property, Plant & Equipment-Gross, less accumulated depreciation. Almost all fixed assets except land are depreciable. Mineral deposits or reserves and standing timber are considered depletable and are important for mining and petroleum firms. In determining the value of a fixed asset, we must remember that their economic life is not unlimited, that eventually, they will wear out or otherwise prove economically useless in their present employment. The accounting reports allow for the loss of value on fixed assets over time by setting up a reserve for depreciation, allowance for depreciation, or accumulated depreciation account. Every fiscal period a previously determined amount is set up as the current charge for depreciation and is subtracted as an expense on the income statement. The matching credit is placed in the allowance for depreciation account, where it accumulates along with the entries from previous periods until (1) the allowance for depreciation equals the depreciable value (original cost less estimated scrap value) of the asset or (2) the asset is sold, lost, or destroyed. On the balance sheet, the allowance for depreciation constitutes a valuation account or reserve; the historically accumulated depreciation is subtracted from the original acquisition cost of the fixed assets, and the balance, called net fixed assets, is added into the sum of the total assets.

The problem of making adequate allowance for depreciation and determining the periodic depreciation charge has caused considerable difficulty for accountants. A commonly used depreciation method presented to the public on the reported books is the "straight-line" method. This technique is quite simple (accounting, among other reasons, for its popularity). The probable useful life of the asset is estimated; the estimated scrap value of the asset is deducted from its original cost in order to obtain its depreciable value; the depreciable value divided by the estimated life gives the yearly depreciation charge. The depreciation charge deducted as an expense remains the same year after year although the net book value of the asset is constantly reduced. For example, assume a warehouse has an estimated life of 20 years, an original cost of $225,000,000, and a $25,000,000 estimated scrap value. The yearly charge will be $10,000,000 per annum or 5 percent of its depreciable value. At the end of ten years, the accumulated allowance for depreciation will be $100,000,000 and the net asset value of the warehouse will be carried at $125,000,000.

Although the straight-line method is popular for public accounting reports, a constant depreciation charge does not reflect the fact that for most fixed assets the loss in economic value is higher in the earlier periods of use. Because of this, the Bureau of Internal Revenue now allows firms to adopt alternative, accelerated depreciation policies (such as the Accelerated Cost Recovery System, ACRS), instituted in 1981 and modified by the Tax Reform Act of 1986. Most industrial equipment is classified as having a seven-year life in ACRS, which allows firms to depreciate annually 14.29%, 24.49%, 17.49%, 12.49%, 8.93%, 8.93%, 8.93%, and 4.45%, respectively, in years one through eight. The great tax advantage of the accelerated depreciation is that it allows the company to defer some of its income tax liabilities to the future, thus providing a greater current after-tax cash flow. Accelerated depreciation is a significant tax saving in present value terms. Currently, most firms fudge their reporting; they use accelerated depreciation for tax reporting and straight line on the reports presented to the public. The difference is captured in an account labeled deferred income taxes which represents the difference between the taxes paid under accelerated depreciation and what the tax would have been if straight depreciation had been used. Presumably, sometime in the future when the accelerated depreciation runs out, the tax charge will rise. However, given the many possible tax and economic changes, the amount of future taxes is uncertain.

Actually reporting would be better, closer to the economic truth, and nothing would be lost if companies used accelerated depreciation on their public books. The sophisticated investor would understand the compensation of increased after-tax cash flow for the decrease in current reported book earnings.

Depreciation allowances are generally based on the original acquisition cost of the fixed asset to the firm. Any subsequent change in the value of the fixed asset—for example, through price level changes—is generally not reflected in the value of the asset on the books nor in the allowable depreciation rate. The depreciation rate is set by the original purchase price and is not changed for the new price level that may exist currently.[4] Thus even if the funds released to the business by its depreciation allowances were actually segregated (which they are not) for the replacement of fixed assets, they would not prove adequate if the replacement or reproduction cost

---

[4]This constitutes "historical cost depreciation" in contrast to, say, "replacement cost depreciation," where allowance might be made for changes in current costs.

of these assets had gone up in the meantime.[5] Many authorities have argued for changes in the basic accounting system that would adjust depreciation rates to movements of the general price level. While there is much in favor of such a reform, many practical difficulties stand in the way. Moreover, many accountants and public finance experts argue that the accountants' job is to measure and keep track of explicit costs and explicit legal claims as best they can, to introduce cost changes that do not reflect an actual transaction might put all accounting on an estimate, guess, and valuation basis.

In any case, much of the argument for current cost depreciation is deflected when accelerated depreciation is used. Because a large part of the depreciation is taken "up front," future changes in the value of the asset are of much less importance.

Akin to the allowance for depreciation is the account called allowance for depletion which appears on the books of mining or extractive companies and other companies, such as lumber firms, engaged in processing natural resources. The accumulated depletion account represents the proportionate cost of the amount of ore, crude oil, etc., which has been removed since the company started operation. It is subtracted on the balance sheet from the original acquisition costs of the company's estimated mineral reserves or resources. For income tax purposes, however, most companies take a percentage depletion allowance. (The rate varies for different types of minerals.) The allowable percentage is applied to the market value of the ore or crude oil and is subtracted from income before computing taxes. Under the law, annual percentage depletion can continue to be taken even if the accumulated depletion already equals the original cost of the oil or mineral reserves.

Other Assets Other assets consist of items such as permanent investments and the so-called intangible assets, i.e., goodwill, franchises, trademarks, patents, and copyrights, and deferred charges.

3.3 Liabilities and Stockholder Equity

The liabilities and stockholder equity section of the balance sheet shows the claims of owners and creditors against the asset values of the business. It presents the various sources from which the firm obtained the funds to purchase its assets and thereby conduct its business. The liabilities represent the claims of people who have lent money or extended credit to the firm; the ownership, capital, net worth, or equity accounts (these terms are interchangeable) represent the investment of the owners in the business.

---

[5]On the other hand, a fall in price levels would enable a firm to increase its physical capacity through the reinvestment of depreciation allowances.

This, the credit side of the balance sheet, is often called the financial section of the balance sheet or the firm's financial structure. It is especially important to the student of finance. Many of the items found here will be discussed briefly since they will be taken up in considerably more detail in other parts of this book.

Current Liabilities The current liabilities are those liabilities, claims, or debts that fall due within the year. Among the more common current liabilities are accounts payable, representing creditors' claims for goods or services, and notes payable or trade acceptances payable, arising out of similar economic transactions. Owed to bankers, obligations to bankers, notes payable to the bank, bank loans payable, or similar accounts show the amounts owing to banks for money borrowed. Usually, these arise from short-term loans, but the amounts due within the year on installment or term loans are also a current liability. Similarly, any portion of the long-term debt, i.e., bonds, mortgages, etc., maturing during the year is also carried in the current liability section.

Accruals, a common group of current liabilities, represent claims that have built up but are not yet due, such as accrued wages, interest payable, and accrued taxes.[6] An item that bulks large for many corporations today is the amount owing on the federal and state corporation profits taxes. It appears as accrued income tax, provision for federal income tax, or other similar titles.

Dividends on the common or preferred stock that have been declared but have not yet been paid are carried among the current liabilities as dividends payable.

The relation of current liabilities to current assets is useful in many types of financial analysis and is especially important in analyzing the short-run credit position of the firm. Thus, the current liabilities are divided into the current assets to obtain the current ratio, and the current liabilities are subtracted from current assets to obtain the firm's net working capital. The larger the current ratio and the larger the net working capital relative to its total operations, the greater is the comparative safety of the firm's short-run financial position. The methods commonly used to judge the safety of the current liability coverage or the adequacy of net working capital vary with the type of firm and industry and with the judgment and analytical ability of the analyst. This subject will be discussed more thoroughly in Chap. 5.

---

[6]Sloan (1996) and Richardson et al. (2001) have shown that accrual's provide information about earnings' quality. Inventory accruals, the primary source of asset accruals, and accounts payables, the primary source of liability accruals, convey information. Richardson et al. (2001) showed that the setting of accruals, asset accruals less liability accruals, is an important source of financial information. Nondiscretionary accruals, accruals associated with sales growth and the level of firm operating activities, conveys information regarding earnings quality. Accrual information influences information attributable to efficiency. Sloan uses FASB 95 to define accruals as the difference between net income and cash flows from operating activities. Sloan (1996) found that stocks with extreme accruals have less persistent earnings and experience mean reversion in next year's earnings.

Long-Term Debt Under the classification of long-term debt, fixed liabilities or funded debt is placed the amount the corporation owes on bond issues, mortgage notes, debentures, borrowings from insurance companies, or term loans from banks. The company may have obtained funds to acquire assets and invest in the business from these sources, and this section of the balance sheet shows the amounts still owing.

There are generally three distinctions between long-term debt and current liabilities. First, the items making up the long-term debt are usually more "formal" than those in the current liability section. A written legal contract or indenture describes the obligation, contains provisions for repayment under different circumstances, and details various devices for protecting the creditors against default and contains other clauses or provisions which might work to the benefit of the debtor company. The long-term debt is also often composed of "securities," or printed certificates issued by the corporation standing for evidence of the ownership of the debt which may be freely traded or negotiated. The second important distinction is that the long-term debt will not mature for at least a year and usually for some time longer than that. Moreover, the current liabilities are generally composed of recurring items, whereas the long-term obligations are incurred only on occasion. Third, the majority of long-term obligations carry some interest charge, whereas most current liabilities do not.

Deferred Credits Somewhere between the liability and equity section of the balance sheet, we often find a category headed deferred credits or perhaps deferred, prepaid, or unearned income. These show a source of funds or assets for which the firm has not as yet performed any service. For example, suppose a company received a cash prepayment for a job on which work is not yet completed. The deferred credit classification does not mean that the firm owes money for this payment but that it owes for completion of the project. Furthermore, if the firm has made this contract on a normal basis, some part of the prepayment will not be covered by services or goods but will revert to the firm as profit. As the contract progresses, the accountants will normally analyze the results to date and apply a proportionate part of the prepayment to expenses, another part to profits (if any), and last leave (among the deferred credits) only that proportion which represents the uncompleted part of the contract.

Among the deferred credits are any rent payments made in advance by the company's tenants. Quite often, an unamortized bond premium may appear. If the company at one time issued bonds bearing a coupon interest rate above that of the going market rate at the time, the bonds would have sold on the market at some premium above their face or stated value. This premium is in a sense, then, compensation to the company for their generous interest rate. The premium is amortized over time as a reduction against the stated interest payment, i.e., it becomes an offset against an expense item. The unamortized portion is carried as a deferred credit since it is a sort of partial prepayment of the company's obligation to pay interest.[7]

Provisions for risks and charges and deferred taxes also are in total liabilities.

Common Equity Common stock, capital surplus, revaluation reserves, retained earnings, and ESOP guarantees constitute the majority of common equity.[8] Those terms and other variants are used interchangeably; they mean approximately the same thing, and the student should learn to identify these terms so that he will not be confused if one or the other is used. This section of the balance sheet contains the items making up the ownership claims against the business. It represents the original investments of the owners plus any earnings they have retained in the business or less any accumulated losses the business may have suffered. Retained earnings change every year by the net income of the firm, less the dividend payment. The firm's retained earnings are the primary source of the firm's cash flow and growth. Common equity is often referred to as the capital section of the balance sheet [Schwartz, 1962].

Preferred Stock The amount shown as preferred stock represents the par or stated value of the various types of preferred stock issued, sold, and outstanding. The class of preferred stock is usually identified by its stated yearly dividends, i.e., the $7 p'f'd., the $5 p'f'd., or perhaps in percentage terms the 41/2 percent or 5 percent. Although the creditors may classify or consider the preferred issues simply as another form of equity, these shares have a prior claim on dividends and usually in case of dissolution have a claim prior to the common stockholders on the assets. They therefore, from the viewpoint of the common shareholders, take on some of the aspects of a creditor claim.

Common Stock The common stock account shows the par value or stated value of the common stock issued, sold, or outstanding. It is often said that this account represents the amount that the stockholders originally put into the business, but this is not likely to be literally true and should be modified by our historical knowledge of the firm's financial affairs. In one sense the capital stock account may represent more than the "original" investment, since common stock may have been issued and sold periodically on the primary market as the firm raised funds to expand and to improve its equity base. In another sense, the capital stock account may represent less than the original investment, for if in time the value of the firm's stock went over par, or over its stated value, and new issues were sold at a higher price, the difference is classified as capital or paid-in surplus. However, the new purchasing stockholders, at least, might well consider this amount part of their original investment.

---

[7]This matter is taken up again in Chap. 9 on Bonds.

[8]They may be called the proprietorship account in a single proprietorship or the partner's equity in a partnership.

Capital Surplus The capital surplus accounts, paid-in surplus, donated surplus, premium on capital stock, or perhaps investment in excess of par value of capital stock represent funds or assets given to the business on behalf of the ownership interests. These funds or assets do not arise out of the "normal" operations, of the firm but out of certain financial transactions. For example, a capital surplus would arise if someone, perhaps but not necessarily a stockholder, were to donate the firm some assets without asking for stock or other legal obligations in return. Most commonly capital surplus arises when a firm floats an issue of its stock at more than par value. After a company has been in operation for a time, the market value of the stock is more likely than not higher than the par value. The amount the company obtains in excess of the par value is classified as paid in surplus or premium on stock issued, etc. A premium on the issue price of either preferred or common stock is considered capital surplus.

Retained Earnings Retained earnings or earnings surplus shows the amount that the firm has reinvested in the business out of earnings that could otherwise have been paid out in common stock dividends. These surpluses differ from capital surplus in that it arises out of accumulated retained earnings and not out of financial trans-actions.[9] If the firm's operations over time show accumulated losses rather than earnings there is, of course, no earned surplus account but an accumulated deficit, which is subtracted from the other capital accounts on the balance sheet. (One year's unsuccessful operation may not create a deficit on the balance sheet, since the losses of the current period may be more than covered by previously accumulated surplus.) The earned surplus accounts are basically derived from this equation: earnings minus losses minus dividends equal earned surplus. For the account to be negative, accumulated losses and dividends over time have to exceed the amounts earned.[10] The retained earnings-to-total-assets ratio is a component of the Altman Z bank-ruptcy prediction model shown in Chap. 5.

Classified among the earned surplus accounts are the so-called appropriated surplus accounts, or surplus reserves. These accounts bear such titles as reserve for contingencies, reserve for plant expansion, or reserve for sinking fund. They are a source of much confusion; they do not represent cash or particular asset any more than the earned surplus account out of which they are derived represents cash or any distinguishable particular asset. These reserves are set up on the books by debiting earned surplus and crediting the reserve; i.e., the unappropriated earned surplus account is reduced, and a surplus reserve account is created by an equal amount. The purpose of these accounts is to warn the reader of the balance sheet that certain events are likely to occur that will reduce either the liquidity of the firm or both the

---

[9] In bank balance sheets, the equivalent to the unappropriated earned surplus account is entitled undivided profits. The bank's surplus account consists of capital surplus plus the retained earnings that are considered permanently committed to the business.

[10] To add to the beginner's confusion, current operating losses are often called operating deficits, or just deficits.

liquidity and the asset holdings of the firm. For example, if the firm is embarked on a program of plant expansion and plans to use some of its present cash or marketable securities to finance the expansion, then the completion of the building program will reduce liquidity but not total assets. If, however, the firm faces a contract renegotiation with the possibility that it may have to relinquish some of its funds, the firm may lose cash without obtaining any other asset to offset it. If the management of the firm has plans for or is worried about losing some of its liquid assets, obviously it cannot distribute them as cash dividends. This is what the appropriated surplus accounts are supposed to show.

The question arises whether the existence of the appropriated surplus accounts means that the rest of the surplus—the unappropriated earned surplus—is available, for dividends. Unfortunately, the answer is no. The unappropriated earned surplus may have accumulated over the years, and the funds it represents may have long since gone into physical plant, inventory, or other operating assets. Although legally possible, it may not be financially possible to payout any large percentage of the "free" surplus without crippling the operations of the company. As long as the company maintains the level of its operations, for all practical purposes, the earned surplus may be as permanently committed to the business as the rest of the capital accounts. In fact, cash is the source of dividends or stock buy backs, not the surplus account.

3.4 Book Value of Common Stock

The book value of the common stock is not directly indicated on the balance sheet, but it is readily derived from the balance sheet data. The book value is the net asset value of a share of common stock as presented by accounting convention on the balance sheet. To obtain this figure, we subtract the total liabilities from total assets shown on the balance sheet, subtract the voluntary liquidation value of the preferred stock plus any accumulated dividends, and divide the remainder by the number of common shares outstanding. Alternately, the book value per share equals the stated or par value of the common shares issued and outstanding, plus all the capital surplus, earned surplus, and surplus reserve accounts, less any liquidation premium or accrued dividends on the preferred shares, divided by the number of common shares outstanding. The book value of common stock did not warrant much analysis in the original Graham and Dodd (1934), but is a major component in the stock valuation analysis of Fama and French (1992) and Chan et al. (1991).

Preferred stock is not included in book value either as a sum or as part of the divisor. If the term book value is used, it is usually understood as referring to the book value of the common stock, since the concept of book value of the preferred is not important or useful. Except in rare instances, the preferred stockholders are not conceived of as having any ownership or interest in the surplus accounts; it is the common shares pro-rata equity in the surplus account which lends meaning to the concept of book value.

Calculating and understanding the concept of book value is not difficult; it shows the net equity per share of the common stock. The book value of a share of stock, however, reflects only the information, which is given formally by the accounting data on the balance sheet. Although book value has some significance in indicating the worth of a share, it cannot give the earning power per share of stock, its market value, its value for control, or its probable future value. Book value is only one of many financial benchmarks.

If a company owns a minor part of another firm, it can be represented by a Minority Interest item. The book value of the subsidiary company's minority stock (i.e., those shares of stock the parent company does not own) will be placed on the balance sheet midway between the liability and capital sections, since this account, usually entitled interest of minority shareholders in consolidated subsidiaries, is somewhat of a hybrid.

The reader has been introduced to the component accounts of the firm's assets, total liabilities, and equity capital. A simple numerical example illustrates the accounts in a balance sheet.

Current Assets		**Current Liabilities**	
Cash	25	Accounts Payable	115
Marketable Securities	75	Short-Term Debt	110
Inventory	150	Total Current Liabilities	175
Total Current Assets	250	Long-Term Debt	525
Net Fixed Assets	750	Stockholder Equity	300
Total Assets	1,000	Total Liabilities and Equity	1,000

The simple firm has assets of $1000 (or $100 MM, if the reader prefers to think of larger numbers). The firm has current assets, cash, marketable securities, and inventory, of $250. The firm's current assets exceed its current liabilities, such as the immediate obligations of the firm could be paid, if payment would be requested, and its current assets, including inventory, could be liquidated at its book value. The bulk of the firm's assets are its net fixed assets, property, plant, and equipment, net of accumulated depreciation. One can think of the firm's fixed assets as including the factory where products are produced. The life of the fixed assets greatly exceeds the life of current assets, which is one year. Most firms with substantial fixed assets, as a percent of total assets, will seek to finance these assets by issuing long-term liabilities. The matching of asset and liability lives is often extremely important, particularly if the firm's creditors demand immediate payment. Management of the simple firm has issued debt of $525, and contributed (only) $300 of equity. That is, the firm has secured financing of its longer-lived assets by issuing debt (bonds, with a fixed maturity, or repayment date) than by issuing stock. Management maintains control of the firm's assets and profitability, as long as interest is paid on its debt and debt principle can be repaid. The concept of debt and equity financing is addressed in Chap. 10. In the world of business, the management of many firms prefer to use debt financing, rather than equity financing, to pay for expansion of the firms' assets. Rates of return on equity investment are normally higher when the firm finances its asset growth using debt capital.

There are over 25 million firms in the US economy, as the reader will remember from Chap. 2. At the time of our first edition, we urged the reader to find balance sheet information from many sources in the business world. Public libraries normally carry the Standard & Poor's *Stock Market Guide* or its *Stock Reports*, the Value Line *Investment Survey*, or Mergent's (formerly Moody's) *Industrial Manual*, which contain balance sheet information on as many as 3000 firms. Now, one can access this data on the internet, at MSN/Money. In order that the reader may follow the discussion more readily, this chapter illustrates the debt and equity financing of assets with the presentation of the balance sheets of three large corporations: International Business Machines (IBM), one of the world's largest computer and software firms; Boeing (BA), a leading aircraft manufacturer and defense contractor; and Dominion Energy (D), the holding company for Virginia Electric and Power Company, VEPCO, a utility firm. IBM is predominately a service and software provider, not the hardware firm of the first edition. Software and services account for the majority of IBM operating profits. VEPCO, an electric utility firm, provides electric power for Virginia and, to a lesser area, North Carolina. We show balances sheets for these firms in Tables 3.1, 3.2, and 3.3, respectively, for the 2017–2020 period, as they appear on MSN.

The reader sees four years of IBM balance sheet financial data on MSN/Money in Table 3.1, shown in millions of dollars. IBM is a firm with current assets that are approximately 25% ($39,165 /$ 155971 $= 25.1\%$) of its total assets in 2020. Goodwill and intangible assets, $73,413, are IBM's largest assets in 2020, most of which is the Goodwill associated with its acquisition of Red Hat in 2019 for approximately $34 billion. Goodwill and intangible assets are 47.1% of IBM's assets in 2020. Net property, plant, and equipment is only $14,726. IBM's liabilities and stockholder equity in Table 3.1 may appear highly unusual; that is, if total liabilities are $135,244 and retained earnings are $162,717, then how can total liabilities and equity total $155,971? The online financial snapshots are correct, but not sufficient for financial analysis. The MSN snapshot does not report Treasury Stock, repurchased equity of the firm, which totaled $169,229 in 2020. We will see Treasury Stock on the FactSet balance sheet reported later in this chapter. IBM's equity of $20,598 is only 13.2% of total liabilities and stockholder equity, which is very low and a firm in the US economy or in its industry. IBM is a highly leveraged firm. Furthermore, IBM's total current assets are approximately equal to its current liabilities. We will discuss the role of current asset management and the current ratio in determining the health of the firm in Chap. 5.

The reader sees four years of Boeing (BA) balance sheet financial data on MSN/Money in Table 3.2, shown in millions of dollars. Boeing has 80% ($121,642/ $152,136) of its assets in current assets, primarily inventories, in 2020. Undelivered planes are inventory.

Boeing's total equity is negative in 2020, reflecting its substantial loss in 2020. Again, Treasury Stock is not included in the MSN Snapshot of Boeing. BA's current assets exceed its current liabilities, and its total liabilities are 118.9% ($170,211/ $152,136) of its total assets. Is this possible? Where has all its stockholder equity gone? We will address these questions with the FactSet presentation of the Boeing balance sheet later in the chapter.

Table 3.1 MSN balance sheet: IBM

Balance sheet				
Values in millions (except for per share items)	2017	2018	2019	2020
Period end date	12/31/2017	12/31/2018	12/31/2019	12/31/2020
Statement source	10-K	10-K	10-K	10-K
Currency code	USD	USD	USD	USD
Assets				
Cash, equity, and short-term investments	-	-	-	-
Receivables	31,630	30,563	24,219	19,186
Inventories	1,583	1,682	1,619	1,839
Deferred current assets	-	-	-	-
Total current assets	**49,735**	**49,146**	**38,420**	**39,165**
Net property, plant, and equipment	11,117	10,792	15,006	14,726
Goodwill and other intangible assets	40,530	39,352	73,457	73,413
Investments and advances	10,884	10,080	9,210	7,735
Other noncurrent assets	-	-	-	-
Total noncurrent assets	**75,621**	**74,236**	**113,766**	**116,806**
Total Assets	**125,356**	**123,382**	**152,186**	**155,971**
Liabilities and shareholders' equity				
Payables	10,670	9,604	7,735	8,209
Accrued expenses, current	7,776	7,073	9,143	11,644
Current debt	6,987	10,207	8,797	7,183
Deferred liabilities, current	-	-	-	-
Other current liabilities	-	-	-	-
Total current liabilities	**37,363**	**38,227**	**37,701**	**39,869**
LT debt and capital lease obligation	-	-	-	-
Deferred liabilities, noncurrent	545	3,696	5,230	5,472
Other noncurrent liabilities	29,886	28,925	34,168	35,548
Total liabilities	**107,631**	**106,453**	**131,201**	**135,244**
Capital stock	446	447	448	449
Retained earnings	153,126	159,206	162,954	162,717
Total equity	**17,594**	**16,795**	**20,841**	**20,598**
Total liabilities and equity	**125,356**	**123,382**	**152,186**	**155,971**
Ordinary shares outstanding	922.2	892.5	887.1	892.7

The reader at this point probably is thinking, are all firms using Treasury Stock such that there is no stockholder equity? Let us look at Dominion Energy, formerly Dominion Resources, a utility firm, where we see substantial equity in its capital structure; see Table 3.3.

Dominion's current assets are lower than its current liabilities, but at least its total liabilities are only 72.4% ($69,444/ $95,905) of its total liabilities and stockholder equity. When did stockholder equity start to disappear? It started with stock repurchasing activities in 1987, but stock repurchases flourished during the 1987–2018 time period. One fact should be very clear. Four years of financial data may not provide enough data for stock valuation.

Table 3.2 MSN balance sheet: BA

Balance sheet				
Values in millions (except for per share items)	*2017*	*2018*	*2019*	*2020*
Period end date	12/31/2017	12/31/2018	12/31/2019	12/31/2020
Statement source	10-K	10-K	10-K	10-K
Currency code	USD	USD	USD	USD
Assets				
Cash, equity, and short-term investments	-	-	-	-
Receivables	11,397	14,364	12,471	10,051
Inventories	61,388	62,567	76,622	81,715
Deferred current assets	-	-	-	-
Total current assets	**85,194**	**87,830**	**102,229**	**121,642**
Net property, plant, and equipment	12,672	12,645	13,684	13,072
Goodwill and other intangible assets	8,132	11,269	11,398	10,195
Investments and advances	4,372	3,749	3,472	2,952
Other noncurrent assets	-	-	-	-
Total noncurrent assets	**27,168**	**29,529**	**31,396**	**30,494**
Total assets	**112,362**	**117,359**	**133,625**	**152,136**
Liabilities and shareholders' equity				
Payables	13,587	14,561	17,382	12,971
Accrued expenses, current	11,671	13,051	20,921	22,128
Current debt	1,335	3,190	7,340	1,693
Deferred liabilities, current	-	-	-	-
Other current liabilities	-	-	-	-
Total current liabilities	**74,648**	**81,590**	**97,312**	**87,280**
LT debt and capital lease obligation	-	-	-	-
Deferred liabilities, noncurrent	2,188	1,736	413	1,010
Other noncurrent liabilities	24,031	22,966	24,238	20,031
Total liabilities	**110,649**	**116,949**	**141,925**	**170,211**
Capital stock	5,061	5,061	5,061	5,061
Retained earnings	49,618	55,941	50,644	38,610
Total equity	**1,656**	**339**	**(8,617)**	**(18,316)**
Total liabilities and equity	**112,362**	**117,359**	**133,625**	**152,136**
Ordinary shares outstanding	591.0	567.6	562.9	582.3

We also show a modified set of balance sheet ratios for our three companies in Tables 3.4, 3.5, and 3.6 for selected years during the 1996–2020 period, drawing data from the FactSet database. The longer time period allows the reader and researcher to more adequately assess the firm's current asset management policy, its use of debt, issuance and repurchase decisions of both debt and stock, and its dividend policy, which affects its retained earnings. In Table 3.4, reported in billions of dollars is that IBM has not built up its cash and other current assets during the 1996–2020 time period. IBM's current assets have fallen to approximately 23.8% ($37.058/$15.971) in 2020 from 50.2% ($40.695/$81.132) in 1996. Intangible assets are quite small, about $10 billion, in 2005, but will rise to $73.413 billion, spurred by the Red Hat acquisition previously mentioned. IBM has become a firm of largely intangible assets.

Table 3.3 MSN balance sheet: D

Balance sheet				
Values in millions (except for per share items)	2017	2018	2019	2020
Period end date	12/31/2017	12/31/2018	12/31/2019	12/31/2020
Statement source	10-K	10-K	10-K	10-K
Currency code	USD	USD	USD	USD
Assets				
Cash, equity, and short-term investments	-	-	-	-
Receivables	1,786	2,080	2,425	2,507
Inventories	1,477	1,418	1,616	1,550
Total current assets	**4,334**	**5,161**	**6,096**	**6,886**
Net property, plant, and equipment	7,176	6,461	-	564
Goodwill and other intangible assets	7,090	7,080	8,080	8,146
Investments and advances	6,964	6,560	7,905	10,238
Other noncurrent assets	-	-	-	-
Total noncurrent assets	**72,251**	**72,753**	**97,727**	**89,019**
Total assets	**76,585**	**77,914**	**103,823**	**95,905**
Liabilities and shareholders' equity				
Payables	-	-	-	-
Accrued expenses, current	848	836	1,284	1,187
Current debt	6,376	4,031	3,311	3,057
Total current liabilities	**9,636**	**7,647**	**9,940**	**10,843**
LT debt and capital lease obligation	-	-	-	-
Deferred liabilities, noncurrent	4,523	5,116	6,277	5,953
Other noncurrent liabilities	12,108	11,959	24,575	18,691
Total liabilities	**57,215**	**55,866**	**69,790**	**69,444**
Capital stock	9,865	12,588	23,824	21,258
Retained earnings	7,936	9,219	7,576	4,189
Total equity	**17,142**	**20,107**	**31,994**	**26,117**
Total liabilities and equity	**76,585**	**77,914**	**103,823**	**95,905**
Ordinary shares outstanding	645	681	838	806

IBM's use of leverage dramatically increased during the 1996–2020 time period. Long-term debt rose from 12.25($9.872/$81.132) in 1996 to 38.4% ($57.929/$155.971). Retained earnings rose from $11.189 in 1996 to $162.717 in 2020, but Treasury Stock, stock repurchased, rose from −0.14 billion in 1996 to −169.33 billion in 2020. While the book value of IBM increased, essentially doubling, during the 1996–2020 time period, its tangible book value became substantially negative due to its Goodwill and intangible assets, and its Provisions for Risks and Charges, primarily its unfunded pension liabilities. IBM's balance sheet data reported in Table 3.4 suggests that IBM's equity was virtually unchanged during the 1996–2020 time period while its debt increased substantially.

We see similar issues with Boeing's long-term balance sheet reported in Table 3.5, with FactSet data. Boeing's equity fell dramatically during the 2004–2020 time period while its debt and Provisions for Risks and Charges, unfunded pension liabilities, increased substantially. Treasury Stock increased faster than retained earnings, creating negative book values and tangible book values.

Table 3.4 FactSet IBM balance sheets, selected years

Ticker	IBM-US								
International Business Machines Corporation									
D-US 25746U109 2542049 NYSE Common stock									
Source: FactSet fundamentals									
GAAP/IFRS balance sheet	*12/2020*	*12/2019*	*12/2015*	*12/2009*	*12/2008*	*12/2007*	*12/2005*	*12/2000*	*12/1996*
Assets									
Cash and short-term investments	14.275	9.009	8.194	13.974	12.907	16.146	13.686	3.722	8.137
Accounts receivables, net	18.472	22.486	27.353	25.65	26.383	27.717	23.29	10.447	16.515
Inventories	1.839	1.619	1.551	2.494	2.701	2.664	2.841	4.765	5.87
Total current assets	37.058	36.524	42.504	48.935	49.004	53.177	45.661	43.88	40.695
Net property, plant, and equipment	14.726	15.006	10.727	14.165	14.305	15.081	13.756	16.714	17.407
Total long-term investments	7.886	9.304	11.706	12.418	12.758	13.543	10.817	15.017	12.341
Intangible assets	73.413	73.457	35.508	22.703	21.104	16.392	11.104	1.63	2.502
Other assets	13.647	12.713	5.228	6.606	3.966	20.725	22.578	5.172	8.187
Total assets	**155.971**	**152.186**	**110.495**	**109.022**	**109.524**	**120.431**	**105.748**	**85.381**	**81.132**
Liabilities and shareholders' equity									
Current									
ST debt and curr. portion LT debt	8.54	10.177	6.461	4.168	11.236	12.235	7.216	10.205	12.957
Accounts payable	4.908	4.896	6.028	7.436	7.014	8.054	7.349	8.192	4.767
Total current liabilities	39.869	37.701	34.269	36.002	42.435	44.31	35.152	36.406	34
Long term									
Long-term debt	57.929	57.981	33.428	21.932	22.689	23.039	15.425	18.371	9.872
Provision for risks and charges	27.214	25.419	22.945	21.712	25.197	18.032	15.847	7.713	3.554
Deferred tax liabilities	5.472	5.23	0.253	0.47	0.27	1.064	1.616	1.623	1.627
Other liabilities	4.761	4.87	5.176	6.151	5.468	5.516	4.61	0.644	10.451
Other liabilities (excl. deferred income)	0.46	1.019	1.405	2.589	2.297	2.456	2.173	-0.622	10.451
Total liabilities	135.245	131.201	96.071	86.267	96.059	91.961	72.65	64.757	59.504
Equity									
Common equity	20.597	20.841	14.262	22.637	13.465	28.47	33.098	20.377	21.375
Common stock par/carry value	56.556	55.895	53.262	41.81	39.129	35.188	28.926	0.378	0.636
Retained earnings	162.717	162.956	146.124	80.9	70.353	60.64	44.734	23.784	11.189
Treasury stock	-169.33	-169.41	-155.51	-81.243	-74.171	-63.945	-38.546	-13.8	-0.135
Total liabilities and shareholders' equity	**155.971**	**152.186**	**110.495**	**109.022**	**109.524**	**120.431**	**105.748**	**85.381**	**81.132**
Per share									
Book value per share	23.07	23.49	14.77	17.34	10.06	20.55	21.03	11.56	10.52
Tangible book value per share	-59.17	-59.31	-22.00	-0.05	-5.70	8.72	13.97	10.63	9.29

Boeing's equity was obviously a result of its plane crashes, but Boeing made very liberal use of debt issuance and stock repurchases during the 1996–2020 time period. These policies are clearly reported in Table 3.5 for BA.

The FactSet Dominion Energy, D, balance sheet data is shown in Table 3.6. Dominion Energy has positive book value and tangible book value during the 1996–2020 time period and has not repurchased stock. D's liabilities to total liabilities and stockholder equity ratio rose from 61.4% ($9.157/$14.906) to 72.4% ($69.444/$95.905) in 2020. Yes, Dominion increased its long-term debt and unfunded pension liabilities, but it did so more conservatively than IBM or Boeing.

The greater use of long-term debt may produce greater interest expenses and lower profitability for Dominion Resources, as we will see in Chaps. 4 and 5. Larger debt ratios imply greater risk in our economy, as we will see in Chap. 10. We address the valuation implications of earnings, dividends, and debt policies in Chaps. 8 and 14.

Table 3.5 FactSet Boeing Company balance sheets, selected years

Ticker	BA-US								
Boeing Company									
D-US 25746U109 2542049 NYSE	Common stock								
Source: FactSet Fundamentals									
GAAP/IFRS balance sheet	12/2020	12/2019	12/2018	12/2014	12/2008	12/2007	12/2006	12/2004	12/1999
Assets									
Cash and short-term investments	25.59	10.03	8.564	13.092	3.279	9.308	6.386	3.523	3.454
Accounts receivables, net	10.051	12.471	14.364	7.623	4.087	4.326	4.377	3.686	-
Inventories	81.715	76.622	62.567	46.756	15.612	9.563	8.105	4.247	6.539
Total current assets	121.642	102.229	87.83	67.785	25.964	27.28	22.983	15.1	15.712
Net property, plant, and equipment	13.072	13.684	12.645	11.007	8.762	8.265	7.675	8.443	8.245
Total long-term investments	3.028	3.472	3.749	4.881	7.185	10.888	12.605	13.502	5.369
Intangible assets	10.924	11.398	11.269	7.988	6.332	5.174	4.745	2.903	2.233
Other assets	3.384	2.159	1.582	0.961	1.422	7.182	2.735	13.861	4.588
Total assets	**152.136**	**133.625**	**117.359**	**99.198**	**53.779**	**58.986**	**51.794**	**53.963**	**36.147**
Liabilities and shareholders' equity									
Current									
ST debt and curr. portion LT debt	1.961	7.592	3.19	0.929	0.56	0.762	1.381	1.321	0.752
Accounts payable	12.928	15.553	12.916	10.667	5.871	5.714	5.643	4.563	4.909
Total current liabilities	87.28	97.312	81.59	56.717	30.925	31.538	29.701	20.835	13.656
Long term									
Long-term debt	62.974	20.94	10.657	8.141	6.952	7.455	8.157	10.879	5.98
Provision for risks and charges	18.545	20.816	19.907	23.984	15.705	8.162	8.806	9.128	4.877
Deferred tax liabilities	1.01	0.413	1.736	-	-	1.19	-	1.09	-
Other liabilities	0.402	2.444	3.059	1.566	1.491	1.637	0.391	0.745	0.172
Other liabilities (excl. deferred income)	0.402	2.444	3.059	1.566	1.491	1.637	0.391	0.00	0.172
Total liabilities	170.211	141.925	116.949	90.408	55.073	49.982	47.055	42.677	24.685
Equity									
Common equity	-18.316	-8.617	0.339	8.665	-1.294	9.004	4.739	11.286	11.462
Common stock par/carry value	5.061	5.061	5.061	5.061	5.061	5.061	5.061	5.059	5.059
Retained earnings	38.61	50.644	55.941	36.18	22.675	21.376	18.453	15.565	10.487
Treasury stock	-52.641	-54.914	-52.348	-23.298	-17.758	-14.842	-12.459	-8.81	-4.161
Total liabilities and shareholders' equity	**152.136**	**133.625**	**117.359**	**99.198**	**53.779**	**58.986**	**51.794**	**53.963**	**36.147**
Per share									
Book value per share	-31.45	-15.31	0.60	12.26	-1.78	11.72	6.01	14.23	13.16
Tangible book value per share	-50.21	-35.56	-19.26	0.96	-10.50	4.99	-0.01	10.57	10.60

The data of these three corporations can be found on the Wharton Research Data Services (WRDS) database. The WRDS data is as-reported data, not restated for a particular time, as is often the case with financial data; restated data makes the history of a particular firm consistent over time for events such as a merger. The investor can make rational financial decisions when he uses data known at that moment in time. Events, such as mergers and divestments, can create vastly different pictures of the firm. We show the ratio of current assets-to-total assets (CATA), current liabilities-to-total assets (CLTA), total debt-to-total assets (TDTA), defined as current liabilities plus long-term debt, and stockholder equity-to-total assets (SEQTA) for selected years during the 1963–2020 period for the three firms in

Table 3.6   FactSet Dominion Energy Inc. balance sheets, selected years

Ticker	D-US								
Dominion Energy Inc.									
D-US 25746U109 2542049 NYSE	Common stock								
Source: FactSet Fundamentals									
GAAP/IFRS balance sheet	*12/2020*	*12/2019*	*12/2015*	*12/2009*	*12/2008*	*12/2007*	*12/2005*	*12/2000*	*12/1996*
Assets									
Cash and short-term investments	0.24	0.269	0.862	1.176	1.563	1.044	3.575	0.635	0.1272
Accounts receivables, net	2.295	2.278	1.2	2.05	2.354	2.13	3.335	-	-
Inventories	1.55	1.742	1.348	1.185	1.166	1.045	1.167	0.327	0.2255
Total current assets	6.886	6.088	4.191	6.817	7.661	6.656	10.129	5.866	1.3205
Net property, plant, and equipment	58.412	69.581	41.554	25.592	23.274	21.352	28.94	14.849	10.5094
Total long-term investments	10.568	8.242	5.774	3.492	3.257	4.068	5.416	2.602	1.9351
Intangible assets	8.146	9.737	3.864	4.047	4.215	4.094	4.917	3.502	-
Other assets	11.893	10.175	3.414	2.606	3.646	2.953	3.258	2.106	1.2309
Total assets	**95.905**	**103.823**	**58.797**	**42.554**	**42.053**	**39.123**	**52.66**	**29.166**	**14.9056**
Liabilities and shareholders' equity									
Current									
ST debt and curr. portion LT debt	3.143	4.161	5.335	2.432	2.474	3.234	3.948	3.573	1.1289
Accounts payable	0.944	1.115	0.726	1.401	1.499	1.734	2.756	1.736	0.4106
Total current liabilities	10.843	9.939	8.12	6.833	7.794	7.746	14.48	7.592	1.9755
Long term									
Long-term debt	34.473	34.266	23.616	15.481	14.956	13.235	14.653	10.101	4.7276
Provision for risks and charges	7.11	7.232	3.086	2.865	3.327	1.722	2.249	0	-
Deferred tax liabilities	5.953	6.277	7.414	4.244	4.137	4.253	4.984	2.879	1.9083
Other liabilities	11.065	12.076	2.959	1.689	1.505	2.476	5.64	0.707	0.5458
Total liabilities	69.444	69.79	45.195	31.112	31.719	29.432	42.006	21.279	9.1572
Equity									
Common equity	23.73	29.607	12.664	11.185	10.077	9.406	10.397	6.992	4.9244
Common stock par/carry value	21.258	23.824	6.68	6.525	5.994	5.733	11.286	5.979	3.4714
Retained earnings	4.189	7.576	6.458	4.686	4.17	3.51	1.55	1.028	1.4379
Treasury stock	0	0	0	0	0	0	0	0	0
Total liabilities and shareholders' equity	**95.905**	**103.823**	**58.797**	**42.554**	**42.053**	**39.123**	**52.66**	**29.166**	**14.9056**
Per share									
Book value per share	29.44	35.33	21.25	18.67	17.28	16.30	14.98	-	-
Tangible book value per share	19.33	23.71	14.77	11.92	10.05	9.21	7.90	-	-

Tables 3.7, 3.8, and 3.9.[11] The reader sees the vast differences among IBM, Boeing, and Dominion Energy with respect to their respective capital structures. The stockholder equity-to-total-assets ratios are very different for the three firms; Dominion Energy has a larger stockholder equity-to-total-assets ratio relative to the corresponding IBM and Boeing ratios during much of the period. IBM and Boeing are far more debt-intensive firms. In 2020, Boeing plunged into a negative position on its stockholder equity-to-total-assets ratio, which was always very low compared to IBM and Dominion Energy, and all WRDS firms, as shown in Table 3.10. It is interesting that the three firms had very similar stockholder equity-to-total-assets ratios in 2004 and 2007. However, post-2007 and the Global Finance Crisis,

---

[11] Because we use current liabilities and long-term debt in the TDTA calculation, and not total liabilities, the TDTA and SEQTA may not sum to 100%. Current liabilities and long-term debt represent the vast majority of total liabilities.

Table 3.7 IBM
WRDS Database, selected years, 1963–2020

Year	CATA	CLTA	TDTA	SEQTA
1963	47.8%	15.0%	32.9%	67.1%
1970	39.9	21.9	28.7	69.6
1980	37.2	24.5	32.3	61.7
1985	49.6	21.7	29.2	60.8
1990	44.4	28.8	42.5	48.9
1995	50.6	39.4	52.0	27.8
2000	49.9	41.2	62.0	23.1
2004	43.0	36.5	50.0	27.3
2007	44.2	36.8	56.9	23.5
2010	42.4	35.8	55.1	20.3
2015	38.5	31.0	61.3	12.9
2017	39.7	29.8	61.6	14.0
2020	25.1	25.6	62.7	13.0

Where current assets-to-total assets (CATA):
Current liabilities-to-total assets (CLTA)
Total debt-to-total assets (TDTA)
And stockholder equity-to-total assets (SEQTA)

Table 3.8 Boeing Company
WRDS Database, selected years, 1963–2020

Year	CATA	CLTA	TDTA	SEQTA
1963	78.5%	43.3%	60.0%	41.8%
1970	67.5	42.4	66.2	30.9
1980	69.1	52.7	54.0	39.0
1985	73.2	47.8	47.9	47.2
1990	60.1	48.8	51.0	47.8
1995	59.6	33.6	44.2	44.8
2000	37.7	43.5	61.5	26.2
2004	27.9	38.6	58.8	20.9
2007	46.2	53.5	66.1	15.3
2010	59.2	51.6	68.4	4.0
2015	72.3	53.4	62.0	6.7
2017	70.6	60.9	71.5	0.4
2020	80.0	57.4	98.8	-12.0

Where current assets-to-total assets (CATA):
Current liabilities-to-total assets (CLTA)
Total debt-to-total assets (TDTA)
And stockholder equity-to-total assets (SEQTA)

Table 3.9 Dominion Resources
WRDS Database, selected years, 1963–2020

Year	CATA	CLTA	TDTA	SEQTA
1963	4.9	4.8	52.0	31.9
1970	4.0	6.0	56.9	31.5
1980	7.9	8.5	53.0	28.7
1985	8.5	12.8	46.5	31.6
1990	7.1	8.5	48.5	33.0
1995	7.9	10.0	44.2	34.1
2000	20.0	25.9	61.6	23.8
2004	15.7	17.7	51.8	25.1
2007	17.0	19.8	53.6	24.7
2010	12.6	13.5	50.3	28.6
2015	7.1	13.8	54.0	21.5
2017	5.6	12.5	53.0	22.4
2020	7.2	11.3	47.3	27.2

Where current assets-to-total assets (CATA):
Current liabilities-to-total assets (CLTA)
Total debt-to-total assets (TDTA)
And stockholder equity-to-total assets (SEQTA)

Table 3.10 All Firms
WRDS Database, selected years, 1963–2020

Year	CATA	CLTA	TDTA	SEQTA
1963	57.1	19.2	38.8	51.7
1970	55.3	21.7	45.7	48.6
1980	56.9	25.9	49.7	41.6
1985	54.7	24.5	46.8	41.5
1990	51.4	24.9	48.4	38.0
1995	53.3	24.2	43.1	35.6
2000	48.2	23.1	44.7	35.6
2005	48.9	22.0	40.6	39.2
2010	46.2	19.9	38.1	40.7
2015	42.0	19.2	42.5	35.9
2020	42.8	17.2	43.5	37.0

Where current assets-to-total assets (CATA):
Current liabilities-to-total assets (CLTA)
Total debt-to-total assets (TDTA)
And stockholder equity-to-total assets (SEQTA)

Dominion has issued far less debt than Boeing and IBM. The stockholder equity-to-total-assets ratio for Dominion Energy is far greater than the corresponding ratio post-2007. Dominion Resources has used far more long-term debt, relative to its current liabilities, than IBM or Boeing, to finance its assets.

3.5 Summary and Conclusions

The corporate balance sheet allows the reader and investor to view the firm at a particular point in time. The reader sees the composition of the firm's capital structure, as well as the composition of current and fixed assets. Management's decisions on capital investment, dividend policy, and reliance on external financing can alter the presentation of corporate balance sheets and several financial ratios that many investors calculate to assess the firm's financial health over time. The balance sheet numbers are hard to analyze without using them in the returns on assets, sales, and net income, in Chap. 4, or in the Altman Z bankruptcy ratios, in Chap. 5.

References

Bernstein, L. (1988). *Financial statement analysis: Theory, application, and interpretation* (4th ed.). Irwin, Chapters 4, 5, 6, 7, and 8.

Brealey, R. A., & Myers, S. C. (2003). *Principles of corporate finance* (7th ed.). McGraw-Hill/Irwin. Chapter 31.

Chan, L. K. C., Hamao, Y., & Lakonishok, J. (1991). Fundamentals and stock returns in Japan. *Journal of Finance, 46*, 1739–1764.

Chan, L. K. C., Lakonishok, J., & Hamao, Y. (1993). Can fundamentals predict Japanese stock returns. *Financial Analysts Journal, 49*, 63–69.

Fama, E. F., & French, K. R. (1992). Cross-sectional variation in expected stock returns. *Journal of Finance, 47*, 427–465.

Fama, E. F., & French, K. R. (1995). Size and the book-to-market factors in earnings and returns. *Journal of Finance, 50*, 131–155.

Graham, B., & Dodd, D. (1934). *Security analysis: Principles and technique* (1st ed.). McGraw-Hill Book Company.

Graham, B., Dodd, D., & Cottle, S. (1962). *Security analysis: Principles and technique* (4th ed.). McGraw-Hill Book Company.

Horngren. (1984). *Introduction to financial accounting* (2nd ed.). Prentice-Hall. Chapter 7.

Lakonishok, J., Shliefer, A., & Vishay, R. W. (1994). Contrarian investment, extrapolation and risk. *Journal of Finance, 49*, 1541–1578.

Palepu, K. G., Healy, P. M., & Bernard, V. L. (2000). *Business analysis & valuation* (2nd ed.). South-Western College Publishing. Chapters 4, 5, 8.

Penman, S. H. (2001). *Financial statement analysis and security analysis*. McGraw-Hill/Irwin. Chapter 7.

Richardson, S. A., Sloan, R. G., Soliman, M. T., & Tuna, I. (2001). *Information in accruals about the quality of earnings*. University of Michigan Working Paper.

Schwartz, E. (1962). *Corporate finance*. St. Martin's Press. Chapter 3.

Sloan, R. G. (1996). Do stock prices reflect information in accruals and cash flows about future earnings? *The Accounting Review, 71*, 289–315.

Standard & Poor's Stock Guide, *May* 2004.

Sunder, S. (1973). Relationship between accounting changes and stock prices: Problems of measurement and some empirical evidence. *Journal of Accounting Research, 11*, 1–45.

Weston, J. F., & Copeland, T. E. (1986). *Managerial finance* (8th ed.). The Dryden Press. Chapters 2, 11, 26.

White, G. I., Sondhi, A. C., & Fried, D. (1998). *The analysis and use of financial statements* (2nd ed.). Wiley, Chapters 2, 4.

Chapter 4
The Annual Operating Statements: The Income Statement and Cash Flow Statement

The balance sheet of a company and the income statement are related. The balance sheet is an accounting snapshot at a point in time. The income, profit and loss, or operating statement is a condensation of the firm's operating experiences over a given period of time.[1] It depicts certain changes that have occurred between the last balance sheet and the present one. The balance sheet (position statement) and the income statement may be reconciled through the retained earnings, or earned surplus, account. If this reconciliation is presented formally, it becomes the surplus statement.

The income statement is very important. The balance sheet depicts how much assets historically have been invested in a firm; the operating statement indicates how successful (whether by efficiency, daring, or chance) the company has been in making a return on the assets committed to it. In this volume, we regard net income, the bottom line of the income statement, as the most important result of a company's operations. The net income, and the resulting earnings per share, calculated as net income divided by the number of common stock shares outstanding, is the statistics that analysts seek to forecast and use in stock valuation models of the firm.

---

[1] Operating statements for internal control can be made up for any feasible time period. Most large firms present quarterly statements for their investors, although the annual results carry the most weight and go into the record books of the financial services, such as *Mergent*'s (formerly *Moody*'s), *Standard & Poor's*, and others, and are available online at MSN/Money, as we will illustrate in this chapter.

J. B. Guerard Jr. et al., *Quantitative Corporate Finance*,
https://doi.org/10.1007/978-3-030-87269-4_4

4.1   Form and Content of the Income Statement

The operating statement takes different forms, depending on the intended audience. The detailed breakdown made for the operating management is usually not presented in the annual report to the general stockholders. Furthermore, the format and the order of items on the report differ according to the tastes and traditions of the management of different firms.

Financial services that gather data on corporations for the investment community use a similar format for all firms to make it easy to compare companies. The form used by the services breaks out most of the important variables that are interesting for investment analysis.

The following discussion explains the major items appearing in the suggested statement and gives something of their significance.

1. *Sales or revenues.* The sales account shows the total gross revenue received by the firm during the period. It includes sales for cash and for credit, whether or not they were collected at the end of the period. The sales figure should be net of allowances made to the buyers for spoiled or poor quality goods or returned shipments.

2. *Direct operating cost or the cost of goods sold.* The direct operating costs, or more commonly called the cost of goods sold, are the amounts spent on material, labor on the goods sold, plus the other costs expended during the period such as selling, administrative, and advertising costs. Items such as the expense for fuel, power, light, local taxes other than those on income, and telephone charges are included in this figure. Although under certain circumstances it is helpful to have separate figures on the cost of goods sold and other costs, this breakdown is generally interesting only to the operating management. The direct operating costs do not, in any case, include such noncash charges as the depreciation of fixed assets, depletion, or the amortization of franchises or patents.

3. *Regular nonoperating income.* This account includes interest income, dividends on investments, and similar items. Income from major subsidiaries should be consolidated on the reported income statement, even if this is not done for tax purposes. Thus, this account does not include the dividends from dominated subsidiary companies. Irregular income, such as that which might occur from the sale of an operating asset at a profit, is presented near the foot of the statement after the results of regular operations are reported.

4. *Earnings before depreciation, interest, and taxes.* This figure represents the gross return on the company's operation. It is the amount by which the revenues exceed the "variable" costs or costs of goods sold. Out of this sum come the funds to satisfy various claimants to a return from the firm and internal funds that can be used to reduce debt or buy new assets according to the company's position after fixed claims are met and the distribution to the owners is decided.

5. *Depreciation (and other noncash charges).* Noncash charges are such items as depletion, depreciation, or amortization of franchises, patents, etc. (these

charges were discussed in Chap. 3). Depreciation is usually by far the most important of these items.

The depreciation account is the estimated capital (i.e., fixed assets) used up during the year. The depreciation charge is based on the cost—not the present value of the assets and the schedule of depreciation charges on an asset once set initially cannot be varied except under special circumstances. The level of depreciation charges is important in setting the amount of corporation profit tax due. It is important in reminding the management that not all the returns coming in are income, some must be considered a return of capital, and dividends policies should be set accordingly.

The annual depreciation charges do not, however, set the amount of fixed assets that will be replaced or new fixed assets purchased. This decision is based on the forecasted future profitability of the replacement or of the new assets, as the reader will see in Chap. 11. If investment in new fixed assets appears to generate an acceptable rate of return, then available internal funds or other sources of funds will be found. If capital expenditures exceed internally generated cash flows, then new debt or equity must be raised to finance the investment. If the new investment in fixed assets does not exceed the amount of depreciation, then the extra funds can be used for something else,[2] such as retiring long-term debt, repurchasing equity, or paying larger dividends.

6. *Earnings before interest and taxes (EBIT).* The EBIT represents the income of the firm after the book charge for depreciation has been made. After this figure is determined, the effect of many past financial decisions comes into play. The amount of interest that will be paid is based on the amount of interest-bearing debt the firm has incurred, and this influences the profits tax. The amount available for the common stockholders is obtained after dividends on the preferred stock are subtracted. These figures are influenced by decisions on alternate methods of financing the firm. The pros and cons of these decisions make up a large part of the subject of corporation finance.

7. *Interest.* This item represents the interest paid by the company on debt. It is reduced by the amortization of any premium on bonds payable and increased by the amortization of bond discounts. Interest expense is deductible before calculating the corporation income tax.

8. *Corporation profits tax.* The current corporate income tax schedule for tax years beginning after December 31, 1992, is as follows:

---

[2]To refer to depreciation charges as a source of funds is the common shorthand usage, which to be frank is somewhat inaccurate. Strictly speaking, only the firm's operations provide funds. If the firm's revenues did not exceed its direct expenses, there would be no funds flow for the period. However, since the depreciation charge is added to retained earnings for the period to obtain the total of reinvested internally generated funds, the custom has grown of referring to depreciation as a source of funds.

Taxable income	Tax rate	Minus $ = $ tax
$0–$50,000	15%	Minus $0 = $ tax
$50,001–$75,000	25%	Minus $5,000 = $ tax
$75,001–$100,000	34%	Minus $11,750 = $ tax
$100,001–$335,300	38%	Minus $16,750 = $ tax
$335,301–$10,000,000	34%	Minus $0 = $ tax
$10,000,001–$15,000,000	35%	Minus $100,000 = $ tax
$15,000,001–$18,333,333	38%	Minus $550,000 = $ tax
> $18,333,334	35%	Minus $0 = $ tax

Source: Small Business, Quickfinder Handbook. www.quickfinder.com

Net income will be substantially higher for firms in the future because the US corporate income tax rate was reduced in H.R. 1, the Tax Cuts and Jobs Act of December 20, 2017, from 35% to 21%, for incomes exceeding ten million dollars and because firms can now deduct all costs of new investment in the year of service, rather than depreciating them over multiple time periods.

9. *Earnings after taxes, or net income.* The earnings after taxes as depicted in our model statement are the amount earned on the total equity of the corporation from regular sources. It is not the dividends paid on the owners' investment nor is it the amount earned on the common stock equity.

10. *Nonrecurring losses or gains (after taxes).* Nonrecurring losses or gains arise out of transactions such as the sale of fixed assets (buildings, land, or equipment), often associated with discontinued operations or the sale of subsidiaries or investments in securities. Losses can also occur because of natural disasters (floods or fires) or because of liabilities on lawsuits. In any case, these gains or losses do not arise out of the normal operations of the business. These items are given separately because they are special or nonrecurring. If a firm sells a plant or subsidiary at a profit, the earnings for a given year are raised, but the earnings generated by the subsidiary will no longer be available in the future. The tax on nonrecurring gains will probably not be at the regular 35 percent rate, for if the transaction is classified as a capital gain the maximum rate is 28 percent. Whether or not nonrecurring losses are fully deductible for tax purposes depends on the circumstances. It is suggested that when nonrecurring gains or losses occur, they be entered net of taxes (a loss would be reduced if there were regular income tax which could be used as an offset) after the regular part of the income report. Details should be provided in footnotes. In our illustrative statement, there are no nonrecurrent items.

11. *Dividends on preferred shares.* The dividends declared on the preferred shares are subtracted from net profits to obtain the earnings available to the common. Although the preferred dividends are not a legally fixed obligation of the firm, they represent a claim senior to any return on common shares, and there can be no calculation of earnings on the common shareholders' investment until they

are accounted for. The preferred dividends are a prior charge from the view of the true residual owners of the firm—the common shareholders. Any accrued past preferred dividends should be cited in a footnote to the statement.[3]

12. *Earnings available to common stock.* This amount represents the accounting profit or earnings accruing to the shareholders of the business after all prior deductions have been made. The term "accounting profits or earnings" is used deliberately. The accounting profits and the "true" or "economic" profits of the firm can differ considerably. (Economic profit is the amount which remains after all imputed costs—such as interest on all the funds employed no matter what the source—are subtracted.) However, the economists differ on their definition of economic profits, and none of these definitions are easily implemented in practice. Thus, the reported accounting profits serve as a useful available measure of the firm's success. Most stock valuation models, such as those presented in Chaps. 8 and 14, rely upon reported earnings. Moreover, under the discipline of the accounting formalities, profits are reported on a sufficiently consistent basis to enable them to be used in the determination of important legal obligations and privileges. But even within the accounting rules, there exist legitimate alternative methods of reporting certain expenses and charges, which can cause considerable variation in the operating results of any year.[4]

13. *Dividends on common stock.* Dividends are shown here as a charge on earnings. In most jurisdictions, however, legally they constitute a charge against surplus, not current earnings, and may be declared as long as there is a sufficient credit balance in the surplus account, even if there are no current profits. Firms may elect to do this. As a practical matter, however, the dividend policies of most firms are conditioned by their current earnings position and not by their retained surplus account, and so from the point of view of functional relationships, the order of accounting presented seems quite correct.[5] Common stock dividends are a voluntary distribution of the profits of the firm and not a legal obligation. Their declaration does not reduce the profits of the firm. Thus, dividends are deducted after the earnings on the common are calculated. Moreover, the profits tax liability of the firm is not affected by either the payment of preferred dividends or common dividends.

14. *Addition to earned surplus for the year.* What is left after all dividends are subtracted from the reported profits are the retained earnings, reinvested earnings, or net addition to surplus, for the year. If expenses exceed revenues, there would be instead an operating loss or deficit for the year. The retained earnings for the period depend on the level of profit and the dividend policy of the

---

[3] The relationship between common and preferred stock is explained more fully in Chap. 8.

[4] See the discussion in Chap. 3 on depreciation and inventory valuation.

[5] For a fuller discussion of dividends and their relationship to earnings and the value of common stock see Chap. 8.

company. These, in turn, are influenced by factors such as the stability and amount of the company's cash flow, the firm's growth prospects, and its need for equity funds either to acquire additional assets or to repay debt.

15. *Earnings per share and cash flow per share.* The amount earned and paid in dividends on the individual stockholder's share is of more direct importance to him than the total amount earned by the firm. The earnings and trend of dividends on the individual shares, in the long run, establish their value in the market.

The earnings per share are obtained by dividing the total earnings available to the common stock by the number of shares of stock outstanding. Often the earnings per share are shown once, reflecting the regular, recurring income, and again, including extraordinary income items. An additional figure, not always available but often useful, is the cash flow per share. It includes earnings available to the common shareholders plus noncash charges divided by the number of shares. This figure shows the gross funds available per share of stock which may be used to repay debt, acquire assets, and pay dividends. An interesting possibility is to subtract required amortization of debt from the cash flow per share and arrive at the figure of "free" or "disposable" cash flow per share. This figure might prove useful in comparing two firms where earnings are similar, but one firm is required to make payments on the principal of its debt.

A simple income statement (in $MM) illustrates its components, as was done for the balance sheet in Chap. 3.

Sales	$722.50
−Cost of Goods Sold	470
−Depreciation	95
Gross Income	157.50
−General, Selling Expenses	72.50
EBIT	85
−Interest Paid	40
Earnings Before Taxes	45
−Taxes	9.45
Earnings After Taxes	35.55

The firm produced sales of $722.5 million by selling its products, and after subtracting its cost of goods sold and depreciation and amortization charges, gross income of $157.5 million is produced.[6] One subtracts general, selling, and administrative expenses from gross income, to create earnings before interest and taxes, EBIT ($85.0 MM for the simple firm). Interest expenses of $40 million must be paid

---

[6]The previous tax rate, pre-2017, produced taxes of approximately $15.8 MM, producing earnings after taxes of $29.2 MM. The increased net income, $6.35 MM, enhances the firm valuation by $63.5 MM, if the firm is valued by analysts with a price-earnings multiple of 10. Christmas came early for the management of many companies, and stockholders, in 2017!

to the firm's bondholders, such that earnings before taxes are $45 million. The firm pays taxes of $9.45 million, leaving earnings after taxes, or net income, of $35.55 million. A net income figure is often used by analysts to value the firm. The net income number by itself may not as useful to an investor unless net income is represented as a percent of sales, total assets, or equity. The ratio of net income relative to its total assets, referred to as a firm's ROA, or return on assets, allows the investor to compare the profitability of the firm relative to all firms or firms within an industry or sector.[7] Comparisons can be made regarding net income relative to a firm's sales, or return on sales, ROS, and equity, referred to as the return on equity, ROE. One prefers to see higher values of ROA, ROS, and ROE. We will calculate these ratios in Chap. 5.

4.2   Retained Earnings vs Dividends

Corporate profits in the year earned are taxed by the federal government at a rate of 21 percent, and any distribution of dividends is taxed as additional income to the recipient at the appropriate personal income tax bracket rate.[8] This is what constitutes the "double taxation of corporation income."

Although the disparity between the corporation profits rate and the personal income tax rate, and the fact that they are applied separately may be economically detrimental to some holders of corporation shares, it may benefit others. For individuals in very high-income brackets, the 21 percent corporation profits tax may be lower than their personal tax rate. If the corporation retains some profits and pays dividends out of these at a later date—for example, at retirement, when the individual's tax bracket may have fallen—there can be a net savings of overall tax. More usual is the use of retained earnings to generate more earnings (and potentially higher dividends) so that, for example, the stock can be sold at a capital gain equivalent to the retained earnings. Retained earnings provide financing for the firm's investment projects, as we will see in Chaps. 8 and 11. Firms pay dividends to provide cash for its investors. Stock valuation is often extremely dependent upon dividends.

---

[7] The return on assets may be expressed as the ratio of net income to total assets, or the ratio of net income to average total assets. Accountants often prefer to express the ROA as earnings before interest and taxes (EBIT) divided by average total assets [White et al., 1997, p. 147]. Alternatively, White, Sondhi, and Fried present the ratio of net income plus after-tax interest cost, relative to average total assets (p.147). These accountants use average sales and stockholder equity in the ROS and ROE calculations. We use the year-end total assets, sales, and stockholder equity in this text.

[8] That is, if the recipient of the income is an individual and not a nontaxable institution or another corporation. An 85 percent dividend received credit is given to corporations for dividends received from another corporation which is nonconsolidated. (To consolidate for tax purposes, the parent must own a minimum of 80 percent of the subsidiary's shares.) The other 15 percent is taxed at the usual corporate income tax rate.

4.3   Income Statement in the World of Business: IBM, Boeing, and Dominion Energy

Income statements can be found for investors and researchers. A source of the most recent 4 years (2017–2020) of income statement data can be found on MSN, Money. We will continue with the examples of IBM, Boeing, and Dominion Energy that we used in Chap. 3. Table 4.1 is a modified income statement for IBM, for the 2017–2020 period, which might be higher for various financial analyses. Revenues, gross profit, operating income, net income, and basic and fully diluted earnings per share (eps) are reported in millions of dollars. Similar income statements are shown for Boeing and Dominion Energy in Tables 4.2 and 4.3. We will stress net income and basic and fully diluted eps as investment variables for stock selection and stock valuation. In Table 4.1, IBM's revenues and net income has fallen during the 2017–2020 time period from $79.14 billion to $73.62 billion and $5.753 to $5.590 billion, respectively.

Boeing revenues decreased substantially, over 60%, during the 2017–2020 time period from $94.01 billion to $58.66 billion, due to crashes with the 737 Max plane and the pandemic and the lack of travel. Boeing's net income fell from $8.452 billion in 2017 to a loss of 11.873 billion in 2020. See Table 4.2. This substantial decrease in

Table 4.1 MSN snapshot of the income statement: IBM

Income statement				
Values in millions (except for per share items)	2017	2018	2019	2020
Period end date	12/31/2017	12/31/2018	12/31/2019	12/31/2020
Statement source	Annual	Annual	Annual	Annual
Total revenue	79,139	79,591	77,147	73,620
Cost of revenue	42,196	42,655	40,659	38,046
Gross profit	36,943	36,936	36,488	35,574
Selling, general, and administrative	19,120	18,217	19,501	19,037
Research and development	5,590	5,379	5,989	6,333
Special income/charges	(199)	(598)	(555)	(2,922)
Operating expenses	67,739	68,249	66,981	68,982
Operating income	11,400	11,342	10,166	4,638
Net interest income	-	-	-	(1)
Other income/expense, net	-	-	-	(1)
Pre-tax income	11,400	11,342	10,166	4,637
Provision for income tax	-	-	-	-
Net income	5,753	8,728	9,431	5,590
Dividend per share	5.90	6.21	6.43	6.51
Tax rate	1.4649	5.1314	5.7545	(21.0050)
Basic EPS	6.17	9.57	10.63	6.28
Diluted EPS	11.98	11.74	10.73	6.26

Table 4.2 MSN snapshot of the income statement: BA

Income statement				
Values in millions (except for per share items)	*2017*	*2018*	*2019*	*2020*
Period end date	12/31/2017	12/31/2018	12/31/2019	12/31/2020
Statement source	Annual	Annual	Annual	Annual
Total revenue	**94,005**	**101,127**	**84,818**	**58,656**
Cost of revenue	76,612	81,490	72,093	63,843
Gross profit	**17,393**	**19,637**	**12,725**	**(5,187)**
Selling, general, and administrative	4,095	4,567	3,909	4,817
Research and development	3,179	3,269	3,219	2,476
Special income/charges	21	75	(7,568)	(296)
Operating expenses	**83,661**	**89,140**	**86,793**	**71,423**
Operating income	**10,344**	**11,987**	**(1,975)**	**(12,767)**
Net interest income	107	91	438	447
Other income/expense, net	107	91	438	447
Pre-tax income	**10,107**	**11,604**	**(2,259)**	**(14,476)**
Provision for income tax	-	-	-	-
Net income	**8,452**	**10,453**	**(636)**	**(11,873)**
Dividend per share	5.68	6.84	8.22	2.06
Tax rate	28.8909	9.8587		
Basic EPS	14.03	18.05	(1.12)	(20.88)
Diluted EPS	11.77	17.85	(1.12)	(20.88)

Table 4.3 MSN snapshot of the income statement: D

Income Statement				
Values in millions (except for per share items)	*2017*	*2018*	*2019*	*2020*
Period end date	12/31/2017	12/31/2018	12/31/2019	12/31/2020
Statement source	Annual	Annual	Annual	Annual
Total revenue	**12,586**	**11,199**	**14,401**	**14,172**
Cost of revenue	6	122	88	53
Gross profit	-	-	-	-
Selling, general, and administrative	-	-	-	-
Research and development	-	-	-	-
Special income/charges	132	226	(2,014)	(2,141)
Operating expe Nses	**8,649**	**8,186**	**12,866**	**12,117**
Operating income	**3,937**	**3,013**	**1,535**	**2,055**
Net interest income	344	885	811	3,033
Other income/expense, net	344	885	811	3,033
Pre-tax income	**3,090**	**2,619**	**869**	**1,411**
Provision for income tax	-	-	-	-
Net income	**2,999**	**2,447**	**1,341**	**(466)**
Dividend per share	3.04	3.34	3.67	3.45
Tax rate	26.5696	18.1749	24.0506	5.8824
Basic EPS	4.72	3.74	1.66	(0.56)
Diluted EPS	3.38	3.12	0.77	1.70

net income sent its stock price crashing, down over 90% at one time! Boeing spent $2.5 billion on research and development expenditures in 2020, down almost a billion dollars from its 2017 level.

Dominion Energy increased its revenues and net income during the 2017 to 2020 time period from $12.59 billion to 14.17 billion, and from $3.0 to a loss of 0.50 billion, respectively. See Table 4.3.

It is clear that the pandemic significantly altered corporate income statements in 2020.

It would seem that IBM is in a virtually no-growth environment, if not a declining growth environment, hardly the case that Guerard and Schwartz reported and discussed in 2007. IBM had revenues of $73.62 billion in 2020, experienced costs of goods sold of $38.05 billion, producing gross profits of $35.57 billion. See Table 4.1. Research and development expenditures of $6.32 billion and general, selling, and administrative expenditures of $19.04 billion, produced operating income of $4.64 billion and net income of 5.59 billion in 2020.

IBM experienced virtually no growth in sales or net income, during the 1996–2020 time period; see Table 4.4. However, although net income rose only from $5.41 billion in 1996 to $5.59 billion in 2020, eps rose from $2.50 to $8.81, a 5.4% average annual growth rate during the 1996–2020 time period. How is this possible? Net income d divided by the number of common stock shares outstanding, to calculate the earnings per share, eps. IBM repurchased a great deal of stock during the 1996 to 2020 time period, driving down its shares outstanding from 2,113.4 million shares to 892.65 million. IBM created an "illusion" of eps growth by repurchasing its stocks; see Table 4.4.

Boeing experienced great sales growth from 1996 to 2015, with sales growing from $22.681 billion to $96.114 billion, an average annual sales growth rate of 7.89%. Net income rose over the corresponding time period from $1.095 to $5.172 billion, some 8.51% annually. However, the two 737 crashes and the 2020 pandemic travel restrictions lowered the 1996 to 2020 average annual growth rates of sales to net income to 4.13% and an undefined average annual eps growth rate.

Dominion Energy sales grew at an average annual rate of 4.48% during the 1996–2020 time period, while its net income grew at an average annual growth rate of 4.41%, respectively. Net income available to common stockholders became negative in 2020 because it discontinued operations; see Table 4.6. The years of 4–8% growth in sales, net income, and eps disappeared with COVID-19 in 2020.

What is important about revenues and net income? Firms that produce higher revenue growth and earnings per share, eps, growth than expected by analysts produce higher stock returns than lower growth firms. It is important to grow revenues and eps, but more important to stockholders who seek to outperform the market that these growth rates exceed expectations, as we will see in Chap. 8. Furthermore, the as-reported, "bottom line" net income data for the three firms will show relative to the firm's total assets (ROA), sales (ROS), and equity (ROE) in the coming chapter. The ROA ratio illustrates the effective use of the firm's assets. That is, a higher ROA for the firm relative to a set of comparable firms, its industry or

Table 4.4   FactSet IBM income statement, selected years

Ticker	IBM-US								
International Business Machines Corporation									
IBM-US 459200101 2005973 NYSE	Common stock								
Source: FactSet Fundamentals									
GAAP/IFRS Income statement									
	12/2020	*12/2019*	*12/2015*	*12/2009*	*12/2008*	*12/2007*	*12/2005*	*12/2000*	*12/1996*
Sales	**73,620**	**77,147**	**81,741**	**95,758**	**103,630**	**98,786**	**91,134**	**88,396**	**75,947**
Cost of goods sold (cogs) incl. D&A	38,768	41,762	41,927	52,678	58,597	57,690	55,065	55,972	45,408
Cogs excluding D&A	32,073	35,703	38,072	47,684	53,147	52,489	49,877	50,977	40,396
Depreciation & amortization expense	6,695	6,059	3,855	4,994	5,450	5,201	5,188	4,995	5,012
Depreciation	4,227	4,209	2,662	3,773	4,140	4,038	4,147	4,513	3,676
Amortization of intangibles	2,468	1,850	1,193	1,221	1,310	1,163	1,041	482	1,336
Gross income	**34,852**	**35,385**	**39,814**	**43,080**	**45,033**	**41,096**	**36,069**	**32,424**	**30,539**
Sg&a expense	25,176	24,768	24,641	25,907	28,389	27,262	24,928	20,790	21,508
Research & development	6,038	5,602	4,980	5,523	6,015	5,754	5,378	5,151	4,654
Other SG&A	19,138	19,166	19,661	20,384	22,374	21,508	19,550	15,639	16,854
EBIT (Operating income)	**9,676**	**10,617**	**15,173**	**17,173**	**16,644**	**13,834**	**11,141**	**11,634**	**9,031**
Nonoperating income - net	-395	1,641	1,651	1,627	1,423	1,966	2,526	617	707
Pretax income	4,659	10,191	15,953	18,143	16,715	14,489	12,226	11,534	8,587
Income taxes	-864	731	2,581	4,713	4,381	4,071	4,232	3,441	3,158
Consolidated net income	5,523	9,460	13,372	13,430	12,334	10,418	7,994	8,093	5,429
Minority interest	22	25	8	5	0	0	0	0	0
Net income	5,501	9,435	13,364	13,425	12,334	10,418	7,994	8,093	5,429
Discontinued operations	89	-4	-58	0	0	0	-24	0	0
Preferred dividends	0	0	0	0	0	0	0	20	20
Net income available to common	**5,590**	**9,431**	**13,306**	**13,425**	**12,334**	**10,418**	**7,970**	**8,073**	**5,409**
Per share									
EPS (recurring)	8.81	11.38	13.75	10.09	9.27	7.52	5.42	4.44	2.50
EPS (basic)	6.28	10.63	13.60	10.12	9.07	7.32	4.98	4.58	-
Basic shares outstanding	890.35	887.24	13.22	1327.16	1359.77	1423.04	1600.59	1763.04	2113.41
Total shares outstanding	892.65	887.11	965.73	1305.34	1339.10	1385.23	1573.98	1762.90	2031.92
EPS (diluted)	6.23	10.56	13.54	10.01	8.93	7.18	4.90	4.44	2.51
Diluted shares outstanding	896.56	892.81	982.70	1341.35	1381.77	1450.57	1627.63	1812.12	2159.42
Total shares outstanding	892.65	887.11	965.73	1305.34	1339.10	1385.23	1573.98	1762.90	2031.92
Dividends per share	6.51	6.43	5.00	2.15	1.90	1.50	0.78	0.51	0.33
Payout ratio	104.41	60.87	36.93	21.48	21.28	20.89	15.93	11.49	12.70
EBITDA									
EBITDA	16,371	16,676	19,028	22,167	22,094	19,035	16,329	16,629	14,043
EBIT	9,676	10,617	15,173	17,173	16,644	13,834	11,141	11,634	9,031
Depreciation & amortization expense	6,695	6,059	3,855	4,994	5,450	5,201	5,188	4,995	5,012

sector or the economy, shows that the firm's assets are employed to generate higher net income and a more effective use of assets. A higher relative ROS indicates that for a given sales volume, the firm generates higher net income than comparable firms. The ROE ratio helps quantify the firm's profitability relative to its use of leverage. Chapter 5 makes extensive use of the ROA, ROS, and ROE ratios.

The online reader sees 4 years of income statements on MSN, its Money tab. Is 4 years of data enough to make an intellect assessment of the company's financial conditions? Probably not. Let us now show selected years of 24-year income statements for IBM, Boeing, and Dominion using the FactSet database service. We show the 20-year income statements in Tables 4.4, 4.5 and 4.6 for our three representative firms. The reader may seem confused at this time in time; IBM's revenues and net income fell, but its eps rose. Is not eps defined as net income divided by common stock shares outstanding? Indeed, it is. IBM eps rose

Table 4.5 FactSet BA income statement, selected years

Ticker	BA-US								
Boeing Company									
BA-US 097023105 2108601 NYSE									
Source: FactSet Fundamentals									
GAAP/IFRS Income Statement									
	12/2020	*12/2019*	*12/2015*	*12/2009*	*12/2008*	*12/2007*	*12/2005*	*12/2000*	*12/1996*
Sales	**58,161**	**76,559**	**96,114**	**68,281**	**60,909**	**66,387**	**54,845**	**51,321**	**22,681**
Cost of goods sold (COGS) incl. D&A	57,148	72,068	82,088	56,540	50,386	53,426	46,211	43,712	19,053
COGS excluding D&A	54,902	69,797	80,255	54,874	48,895	51,940	44,708	42,233	18,062
Depreciation & amortization expense	2,246	2,271	1,833	1,666	1,491	1,486	1,503	1,479	991
Depreciation	1,929	1,940	1,609	1,459	1,325	1,334	1,412	1,317	-
Amortization of intangibles	317	331	224	207	166	152	91	162	-
Gross income	1,013	4,491	14,026	11,741	10,523	12,961	8,634	7,609	3,628
SG&A expense	7,286	6,427	6,856	9,870	6,852	7,381	6,433	3,776	2,274
Research & development	2,476	3,219	3,331	6,506	3,768	3,850	2,205	1,441	1,200
Other SG&A	4,810	3,208	3,525	3,364	3,084	3,531	4,228	2,335	1,074
EBIT (Operating income)	-6,273	-1,936	7,170	1,871	3,671	5,580	2,201	3,517	1,354
Nonoperating income - net	856	1,201	411	342	523	781	969	484	287
Pretax income	-14,476	-2,259	7,155	1,731	3,995	6,118	2,819	2,999	1,363
Consolidated net income	-11,941	-636	5,172	1,335	2,654	4,058	2,562	2,128	1,095
Minority interest	-68	0	0	0	0	0	0	0	0
Net income	-11,873	-636	5,172	1,335	2,654	4,058	2,562	2,128	1,095
Discontinued operations	0	0	0	0	0	0	0	0	0
Preferred dividends	0	0	0	0	0	0	0	0	0
Net income available to common	**-11,873**	**-636**	**5,172**	**1,335**	**2,654**	**4,058**	**2,562**	**2,128**	**1,095**
Per share									
EPS (recurring)	-12.38	-0.13	7.59	2.01	3.64	5.30	3.24	2.89	1.59
EPS (basic)	-20.88	-1.12	7.53	1.88	3.67	5.34	3.25	2.48	1.60
Basic shares outstanding	568.60	565.40	686.90	709.60	722.60	759.30	788.50	859.50	687.32
Total shares outstanding	582.32	562.91	666.62	755.85	726.60	768.04	800.17	836.33	694.73
EPS (diluted)	-20.88	-1.12	7.44	1.87	3.64	5.26	3.19	2.44	1.60
Diluted shares outstanding	568.60	565.40	695.00	713.40	729.00	772.50	802.90	871.30	687.32
Total shares outstanding	582.32	562.91	666.62	755.85	726.60	768.04	800.17	836.33	694.73
Dividends per share	2.06	8.22	3.64	1.68	1.60	1.40	1.00	0.59	0.56
Payout ratio	-	-	48.91	89.78	43.95	26.64	31.34	24.18	35.11
EBITDA									
EBITDA	-4,027	335	9,003	3,537	5,162	7,066	3,704	4,996	2,345
EBIT	-6,273	-1,936	7,170	1,871	3,671	5,580	2,201	3,517	1,354
Depreciation & amortization expense	2,246	2,271	1,833	1,666	1,491	1,486	1,503	1,479	991

substantially because its number of common stock shares outstanding substantially during the 24-year time period. The repurchasing of common stock creates an illusion of earnings.

growth. For most firms, as net income rises, so does its eps. The growth in net income and eps may not be proportional, because of stock repurchases. Does this eps growth "fool the market"? Stock returns suggest yes. We will further examine this issue in Chap. 8 on stock valuation.

4.4   Annual Cash Flow Statement

The income statement provides a picture of the firm's operations during the past year. The "bottom line" of the income statement is the firm's net income or after-tax profits. The firm's sources of cash flow from its operations are positive net income, depreciation and other noncash expenses, net decreases in its current assets and net

Table 4.6 FactSet D income statement, selected years

Ticker	D-US								
Dominion Energy Inc									
D-US 25746U109 2542049 NYSE	Common stock								
Source: FactSet Fundamentals									

GAAP/IFRS Income Statement	12/2020	12/2016	12/2010	12/2009	12/2008	12/2006	12/2001	12/1997	12/1996
Sales	**14,147**	**11,407**	**14,598**	**14,059**	**16,290**	**16,482**	**10,558**	**7,263**	**4,854**
Cost of goods sold (COGS) incl. D&A	5,513	4,427	7,548	8,031	8,866	9,282	5,335	2,372	2,554
COGS excluding D&A	2,677	2,578	6,290	6,712	7,675	7,543	4,013	1,466	1,859
Depreciation & amortization expense	2,836	1,849	1,258	1,319	1,191	1,739	1,322	906	694
Depreciation	2,767	1,776	1,151	1,164	1,096	1,633	-	-	-
Amortization of intangibles	69	73	107	155	95	106	-	-	-
Gross income	**8,634**	**6,980**	**7,050**	**6,028**	**7,424**	**7,200**	**5,223**	**4,891**	**2,300**
SG&A expense	3,588	3,064	3,171	3,340	3,257	3,280	2,246	1,238	808
Research & dev elopment	-	-	-	-	-	-	-	-	-
Other SG&A	3,588	3,064	3,171	3,340	3,257	3,280	2,246	1,238	808
EBIT (Operating income)	**4,175**	**3,320**	**3,347**	**2,197**	**3,668**	**3,345**	**2,582**	**1,081**	**1,208**
Nonoperating income - net	757	549	3,184	1,095	-100	174	184	-280	10
Pretax income	1,411	2,867	5,037	1,916	2,715	2,489	1,012	727	727
Income taxes	83	655	2,057	612	879	920	370	233	213
Consolidated net income	1,328	2,212	2,980	1,304	1,836	1,569	642	494	515
Minority interest	-149	89	17	17	0	6	0	47	0
Net income	1,477	2,123	2,963	1,287	1,836	1,563	642	447	515
Discontinued operations	-1,878	0	-15	0	-2	-183	0	0	0
Preferred dividends	65	0	0	0	0	0	98	48	43
Net income available to common	**-466**	**2,123**	**2,948**	**1,287**	**1,834**	**1,380**	**544**	**399**	**472**
Per share									
EPS (recurring)	1.32	3.47	5.70	2.74	3.16	1.96	2.13	1.08	1.32
EPS (basic)	-0.56	3.44	5.01	2.17	3.17	1.98	1.09	-	1.33
Basic shares outstanding	831.00	616.40	588.90	593.30	577.80	699.40	500.40	370.40	356.60
Total Shares outstanding	806.00	628.00	581.00	599.00	583.00	698.00	-	-	-
EPS (diluted)	-0.56	3.44	5.00	2.17	3.16	1.96	1.08	1.08	1.33
Diluted shares outstanding	831.00	617.10	590.10	593.70	580.80	703.20	505.00	370.40	356.60
Total shares outstanding	806.00	628.00	581.00	599.00	583.00	698.00	-	-	-
Dividends per share	3.45	2.80	1.83	1.75	1.58	1.38	1.29	1.29	1.29
Payout ratio	-	81.40	36.63	80.65	50.00	70.32	120.00	120.00	97.36
EBITDA									
EBITDA	7,011	5,169	4,605	3,516	4,859	5,084	3,904	1,987	1,902
EBIT	4,175	3,320	3,347	2,197	3,668	3,345	2,582	1,081	1,208
Depreciation & amortization expense	2,836	1,849	1,258	1,319	1,191	1,739	1,322	906	694

increases in its current liabilities (funds from operating activities), decreases in its long-tern (fixed) assets (funds from investing), and issuing new debt or equity (funds from financing). The firm uses its cash flow to pay dividends, engage in capital expenditures, pursue research and development (R&D) expenditures repurchase debt and equity, decrease its current liabilities, or increase its current and/or fixed assets. The firm's sources of funds must equal its uses of funds.[9] Stockholders prefer to see its firms cash flow derived from profits, not depreciation, because depreciation is an expense that serves to provide the firm with cash flow to replenish its capital investment.

During the 2017–2020 time period, IBM generated a very large net income, producing very high returns on equity, reported in Chap. 5, and large cash flow from operating activities. IBM's net income and depreciation, $9.82 billion in 2020, provided more than one-half of its $18.197 billion in total cash flow from operations; see Table 4.7. Again, the reader sees 4 years of financial data presented on MSN/Money. IBM spent its operating funds on capital expenditures, $2.618 billion, dividends, $5.80 billion, and $302 million on common stock repurchases. The primary differences difference between cash flow from operations, $18.197 billion, and financing activities cash flow, $8.314 billion, are composed of dividends paid, debt and stock repurchases, and capital expenditures. During 2017–2020, IBM spent a very large portion of its cash flow from operations to pursue debt and stock repurchases and pay dividends and engage in capital expenditures. We will use the sources and uses of funds, from the cash flow statement in valuation, in Chap. 8, and as an effective portfolio constraint in Chap. 14.

In 2020, in light of the Max 737 crashes and the travel bans of the pandemic, Boeing's net income and depreciation, $ -11.941 and $2.246 billion, respectively, were responsible for much of Boeing's $ -18.410 billion in total cash flow from operations; see Table 4.8. Boeing spent its operating funds on capital expenditures, $1.303 billion, dividends, $1.158 billion, and increasing its inventory, $12,391 and 11.002 billion, respectively, in 2019 and 2020, due to the grounding of the Max 737. All in all, it was by far the worst year in Boeing's history of financial operations. One can see from Table 4.8 how successful Boeing was in generating net income and cash flow from operations, $8.458 billion and $13,356 billion, in 2017, and $10.460

---

[9]Dtrina and Largay (1985) examined cash flow reporting with regard to "cash flow from operations" (CFO). Problems can develop because of ambiguity in terms of the definition of "operations," the measurement of the current position of long-term leases (noted in Chap. 9), diversity in reporting practices, and reclassification of current and noncurrent accounts. Krishnan and Largay III (2000) reported that past direct method cash flows offer better predictions of future operating cash flows than indirect method cash flows information. The direct method of presenting cash flows OCF for time t is predicted to be:

$$OCF_t = \int (CSHRD_{t-1}, CSHPD_{t-1}, INTRD_{t-1}, INTPD_{t-1}, TXPD_{t-1})$$

where

- $CSHRD_{t-1}$ = Cash Received from Customers at time t-1,
- $CSHPD_{t-1}$ = Cash Paid to Suppliers and Employees at time t-1,
- $INTRD_{t-1}$ = Interest Received at time t-1,
- $INTPD_{t-1}$ = Interest Paid at time t-1,

and

- $TXPD_{t-1}$ = Taxes Paid at time t-1.

Table 4.7 MSN snapshot of the CF statement: IBM

Cash flow				
Values in millions (except for per share items)	2017	2018	2019	2020
Period end date	12/31/2017	12/31/2018	12/31/2019	12/31/2020
Statement source	10-K	10-K	10-K	10-K
Net income	**5,753**	**8,728**	**9,431**	**5,590**
Operating gains/losses	-	-	-	-
Depreciation, amortization, and depletion	3,021	3,127	4,209	4,227
Deferred taxes	(931)	853	(1,527)	(3,203)
Other non-cash items	534	510	679	937
Change in receivables	1,297	1,006	502	5,297
Change in inventories	18	(127)	67	(209)
Change in pay/accrued exp	47	126	(503)	138
Change in other current assets	-	-	-	-
Change in other current liabilities	-	-	-	-
Change in other working capital	6,812	553	1,225	8,248
Cash flow from operating activities	**16,723**	**15,247**	**14,771**	**18,197**
Purchase/sale of prop, plant, equip: net	(3,229)	(3,395)	(2,286)	(2,618)
Purchase/sale of business, net	(701)	(139)	(32,630)	(336)
Purchase/sale of investments, net	-	-	-	-
Sale of investments	3,910	6,487	3,961	5,618
Other investing changes, net	(2,028)	(504)	6,720	475
Cash flow from investing activities	-	-	-	-
Issuance/payments of debt, net	-	-	-	-
Issuance/payments of common stock, net	(4,533)	(4,614)	(1,633)	(302)
Cash dividends paid	(5,506)	(5,666)	(5,707)	(5,797)
Other financing changes, net	174	112	98	92
Cash flow from financing activities	-	-	-	-
Cash, equivalents, start of period	8,073	12,234	11,604	8,314
Cash, equivalents, end of period	12,234	11,604	8,314	13,675
Change in cash	**4,161**	**(630)**	**(3,290)**	**5,361**
Free cash flow	26,002	24,877	23,385	27,224
Effect of exchange rate changes	937	(495)	(167)	(87)

billion and $15.322 billion, respectively in 2018. The two great years of 2017 and 2018 were followed by disasters in 2019 and 2020, creating great volatility in Boeing's stock returns, which the reader will see in Chap. 14.

During the 2017–2020 time period, Dominion Energy generated significant and substantial cash flow from operating activities and free cash flow. In 2017, Dominion Energy's net income and depreciation, $3.120 billion and 2.202 billion, respectively, were approximately equal provided the vast majority of its $4.502 billion in total cash flow from operations; see Table 4.9. In 2017, Dominion Energy spent its operating funds on capital expenditures, $5.504 billion, dividends, $1.931 billion, and issued a total of $3.61 billion of new debt and equity, respectively. Dominion repurchased $2.873 billion of common stock in 2020, whereas it issued common stock in 2017–2019 totaling $7.86 billion.

Four years of financial data is hardly sufficient for financial analysis and stock valuation. Let us proceed to look at selected years during the 1996–2020 time period of FactSet's cash flow statements for IBM, BA, and D.

Table 4.8 MSN snapshot of the CF statement: BA

Cash flow				
Values in millions (except for per share items)	2017	2018	2019	2020
Period end date	12/31/2017	12/31/2018	12/31/2019	12/31/2020
Stmt source	10-K	10-K	10-K	10-K
Net income	**8,458**	**10,460**	**(636)**	**(11,941)**
Operating gains/losses	-	-	-	-
Depreciation, amortization, and depletion	2,047	2,114	2,271	2,246
Deferred taxes	-	-	-	-
Other non-cash items	497	446	796	1,919
Change in receivables	(743)	(2,681)	(472)	(575)
Change in inventories	(1,403)	568	(12,391)	(11,002)
Change in pay/accrued exp	130	2	1,600	(5,363)
Change in other current assets	(19)	98	(682)	372
Change in other current liabilities	-	-	-	-
Change in other working capital	2,252	2,284	(4,629)	(17,335)
Cash flow from operating activities	**13,346**	**15,322**	**(2,446)**	**(18,410)**
Purchase/sale of prop, plant, equip: net	(1,739)	(1,722)	(1,834)	(1,303)
Purchase/sale of business, net	(324)	(3,230)	(455)	-
Purchase/sale of investments, net	-	-	-	-
Sale of investments	3,607	2,898	1,759	20,275
Other investing changes, net	6	(11)	(13)	(18)
Cash flow from investing activities	-	-	-	-
Issuance/payments of debt, net	-	-	-	-
Issuance/payments of common stock, net	(9,236)	(9,000)	(2,651)	-
Cash dividends paid	(3,417)	(3,946)	(4,630)	(1,158)
Other financing changes, net	(132)	(222)	(256)	(173)
Cash flow from financing activities	-	-	-	-
Cash, equivalents, start of period	8,869	8,887	7,813	9,571
Cash, equivalents, end of period	8,887	7,813	9,571	7,835
Change in cash	**18**	**(1,074)**	**1,758**	**(1,736)**
Free cash flow	18,633	21,059	4,145	(15,949)
Effect of exchange rate changes	80	(53)	(5)	85

During the 1996–2020 time period, IBM produced net income totaling $246,737 billion and had depreciation, depletion, and amortization expenses of $124.934 billion. Net operating cash flow of $382.871 billion was generated by IBM. IBM spent $123.529 billion on capital expenditures, paid dividends of $70.461 billion, repurchased stock of $170.944 billion, and issued long-term debt of $36.156 billion. IBM's large dividend payments, repurchases of common stock, and large capital expenditures generated net financial cash flow of $-197.954 billion during the 1996–2020 time period. In the King's English, IBM generated tremendous net operating cash flow but outspent these funds rather substantially on dividends, stock repurchases, and capital expenditures. IBM's free cash flow, defined as net income plus depreciation minus capital expenditures, totals $274.062 billion during the 1996–2020 time period. The reader is referred to Table 4.10 for selected years of the FactSet IBM cash flow statement.

Table 4.9 MSN snapshot of the CF statement: D

Cash flow				
Values in millions (except for per share items)	2017	2018	2019	2020
Period end date	12/31/2017	12/31/2018	12/31/2019	12/31/2020
Statement source	10-K	10-K	10-K	10-K
Net income	**3,120**	**2,549**	**1,376**	**(550)**
Operating gains/losses	-	-	-	-
Depreciation, amortization, and depletion	2,202	2,280	2,977	2,836
Deferred taxes	(3)	517	216	(324)
Other non-cash items	(42)	177	661	(80)
Change in receivables	(103)	(110)	(71)	(238)
Change in inventories	15	(29)	(90)	39
Change in pay/accrued exp	(89)	67	(225)	35
Change in other current assets	(71)	(247)	195	212
Change in other current liabilitie S	-	-	-	-
Change in other working capital	(774)	(241)	(886)	(796)
Cash flow from operating activities	**4,502**	**4,773**	**5,204**	**5,227**
Purchase/sale of prop, plant, equip: net	(5,504)	(4,254)	(4,980)	(6,020)
Purchase/sale of business, net	(405)	(151)	(341)	(311)
Purchase/sale of investments, net	-	-	-	-
Sale of investments	1,831	1,804	1,712	4,278
Other investing changes, net	308	23	109	12
Cash flow from investing activities	-	-	-	-
Issuance/payments of debt, net	2,308	753	(4,855)	3,698
Issuance/payments of common stock, net	1,302	2,461	4,097	(2,921)
Cash dividends paid	(1,931)	(2,185)	(2,983)	(2,873)
Other financing changes, net	(269)	(274)	1,842	(446)
Cash flow from financing activities	-	-	-	-
Cash, equivalents, start of period	322	185	391	269
Cash, equivalents, end of period	185	391	269	247
Change in cash	**(137)**	**206**	**(122)**	**(22)**
Free cash flow	11,937	11,212	13,167	14,120
Effect of exchange rate changes	-	-	-	-

During the 1996–2020 time period, Boeing produced net income totaling $64.419 billion and had depreciation, depletion, and amortization expenses of $42.483 billion. Net operating cash flow of $122.861 billion was generated by BA. Boeing spent $41.100 billion on capital expenditures, paid dividends of $35.996 billion, repurchased stock of $63.779 billion, and issued long-term debt of $51.40 billion (the majority, $36.250 billion in 2020). As with IBM, Boeing's large dividend payments, repurchases of common stock, and large capital expenditures generated net financial cash flow of $-45.753 billion during the 1996–2020 time period. BA's free cash flow, defined as net income plus depreciation minus capital expenditures, totals $84.813 billion during the 1996–2020 time period. The reader is referred to Table 4.11 for selected years of the FactSet BA cash flow statement.

Table 4.10 FactSet snapshot of the CF statement: IBM

Ticker   IBM
International Business Machines Corporation
IBM-US 459200101 2005973 NYSE   Common stock
Source: FactSet Fundamentals

GAAP/IFRS cash flow	12/2020	12/2019	12/2015	12/2009	12/2008	12/2007	12/2005	12/2000	12/1996
Operating activities									
Net income/starting line	5,590	9,431	13,190	13,425	12,334	10,418	7,934	8,093	5,429
Depreciation, depletion, & amortization	6,695	6,059	3,855	4,994	5,450	5,201	5,188	4,995	5,012
Other funds	866	-417	1,020	163	321	624	-449	-792	-1,356
Funds from operations	9,948	13,546	19,452	20,355	20,005	16,983	14,858	12,325	9,096
Changes in working capital	8,249	1,224	-2,444	418	-1,193	-889	56	-3,051	1,179
Net operating cash flow	18,197	14,770	17,008	20,773	18,812	16,094	14,914	9,274	10,275
Investing activities									
Capital expenditures	-3,230	-2,907	-4,151	-4,077	-4,887	-5,505	-4,634	-6,181	-6,178
Net assets from acquisitions	-336	-32,630	-3,349	-1,194	-6,313	-1,009	-1,482	0	-716
Sale of fixed assets & businesses	691	1,613	-31	730	421	847	2,039	1,619	1,314
Purchase/sale of investments	-628	268	-231	-2,005	1,510	992	-346	314	-143
Net investing cashflow	-3,028	-26,936	-8,159	-6,729	-9,285	-4,680	-4,463	-4,248	-5,723
Financing activities									
Cash dividends paid	-5,797	-5,707	-4,897	-2,860	-2,585	-2,147	-1,250	-929	-706
Change in capital stock	0	-1,361	-4,609	-7,429	-10,578	-18,828	0	-6,073	-5,005
Repurchase of common & preferred stk.	0	-1,361	-4,609	-7,429	-10,578	-18,828	0	-6,073	-5,005
Sale of common & preferred stock	0	0	0	0	0	0	0	-	0
Issuance/reduction of debt, net	-3,714	16,284	19	-7,463	-2,444	12,112	609	643	1,759
Change in current debt	-853	-2,597	101	-651	-6,025	1,674	-232	-1,400	-919
Change in long-term debt	-2,861	18,881	-82	-6,812	3,581	10,438	841	2,043	2,678
Issuance of long-term debt	10,504	31,825	5,540	6,683	13,829	21,744	4,363	9,604	7,670
Reduction in long-term debt	-13,365	-12,944	-5,622	-13,495	-10,248	-11,306	-3,522	-7,561	-4,992
Net financing cash flow	-9,721	9,042	-9,166	-14,700	-11,834	-4,740	-7,147	-6,359	-3,952
Free cash flow	15,579	12,484	13,429	17,326	14,641	11,464	11,072	3,658	4,392

Table 4.11 FactSet snapshot of the CF statement: BA

Ticker   BA-US
Boeing Company
BA-US 097023105 2108601 NYSE   Common stock
Source: FactSet Fundamentals

GAAP/IFRS Cash Flow	12/2020	12/2019	12/2015	12/2009	12/2008	12/2007	12/2005	12/2000	12/1996
Operating activities									
Net income/starting line	-11,941	-636	5,176	1,312	2,672	4,074	2,572	2,128	1,095
Depreciation, depletion, & amortization	2,246	2,271	1,833	1,666	1,491	1,486	1,503	1,479	991
Other funds	8,620	548	559	679	366	1,374	1,795	852	133
Funds from operations	-1,075	2,183	7,568	3,657	4,529	6,934	5,870	4,459	2,219
Changes in working capital	-17,335	-4,629	1,795	1,910	-4,902	2,675	1,118	1,483	4
Net operating cash flow	-18,410	-2,446	9,363	5,603	-401	9,584	7,000	5,942	2,223
Investing activities									
Capital expenditures	-1,303	-1,961	-2,450	-1,186	-1,852	-1,731	-1,547	-932	-762
Net assets from acquisitions	0	-455	-31	-679	-964	-75	-172	-5,727	0
Sale of fixed assets & businesses	296	805	42	27	34	59	1,760	169	0
Purchase/sale of investments	-17,341	101	554	-1,588	4,670	-1,893	-141	0	0
Net investing cash flow	-18,366	-1,523	-1,846	-3,834	1,888	-3,822	-98	-7,628	300
Financing activities									
Cash dividends paid	-1,158	-4,630	-2,490	-1,220	-1,192	-1,096	-820	-504	-379
Change in capital stock	36	-2,593	-6,352	-40	-2,893	-2,566	-2,529	-2,221	214
Repurchase of common & preferred stk.	0	-2,651	-6,751	-50	-2,937	-2,775	-2,877	-2,357	0
Sale of common & preferred stock	36	58	399	10	44	209	348	136	214
Issuance/reduction of debt, net	36,250	13,218	861	5,410	-725	-1,366	-1,378	2,067	-822
Change in current debt	0	0	0	0	0	0	0	0	0
Change in long-term debt	0	0	0	0	0	0	0	2,067	-822
Issuance of long-term debt	0	0	0	0	0	0	0	2,687	0
Reduction in long-term debt	0	0	0	0	0	0	0	-620	-822
Net financing cash flow	34,955	5,732	-7,920	4,134	-5,202	-4,884	-4,657	-658	-1,878
Free cash flow	-19,713	-4,280	6,913	4,417	-2,075	7,853	5,453	5,010	1,461

Table 4.12   FactSet snapshot of the CF statement: D

Ticker   D-US
Dominion Energy Inc
D-US   25746U109   2542049   NYSE   Common stock
Source: FactSet Fundamentals

GAAP/IFRS cash flow

	12/2020	12/2016	12/2010	12/2009	12/2008	12/2006	12/2001	12/1997	12/1996
Operating activities									
Net income/starting line	-550	2,212	2,825	1,304	1,834	1,380	642	447	472
Depreciation, depletion, & amortization	2,836	1,849	1,258	1,319	1,191	1,739	1,322	906	694
Other funds	4,061	-173	-3,121	1,126	342	183	43	-87	-234
Funds from operations	6,023	4,613	1,644	3,255	3,636	3,812	2,208	1,280	999
Changes in working capital	-796	-486	181	531	-977	193	195	-18	33
Net operating cash flow	5,227	4,127	1,825	3,786	2,659	4,005	2,403	1,262	1,032
Investing activities									
Capital expenditures	-6,331	-6,125	-3,422	-3,837	-3,554	-4,052	-2,168	-773	-484
Net assets from acquisitions	0	-4,381	0	0	-21	-91	-2,215	-2,099	-291
Sale of fixed assets & businesses	0	0	4,191	0	343	393	141	123	0
Purchase/sale of investments	3,403	-82	-435	-76	-337	-86	-74	-19	-410
Net investing cash flow	-2,916	-10,703	419	-3,695	-3,490	-3,494	-4,193	-2,880	-1,293
Financing activities									
Cash dividends paid	-2,873	-1,727	-1,093	-1,056	-916	-970	-649	-478	-460
Change in capital stock	-2,921	3,124	-826	456	240	-61	992	426	170
Repurchase of common & preferred stk.	-3,080	0	-900	0	0	-540	0	0	0
Sale of common & preferred stock	159	3,124	74	456	240	479	992	426	170
Issuance/reduction of debt, net	3,907	4,858	-311	513	1,309	530	1,552	1,806	600
Change in current debt	209	-1,254	91	-735	273	713	-1,620	-99	135
Change in long-term debt	3,698	6,112	-402	1,248	1,036	-183	3,172	1,905	465
Issuance of long-term debt	6,577	7,722	1,090	1,695	3,290	2,450	7,365	6,282	802
Reduction in long-term debt	-2,879	-1,610	-1,492	-447	-2,254	-2,633	-4,193	-4,377	-337
Net financing cash flow	-2,333	6,230	-2,232	-112	615	-515	1,916	1,829	305
Free cash flow	-1,104	-1,998	-1,597	-51	-895	-47	235	489	548

During the 1996–2020 time period, Dominion Energy, D, produced net income totaling \$32.288 billion and had depreciation, depletion, and amortization expenses of \$38.032 billion. Net operating cash flow of \$73,143 billion was generated by D. Dominion spent \$89.288 billion on capital expenditures, paid dividends of \$29.831billion, repurchased stock of \$13.737 billion, and issued long-term debt of \$24.346 billion, including 3.91 billion in 2020. Dominion generated cash flow in excess of large dividend payments, repurchases of common stock, and large capital expenditures such that D generated net financial cash flow of \$1989.1 billion during the 1996–2020 time period. BA's free cash flow, defined as net income plus depreciation minus capital expenditures, was \$-16.145 billion during the 1996–2020 time period. The reader is referred to Table 4.12 for selected years of the FactSet BA cash flow statement.

IBM, Boeing, and Dominion Energy had different mixtures of net income and depreciation on its income and cash flow statements. The firm spent the funds in different mixtures of capital expenditures, research and development expenditures, dividends, common stock repurchases, and long-term debt issuances. What should the reader take away from income statements and cash flow statements? First, 2 years of annual report data is rarely sufficient to assess a company and its capacity to generate revenues and net income. Third, the reader must be aware of alternative sources of financial data, such as MSN, FactSet, and Compustat.

4.5 US Firm Cash Flow Analysis, 1971–2020

In this chapter, we specifically examined the income statement and cash flow statements of IBM, Boeing, and Dominion Energy. IBM, Boeing, and Dominion Energy had different mixtures of net income and depreciation on its income and cash flow statements. The firm spent the funds in different mixtures of dividends, common stock repurchases, and long-term debt issuances. How might these data look for all firms? Let us examine the net income, dividends, equity issued, equity repurchased, long-term debt issued, and long-term debt repurchased for all US firms listed on US stock exchanges, 1971–2020. The database consists of over 530,000 firm-years of data. The source of our data is the Compustat database. US firms totaled $40,155.392 billion of net income, ni; paid dividends, div, of $25,600.023 billion; repurchased common stock, prstkc, of $15,694.445 billion, issued common stock of $598.123 billion; issued long-term debt, dltis, of $170,046.276 billion; and repurchased long-term debt, dltr, of $155.933 billion. Firms paid 63 percent of their net income to stockholders as dividends; repurchased large amounts of stocks, about 62% of what firms paid dividends; and increased net debt (long-term debt issued exceeded long-term debt repurchased) and decreased net equity (common stock issued was less than common stock repurchased) (Table 4.13).

Table 4.13 WRDS cash flow 1971–2020, $MM

Variable	Sum ($MM)	N
ni	$40,155.392	531306
div	25,600.023	531306
prstkc	15,694.452	531306
sstk	598.123	531306
dltis	170,463.276	531306
dltr	155,933.084	531306
NetEquityRep.	14,530.192	531306
NetdebtIssued	15,963.329	531306

4.6 Summary

The authors introduced the reader to operating statements of companies, the annual income statement, and cash flow statements. We specifically examined IBM, Boeing, and Dominion Energy, firms introduced to the reader in Chap. 3. The reader learned about revenues, gross and net income, dividends, common stock issuance and repurchases, and long-term debt issuances and repurchases. We will use these data for common stock valuation in Chap. 8. In this next chapter, we introduce the reader to ratio and concepts of firm efficiency using the data reported in Chaps. 3 and 4.

References

Bernstein, L. (1988). *Financial statement analysis: Theory, application, and interpretation* (4th ed.). Irwin, Chapters 9, 10, 11.

Bradley, I. F. (1959). *Fundamentals of corporation finance* (rev ed.). Holt, Rinehart and Winston, Chapters 4, 13.

Brealey, R. A., & Myers, S. C. (2003). *Principles of corporate finance* (7th ed.). McGraw-Hill/ Irwin, Chapters 31.

Dtrina, R. E., & Largay, J. A., III. (1985). Pitfalls in calculating cash flow from operations. *The Accounting Review, 40*, 314–326.

Hunt, P., Williams, C. M., & Donaldson, G. (1961). *Basic business finance* (rev ed.). Richard D. Irwin, Ch. 8.

Guerard, J. B., Jr., & Schwartz, E. (2007). *Quantitative corporate finance*. Springer.

Guerard, J. B., Jr., & Vaught, H. T. (1989). *The handbook of financial modeling*. Probus, Chapters 1, 2, 4.

Johnson, R. W. (1959). *Managerial finance*. Allyn and Bacon, Ch. 5.

Krishnan, G. V., & Largay, J. A., III. (2000). The predictive ability of direct method cash flow information. *Journal of Business Finance and Accounting, 27*, 215–245.

Paton, W. A., & Dixon, R. (1958). *Essentials of accounting*. Macmillan, Chapters. 5, 6, 18.

Penman, S. H. (2001). *Financial statement analysis and security analysis*. McGraw-Hill/Irwin, Chapters 7, 8, 9.

Samuelson, P. A. (1961). *Economics* (5th ed.). McGraw-Hill, Ch. 5 (appendix).

Schultz, W. J., & Reinhardt, H. (1955). *Credit and collection management*. Prentice-Hall, Chapters. 8, 9, 10, 11.

Schwartz, E. (1962). *Corporate finance*. St. Martin's Press, Chapter 4.

Van Horne, J. C., & Wachiwicz, J. M. (2001). *Fundamentals of financial management* (Eleventh ed.). Prentice-Hall. Chapter 6.

Weston, J. F., & Copeland, T. E. (1986). *Managerial finance* (8th ed.). The Dryden Press, Chapter 2.

White, G. I., Sondhi, A. C., & Fried, D. (1997). *The analysis and use of financial statements* (2nd ed.). Wiley, Chapters 2, 3.

Chapter 5
Financing Current Operations and Efficiency Ratio Analysis

Current financing encompasses managing and utilizing current assets and incurring and repaying current debt. The current assets of a firm differ from fixed assets; these differences are not abrupt but represent a continuum. The current assets (cash, receivables, inventory, etc.) support the short-run operations of the business. Current assets are what the classical economists called "circulating capital." Within the current asset grouping, however, some items remain in the firm's possession longer than others.

5.1 Working Capital Concepts

The reader was introduced to the firm's annual cash flow statement in the previous chapter. This chapter integrates current asset management, sources and uses of funds, and ratio analysis. In financial terminology, the total current assets, primarily cash, and other items likely to be turned into cash within the year, are called the gross working capital of the firm. The current assets of the firm, less the current liabilities, are the net working capital of the firm. When the term working capital is used, net is usually intended.

The net working capital of the firm is an important aspect of its financial strength. It represents the current assets, not offset by current creditor claims, which can be used to meet unexpected expenditures, absorb irregularities in the firm's cash flow, and cushion any short-run interruption, seasonality, or lumpiness in the flow of funds through the firm.[1] In addition, adequate working capital allows

---

[1] Although only a fraction of net working capital is likely to be in cash, an adequate amount of net working capital assures a substantial short-term cash flow and serves as a good base for short-term credit.

the firm to make especially opportune purchases and to take advantage of immediate expansion possibilities. Net working capital is not to be confused with cash, but it functions similarly as a sort of generalized short-term working balance for the firm.

How much net working capital a firm should carry depends on many factors. Generally, the net working capital a firm carries depends on its sales volume, its needs for gross circulating capital relative to its sales volume, and the stability of its operations. A firm can increase its rate of profit by economizing on the use of net working capital, but this has the effect of increasing its short-term financial risk. A firm can minimize its net working capital requirements; however, if any of the following conditions exists: (1) its flow of cash receipts tends to be stable and predictable; (2) its current assets tend to be marketable, stable in value, or short term and liquid; (3) the amount of long-term debt is small for the type of firm.

On the other hand, net working capital needs are higher for firms with the following characteristics: (1) those with erratic or irregular cash flows or cash demands, (2) those carrying slow-moving inventories which do not have active trading markets or whose market prices tend to fluctuate, and (3) those firms that by the nature of their business extend considerable credit to dubious risks. Since net working capital is the current assets minus the current debt, net working capital (by definition) must be supplied by equity and/or long-term debt. Thus, net working capital is financed from permanent or semipermanent sources. Working capital is not a static item, however, but expands and contracts with the level of the firm's operations. (Gross working capital is likely to show much wider proportionate changes.) Only on occasion does an established firm go out to the capital markets for additional working capital; normally, working capital is increased or depleted by internal operations of the business.

The main financial items affecting the level of working capital fall within two categories. Those that increase net working capital funds are: (1) net operating profits, (2) current depreciation charges insofar as they are covered by revenues over variable costs, (3) sales of fixed or other assets for current funds, and (4) issuing new bonds or stocks if the funds obtained are not all used to acquire fixed assets. Financial events that decrease net working capital funds are (1) operating losses, (2) payment of dividends, (3) purchase of fixed or other noncurrent assets, (4) repayment of long-term debt, and (5) buybacks, i.e., repurchasing the corporation's own shares with available funds.

Ratio analysis is an alternative to the flow of funds method of working capital analysis, although the two can be used to supplement each other. Ratio analysis is older and possibly the more popular approach than the flow of funds method of management and is the most readily available to credit managers of other companies, or other outsiders. A person within the firm sometimes finds other analytical tools more useful.

Ratio analysis consists of studying ratio or percentage relationships of meaningful financial data. The results are compared (1) with standard ratios—i.e., the averages of similar firms—(2) with the firm's ratios in previous years, or (3) with some implicit standards existing in the mind of the analyst. In the hands of a skilled practitioner, both "external analysis" (comparisons to standard ratios) and internal analysis (i.e., trends and relationships of the ratios within the company) can be revealing.

5.2 Snapshots of Financial Ratios

Many financial textbooks start with a presentation of financial statements and ratios. Financial ratios of liquidity, profitability, growth, capital structures, and asset efficiency are useful for comparing firms. The comparison of firms may be within the industry or across databases, as we do with the WRDS database. If one goes on MSN/Money, as we reported earlier in Chaps. 3 and 4 to obtain four years of financial balance sheet and income statement data, one can obtain current ratios as well as an estimated beta. The "key statistics" information on companies is very useful. One need only read Burton Malkiel's *A Random Walk Down Wall Street*, in any of his 15+ editions to see the benefits of the use of earnings, book value, and cash flow. We will use these data in our stock selection modeling of Chap. 14. As with any data, one must be vigilant, or as they say downhome, "one always gets what one pays for." We download and reprint the MSN/Money snapshot of IBM in June 2021. One sees the $73,62 billion of revenues discussed in Chap. 4 and the market capitalization of $132.11 billion and estimated beta of 1.22 (an "aggressive" asset beta exceeding 1.0) that is addressed in Chap. 14. The price-to-forecasted earnings (PEG), book value, cash flow, and sales are readily available. The return on capital, equity, and assets are presented in the snapshot and we will present similar data this chapter for the returns on sales (ROS), assets (ROA), and equity (ROE) in the next section of this chapter for comparisons with the WRDS database of some 10,000 firms in 2020.

A negative PEG ratio, such as one sees for IBM, is never good. Moreover, there are no benchmarks for the return ratios on MSN in the snapshot, such as we will present for ROS, ROA, and ROE calculations. There are industry comparisons for IBM relative to technology firms and IBM fares very poorly in these calculations, having far lower sales and net income growth (see Table 5.1B) than its industry. Its year-over-year net income growth of -41.44 % is far (far) lower than its industry benchmark of 13.26%

Table 5.1 MSN snapshot ratios: IBM

Key Statistics	
Revenue	73.62B
Net Income	5.59B
Market Cap.	132.11B
Enterprise Value	177.12B
Net Profit Margin %	7.62
PEG (Price/Earnings Growth) Ratio	-1.64
Beta	1.22
Forward P/E	13.52
Price/Sales	0
Price/Book Value	6.39
Price/Cash flow	10.7
EBITDA	17.30B
Return on Capital %	3.48
Return on Equity %	25.43
Return on Assets %	3.48
Book Value/Share	23.94
Shares Outstanding	893.52M
Last Split Factor (Date)	2:1 (5/27/1999)
Last Dividend (Ex-Date)	1.63 (2/9/2021)
Dividend Declaration Date	1/26/2021

5.1 B Growth Rate (%)	Company	Industry
Sales (Revenue) Q/Q (Last Year)	0.90	11.80
Net Income YTD/YTD (Last Year)	-41.44	13.26
Net Income Q/Q (Last Year)	-26.42	6.57
Sales (Revenue) 5-Year Annual Average	-2.07	-
Net Income 5-Year Annual Average	-15.93	-
Dividends 5-Year Annual Average	5.42	-

Table 5.1.C Profit Margins (%)	Company	Industry
Gross Margin	48.32	50.96
Pre-Tax Margin	6.30	13.81
Net Profit Margin	7.62	11.10
Average Gross Margin 5-Year Annual Average	47.37	50.74
Average Pre-Tax Margin 5-Year Annual Average	12.81	15.35
Average Net Profit Margin 5-Year Annual Average	12.60	12.42

Moreover, IBM produced, in 2020, far lower profitability than its industry. Its net profit margin of 7.62% is substantially lower than its industry benchmark of 11.10%. Its five-year average net profit margin (on sales) is approximately the industry average.

An undefined PEG ratio, such as one sees for Boeing, BA, is not good either. The reader is referred to Table 5.2. Boeing lost $11.87 billion in 2020, as we noted in Chap. 4.

There are industry comparisons for BA and Boeing fares very poorly in these calculations, having far lower sales and net income growth; see Table 5.2B than its industry. Its year-over-year net income growth of -1766.82 is far (far) lower than its industry benchmark of 9.45%.

Boeing produced, in 2020, far lower profitability than its industry. Its net profit margin of -20.36% is substantially lower than its industry benchmark of -.68%. Its five-year average net profit margin (on sales) of 2.34% is lower than the industry average of 4.37%.

Dominion Energy, D, is a very stable utility firm with a very low beta (0.37) which indicates that D is a "defensive" asset. A negative PEG ratio, such as one sees for D, is not good.

Table 5.2 MSN snapshot ratios: BA

Key Statistics	
Revenue	58.66B
Net Income	-11.87B
Market Cap.	147.75B
Enterprise Value	189.63B
Net Profit Margin %	-20.36
PEG (Price/Earnings Growth) Ratio	-
Beta	1.63
Forward P/E	-192.58
Price/Sales	0
Price/Book Value	-
Price/Cash flow	-
EBITDA	-4.24B
Return on Capital %	-8.09
Return on Equity %	-
Return on Assets %	-8.09
Book Value/Share	-30.9
Shares Outstanding	584.81M
Last Split Factor (Date)	2:1 (6/9/1997)
Last Dividend (Ex-Date)	0.00 (-)
Dividend Declaration Date	3/24/2020

Table 5.2B Growth Rate (%)	Company	Industry
Sales (Revenue) Q/Q (Last Year)	-10.00	-6.07
Net Income YTD/YTD (Last Year)	-1,766.82	79.45
Net Income Q/Q (Last Year)	14.49	-116.24
Sales (Revenue) 5-Year Annual Average	-9.40	-
Net Income 5-Year Annual Average	-	-
Dividends 5-Year Annual Average	-11.66	-

Table 5.2C Profit Margins (%)	Company	Industry
Gross Margin	-8.84	19.02
Pre-Tax Margin	-24.68	-0.10
Net Profit Margin	-20.36	-0.68
Average Gross Margin 5-Year Annual Average	13.66	21.84
Average Pre-Tax Margin 5-Year Annual Average	2.49	5.90
Average Net Profit Margin 5-Year Annual Average	2.34	4.37

There are industry comparisons for D relative to utility firms and D fares very well in these calculations, having far higher net income growth (see Table 5.1B) than its industry. Its year-over-year net income growth of 125.92 % is far higher than its industry benchmark of 13.54%; see Table 5.3B.

Furthermore, D's five-year average net profit margin of 13.44% greatly exceeds the utility form average of 8.52%. The reader is referred to Table 5.3C. The MSN/Money snapshots are useful, but only a first step.

Table 5.3 MSN snapshot ratios: D

Key Statistics	
Revenue	14.17B
Net Income	-466.00M
Market Cap.	61.58B
Enterprise Value	102.67B
Net Profit Margin %	9.37
PEG (Price/Earnings Growth) Ratio	-3.79
Beta	0.37
Forward P/E	19.75
Price/Sales	0
Price/Book Value	2.6
Price/Cash flow	15.06
EBITDA	7.03B
Return on Capital %	2.54
Return on Equity %	10.05
Return on Assets %	2.54
Book Value/Share	33.13
Shares Outstanding	806.52M
Last Split Factor (Date)	2:1 (11/20/2007)
Last Dividend (Ex-Date)	0.63 (6/3/2021)
Dividend Declaration Date	5/5/2021

Table 5.3B Growth Rate (%)	Company	Industry
Sales (Revenue) Q/Q (Last Year)	-13.92	1.79
Net Income YTD/YTD (Last Year)	125.92	13.54
Net Income Q/Q (Last Year)	457.04	7.32
Sales (Revenue) 5-Year Annual Average	3.94	-
Net Income 5-Year Annual Average	-5.75	-
Dividends 5-Year Annual Average	5.90	-

Table 5.3C Profit Margins (%)	Company	Industry
Gross Margin	-	33.03
Pre-Tax Margin	9.96	10.01
Net Profit Margin	9.37	7.80
Average Gross Margin 5-Year Annual Average	-	34.86
Average Pre-Tax Margin 5-Year Annual Average	16.94	10.46
Average Net Profit Margin 5-Year Annual Average	13.44	8.52

5.3   The Calculations of Financial Ratios and Their Implications for Stockholder Wealth Maximization

We have stressed the net income, or profits, of firms in this monograph. The as-reported "bottom line" net income data for the three firms will show relative to the firm's sales (ROS), total assets (ROA), and equity (ROE) . The ROA ratio illustrates the effective use of the firm's assets. That is, a higher ROA for the firm relative to a set of comparable firms, its industry or sector or the economy, shows that the firm's assets are employed to generate higher net income and a more effective use of assets. A higher relative ROS indicates that for a given sales volume, the firm generates higher net income than comparable firms. The ROE ratio helps quantify the firm's profitability relative to its use of leverage. We report return on sales, assets, and equity for IBM, BA, and D during the 1970–2020, for selected years. See Table 5.4. IBM and Boeing have produced ROEs exceeding 20% annually, in the selected years. IBM and BA are highly levered firms and leverage magnifies the profitability ratios. Furthermore, as IBM and BA have repurchased stock during the 1987–2020 time period, a decreasing equity, the denominator, drives up the ROE. D has the largest ROS return, much higher than IBM or BA. Finally, IBM has a much higher ROA than D or BA. The point is that firm profitability is very important, but it is only an input to firm valuation or its relative expected return. Let us proceed to include other categories of ratios in our analysis.

5.3.1   The Dupont Analysis Return on Invested Capital

One of the first profitability margins was the DuPont Analysis return on invested capital. The "DuPont System" rate of return, dating back to the early twentieth century, and the previously discussed Altman Z Model. The DuPont system, or measure, takes its net operating income divided by sales and multiplies it by the ratio of sales to investment, producing a return on investment, ROI.

$$\frac{NOI}{Sales} \text{ x } \frac{Sales}{Investment} = ROI \tag{5.1}$$

Stockholders should invest in firms with higher ROIs, and management could seek to maximize the DuPont ROI to maximize its stock price. The DuPont analysis uses information inherent in its return on sales and sales turnover ratios.

Pierre DuPont and Donald Brown, a DuPont employee, developed the DuPont return on investment relationship to access the firm's financial performance. General Electric calculated profitability by dividing earnings by sales (or costs). However, this calculation ignored the magnitude of invested capital. In 1903, Pierre DuPont created a new general ledger account for "permanent investment," where capital expenditures were charged at cost. The DuPont Corporation executive committee

Table 5.4 Profitability ratios of IBM, BA, and D
1970–2020, selected years

fyear	tic	cusip	ROS	ROA	ROE
1970	IBM	45920010	0.1356	0.1192	0.1711
1975	IBM	45920010	0.1378	0.1281	0.1743
1980	IBM	45920010	0.1359	0.1334	0.2165
1987	IBM	45920010	0.097	0.0826	0.1374
1990	IBM	45920010	0.0872	0.0687	0.1405
1993	IBM	45920010	-0.1292	-0.0999	-0.4104
2000	IBM	45920010	0.0916	0.0916	0.3924
2007	IBM	45920010	0.1055	0.0865	0.3659
2010	IBM	45920010	0.1485	0.1307	0.6436
2019	IBM	45920010	0.1222	0.062	0.4525
2020	IBM	45920010	0.0759	0.0358	0.2714
	average		*0.0916*	*0.0763*	*0.2323*

Fyear	tic	cusip	ROS	ROA	ROE
1970	D	25746U10	0.1925	0.0394	0.098
1975	D	25746U10	0.1497	0.04	0.0902
1980	D	25746U10	0.114	0.0372	0.0952
1987	D	25746U10	0.1474	0.0496	0.1295
1990	D	25746U10	0.1426	0.0458	0.115
1993	D	25746U10	0.126	0.0419	0.1063
2000	D	25746U10	0.0471	0.0149	0.0581
2007	D	25746U10	0.163	0.0653	0.2644
2010	D	25746U10	0.1848	0.0656	0.2291
2019	D	25746U10	0.0819	0.0131	0.0424
2020	D	25746U10	-0.0283	-0.0042	-0.0154
	average		*0.1201*	*0.0371*	*0.1103*
1970	BA	9702310	0.006	0.0084	0.0273
1975	BA	9702310	0.0205	0.0427	0.0756
1980	BA	9702310	0.0637	0.1012	0.2594
1987	BA	9702310	0.0313	0.0382	0.0963
1990	BA	9702310	0.0502	0.0949	0.1986
1993	BA	9702310	0.0489	0.0608	0.1358
2000	BA	9702310	0.0415	0.0506	0.1931
2007	BA	9702310	0.0614	0.0691	0.4525
2010	BA	9702310	0.0514	0.0482	1.1956
2019	BA	9702310	-0.0075	-0.0048	0.0738
2020	BA	9702310	-0.2024	-0.078	0.6482
	average		*0.015*	*0.0392*	*0.3051*

was presented with monthly sales, income, and return on invested capital on the firm's thirteen products in 1904 [Chandler (1977)]. Donald Brown contributed to the DuPont analysis pointing out that as sales volume rose, the return of invested capital rose, even if prices remained constant. Brown's "turnover" analysis was defined as sales divided by total investment. The multiplication of turnover by the ratio of earnings to sales produced the DuPont return on invested capital, still in use by the DuPont Corporation and most American firms. Total investment includes working capital, cash, inventories, and accounts receivable, and permanent investment, bonds, preferred stock, and stocks. The DuPont return on invested capital combined and consolidated financial, capital, and cost accounting. The DuPont return on total investment helped DuPont develop many modern management procedures for creating operating and capital budgeting and making short-run and long-run finan- cial forecasts.

Innumerable ratios can be developed, since the financial accounts can be placed in almost unlimited combinations. For most purposes, however, about thirteen popular ratios suffice for whatever can be learned from this method about the firm's current financial position.[2] In many cases, only six to ten of these ratios are needed for an understanding of the problem. If special areas seem to warrant additional attention, it is not difficult to develop other ratios.

Guerard and Schwartz (2007) calculated thirteen ratios that were

most generally used in financial analysis. The first six are most relevant for current analysis. The remaining seven reveal more general relationships.

Current Analysis Ratios

1. Current ratio, CR
2. Acid test, AT
3. Sales/receivables, SR
4. Sales/inventory, SI
5. Sales/net working capital, SNWK

General Analysis Ratios

6. The financial structure ratios—total debt/assets, TDTA
7. Sales/total assets, SA
8. Net profit/total assets, ROA
9. Net profit/tangible net worth, ROTNW
10. Net profit/equity, ROE
11. EBIT/interest, TIE

Composite Firm Relative Valuation Ratios

12. DuPont Analysis, DuPontA
13. Altman Z Model, NewZ

---

[2] As additional ratios are used, one soon discovers that the same information is being presented in a different form.

These thirteen ratios are still used by many analysts; however, we will restrict ourselves to a small set of ratios with more demonstrated predictive power with regard to stock prices within the WRDS universe.

1. Current ratio, CR
2. Total debt ratio, TDTA
3. Sales/total assets, SA
4. Net profit/equity, ROE
5. Net profit/total assets, ROA
6. Net profit/total sales, ROS
7. Altman Z Model, NewZ

Having discussed TDTA in Chap. 3 and ROS, ROA, and ROE in Chap. 4 and previously, let us proceed to examine additional variables.

5.3.2   Current Ratio

The current ratio is obtained by dividing the current liabilities into the current assets. It indicates how many times current liabilities are covered by gross working capital. The higher the current ratio, the more conservative the current financial position of the firm. A two-to-one ratio is a rule-of-thumb benchmark indicating a minimum level of the working capital position. Other circumstances must always be considered; no financial analysis can proceed rigidly. A ratio below two does not necessarily make the firm unsafe nor does a current ratio well over two ensure financial soundness. Much depends upon the collectibility and time structures of the firm's receivables and the type and quantity of inventory the firm carries. The public often have a current ratio of one to one or below.[3] In an electric utility company, for example, the low current ratio is possible because of its minimum inventory requirements and the stability of its revenues and cash flow.[4]

---

[3] This means that these companies, in effect, carry no net working capital.

[4] For completeness, we will relegate several widely used ratios to footnotes.

Acid test. The acid test, or quick ratio, is obtained by dividing current liabilities into the firm's net receivables and cash. This ratio highlights the firm's short-term liquidity position. The rule-of-thumb measure of a satisfactory acid test ratio is one to one. From an obverse point of view, the acid test ratio tends to indicate the amount of inventory in the working capital position of the firm. For example, if the current ratio is 3 to 1 and the acid test is only .85 to 1, the inventory account probably constitutes a heavy proportion of the current assets.

Sales to receivables or the receivable turnover ratio. This ratio is obtained by dividing credit sales by the outstanding trade accounts and trade notes receivable and indicates the collectibility and current condition of the receivables. The higher the sales to receivables ratio, the more current are the receivables. A variant of this ratio is to divide the turnover rate into 360 (representing the approximate number of days in the year). The resulting figure gives the number of days it takes to collect an average account. This figure can be compared to the usual terms (or allowable credit time)

5.3.3 Sales to Total Assets

The ratio of sales to total assets indicates how intensively the total assets are used in production. A ratio of low sales to total assets in comparison with similar firms or with previous periods gives some indication of idle capacity, i.e., excess assets compared to the level of operations.

Inter-industry comparisons of this ratio are not very useful. A wholesale distributor, for example, with no processing costs, a small margin, and a large turnover of goods shows a relatively high volume of sales to total assets. A better ratio to measure the basic concept of the rate of utilization of capital would be value

---

granted by the firm to ascertain whether the average account is collected in a period close to the credit terms

Sales to inventory: approximate inventory turnover. This ratio is obtained by dividing the inventory into the sales figure. The result is useful in analyzing how rapidly the firm's inventory is sold. A slow turnover—relative to the type of business or its own previous performance—may indicate that the firm is overstocked or that the inventory contains too many old or out-of-style items or that the management is speculating in inventory. Again, as in the case of the receivable turnover, the inventory turnover figure can be divided into 360 days to get an average of how many days it takes for a given dollar amount of merchandise to be turned into an equivalent amount of sales.

Many analysts prefer to reserve the term "inventory turnover" for the ratio of inventory divided into the cost of goods sold. The ratio then indicates the true physical turnover of the inventory. Since the sales figure contains the gross markup (profit) over cost, the sales over inventory ratio overstates the actual physical turnover of the goods. The higher the customary gross margin, or markup over cost in the sales figure, the better the inventory into sales ratio appears in comparison to the true turnover ratio. Unfortunately, the figure for the cost of goods sold is not always as available as the amount of sales. Thus, the standard ratios are more often based on sales. Ratio analysis, in any case, is not an exact science but is based on historical or intra-industry comparisons. The ratio serves its purpose as long as it depicts a logical relationship and comparisons using it are made on a consistent basis.

Sales/net working capital: working capital turnover. The net working capital turnover is obtained by dividing net working capital into the annual sales. This ratio is double-edged; a high ratio can indicate either efficiency or risk. A low turnover may indicate managerial inefficiency in moving goods and collecting receivables, or it may indicate excessively conservative management—a tendency to hold redundant idle funds or a failure to use a reasonable amount of available current credit. The other edge of the ratio appears if the turnover is too high in contrast with the industry norm. It may not necessarily indicate efficiency but a tendency to take on undesirable levels of risk. An especially high net working capital turnover can indicate overtrading on current account—an attempt to carry a heavy volume of business on an inadequate current capital base. Such speculative striving on the part of the management can be dangerous to both owners and creditors.

It should be obvious that the ratios are not to be used singly but in a composite manner to fill out a financial portrait of the company. Thus, the position of the net working capital turnover can be checked against the other current operating ratios. For example, a firm with a high net working capital turnover and ordinary inventory and receivable turnovers whose current ratio is tight is likely to be overtrading. A firm with a low net working capital turnover, a normal cash cycle, and a very high acid test ratio may be holding excessive idle funds. These are only two possibilities. An experienced analyst may be able to rough out normal relationships in his head while scanning the financial figures. His instinct may lead him quickly to any items that are out of line and suggest the few ratios necessary to highlight the potential trouble spot.

added to total asset—i.e., something approaching a capital coefficient. Unfortunately, possibly because of statistical difficulties, value-added ratios are not commonly used in financial analysis.

5.3.4 The Altman Z Model

A well-known and well-respected composite model to access the firm's health is the Altman Z Model, which is useful to identify potential bankrupt firms. The Altman Z Score used 5 primary ratios in its initial 1968 version.

$$Z = .012\,X_1 + .014\,X_2 + .033\,X_3 + .006\,X_4 + .999\,X_5 \tag{5.2}$$

Where $X_1 = \dfrac{\text{Current Assets} - \text{Current Liabilities}}{\text{Total Assets}}$

$X_2 = \dfrac{\text{Retained Earnings}}{\text{Total Assets}}$

$X_3 = \dfrac{\text{EBIT}}{\text{Total Assets}}$

$X_4 = \dfrac{\text{Market Value of Equity}}{\text{Book Value of Debt}}$

$X_5 = \dfrac{\text{Sales}}{\text{Total Assets}}$

The Altman Z Score used a liquidity, past profitability, (present) profitability, leverage, and sales turnover ratios to produce a single score. An Altman Z Score of less than 2.67 implied that the firm was not healthy. An Altman Z Score exceeding 2.67 implied financial health. The Altman Z Score successfully predicted impending bankruptcy for 32 of 33 firms (97%) in the year prior to bankruptcy, for his initial sample. The model correctly predicted 31 of 33 (94%) non-bankrupt firms in this sample for the year prior to bankruptcy.

Altman modified his equation in 2000 to become:

$$Z = .717X_1 + .847X_2 + 3.107X_3 + .420X_4 + .998X_5 \tag{5.3}$$

where X_4 is now book value of equity relative to its book value of debt. The new critical level is 2.0. The ratios of companies can be compared to all firms, companies in a sector, or broad segments of the economy, or specific industry groups. IBM maintains fewer liquid assets, as measured by its current ratio, than most companies in the WRDS universe. IBM uses greater leverage than the WRDS universe. IBM's greater leverage and profitability, as shown in Table 5.4, produced substantially higher ROE and DuPont ratio values than most US companies.

In 2020, the IBM Altman Z is 1.429, exceeding the WRDS universe median value of 1.04. IBM did not declare bankruptcy in 1993, when its Altman Z ratio collapsed to 1.09, far below the critical level of 1.00.

The seven ratios that are used in much of this text are reported in Table 5.5 for IBM, Boeing, and Dominion Energy, for selected years 1970–2020. What rends "jump off the page" at the reader. First, firm liquidity is falling, as evidenced by the falling current ratio, CR, for the three firms. Total debt to total assets, TDTA, is rising for all firms during the 1970–2020 time period. Firm profitability, as measured by ROS, the return on sales, has a general declining trend for IBM, BA, and D. The ROE is highly volatile. More importantly, the Altman Z Score, an overall measure of financial health and reliable tool to differentiate bankrupt and non-bankrupt firms, is declining for all three firms, falling below 2.0, a critical level. The reader is refereed to Table 5.5.

5.4   The Time Series of WRDS Ratios in the United States, 1970–2020

Is there a consistent pattern of movement in financial ratios over the 1970–2020 period? Yes. For all firms listed on the WRDS database, firms have substantially lowered their liquidity over the 55-year period, the median current ratio falling from 2.065 in 1970 to 1.859 in 2020. Sales efficiency has fallen from 1.816 in 1962 to 0.290 in 2020. The sales efficiency decline is very dramatic. The median debt-to-assets ratio has risen, from 0.449 in 1970 to 0.616 in 2020 (Table 5.6).

The median on equity has fallen from 9.8 percent in 1962 to 2.4 percent in 2030. The falling liquidity, sales efficiency, and return on equity have driven the median Altman (new) Z from 2.097 in 1988 to 0.950 in 2017. The Altman Z statistic has fallen dramatically during the 2017–2020 time period. This should be a surprise to very few investors and market watchers. The reader will not be surprised to see bankruptcies rise during the 1963–2017 period, as we will show in Chap. 19.

5.5   Industry Production of Financial Ratios

Industry has created massive databases for ratio calculations. In Table 5.7, we present many (the vast) majority of FactSet profitability, valuation, per share, asset turnover, DuPont analysis, operating efficiency, operating cycle, liquidity, coverage and leverage ratios. These ratios have been standard tools of financial analysis as noted in Weston and Copeland (1986), Van Horne (2002), and Brealey and Myers (2003). In Table 5.7, we see that IBM generated very large Profitability and DuPont ratios, had very low liquidity ratios, and used high degrees of leverage during the 1996–2020 time period, for selected years.

Table 5.5   A more inclusive set of ratios, including the Altman Z Score
1970–2020, selected years

fyear	tic	cusip	CR	SNWK	TDTA	ROA	ROE	ROS	TIE	NewZ
1970	IBM	45920010	1.8056	4.9631	0.3035	0.1192	0.1711	0.1356	37.9391	2.9964
1975	IBM	45920010	2.413	3.0381	0.2649	0.1281	0.1743	0.1378	54.6732	3.4096
1980	IBM	45920010	1.5208	7.712	0.3839	0.1334	0.2165	0.1359	17.6616	2.8092
1987	IBM	45920010	2.3189	3.073	0.3992	0.0826	0.1374	0.097	12.5073	2.482
1990	IBM	45920010	1.5398	5.0585	0.5109	0.0687	0.1405	0.0872	7.6293	2.0449
1993	IBM	45920010	1.1826	10.3629	0.7567	-0.0999	-0.4104	-0.1292	0.2335	1.0938
2000	IBM	45920010	1.2053	11.8271	0.7666	0.0916	0.3924	0.0916	15.7978	1.8217
2007	IBM	45920010	1.2001	11.1409	0.7636	0.0865	0.3659	0.1055	10.0349	1.7744
2010	IBM	45920010	1.1862	13.2209	0.7958	0.1307	0.6436	0.1485	20.2004	2.0978
2019	IBM	45920010	1.0191	107.2976	0.8621	0.062	0.4525	0.1222	6.2747	1.5453
2020	IBM	45920010	0.9823	-104.574	0.8671	0.0358	0.2714	0.0759	4.9794	1.4292
fyear	tic	cusip	CR	SNWK	TDTA	ROA	ROE	ROS	TIE	NewZ
1970	D	25746U10	0.6793	-10.7339	0.5977	0.0394	0.098	0.1925	2.476	0.7174
1975	D	25746U10	0.964	-107.999	0.557	0.04	0.0902	0.1497	1.8054	0.8685
1980	D	25746U10	0.9366	-60.6811	0.6092	0.0372	0.0952	0.114	.	.
1987	D	25746U10	1.0204	194.6826	0.6167	0.0496	0.1295	0.1474	3.2842	1.0089
1990	D	25746U10	0.8402	-23.7399	0.6013	0.0458	0.115	0.1426	2.4604	0.9734
1993	D	25746U10	0.8618	-26.5981	0.6064	0.0419	0.1063	0.126	3.0182	0.9474
2000	D	25746U10	0.7727	-5.365	0.7444	0.0149	0.0581	0.0471	1.4932	0.6076
2007	D	25746U10	0.8593	-14.3798	0.7523	0.0653	0.2644	0.163	3.2335	0.8575
2010	D	25746U10	0.9354	-40.7426	0.7138	0.0656	0.2291	0.1848	4.031	0.91
2019	D	25746U10	0.6125	-4.3033	0.6722	0.0131	0.0424	0.0819	3.0038	0.5398
2020	D	25746U10	0.6351	-3.5815	0.7241	-0.0042	-0.0154	-0.0283	2.7209	0.4281
fyear	tic	cusip	CR	SNWK	TDTA	ROA	ROE	ROS	TIE	NewZ
1970	BA	9702310	1.5901	5.6	0.6913	0.0084	0.0273	0.006	0.5762	1.9362
1975	BA	9702310	1.9668	6.4378	0.4354	0.0427	0.0756	0.0205	5.2422	3.2554
1980	BA	9702310	1.3106	9.7008	0.6097	0.1012	0.2594	0.0637	60.027	2.534
1987	BA	9702310	1.3184	6.8275	0.6031	0.0382	0.0963	0.0313	7.7407	1.9293
1990	BA	9702310	1.2297	16.8468	0.5221	0.0949	0.1986	0.0502	54.6429	2.9591
1993	BA	9702310	1.4048	9.621	0.5522	0.0608	0.1358	0.0489	8.9471	2.2291
2000	BA	9702310	0.8674	-21.1633	0.7378	0.0506	0.1931	0.0415	6.8558	1.8371
2007	BA	9702310	0.865	-15.5911	0.8474	0.0691	0.4525	0.0614	9.7023	1.6988
2010	BA	9702310	1.1463	12.4215	0.9583	0.0482	1.1956	0.0514	6.7859	1.3667
2019	BA	9702310	1.0505	17.2499	1.0621	-0.0048	0.0738	-0.0075	6.8651	0.9914
2020	BA	9702310	1.3937	1.707	1.1188	-0.078	0.6482	-0.2024	-4.1325	0.4287

In Table 5.8, the readers see that Boeing generated very large Profitability ratios, had very low liquidity ratios, and used seemingly outrageous degrees of leverage during much of the 1996–2020 time period. Obviously, the plane crashes of 2019 and the pandemic of 2020 crushed Boeing.

Table 5.6 WRDS universe median values, 1962–2020, selected years

Fyear	CR	SA	TDA	DuPontA	ROE	New Altman Z
1970	2.065	1.816	0.449	0.063	0.098	
1975	1.879	1.262	0.487	0.062	0.102	
1980	1.765	1.224	0.486	0.078	0.125	
1985	1.870	0.996	0.452	0.056	0.093	
1990	1.578	0.957	0.460	0.047	0.084	2.097
1995	1.688	0.859	0.409	0.053	0.088	2.183
2000	1.630	0.629	0.402	0.031	0.065	1.765
2005	1.755	0.573	0.357	0.050	0.099	1.705
2010	1.797	0.500	0.338	0.039	0.066	1.635
2015	1.620	0.362	0.378	0.029	0.056	1.313
2017	1.724	0.353	0.375	0.063	0.044	1.389
2020	1.859	0.290	0.616	0.000	0.024	0.950

In Table 5.9, we view Dominion Energy, D, with regard to Factset ratios. D generated very large profitability ratios, had very low liquidity ratios, and used high degrees of leverage during the 1996–2020 time period, for selected years.

5.6 Limitations of Ratio Analysis

Although ratio analysis is an extremely versatile tool, applying it can be dangerous if its limitations are not understood. Ratios differ among industries, as we have shown. Ratio analysis may be useless if the analyst does not have a feeling for the normal differences among different industries. Furthermore, if the firm has misled the financial community, such as WorldCom and Enron, in its financial disclosures, then ratio calculation and analysis using incorrect data cannot be correct. We will examine bankruptcy in Chap. 19 and calculate the WorldCom in 2001.

The user of ratios should be aware of seasonal factors. Usually, according to the results of the ratios, a firm appears in its best financial health at the seasonal low point of its annual operations. Inventory and collectibles are at minimum levels; funds normally invested in these assets have been released and are generally applied to reduce current obligations. These developments tend to improve almost all the working capital ratios. Most firms draw up their annual statements as of December 31. For many industries, the end of the normal calendar year is as good a point as any to present the financial statements. In industries whose seasonal low point is not at December 31, some firms set the closing of their books at the end of their "natural"

Table 5.7A   FactSet snapshot ratios: IBM
$151.28
International Business Machines Corporation
IBM-US 459200101 2005973 NYSE Common stock
Source: FactSet Fundamentals
All figures in millions of US Dollar except per share and labeled items.

	DEC '20	DEC '16	DEC '10	DEC '09	DEC '08	DEC '06	DEC '01	DEC '97	DEC '96
Profitability									
Gross Margin	47.3	46.8	45.4	45.0	43.5	41.2	37.0	39.0	40.2
SG&A to Sales	34.2	31.4	26.8	27.1	27.4	27.8	26.2	27.4	28.3
Operating Margin	13.1	15.4	18.7	17.9	16.1	13.4	10.8	11.6	11.9
Pretax Margin	6.3	15.4	19.8	18.9	16.1	14.6	12.8	11.5	11.3
Net Margin	7.5	14.9	14.9	14.0	11.9	10.3	9.0	7.8	7.1
Free Cash Flow Margin	21.2	16.8	15.4	18.1	14.1	11.7	10.0	2.6	5.8
Return on Assets	3.6	10.4	13.3	12.3	10.7	9.0	9.0	7.5	6.9
Return on Equity	26.6	73.1	64.9	74.4	58.8	30.6	34.9	29.4	24.6
Return on Common Equity	26.6	73.1	64.9	74.4	58.8	30.6	35.1	29.7	24.8
Return on Total Capital	11.0	21.5	37.1	35.7	30.0	22.9	18.6	20.0	20.4
Return on Invested Capital	7.0	23.6	33.2	33.3	28.1	20.7	19.7	18.7	17.0
Cash Flow Return on Invested Capital	23.1	33.7	43.7	51.5	42.9	33.1	36.3	27.3	32.1
Valuation									
Price/Sales	1.5	2.0	1.9	1.8	1.1	1.7	2.5	1.3	1.1
Price/Earnings	20.2	13.4	12.7	13.1	9.4	15.9	27.8	17.4	14.8
Price/Book Value	5.5	8.6	7.8	7.5	8.4	5.1	8.8	5.2	3.6
Price/Tangible Book Value	-2.1	-6.9	-32.3	-2,588.9	-14.8	10.9	9.8	5.7	4.1
Price/Cash Flow	6.2	9.4	9.7	8.5	6.2	10.0	15.0	11.9	8.0
Price/Free Cash Flow	7.2	11.9	12.3	10.1	7.9	14.2	24.9	51.0	18.6
Dividend Yield (%)	5.2	3.3	1.7	1.6	2.3	1.1	0.5	0.7	0.9
Enterprise Value/EBIT	17.0	15.5	10.6	10.7	8.0	13.0	24.7	13.3	10.2
Enterprise Value/EBITDA	10.1	11.4	8.4	8.3	6.1	9.2	16.2	8.6	6.5
Enterprise Value/Sales	2.2	2.4	2.0	1.9	1.3	1.7	2.7	1.5	1.2
Total Debt/Enterprise Value	0.4	0.2	0.1	0.1	0.3	0.1	0.1	0.2	0.2
Per Share									
Sales per Share	82.11	83.36	77.58	71.39	75.00	58.85	48.48	38.83	35.17
EPS (recurring)	8.81	13.17	11.50	10.09	9.27	6.30	4.51	3.00	2.50
EPS (diluted)	6.23	12.38	11.52	10.01	8.93	6.11	4.35	3.01	2.51
Dividends per Share	6.51	5.50	2.50	2.15	1.90	1.10	0.55	0.39	0.33
Book Value per Share	23.07	19.29	18.77	17.34	10.06	18.92	13.70	10.10	10.52
Cash Flow per Share	20.30	17.69	15.19	15.49	13.61	9.67	8.05	4.38	4.76
Total Shares Outstanding (M)	892.65	945.87	1,227.99	1,305.34	1,339.10	1,506.48	1,723.19	1,936.18	2,031.92
Asset Turnover Analysis									
Total Assets	0.5	0.7	0.9	0.9	0.9	0.9	1.0	1.0	1.0
DuPont Analysis									
Asset Turnover (x)	**0.5**	**0.7**	**0.9**	**0.9**	**0.9**	**0.9**	**1.0**	**1.0**	**1.0**
x Pretax Margin (%)	6.3	15.4	19.8	18.9	16.1	14.6	12.8	11.5	11.3
= Pretax Return on Assets (%)	**3.0**	**10.8**	**17.7**	**16.6**	**14.5**	**12.7**	**12.8**	**11.1**	**10.9**
x Tax Rate Complement (1-Tax Rate) (%)	118.5	96.4	75.2	74.0	73.8	70.7	70.5	67.5	63.2
= Return on Assets (%)	**3.6**	**10.4**	**13.3**	**12.3**	**10.7**	**9.0**	**9.0**	**7.5**	**6.9**
x Equity Multiplier (Assets/Equity)	7.4	7.0	4.9	6.1	5.5	3.4	3.9	3.9	3.6
= Return on Equity (%)	**26.6**	**73.1**	**64.9**	**74.4**	**58.8**	**30.6**	**34.9**	**29.4**	**24.6**
x Earnings Retention (1-Payout) (%)	-4.4	55.6	78.3	78.5	78.7	82.0	87.4	87.1	87.3
= Reinvestment Rate (%)	-1.4	40.8	51.0	58.5	46.5	25.1	30.5	25.6	21.4
Note: EBIT Return on Assets (%)	6.3	10.8	16.8	15.7	14.5	11.7	10.9	11.2	11.4
Note: Interest as % Assets	0.8	0.7	0.4	0.5	0.6	0.3	0.3	0.9	0.9

year. The analyst must be aware that the statement date often presents a better financial picture of the firm than that which prevails during the operating year. He must also be careful in making comparisons to allow for differing seasonal patterns among firms and industries and to take into account the use of possible different statement dates.

An attempt to solve the problem of seasonal or other variation in the statements has been made by taking such items as inventory or accounts receivable as an average between two statement figures. As in so many attempts to be extremely accurate, the slight improvement in results does not seem worth the additional effort. The figures may be difficult to obtain, and if the figures used to obtain the average

Table 5.7B FactSet snapshot ratios continued: IBM

	DEC '20	DEC '16	DEC '10	DEC '09	DEC '08	DEC '06	DEC '01	DEC '97	DEC '96
Operating Efficiency									
Revenue/Employee (actual)	196,163	192,855	215,298	218,737	236,555	231,718	234,236	291,348	315,637
Net Income/Employee (actual)	14,658	28,670	31,977	30,666	28,155	23,865	21,068	22,611	22,563
Assets/Employee (actual)	415,590	283,470	244,578	249,036	250,009	261,651	234,378	302,447	337,186
Receivables Turnover	3.4	2.8	3.6	3.5	3.7	3.6	3.0	3.3	3.3
Inventory Turnover	22.4	27.4	22.0	20.3	21.8	19.0	11.9	8.7	7.4
Payables Turnover	8.0	7.0	7.1	7.3	7.8	7.0	7.0	9.5	9.7
Asset Turnover	0.5	0.7	0.9	0.9	0.9	0.9	1.0	1.0	1.0
Working Capital Turnover	-	10.5	13.2	7.4	15.8	20.0	11.7	11.4	11.3
Operating Cycle - Days									
Days of Inventory on Hand	**16.3**	**13.3**	**16.6**	**18.0**	**16.7**	**19.2**	**30.6**	**41.9**	**49.0**
+ Days of Sales Outstanding	107.6	132.0	100.5	103.6	99.2	102.4	122.7	109.2	111.9
= Operating Cycle	**123.9**	**145.3**	**117.1**	**121.6**	**115.9**	**121.5**	**153.3**	**151.2**	**160.9**
- Days of Payables Outstanding	45.9	52.5	51.1	50.3	46.9	52.0	51.9	38.6	37.7
= Net Operating Cycle	78.0	92.8	66.0	71.3	69.0	69.5	101.5	112.6	123.2
Liquidity									
Current Ratio (x)	0.9	1.2	1.2	1.4	1.2	1.1	1.2	1.2	1.2
Quick Ratio (x)	0.9	1.2	1.1	1.3	1.1	1.0	1.1	1.1	1.0
Coverage									
Net Debt/EBITDA	2.9	2.0	0.7	0.5	1.0	0.7	1.5	1.4	1.0
EBIT/Interest Expense (Int. Coverage)	7.3	15.8	39.2	33.6	24.2	42.3	34.3	12.0	12.1
Leverage									
LT Debt/Total Equity	281.2	189.9	94.8	96.9	168.5	48.3	67.6	69.1	45.6
LT Debt/Total Capital	66.5	57.4	42.3	45.0	47.9	26.9	31.4	29.3	22.2
LT Debt/Total Assets	37.1	29.5	19.3	20.1	20.7	13.3	18.6	16.8	12.2
Total Debt/Total Assets (%)	42.6	35.9	25.2	23.9	31.0	22.0	31.6	33.0	28.1

All figures in millions of US Dollar except per share and labeled items

come out of about the same seasonal position, the averages still do not represent the intra-operational relation of the accounts. Moreover, standard ratios used for comparison are not generally based on averages. Only where there has been a drastic change in the level of the firm's activity does the use of averages seem justified, and in such cases, ratios may not be the most fruitful method of analysis.

The various tools of working capital analysis are useful to the internal management in forecasting the need for short-term funds and in assuring the safety and short-run liquidity of the firm. Investors may, of course, also be interested in the working capital relationships. Of all the techniques of short-term financial analysis, however, ratios are most commonly applied as a guide to decisions on the granting of credit. The most consistent users of ratio analysis are likely to be found in the credit departments of suppliers and in the loan departments of commercial banks.[5]

When used in granting credit, ratio analysis helps classify the credit customer into different grades of risk. The grantor of credit desires some inkling of the chance of loss—the percentage of the accounts he may be unable to collect—on each class of customer of different financial strength.[6] An experienced and perceptive credit analyst can roughly grade various credit applicants. The credit department might roughly predict that 0.5 percent of the total credit sales made over the year to firms having certain financial and business characteristics are likely to be uncollected. For

---

[5]Ratio analysis and related techniques are often called credit analysis.

[6]The banks, which have a lower margin for risk, obviously have to restrict their credit more severely than suppliers. The following discussion applies mainly to trade credit.

Table 5.8A   FactSet snapshot ratios: BA
$247.28
Boeing Company
BA-US 097023105 2108601 NYSE Common stock
Source: FactSet Fundamentals

	DEC '20	DEC '16	DEC '10	DEC '09	DEC '08	DEC '06	DEC '01	DEC '97	DEC '96
Profitability									
Gross Margin	1.7	14.6	19.4	17.2	17.3	18.0	16.2	11.3	16.0
SG&A to Sales	12.5	8.7	12.1	14.5	11.2	12.1	7.4	9.0	10.0
Operating Margin	-10.8	5.9	7.3	2.7	6.0	5.9	8.1	2.3	6.0
Pretax Margin	-24.9	5.9	7.0	2.5	6.6	5.2	6.1	-0.7	6.0
Net Margin	-20.4	5.2	5.1	2.0	4.4	3.6	4.9	-0.4	4.8
Free Cash Flow Margin	-33.9	8.3	2.8	6.5	-3.4	9.5	4.7	1.5	6.4
Return on Assets	-8.3	5.3	5.1	2.3	4.7	3.9	6.3	-0.5	4.5
Return on Equity	-	136.8	135.3	320.1	68.8	27.9	25.9	-1.5	10.5
Return on Common Equity	-	136.8	135.3	320.1	68.8	27.9	25.9	-1.5	10.5
Return on Total Capital	-18.7	40.7	30.9	17.5	31.3	20.2	22.0	6.0	9.9
Return on Invested Capital	-41.7	38.4	23.2	13.3	24.0	13.2	14.0	-1.0	8.1
Cash Flow Return on Invested Capital	-64.6	82.5	20.7	56.0	-3.6	44.8	18.9	12.4	16.4
Valuation									
Price/Sales	2.1	1.1	0.8	0.6	0.5	1.1	0.6	1.0	1.6
Price/Earnings	-	20.5	14.7	28.9	11.7	31.3	11.4 -		33.4
Price/Book Value	-6.8	117.6	17.3	19.2	-24.0	14.8	2.9	3.7	3.4
Price/Tangible Book Value	-4.3	-13.6	-9.3	-8.1	-4.1	-11,678.6	7.1	4.5	4.4
Price/Cash Flow	-6.6	9.5	16.5	6.9	-77.6	9.3	8.4	22.6	16.5
Price/Free Cash Flow	-6.2	12.7	26.6	8.7	-15.0	12.0	11.7	67.0	25.1
Dividend Yield (%)	1.0	2.8	2.6	3.1	3.7	1.4	1.8	1.1	1.1
Enterprise Value/EBIT	-26.2	17.3	10.6	22.8	9.6	20.1	9.0	47.2	26.4
Enterprise Value/EBITDA	-40.8	12.9	7.8	12.1	6.8	14.1	6.6	19.7	15.2
Enterprise Value/Sales	2.8	1.0	0.8	0.6	0.6	1.2	0.7	1.1	1.6
Total Debt/Enterprise Value	0.4	0.1	0.2	0.3	0.2	0.1	0.3	0.1	0.1
Per Share									
Sales per Share	102.29	147.12	86.40	95.71	83.55	78.12	70.18	47.21	33.00
EBIT (Operating Income) per Share	-11.03	8.62	6.31	2.62	5.04	4.62	5.69	1.08	1.97
EPS (recurring)	-12.38	7.67	4.64	2.01	3.64	3.34	4.20	-0.18	1.59
EPS (diluted)	-20.88	7.61	4.45	1.87	3.64	2.84	3.41	-0.18	1.60
Dividends per Share	2.06	4.36	1.68	1.68	1.60	1.20	0.68	0.56	0.56
Dividend Payout Ratio (%)	-	57.3	37.8	89.8	43.9	42.3	20.0 -		35.1
Book Value per Share	-31.45	1.32	3.76	2.82	-1.78	6.03	13.57	13.31	15.75
Tangible Book Value per Share	-50.21	-11.42	-7.00	-6.70	-10.50	-0.01	5.49	10.85	12.18
Cash Flow per Share	-32.38	16.33	3.97	7.85	-0.55	9.52	4.60	2.16	3.23
Free Cash Flow per Share	-34.67	12.27	2.45	6.19	-2.85	7.39	3.31	0.73	2.13
Diluted Shares Outstanding (M)	568.60	642.80	744.30	713.40	729.00	787.60	829.30	970.10	687.32
Basic Shares Outstanding (M)	568.60	635.50	738.10	709.60	722.60	771.00	816.20	970.10	687.32
Total Shares Outstanding (M)	582.32	617.15	735.26	755.85	726.60	788.74	797.89	973.48	694.73
Asset Turnover Analysis									
Cash & ST Investments	3.3	8.6	5.9	9.4	9.7	10.0	70.8	8.8	5.0
Receivables	5.2	10.4	10.8	11.2	10.1	10.9	9.6	16.6	11.9
Inventories	0.7	1.8	2.5	3.5	4.0	6.3	7.1	5.1	2.7
Current Assets	0.5	1.4	1.7	2.2	2.3	2.7	3.6	2.7	1.6
Fixed Assets	4.3	7.6	7.3	7.8	7.2	7.6	6.7	6.0	3.4
Total Assets	0.4	1.0	1.0	1.2	1.1	1.1	1.3	1.4	0.9
DuPont Analysis									
Asset Turnover (x)	0.4	1.0	1.0	1.2	1.1	1.1	1.3	1.4	0.9
x Pretax Margin (%)	-24.9	5.9	7.0	2.5	6.6	5.2	6.1	-0.7	6.0
= Pretax Return on Assets (%)	-10.1	6.0	6.9	3.0	7.1	5.7	7.9	-1.1	5.6
x Tax Rate Complement (1-Tax Rate) (%)	-	87.9	73.5	77.1	66.4	69.1	79.3	-	80.3
= Return on Assets (%)	-8.3	5.3	5.1	2.3	4.7	3.9	6.3	-0.5	4.5
x Equity Multiplier (Assets/Equity)	-10.6	25.8	26.7	138.9	14.6	7.1	4.1	2.7	2.3
= Return on Equity (%)	-	136.8	135.3	320.1	68.8	27.9	25.9	-1.5	10.5
x Earnings Retention (1-Payout) %	-	42.7	62.2	10.2	56.1	57.7	80.0 -		64.9
= Reinvestment Rate (%)	-	59.7	84.1	27.6	37.9	15.8	20.5	-6.2	6.9
Note: EBIT Return on Assets (%)	-4.4	6.0	7.2	3.2	6.5	6.5	10.4	3.2	5.5
Note: Interest as % Assets	1.4	0.3	0.8	0.5	0.4	0.5	1.3	1.3	0.5

another group of customers, perhaps the loss rate in uncollected accounts might amount to 15 percent of the annual sales: Presumably, there are all sorts of classes or grades in between.[7]

---

[7] If the loss experience in a given credit grade is to be fairly predictable, the classes have to be made quite large. Very fine grades might be useful for the purpose of finding the exact credit cutoff point. But they decrease the predictability of the results in each class.

Table 5.8B FactSet snapshot ratios continued: BA

	DEC '20	DEC '16	DEC '10	DEC '09	DEC '08	DEC '06	DEC '01	DEC '97	DEC '96
Operating Efficiency									
Revenue/Employee (actual)	412,489	628,379	400,660	434,634	375,518	399,545	309,564	192,437	158,608
Net Income/Employee (actual)	-84,206	32,505	20,629	8,498	16,363	14,325	15,032	-748	7,657
Assets/Employee (actual)	1,078,979	597,987	427,196	394,990	331,560	336,325	257,144	159,702	187,685
Receivables Turnover	5.2	10.4	10.8	11.2	10.1	10.9	9.6	16.6	11.9
Inventory Turnover	0.7	1.8	2.5	3.5	4.0	6.3	7.1	5.1	2.7
Payables Turnover	4.4	7.0	8.0	8.9	9.7	9.4	9.9	9.3	5.8
Asset Turnover	0.4	1.0	1.0	1.2	1.1	1.1	1.3	1.4	0.9
Working Capital Turnover	1.7	7.7	12.4	28.5 -	-	-		9.0	3.5
Operating Cycle - Days									
Days of Inventory on Hand	505.6	204.3	145.2	105.0	91.2	58.0	51.3	71.4	132.9
+ Days of Sales Outstanding	70.7	35.1	33.7	32.6	36.2	33.4	38.0	22.0	30.7
= Operating Cycle	576.3	239.4	178.9	137.6	127.4	91.5	89.4	93.4	163.6
- Days of Payables Outstanding	83.5	52.3	45.6	40.9	37.5	38.8	36.7	39.2	62.9
= Net Operating Cycle	492.8	187.1	133.2	96.7	90.0	52.6	52.7	54.2	100.6
Liquidity									
Current Ratio (x)	1.4	1.2	1.1	1.1	0.8	0.8	0.8	1.4	1.7
Quick Ratio (x)	0.5	0.4	0.5	0.6	0.3	0.5	0.5	0.7	0.9
Cash Ratio	0.3	0.2	0.3	0.3	0.1	0.2 -	-		
Cash & ST Inv/Current Assets	21.0	16.0	25.9	31.8	12.6	27.8	3.9	26.7	34.9
CFO/Current Liabilities	-21.1	20.9	8.3	17.0	-1.3	25.2	18.6	14.8	25.7
Coverage									
Net Debt/EBITDA	-	0.0	0.3	0.5	0.8	0.6 -	-	-	-
Net Debt/(EBITDA-Capex)	-7.4	0.0	0.4	0.7	1.3	0.9 -	-	-	
Total Debt/EBITDA	-	1.3	1.9	3.7	1.5	1.8	1.9	2.7	1.7
EBIT/Interest Expense (Int. Coverage)	-2.8	11.6	8.3	4.4	12.2	12.6	6.5	1.9	7.7
EBITDA/Interest Expense	-1.9	24.3	12.5	10.4	25.6	21.6	9.9	4.9	16.2
Fixed-charge Coverage Ratio	-2.8	11.6	8.3	4.4	12.2	12.6	6.5	1.9	7.7
CFO/Interest Expense	-8.5	34.3	5.7	16.5	-2.0	31.2	5.9	4.1	15.3
Cash Dividend Coverage Ratio	-0.9	2.7	4.8	3.0	3.8	5.6	8.5	4.6	5.9
LT Debt/EBITDA	-15.6	1.3	1.8	3.5	1.3	1.6	1.7	2.4	1.7
Net Debt/FFO	-36.6	0.0	0.3	0.5	0.9	0.6 -	-	-	
LT Debt/FFO	-58.6	1.3	1.9	3.3	1.5	1.5	2.2	2.4	1.8
FCF/Total Debt	-0.3	0.8	0.1	0.3	-0.3	0.6	0.2	0.1	0.4
CFO/Total Debt	-0.3	1.1	0.2	0.4	-0.1	0.8	0.3	0.3	0.6
Leverage									
LT Debt/Total Equity	-343.8	1,171.1	414.8	574.1	-537.2	172.1	100.4	47.3	36.4
LT Debt/Total Capital	135.1	88.8	75.5	81.2	111.8	57.1	47.1	30.9	26.7
LT Debt/Total Assets	41.4	10.6	16.7	19.7	12.9	15.7	22.5	16.1	14.8
Total Debt/Total Assets (%)	42.7	11.1	18.1	20.8	14.0	18.4	25.4	18.0	14.9
Net Debt/Total Equity (%)	-214.8	-9.4	68.8	79.9	-327.1	66.5 -	-	-	
Total Debt/Equity (%)	-	1,218.1	449.1	607.3 -		201.3	113.3	52.9	36.5
Net Debt/Total Capital	84.4	-0.7	12.5	11.3	68.1	22.1 -	-	-	
Total Debt/Total Capital	139.3	92.4	81.8	85.9	120.8	66.8	53.1	34.6	26.7

All figures in millions of US dollar except per share and labeled items

The credit department has to decide which grades of risks may be extended credit. Credit losses can be minimized by selling only to the financially strongest customers, but this would cause a loss of sales volume. The credit department's objective is to minimize losses, but management wishes also to maximize business volume. The problem can be brought near solution by the marginal approach. Credit is refused to that class of firms where the potential rate of loss is too large. All other applicants are extended normal credit terms. It is not necessary to grade the intra-marginal credit risks. The credit department needs only two basic working grades. Their major task is at the margin where it must be decided whether a customer's financial position is too shaky to extend him credit profitably.

Where should the credit line be drawn? The amount of risk a creditor can take profitably depends largely on the operating and financial position of the credit-granting firm itself. A firm with a strong net working capital position may be able,

Table 5.9A FactSet snapshot ratios: D
$77.32
Dominion Energy Inc
D-US 25746U109 2542049 NYSE Common stock
Source: FactSet Fundamentals

	DEC '20	DEC '16	DEC '10	DEC '09	DEC '08	DEC '06	DEC '01	DEC '97	DEC '96
Profitability									
Gross Margin	61.0	61.2	48.3	42.0	16.6	13.7	43.3	61.3	47.4
SG&A to Sales	25.4	26.9	21.7	23.8	20.0	19.9	21.3	17.0	16.6
Operating Margin	29.5	29.1	22.9	15.6	22.5	20.3	24.5	22.7	24.6
Pretax Margin	10.0	25.1	34.5	13.6	16.7	15.1	9.6	10.0	15.0
Net Margin	10.4	18.6	20.3	9.2	11.3	9.5	6.1	6.2	10.6
Free Cash Flow Margin	-7.8	-17.5	-10.9	-0.4	-5.5	-0.3	2.2	6.7	11.3
Return on Assets	1.5	3.3	6.9	3.0	4.5	3.1	2.0	2.5	3.6
Return on Equity	5.1	15.6	25.0	11.8	18.4	13.1	7.7	7.5	9.1
Return on Common Equity	5.3	15.6	25.6	12.1	18.8	13.4	7.1	8.0	9.8
Return on Total Capital	6.1	7.0	11.3	7.7	13.6	10.8	11.0	12.1	10.7
Return on Invested Capital	2.3	5.2	10.8	4.9	7.6	5.9	3.2	3.8	5.0
Cash Flow Return on Invested Capital	8.2	10.2	6.6	14.5	11.0	15.0	12.0	10.6	10.0
Valuation									
Price/Sales	4.4	4.1	1.7	1.6	1.3	1.8	1.4	1.1	1.4
Price/Earnings	-	22.3	8.6	17.9	11.3	21.4	28.0	19.8	14.5
Price/Book Value	2.6	3.3	2.1	2.1	2.1	2.3	-	-	-
Price/Tangible Book Value	3.9	6.3	3.0	3.3	3.6	3.7	-	-	-
Price/Cash Flow	12.0	11.5	13.8	6.1	7.8	7.4	6.3	6.2	6.7
Price/Free Cash Flow	-56.6	-23.7	-15.8	-453.1	-23.3	-627.2	64.6	16.1	12.5
Dividend Yield (%)	4.6	3.7	4.3	4.5	4.4	3.3	4.3	6.1	6.7
Enterprise Value/EBIT	24.1	25.6	12.5	18.3	10.1	14.2	12.1	16.6	11.1
Enterprise Value/EBITDA	14.4	16.4	9.1	11.5	7.6	9.3	8.0	9.1	7.1
Enterprise Value/Sales	7.1	7.5	2.9	2.9	2.3	2.9	3.0	2.5	2.8
Total Debt/Enterprise Value	0.4	0.4	0.4	0.4	0.5	0.4	0.5	0.5	0.4
Per Share									
Sales per Share	17.02	18.48	24.74	23.68	28.05	23.44	20.91	19.61	13.61
EBIT (Operating Income) per Share	5.02	5.38	5.67	3.70	6.32	4.76	5.11	2.92	3.39
EPS (recurring)	1.32	3.47	5.70	2.74	3.16	1.96	2.13	1.08	1.32
EPS (diluted)	-0.56	3.44	5.00	2.17	3.16	1.96	1.08	1.08	1.33
Dividends per Share	3.45	2.80	1.83	1.75	1.58	1.38	1.29	1.29	1.29
Dividend Payout Ratio (%)	-	81.4	36.6	80.6	50.0	70.3	120.0	120.0	97.4
Book Value per Share	29.44	23.26	20.65	18.67	17.28	18.50	-	-	-
Tangible Book Value per Share	19.33	12.08	14.14	11.92	10.05	11.44	-	-	-
Cash Flow per Share	6.29	6.69	3.09	6.38	4.58	5.70	4.76	3.41	2.89
Free Cash Flow per Share	-1.33	-3.24	-2.71	-0.09	-1.54	-0.07	0.47	1.32	1.54
Diluted Shares Outstanding (M)	831.00	617.10	590.10	593.70	580.80	703.20	505.00	370.40	356.60
Basic Shares Outstanding (M)	831.00	616.40	588.90	593.30	577.80	699.40	500.40	370.40	356.60
Total Shares Outstanding (M)	806.00	628.00	581.00	599.00	583.00	698.00	-	-	-
Asset Turnover Analysis									
Cash & ST Investments	55.6	18.1	14.8	10.3	12.5	6.2	15.5	21.1	47.4
Receivables	5.5	7.4	6.6	5.9	6.6	5.2	4.8	7.7	7.3
Inventories	3.3	3.1	6.4	6.8	8.0	8.2	11.8	10.4	11.2
Current Assets	2.2	2.7	2.4	1.9	2.3	1.8	1.9	4.0	4.0
Fixed Assets	0.2	0.2	0.6	0.6	0.7	0.6	0.6	0.6	0.5
Total Assets	0.1	0.2	0.3	0.3	0.4	0.3	0.3	0.4	0.3
DuPont Analysis									
Asset Turnover (x)	**0.1**	**0.2**	**0.3**	**0.3**	**0.4**	**0.3**	**0.3**	**0.4**	**0.3**
x Pretax Margin (%)	10.0	25.1	34.5	13.6	16.7	15.1	9.6	10.0	15.0
= Pretax Return on Assets (%)	**1.4**	**4.4**	**11.8**	**4.5**	**6.7**	**4.9**	**3.2**	**4.1**	**5.0**
x Tax Rate Complement (1-Tax Rate) (%)	94.1	77.2	59.2	68.1	67.6	63.0	63.4	67.9	70.8
= Return on Assets (%)	**1.5**	**3.3**	**6.9**	**3.0**	**4.5**	**3.1**	**2.0**	**2.5**	**3.6**
x Equity Multiplier (Assets/Equity)	3.4	4.8	3.6	3.9	4.1	4.3	3.8	3.0	2.5
= Return on Equity (%)	**5.1**	**15.6**	**25.0**	**11.8**	**18.4**	**13.1**	**7.7**	**7.5**	**9.1**
x Earnings Retention (1-Payout) (%)	-	18.6	63.4	19.4	50.0	29.7	-20.0	-20.0	2.6
= Reinvestment Rate (%)	-5.0	2.9	15.9	2.3	9.2	5.0	-1.3	-1.3	0.2
Note: EBIT Return on Assets (%)	4.2	5.1	7.8	5.2	9.0	6.6	8.1	6.2	8.4
Note: Interest as % Assets	1.3	1.4	2.1	2.1	2.0	2.1	2.9	3.1	2.6

Table 5.9B FactSet snapshot ratios continued: D

	DEC '20	DEC '16	DEC '10	DEC '09	DEC '08	DEC '06	DEC '01	DEC '97	DEC '96
Operating Efficiency									
Revenue/Employee (actual)	817,746	704,136	923,924	785,419	905,000	941,829	617,427	469,854	434,401
Net Income/Employee (actual)	85,376	131,049	187,532	71,899	102,000	89,314	37,544	28,924	46,062
Assets/Employee (actual)	5,543,642	4,420,370	2,709,937	2,377,318	2,336,278	2,815,371	2,009,883	1,306,294	1,333,954
Receivables Turnover	5.5	7.4	6.6	5.9	6.6	5.2	4.8	7.7	7.3
Inventory Turnover	3.3	3.1	6.4	6.8	8.0	8.2	11.8	10.4	11.2
Payables Turnover	5.2	5.3	5.1	5.6	5.6	3.8	3.2	4.4	6.8
Asset Turnover	0.1	0.2	0.3	0.3	0.4	0.3	0.3	0.4	0.3
Operating Cycle - Days									
Days of Inventory on Hand	109.0	118.4	56.8	53.4	45.5	44.6	30.9	35.1	32.7
+ Days of Sales Outstanding	66.5	49.2	55.3	61.5	55.1	69.9	75.3	47.1	50.3
= Operating Cycle	175.4	167.6	112.1	114.9	100.6	114.5	106.2	82.2	83.0
- Days of Payables Outstanding	70.6	68.4	71.9	65.7	65.7	97.0	114.8	83.7	53.5
= Net Operating Cycle	104.8	99.2	40.3	49.2	34.9	17.5	-8.6	-1.5	29.5
Liquidity									
Current Ratio (x)	0.6	0.5	0.9	1.0	1.0	0.7	0.7	0.6	0.7
Quick Ratio (x)	0.5	0.3	0.7	0.8	0.8	0.6	0.6	0.6	0.6
Cash Ratio	0.0	0.0	0.1	0.2	0.2	0.2	-	-	-
Cash & ST Inv/Current Assets	3.5	9.4	14.8	17.3	20.4	21.4	13.5	24.7	9.6
CFO/Current Liabilities	48.2	50.9	31.6	55.4	34.1	35.7	32.1	34.8	52.3
Coverage									
Net Debt/EBITDA	5.2	6.7	3.7	4.8	3.3	3.5 -			
Net Debt/(EBITDA-Capex)	55.0	-36.3	14.2	-52.1	12.2	17.3 -			
Total Debt/EBITDA	5.3	6.8	3.8	5.1	3.6	3.9	4.2	4.6	3.1
EBIT/Interest Expense (Int. Coverage)	3.0	2.9	3.3	2.3	3.9	2.9	2.5	1.7	3.1
EBITDA/Interest Expense	5.4	5.3	5.1	4.0	5.7	4.9	3.9	3.2	4.9
Fixed-charge Coverage Ratio	2.8	2.9	3.3	2.3	3.9	2.9	2.2	1.5	2.7
CFO/Interest Expense	4.0	4.2	2.0	4.3	3.1	3.9	2.4	2.0	2.7
Cash Dividend Coverage Ratio	2.1	2.7	1.5	3.1	4.0	3.9	3.4	2.7	2.2
LT Debt/EBITDA	4.9	5.8	3.4	4.4	3.1	2.9	3.4	3.6	2.5
Net Debt/FFO	6.2	7.5	10.2	5.1	4.4	4.7 -			
LT Debt/FFO	5.7	6.6	9.6	4.8	4.1	3.9	6.0	5.6	4.7
FCF/Total Debt	0.0	-0.1	-0.1	0.0	-0.1	0.0	0.0	0.1	0.1
CFO/Total Debt	0.1	0.1	0.1	0.2	0.2	0.2	0.1	0.1	0.2
Leverage									
LT Debt/Total Equity	132.0	207.0	128.6	135.3	144.7	112.3	151.4	117.7	82.2
LT Debt/Total Capital	54.1	60.8	52.7	52.7	53.9	45.1	52.5	47.0	40.7
LT Debt/Total Assets	35.9	42.2	36.8	36.4	35.6	30.0	38.6	35.6	31.7
Total Debt/Total Assets (%)	39.2	49.0	41.2	42.1	41.4	39.8	47.9	45.5	39.3
Net Debt/Total Equity (%)	143.1	237.5	137.4	146.3	153.5	135.7 -			
Total Debt/Equity (%)	144.0	240.3	144.0	156.6	168.7	148.8	188.1	150.2	101.9
Net Debt/Total Capital	58.6	69.8	56.3	57.0	57.1	54.5 -			
Total Debt/Total Capital	59.0	70.6	59.0	61.0	62.8	59.8	65.3	60.0	50.5

All figures in millions of US Dollar except per share and labeled items

from a financial point of view, to extend credit to poorer accounts than a firm with a narrower working capital base.[8] The supplier with a strong financial cushion can absorb occasional losses on credit sales, whereas a weaker supplier may fear that a bunch of credit losses would push his own firm close to the brink.

Pure economic analysis reveals that the firm should sell up to that grade of credit risk where the probable amount of loss on uncollected accounts just equals the difference between the sales price and the firm's marginal cost. Of course, in actual operations, most firms have no precise measure of marginal costs but have adopted a variety of devices which, perhaps unconsciously, roughly approximate the results obtainable by the marginal theory.

---

[8] In some cases where immediate financial risks are high but the possibilities of good profits exist in the buying firm's activities, a supplier in a strong financial position has on occasion purchased some of its customer's common stock as well as sold to it on credit. This gives the supplier an opportunity to obtain possible compensating gains to offset the credit risk. The use of convertible securities is an analogous situation.

From the view of applied economics, a firm having open or extra capacity may profitably sell on credit up to that grade or class of risks where the chance of loss equals the margin of profit over the estimated variable costs at the going rate of production. For example, if at the current rate of output the variable costs on a product were 60 percent of the selling price and the credit loss in selling to a particular grade of customer was estimated at 30 percent of the sales, sales to this group of customers would cover out-of-pocket costs and return an additional 10 percent to cover fixed costs and possibly profits. To be precise, the seller could actually grant credit to the marginal customers up to the exact point where the predicted rate of credit losses (e.g., 40 percent) equaled the margin of profit over the variable costs at the given level of output.[9] However, credit acceptance tightens severely when the firm is operating at or near capacity.

Ideally, the credit managers' decision is mainly concerned with those cases where the financial characteristics of the customer (as determined by ratio analysis, etc.) are such as to make him appear to be close to the marginal credit risk. All firms which appear obviously stronger are extended credit, and weaker firms are denied credit terms. At the managerial level, there is often some tension between the sales manager whose goal is to maximize sales and the credit manager who wishes to minimize credit losses. As one observer noted, "If the credit manager and the sales manager do not get into an occasional fight, one or the other is not doing his job."

5.7   A Summary of Ratio Analysis

The financial community has long calculated ratios to assess the liquidity, profitability, leverage, and efficiency of firms. Ratio calculations and analysis to summarize financial information found on the balance sheet and income statement. The investor can often easily assess the financial health of a firm by calculating the ratios introduced in this chapter. Ratio analysis can be extremely useful in screening potentially poorer performing stocks, identifying problem firms, if not, bankrupt firms, using the Altman Z Score model. The reader must be careful to compare firm ratios with its competitor firms.

---

[9]No net profit might be made out of selling to this group, but we presume there would be profits on the sales to all the firms that were intra-marginal as far as credit loss probabilities are concerned.

References

Altman, E. I. (1968). Financial ratios, discriminate analysis and the prediction of corporate bankruptcy. *Journal of Finance, 23*, 589–609.

Baumol, W. J. (1952). The transactions demand for cash: An inventory theoretic approach. *The Quarterly Journal of Economics, 65*, 545–556.

Bernstein, L. (1988). *Financial statement analysis: Theory, application, and interpretation* (4th ed.). Irwin. Chapter 4.

Brealey, R. A., & Myers, S. C. (2003). *Principles of corporate finance* (7th ed., p. 29). McGraw-Hill/Irwin. Chapters 27.

Carleton, W. T., & McInnes, J. M. (1982). Theory, models, and implementation in financial management. *Management Science, 28*, 957–978.

Chandler, A. D., Jr. (1977). *The visible hand: The managerial revolution in American Business.* Cambridge: The Belknap Press of Harvard University Press. Chapter 13.

Guerard, J. B., Jr., & Schwartz, E. (2007). *Quantitative corporate finance.* Springer.

Guerard, J. B., Jr., & Vaught, H. T. (1989). *The handbook of financial modeling* (p. 4). Probus. Chapters 1, 2.

Hunt, P., Williams, C. M., & Donaldson, G. (1961). *Basic business finance*, rev. ed., Richard D. Irwin. Chapter. 8.

Johnson, H. T., & Kaplan, R. S. (1991). *Relevance Lost: The Rise and Fall of Management Accounting.* Harvard University Press. Chapter 4.

Maness, T. S., & Zietlow, J. T. (2005). *Short-term financial management* (3rd ed.). South-Western. Chapter 15.

Mao, J. C. T. (1969). *Quantitative analysis of financial decisions.* The Macmillan Company. Chapters 13 and 14.

Miller, M. H., & Orr, D. (1966). A model of the demand for money by firms. *The Quarterly Journal of Economics, 80*, 413–435.

Pogue, G. A., & Bussard, R. N. (1972). A linear programming model for short-term financial planning under uncertainty. *Sloan Management Review, 13*, 69–99.

Schultz, W. J., & Reinhardt, H. (1955). *Credit and collection management.* Prentice-Hall. Chapters. 8, 9, 10, 11.

Schwartz, E. (1962). *Corporate finance.* St. Martin's Press. Chapter 13.

Stone, B. K. (1972). The use of forecasts and smoothing in control-limit models for cash management. *Financial Management*, 72–84.

Standard & Poor's *500 Guide.* (2003). Edition. McGraw-Hill.

Vander Weide, J. H., & Maier, S. F. (1985). *Managing corporate liquidity: An introduction to working capital management.* Wiley. Chapter 10.

Van Horne, J. C. (2002). *Financial management & policy* (12th ed.). Prentice Hall.

Weston, J. F., & Copeland, T. E. (1986). *Managerial finance* (8th ed.). The Dryden Press. Chapters 8, 12.

White, G. I., Sondhi, A. C., & Fried, D. (1997). *The analysis and use of financial statements* (2nd ed.). Wiley. Chapter 4.

Chapter 6
Financing Current Operations and the Cash Budget

A firm with increasing sales volume needs increased current assets to service the new level of activity. Given normal inventory turnover, higher sales necessitate a higher level of stocks. Similarly, greater sales levels enlarge the average amount of receivables the firm carries since additions to the accounts come in faster than old accounts are collected. The corporation officers responsible for working capital management must decide how to finance the required increase in current assets.

A temporary or seasonal rise in activity may be financed with short-term or "temporary" sources. The additional gross working capital required can be acquired by an increase in suppliers' credit, negotiating a bank loan, or selling open market commercial paper. The use of short-term financing neither adds to nor detracts from the absolute amount of a firm's net working capital. The working capital ratios, however, decline, indicating an increase of risk in the firm's current position, as we discussed in the previous chapter. Table 6.1 illustrates the effect on the working capital position of a firm when an increase in inventory is financed with short-term debt, specifically with a bank loan. The current and acid test ratios decline; the net working capital turnover could rise to what might be a dangerous level. The working capital position of the company would be considered rather risky if it maintained these proportions in its current accounts for any length of time.

However short-term financing is comparatively cheap and flexible and may be used with relative safety under any of the following conditions:

1. The growth in sales is temporary or seasonal.

   Under these circumstances, at the end of the peak period, sales decline, with a resulting decline in the level of inventory required. As sales slacken, more old receivables are collected than is added in new accounts. The funds released from inventory and receivables can be used to repay current liabilities and restore the firm to its original (or, if there are profits, a better) working capital position. Nevertheless, even if the need for more current assets is expected to be temporary, using much additional short-term financing is dangerous if the starting level of net working capital is inadequate.

© The Author(s), under exclusive license to Springer Nature Switzerland AG 2022
J. B. Guerard Jr. et al., *Quantitative Corporate Finance*,
https://doi.org/10.1007/978-3-030-87269-4_6

Table 6.1 Effect of a bank loan on the working capital ratios

Working capital position before bank loan			
Cash	$50,000	Accounts payable	$100,000
Receivables	100		
Inventory	100,000		
Total current assets	$250,000	Total current liabilities	$100,000
Going level of sales		$1,500,000	
Net working capital	= $150,000		
N.W.C. turnover	= 10		
Current ratio	= 2.5		
Acid test ratio	= 1.5		
Working capital position after bank loan			
Cash	$50,000	Accounts payable	$100,000
Receivables	100	Bank loan	100,000
Inventory	200,000		
Total current assets	$350,000	Total current liabilities	$200,000
Going level of sales	$2,000,000		
Net working capital	= $150,000		
N.W.C. turnover	= 13.33		
Current ratio	= 1.75		
Acid test ratio	= .75		

2. The firm holds substantial net working capital.

   If the firm has a strong working capital position, it may be able to use some additional short-term financing with impunity. The resulting changes in the working capital ratios would merely bring them down to the standard or "normal" range. Within limits, the firm can safely finance an increase in operations without using any new long-term financing.

3. The operating margin over variable costs is high.

   A good profit margin over out-of-pocket costs may permit the firm to maintain its working capital from the operating profits on the increased sales. Whether this can be done depends on the comparative rate of fund inflow, the rate at which new assets are required, and the path of the rising sales curve. If the latter shows a short rise and then plateaus, the high profits may soon restore the working capital position that existed before the rise. If, however, there is no period of consolidation, profits alone may not be sufficient to furnish new net working capital at the desired rate. Relying on profits during a period of growth as the source of net working capital constrains the firm to careful budgeting of new fixed asset acquisitions. Moreover, such a policy is likely to keep dividend payouts niggardly.

Paradoxically, a firm finds the most pressure against its current financial position during periods of rising activity (and periods of sharp losses). In periods of mildly declining activity, the funds released by the decrease in current assets can be used to repay current debt. As long as the current ratio exceeds 1:1, an equal decrease in current assets and current liabilities improves the working capital ratios. Thus, unless significant losses are experienced in realizing the current assets, the firm develops increased operating liquidity.[1]

On the other hand, the conditions for the financial ruin of a firm are often set during the upgrade. If financial planning is lax and if the management fails to provide adequate working capital, the company's current position can be weakened as its sales increase. A company whose current and acid test ratios are low and whose volume of business is high relative to its net working capital is "overtrading." A relatively minor economic reverse—the failure of a large account receivable or a drop in inventory values—can bring the firm to financial disaster. The reader is reminded of the Altman Z Score calculated in the previous chapter and its predictive power in identifying potential distressed firms.

6.1   Sources of Short-Term Financing

Almost all firms make some use of short-term financing. It is comparatively cheap and relatively available. It may be obtained quickly and conveniently; some kinds of short-term financing are supplied practically automatically. (Among these are trade credit, and expense and tax accruals.) The major types of short-term credit are trade credit, bank credit, open market commercial paper, finance companies, and factors. The order in which these are listed is largely based on the prevalence and volume of the type of credit as used by the nonfinancial firm. Trade credit is part of almost every firm's financial structure because credit is customarily given on almost every commercial sale. Bank credit is available to sound firms of all sizes and is used frequently. The use of commercial paper is restricted to large firms of good standing. Commercial paper is one of the credit instruments traded on the "money market."[2]

---

[1] In the early stages of a sales downturn, the firm may accumulate unintended inventory. The firm will cut current purchases to restore the inventory to a more "normal" relationship to sales. Adequate net working capital is important to bring the firm through just such transitional situations.

[2] In the broadest sense, the money market may be defined as the broad complex of institutions extending, dealing in, and using short-term money credit. It refers both to the dealers in short-term credit obligations and the debt instruments and types of loans customary on the market. The operations of the money market set the structure and level of the short-term interest rates. The center of its activities in the United States is New York City, although its eventual scope is nationwide and even international. Among the major suppliers of funds to the money market are the banks, insurance companies, other financial institutions, and corporations having temporary excess funds. The important rates set on the money market are the rates on short-term treasury bills, commercial paper, bankers' acceptances, prime commercial loans, and margin loans to brokers on security collateral. The rates on the capital and money markets are related.

The total volume of credit supplied by the finance companies and factors is quite large, but although its use is spreading, it is not widely used outside a few industries.

6.2 Trade Credit

Trade credit, also called suppliers' credit, open book account, or ledger credit, develops because most commercial purchases need not be paid for immediately. During the period the purchasing firm has the goods but has not yet made payment, some part of its operating assets are being financed by its "creditor." On the books of the purchaser, the balance of not-yet-paid suppliers' bills is carried as accounts payable.

The length of time the purchaser has before payment and the discount, if any, for prompt or early payment are called the "terms." Credit terms vary according to the customs prevailing in different industries or lines of trade. One type of term, which is growing more common, is the single statement or end-of-the-month billing terms. A popular single-statement term is customarily written "2/10 e.o.m. n/30." All purchases made during the current month appear on the statement of the first day of the following month. If the purchaser pays the bill within 10 days after the end of the month (e.o.m.) in which he made the purchase, he may take 2 percent off the face of the bill as a cash discount or reward for prompt payment. If the purchaser fails to pay during the first 10 days of the month, the gross amount is due at the end of the month (i.e., net/30).

Single-statement terms save the seller booking costs and billing problems since he can consolidate the billing for sales during a month on a single statement. The cash discounts are actually less a reward for prompt payment than a penalty to the slow payer for the worry and extra risk put on the seller.

Open book credit is a cheap source of funds if the bills are paid within the 10-day end-of-the-month discount period. A purchaser who buys at the beginning of the month has 40 days in which to pay his bill and still receive his 2 percent discount. (As a matter of fact, the buyer may have upwards of 45 days of free credit, since a purchase made after the twenty-fifth of the month usually does not appear on the statements until the end of the following month.) If the discount credit period passes, however, the purchaser should consider the loss of the discount a cost of receiving credit for only the additional 20 days from the tenth until the end of the month. This loss of the 2 percent discount comes to a high rate when reckoned on an annual basis. The payment of 2 percent for use of funds for 20 days is equivalent to an annual simple interest charge of 36 percent.[3]

A firm that foresees difficulty in meeting the cash discount period on its payables should, if it can, arrange a bank loan or alternate source of credit. (The cost of a bank loan is hardly likely to be more than 6–10 percent per annum, and perhaps the cost of

---

[3]I.e., 20 days goes in 360 days 18 times ($18 \times 2\% = 36\%$).

factoring may run 12 percent. The cost in lost cash discounts may amount to 36 percent per annum or even more in some lines of trade.) However, some firms do not consider the lost cash discount as amounting to the equivalent of as much as 36 percent a year because in practice they delay payment beyond the whole term period. Of course, the firm is breaking the credit terms and is considered a delinquent or slow payer. Creditors may forgive such behavior if it happens infrequently. But frequent abuse of credit terms becomes known in the trade and the habitually delinquent firm may find its credit curtailed. It will not receive the best service, and some suppliers will not deal with it at all except on a cash basis.

6.3 Bank Credit

Types of Bank Loans Although commercial banks place funds in government and corporation bonds, purchase mortgages, and make personal loans, their prime function—one they are uniquely equipped to perform—is to make commercial loans.[4] The classical business or commercial loan is a short-term (30-to-90-day) grant of credit to a business unit to enable it to cover temporary gross working capital needs. A forecasted or planned event—i.e., a seasonal tapering off in activity that releases funds from inventory and receivables, or the development of long-term financing—will provide the firm with the money to repay the loan.

Although it is general bank policy to assume that all commercial loans will be paid out of the business' cash-generating capacity, there are two classes of loans, secured and unsecured. The unsecured loan is granted merely on the general credit of the firm; the secured loan is backed by a lien against a firm asset, by the pledge of the credit of some person in addition to that of the business, or by a lien against some personal assets of those connected with the business.

One type of secured loan is based on a lien against inventory. Inventories constitute good collateral for loans when the goods are homogeneous in nature and unspecialized, and where there is a going market (preferably organized) and the market price of the goods is relatively stable. Sometimes banks make loans against warehouse receipts. The inventory is held in a public warehouse and the goods may not be withdrawn from the warehouse by the firm without the counter-signature of the properly designated bank official. As any part of the goods is withdrawn, a proportionate amount of the original loan is repaid.

Some bank loans are secured by a lien against the firm's outstanding receivables. If the firm fails to repay its loan, the bank may try to satisfy its claim by collecting the accounts receivable for itself. This is, of course, a last resort; before granting the loan, the bank tries to ascertain whether the borrowing firm can repay out of its

---

[4] As of December 31, 1960, commercial loans totaled $42.4 billion and constituted 80 percent of aggregate earning assets of all commercial banks in the United States. As of December 2002, commercial loans totaled $965.3 billion.

regular operations. The asset pledged as security is only an additional safety feature which may help to reduce the bank's losses if the loan is not repaid as expected. The bank, however, may set the terms of the loan so the company will have to pay back part of the loan as the pledged accounts are paid. This keeps the proportion of bank debt to the value of the assets pledged constant over the life of the loan.

Many business loans are backed by the pledge of personal assets of the owners as collateral. The assets are usually liquid, that is, marketable securities, cash values of insurance policies, savings accounts, etc. In a personal proprietorship, these personal assets of the owners may furnish eventual backing for business debts, but only after the personal creditors are satisfied; moreover, all the general business creditors may look to the net personal assets. However, the direct pledge of personal assets as collateral gives the bank the first claim against these assets. The personal holdings of corporation owners are not normally within the reach of the business creditors. In a smaller company, this is not always an advantage, for it can restrict the amount of credit available to the firm. Often, then, the principal officers or major stockholders of a closely held corporation may pledge personal securities, savings, or other assets to help obtain a bank loan for their company.

Some loans are made on a cosigner basis. Some party other than the one primarily liable assumes contingent liability on the debt or guarantee the loan in case payment is not made as agreed. Friends and business acquaintances of the owner sometimes cosign a promissory note along with the proprietor of a small business. A supplier may cosign to help a new but growing customer establish bank credit. A parent company may sign or endorse the note of a subsidiary. In a family or closely held corporation, the principal officer may sign personally as well as in his capacity of agent for the corporation. In so doing, he waives his limited liability and pledges his personal credit at least for the amount of the particular debt. In general, as far as the bank is concerned, the effect of a cosigner loan is to bring in the credit and resources of another to help buttress the inadequate position of the original applicant for the loan.

A firm usually prefers to receive an unsecured loan rather than any of the varieties of secured loans. Receiving an unsecured loan is an implicit acknowledgment of a high credit rating. Moreover, although a secured loan does not change the overall situation of the other creditors, as shown by ratio analysis, from what would exist if an unsecured loan had been obtained, it does segregate a group of assets for the prior benefit of the secured creditor in the event financial difficulties should occur. Less desirable assets are thus left for the general creditors, and the rate of loss they could suffer may be higher than that of the secured creditors. If the general trade creditors know that a solid asset is pledged for a bank loan, they may be cautious in extending additional credit to the firm.

The foregoing analysis does not imply that incurring a bank loan normally weakens the general creditors' position. Ordinarily, the proceeds of the bank loan are used to reduce accounts payable or accruals. The remaining claims are in no worse position than before since the total debt is the same—only some of it has been transformed into bank debt.

Cost of Bank Credit   The cost of bank credit varies with the size and credit standing of the firm and according to the general level and the shape of the market structure of the interest rate. The prime rate is the lowest commercial loan borrowing rate and is set by the banks in the large cities as the charge for their biggest and soundest customers. The average interest cost is lower on a large loan than on a smaller one, because administrative, bookkeeping, investigating costs, etc. are more or less fixed and do not vary with the size of the individual loan. Also, credit risk tends to be higher for smaller concerns. For a given class of customers, the commercial loan rate is generally lower than the capital market rate (effective bond interest rate), because the bank loan rate belongs to the class of short-term interest rates, and in a normal market the interest rate curve is lower at the shorter maturities.[5] The commercial loan rate tends to be low, of course, during periods of slack demand for funds and when the monetary authorities are pursuing an easy money policy; the rate is higher during times of credit stringency or tight money.

Not only is the commercial loan rate usually a notch below the effective capital market rate, but the use of bank borrowing in lieu of a bond or debenture issue avoids flotation costs, the legal and administrative costs of drawing up the indenture, and the administrative costs (if the issue requires it) of registering with SEC. Further-more, a bank loan may be more closely tailored to the firm's requirements in contrast to a capital market issue; a bond issue, for example, is often floated ahead of the company's need for funds, thus requiring the firm to pay interest on the debt in the meantime. A seasonal inflow of funds can be used to extinguish bank debt and reduce the interest costs when the borrowed capital is not needed. If, however, the debt is a debenture or bond, the firm is not likely to make retirements out of seasonal funds, and interest payments continue. Of course, the net interest cost is somewhat reduced by the earnings on the short-term investments the firm can make in the interim.

On the other hand, although long-term capital may be costlier, its advantages over bank credit lie in its semipermanence—the lack of continuous current pressure on the firm gives management the opportunity to forecast and plan further into the future. Thus, many firms may prefer to fund (i.e., use long-term sources) as much of their current debt as possible, relying on bank credit for perhaps a small part of the peak of their seasonal needs. Such firms may have excess funds off-season, but capital costs can be reduced by placing seasonally redundant funds in such eminently secure short-term money market investments as treasury bills, tax certificates, marketable commercial paper, etc.

The interest cost on a commercial loan is usually computed on a simple interest basis for the life of the loan. Thus, if the interest rate is 5 percent per annum, the interest charge on a $100,000 loan for 60 days will be $833.33 (i.e., 5/6 percent of $100,000 for 60 days equals 5 percent interest on an annual basis). Where the amount of loan outstanding varies from day to day, the charge may be set up on a per-diem basis so that the lender pays only for the funds he actually uses. In making

---

[5] See Chap. 9 for a discussion of this phenomenon.

most loans, however, the bank discounts the note (i.e., the negotiable legal evidence of debt) and thereby earns a rate slightly above that of simple interest. When a note is discounted, the bank subtracts the interest for the term of the loan in advance and extends to the borrower the difference between the face amount of the note and the interest. In other words, the borrowers have the use of slightly less than the face value of their loans, and the bank retains the discounts which makes the interest charge minutely higher. The extra amount brought in by discounting rather than straight interest charges is relatively small.

Another device used by banks to vary the effective interest charge is the compensating balance. Traditionally the compensating balance has been set at 20 percent, which means that a successful applicant for bank accommodation is requested to keep 20 percent of his loan on deposit at the bank over the life of the loan. Although this requirement has been often cited as a safety measure, its actual effect is to raise the rate of interest on the usable portion of the loan.[6] To the extent the compensating balance exceeds the firm's normal cash balance, the effective interest cost on the loan exceeds the nominal rate. The use of the compensating balance is no longer a fixed custom. During times of easy money when the banks have considerable free reserves, no compensating balance is requested of prime borrowers. At other times, the percentage of compensating deposit required varies with conditions on the money market and the general credit standing of the borrower. Although a compensating balance is no longer an invariable condition, it is generally expected that the borrower will keep his normal working balance at the bank that accommodates his credit needs. Usually, a bank tries to meet the credit needs of its own depositors before financing other applicants.

Line of Credit The term line of credit is used in three different but related meanings. Suppliers may set a line of credit or credit limit for individual customers. This is the amount of outstanding trade credit they are willing to allow before cutting off further sales or reviewing the account. If the customer knows what the amount is, it is called an explicit line of credit; an implicit credit limit exists when the amount set by the suppliers' credit manager is not made known to the customer.

An open or general line of credit exists when a large corporation makes an agreement with its banks for a specified amount which the company may borrow at any time during a specified period, perhaps a year. The line of credit is subject to review and renewal at the end of the period. The firm is charged interest at the going commercial loan rate only for the time any loan is outstanding and only for the actual amount of the line of credit used. The firm usually pays an additional fee for the whole line of credit over the year—whether used or unused—since the bank obviously has to restrict its investment policies to have the standby credit available.

A seasonal or temporary line of credit with a bank provides a company with funds for its peak needs. The firm has only to pay interest on the funds it actually borrows,

---

[6]On a loan granted with a compensating balance, the bank is able to keep some proportion of its reserves otherwise committed to the loan. This savings in reserves can be lent to another borrower with a subsequent increase in the bank's earning power.

but the bank is morally committed to having the funds available. The amount of a
seasonal line of credit is generally determined by drawing up a cash budget (i.e., a
short-term cash forecast of the firm's operations). This not only enables the negoti-
ators to project the firm's likely maximum cash needs, but it also provides a
quantitative framework to ascertain how the firm will develop the free funds to
repay the bank debt after the operational peak is over.

Borrowing in Anticipation of Permanent Financing   Bank credit is a simple and
quick method of raising funds. Thus, it is often used as a temporary method of
financing until the firm determines how much permanent financing it needs, decides
upon the best type of security to float, or organizes the floating of an issue. Thus, a
utility firm may use bank financing to begin an addition to the plant, and after the
final cost is determined, float new bonds to retire the bank debt. A firm with access to
the capital markets may be caught in a sudden wave of expansion and finance its
increased activity with bank borrowings until it can organize a long-term security
issue. The bank may lend money to recognized companies that already have issues
traded on the secondary markets, with the full knowledge that the debt is to be retired
by the issue of additional stocks or bonds.[7] In some instances, the security issue may
be an afterthought. Both parties may have thought the debt would be paid out of the
cash flow and underestimated the time period for which the new financing was
required. When it becomes apparent that the firm cannot clear up its bank debt
without curtailing profitable opportunities and the banks are reluctant to supply long-
term capital, the management will seek other financing to repay the bank. If the bank
debt is repaid out of the proceeds of a bond issue, the whole process is called funding
the debt.

The sequence of bank credit to permanent financing may work quite well for the
firm already established in the capital markets. The pattern also describes a major
reason why a firm may venture out with its first public issue. But for many firms, the
link between the money markets and the capital market is broken. Because they are
small, or new, or for other reasons, these firms cannot make the transition to the
capital markets easily or cheaply. The company may have outgrown the resources of
its owners, and it may have opportunities that the commercial bankers recognize, but
which they are unwilling to finance because funds would be committed over too long
a term.[8]

---

[7]This seems to establish another advantage to having a publicly traded issue. The firm not only has
access to the capital markets but has a better rating in the money market in addition.

[8]The small business investment companies growing out of the Small Business Act of 1959 were
hopefully designed to fill this gap in our financial institutions by providing capital to intermediate-
size firms.

6.4   Other Forms of Short-Term Financing

Open Market Commercial Paper   Open market commercial paper is the term for the short-term negotiable notes with maturities from about 1 to 9 months that strong corporations float on the money markets. Open market paper is moved onto the market by commercial paper dealers (who sell the issue to insurance companies), pension funds, business corporations, and smaller commercial banks for the short-term investment of their liquid reserves.[9] It is an important fundraising device for large commercial finance companies.[10] In total, the amount of finance company commercial paper floated is about double the amount floated by all other kinds of companies combined. The rates on commercial paper are generally below the prime loan rate but somewhat above the prevailing rate on treasury bills of the same maturity. The advantages to the lender in buying commercial paper are its ready availability in convenient maturities and its minimal administrative detail and costs. The advantage to the borrower is that the wide distribution his paper may obtain in what is essentially a nationwide market frees him from relying on local bank resources and provides him with lower net interest rates. On the other hand, the use of commercial paper does not establish personal rapport with the bankers, who may sometimes advise and support the firm through temporary hard times. Neither does the use of commercial paper lend itself to easy change of terms and renewal as is sometimes possible with a bank loan. The total volume of commercial bank loans far exceeds the funds raised by the issuing of commercial paper, although the latter is by no means negligible.

The large commercial finance companies are the department stores of the financial world. Initially adjuncts or subsidiaries of the automobile companies, their major business still consists in discounting dealer sales paper and financing dealer automobile inventories. Branching out into other financing areas, the large companies usually have a personal finance subsidiary and an account receivable factoring subsidiary. They finance major furniture and appliance dealers both for inventory and on installment receivables, and they may do some direct sales financing themselves. In many activities, the finance companies compete with commercial banks. Although their charges may be higher (the finance companies often benefit from closer contact with the dealers), they have a specialization that enables them to act efficiently and quickly, and the willingness to take more risk than the bank. A large percentage of the funds which the finance houses dispense are raised through the banking system either by the establishment of lines of credit with large banks or by

---

[9]The spread received by the dealers for floating the notes is very low, but since the issues are large and the issuing firms have the highest credit standing, marketing problems are not difficult.

[10]In 1960 about $3.5 billion of finance company commercial paper outstanding was placed with insurance companies, corporations, or other large buyers without passing through the dealers and the open market. By December 2002, the commercial paper market had grown to $1321.5 billion, of which $1167.5 billion was finance and $154 billion was nonfinance. Source: U.S. Federal Reserve (www.federalreserve.gov).

the sale of open market paper, a large percentage of which is picked up by the banks outside of New York.

Factoring is an important source of financing in the textile and garment trades and also appears in leather goods, sporting goods, and lumber trades, among other industries. Many factoring companies are today subsidiaries of the major finance companies. Factoring arrangements are based on an advance of funds by the factor against an assignment of trade receivables. The factor's relation with the company is governed by an underlying agreement which can be varied to meet different circumstances. Generally, the factoring agreement is not a one-shot deal but a continuing arrangement; the company's debt to the factor is decreased as the assigned receivables are collected, but new debt is created as the receivables arising out of new sales are surrendered to the factor for immediate funds. Usually, the factor provides cash up to about 75 percent of the firm's accounts receivable. In certain cases, however, the firm may be given the privilege of assigning up to 85 percent of its accounts if it pays a higher financing charge for the last 10 percent.

Factoring arrangements are either recourse or non-recourse. In a non-recourse factoring agreement, the factor buys the accounts outright and cannot collect from the selling company for any losses on defaulted accounts. Of course, the factor advances only about 75 percent of the face of the accounts, and this gives him some margin to cover losses.[11] Moreover, in non-recourse factoring, the factor has the right to reject accounts that do not meet his credit standards. In factoring with recourse, the factor does not take the loss on bad accounts; the borrowing firm makes good on all non-collected receivables. The financing costs are generally lower on the recourse agreement.

Factoring agreements are also classified as being on a notification or non-notification basis. In a non-notification arrangement, the firm's customers are not necessarily told of the factoring agreement, and they continue to send their payments to the office of the selling firm. Here the receipts are segregated, and the proper share is turned over to the factor. Borrowing companies generally prefer a non-notification basis. When the accounts are on a notification basis, the customers are instructed to send their payments directly to the factor. The factor, of course, applies these receipts against the selling firm's indebtedness. Under a notification, non-recourse basis, the factor often in effect operates the selling firm" credit department. He makes the initial credit analysis on a sale, takes charge of collections, and follows up on bad accounts.

Factoring charges are usually based on the firm's daily balance. Depending on the level of the general interest rate, the specific factoring agreement, and the credit standing of the selling firm and its customers, factoring rates may run from something below 1 percent to 3 percent per month. These relatively high charges, however, compare favorably with the customary cash discounts running as high as 8/10 n/60 in some trades where factoring is common. Factoring costs, of course, run

---

[11] When the account is collected in full the 25 percent coverage belongs to the seller; it is either returned to him or used to reduce his balance at the factors.

higher than the bank loan rates, but a factor takes risks the bank will not, and many of the services a factor performs are not available at the banks.

Factors are heavily used in industries where entry of new firms has been comparatively easy and where many companies are short of working capital. Thus, although the use of factoring does not weaken the financial position of the firm, its customary use by companies with inadequate initial capital has sometimes led to the mistaken notion that factoring is a "last resort" type of financing. In spite of this psychological drawback, factoring is a useful and convenient method of speeding up the cash cycle for many business firms.

6.5 Quantitative Working Capital Models: Cash Management

There are several well-known quantitative working capital models. The first model, developed by Baumol in 1952, optimized the firm's cash balance by minimizing total costs per year of maintaining cash balances. Total costs, TC, were composed of transactions costs and an opportunity cost of maintaining additional cash balances.

$$TC = t\left(\frac{Cu}{C}\right) + i\left(\frac{C}{2}\right) \tag{6.1}$$

where

- Cu = cash usage for the year
- t = transactions costs per sale or purchase of marketable securities
- i = annual interest rate on marketable securities
- and C = cash balance

One minimizes the costs of selling and buying securities and holding cash on which one could earn interest. Thus, to minimize one's total costs, one takes the derivative of TC with respect to its cash balance and sets the derivative to zero.[12] The optimal cash level, C^*, is:

$$C^* = \sqrt{\frac{2tCu}{i}} \tag{6.2}$$

The optimal cash balance is positively associated with total cash usage and negatively associated with the interest rate, in the Baumol formulation. If a firm

---

[12] $dTC/dQ = i/2 - tCu/Q^2 = 0,$
$i/2 = tCu/Q^2$
$Q = \sqrt{2\ tCu/i}$

uses $100 million of cash in a given year, the interest rate is 5 percent, and its transaction cost is $50, then its optimal cash balance is:

$$C^* = \sqrt{\frac{2(\$50)\$100,000,000}{.05}}$$

$$C^* = \sqrt{\frac{10,000,000,000}{.05}}$$

$$C^* = \$447,213.60$$

The optimal cash balance for the firm should be $447,213.60.

A second cash management model is the Miller-Orr (1966) control limit model that built upon the Baumol formulation. The firm earns interest on its savings deposits and has a noninterest-bearing cash balance. The firm transfers M dollars from its savings deposit into its cash account. The average cash balance is $\frac{M}{2}$, or m. Cash usage to a minimum acceptable level, zero, occurs every L days. Cash transfers occur every M/m days. In the Miller-Orr framework,

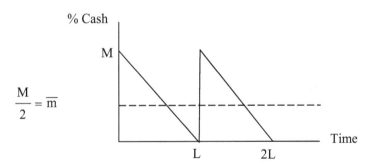

Miller and Orr defined v as the annual interest rate and γ as the transaction cost. The Baumol formulation could be written as:

$$M^* = \left(\frac{2\gamma m}{v}\right)^{.5} \tag{6.3}$$

The optimal period of M* transfers is:

$$L^* = \left(\frac{2\gamma}{m^*}\right)^{.5} \tag{6.4}$$

If p is the probability that the cash balance will increase by m dollars, then 1−p, or q, is the probability of a cash decrease. The expected costs of maintaining the cash balance over a T-day planning horizon is:

$$E(dc) = \gamma \frac{E(N)}{T} + vE(M) \tag{6.5}$$

where

- $E(dc)$ = expected daily costs
- $E(N)$ = expected number of cash transfers
- $E(M)$ = expected (average) daily cash balance

Miller and Orr further assumed that a firm maintained an upper bound on its cash balance of h, a balance level of z, and its transfer of lump sum cash would be $(h - z)$ dollars. Miller and Orr converted the expected value of the cash duration, $Z(h - z)$, into time (days by dividing by t, the number of operating days). The expected value of the cash duration in days is:

$$= \frac{Z^1 \left(h^1 - Z^1\right)}{m^2 t} \tag{6.6}$$

where

- $Z^1 = Z - m$
- and $h^1 = h - m$

The mean cash balance is $\frac{h+z}{3}$.

The cash transfer, Z, is $h - Z$.[13]

Miller and Orr found that the optimal cash transfer was a positive function of cash transfer cost, daily cash balances, and an inverse function of the daily interest rate.

$$z^* = \left(\frac{3\gamma m^2 t}{4v}\right)^{.333} \tag{6.10}$$

- $Z^* = 2z^*$
- and $h^* = 3z^*$

---

[13] Thus Miller and Orr minimized total costs:

$$\min \, \varepsilon(dc) = \frac{\gamma m^2 t}{zZ} + \frac{v(Z + 2z)}{3} \tag{6.7}$$

$$\frac{\partial \varepsilon(dc)}{\partial z} = -\frac{\gamma m^2 t}{Z^2 z} + \frac{2v}{3} = 0 \tag{6.8}$$

$$\frac{\partial \varepsilon(dc)}{\partial Z} = -\frac{\gamma m^2 t}{Z^2 z} + \frac{\gamma}{3} = 0 \tag{6.9}$$

The Miller-Orr model allows the corporate treasurer to invest excess cash when cash balances reach the upper control limit, h, returning the cash balance to its mean cash balance.

6.6   The Cash Budget

Short-term borrowing is generally arranged on the anticipation that the normal operations of the business will create enough "cash throw-off" to retire the debt. The cash budget is a device for making the assumptions behind this anticipation explicit. The cash budget is a short-term forecast of the firm's cash operations. It shows the items that will absorb funds as activity rises, and it indicates how cash will be made available to retire current obligations after the peak need is over.[14]

Since the cash budget estimates the firm's short-term needs for cash, it is extremely useful in arranging bank loans and seasonal lines of credit. However, our purpose in presenting the cash budget here transcends its practical value as a tool of financial planning. The cash budget serves as a clear expository device depicting the absorption and release of funds as the pulse of business operations surges upward and then slackens.

The following information or estimates, broken down by the relevant time periods, are needed to construct a cash budget:

1. A forecast of sales, broken down into the probable proportions of cash and credit sales
2. An estimate of purchases coupled with information on customary credit terms
3. The estimation and construction of a function relating the volume of cash collections to the amount of outstanding receivables
4. An estimate of out-of-pocket charges (or cash operating expenses) for the period ahead

For illustration suppose we construct a cash budget to apply to the following problem: You are the treasurer of Goldman Dress Manufacturing, Inc. You are about to negotiate a credit limit at the bank for your seasonal operating needs.

It is estimated that about 80 percent of the accounts receivables of the preceding month are collected each month. About 90 percent of the collections obtain the cash discount.

- Variable expenses are 20 percent of each month's sales.
- Staff salaries and other administrative expenses run $10,000 per month.
- Depreciation on equipment is $5000 per month.

---

[14]If the cash budget indicates that the firm will require funds for a longer term, the management should, if it can, arrange more permanent financing.

Table 6.2 Cash budget of Goldman Dress Manufacturing, Inc.

	1st month	2nd month	3rd month	4th month	5th month
Accounts receivable, beginning of the month	50,000	110,000	172,000	284,000	256880
Gross collections for the month (80% of the above)	40,000	88,000	137,600	227,520	205,500
	10,000	22,000	34,400	56,880	51,380
Sales (estimated per schedule B)	100,000	150,000	250,000	200,000	n.a.
Accounts receivable–end of the month	110,000	172,000	284,400	256,880	n.a.
Cash receipts:					
Net cash collection after discounts					
(98.2% of the gross collections)	39,280	86,416	135,123	223,425	201,805
Estimated cash disbursements:					
Payment of trade accounts–less discounts	36,800	138,000	184,000	23,000	23,000
Variable expenses (est. for mo. 5)	20,000	30,000	50,000	40,000	30,000
Administrative expenses	10,000	10,000	10,000	10,000	10,000
Total disbursements	66,800	178,000	244,000	73,000	63,000
Net cash gain or (outflow) for the month	(27,520)	(91,584)	(108,877)	150,425	138,805
Starting cash 1st month	10,000-				
Cumulative cash position (deficit)	(17,520)	(109,104)	(217,981)*	(67,556)	71,249

*Peak forecasted cash needs. Probable line of credit around $230,000.

Table 6.2 depicts the construction of a cash budget based on the information given on the Goldman Dress Manufacturing Company. Although all the figures on the budget appear solid, it is important to remember that the whole schedule is built on a forecast of short-term sales, an estimated purchase schedule related to these sales, and a cash collection estimate based on past experience of the relation of cash payments to the receivable balances. If any of these factors deviate from expectations, the budget will be more or less imprecise. Of course, a budget is not sacred and is subject to revision to conform to developing facts. It is a guide to intelligent action, not a fixed concept.

According to the Goldman Dress Manufacturing, Inc. budget, the peak need for cash will reach approximately $218,000. This estimate is based on the assumption that payments for the purchases and receipts from the receivables will come in about the same time during the month. For some companies, payments are customarily concentrated at the beginning of the month and receipts come in evenly throughout the month. It is possible to have an intra-monthly peak which, though it exists but a short time, must be covered. In such a case, the time periods upon which the budget is constructed should be made shorter than a month so that the peak need for cash is clearly indicated.[15]

The information given in the Goldman Dress Manufacturing Company example included depreciation charges of $5000 per month. The reader might note that this item is extraneous and was not used in constructing the cash budget.

Although data were not specifically given for the fifth month, the major cash flow items affecting the fifth month were implicitly contained in the information

---

[15] A budget can be made to cover the first third of every month and then the last two-thirds. This is useful, if there is a peak need around the tenth of the month.

presented. The estimated cash flow for the fifth month is particularly interesting, for it shows the company well out of temporary debt with cash holdings of $71,000— approximately $60,000 more than at the start of the season. This does not necessarily mean that the company will make a profit for the year's operation, or even for the season. Since there are other expenses or incomes which may be noncash in nature, the cash flow for any short-term period seldom corresponds to the usual definition of profit or loss. However, in constructing a budget even for so simplified a case as the Goldman Company, we can observe how the slackening activity led to a net cash inflow more than sufficient to repay the indebtedness built up at the height of the seasonal operations. Thus we have an illustration of the argument presented at the beginning of this chapter.

If a cash budget shows that the firm's increasing activity is long run so that the repayment of short-term borrowing might prove difficult, the management may consider raising longer-term capital-funding the short-term debt. Of course, it must determine whether moving to this new, more permanent level of activity is profitable. A firm does not have to increase its activities if there is no economic gain in it. If the firm decides to use more permanent financing, the factors of cost, risk, and the existing financial structure will determine the type of issue.

The cash budget problem has been presented here in isolation. Using the information given, we determined that the Goldman Company should apply for a seasonal line of credit with a maximum (allowing for possible minor miscalculations) of about $230,000. In the context of our little example, the other circumstances of the Goldman Manufacturing Company's financial position can be assumed to support this conclusion as the best course of action. On the other hand, suppose the financial structure of the Goldman Company was already risky. Then the firm should try to arrange additional long-term financing carry a bigger net working capital balance and reduce the amount of its necessary seasonal short-term borrowing. If this procedure were to lead to an unduly heavy cash balance in the off-season, the excess funds can earn a return (albeit low) by investment in the money market.[16]

The cash budget—like other tools of financial analysis—is not an end in itself but merely a guide to understanding. Comprehending the financial problems of a firm may require the use of more than one method of analysis. When, however, these methods overlap or duplicate the same information, the simplest tool adequate to the problem is preferable.

---

[16] Many large corporations are suppliers of funds to the money market, investing temporarily idle cash in commercial paper, bankers acceptances, etc.—but above all in short-term US Treasury notes and bills. In the past, short-term corporate investment funds often went into "call loans" to stock brokers. In 1929, an exceptionally large amount of these corporation funds helped support the excesses of the stock market by supplying cash for margin loans to speculators. Member banks were prohibited from placing "call loans" for nonbank lenders in 1933.

References

Altman, E. I. (1968). Financial ratios, discriminate analysis and the prediction of corporate bankruptcy. *Journal of Finance, 23*, 589–609.

Baumol, W. J. (1952). The transactions demand for cash: An inventory theoretic approach. *The Quarterly Journal of Economics, 65*, 545–556.

Bernstein, L. (1988). *Financial statement analysis: Theory, application, and interpretation* (4th ed.). Irwin. Chapter 4.

Brealey, R. A., & Myers, S. C. (2003). *Principles of corporate finance* (7th ed.). McGraw-Hill/Irwin. Chapters 27, 29.

Carleton, W. T., & McInnes, J. M. (1982). Theory, models, and implementation in financial management. *Management Science, 28*, 957–978.

Chandler, A. D., Jr. (1977). *The visible hand: The managerial revolution in American business.* The Belknap Press of Harvard University Press. Chapter 13.

Guerard, J. B., Jr., & Schwartz, E. (2007). *Quantitative corporate finance.* Springer.

Guerard, J. B., Jr., & Vaught, H. T. (1989). *The handbook of financial modeling.* Probus, Chapters 1, 2, 4.

Hunt, P., Williams, C. M., & Donaldson, G. (1961). *Basic business finance* (rev. ed.). Richard D. Irwin. Chapter. 8.

Johnson, H. T., & Kaplan, R. S. (1991). *Relevance lost: The rise and fall of management accounting.* Harvard University Press. Chapter 4.

Maness, T. S., & Zietlow, J. T. (2005). *Short-term financial management* (3rd ed.). South-Western. Chapter 15.

Mao, J. C. T. (1969). *Quantitative analysis of financial decisions.* The Macmillan Company. Chapters 13 and 14.

Miller, M. H., & Orr, D. (1966). A model of the demand for money by firms. *The Quarterly Journal of Economics, 80*, 413–435.

Pogue, G. A., & Bussard, R. N. (1972). A linear programming model for short-term financial planning under uncertainty. *Sloan Management Review, 13*, 69–99.

Schultz, W. J., & Reinhardt, H. (1955). *Credit and collection management.* Prentice-Hall. Chapters. 8, 9, 10, 11.

Schwartz, E. (1962). *Corporate finance.* St. Martin's Press. Chapter 13.

Standard & Poor's *500 Guide.* 2003 Edition. McGraw-Hill.

Stone, B. K. (1972). The use of forecasts and smoothing in control-limit models for cash management. *Financial Management*, 72–84.

Vander Weide, J. H., & Maier, S. F. (1985). *Managing corporate liquidity: An introduction to working capital management.* Wiley, Chapter 10.

Weston, J. F., & Copeland, T. E. (1986). *Managerial finance* (8th ed.). The Dryden Press, Chapters 8, 12.

White, G. I., Sondhi, A. C., & Fried, D. (1997). *The analysis and use of financial statements* (2nd ed.). Wiley. Chapter 4.

Chapter 7
Capital and New Issue Markets

The essential advantage of the corporation over other forms of business organization lies in its ability to bring together large amounts of capital. This is especially true of those corporations, which are *publicly held* or *widely held*.[1] The small *family-held*, *closely held*, or *closed* corporation has basically the same financial sources as other business forms, i.e., trade credit, the banks, and the resources of family and friends willing to invest in the business. A widely held corporation, although it will also finance through trade creditors and banks, has the additional ability to tap the flow of investment funds available in the security markets, the market for stocks and bonds. The security market, alternately called the investment, financial, or capital market, comprises the group of institutions through which old corporate securities are traded and new issues of securities are floated. These markets have provided a significant part of the capital that has nourished the growth of corporate enterprise. Thus, to understand the birth of new corporations and the growth of established firms, it is necessary to understand the functions of the security markets.

The capital market is divided by function into two aspects: the secondary or old issues market and the primary or new issues market. The secondary market makes it convenient for an owner of securities—stocks or bonds—to sell them to someone else. The primary market is concerned with floating new issues, with raising funds for existing corporate expansion or the promotion and creation of new publicly held firms.

---

[1] English practice distinguishes between "public corporations" (basically our widely held corporations) and "private corporations."

© The Author(s), under exclusive license to Springer Nature Switzerland AG 2022   119
J. B. Guerard Jr. et al., *Quantitative Corporate Finance*,
https://doi.org/10.1007/978-3-030-87269-4_7

7.1 The Secondary Markets

The secondary markets consist of the organized stock exchanges—the so-called listed exchanges—and the group of security dealers and traders composing the over-the-counter market organized as the National Association of Security Dealers Automated Quotation (i.e., NASDAQ). Stock exchanges have changed since Guerard and Schwartz (2007). In the United States, the organized exchanges include the famous New York Stock Exchange, NYSE, the much smaller American Stock Exchange (AMEX, also in New York City) and the various smaller regional exchanges. NYSE American, formerly known as the AMEX, and more recently as NYSE MKT, is an American stock exchange situated in New York City.[2] NYSE Euronext acquired AMEX on October 1, 2008, with AMEX integrated with the Alternext European small-cap exchange and renamed the NYSE Alternext US. In March 2009, NYSE Alternext US was changed to NYSE Amex Equities. On May 10, 2012, NYSE Amex Equities changed its name to NYSE MKT LLC. Following the SEC approval of competing stock exchange IEX in 2016, NYSE MKT rebranded as NYSE American and introduced a 350-microsecond delay in trading, referred to as a "speed bump," which is also present on the IEX.

In 2019, some 2400 companies were listed on the NYSE with a market capitalization of $22.9 trillion. The NYSE domestic companies met the following criteria for listing[3]:

1. Total number of shareholders of 2200 or 2000 round-lot holders.
2. 1 100,000 shares of stock outstanding
3. Market value of public companies of at least $200,000,000.
4. An affiliated company standard in which the stockholders (1) requirement is met and market capitalization of entity exceeds $500,000,000.
5. Aggregate pre-tax earnings over last three fiscal years of $10,000,000 as:

    (a) Most recent fiscal years pre-tax earnings of $2,500,000 and previous two fiscal years pre-tax earnings of $2,000,000 (each); or.
    (b) Aggregate last three fiscal years operating cash flow of $25,000,000, if global market value exceeds $500,000,000 and revenues exceed $100,000,000 in last 12 months; or.
    (c) Average global market capitalization exceeds $1,000,000,000 and last fiscal years revenues exceed $100,000,000.

The number of companies listed for trading on the regional exchanges is considerably less. Listed companies are considered to have an active following or market. However, the securities of any company which develops a "following" may be traded at some time or other. Thus the over-the-counter market may deal

---

[2] AMEX was previously a mutual organization, owned by its members. In December 2002, some 796 companies were traded on the American Stock Exchange.

[3] Source:www.nyse.com. Rule 102.01C

sporadically with the securities of more than seventy thousand companies. The shares of most major well-known companies are listed on the New York Stock Exchange, but some prominent newer companies are traded on the NASDAQ which has grown enormously in recent years including such notable firms as Microsoft and Apple Computers. The relative out-performance of the NASDAQ during the 1985–2000 period gave rise to its new Tower at Times Square and coverage on finance news. The world's largest electronic exchange lists over 3600 companies, and its informal over-the-counter market handles the largest volume of transactions in both government and corporation bonds.

The organized stock exchanges are strictly secondary markets: they do not float or sell any new issues. The over-the-counter market straddles both functions; the firms on it handle the trading of previously issued securities and also take part in the sales effort on new issues. The function of the organized exchanges and the NASDAQ should not be misunderstood. The monies invested for the purchase of existing stocks do not go to the underlying company. They pass from the individual buyers to the individual sellers. The stock exchange does not set the price of securities. The "market," an abstraction for the freely competing buyers and sellers bidding for and offering various securities, sets the price on each individual security.

A letter published in the *Wall Street Journal* some years ago defines the contribution of the secondary market to the economy:

> A stock exchange is a second-hand market; that is, it enables those who have "old" securities to sell them to others. Such change of ownership obviously has no effect on the amount of capital assembled by the company whose securities are traded. The great economic significance of the stock exchanges in this context is that if such a splendidly functioning second-hand market were not available for disposing of securities once purchased, there would be a greater difficulty than there is in selling new issues and, thus, in accumulating large aggregates of capital.[4]

The secondary market furnishes the desirable quality of liquidity or marketability to investments in the stocks and bonds of traded corporations. The market will accept a somewhat lower rate of return on liquid or marketable investments over those that have no market. The cost of capital is lower for corporations that are established on secondary markets. Equivalently we can say that the issuing corporation can acquire funds at a lower cost if its securities have access to the secondary market.

The secondary market provides some other useful function in a modern economy. By setting the levels of prices on already issued securities, it indicates the yields, interest rates, and price-earnings levels that must be placed on new securities to float them successfully on the primary markets. Thus the secondary market acts as an indicator or measuring gauge for the costs of new capital funds. It also helps to ration capital. Since a relatively higher price is set on the securities of industries whose prospects seem promising and a lower price placed on the securities of industries

---

[4]Quoted from a letter by Leverett Lyon appearing in the *Wall Street Journal*, March 20, 1955.

whose future seems less bright, the secondary markets help direct the flow of capital funds toward the more favored areas of the economy.[5]

The secondary market facilitates the mobility of funds. It harnesses the resources of investors who have somewhat more aversion to risk in support of the primary market. This occurs because some of the overflow of funds out of the secondary market helps to furnish money for the new issue (primary) market.[6] An idealized schematic explanation of this statement would take three steps:

1. Primary market. The venturesome investor puts his funds into a new enterprise.
2. The over-the-counter market (part of the secondary market). The new company having become fairly well established, our risk-taking investor sells its securities. He takes the funds released by the sale and searches for another new venture.
3. The organized exchange. After some seasoning on the over-the-counter market, the firm's securities are listed on an organized exchange. The securities are gradually taken by relatively more conservative investors. The funds from these sales are placed in the securities of newer and smaller companies traded over the counter. As the securities are passed up the ladder to somewhat more conservative investors, funds filter down to the new issue market. The secondary market enables the providers of true risk capital to withdraw their funds from successful ventures and seek out new enterprises.

The secondary market performs its functions well when it is in a more or less "normal" range. When the market is in a speculative fever or seized in a wave of excessive pessimism, its malfunctioning probably adds an extra disequilibrating factor to the aggregate problem.

7.2 The Primary Market

The primary market is the complex of investment bankers, over-the-counter houses, and security dealers who combine to sell new issues to individual investors and institutions.

Although it is smaller in dollar volume than the trading markets, the primary market from an aggregate economic view is more important. On the secondary market, if one investor sells his securities to another, no new funds necessarily enter the economy, no new business assets appear, and the underlying company

---

[5] In a price or market economy, the process of allocation starts when increased demand raises profits or rates of return in a given industry or segment of the economy, or decreased demand lowers profits. However, the security markets are quick to recognize such portents and mark prices up or down accordingly.

[6] If *net* new funds are invested in the secondary markets, some portions of old portfolios— securities—will have been liquidated, turned into cash. Since it is unlikely that all of these funds will be spent on consumption, a large portion must be reinvested in the primary market, i.e., net new securities.

does not grow nor does the productive capacity of the economy increase. But when a new issue is sold on the primary market, the company making the sale raises new funds from the investors. These funds may be used to pay off old debt, but more likely they will be used to acquire new assets: new inventory, plant, or equipment. So used, these funds re-enter the income stream, increasing the gross national product. Moreover, these new assets constitute real capital formation, increasing the productive capacity of the firm and the economy. The primary market is a part of the machinery through which aggregate savings are tapped to create net new investment and economic growth. If the savings of the economy do not reenter the income stream at the proper rate, the rate of economic growth will be retarded.

The investment banking firms and the new issues department of the brokerage firms are an integral part of the primary market. The new issue firms help devise the legal content of the new issue, agree on the interest rate or price that can be set on the securities, and then take over the sale of the issue on the market. (The investment firm that finds the issue and guides it through the primary market is called the originating house.) When a security firm floats an issue of securities on a firm commitment or underwriting basis, it purchases the whole issue directly from the corporation before it is placed for sale on the market. US firms have moved from using rights issues to using the firm commitment method of selling seasoned public offerings (SPOs) during the 1950–2003 period, and 60% of IPOs are sold using the firm commitment method (Eckbo & Masulis, 1995). The funds to enable the investment bankers to carry the securities (their inventory) until they are sold are generally obtained from the commercial bankers. The price the new issue firms pay to the issuing corporation is somewhat less than the buying price set for the market. This difference, the charge, or "spread," compensates the security firms for their advice, their selling services, and for the risks they assume.

There are some 8–10 investment firms who handle most of the originating and underwriting business. Such firms as Goldman Sachs, Merrill Lynch, and Morgan Stanley are among the largest underwriters. Although these firms are rivals, they will cooperate and form groups or syndicates of various sizes to float large issues. Many possible combinations of bankers and dealers and organizational structures can be used in floating an issue of securities; there are, however, three recognizable functions performed for most security issues: originating, underwriting, and selling. (On a small issue, one or all of these functions may be combined.)

7.3 The Originating House

The originating house advises the issuing firm on the features, terms, and prices of the security to be floated. It sets the spread it will take for floating the issue. It undertakes to get together the underwriting group. The spread is divided among the underwriters and sellers. In addition to the originating house, a relatively small fee for being the sponsor of the whole issue obtains its pro-rata share of the spread, usually a major allotment, for its participation in the underwriting. (The largest share of the spread usually goes to the selling group since they generally put in the most effort and time.) The originating house also receives its proportion of the spread on the amount of the securities that they hold out to sell themselves. Thus, the competition to originate issues rests on the advantages in functioning and obtaining returns on three levels, originating, underwriting, and selling.

7.3.1 The Underwriting Group

The underwriting group, of which the originating house is a part, agrees to purchase the total issue at a fixed price. Each house assumes a part of the issue. The members of the underwriting group are on a firm commitment basis. Thus, they may take a loss if the issue moves slowly or if the price at which the market will take the securities drops below the spread; in short, they risk a loss if they are not able to sell all of their quota at the price agreed upon with the issuing corporation.

7.3.2 The Selling Group

The selling group is composed of the underwriters plus other firms invited to help sell the issue. Usually, a concerted drive is made to sell the issue (close the books) as quickly as possible since the risk to the underwriters increases as the selling period lengthens. Generally, the selling group is on a "best-effort" basis. This means they get a commission on what they sell; they are not, however, forced to buy the unsold part of their quota, but may return it to the underwriters. Of course, a firm in the selling group that fails to place its quota too often is likely to find itself uninvited to participate in later flotations.

7.3.3 Other Aspects of Investment Banking

Best Effort vs. Firm Commitment or Underwriting Basis
Sometimes the securities of a new, lesser-known, or weaker firm are floated on a "best-effort" basis. The investment banker has no commitment to purchase the issue at a fixed price. He does his best to sell the issue, and he gets a commission on all he sells. In contrast to the underwriting agreement, the investment bankers take no risk of loss since they may return any unsold securities to the issuing corporation.

7.4   Initial Public Offerings (IPOs)

An interesting aspect of investment banking is the market for IPOs, the initial public offering of the shares of new companies. In the 1990s, the floatation of IPOs in the "hi-tech" industries was a source of considerable income for investment banking firms. IPOs have been sold on a firm commitment or underwriting basis rather than floated on a "best-effort" basis during the 1980–2003 period. The investment bankers gave no commitment to purchase the issue at an agreed-upon fixed price.

Profits to the security firms on a successful IPO could be quite large. The spread between the market offering price and the return to the issuing companies could be as high as 20% (Ibbotson & Jaffee, 1975; Ritter, 1984; and Chen & Ritter, 2000). In the boom period for new issues, it has been apparently common practice for the IPO shares to be underpriced. There is a pronounced underpricing differential between IPOs with an offering price below $3.00 and those with an offering price exceeding $3.00. Ibbotson et al. (1994) report that IPOs with offering prices less than $3.00 produced returns of 42.8% for the 1975–1984 period, whereas the corresponding IPOs with offering prices exceeding $3.00 produced an average return of 8.6%. In the trading hours shortly after the issue was placed on the market, the market price might rise 50–100% over the original offer price. It seemed that the managers of the corporation were not opposed to these phenomena. Although the funds going directly to the corporation were somewhat less than they might be, the pro-forma wealth in their own holdings of shares rose with the market. In the 3 years following IPOs during the 1975–1984 period, Ritter (1991) found that IPOs substantially underperformed matched firms on a risk-adjusted basis. (A profitable investor should only be a short-term holder in IPOs.) A recent study of more than 1100 IPOs, raising between the $20 million and $80, launched during the 1995–1998 period found that 90% of the issuers paid gross spreads on underwriting discounts of exactly 7% (Chen & Ritter, 2000). The IPO gross spreads in many developed nations were approximately one-half the US spreads, calling into question the competitiveness of the US IPO market. The largest five investment banks underwrote 60% of the deals during the 1999–2000 period, and the largest 15 banks accounted for 95% of the deals (Fang, 2005). Fang found evidence that the larger, more reputable banks had more stringent underwriting standards, particularly for junk bonds, and produced higher net proceeds for their clients, despite charging higher fees.

New corporations start out by selling common shares. A corporation is formed by stockholders, the owners of the firm. Most corporations are relatively small, are held privately, and remain so throughout their economic life. However, an interesting juncture in a firm's life is when a decision is made to "go public." The firm engages in an initial public offering (IPO). It sells shares on the market to outside stockholders. In the promotion of a new venture, the inside shareholders may decide on an IPO to finance the enterprise. They find an investment banker willing to form a syndicate to float shares to the general public. A new public issue requires a published prospectus presenting a detailed study of the proposed new business, which must obtain the approval of the SEC. The prospectus contains estimates of potential revenues, etc. and projects the necessary initial investment. The prospectus must specifically acknowledge that all details are estimates and that the whole proposal is subject to risk. Of course, a completely new enterprise has no past record of revenues and profits.

7.4.1   Expansion of a Privately Held Firm into a Public Corporation

A new business generally starts out as a proprietorship or a partnership. If it does take the corporate form, the initial investors are usually a group of relatives, associates, or friends. Such a new corporation is classified as closely or family held. If the firm has growth or expansion plans, it may invite the public to participate in the firm by selling shares through flotation of stock on the market.[7] If the publicly held shares exceed that of the internal group, it is now a publicly held company.

The flotation of new stock may secure equity capital in a larger amount and at a more rapid rate than the retained earnings of the firm or the private resources of the initial owners can or are willing to supply. If capital is required to take advantage of a profitable opportunity to expand, a sale of new common stock not only supplies new equity funds, but by buttressing the equity base, enables the firm to borrow additional debt on better terms. Alternatively, a firm that has financed past expansion with debt capital might wish to float a stock issue to secure additional risk-bearing equity capital in order to place its financial affairs on a firmer footing.

The owners of a firm may go to the market with a capital stock flotation for reasons other than the raising of additional funds. Having some part of the stock traded on the market establishes a going price. This makes it easier to adjust the ownership value of different individual owners and to settle estates. The original owners may no longer be interested in the direct management of the company or in having so much invested in one company. A public flotation of new stock brings outside funds and participation into the company. By allowing a market to develop for the company's shares, the original holders can dispose of some of their present shares and diversify their holdings. A company with shares on the market can also make more appealing offers in the way of stock option or stock bonus plans to attract managerial talent. In the initial flotation of a closely held company, shares are often sold both for the company and for the account of the original owners. This serves the dual purpose of increasing the corporation's equity base and decreasing the participation of the original shareholders.

Floating the stock of a privately held company that wishes to go public is not as difficult as selling the stock of an entirely new venture. The SEC prospectus can show a record of past earnings and perhaps even dividends on which to base the

---

[7]If the firm is legally a single proprietorship or partnership, it must, of course, first organize as a corporation before it can sell shares.

market price of the shares. Potential investors can compare the firm's past performance and growth possibilities with that of similar companies. The price for the new issue—based on such criteria as price-earnings ratios, dividend yields, financial structure, earnings growth per share, and asset structure—can be set competitively with other stocks already on the market.

7.4.2   The Problem of Control

A firm contemplating a market flotation may consider the problem of control. Fear of losing the advantages of controlling the management of the company has prevented some firms from selling stock to the public even though from a purely financial point of view it would be profitable to both the original and the new shareholders. On the other hand, if the original ownership group has lost interest in the direct management of the firm, the problem arises of how to obtain and maintain loyal and competent management. In normal times, there is a shortage of good managerial and executive talent, rather than the reverse.

Going public creates marketable shares which can be used to help solve the agency-principal problem where the principals, the owners (stockholders) of the firm, need to set up a reward or compensation structure which will induce the managers (the agents) to operate as fully as possible in the owner's interests. (In simpler times, the problem was resolved by offering the manager the owner's daughter and/or a share in the business.) In the corporate world, part of the agent's compensation can be in the form of grants of shares or stock options. Presumably, if the managers hold shares, they will operate to maximize the share value, thus increasing their own wealth and that of the rest of the stockholders. In actuality, the solution has not always worked that well, especially in cases where the holding period for the managerial shares was very short and the agents operated to push up the short-run price of the stock at the cost of the long-run health of the company and a poor outcome for the longer-term holders of the stock.

If there is a strong desire to maintain control, the holding group may issue only a minority of the common stock. (The company is then classed as closely or family-controlled rather than closed or family held.) Close control, however, often breaks down over time as the ties among the original founding group become tenuous. At this point, a strong outside interest may link up with some faction or descendants of the original holders and acquire control.

7.4.3   Promotion of a Subsidiary by Parent Corporations

A firm venturing into a new field that differs from its main line of business may find it strategic to do so by forming a subsidiary. The parent company lends organizational and managerial talent to the new company and helps in the financing, both with credit and by purchasing all or a significant part of the new company's shares. If the parent company owns all of the subsidiary company's common stock, the subsidiary is called 100% owned or wholly owned. Alternatively some of the subsidiary's shares may be floated on the market and the firm organized with some public participation.[8]

Why should a company go through the trouble of promoting a subsidiary rather than simply creating a division of the main firm? The reasons are legal, organizational, and financial. From the legal point of view, the subsidiary may be able to obtain a broader charter or operate in a different jurisdiction or country than the parent company. From the organizational point of view, the subsidiary allows for decentralization of administration. The parent company can endow the management of the subsidiary with some independent responsibility for the economic success of the new venture. Moreover by organizing a separate subsidiary, the firm is able to limit its financial responsibility in the new venture. If all goes well, the parent firm benefits almost as much from the subsidiary's prosperity as if it were a division or branch. In the event of losses or operating difficulties, the parent company can minimize its losses by cutting the subsidiary loose. At worst, the parent company is not likely to lose more than its original investment unless it has endorsed all or part of the subsidiary's debt.

When the parent company organizes a subsidiary with "public participation," it purchases or acquires a large part of the new subsidiary's stock, while the rest of the shares are sold to the public. This limits the investment at risk of the parent company, but still gives the new subsidiary an adequate equity base. Moreover, if the new company starts out with some public ownership of its stock, its shares will develop investor acceptance, and it will be easier to float new stock to finance future expansion plans. A political advantage in having public participation lies in reducing the onus for demonstrating "monopolistic expansion tendencies." If the subsidiary is located in a foreign country, it is often useful to have a group of native stockholders interested in the success of the venture.

The subsidiary as a new legal entity or firm might be considered a new venture. However, it is easier for a subsidiary to carry out a successful public flotation of its stock than it would be for an entirely new company. A firm partly financed by a parent company possesses considerable advantages; probably it has a seasoned management, and its assets are likely to be in working order from the beginning. In short, the parent company may bestow considerable operating and credit facilities on its offspring.

---

[8]None of these arrangements need be perpetual. Companies have sold off all or part of their investment in a subsidiary or "spun off" the shares to their own stockholders. In other instances, a parent company has purchased the public's holding in a subsidiary.

7.5   Formation of a Joint Subsidiary by Two or More Parent Companies

Sometimes two or more parent companies form a new venture because the new area of operation falls somewhere between the usual operations of both companies. Each company limits its overall financial participation in the new company, but each may contribute technical or economic know-how in the area it knows best.

Of special interest is the formation of joint ventures linking firms in different countries. Using one country's firm's processes or products, the other country's firm may furnish local manufacturing facilities and/or knowledge of local markets. For example, if a foreign firm locates a joint subsidiary in the United States, the American parent may furnish the production facilities or knowledge of distribution channels while the foreign parent brings a new patent, products, or processes.

Of course, any of these joint subsidiaries may involve public participation. If the subsidiary is successful and grows, sometimes one or both of the parent companies may divest itself of its holdings either by selling its shares in the market or by "spinning them off," i.e., giving its holdings in the subsidiary to its own share-holders. This may be done if there is an antitrust problem or if the combined management becomes too unwieldy. In a sense, the parent corporation performs the role of the "classical promoter." Another possibility is an eventual reshuffling of interests between the parent companies so that one buys out the interest of the other and makes the offspring into a wholly owned subsidiary.

7.6   The SEC and the Flotation of New Issues

From 1932 to 1934, the Senate Committee on Banking and Currency, whose chief counsel was Ferdinand Pecora, held a series of investigations on the operations of the securities business in the 1920s. These hearings disclosed considerable evidence that many areas of the security market operations fell short of normal notions of honesty and that in the promotion of many stock and bond issues there had been a consid-erable disregard of the interests of the investors. No doubt the reaction to the hearings were partially colored by the attitudes of the Depression decade to the prodigal and profligate twenties; nevertheless, the legislation that resulted was well considered and constructive. More recently, the market plunge of 2002 brought about a demand and rules for increased surveillance.

The provisions of the Securities Act of 1933 as amended require the preregistra-tion of all public issues over a set dollar amount that is not the obligations of the federal, state, or local governments or their instrumentalities.[9] A registration

---

[9]Of course, the exemption of issues from federal pre-registration, which are under the minimum amount, does not excuse the promoters of such securities (nor, as a matter of fact, the promoters of registered securities) for violations of pertinent state statutes nor from the penalties of the general antifraud provisions of the common law.

statement containing exhaustive details on the new issue—reports by accountants and engineers, a description of the company's properties and business, a description of the securities offered for sale, past or projected earnings, liabilities incurred and assets acquired, the names of the promoters and their interest in the firm, the names of the officers, and the names and addresses of the major stockholders—must be filed with the SEC. The registration statement approximates the size of a "small telephone book." The SEC puts in a "stop order," holding up the issue if any item on the registration statement appears questionable. If the corporation fails to amend the statement, the commission may refuse to grant permission to float the issue. Nevertheless, the commission does not vouch for the accuracy of every item in the registration statement, and the law specifically forbids issues to state or imply that the SEC has approved the issue. Since the registration statement is very lengthy, the issuing corporation draws up a prospectus, a booklet of about ten to twenty pages summarizing the most pertinent data in the registration statement. The prospectus, which must be given to all prospective purchasers, is full of information useful to a thoughtful investor. The Securities Act of 1933 prohibited deceit, misrepresentations, and other fraud in the sale of securities but could not guarantee the information provided was accurate.[10]

Although fines and imprisonment may be enforced against violators of the commission's orders, the heart of the enforcement of the Securities Act lies in the potential individual liability which exists against the (1) corporation; (2) its chief officers; (3) all of its directors, and major stockholders (who may control directors); (4) underwriters and sellers; and (5) all experts whose statements are included, for any omission of a material fact or misstatement appearing on either the prospectus or registration statement. Any purchaser of the issue may bring civil suit for damages against any or all of these groups if he suffers a loss and an inaccurate statement appears on the prospectus or registration statement.

The registration statement has been criticized for its length and cost, since very few investors need (or indeed are willing) to study and evaluate any more information than that presented in the prospectus. The statement is put on file, however, and prospective investors may request photostatic copies. The institutional investors (banks, insurance companies, etc.), who constitute the backbone of the bond market, make good use of the details on the registration statement, and perhaps the mutual funds, the managers of pension funds, the brokerage houses, and very large and sophisticated individual investors may find it valuable to study the quantity of data available in the registration statement. Because these groups constitute a sort of leadership among investors, the results of their evaluations can affect the acceptance of the issue by the capital markets. Of course, if they behave badly, the result may be a speculative bubble which could upset the normal operations of the market for a considerable time.

Risky, foolish, or ill-conceived companies may still legally sell their securities so long as they reveal the truth in their registration statement and prospectus. (A prospectus, for example, may reveal that an overwhelming proportion of the

---

[10]www.sec.gov/about/

company's stock is going to the organizers for assets of nebulous value, whereas the outside investors are expected to furnish all the cash and shoulder almost all the risk for a minority share of the prospective gains. Nevertheless, all the facts are revealed, and the issue is perfectly legal.) The Securities Act of 1933 as amended has been called the "Truth in Securities Act."

The reforms brought about by the Securities Act and the power granted to the SEC has entailed some economic costs. The expenses of drawing up the registration statement and prospectus may weigh heavily on the issues of small companies. The fear of the many-sided penalties may preclude the promotion of some risky but justifiable ventures. Nevertheless the greater the reliance investors can repose in the facts offered in the new issue market increases their willingness to risk their capital. Investor confidence is an important part of the dynamic of a relatively free enterprise system. Without it, a large source of risk capital dries up, and the formation of new enterprise increases in productive capacity, and economic growth can slow to a deleterious level.

7.6.1   Secondary Floatations

Investment bankers on occasion underwrite the sale of a large block of securities for an estate, a large individual owner, or an endowed institution. These secondary flotations or secondary offerings differ from a regular primary issue in that the proceeds of the sale do not go to the corporation but to the original holders of the securities. Selling a large block of securities through the regular market channels might depress the price; moreover, if the securities are sold slowly over time, the sellers assume the risk of changes in the market price in the interim. Often, then, the underwriting function, the marketing facilities, and the skills of the investment bankers are used to sell the securities at a determined price at an immediate time.

7.7   Issuing Securities Through Rights

Unless the shareholders waive their rights, an established corporation must allow its common stockholder a privileged subscription position on new common stock issues. When a company issues new stock under the conditions of privileged subscription or preemptive rights, the present common shareholders are given the first opportunity to buy the new shares in proportion to their present holdings.[11] If an issue of new stock is made through privileged subscription, the new shares are usually priced under the going market price. The privilege of buying the new shares

---

[11] Preemptive rights are not confined to issues of common stock but may be extended to the issue of other types of securities if these are *convertible* to common stock.

at this bargain price attaches to each outstanding share. It is represented by a negotiable "right" which can be either exercised or sold before the subscription date to be used by someone else to purchase the new shares. The intrinsic value of a right on the market can be obtained through the use of a simple formula, namely:

$$VR = \frac{MPOS - PSP}{N + 1}$$

where VR is the value of the right, MPOS is the market price of the existing shares, PSP is the privileged subscription price, and N is the number of old shares required to buy one new share.

For example, suppose a firm with 6,000,000 outstanding shares floats a new issue of 600,000 shares giving preemptive rights to the existing stockholders. The going price of the stock on the market is $36.60, and the subscription price is $30.00.

$$\text{Value of a Right} = \frac{\$36.60 - \$30.00}{10 + 1} = \frac{\$6.60}{11} = \$0.60$$

The intrinsic value of a right is $0.60, and it requires ten of these rights plus $30 to buy a new share. If a shareholder sells a right he will get $0.60 for it, representing a compensation for the dilution of his equity, and the market price of his stock drops to $30.00 since the stock is now quoted *ex-rights*. Conversely, a prospective new shareholder can buy ten rights (10 X $0.60) and use them to buy a new share, which, including the payment for the rights, costs him $36.00 ($0.60 less than the present price); however, the new share is also ex-rights. If the original holder exercises his rights himself, he acquires one new share at $30 for every ten old shares, valued at $36.60, he presently holds. The average value of the eleven shares works out exactly to $36.00. Because the value of a right is considered a return of capital, the stockholder who sells his rights need not pay personal income tax on the proceeds; however, the sale is used to reduce his original acquisition costs so that if he later sells his securities at a profit, his taxable capital gains will be that much larger.

Because the market value of rights moves up or down proportionately more than the value of the underlying shares, generally rights tend to sell at a slight premium. Since they have a more volatile price movement, they give the speculator a "greater kick" for his investment, and there are those who will pay for the privilege of gambling.

Although a company could manage to sell a new issue of stock by itself by pricing the new shares below the market and using rights, most firms prefer to enlist the aid of an investment banker to ensure the success of the issue. In fact, the use of rights costs substantially less (about one-third) than the cost of using an investment banker (Smith, 1977). The investment banker makes a "standby pledge" by which he is obligated to purchase all shares not subscribed to by the rights holders. The period between the issue of the rights and the subscription deadline may be as long as 30 days. It is possible in this time for the market price of the stock to drop below its subscription price; the banker would then suffer a loss under his standby commitment. Ordinarily, however, since the new issue has been floated below market, all rights (except those of some few stockholders who may not be fully informed or wait too long or are negligent) are exercised. Under these circumstances, picking up the few neglected rights furnishes an additional source of income to the investment bankers. In addition to the standby agreement, the underwriting banker on a privileged subscription usually administers and records the exercise of the rights and the accounting for the new shares. The costs of rights financing are often approximately 30% of the cost of using an underwriter.

7.8   Stock Tenders

Technically, a tender is a pledge of the willingness of a holder to sell his securities in response to an open bid by a corporation or an individual to buy outstanding securities at a designated price. In financial language, however, the whole transaction—bid and response—is likely to be called a tender. Sometimes a firm may bid for bond tenders when it wishes to pick up bonds for retirement. The stock tender, however, is probably more common and certainly more interesting.

Stock tenders may be thought of as a reverse flotation. On the completion of a successful tender request, shares formerly held in the market are concentrated with one holder or retired. A bid for tenders may be used for the following purposes:

1. By a corporation to retire some of its own shares.
2. By one group of shareholders to buyout dissident shareholders.
3. By a corporation to acquire another corporation.
4. By a corporation to extend its holdings in a subsidiary, possibly with the view of acquiring enough shares to enable it to consolidate its income statement for tax purposes.
5. By an individual or group of individuals to acquire enough shares to obtain control of a corporation.

During the past 50 years, most tender offers have been merger related.

The group making the offer to buy will designate a transfer agent—usually a large bank—to register the shares tendered for sale according to the terms of the offer. The price at which the stock will be purchased is fixed somewhat above the market price. Once the offer is made, however, the market price of the stock will not fall below the tender price as long as the tender is still open. A date is set for the termination of the tender. Most requests for tenders run about 2 weeks to a month. A tender offer may be open, in which case all stock tendered before the termination date will be purchased. The terms may, however, set an upper limit; e.g., only the first 100,000 shares submitted will be picked up. When a definite number of shares is required for the purchasing group's purpose, there may be an all or nothing tender. The terms will state, for example, that a minimum of 100,000 shares must be submitted or none of the shares will be bought. An all or nothing tender offer can be combined, of course, with an upper limit. The tender price can be stated in cash, in an exchange of securities, or in a combination of both.

The purposes of a request for tenders could be accomplished by purchasing the desired shares on the open market. By using the tender device, however, the price paid per share can be limited. Perhaps some of these shares could have been bought for less on the open market, but the purchasing activity in the shares might raise their price so that the average cost of the shares becomes greater than they would be if purchased by tender.

One interesting use of a request for the submission of tenders is the retirement of some of its outstanding common shares by the corporation itself. Such a tender is akin to a partial liquidation and might be employed if the corporation has redundant liquid funds which are not contributing much to its earnings. The tender accomplishes a reduction in the firm's ownership capital with a subsequent increase in the earnings of the remaining shares. Perhaps a partial liquidating dividend would be the fairer way to accomplish the same purpose. However, the buyback or tender limits the personal tax to the capital gains rate on the shares.

7.9   Costs of Floating an Issue

The major cost of a security is its servicing charge – the amount of interest or the earnings yield the firm must make overtime to entice the market to buy the issue. The actual cost of floating the issue – the legal expenses, fees, and the spread of the investment banker – is usually a relatively minor consideration and only rarely affects the type or timing of the issue.

In general, the costs of flotation depend on the expected risks and difficulties of selling the securities. This is affected by the type of underwriting agreement, the class of security, the size of the issue, and the security market seasoning and reputation of the issuing company. Equity floatation costs are considerably greater than debt floatation costs. Straight debt costs less to issue than convertible debt; see Table 7.1.

1. Issues of greater price stability cost less to float than issues, which are by nature, somewhat more volatile. Debt securities (bonds and debentures) are relatively inexpensive to float compared to stocks, and preferred stocks are floated for a cheaper percentage than common shares.
2. Issues of seasoned, well-known companies can be floated for a relatively lower spread than those of new or more speculative companies.
3. An issue of common shares through the use of rights and a standby arrangement with the investment bankers will cost less than an underwritten flotation price with rights waived.
4. Since the bankers incur certain costs, such as advertising, which do not vary proportionately with the size of the issue, larger issues are generally handled at

Table 7.1 Direct costs as a percentage of gross proceeds for equity (IPOs and SEOs) and straight and convertible bonds offered by Domestic Operating Companies, 1990–1994

Proceeds (millions of dollars)	Equity IPOs GS[a]	E[b]	TDC[c]	SEOs GS	E	TDC	Bonds Convertible Bonds GS	E	TDC	Straight Bonds GS	E	TDC
$ 2-9.99	9.05%	7.91%	16.96%	7.72%	5.56%	13.28%	6.07%	2.68%	8.75%	2.07%	2.32%	4.39%
10-19.99	7.24	4.39	11.63	6.23	2.49	8.72	5.48	3.18	8.66	1.36	1.40	2.76
20-39.99	7.01	2.69	9.70	5.60	1.33	6.93	4.16	1.95	6.11	1.54	0.88	2.42
40-59.99	6.96	1.76	8.72	5.05	0.82	5.87	3.26	1.04	4.30	0.72	0.60	1.32
60-79.99	6.74	1.46	8.20	4.57	0.61	5.18	2.64	0.59	3.23	1.76	0.58	2.34
80-99.99	6.47	1.44	7.91	4.25	0.48	4.73	2.43	0.61	3.04	1.55	0.61	2.16
100-199.99	6.03	1.03	7.06	3.85	0.37	4.22	2.34	0.43	2.76	1.77	0.54	2.31
200-499.99	5.67	0.86	6.53	3.26	0.21	3.47	1.99	0.19	2.18	1.79	0.40	2.19
500 and up	5.21	0.51	5.72	3.03	0.12	3.15	2.00	0.09	2.09	1.39	0.25	1.64
Average	7.31%	3.69%	11.00%	5.44%	1.67%	7.11%	2.92%	0.87%	3.79%	1.62%	0.62%	2.24

Source: reprinted with permission from the *Journal of Financial Research*, Vol. 19, No. 1 (Spring 1996), pp. 59–74, "The Costs of Raising Capital," by Inmoo Lee, Scott Lochhead, Jay Ritter, and Quanshui Zhao

Note:

[a]GS—gross spreads as a percentage of total proceeds, including management fee, underwriting fee, and selling concession

[b]E—other direct expenses as a percentage of total proceeds, including management fee, underwriting fee, and selling concession

[c]TDC—total direct costs as a percentage of total proceeds (total direct costs are the sum of gross spreads and other direct expenses)

lower percentage underwriting and selling fees than smaller issues of the same quality.

5. The split between the selling fees, underwriting fees, and the originating house or managing house charges depends on the expected risks and difficulties in selling the issue. Although the underwriters take the risks, the sellers may have to incur more expenses. Their share of the spread must be large enough to encourage them to sell the issue.

7.10 Regulation of the Capital Markets

Under the New Deal administration of the 1930s, Congress passed a number of laws designed to eliminate questionable practices and to enforce a higher level of business morality in the investment markets. Over time as conditions developed, additional acts and regulations have been added to the mix.

Securities Act of 1933 As previously discussed, the Securities Act of 1933 mandates new issues to file full information, registration statements, and prospectuses, with the SEC, as was stated earlier in this chapter. The SEC is composed of five commissioners, appointed by the president, confirmed by the Senate, and whose term is 5 years. No more than three commissioners can belong to the same political party, reducing potential partisanship. The president designates one of the

Commissioners as Chairman, the top executive in the SEC. President Franklin D. Roosevelt appointed Mr. Joseph P. Kennedy as the first SEC Chairman. The current chairman is Mr. William Donaldson, and Cynthia Glassman, Harvey Goldschmid, Paul Atkins, and Roel Campos. The Commissioners interpret Federal Securities laws, amend existing rules, propose new rules in response to changing market conditions, and enforce rules and laws. The SEC is composed of four divisions: (1) corporate finance, which reviews registration statements for newly offered securities as well as 10-K (annual) and 10-Q (quarterly) filings, annual reports and proxies, and merger and acquisition filings; (2) market regulation, which establishes and maintains standard for fair, orderly, and efficient markets, including monitoring the financial integrity program for broker-dealers, reviewing self-regulatory organizations (SROs), and surveilling markets; (3) investment management, overseeing and regulating the $15 trillion investment management industry, including mutual funds and investment advisors; and (4) enforcement, which investigates possible violations of security laws. Enforcement obtains evidence of possible violations. The most common SEC investigations include: insider trading, misrepresentation of information about securities, manipulating market prices of securities, sale of securities without proper registration, and stealing customer's funds or securities. The Act had the objective of enforcing full and truthful disclosure of all pertinent facts to prospective investors on the primary market.

The Securities Exchange Act of 1934   The Securities Act of 1933 was followed by the Securities Exchange Act of 1934, which established the Securities Exchange Commission, the SEC. The SEC was empowered to register, regulate, and oversee brokerage firms, transfer agents, clearing agencies, as well as the self-regulatory organizations (SROs), which include the NYSE and AMEX. The Securities Exchange Act of 1934 required annual (and other periodic) reports of firms with assets exceeding $10 MM, and more than 500 owners. The Securities Exchange Act governs proxy solicitations, tender offers involving direct purchase or a tender offer of more than 5% of a company's securities. Fraudulent insider trading was prohibited by the Securities Exchange Act of 1934. The existence of the SEC has not prevented all losses on questionable securities. Unscrupulous promoters and security salesmen may still take advantage of investors simply by risking the fines, prison terms, and civil suits which false statements may entail. The SEC does not guarantee the profitability of any venture. An investor making a bad choice will lose money on his purchase of securities.

Trading on the exchange has been regulated in the following respects:

1. The Federal Reserve Board is given authority to set initial margin requirements (the amount required of the purchaser in cash as a percentage of the market price of stock) and to regulate the extension of credit on the market.[12] (Low margin

---

[12]In the feverish era of the 1920s, initial margin went as low as 10%.

requirements create a weak market. If the stocks start to drop, the brokers may strengthen the downtrend by selling the margined shares in order to protect their loans.)

2. Brokers' indebtedness to all persons may be regulated and such indebtedness may not exceed twenty times the net capital of the firm.
3. Brokers may not lend customers' securities without their consent. The lending of securities is used to effectuate a short sale. A short sale is made by someone who expects the price to go down. He borrows the stock and sells it at the going market price. When (or if) the price falls, he repurchases the stock (covers his sale) and returns it to the lender. The difference between the repurchase price and the price at which the shares were originally sold is the short sellers' profit.
4. The manipulation of stock prices by wash sales, excessive activity, and misleading statements is made unlawful.
5. Options (puts, calls, etc.) and short selling are regulated by the commission. Short sales can only be made on the "up-tick"—the stock must have moved up a fraction of a trading point since its last regular sale. (This rule handicaps a "bear raid," where a speculative group hammers down the price of a stock by continuously offering more shares as it motes downward.) We discuss option pricing and strategies in Chap. 16.
6. The commission was authorized to supervise the activities of the "specialists," members of the exchange who act as a broker on certain transactions and in other instances "create a market" (act as dealers) on companies stocks assigned them by the exchange. The specialist is a critical part of the stock exchange's machinery; it is doubtful if a manipulative "pool" could be successful without enlisting the sub rosa aid of the specialist.

Banking Act (Glass Steagall Act) of 1933 One of the clauses of the Banking Act of 1933 was a requirement that deposit bankers divest themselves of investment banking affiliates. (Commercial banks were prohibited from floating or dealing in any securities except federal and state and local government obligations.) Many of the investment banking affiliates were organized as subsidiaries by the big city commercial banks in order to participate in the profitable security issue business of the 1920s. The history of these investment banking affiliates did not in the end reflect very favorably on the acumen a financial judgment of their organizers and management.[13]

The security affiliates of commercial banking firms represented a potential conflict of interest; they violated the general legal concept of the "arms-length transaction" which is enforced wherever fiduciary (or quasi-fiduciary) relationships are involved. Possible conflicts of interests between the commercial bank, its depositors, and its investment banking subsidiary could arise in the following instances:

---

[13] See F.A. Bradford, *Money and Banking*, Longmans, Green, 1949, p. 800.

1. If a depositor asked his banker for advice on investments, the banker might be tempted to recommend an issue being floated by the security affiliate. Indeed, the list of depositors of the commercial bank often served as a prime source of prospects for the affiliate's sales.
2. When the commercial bank bought securities for its portfolio, it might choose an inferior issue floated by the affiliate. (And the high quality of the bank's investment is one aspect of the safety of its deposits.)
3. If the bank felt that one of its borrowers was stretching out his repayment, they might have their affiliate float an issue for him, the proceeds to repay the loan. Here the commercial bank would get rid of a potentially embarrassing asset by shifting it to the outside investors. This is a subtle violation of the implicit investment banking code—that the investment banker lends his name to the general soundness of the issue considering its description and class.

Glass Steagall Act Amended   In 1999, the Gram-Leach-Bliley Act (GLB) severely amended the original Glass Steagall Act. Large financial conglomerates, deregulated as financial holding companies (FHL), are allowed to have an investment banking subsidiaries as long as a "fire wall" exists between the activities of investment banking (new issues) and the regular banking or other activities of the conglomerate. The result has been a merger of all types of banks, stock brokerage firms, and insurance companies offering a department store of financial services. Each sector is supposed to be overseen by its own traditional functional regulator.

Retail Brokerage Houses   Of course the problem of conflicts of interest between different activities in the financial industry are endemic. A recent area of concern has developed in the brokerage industry where the brokerage firms also had large investment banking operations. Here a conflict of interest arises when the retail broker advises his clients to purchase a company's new issues unwritten by his firm.

Public Utility Act of 1935   The main purpose of this Act is breaking up and regulating holding companies in the gas and electric utility industries. In addition, however, the Public Utility Act of 1935 gave the Securities and Exchange Commission considerable powers over the financial affairs of the holding companies and operating subsidiaries.[14] The SEC must approve the terms and types of new utility issues and it sets broad bounds on permissible financial structures for the companies.

The SEC, in its authority under the 1935 Act, encouraged the use of competitive bidding in floating utility company securities. In 1941, it imposed a competitive bidding requirement on all public utility issues. (The ICC has required competitive bidding on the issue of railroad equipment trust certificates since 1926, and it imposed this requirement on all railroad securities in 1944. Municipal issues are almost always floated under competitive bids.)

---

[14]This authority over financial affairs should not be confused with the rate-regulating function given to the various state Public Utility Commissions and to the Federal Power Commission for interstate business.

Most corporation issues are still floated on a negotiated basis. The originating house advises the company on the features and terms of the issue and negotiates the spread and fees. Under competitive bidding, the issuing firm prepares all the terms of the issue except the coupon rate (stated interest rate) on bonds and the price. The investment bankers then submit bids citing the stated interest rate they will put on the issue, the price to the issuing firm, and the price to the market. The investment banker whose bid shows the lowest effective interest lost to the issuer (function of the rate and the price) wins the privilege of floating the issue and managing the underwriting syndicate. On a competitive bid issue, the firm performing the function corresponding to the originating house is called the managing firm.

Competitive bidding was pushed by some Midwestern investment bankers who felt that eastern bankers had established relationships with the corporations which gave them an unfair share of the new issue market, but their complaints found sympathetic ears at the government commissions. It was argued that the connections developed by the negotiated type of offering lead to investment banker domination of some firms. (The corporation became more interested in deals which furnished securities to be marketed than in company operations.) The most pertinent argument was that competitive bidding would lower flotation costs.

The old-line investment bankers originally opposed competitive bidding. They argued that it deprived the issuing firm of the private advice, market knowledge, and competent guidance of the investment banker in selecting the most advantageous type of security and setting up the best possible terms and provisions before the issue of the security. There may be some force to this argument when applied to industrial issues. However, the features of the issues commonly subject to competitive bidding—utilities, municipals, and equipment trust certificates—are today actually quite uniform and standardized. There may not be any pressing need for the special expertise of the investment banker.

The Maloney Amendment, 1938   This amendment to the Securities Exchange Act gave the SEC some control over the secondary market activities of the over-the-counter dealers. (The primary market activities of these dealers falls under the general purview of the Securities Act of 1933). The Maloney Amendment provides a system of self-regulation for the over-the-counter traders promulgated and administered by the National Association of Security Dealers. The Securities Exchange Commission passes on the rules, supervises their administration, and reviews the disciplinary decisions of the Association.

The Investment Company Act of 1940[15]   This Act largely unchanged to this day, provides for the federal registration of both the closed-end and open-end investment companies. (Open-end companies—mutual funds—stand ready to redeem their own shares at the pro-rata underlying portfolio value. An investor in a closed-end

---

[15]This act follows many of the provisions suggested by Henry Simon in his pamphlet "A Positive Program for Laissez Faire," University of Chicago, 1935, reprinted in *Economic Policy for a Free Society*, University of Chicago Press, 1948.

company must sell his shares on the market.) Investment companies are financial intermediaries; they place the funds given to them by their own investors into securities of operating companies. People invest with them because they may provide better diversification and better portfolio management than the investor could achieve himself.

Although investment companies had a long history of successful operations dating from Scotland in the 1870s, they fell into disrepute in the United States in the 1930s. The Investment Company Act provided that companies who announced their investment policy clearly and adhered to it could register with the commission. Registration gave the company status in the eyes of the investor and removed some of the onus of the 1929 debacle. Among other things, a registered company must practice proper diversification of its portfolio, limit the issue of its prior claim securities and the use of margin trading, and renounce any holding company ambitions. If a registered company pays out 90% of its income and capital gains to its own shareholders, it needs pay no federal income tax. The effects of the Investment Company Act must be considered salutary. The investment companies, especially mutual funds, have experienced vigorous growth over the last 60 years and brought a healthy flow of savings into the capital markets.

Sarbanes-Oxley Act of 2002   President Bush signed the Sarbanes-Oxley Act of 2002, on July 30, 2002, which enhanced corporate responsibility and financial disclosure. The Act created a "Public Company Accounting Oversight Board" (PCAOB), to oversee activities in the auditing process.[16] The SEC adopted requiring mutual fund CEOs and CFOs verification of stockholder reports, disclosure of codes of ethics, audit independence, and listing standards for audit committees. The audit independence ruing prohibited auditors from providing certain non-audit services related to conflict of interests standards. Exchanges and national security associations were prohibited from listing securities of any security not in compliance with certain audit committee requirements. Disclosure of the firm's adoption of a code of ethics was required, as was the presence of one "financial expert" serving on the audit committee.[17] Investment companies were encouraged by rule amendments to disclosure more balanced information to potential investors with respect to past performance. In December 204, the SEC established an advisory committee to examine the impact of the Sarbanes-Oxley Act on smaller public companies. The advisory committee, known as the Securities and Exchange Commission Advisory Committee on Smaller Public Companies, will concentrate on (1) the frameworks for internal control over financial reporting for smaller companies; (2) corporate disclosure and reporting requirements for smaller public firms, including differential requirements based on market capitalizations; and (3) accounting standards applicable to smaller firms.[18] There could well be a cost to the public by the Sarbanes-Oxley

---

[16] www.sec.gov/about

[17] SEC Annual Report, 2003, pp. 44–52.

[18] www.sec.gov/news/press/2004-174

Table 7.2 Gross proceeds of new securities of all corporations offered for sale for selected years 1935–2017 (in millions of dollars)

Calendar Year	Gross Proceeds from Issues of Bonds, Debentures and Notes	Gross Proceeds from Issues of Common Stock
1935	$2,224	$ 22
1940	2,386	108
1945	4,855	397
1950	4,919	811
1955	7,420	2,185
1960	8,081	1,664
1971	37,809	10,146
1975	97,088	18,819
1980	179,633	44,254
1985	357,436	83,347
1990	696,799	53,944
1995	1,197,158	97,636
2000	2,748,933	392,684
2003	2,645,478	172,294
2010	6,038,082	19,448
2019	18,362,877	27,584
2020	15,167,057	37,761

Source: Securities and Exchange Commission, 27th Annual Report, Compustat Database

Act. In 2002, some 67 firms delisted their common stock, whereas, in 2003, 198 firms delisted common stock, or "went dark." The costs of compliance may be too high for many firms.[19]

7.11   The Capital Market as a Source of Funds

Financial institutions, the financial intermediaries, far exceed individual suppliers of funds. However, the institutions are not an autonomous source of funds; they merely reinvest the moneys of their individual policy holders and savers. The amount of funds shown supplied by individuals on their own account appears to be relatively small. This is a strategic item, however, because although the data is not available, it is generally presumed that this is the largest source of financing for new enterprises.

In Table 7.2, we have the breakdown of corporation use of the capital market according to the type of issues. The low interest rates that have prevailed since the 1930s and the income tax deductibility of interest seem to have encouraged debt issues over stock. A larger proportion of debt is retired every year than common stock, and on a net basis, the issuance of debt has been five times the issuance of

---

[19]C. Leuz, "Why Firms go Dark: Causes and Economic Consequences of Voluntary SEC Deregulations," working paper, The Wharton School, University of Pennsylvania. See also R.E. Verrecchia) and C. Leuz "The Economic Consequences of Increased Disclosure." *Journal of Accounting Research* (2000)

Table 7.3 Internal and external capital financing, selected years ($billion)

	Internal		External			
Year	Retained Earnings	Depreciation	Debt Issued	Stock Issued	NetDebt Issued	NetEquity Repurchased
1971	$307.107	44.62	36.21	0.00	16.10	1.53
1980	957.10	141.05	173.02	9.28	90.14	-0.99
1987	1,740.48	326.63	437.01	4.62	151.85	62.17
1990	1,971.61	380.50	666.30	4.85	154.02	50.36
2000	4,871.93	1,090.12	2,734.20	13.23	1,012.27	234.02
2007	9,427.05	1,386.33	5,007.70	1,113.00	1,817.72	1,092.79
2010	9,193.87	1,544.47	6,038.00	19.45	-1,105.63	502.68
2018	14,242.77	1,989.20	15,337.80	1,189.90	363.25	1,166.21
2020	14,367.32	2,042.06	15,167.00	36.76	791.01	684.92

Source: WRDS database
Source: Standard & Poor's, Compustat
Net debt = debt issued – debt repurchased
Net equity = equity issued – equity repurchased

equity since 1971. Moreover, the use of equity is understated because the largest part of equity finance arises from the retention of earnings rather than new issues. In general, Table 7.2 shows a sharp increase in corporate use of the capital markets since 1975. This has been a true "growth industry."

In the late 1930s and the early 1940s, financial writers were predicting the demise of the capital markets as a source of long-term financing. Firms were doing most of their long-term financing internally through depreciation and retained earnings and were divorcing themselves from any reliance on external financing, i.e., the capital markets. Even as they wrote, however, there were indications of a recovery in the capital markets.

Table 7.3 depicts the relative importance of internal financing and external financing sources. Internal financing looms very large in financing firm growth. The capital markets allow firms to raise capital to support capital expenditures and generate future profits. Nevertheless, firms generate most of their sources of funds from retained earnings, net income less dividends. The capital markets, as we will see in Chap. 8, evaluate the effectiveness of the firm's retained earnings policy by establishing a price-earnings multiple for the security. A higher price-earnings multiple often shows confidence in management, among other things. Nevertheless, although depreciation is counted as one of the sources of funds, it is part of the gross savings of the economy, not net. Over time, depreciation tends to approximate capital consumption. Fixed assets must be replaced eventually. Corporations earned net income of $37,005.19 million during the 1950–2017 period. The corporations paid dividends of $27,505.91 million. Amortization and depreciation charges totaled $37,681.25 million during the corresponding period. Cash flow, net income plus depreciation, totaled approximately $74.8 billion. Corporations raised $90,765.43 million of new debt issues, and $10,617 million of new equity issues;

Table 7.4 Industry statistics: public offerings and mutual fund assets

	1980	1985	1990	1995	1998
Public Offerings: IPOs Registered with SEC	1.3	90.0	50.0	122.0	257.0
Public Offerings (excluding Private Placements)	58.0	133.0	309.0	705.0	1,819.0
Managed Funds: Assets under Management by Investment Companies	235.0	591.0	1,350.0	2,879.0	5,180.0
Mutual Funds	135.0	495.0	1,065.0	2,812.0	5,525.0

Source: SEC, 1999 Annual Performance Report

however, if one calculates net debt and equity issues, defined as new debt issues less debt repurchased, and new equity issues less equity repurchased, one finds that nets debt issues totaled $26,466.74 million during the 1950–2017 period, whereas net equity issues were $ -6341.6 million. Thus, new debt issues exceeded new equity issues by an 8.5:1 ratio. Moreover, the internal generated cash flow totaled $52.30 billion, whereas net debt and equity issues totaled $14.56 billion. Internal financing is the greatest source of corporate financing. Internal funds, defined as retained profits and depreciation, exceeds external funds, defined as net debt issues plus net equity issues, during much of the 1950–2017 period. Depreciation has become the largest source of internal funds; see Table 7.3. Depreciation has exceeded retained profits during the 1971–2018 period. Net debt issues dominate net equity issues. External financing rose relative to internal financing during the late 1990s and early 2000s, but still fell short relative to internal financing because of depreciation. The capital market, especially in recent years, is a significant supplier of savings for net capital formation. The relative insignificance of IPOs is shown in Table 7.4, where IPOs, public offerings, and mutual fund assets are shown. Mutual fund assets dwarf IPOs and are 2–3 times the amount of public offerings during the 1980–1998 period. The post-IPO performances of the 2019 Tech IPOs has not been stellar (Table 7.5).

7.12   The Debate on the Optimal Organization of the Capital Market

In the European continental capital markets and in Japan, the various financial functions are quite integrated. Thus, it is quite common for the commercial banks to sell insurance to their clients and perform the brokerage function, buying and selling shares and bonds for their customers. Until recently, in the United States and Great Britain, there has been a tendency to keep various financial institutions independent so that for example the transactions between a bank and an insurance

Table 7.5 Post-IPO stock performance for tech companies

COMPANY	IPO DATE	OFFER PRICE	PRICE AS OF 13-DEC-2019	PRICE RETURN
Lyft, Inc	28-Mar-2019	$72.00	$46.75	(35.1%)
Peloton Interactive, Inc.	25-Sep-2019	29.00	31.53	8.7%
Pinterest, Inc.	17-Apr-2019	19.00	17.45	(8.2%)
Slack Technologies, Inc.	19-Jun-2019	26.00	21.39	(17.7%)
Uber Technologies, Inc.	09-May-2019	45.00	28.36	(36.7%)

company would be at arm's length. The growth of holding company financial companies owning insurance companies and brokerage company subsidiaries, and brokerage companies offering a variety of banking and insurance options have begun to blur the boundaries.

Putting activities under one roof may increase "transaction efficiency." It can lower transaction costs. However, as shown in the recent case of Merrill Lynch's financial advisors lying in bed with its new issue branch and promoting risky new issues to their customers, a source of abuse can arise when the presumed neutral financial advisor and the seller wear the same uniform. The advantages of transaction efficiency can soon evaporate if the external investor begins to doubt that he is getting the benefit of an "arms length transaction" and operating in an unbiased market. A loss of investor confidence raises the risk premium on all financial sources and taxes the economy across the board.

7.13   Capital Markets and Long-Term Economic Growth

The capital market is an institution of tremendous importance. It may be enlightening to sketch its influence on three levels: how it affects investors, corporation management, and the economy as a whole. From the viewpoint of the investor, the prices shown by the securities market is the considered judgment of his peers on the success and potential success of the company in which he has put his funds. The attitude he will have toward the management, his willingness to put more money in the company or similar companies, will be affected by the verdict of the investment market.

The securities markets play a significant role in the managerial and financial decisions of large publicly held corporations. The security markets affect the stockholder's attitude toward the corporation and its management. The verdict of the securities market reflects what sophisticated investors think of the success and potential of the company. From the operational viewpoint of the corporation's financial management, the security or capital funds markets are a source of long-term capital. Through the machinery of the capital markets and the investment bankers, the corporation can obtain funds from long-term debt issues (bonds or debentures) or by increasing its equity (floating more stock). The financial management of the corporation must understand these markets and the relations between the primary and secondary markets if it is to make careful, rational decisions on problems of long-term financing.

From the view of the economy as a whole, the activities of the investment markets play a major role in capital formation and economic growth. The tradition of investing savings in securities is quite strong in economically developed countries; in other parts of the world, savings are too often placed in precious metals, jewelry, or land rather than invested in productive enterprises.

The investment markets are in nature international as well as national. The securities of many large foreign corporations are listed on the American exchanges, and efficient machinery exists for buying foreign securities from foreign exchanges or co-listings. The vast apparatus of the financial market serves an essential service in a free enterprise economy of allocating a large portion of new capital formation among various firms, countries, organizations, and governments.

References

Asquith, P. (1983). Merger bids, uncertainty, and stockholder returns. *Journal of Financial Economics, 11*, 51–83.

Asquith, P., & Mullins, D. (1983). The impact of initiating dividend payments on shareholder wealth. *Journal of Business, 56*, 77–96.

Beatty, R. P., & Ritter, J. R. (1986). Investment banking, reputation, and the underpricing of initial public offerings. *Journal of Financial Economics, 15*, 213–232.

Bhagat, S. (1983). The effect of pre-emptive right amendments on shareholder wealth. *Journal of Financial Economics, 12*, 289–310.

Booth, J. R., & Smith, R. L., III. (1986). Capital raising, underwriting and the certification hypothesis. *Journal of Financial Economics, 15*, 261–281.

Bosland, C. C. (1949). *Corporate finance and regulation* (p. 12). Ronald Press. Chs. 10.

Bradley, M., & Wakeman, L. M. (1983). The wealth effects of targeted share repurchases. *Journal of Financial Economics, 11*, 301–328.

Bradley, M., Desai, A., & Kim, E. H. (1983). The rationale behind interfirm tender offers: Information or synergy? *Journal of Financial Economics, 11*, 183–206.

Brealey, R., Myers, S., & Allen, F. (2006). *Principles of corporate finance* (8th ed.). MccGraw-Hill/Irwin. Chapters 15 and 18.

Brickley, J. (1983). Shareholder wealth, information signaling and the specially designated dividend: An empirical study. *Journal of Financial Economics, 12*, 187–209.

Charest, G. (1978). Dividend information, stock returns, and market efficiency – II. *Journal of Financial Economics, 6*, 297–330.

Chen, H. C., & Ritter, J. R. (2000). The seven percent solution. *Journal of Finance, 55*, 1105–1131.

Dann, L. (1981). Common stock repurchases: An analysis of returns to bondholders and stockholders. *Journal of Financial Economics, 13*, 157–186.

Dann, L. Y., Mayers, D., & Raab, R. J., Jr. (1977). Trading rules, large blocks and the speed of Price adjustment. *Journal of Financial Economics, 4*, 3–22.

Dewing, A. S. (1953). *The financial policy of corporations* (5th ed., p. 35). Ronald Press. Chs. 33, 34.

Dodd, P. (1980). Merger proposals, management discretion and stockholder wealth. *Journal of Financial Economics, 8*, 105–138.

Dodd, P., & Ruback, R. S. (1977). Tender offers and stockholder returns: An empirical analysis. *Journal of Financial Economics, 5*, 351–374.

Dodd, P., & Warner, J. B. (1983). On corporate governance: A study of proxy contests. *Journal of Financial Economics, 11*, 401–438.

Dyl, E. A., & Joehnk, M. D. (1976). Competitive versus negotiated underwriting of public utility debt. *Bell Journal of Economics, 7*, 680–689.

Eckbo, B. E., & Masulis, R. W. (1995). Seasoned equity offerings: A survey. In R. A. Jarrow, V. Maksimovic, & W. T. Ziemba (Eds.), *Handbooks in operations research and management: Finance* (Vol. 9, pp. 1017–1072). Elsevier.

Fama, E. F. (1980). Agency problems and the theory of the firm. *Journal of Political Economy, 88*, 288–307.

Fama, E. F., & Jensen, M. C. (1985). Residual claims and investment decisions. *Journal of Financial Economics, 14*, 101–119.

Fang, L. H. (2005). Investment Bank reputation and Price and quality of underwriting services. *Journal of Finance, 60*, 2729–2762.

Freund, W. C. (1958, May). An appraisal of the source sand uses of funds: Approach to the analysis of financial markets. *Journal of Finance, 13*, 275–294.

Guerard, J. B., Jr. (2010). The corporate sector as a net exporter of funds. In J. R. Aronson, H. L. Parmet, & R. J. Thorton (Eds.), *Variations in economic analysis: Essays in honor of Eli Schwartz*. Springer.

Hansen, R. S., & Pinkerton, J. M. (1982). Direct equity financing: A resolution of a paradox. *Journal of Finance, 37*, 651–665.

Hite, G. L., & Owers, J. E. (1983). Security Price reactions and around corporate spin-off announcements. *Journal of Financial Economics, 12*, 409–436.

Ibbotson, R. (1975). Price performance of common stock new issues. *Journal of Financial Economics, 2*, 235–272.

Ibbotson, R. G., & Jaffe, J. F. (1975). 'Hot Issue' Markets. *Journal of Finance, 30*, 1027–1042.

Ibbotson, R. G., & Ritter, J. R. (1995). Initial public offerings. In R. A. Jarrow, V. Maksimovic, & W. T. Ziemba (Eds.), *Handbooks in operations research and management: Finance* (Vol. 9, pp. 903–1016). Elsevier.

Ibbotson, R., Sindelar, J., & Ritter, J. (1994). The Market's problem with the pricing of initial public offerings. *Journal of Applied Corporate Finance, 6*, 66–74.

Investment Bankers Association. (1949). *Fundamentals of investment banking* (pp. 1027–1042). Prentice-Hall.

Jensen, M. C., & Meckling, W. H. (1976). Theory of the firm: Managerial behavior agency costs and ownership structure. *Journal of Financial Economics, 3*, 305–360.

Logue, D. E., & Jarrow, R. A. (1978). Negotiation vs. competitive bidding in the Sale or securities by public utilities. *Financial Management, 7*, 31–39.

Mandelker, G., & Raviv, A. (1977). Investment banking: An economic analysis of optimal underwriting contracts. *Journal of Finance, 32*, 683–694.

Masulis, R. M. (1980). Stock repurchase by tender offer: An analysis of the causes of common stock Price changes. *Journal of Finance, 35*, 305–319.

Masulis, R. (1983). The impact of capital structure change on firm value: Some estimates. *Journal of Finance, 38*, 107–126.

Mikkelson, W. (1981). Convertible calls and security returns. *Journal of Financial Economics, 9*, 237–264.

Mikkelson, W. H., & Partch, M. M. (1985). Stock Price effects and costs of secondary distributions. *Journal of Financial Economics, 14*, 165–194.

Miller, M. (1977). Debt and taxes. *Journal of Finance, 32*, 261–276.

Modigliani, F., & Miller, M. (1958). The cost of capital, corporation finance and the theory of investment. *American Economic Review, 48*, 261–297.

Modigliani, F., & Miller, M. (1963). Corporate income taxes and the cost of capital: A correction. *American Economic Review, 53*, 433–443.

Myers, S. (1977). Determinants of corporate borrowing. *Journal of Financial Economics, 5*, 147–175.

Myers, S. (1984). The capital structure puzzle. *Journal of Finance, 39*, 575–592.

Myers, S. C., & Majluf, N. S. (1984). Corporate financing and investment decisions when firms have information that investors do not have. *Journal of Financial Economics, 13*, 187–221.

Phillips, S. M., & Smith, C. W., Jr. (1980). Trading costs for listed options: The implications for market efficiency. *Journal of Financial Economics, 8*, 179–201.

Ritter, J. R. (1984). The 'hot issue' market of 1980. *Journal of Business, 57*, 215–240.

Ritter, J. R. (1991). The long-run performance of initial public offerings. *Journal of Finance, 46*, 3–27.

Rock, K. (1985). Why new issues are underpriced. *Journal of Financial Economics, 34*, 135–151.

Schipper, K., & Smith, A. (1983). Effects of Recontracting on shareholder wealth: The case of voluntary spin-offs. *Journal of Financial Economics, 12*, 437–467.

Scholes, M. (1972). Market for securities: Substitution versus Price pressure and the effects of information on share prices. *Journal of Business, 45*, 179–211.

Smith, C. (1977). Alternative methods for raising capital: Rights versus underwritten offerings. *Journal of Financial Economics, 5*, 273–307.

Smith, C. (1986). Investment banking and the capital acquisition process. *Journal of Financial Economics, 15*, 3–29.

Stock Exchange Practices, Hearings Before Subcommittee of the Committee on Banking and Currency, U.S. Senate, 72d and 73d Congress, 1832-34.

Chapter 8
The Equity of the Corporation: Common and Preferred Stock

This chapter deals mainly with the financial function of stockholding, i.e., the supplying of risk capital and the expected rewards thereof. The shareholders' expected return is the supply cost of equity capital. Only if the firm is able to give its shareholders at the minimum the "normal" rate of return on risk capital can the company be considered an economic success. Thus, a large part of the discussion is centered on the behavior of the investment markets. This follows from the assumption that the major objective of financial management is to maximize the long-run value of the common stock. If management is to develop financial strategies aimed at maximizing the long-run value of the common stock, it must understand the rationale of the investment markets. It is this market that measures relative risk and provides approximations of the rates of return on different classes of risk capital.

8.1 Common Stock

Common Stock as Risk Capital The basic function of the shareholder is to furnish risk capital. All the creditors and investors in the firm take some risk, but because the claims of the common shareholders are subordinated to those of all the others, the common stock absorbs the initial brunt of any downturn.

The owner of the common shares of the corporation receives no definite rate of return. The corporation owes no contractual obligation to pay any fixed amount to its owners. The return to the shareholders depends on how much is left after bank debts, bonds, and other obligations which have been acquired at a fixed rate have been serviced. As long as the earnings of a business are subject to external variation, at least one class of capital suppliers must be content with a variable return. A firm could not promise fixed returns to the short-term creditors and bond owners, if there were no equity holders willing to take whatever earnings are left or to absorb losses.

J. B. Guerard Jr. et al., *Quantitative Corporate Finance*,
https://doi.org/10.1007/978-3-030-87269-4_8

The equity holders bear the initial risks of operation; the debt holders' return is not usually endangered until the earnings of the shareholders shrink or turn into losses.

The investments of the creditors are safe until the equity can no longer absorb the loss of asset value. The equity of a corporation is a shield protecting the debt holders. Other things being equal, the proportion of ownership capital to total assets affects both the safety margin of earnings before the fixed interest charges and the asset safety margin of the debt holders.

Shareholders bear the initial economic risks of the firm, but they are rewarded with the residual earnings left after all operating expenses, financial charges, and taxes are deducted; in a successful firm, the residual earnings exceed the fixed returns promised to the creditors. The reader is referred to the income statements of IBM, Dominion Resources, and DuPont discussed in Chap. 4. The net income of a corporation is most often expressed as a percent of sales, total assets, or stockholder equity, as the reader saw in Chaps. 4 and 5. The earnings of the common stockholders are calculated only after the expenses, interest charges, and profits taxes are deducted. Though the shareholders' reward comes last, they have one great advantage; their return is not contractually fixed or limited. As legal owners of the company, entitled to all the earnings left after the creditors and the preferred shareholders receive their fixed return, they receive the rewards of successful operation and company growth. If sales increase, if profitable new products are introduced, or cost-saving techniques are successfully applied, the increased returns redound mainly to the benefit of the stockholders. Those who take the larger end of the risk do not necessarily lose. When both dividends and appreciation of capital are taken into account, a "long-run" investment in a diversified portfolio of common stocks has generally proven more profitable than investment in any other type of security.[1] Various statistical tests lend support to the historical validity of this conclusion, which has been called "the common stock theory of investment."

Of course, the superior performance of common shares in the past is no guarantee of gains in the future. The long-run nature of the test periods (12–20 years, or even 54 years as in the WRDS database) is no comfort to the short-term speculator or to those who wish or need an assured amount of money at a given time, and the selection of an improperly diversified or over-priced portfolio can surely lead to losses. Nevertheless, the data indicates that in general those who have risked funds in the ownership of corporation shares have not gone unrewarded.

8.2   Rewards of Common Shareholders

The returns to the outside shareholder of a corporation fall into three categories: dividends, earnings, and capital appreciation on the market price of the stock. These categories are not independent; the emergence of capital appreciation depends

---

[1] See Chelcie C. Bosland, *The Common Stock Theory of Investment*, Ronald Press, 1937.

heavily on the future course of earnings. The two explicit measures of return are the dividend yield, current dividends divided by the market price, and the price/earnings (P/E) ratio, current earnings divided into the market price (the reciprocal is the earnings rate).[2] Implicit in the valuation of shares is g, the expected growth rate of earnings.

Buybacks   The two cash returns that shareholders may expect to receive from their holdings are buybacks and dividends. Buybacks are a program where the corporation periodically uses a proportion of its cash flow earnings to buy back its shares from the open market. A buyback reduces the amount the firm reinvests in the business and is a full or partial substitute for dividends. Currently, the cash amount of buybacks exceeds the amount of dividends. Under past tax policy where dividends were fully taxable at the personal level but capital gains, if any were capped at 20%, buybacks had a tax advantage. (Hopefully, an eventual sensible tax reform would allow a full or partial deduction of dividends at the corporate level. This would do much to restore dividends as the main cash payout program of the corporations.)

Buybacks and dividends seem at first glance to be different creatures. However, at the internal corporate level, the two have the same effect; they reduce the amount of cash available within the corporation and they reduce the proportion of equity in the financial structure. For the shareholder, receiving a dividend gives him cash, and all else remaining equal, the value of his share holdings remains the same. In case of a buyback, the "universal shareholder" receives cash for a portion of his shares and now holds fewer shares. However, the divisor (the number of shares) against total earnings has been reduced, earnings per share increases, and thus the price of each of the remaining shares should rise. Fewer shares times a higher price per share, all else being equal, the value of the portfolio should stay the same, just as in the case of dividend payouts.

Equity repurchases (buybacks) are generally motivated by several factors. First, equity repurchases, in contrast with dividends, serve to increase the earnings per share and thus lead to a possible increase in the stock price. Equity buybacks increase the financial leverage of the firm and are often used in mergers and acquisitions and stock option plans. Buybacks may be viewed as flexible dividends, by investors who do not want cash from the sale (Bierman, 2001a, b). Equity repurchases may be a good use of corporate cash and can be viewed as an investment. Buybacks enhance the value of the book value of the remaining shares, particularly if the shares are repurchased at a price less than the current book value. Bierman (2001a, b) holds that the primary advantages to share buybacks are:

---

[2]Price/earnings ratios are given as multipliers. They are the number of times the going market price of the stock exceeds annual earnings per share. The price/earnings ratio equals $\frac{\text{market price}}{\text{annual earnings}}$. Current dividend yield, on the other hand, is obtained by dividing annual dividends by the market price. It is therefore stated as a percentage. No special reason can be adduced as to why these two relationships are customarily presented in these different manners. It does, however, help to prevent unnecessary confusion on the security markets.

1. The tax shield that occurs when the firm repurchases stock. A portion of the transaction is regarded as a return to stockholders capital and not taxed, generating increased after-tax cash flow, whereas the increase in stock prices generates capital gains, taxed at 20%, as opposed to dividends taxed as ordinary income, at a rate of 35,000.
2. The lower transactions costs for investors who do not sell, and increase their "investment" in the firm.
3. A higher stock price than paying an equivalent dividend.
4. Hiding dilution effects of issuing stock options.
5. Supporting the stock price.
6. Provides a signal that management believes the price is too low.
7. Often increases earnings per share.

Of course, one important difference remains. Dividends go willy nilly to all the shareholders; during a buyback, some shareholders may sell varying proportions of their holdings and others may hold on to all their shares.

Dividends If a firm is making some earnings, has cash resources, and is reasonably well established, the management often endeavors to set up some sort of dividend policy. Generally, dividends are paid out of current earnings and constitute an interim sharing of the profits of the enterprise. The proportion of earnings distributed as dividends is the payout rate.

The board of directors is not legally obligated to declare dividends. Failing to pay dividends on stock is not an act of bankruptcy, as is the failure or inability to pay the interest on debt. This flexibility of return—the ability to forego, postpone, or cut dividends—makes capital stock the safety element in a firm's financial structure.

For a purely dividend-paying company, the expected flow of dividends sets the basic value of a company's shares. If a stock has a current dividend of $2, a forecasted growth rate of 3.0% per annum, and the market desires an overall rate of return of 9%, the stock would have a market price of about $33.33 ($\frac{\$2.00}{(.09-.03)} = \$33.33$) and a current yield of 6.0%.

Various aspects of risk raise the yield rate on common shares and reduce the relative market price. On the other hand, the prospect of a growing stream of earnings and dividends raises the market price and reduces the present yield. The expectation of growth in earnings and dividends will lower current dividend yields below the interest yield on sound bonds.

Thus the expression:

$$\$1 + \frac{1}{(1.08)} + \frac{1}{(1.08)^2} + \frac{1}{(1.08)^3} \cdots \frac{1}{(1.08)^n}$$

carried to a large number of times, eventually sums to $12.50, but the unlimited progression of

$$\$1 + \frac{1(1.08)}{(1.08)} + \frac{1(1.08)^2}{(1.08)^2} + \frac{1(1.08)^3}{(1.08)^3} \cdots \frac{(1.08)^\infty}{(1.08)^\infty}$$

has no finite number. Presumably, a stock whose dividend is expected to grow at a rate equal to or above the going yield rate for an indefinitely long period could fetch an infinitely high price on the market. But, long-run dividends based on earnings cannot grow above the yield rate forever. Nevertheless, prospective growth will raise the stock price, raise the P/E ratio, and lower present yields.

8.3   The Corporate Sector: A Net Exporter of Funds

There is a "balanced economic growth model" which carries the assumption that the amount of income from capital equals the total of aggregate savings and real investment and that perforce the rate of return on capital and the growth rate of the economy are equal. However, the cash flow generated by the operating assets exceeds the amount of possible net real investment, and it necessarily follows that at the end of the fiscal period, the overall corporate sector will have more funds than it can desirably reinvest in the business. In short, the data shows that if in any given year the growth rate is 4.0% and the aggregate after-tax earnings on the real value of *corporate equity* is 8.0%, then the corporate sector one way or another will distribute about 50% of its equity earnings or profits to the rest of the economy. The outflow of cash in the form of dividends, interest paid on debt, buybacks, and repayment of debt has substantially exceeded new funds raised on the capital markets, primarily raised by issuing debt. The reader is referred to Table 8.1 for an initial examination of this issue, as was presented in Guerard and Schwartz (2007).

Table 8.1 The corporate sector net exports of funds
1971–2000 ($MM)

Fund Sources/ Uses	Year						
	1971	1975	1980	1985	1990	1995	2000
Debt Issued	$35,762.8	84,427.6	144,809.8	299,454.4	625,378.4	1,156,219.5	2,204,155.7
Equity Issued	9,262.5	14,611.9	37,132.9	76,050.7	51,664.4	118,036.6	360,740.4
Debt Repurchased	19,907.5	45,473.0	75,759.6	118,685.9	485,998.4	838,614.1	1,546,809.7
Equity Repurchased	1,585.3	1,501.3	7,959.1	51,989.7	52,183.7	88,825.8	230,042.4
Net Debt	15,855.2	38,954.6	69,050.2	111,798.5	139,380.0	317,605.4	657,346.0
Net Equity	7,667.2	13,110.9	29,173.8	24,061.1	-519.4	29,210.8	130,698.0
Dividends	23,743.6	33,438.9	67,759.3	100,397.8	163,604.3	228,352.1	303,767.2
Interest Paid	20,331.2	44,028.2	97,911.8	161,184.8	338,105.4	357,941.9	634,214.4
Net Exports	20,542.3	25,401.9	67,447.1	56,753.1	362,849.0	239,477.8	149,938.3

In 2003, the corporate sector exported over \$350 billion of funds. The surplus of funds over any possible reasonable capital investment policy is the rationale behind the cash buyback of shares and the payment of dividends. In short, perforce, the corporate sector is a net exporter of funds to the rest of the economy.[3]

Earnings   The investor's cash returns, buybacks and dividends, are the fruit of the company's earnings. The level of current and projected earnings impacts dividend policy and potential buybacks. A forecast of dividend levels can be obtained by studying past payout rates and attempting to predict future earnings levels. Dividend policies show a high correlation of dividends with earnings, although, as earnings rise, there is usually some lag before former payout levels are resumed. Buybacks are more difficult to predict. In the past, before the prevalence of buybacks, if a company had a "normal" dividend payout policy, the prospective investor would not be far wrong if he concentrated on the trend of earnings in evaluating the worth of a share of stock.

The amount left after common dividends are paid represents retained earnings or earnings reinvested in the firm. Retained earnings add to the equity of the common shareholders, and the amount of correlated funds can be used to finance additions to the operating assets of the corporation or to retire debt.

Additional assets, properly employed, should add to the future earnings of the corporation. An increase in assets financed by ownership capital as opposed to debt improves the credit standing of the firm and enables it to acquire debt funds at a relatively lower rate. Funds represented in earnings should increase the future profits of the shareholders and eventually result in buybacks or higher dividends. It is not the increment in the book value of the shares, but a hoped-for sequence of increased earnings that makes retained earnings of value to the shareholder.

Capital Appreciation   If the shareholder sells his shares at a higher price than he paid for them, the difference is called capital appreciation or capital gains. The reverse can be true and the sale of securities may show a capital loss. However, a long-term investment in a portfolio of common shares almost always outperforms a comparable investment in other financial assets. Essentially as one might expect, there is a net return for bearing risk.

Capital appreciation develops from growth in earnings or from a decline in the basic discount rate or a favorable reevaluation of the company's risk.

If the market reevaluates a stock favorably, assigning it plus factors for future growth, increased dividends and buybacks and greater stability all at once, the stock can show a rapid rise in price in a brief span of time. However, the market for common stock is notoriously volatile, and a wave of optimism as to the future of the economy and the share of corporate earnings can send the general level of stock prices unsustainably high. Widespread pessimism can have the opposite effect.

---

[3] See E. Schwartz and R.J. Aronson, "The Corporate Sector: A Net Exporter of Funds," *Southern Economic Journal*, October 1966.

8.4 Corporate Exports and the Maximization of Stockholder Wealth

Corporations seek to maximize stockholder wealth. Brealey et al. (2006) stated that the goal of corporate finance is to implement corporate decisions to enhance stockholder wealth. We assume that the market mechanism produces the most effective measurement of corporate strategies. Schwartz and Aronson (1966) held that the rate of return on capital and the growth rate of the economy are equal. If corporate cash flows exceeded real investment opportunities, then the firm should use distribute the surplus funds in the form of dividends, interest paid on debt, buybacks, and repayment of debt. Schwartz and Aronson reported that dividends and interest paid substantially exceeded new funds raised on the capital markets. Schwartz and Aronson referred to the excess of corporate cash flows exported as "corporate exports." Guerard and Schwartz (2007) updated the analysis and confirmed the dramatic growth in corporate funds exported during the 1971–2005 period using a universe of publically traded firms in the United States with total assets exceeding $200 million. Dividends paid and interest paid rose substantially, as did stock repurchases.

We specifically look at dividend, debt and equity issuance and buyback strategies, and the implied leverage decisions of firms. Fama and Babick (1968) reported that dividend policies show a high correlation of dividends with earnings, although, as earnings rise, there is usually some lag before former payout levels are resumed. Buybacks are more difficult to predict. In the past, before the prevalence of buybacks, if a company had a "normal," or consistent, dividend payout policy, the prospective investor would not be far wrong if he concentrated on the trend of earnings in evaluating the worth of a share of stock. A consistent dividend policy means the company pays reasonably consistent ratio of dividends divided by earnings. Equity repurchases, whether privately negotiated or via a tender offer generally specify the number of shares the firm seeks to repurchase, the tender price at which it will repurchase shares, and the expiration date of the tender offer, which the firm may extend. Dann (1981) examined 143 cash tender offers to repurchase equity during the 1962–1976 period, made by 122 different firms, and reported a 22.46% tender offer premium, relative to the previous day of the announcement (20.85% relative to the 1 month period before the announcement).

Large excess returns are associated with share buybacks, particularly for stocks in the 1970s. Firms repurchase equity to enhance stockholder wealth. There were marginal debt effects, and approximately 95% of the enhanced value accrued to stockholders. Lakonishok and Vermaelen (1990) reported excess returns to repurchases continuing through 1986, but at a (slightly) diminished rate.[4] The level of current and projected earnings impacts dividend policy and potential

---

[4]Lakonishok and Vermalen reported premiums of 21.79%, 24.09%, and 18.54% on tender offers during the 1962–1986, 1962–1979, and 1980–1986 periods, respectively. They also reported cumulative abnormal returns of 12.54%, 14.58%, and 9.78% to non-tendering stockholders during the corresponding periods. Smaller firms produced the highest abnormal returns.

buybacks; see Bierman (2001a, b, 2010).[5] Deutsche Bank (2014) reports modest long-term buyback premia. A recently published study, Fu and Huang (2016) studied 14,309 stock repurchases over the 1984–2012 period and reported that long-run abnormal returns from stock repurchases for the 200–2012 period became negative. We find no such results with corporate exports. Moreover, we find no such results over the 2003–2014 time period with global firms' stock buybacks.

Debt is issued to finance capital expenditures and dividend payments.[6] Retained earnings add to the equity of the common shareholders and represent the majority of funds can be used to finance additions to the operating assets of the corporation or to retire debt. An increase in assets financed by ownership capital as opposed to debt improves the credit standing of the firm and enables it to acquire debt funds at a relatively lower rate. Debt and internally generated funds are the primary sources of aggregate investment outlays.

New debt issues exceed new equity issues by a multiple exceeding eight times, a result consistent with Dhrymes and Kurz (1967), and Guerard et al. (1987). New debt issued is identified with the financing capital expenditures in the simultaneous equation modeling. Funds represented in earnings should increase the future profits of the shareholders and eventually result in buybacks or higher dividends. It is not the increment in the book value of the shares, but a hoped-for sequence of increased earnings that makes retained earnings of value to the shareholder.

Let's define a variable to designate the net corporate export of funds, CE, of the corporate sector, as defined in Guerard and Schwartz (2007) and Guerard et al. (2014).

$$CE = \text{Dividends Paid} + \text{Interest Paid} + \text{Net Equity Repurchased} \\ - \text{Net Debt Issued}$$

In 1972, the corporate sector exported over $32 billion of funds and by 2014, the corporate sector funds exported to grown to over $ 145 billion. See Table 8.2. The surplus of funds over any possible reasonable capital investment policy is the rationale behind the cash buyback of shares and the payment of dividends. In short, the US corporate sector continues to be a net exporter of funds to the rest of the economy and has risen almost consistently throughout the 1972–2014 period. The steady growth of dividends paid and equity repurchased as led to a great growth

---

[5]A forecast of dividend levels can be obtained by studying past payout rates and attempting to predict future earnings levels. The amount left after common dividends are paid represents retained earnings or earnings reinvested in the firm. Firms target a target debt ratio by choosing between debt and equity repurchases rather than choosing between debt and equity issuances; see Bierman (2001a, b). Repurchases and dividends are motivated by contracting future investment opportunities, see Bierman (2010).

[6]Dhrymes and Kurz (1967) proposed an explicit link among these dividend, investment, and debt issuance decisions and econometrically implemented using sample consists of 181 industrial and commercial firms for which a continuous record exists over the period 1947–1960. Guerard et al. (1987) updated the Dhrymes and Kurz analysis and reported: (1) a strong interdependence is evident between the investment and dividend decisions, and (2) a strong interdependence is evident between the investment and new debt financing decisions.

Table 8.2 Corporate exports components, $ MM
All Compustat companies, and IBM, selected years

Year	Number of Companies	Long-Term Debt Issued	Long-Term Debt Repurchased	Stock Dividends	Equity Repurchased	Equity Issued	Net New Debt	Corporate Exports
1972	4,508	35,435.7	20,728.3	30,444.9	2,113.7	8,679.8	14,707.4	32,123
1975	6,757	93,832.6	47,912.0	41,668.1	15,238.0	18,180.1	459,320.6	73,713
1980	6,889	173,022.6	82,885.2	82,523.2	8,295.3	42,367.2	90,137.4	207,862
1987	9,225	437,062.9	329,071.9	154,007.9	71,264.0	102,673.0	107,991.1	595,483
1990	9,571	666,371.5	512,349.7	192,196.8	55,208.9	57,453.2	154,021.1	877,259
1995	12,491	1,231,890.1	898,656.4	263,986.6	96,141.0	132,137.5	333,236.6	876,265
2000	12,092	2,724,008.9	1,721,730.0	367,292.3	247,253.7	411,075.7	1002,281.5	1289,183
2008	10,884	6,085,805.7	5,138,659.0	776,022.6	703,572.4	891,732.3	947,147.0	3091130
2010	11,088	6,036,302.0	7,142,278.4	727,427.8	521,110.3	541,200.9	-1,105,976.4	4,210,335
2014	10,921	6,703,998.7	6,000,589.7	979,835.3	824,651.0	497,677.5	703,409.0	2,758,859
1972	IBM	201.6	98.3	626.2	0.0	270.6	103.3	37.4
1975	IBM	29.7	870.4	968.9	0.0	284.5	-40.7	787.4
1980	IBM	604.0	94.0	484.0	484.0	422.0	510.0	1,885.0
1987	IBM	408.0	719.0	2,654.0	1,425.0	133.0	-311.0	4,874.0
1990	IBM	4,676.0	4,184.0	2,774.0	491.0	135.0	1,592.0	2,874.0
1995	IBM	6,636.0	9,460.0	634.0	10.0	0.0	-4,110.0	4,346.0
2000	IBM	9,604.0	756.0	929.0	6,093.0	0.0	2,043.0	5,696.0
2008	IBM	4,363.0	3,522.0	1,250.0	6,506.0	3,774.0	841.0	7,676.0
2010	IBM	8,055.0	6,522.0	3,177.0	15,376.0	3,774.0	1,533.0	14,174.0
2014	IBM	8,180.0	4,644.0	4,265.0	13,679.0	709.0	3,536.0	14,729.0
2020	IBM	10,154.0	13,365.0	5,797.0	302.0	0.0	3,211.0	10,705.0

Source: WRDS Database

in corporate exports.[7] IBM's corporate exports have risen substantially with its dividend payments, great reduction in common stock due to equity repurchases.

---

[7] An analysis of the CE components reveals that interest paid has risen faster than dividends paid during the 1971–2014 time period. Net debt issues have risen at an undiminished rate, with the notable exception of 2001–2005 and 2009. Stock repurchases rose substantially following the crash of October 1987. Net equity repurchases increased substantially in the 2002–2006 time period and fell dramatically with the financial crisis. To illustrate the corporate fund generation process, Guerard (2010) showed the ten largest and smallest corporate exporter firms in 1983; the largest corporate exports firms that included AT&T, IBM, and several of the large oil companies dominated positive corporate exports in 1983 as they paid large dividends and interest and generally repurchased more debt than was issued (which made a great deal of sense given the level of interest rates in 1983). A similar process occurred in 2006 as Microsoft, Pfizer, and the oil companies dominated the largest corporate exporting firm (IBM fell to only the 24th largest exporter in 2006). Dividends paid exceeded equity repurchased of the Compustat firms from 1971 to 2014, although equity repurchases have risen relatively to dividends since 1982. We examine the 1971–2014 period because Compustat does not maintain debt and equity issuance, and repurchases, prior to 1971. Guerard et al. (2014) created mean-variance efficient portfolios and tilt the portfolios to purchase stocks with the largest corporate exports. We believe, and will show, that buying stocks on a forecasted earnings acceleration variable basis can be complemented with buying companies that generate positive and corporate exports. The portfolios produce statistically significant outperformance relative to the universe benchmark. We report that portfolios with the largest corporate exports stocks produce higher returns, relative to risk, when the stocks that are not identified as undervalued by a stock selection model are excluded from the portfolios.

Equity repurchases have substantially exceeded dividends for IBM in the post-2000 time period.

IBM's low growth, extensive leverage, and high profitability make it a very interesting study.

8.5 Definitions of the Value of Common Shares

Par Value Par value is basically a legal concept that sets the minimum initial price at which the company's shares are sold at the time of organization. (Par value can be changed through reorganization, through recapitalization, or through stock splits.) Any initial sale of stock at a discount from par presumably involves a legal obligation which the creditors can press against the holder of the stock should the company ever fail. In some jurisdictions, the issue of stock at a discount from par is illegal; in any case, it is extremely rare. (This is why almost all stock certificates bear the notation "fully paid and nonassessable.") The dollar amount of stocks sold at par comprises the capital stock account. In most jurisdictions, no dividends can be declared unless there is a positive balance in the surplus account sufficient to cover the dividend. If the books show no surplus, no dividends can be declared under this rule even though there might be considerable current earnings. In short, the general rule "prohibits dividends out of capital."

Historically, par value was designed to serve a twofold purpose. It set up a sum of paid-in capital to act as a trust fund for the creditors of the corporation in lieu of the unlimited personal liability of the owners. It was also to act as assurance to a subscriber for the original shares that other shareholders were purchasing their shares at an equal price. In practice, par value has served neither of these two functions very well. Credit ratings are more likely to be based on earnings potential, safety and liquidity of assets, and total financial structure than the presumed last line of defense shown by the capital stock account. Furthermore, any attempt to favor stockholders by giving them dividends at a time when the firm appeared unable to meet its credit obligations would constitute an act of "involuntary bankruptcy" and could be halted by invoking bankruptcy proceedings. Both aspects of par value, including the assurance that all are venturing equally in the new company, are considerably vitiated by the fact that the directors have wide discretion in valuing noncash assets exchanged for common stock. Essentially, the investor in a new venture must estimate and study all of the potential rewards and risks and not rely upon the fictitious safety of par value.

Because of the mistaken impressions par value has on occasion given to the investor, and because of its disadvantages in the initial issue, there has been a large growth in the employment of no-par value stock. With an issue of no-par stock, the directors have some discretion as to how much is to be allocated to the capital stock account and how much to the paid-in surplus account. A great advantage of no-par is that if the initial operations of the company do not go well, additional equity capital may be raised by selling new shares at a lower price than the original issue. With par value shares, this might be difficult to accomplish because it would be illegal to sell the second flotation at a discount from par. However there have been some tax difficulties with no-par shares; in New York State, for example, these shares have been given a constructive value of $100 par for purposes of calculating the transfer tax on their sale. Mainly because of this problem, the most popular issue today, replacing no-par stock, is low par stock. These shares may have par values such as $1.00, $3.00, or $5.00 and are usually floated at a price considerably higher than their par value. This stock has most of the advantages of no-par stock without being subject to possible extra burdens under the franchise or transfer tax.

8.5.1 Book Value

The book value is the net balance sheet asset value per share. It is obtained by subtracting liabilities and preferred stock from the accounting value of the total assets and dividing the remainder by the number of common shares outstanding. Book value may be used in certain financial comparisons. It is sometimes useful to see which of a set of companies earns a greater percentage on its book value. Because book value is increased with additions to surplus (i.e., retained earnings), tracing the course of book value over time shows how much net internal financing the company has used. (This historical analysis becomes difficult if any stock dividends have taken place.) The usefulness of book value for financial analysis is reduced by the time changes in price levels and variations in depreciation policies, which results in a wide difference between the balance sheet accounting value of the company's assets and their current economic value.

When the market is pessimistic or depressed, stocks may tend to sell below book value; in a buoyant or optimistic market, shares are generally priced above book value. Growth companies, which show a strong upward trend in earnings, may sell at a considerable multiple of their book value. The statistical association between equity holding period returns and book value of stockholder equity will be calculated in Chap. 14.

8.5.2 Market Value

The relative market price of a corporation's shares is an index of the firm's success. To the so-called value investor, the market price of the stock relative to its prospective yield is the vital datum.

Market price is set by the investors' projection of the firm's potential earnings discounted by the going rate of return for the type of firm. The market price relationship is usually summed up in the going price/earnings ratio. A relatively high P/E ratio indicates that the market has hopes for growing earnings from the firm, and a low P/E ratio suggests fear for the future. The price/earnings ratio sometimes rises with a drop in corporation earnings which the investors believe to be temporary.[8]

8.5.3   Intrinsic or "Normal" Value

Intrinsic value is an idealized concept. It is the price a stock would sell at if the market was supposedly not subject to psychological aberrations.[9] Under the method given by Graham and Dodd, intrinsic value is approximated by studying the past level of earnings, earnings trends, and the pattern of dividends. (Buybacks did not appear in the Graham and Dodd model.) A capitalization rate based on standard levels modified for various qualitative factors affecting the particular firm is applied to a weighted combination of earnings and dividends. The result is an intrinsic or normal value.

Graham and Dodd's basic formula was:

$$\text{Intrinsic Value} = M\left(\frac{E}{3} + D\right) \pm \text{asset factor.}$$

M represents a "normal capitalization multiplier" which ran from 8 to 15 for ordinary industrials—with lower multipliers given for highly leveraged and volatile companies and the higher multipliers reserved for stable and blue chip companies. E is the projected earnings per share, generally based on the average earnings of the last 5 years corrected for any pronounced trend. The figure for earnings is divided by 3, and then D, the projected dividend, is added to the multiplicand. The amount for D is generally based on the dividends of the last 5 years. When a company's payout rate approximated 66%, M approaches the pure earnings multiplier. The asset correction is used to reduce the intrinsic value if it exceeds book value by too large an amount; a plus correction is made if intrinsic value based on earnings falls below the rule-of-thumb "liquidating value" per share.

Graham and Dodd admitted the difficulty of applying the formula to very new companies, to high-risk situations, to highly leveraged companies, and to strong growth companies.

According to the intrinsic value theorists, share prices at any time may swing well above or below intrinsic value, but they tend to return to it over the long run. Although long-term economic growth and stability have raised expectations and raised the P/E multipliers above those proposed by Graham and Dodd, and although buybacks have vitiated the emphasis on dividends, there is still a sense in which the

---

[8]Thus, if a stock had annual earnings of $6 and sold for $60, its P/E was 10 times; if earnings should drop to $5 and the stock price drops to $55, still the P/E ratio would have risen to 11 times.

[9]For the basic statement of this concept, see Graham Dodd, *Security Analysis*, McGraw-Hill, 1951, pp. 410–11.

intrinsic value concept influences managerial decisions. The determination of the cost of equity financing cannot be based on current market data unless the company's shares are selling within some "normal range." A growth firm whose stock is selling at a relatively high price cannot use the explicit current market rate of return as its cost of equity funds but must develop an implicit cost based on some normative valuation.

Liquidation Value If all the assets of a company were sold off, all creditors paid in full, and the preferred stockholders paid the voluntary liquidation price, the remainder per share is the liquidation value. In practice, the exact liquidation value of the shares cannot be known unless this process (which unfortunately is not reversible) was carried out. Rough estimates of the liquidation value can be made by experienced appraisers estimating the realizable value of the assets. The accuracy of such estimates varies considerably for different types of assets. In general, the receipts from current assets and land can be estimated more accurately than the amount which might be recovered from plant and equipment.

A crude index of liquidation value has been developed by the investment analysts. This measure is called "net current assets" per share and is obtained by subtracting all liabilities and preferred stock from the current assets and dividing the remainder by the common shares. No value is given to plant and equipment, but on the other hand, it is assumed that no loss will be incurred in the liquidation of the current assets. Net current asset value is in a sense a minimal liquidation value.

When the shares of a company sell for any considerable time below their liquidation value, it indicates either extremely poor management or that the capital of the business is employed at a rate of return lower than that obtainable in other industries.

The company may be ripe for seizure by a new management group. The new managers may try to raise the rate of return, merge with another company, or engage in a partial or complete liquidation of the firm. From an overall economic viewpoint, the demise of such a company is not distressing; presumably, the resources released could be more productively employed elsewhere.

8.6   Stock Splits and Stock Dividends

A stock split is a financial maneuver by which the book value of each share is reduced but the holders of outstanding stock receive certificates for new shares pro-rated on their present holdings. After a stock split of 2 for 1, a holder of 100 shares of a corporation holds 200 shares; after a split of 3 for 2, she holds 150 shares. The book value of the shares is reduced proportionately by the split, as are the earnings on the new shares. Nothing on the balance sheet changes; the surplus and capital stock accounts have the same totals, but in the information attached to the balance sheet, the number of outstanding shares is increased. A stock split gives the share owners nothing of value that they did not have previously. Nevertheless, the investment market tends to react favorably to a split. Rumors of an impending split generally send the price of the stock upward. A potential favorable effect of the stock split may be to bring the price of the shares down into a more marketable range. The reduced base price draws increased trading interest and adds a slight increment to the market price. Another presumed favorable aspect of a split is the "announcement effect," the assumption that the split implies an optimistic management view of the prospects of the firm.

A stock dividend is the distribution of new shares to the present stockholders, reducing the surplus account and crediting the capital stock account on the corporation's books.[10] Its effects are similar to those of a stock split, although the accounting formalities differ slightly. A stock dividend runs only from about 2% to 10% of the shares in contrast to the much larger division of the stock split. In a stock dividend, the par value of the shares is left unchanged. However, a larger number of shares is now divided into the same total equity, and the book value of the stock is lower. The earnings per share is also reduced proportionately. The stock dividend brings the stockholder nothing of objective value except as it may raise the effective cash dividend rate. The usual stock dividend is under 10%, and generally the cash dividend rate on the shares remains unchanged. Thus a 10% stock dividend is usually equivalent to a 10% rise in the cash dividend and probably has a positive effect on the value of the investor's portfolio.

Federal tax laws do not consider a stock split or a stock dividend personal income. The stock split or stock dividend is used to reduce the acquisition cost of the shares, so that, in case of any partial liquidation of the portfolio, capital gains or losses may be properly calculated.

8.7   Stock Prices and Dividends: An Example of Valuation

Stock values are determined, to a great extent, by the expectations of earnings and dividends growth. Let us examine the case of Dominion Resources, D, the holding company of Virginia Electric & Power Company (VEPCO), as of October 2000. The current price of D was $53.25. The respective dividend and earnings per share data were:

	1999	1998	1997	1996	1995	1994	1990
Dividends	$2.58	2.58	2.58	2.58	2.58	2.55	2.32
Earnings	$2.81	2.75	2.15	2.65	2.45	2.81	2.92
Payout Ratio	.92	.94	1.20	.97	1.05	.91	.78

---

[10]In rather old-fashioned parlance a stock dividend was described as "capitalizing earnings."

The reader notes that Dominion Resources is maintaining its dividend per share during the 1995–1999 period, despite the volatility of its earnings per share. How can one determine the fair market value of Dominion Resources stock, P? Let us use the Williams/Gordon discounted dividends stock valuation equation.

$$P = \frac{E(DPS)}{k - g}$$

where

$E(DPS)$ = dividend per share
k = required rate of return and
g = growth rate

Gordon assumed in his model that dividends and earnings growth is equal. This is not a reasonable assumption for Dominion Resources during the 1990–1999 period.

$$g_{eps} = \sqrt[9]{\frac{eps_{1999}}{eps_{1990}}} = \sqrt[9]{\frac{\$2.81}{\$2.92}} = -.004$$

$$g_{DPS} = \sqrt[9]{\frac{DPS_{1999}}{DPS_{1990}}} = \sqrt[9]{\frac{\$2.58}{\$2.32}} = .012$$

The 9-year compounded earnings per share growth is -0.4%, whereas dividends grew at 1.2%. One will use the nine-year historic dividend per share growth to price the stock.

What should one use for the required rate of return? We can use the required rate of return, k, from the Capital Asset Pricing Model, which we develop in Chap. 14. The required rate of return on equity is an opportunity cost that the investor incurs by purchasing the stock. That is, the stock must produce a return that is consistent with its level of risk. The return of a stock is determined by its dividend payments and its stock price appreciation. Investors must be compensated for bearing risk, and risk is measured by the stock's beta, its measure of systematic risk. The S&P 500 beta for D is 0.20, a defensive asset in that if the market is expected to rise by 10%, then Dominion Resources stock return is expected to rise by only 2%. We will estimate a beta in Chap. 14.

The 90-day Treasury bill yield is 3.5% in October 2001, and the expected market equity premium of stocks relative to Treasury bills historically is 8.8%. The required return on return for Dominion Resources is 5.26%.

$$k = R_F + [E(R_M) - R_F]\beta;$$

$$k = .035 + (.088)(.20) = .0526$$

The market's required rate of return is 5.26%; we should purchase Dominion Resources if we expect the stock to earn more than 5.26%. An alternative strategy, yielding the same answer, is to price Dominion Resources shares and compare its fair market value to its current stock price:

$$P = \frac{\$2.58(1.012)}{.0526 - .012} = \$64.29$$

In October 2001, one could have determined a fair market value for Dominion Resources of $64.29, far exceeding its current stock price of $53.25. One should purchase Dominion Resources.[11]

The current date, as the authors wrote this chapter, is August 23, 2004. The current three-month Treasury Bill yield is 1.41%, the current Dominion Resources beta is 0.23, and the 10-year growth in earnings per share for Dominion Resources is 0.65% (the ninth root of $2.98 in 2003 divided by $2.81 in 1994), and Dominion Resources pays a $2.58 dividend (still). The current fundamental value of Dominion Resources is:

$$k = .014 + .088\,(.23) = .032$$

$$P = \frac{\$2.58(1.0066)}{.032 - .0065} = \$100$$

The current price of Dominion Resources is $64.58.[12] One should purchase Dominion Resources as it is substantially undervalued.

What has D performed since 1982, when purchased, as of December 2019?

---

[11] One of the authors used Dominion Resources as an example of security valuation for an Advanced Financial Management class at Rutgers University in October 2001. Dominion Resources has outperformed the market, the S&P 500, by approximately 3 percent over almost 3 years, the relative outperformance reaching 15 percent in October 2003. One might smirk at such a return; however, one must remember that the beta of Dominion Resources is approximately 0.20, creating a more reasonable market-adjusted excess return. Dominion Resources has cumulatively underperformed the market in 1 month of the 36 months (which one of my former Penn students noticed with an interesting E-mail).

[12] On August 10, 2006, a final version of this chapter was complete in the original edition. The price of Dominion Resources was $ 78.90. D, with a current beta of 0.44, had outperformed the market (S&P 500 Index) by over 500 basis points during the past 2 years.

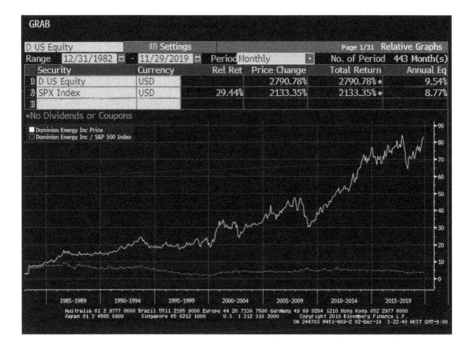

Dominion Resources, now Dominion Energy, has outperformed the S&P 500 by over 80 basis points, annually, with a beta of about 0.26, one-quarter the risk of the market! An investment of $600 for 50 shares now exceeds $64,000. Have you any Grey Poupon? But of course. Dominion Energy closed on July 7, 2021, at a price of $75.64, the stock returning the approximate return on the S&P 500 since June 1981, with a beta of 0.32. D has maintained a great risk-adjusted rate of return for stock.

One could calculate dividend discount models for all exchange-traded stocks in the United States (and globally) and rank the stocks relative to their stock price deviation from fair market value. Alternatively, one could use a multifactor stock selection model to quantify the purchase decision of stocks.[13] The authors prefer the latter method that is developed in Chap. 14.

8.7.1   Noncash-Paying Growth Shares

The value of any corporation stock is its current or its potential future cash throw-off either in the form of dividends, cash buybacks, or purchase by another company. At any point in time, there are shares on the market that currently pay no dividend and show no buyback policy but still command a positive price. The value of such shares is based on an expectation or forecast that sometime in the future the shares will yield

---

[13] See Bierman (2010) for an interesting analysis of stock buybacks and dividends.

a cash throw-off. If the shares are those of a *super-growth company,* one whose periodic earnings grow at a rate in excess of the normal capitalization rate, the current market price will be quite high.

Super growth cannot be infinite; sometime in the future, the firm's growth rate will slow to some norm, the company will begin to throw off cash either in the form of buybacks or dividends, and the price of the shares will be evaluated on the basis of the earnings reached at this time. The current market price of the shares is the present value of this future price. An example of the valuation of a noncash paying, fictional growth company, Crampton Inc., is presented below. The formula for the valuation of the shares follows:

$$\text{Current price} \approx \frac{E(1+g)^n \times M}{(1+i)^n}$$

where

E equals the current earnings
g is the expected average super-growth rate
M is a "normal" market multiplier
n is the forecasted period of super growth
i is a risk adjusted desired rate of return

The data for the fictional Crampton Corporation:

- Crampton's current earnings per share is $1.00.
- The current growth rate of earnings is 30% per annum.[14]
- The forecasted growth period is 20 years.
- The risk-adjusted rate of return is 9.0%.

The final value of the shares is obtained thusly:

- At the end of 20 years, earnings compounded at a growth rate of 30% p.a., = $190.
- The future price of the normalized shares with earnings capitalized at 9% (M = 11.11) equals $2111.
- The present value of a Crampton share, $2111, discounted at 9% for 20 years, equals $376.

Under this supposed scenario, the P/E ratio could be as high as 376 times current earnings. Of course, financial disaster follows if the growth period is shorter than 20 years and/or the growth rate does not reach 20%. On the other hand, much money can be made if the reverse is true or if the investor jumps in at a lower price before the market recognizes Crampton's potential.

---

[14]To make calculations easier, we have set up an average growth rate. In fact, we are likely to have a curvilinear growth path. This does not vitiate the basic concept.

An argument sometimes presented against any valuation of long-term deferred returns is that an investor would be irrational to buy an investment whose cash payout is beyond her expected life or planning horizon. The argument does not hold as long as a market exists containing sufficient purchasers with differing and/or longer horizons. As time passes and the payoff period becomes closer, the price of the investment moves continuously upward. The analogy can be made to that of a forest, which is growing at a rate in excess of the financial rate, will not be harvested for 50 years. One may buy the forest at a current price given the knowledge that there will be other buyers in the interim willing to buy and hold at all intervals before the harvest.

8.7.2 Valuing a Dividend-Paying Super-Growth Stock

A growth company may present the investor with a stream of dividends and earnings growth in excess of the normal yield rate. A solution of the problem is presented for the fictional Shoreham Corporation. In evaluating Shoreham shares, it is necessary to obtain the present value of both the stream of cash dividends and the present value of the future price of the stock.

**Illustration of the Valuation of a Dividend Paying
Growth Stock, Shoreham Inc.**

A. Data:

Current earnings = $5.00
Current dividend = $2.00

B. Assumptions:

1. Earnings per share will grow at the rate of 20.0% per annum.
2. The risk adjusted desired rate of return is 10% per annum.
3. The expected growth period, n, is 10 years.
4. At the end of 10 years, Crampton shares will have a "normal valuation" at a P/E of 11 times current earnings.
5. Dividends will be 40% of annual earnings during the growth period and will necessarily grow at 20% per annum.

C. The basic valuation equation is:

$$PV = \frac{D(1+g)}{(1+i)} + \quad \cdots \quad + \frac{D(1+g)^{10}}{(1+i)^{10}} + \frac{11 \times (E)(1+g)^{10}}{(1+i)^{n}}$$

or the value of the increasing dividend stream for 10 years, plus the present value of the price of the shares at the end of the super-growth period 10 years from date.

D. 1. The value of the dividends is the value of an annuity: $s_{\overline{n}|}{}^{e}$
where e equals g-i, (20%-10%) or 10.0%, and n = 10 years.

Thus: $s_{\overline{10}|}{}^{10\%}$	=	15.937
	x	$2.00
The PV of the Dividends	≈	$31.87

2. The value of the future share price is obtained by multiplying the future earnings by 11 and bringing this price back to present value.

Future earnings equal $E(1+g)^n$ or $\$5(1.20)^{10}$

$(1.20)^{10}$	=	6.192
	×	$5.00
The amount of future earning	=	$30.96
Times the future P/E factor	×	11
The future price of the shares	≈	$340.56
Discounted at the present value factor of $1/(1.10)^{10}$	×	.3855
The present value of the future price of the shares	=	$131.29

The current value of Crampton Shares equals the pv of the dividends plus the pv of the future price of the shares

pv of dividends		$ 31.87
pv of future price		131.29
Current value	=	$163.16

E. Note that on the basis of the current value, the

P/E ratio, $163.26/$5	=	32.7
Or earnings to price $5/163.26	=	3.06%
The current dividend yield, $2/163.26	=	1.2%

The growth stock has a high current price-to-earnings multiple and low current yield.

8.7.3 Super Growth Cannot Be Infinite

An investment that forecasts earnings growth for an infinite period in excess of the market's normal rate of return would have an infinite value.[15] However, in actuality, a super-growth period cannot be infinite, and therefore a growth investment has a defined value.

Three constraining factors prevent the growth period from running for an indefinite length. First, it is apparent that no option can keep generating a return that grows forever at a rate that exceeds the growth rate of the economy. A super-growth enterprise that grows at a rate exceeding that of the GDP (gross domestic product) becomes a larger and larger part of the GDP; it becomes a bigger and bigger proportion of the total economy. Taking this proposition to the logical extreme, when the super-growth enterprise becomes the total GDP, it cannot grow faster than the GDP. More realistically, we may note that as the enterprise becomes a larger part of the economy, its rate of growth is constrained by the limits of the system in which it operates.

The second limiting factor involves the relation of the level of the interest rate to available super-growth projects. Theoretically, a super-growth investment of infinite duration has an infinite value. However, if there were a prevalence of super-growth options, the demand for capital would take a quantum leap. But savings, the supply of capital, is not infinitely elastic. Presumably, in any society, people must consume some resources, eat, clothe themselves, and find shelter. They cannot save all income. The price for savings, i.e., the interest rate, must rise, and equilibrium is reached when the interest rate became higher than the rate of growth. Last, we may take a look at the actual financial side of a super-growth process. Presumably, a super-growth situation requires an increase in net capital per period which is in excess of the investment in the previous period and which requires that in each period the return on the total capital continues at a rate in excess of the cost of capital. Although given some initial superior advantage in technology or product, an

---

[15] If the growth rate exceeds the prevailing discount rate and is of unlimited length, the theoretical present value of the investment option is infinite. This result at the extreme has led to much philosophical discussion in a variant form of the "St. Petersburg Paradox" as to whether any value can be infinite.

enterprise might enjoy super growth for a considerable number of years; such growth is not likely to go on indefinitely on an increasing capital base. When in the particular situation, the enterprise begins to evidence a declining marginal return to capital, the period of super growth must begin to slow and halt.

8.7.4   The Paradox of the Low Current Return on Growth Options

The existing current return on a growth stock appears meager on the basis of market valuation. Relative to the price, current earnings are very low.[16] The explanation is that the discounting of a projected rising stream of return at a normal desired market rate of return produces a very high present value. It is not that the existing earnings are low but the high current price that makes them seem so.

Would purchasers of the shares at the current high price lose return if they do not hold the stock to the end of the growth period? Not so. The current earnings yield is low, but the price of the stock over time will rise toward its future price, resulting in capital gains over each holding period. The periodic capital gains yields the normal rate of return. If actual events, expectations, and market discount rates do not change, every owner for every holding period will obtain the same overall rate of return.

8.8   Risk and Returns to Growth Investments

The volatility of the market price growth stock shares is higher than that for standard shares. Not only must an estimate of the rate of growth be used in the evaluation, but the number of years of forecasted growth, n, is a highly strategic variable. The market price is highly sensitive to n, and a change in the forecast number of the remaining years of super growth can bring about wide swings in the price of the shares. Investors who come in early, before the market recognizes the facts of potential growth, may do very well indeed. Even later investors may garner well above a normal return if, after a period, the forecast of the growth period inherent in the price of the shares is extended. On the other hand, if the expected length of the growth period is shortened, the price will drop sharply.

A bit of analytical advice to potential purchasers of growth shares goes as follows. Given the current market price of the shares, and given an estimate of the percentage rate of growth and a reasonable risk-adjusted desired rate of return (discount rate), the number of years of growth necessary to justify the investment will fall out of the equation. If the number of years of sustained growth implicit in the current market

---

[16] A similar phenomenon also holds for real estate in a growth area where the current short-term rents appear to be low relative to the market price of the properties.

price seems reasonable, the investment may be worthwhile. If the required years of growth seem too long, the investment is dangerous.

8.8.1   The Cost of Capital to a Growth Firm

Because the P/E ratio on a growth share can go very high, some analysts have argued that the cost of capital to a growth company is very low and that, for example, in the Shoreham case, the company could invest in projects that at the margin yielded a return on equity of a mere 3.06%. This is preposterous. The marginal investment would not cover the normal cost of capital. If this strategy were followed, the market would soon note that the relative returns generated by the company were falling and the Shoreham share price would plunge downward.

The cost of capital to a growth firm is the same as that for normal enterprises in its class. Real capital investments at the margin must yield the norm of the market rate of return. However, the returns made on the intramarginal projects of the growth company are well above the normal rate, so that the *average rate of return* is significantly above the cutoff point. This excess rate of return is sustained because increases in product demand and/or decreases in production costs are continuous over the length of the growth period.

The value of a growth stock is raised by factors outside the financial system. The rise in the value of the stock does not lower the cost of capital.

8.8.2   The Cost of Common Stock Financing: The Norm

Assuming the firm is operating in the range of an acceptable financial structure, the cost of common equity capital is the "normal" earnings yield for shares of the given industry. The earnings yield is the inverse of the price/earnings ratio corrected for projected growth or decline. A stock with average earnings of $4, an estimated growth rate of 3.0% and a "normal" price of $48 has a P/E ratio of 15 and a current yield of 6%. Given the projected growth rate, the minimum rate of return after taxes that the company must make on new equity capital is 9.0%. Investing new equity funds in projects earning less than 9.0% lowers the return for both new and old shareholders. In a growth situation, where the investors believe earnings are likely to increase for some indefinite period, the ratio of current earnings may be very low. Here the cost of common stock financing must be imputed from some normal concept.[17]

---

[17] It may be of interest that Adam Smith wrote that the normal rate of profit on ownership capital would be about double the interest rate on sound commercial loans. A. Smith, *The Wealth of Nations*, Modern Library, E. Cannan, Ed. 1937, Ch. IX, p. 97.

The cost of any one source of financing cannot be determined singly. Its interactions with other types of financing must be considered. An increase in equity capital decreases the firm's financial risk and thus lowers the costs of all sources of funds. On the other hand, equity in itself is the most expensive source of funds. Determining the best position between the reduction of risk and the cost of equity funds is the crux of the problem of reaching the "optimum" financial structure. The cost of capital to the firm is the weighted sum of the costs of each source of funds in the optimum mix.[18] However, the basic rule is that unless it is extremely desirable to reduce financial risk, new equity financing should not be used if the effect is to lower the expected earnings per share.

The management must consider financial risk and the cost of funds if it is to succeed in its basic obligation of maximizing the long-run economic position of the common shareholders. The management must not overuse debt financing. Maximizing short-run earnings is not optimum if the shareholders are subject to inordinate financial risk. In making investment decisions, the firm must consider the alternate returns that stockholders could get on their funds elsewhere. The costs of financing new assets must not exceed the projected return on these assets, and neither should the shareholders be subjected to undue risks. These considerations apply to those decisions involving the management of assets and financial decisions, which are under the control of the management. A decline in the value of shares because of unfavorable developments in the general external economy cannot be blamed on management.

8.9 Preferred Stock

Preferred stock is a form of equity or ownership capital. Its claims to a return out of earnings or to a share of the assets in case of liquidation come after the creditors but before those of the common shares. From the viewpoint of the common shareholder, the preferred stock issue performs a function resembling a form of debt—albeit a debt subordinated to the rest of the liabilities. However, the preferred has one basic equity feature: its dividend is not an inflexible claim; it can be postponed or passed without the firm having committed an act of bankruptcy. The equity nature of the preferred shares is recognized in the current corporation income tax law; preferred dividends are not deducted before calculating net income for tax purposes.

---

[18] See Chap. 10.

8.9.1 *Features of Preferred Stock*

Priority on a Stated Dividend The dividend on a preferred is stated on a fixed basis. The dividends on preferred stocks must be paid before any dividends on common can be paid. A corporation has to be in a rather tight financial position before it passes the preferred dividend. The fixity of the preferred stock dividend eliminates any significant gain for the ordinary run of preferred stock from an increase in the prosperity and earnings of the corporation but sets a basic value in the ordinary run of affairs.

A preferred stock has the features of a perpetual annuity. The market price is obtained by dividing the stated dividend by the going rate of return. Thus if the market rate of return on a $5 dividend preferred is 6.0%, the stock sells at approximately $83.33. The market price rises or falls with the level of the interest rate.

Cumulative Dividends Most preferred shares have a cumulative dividend feature. If dividends are passed, they accumulate, and no dividends can be paid on common stock until all arrearages are paid up. The cumulative features deter management from passing the preferred dividend.

A preferred stock carrying passed dividends may sell below par as long as the possibility of the arrearages being paid appears remote. If the economic position of a firm takes a turn for the better and payments of some amounts of the passed dividends are imminent, the price of the preferred stock will rise to somewhere between its long-run investment value plus a discounted value of the accrued past dividends.

Preference to Assets Preferred stock has a priority over the common in case of liquidation. In case of dissolution, the funds from the sale of the assets go first to the creditors and then to the preferred shareholders, and only then is any remainder distributed to the common stockholders. If the funds are insufficient to pay the liquidation value of the preferred stock, the common shares receive nothing. However in involuntary liquidation, through failure or bankruptcy, most of the creditors are probably not paid in full, and both the preferred and the common stock are worthless.

The preferred stock generally has two values: one for "voluntary" liquidation and one for "involuntary" liquidation. Liquidation is involuntary in the case of bankruptcy. The involuntary liquidation value of the preferred generally consists of the par value of the preferred shares plus any accumulated dividends. Voluntary liquidation occurs when the stockholders themselves vote to go out of business. Usually the voluntary liquidation value of the preferred has a premium over par (e.g., a $100 par preferred may have a voluntary liquidation value of $120).

Most companies attach a call option on a preferred issue. This enables the company to buy back (i.e., call) its preferred stock from the holders at a fixed maximum price. The call price is usually set at a premium over par. The voluntary liquidation price and the call price are usually the same.

Nonvoting The preferred shares are generally nonvoting on ordinary matters of corporate policy, but usually have voting rights on matters directly affecting their own position. Many preferred stocks are given the full right to vote on corporation matters if preferred dividends have been passed.

8.9.2 Rational for Preferred Stock Financing

If the going dividend yield on preferred is 6% and the tax rate is 35%, the pre-tax cost of preferred stock financing is 9.23%. The cost of preferred stock financing is relatively high compared to that of tax-sheltered debt. On the other hand, as a subordinated form of financing—bearing some of the risks of equity—it helps to support the debt structure. Thus, if the firm can net 10% on its equity capital after interest and taxes (about 15.4% before a 35% tax), an issue of a 6% preferred can be to the advantage of the common shareholders. The 4% earned over the preferred dividend becomes part of the return of the common shareholders. *The preferred is a form of leverage within the equity section of the financial structure.* Nevertheless, the corporate income tax makes preferred stock financing expensive compared to debt. At present, preferred stock is not very common on the new issue market.

8.9.3 Convertible Preferred

A number of preferred issues have a convertible feature. At the option of the holder, the preferred shares may be turned in for common stock at an agreed ratio or price.[19] The conversion option may be open indefinitely or it may expire after a stated date.

Charter provisions and indentures are generally drawn up to protect the holder of a convertible security against dilution of the value of the convertible privilege. The conversion ratio is adjusted for any common stock splits or dividends. For example, a conversion ratio of 2 for 1 will be raised to 4 for 1 if the common shares are split 2 for 1.

Rationale of Convertible Preferred The logic of issuing convertible preferred rests on the possibility of the firm obtaining, for the time being, equity funds at a lower cost than that desired on the common equity. The conversion feature induces investors to supply the firm with equity capital at a reasonable rate. The convertible feature gives the purchasers the advantages of a hedged bet; they have some of the safety of return given by the preference feature of the shares, and the conversion

---

[19] Since a convertible issue, whether preferred stock, debenture, etc., can be turned into common stock, possibly changing the future sharing in earnings, assets, and voting strength, preferred subscription privileges (the pre-emptive right) is often given to the common stockholders on the initial flotation of any convertible security.

feature allows them to share in the gains if the earnings and market value of the common shares rises.

Convertible preferred shares are sometimes used to make merger terms more attractive. The stockholders of the firm being wooed into the merger are given convertible preferred shares of the acquiring firm in return for the common stockholdings in their present company. The new group of stockholders has some protection against a decline in the fortunes of the acquiring company plus a chance to share in its possible success.

The advantages to one side on a convertible issue are of some disadvantage to the other. If the enterprise is successful, the potential gain to the common is diluted as the preferred shareholders turn their shares into common. The rise in common share earnings is delayed, and there is a drag on the market price of the common until all the preferred is converted. On the other hand, the initial flotation of the converted preferred gives the firm equity funds relatively cheaply, and in the interim before conversion, the amount earned on these funds over and above the preferred dividend benefits the common shareholders. Perhaps the increased equity funds acquired at a strategic time enable the corporation to obtain additional debt funds at better terms, and, of course, the common stockholders benefit from all returns in excess of the cost of the financing mix. Thus, the decision to float a convertible issue rests on a favorable weighing of present benefits against a potential future detriment.

Effects of Conversion Features on the Market Price   The convertible feature on a security is an option to buy the common stock at a fixed price, and the option fetches a price in the market. The more optimistic the market, the higher the price of the option. The amount at which the market prices the option can be inferred from the difference between the market price of the preferred and its pure investment or the value, where the investment value is the price the stock would sell for as a pure preferred stock and the conversion value is the amount the preferred is worth if turned into common stock at the going price.

Table 8.2 gives some examples of possible relationships between these various prices. The preferred illustrated in Table 8.2 has a dividend of $5.50 annually and a conversion rate of 2 for 1 on the common. If the normal yield for a straight preferred stock were 5.5%, the stock would have an investment value of $100. In example 1 in Table 8.3, the conversion value of the preferred is $80, but the market will pay $110 for the preferred; the option price is $10. However, not only is the $10 option premium at risk if the common should fall significantly but in the interim, the conversion option price reduces the current yield on the preferred from 5.5% to 5%. The price of the preferred in example 2 illustrates an optimistic market. The premium for the convertible option is $10 over the conversion price; a current purchaser of the preferred could not convert except at a loss. Although the option price is still only $10, the amount put at risk over the investment value is $30 per share and the current yield is down to 4.23%. In example 3 the conversion value of the preferred is $160, but the option premium has dropped to $2. In a rational market, the conversion option price goes down if the conversion value rises high enough. A

Table 8.3 Possible market prices on a $5.50 dividend convertible preferred stock
(convertible into common at a rate of 2 for 1)
(assume a normal preferred stock yield of 5½%)

	Market Price Common	Investment Value Preferred	Conversion Value Preferred	Market Price Preferred	Premium for Conversion Option
1. Preferred has no positive conversion value	$40	$100	$80	$110	$10
2. Preferred has a slight positive conversion value	60	100	120	130	10
3. Preferred has a considerable conversion value	80	100	160	162	2

purchaser of the preferred would have $62 at risk. He no longer has any significant comparative safety as against a direct investment in the common. Moreover, the yield on the current price of the preferred is now only 3.4%. In short, as the conversion value rises, the advantage of the hedged position diminishes.

Call Prices and Forced Conversion   If the convertible preferred has a call feature, the firm may force conversion by calling in the preferred. If, for example, the preferred is convertible to common 1 for 1, the market price of the common is $70, and the call price of the preferred is $60, an investor has no choice but to change his preferred into common if the preferred is called.

The reasons for a forced conversion vary. The preferred may have a restrictive covenant—such as limitation on bank borrowing or senior debt—which put difficulties in the way of new financing. Or the company might wish to simplify its capital structure preliminary to raising funds on a new security issue. There does not seem to be any obvious economic gain in forcing conversion, for if the common has risen so high as to make this operation possible, then the return on the common must be quite good, and the risk in carrying the preferred cannot be very large.

8.9.4   Protective Features on Preferred Shares

The indentures on preferred issues may carry features designed to give additional protection to the preferred stockholders. These provisions are similar to the protective clauses found on debt issues. There may be provisions relating to the maintenance of minimum working capital ratios before common stock dividends can be paid. There may be limitations on the amount of debt the firm may carry. Minimum working capital ratios and financial structure ratios may be required before additional senior debt can be incurred.

A preferred issue may have a sinking fund provision. The sinking fund is used to retire some fixed portion of the preferred stock annually. By reducing the amount of the preferred stock outstanding in a systematic manner, the position of remaining shares is strengthened. The sinking fund provision usually allows the firm to buy the required amount of preferred shares either at the market price or at a predetermined sinking fund call price.[20]

8.9.5   Floating New Common Equity Issues

Very seldom does an ongoing, mature company issue major amounts of additional capital stock. Usually, any increase in equity capital is provided by retained earnings. New equity may be required in a new company when the inflow of retained earnings does not provide a sufficient equity base to finance the growth of the firm. Or firm may issue new equity when special investment opportunity opens up, or the firm wishes to finance the acquisition of another company. The firm could finance the expansion solely with debt, but if the proportion of debt to equity grows too large, the firm's financial risk increases. Thus a large package of new debt finance often includes an accompanying stock issue.

Rights   As a matter of precaution, to make sure the flotation is successful, the announced price on a new issue of shares is usually set below the market price of the existing stock. To compensate the old shareholders for the potential dilution in the value of their holdings, they are given rights, i.e., options to purchase the new issue at the discounted market price. A market develops for the rights, and the existing shareholders may exercise the rights to purchase additional shares or sell the rights on the market.[21] In either case they are compensated for allowing in new shares at a reduced price. In case where there are major shareholders who exercise some degree of control, exercise of the rights allows them to maintain their proportional share of the voting common.

8.10   Advantage of New Share Financing

The main advantage of new capital stock financing is that the shares require no fixed rate of return and there is no maturity date on the principle. Although the new shareholders will share equally with existing shareholders in the company's success, they do not constitute a legal liability if the company does not do as well as expected.

---

[20] The sinking fund call price is usually lower than the general call price.

[21] Because the rights need not be exercised immediately and provide an option to buy at a fixed price, they generally develop a speculative premium over their current exercise value. Of course, the rights will go to their net value when the rights expire.

If the firm experiences temporary economic setbacks, it need not pay anything to the shareholders until conditions improve. For a new firm, which may suffer operating losses before establishing a profitable level of operations, a heavy equity base if obtainable increases the chance of surviving the lean times.

If the corporation is to justify new common stock financing, it must earn the going rate of return on equity on the new issue. If it does not reach this rate of return, the value of its present stockholders will be drawn down. However, additional equity financing reduces internal risk and lowers the cost of additional debt financing and the overall cost of funds.

References

Bierman, H., Jr. (2001a). Valuation of stocks with extraordinary growth prospects. *Journal of Investing, 10*, 23–26.

Bierman, H., Jr. (2001b). *Increasing shareholder value: Distribution policy, a corporate finance challenge*. Kluwer Academic Publishers.

Bierman, J. H. (2002, Summer). The P/E ratio: An important but imperfect tool. *Journal of Portfolio Management, 28*.

Bierman, H., Jr. (2010). My favorite two corporate finance puzzles. In J. R. Aronson, H. Parmet, & R. L. J. Thornton (Eds.), *Variations in economic analysis: Essays in Honor of Eli Schwart*. Springer.

Brealey, R. A. (1969). *An introduction to risk and return from common stock*. The M.I.T. Press.

Brealey, R., Myers, S. C., & Allen, F. (2005). *Principles of corporate finance*. McGraw-Hill/Irwin.

Copeland, T., Koller, T., & Murrin, J. (1994). *Valuation: Measuring and managing the value of companies* (2nd ed.). Wiley. Chapter 5.

Damon, W. W., & Schramm, R. (1972). A simultaneous decision model for production, marketing and finance. *Management Science, 18*, 161–172.

Dann, L. (1981). Common stock repurchases: An analysis of returns to bondholders and stock-holders. *Journal of Financial Economics, 9*, 115–138.

Dhrymes, P. J. (1974). *Econometrics: Statistical foundations and applications*. Springer.

Dhrymes, P. J., & Kurz, M. (1967). Investment, dividends, and external finance behavior of firms. In R. Ferber (Ed.), *Determinants of investment behavior*. Columbia University Press.

Dhrymes, P. J., & Kurz, M. (1964). On the dividend policy of electric utilities. *Review Economics and Statistics, 46*, 76–81.

Dhrymes, P. J., & Guerard, J. B. Jr. (2005). Investment, dividend, and external financing behavior of firms: A further examination. Unpublished.

Fama, E. F. (1974). The empirical relationship between the dividend and investment decisions of firms. *American Economic Review, 63*, 304–318.

Fama, E. F., & Babiak, H. (1968). Dividend policy: An empirical analysis. *Journal of the American Statististical Association, 63*, 1132–1161.

Fisher, L., & Lorie, J. H. (1968). Rates of return on investments in common stock: The year-by-year record, 1926–1965. *Journal of Business, 41*, 290–306.

Grabowski, H. G., & Mueller, D. C. (1972). Managerial and stockholder welfare models of firm expenditures. *Review of Economics and Statistics, 54*, 9–24.

Guerard, J. B., Jr., Bean, A. S., & Andrews, S. (1987). R&D management and corporate financial policy. *Management Science, 33*, 1419–1427.

Guerard, J. B., Jr., & Schwartz, E. (2007). *Quantitative corporate finance*. Springer.

Gordon, M. J. (1959). Dividends, earnings, and stock prices. *Review of Economics and Statistics, 41*, 96–105.

Gordon, M. J. (1962). The savings investment, and valuation of a corporation. *Review of Economics and Statistics, 44*, 37–51.

Graham, B., Dodd, D., & Cottle, S. (1962). *Security analysis* (4th ed.). McGraw-Hill Book Company.

Guerard, J. B., Jr., Markowitz, H. M., & Xu, G. (2014). The role of effective corporate decisions in the creation of efficient portfolios. *IBM Journal of Research and Development, 59*(6), 1–11.

Higgins, R. C. (1972). The corporate dividend-saving decision. *Journal of Financial and Quantitative Analysis, 7*, 1527–1541.

Jalilvand, A., & Harris, R. S. (1984). Corporate behavior in adjusting to capital structure and dividend targets: An econometric study. *Journal of Finance, 39*, 127–145.

Kuh, E. (1963). *Capital stock growth: A micro-econometric approach*. North-Holland.

Lakonishok, J., & Vermaelen, T. (1990). Anomalous Price behavior around repurchase tender offers. *Journal of Finance, 45*, 455–477.

Lintner, J. (1956). Distributions of incomes of corporations among dividends, retained earnings and taxes. *American Economic Review, 46*, 97–118.

Lintner, J. (1962b). Dividends, earnings, leverage, and the supply of capital to corporations. *Review of Economics and Statistics, 44*, 235–256.

Lintner, J. (1962). Dividends, earnings, leverage, and the supply of capital to corporations. *Review of Economics and Statistics, 44*, 235–256.

Meyer, J. R., & Kuh, E. (1957a). *The investment decision*. Harvard University Press.

McCabe, G. M. (1979). The empirical relationship between investment and financing: A new look. *Journal Financial and Quantitative Analysis, 14*, 119–135.

McDonald, J. G., Jacquilla, B., & Nussenbaum, M. (1975). Dividend, investment, and financial decisions: Empirical evidence on French firms. *Journal of Financial and Quantitative Analysis, 10*, 741–755.

Meyer, J. R., & Kuh, E. (1957b). *The investment decision*. Harvard University Press.

Miller, M., & Modigliani, F. (1961). Dividend policy, growth, and the valuation of shares. *Journal of Business, 34*, 411–433.

Montgomery, A. L., Zarnowitz, V., Tsay, R. S., & Tiao, G. C. (1998b). Forecasting the U.S. unemployment rate. *Journal of the American Statistical Association, 93*, 478–493.

Mueller, D. C. (1967). The firm decision process: An econometric investigation. *Quarterly Journal of Economics, 81*, 58–87.

Peterson, P., & Benesh, G. (1983). A Reexamination of the empirical relationship between investment and financing decisions. *Journal of Financial and Quantitative Analysis, 18*, 439–454.

Schwartz, E., & Aronson, R. J. (1966). The corporate sector: A net exporter of funds. *Southern Economic Journal, 33*, 252–257.

Schwartz, E., & Aronson, R. J. (1967). The corporate sector: A net exporter of funds: Reply. *Southern Economic Journal, 34*, 153–154.

Switzer, L. (1984). The determinants of Industrial R&D: A funds flow simultaneous equation approach. *Review of Economics and Statistics, 66*, 163–168.

Tinbergen, J. (1938a). Statistical evidence on the accelerator principle. *Econometrica*.

Tinbergen, J. (1938b). *Statistical testing of business cycles*. The League of Nations.

Williams, J. B. (1937). *The theory of investment value*. Harvard University Press.

Zarnowitz, V. (2001b). The old and the new in the U.S. economic expansion. *The Conference Board. EPWP*, #01-01.

Zarnowitz, V. (1992c). *Business cycles: Theory, history, indicators, and forecasting*. University of Chicago Press.

Zarnowitz, V., & Ozyildirim, A. (2001b). On the measurement of business cycles and growth cycles. *Indian Economic Review, 36*, 34–54.

Chapter 9
Long-Term Debt

Long-term debt is the term given to those obligations the firm does not have to pay for at least a year. They are also called funded debt or fixed liabilities. Items that may be classed as long-term debt are bonds, debentures, term loans, or, in small firms, mortgages on buildings. The portion of the long-term debt due within the current year is carried in the current liability section of the balance sheet. Firms in the United States issue far more debt than equity shares. In most years during the 1963–2003 period, firms have issued six to eight times more debt than equity. (Of course, most increase in equity is through retained earnings.) Issuing debt raises capital for firm growth and expansion without possibly lessening current stockholder control. The floatation costs are less on debt than on equity, and the cost of debt is less than the expected shareholder return on equity.

9.1 Bonds

A bond is a formal long-term promissory note. It is a promise made by the company to pay a fixed sum of money at a determinable future date with a stipulated interest rate. A bond issue is composed of a number of notes, which may be owned by different holders. Since bonds are negotiable instruments, they are readily transferred from one investor to another, and bonafide holders for value are given the usual protections of the negotiable instrument laws against counterclaims.

The major provisions of a bond issue are listed in the bond indenture. This legal document describes the agreements and pledges made between the corporation and the bondholders (creditors). The indenture is usually quite long, and it is impractical to attach a copy to each bond certificate. Each bond, however, carries a reference to the indenture, and the bond owner is presumed to have access to the indenture; at least he can learn of its major features in the original prospectus, written up at the initial issue of the security.

Because in most cases it would be impractical for an individual bond owner to enforce his rights under the indenture and because it would be a matter of some difficulty for the company to deal with each bondholder separately, the actual enforcement of the indenture is left to a trustee.[1] It is the trustee's duty to ensure that the firm lives up to the provisions in the indenture protecting the bondholder. Because congressional hearings in the 1930s indicated that some trustees had not always been fully vigilant in enforcing the indenture provisions and even that some trustees had interests in the corporation in conflict with the interests of the bondholders, the Trust Indenture Act was passed in 1940. This law establishes a high standard of conduct for the trustees and prohibits individuals or firms who might have conflicts of interest with the bondholders from acting as trustees on the indenture.

9.1.1   Mortgage Agreement

The usual bond issue is generally backed by a mortgage, basically, a lien given to the creditor against some of the firm's assets. This lien gives the creditors a provisional title against the assets which can be exercised if the debtor does not pay his notes as agreed. If the lien is not paid, the mortgaged property may be seized and auctioned off to satisfy the debt. If the property brings more than the mortgage at the sale, the difference goes back to the owner; if the property is not sold for an amount sufficient to satisfy the debt in full, the unpaid portion remains as an obligation of or a general claim against the debtor. The mortgage is given by a company as a protective feature of the indenture to provide additional security for all the owners of the bond issue. A mortgage usually contains a clause requiring minimum maintenance standards on the property and provides for the compulsory retirement of a proportionate part of the outstanding debt if some part of the underlying property is disposed of.

Many bond mortgages contain an "after-acquired" clause. This clause requires that all new property coming under the ownership of the firm fall under the lien of the existing mortgage. If the indenture is closed, that is, no new debt can be issued under the cover of the existing mortgage, it could mean that on the purchase of new assets the firm could offer prospective new lenders only a second mortgage, that is, only a subsidiary lien on the new properties. In certain instances, these circumstances have put an unfortunate rigidity on the corporation's financial plans. However, in most cases where the mortgage is closed, the lien is restricted to definite property and does not flow over (at least as a prior lien) to other assets.

In most existing bond issues where the after-acquired clause is part of the indenture, the mortgage is conditionally open. Additional debt can be created under the existing mortgage up to 60%, for example, of the amount of new assets

---

[1]The trustee is compensated with an annual fee. The business of being a trustee for major corporation issues is concentrated among the larger commercial banks.

acquired by the firm. The corporation thus has considerable flexibility in financing new property without diluting the underlying protection of the old bondholders.

The importance of mortgage provisions to the security of the bondholders has been exaggerated in the past. Earnings and cash flow are more relevant to the safety of the interest payment and the ultimate repayment of the bonds than the pledge of property. The property itself is not likely to have much dollar value unless it can generate earnings; it must show economic usefulness in our competitive society if it is to bring any cash on the market. On the other hand, there has been a recent tendency to deprecate any value to the bondholders in a mortgage provision. But a mortgage claim has considerable worth in the event of ultimate financial disaster. In bankruptcy, the mortgage lien and its priority have tremendous importance in determining the amount and value of the securities the bondholders will get if the firm is reorganized or the amount of cash they will receive in settlement of their claim if the bankrupt firm is liquidated. Because the mortgage pledge is generally placed mainly against fixed assets, it provides additional security to the bondholders even within the "going concern" concept. Because depreciation is a charge against income, but not an out-of-pocket expense, a firm having a considerable amount of depreciable fixed assets has a relatively larger cash flow than a firm with similar earnings but relatively little fixed assets.[2] The inflow of cash is one criterion of a firm's ability to carry debt; a firm with earnings and substantial fixed assets is obviously in a safer position to incur funded debt.

9.1.2 Subordinate Mortgages

Bonds are sometimes secured by a second or third mortgage if a prior mortgage already exists on the property. In case of default, the pledged assets are used first to satisfy the first mortgage holder's claims, and any remaining value goes to the second mortgage. The holders of any part of an unsatisfied mortgage claim become general creditors.

If the prior bond issue is small compared to the total value of the assets pledged, a secondary lien may actually have considerable strength. If a property valued at $1,000,000 carried a first mortgage of $200,000 and a second mortgage of $300,000, the position of the second mortgage bonds might be better than a first mortgage of, let us say, $700,000 on a property of similar value. Bonds backed by a secondary lien, however, are seldom called second mortgage bonds but tend to bear titles like "Improvement Bonds, 1990;" "General Consolidated and Refunding Bonds, 1995;" etc. Many railroad bond issues have extremely complicated mortgage positions. They may have first mortgage claims over one section of trackage,

---

[2]The depreciation charges, although they reduce reported accounting earnings are not a current out-of-pocket expense. See Chaps. 3 and 4.

secondary liens over another, and tertiary claims over still another part of the company's property. These complexities tend to add difficulties to the problem of reorganization in case of failure.

Many bond issues, besides holding mortgages against the firm's fixed assets, may be given a lien against stocks or bonds the borrowing firm carries as investments. (A bond issue backed mainly by a contingent claim against such securities is a collateral trust bond.) The pledge of securities does not always add safety to a bond issue. The stocks and bonds are often those of firms or enterprises whose economic position is closely tied to the borrowing company; if the borrowing corporation falls on hard times, the issues of satellite companies may not be too valuable.

9.1.3   Net Working Capital Maintenance Requirements

Some indentures require a firm to keep its net working capital above a stipulated proportion of the funded debt. If the firm suffers losses, this proportion may be difficult to maintain. If interest payments and the other indenture provisions are met, however, the trustees often ask the bondholders to relax the working capital requirement temporarily, rather than force failure on the firm. Nevertheless, although working capital maintenance provisions may add little to the debt holder's security in a downturn, in normal periods they warn the management of the dangers of drawing down working capital with excessive dividends or by overinvesting in fixed assets.

Dividend restrictions help safeguard the bond issue by preserving the debt-to-equity ratio in the firm's financial structure. Usually, dividends are restricted to earnings made after the bond issue, perhaps plus some arbitrary amount. Dividends may be limited to a fraction (e.g., two-thirds of all subsequent earnings) or prohibited entirely for a definite period of time. The general effect of these provisions is to keep assets within the firm unless there are operating losses.

Most bond indentures provide for a periodic repayment of at least a substantial proportion of the bonds before maturity. The required amount of bonds can be repurchased on the market (at market price) or called in by the use of the sinking fund call option. The call option gives the sinking fund trustees the right to buy back a given portion of bonds at a previously fixed price. The periodic retirement of part of the issue strengthens the position of the remaining outstanding bonds.

9.1.4   Restrictions on Creating New Debt

If a firm incurs new obligations at too heavy a rate (especially if the new creditors have the same or higher priority than the present creditors), the position of the existing debt holders is eroded. Thus restrictions are generally placed on the amount and types of new issues. Yet the firm must have some flexibility in its financial operations; if restrictions are unreasonable, it may turn to a more liberal competitive source to obtain its funds. Therefore most indentures do not prohibit the firm from borrowing new funds outright but attempt to set bounds of safety before additional debt can be floated. Most indentures have a number of conditions, all of which must be satisfied simultaneously.

In first mortgage bonds, one likely condition is to permit the issue of new bonds up to some percentage of new fixed assets acquired. New bonds are also allowed to replace any part of the issue which may have been retired outside of the required sinking fund. Other common restrictions require the firm to have met certain minimum interest charge coverages over perhaps the last five years before new fixed debt obligations can be created. Often, especially for industrial firms, net working capital has to bear a given percentage to the total proposed funded debt.

Indentures on junior debt claims of lower priority may limit the total bank debt that can be created and restrict the amount and kind of senior debt that the firm may incur.

9.2 Other Types of Long-Term Debt

A debenture is a bond unsecured by a lien or mortgage against the firm's assets. It is a general obligation of the corporation and holds the same position as any other general creditor. A debenture issued after other secured long-term debt is already outstanding, is weakened by the priority of claim of the senior issue, and has to bear a higher yield in order to find acceptance in the capital markets. Commonly, however, the debentures constitute the sole long-term debt issue of industrial firms; if the company is strong, the debenture's financial position may be quite secure and the interest rate comparatively low.

A type of debenture that has become quite popular in financing commercial finance companies is the subordinated debenture. A major clause of the indenture provides that the claims of the debenture rank behind those of bank loans the company is carrying or may incur. A subordinated debenture issue adds to the safety of the bankers' loans since it provides additional risk-bearing funds in the firm's financial structure.[3] It serves the same function as preferred stock. From the view of the company, the subordinated debentures allow the firm to negotiate more bank borrowing without putting new equity capital immediately into the firm. Moreover, the interest paid on the debentures is deductible for tax purposes, whereas the dividends on a preferred stock issue are not.

---

[3]To provide a little "sweetening" to attract the investor, the subordinated debentures are often convertible.

9.2.1 Term Loans

A company may negotiate a loan with a bank, a group of banks, or an insurance company that may last from one to perhaps 15 years, usually 3–10 years. These loans, amortized periodically, are designated term loans, whereas the ordinary commercial loan is hardly ever made for more than 90–120 days. The term loan is often used to help finance the purchase of equipment and is customarily secured by a chattel mortgage[4] against new machinery, trucks, etc. They were quite popular in the immediate postwar period when banks were diligently seeking additional ways to put their funds to work. The development of the term loan has provided the medium-sized business with one source of comparatively available, reasonably priced intermediate-term funds.

9.2.2 Private Placements

When a company sells or "places" all of a given issue to a single buyer, for example, an insurance company, a large bank, or a group of insurance companies or banks, rather than sell its securities to the investors at large in the capital market, the transaction is called a private placement.

9.2.3 Equipment Trust Certificates

The equipment trust certificate has been used for some time in financing railroad rolling stock (freight and passenger cars and locomotives) and now appears in the financing of airline equipment. The basic legal provision of the equipment trust certificates indenture leaves the legal title to the newly purchased aircraft in the possession of the trustee on the issue. The trustee (usually a bank) leases the equipment to the airline. The railroad company puts up about 20% of the cost of the new airplanes and agrees to pay an annual or semiannual rent sufficient to cover the interest on the outstanding trust certificates and retire the principal of the certificates, usually in 15 years. The certificates, which have been floated to obtain the money for the equipment, are issued in series so that some part comes due every year; when the last issue is retired (i.e., when the last rental payment is made), the airline exercises its option to buy, and the title to the planes is released to the airline.

The holders of equipment trust certificates are not legally creditors of the corporation but participants in a leasing agreement. Legally their position is that of beneficial owners (through a trust arrangement) of an identifiable lot of aircraft.

---

[4]A chattel mortgage is placed against personality, movable or personal property as distinguished from realty, that is, land and building.

This actually makes their credit position quite strong. Should the airline fail to meet its rental payments (i.e., the principal and interest due on the certificates), the trustee on behalf of the certificate holders can take possession of the planes and sell them to another line. There have been very few cases of default on equipment trust certificates; even in case of bankruptcy and reorganization, the trustee in bankruptcy will keep up the rental payments rather than lose the use of the aircraft, without which the operation of the airline becomes impossible. Equipment trust certificates tend to bear comparatively low interest rates, because of their rather special position, which gives them considerable safety.

9.3   Long-Term Lease

Some firms acquire fixed assets by renting them rather than purchasing and financing them. The lessee (user) makes periodic payments, monthly, quarterly, or semiannual, payments, to the lessor (owner) for the exclusive rights to use the asset. A long-term lease lasts from 10 to 20 years. One form of long-term lease is sale-and-lease-back. A firm sells a building or plant to another company, sometimes a subsidiary or more frequently an insurance company, to obtain immediate funds. It then commits itself to renting the property on a long-term basis. Typically lease payments are made in advance. Leases may be operating, or financial. An operating lease is a short-term obligation, usually for a period less than the economic life of the asset, and is cancelable. Financial leases are longer-termed obligations, approximating the economic life of the asset, and are noncancelable. Sale-and-lease-back arrangements usually allow for repurchase at a fixed price after the rental term is over.

Traditionally leases were not disclosed and were referred to as "off-balance-sheet" financing. Financial Accounting Standards Board (FASB) Statement No. 13, adopted in 1981, requires capitalization on the balance sheet of leases, known as capital leases, meeting any of the following criteria:

1. The lease transfers the title to the lessee by the end of the lease period.
2. The lease contains an option to purchase the asset at a bargain price.
3. The lease period is equal to or exceeds 75% of the economic life of the asset.
4. The present value of the minimum lease payments equal to exceeds 90% of the fair value of the leased property, at the beginning of the lease.

The absence of all four conditions creates an operating lease that is cancelable and not capitalized. The capitalized value of a capital lease is the present value of the minimum lease payments over the lease period. The discount rate used to determine the present value of the lease payments is the lessee's cost of borrowing or the lessor's implicit interest rate. Operating leases can be disclosed in financial foot-notes. Five years of noncancelable operating leases must be disclosed, annually, as well as the total future minimum lease payments.[5]

A capital lease must be amortized over the lease payment. The lease is amortized as the firm depreciates its assets, and the annual interest implied in the lease payment is treated as an expense. The interest expense reduces net income. In an operating lease, only the lease payment itself is deductible as an expense.

Logically, it would seem that long-term leasing or outright purchase of an asset should be considered alternative forms of financing. Renting requires little immedi-ate outlay of funds, but the future cash flow from the project is reduced by the rental payments. The rental payments must contain all sorts of implicit capital costs since the owner of the asset will expect to earn a fair return on his funds. Obviously, the implicit interest rate on a long-term lease will be higher than the cost of "loan" capital to the firm. Whether renting or purchasing assets is better (for those firms which have easy access to the capital markets) depends upon which method shows a higher rate of return.[6] If no special hindering factors exist, purchase rather than rental is economically justified if the firm can make a going rate of return on the extra funds put into the purchase.

Rental saves the firm from the eventual burden of holding obsolescent equipment. This argument may be true where the rate of technical improvement is rapid, the equipment mobile, and the rental period short. In this case, after the lease is over, the company can obtain new equipment if the old proves unserviceable or better designs are available; however, one must assume that owner of the equipment will require high enough rental payments—on the average—to cover the possibilities of rapid obsolescence. In the usual long-term rental contract on real property, which often runs for 20 or 25 years and where the rent payment amortizes the whole cost of the property over this period, potential losses from obsolescence or changes in the economic value of the location can hardly be shifted from the renter to the owner. As a matter of fact, if the property declines in use value, a renter may be in a more difficult position because he finds it harder to adapt the property to new uses than would an owner.

---

[5]In 2003, the S&P 500 Index firms disclosed $482 billion of off-balance sheet operating lease commitments in their footnotes. J. Weil, "Lease Accounting: Accounting Still has an Impact," *Wall Street Journal,* September 22, 2004, A5-A6.

The operating lease commitments were approximately 8% of reported debt of the S&P firms. For many firms, operating lease obligations are much higher. When UAL, the parent firm of United Airlines, files for bankruptcy in December 2002, its assets were $ 24.5 billion, its liabilities $ 22.2 billion, and its noncancelable operating lease commitments were $24.5 billion (for aircraft). US Airways had substantial operating obligations at the time of its bankruptcy.

[6]One must discount the difference between the forecasted after-tax cash flow for the project under a purchase plan and under a rental arrangement, in order to obtain the implicit rate of return on the additional funds required under an outright purchase plan. In general, the rate of return is obtained mathematically by finding the discount factor which equates the cash flow plus any remainder value of the assets with the original investment.

Renting saves taxes. The rental payments in capital leases include both a charge for interest on the funds invested in the property plus the amortization or repayment of the principal amount of these funds. The whole of the rent is deductible for tax purposes; on a bond issue only the interest is a deductible expense, and repayment of the principal is not. This argument oversimplifies the problem of taxes; however, if the property is owned by the firm, depreciation charges may be deducted before calculating taxable income. The tax advantages, if any, can be found by calculating the rate of return under ownership or rental using the net cash flow after taxes. In many cases, the rental program offers no net tax savings.

A post-inflationary situation can encourage sale-and-lease-back arrangements for the tax advantage. At the end of the period, many companies may have assets of considerable market value which have been written off on the corporation's books so that no more depreciation charges are allowable. The possibilities of a sale-and-lease-back arrangement for tax savings now arise. The firm sells its property to an insurance company (sometimes to its own special real estate subsidiary) and agrees to rent the property back at a rate which would cover the going interest rate plus repayment of the value of the property. The firm pays, out of the receipts for the sale of the property, the capital gains tax on the profits of the sale, but it saves out of future taxes on income the amortization on the principal value.[7]

In general, the sale-and-lease-back arrangement is worthwhile if the present value of the savings to be obtained out of future taxes exceeds the amount of the immediate capital gains tax. In a bankruptcy, the lessor's position may be better than a traditional creditor, as the lessor owns the asset and can retrieve it upon the lessee default.

9.4   Lease Accounting: Recent Changes

In 2016, FASB introduced a new standard that will require organizations that lease assets— referred to as "lessees"—to recognize on the balance sheet the assets and liabilities for the rights and obligations created by those leases.

Under the new guidance, a lessee will be required to recognize assets and liabilities for leases with lease terms of more than 12 months. Consistent with Generally Accepted Accounting Principles (GAAP), the recognition, measurement, and presentation of expenses and cash flows arising from a lease by a lessee primarily will depend on its classification as a finance or operating lease.

However, unlike current GAAP—which requires only capital leases to be recognized on the balance sheet—the new ASU will require both types of leases to be recognized on the balance sheet. The ASU also will require disclosures to help investors and other financial statement users better understand the amount, timing,

---

[7]These savings accrue because no deduction would be left for depreciation were the firm to keep the title to the property.

and uncertainty of cash flows arising from leases. These disclosures include qualitative and quantitative requirements, providing additional information about the
amounts recorded in the financial statements.

The accounting by organizations that own the assets leased by the lessee—also
known as lessor accounting—will remain largely unchanged from current GAAP.
However, the ASU contains some targeted improvements that are intended to align,
where necessary, lessor accounting with the lessee accounting model and with the
updated revenue recognition guidance issued in 2014.

Off-balance sheet accounting of operating leases can skew various financial ratios
positively and can mask underlying contractual liabilities for a firm that are not
recognized on the balance sheet. The latest guidance on FASB is a step in the right
direction that helps to level the playing field and remove the arbitrariness that was
inherent in the balance sheet treatment of operating and capital leases.

9.5   The Cost of Debt Capital

The explicit cost of a long-term debt issue is the interest rate the firm pays on the net
funds obtained by the bond issue. But there are costs other than the explicit interest
rate which a firm must consider before floating a debt issue. Debt increases the
financial risk of the firm and thus raises the cost of equity capital. Further, every
additional amount of debt has its effect on the interest rate that the firm will
eventually have to pay on its overall debt and thus may have a marginal cost
measurably higher than the actual rate at which the bonds are floated.

A firm also incurs intangible "managerial" costs on a bond or debt issue. It must
meet cash drains periodically as the issue is amortized, and the indenture on the bond
issue usually carries restrictive clauses that may at some time hamper managerial
freedom or flexibility.

Of course, some of the problems and costs of incurring long-term debt exist with
other types of debt financing. For example, a firm may have the choice of using more
short-term debt (economizing on net working capital) or using additional long-term
debt. Long-term debt may have a higher explicit interest cost, but heavy short-term
debt brings with it more implicit costs and difficulties. If open-book creditors are not
paid on time, the costs of lost cash discounts can come to an astounding rate.[8] Heavy
short-term debt may force the firm to forego immediately profitable buying opportunities or hurriedly to arrange financing at inopportune times. Nor is the notion that
short-term debt is passable but fixed debt is risky or dangerous a valid one. Long-
term debt has its risks, but trying continually to balance and meet current

---

[8] Suppose a firm is given customary terms of 2/10 n/30 e.o.m. This means the firm may deduct 2%
from the monthly statement if it pays within the first 10 days of the month. If it misses this period,
the gross amount is due 30 days from the statement date. The 2% discount lost if the firm does not
pay within 10 days of the monthly statement amounts to 36% on an annual basis.

indebtedness is riskier. If current debt is defined as those obligations due within the year, it is a truism that a firm fails by being unable to meet current obligations, not by being unable to meet its funded debt.

The intangible costs of all debt issues are extremely important; however, the going interest rate (the explicit cost) is an index of certain aspects of these intangible costs. Furthermore, the interest rate is an observable, measurable phenomenon, whereas the implicit costs are not.

The interest rate a firm has to pay on any debt issue is conditioned by four major factors: (1) the basic level of prevailing market interest rates; (2) the length to maturity of the issue; (3) the particular type of issue, its provisions, and its security position, relative to other financing that the firm is already carrying or may incur; (4) the general credit standing of the firm.

How can one measure the credit quality of a firm or a firm's particular issue? The marketplace relies upon several bond rating services to assess risk. Mergent's and Standard and Poor's have long histories of rating debt. Standard and Poor's Corporation uses the following debt ratings:

Rating	Financial strength or condition
AAA	Extremely strong to repay debt and interest
AA	Strong capacity to repay principal and interest
A	Strong capacity to serve debt but more volatile than AA debt
BBB	Adequate debt-serving capacity but more volatile than A debt
BB	Bonds have long-term uncertainty but adequate short-term debt-serving capacity
B	Larger volatility than BB bonds
CCC	Current vulnerability to default and dependent on favorable economic conditions to service debt
CC	Debt subordinated to CCC senior debt
C	Same as CC
D	Debt is in default

All these factors are significant, although some of them may be objectively determined and others are subjective. Standard and Poor uses the following criteria to partially determine debt ratings:

Adjusted	Key	Industrial		Financial		Ratios	
US industrial long-term debt							

Three-year (1998–2000) medians	AAA	AA	A	BBB	BB	B	CCC
EBIT int. cov. (x)	21.4	10.1	6.1	3.7	2.1	0.8	0.1
EBITDA int. cov. (X)	26.5	12.9	9.1	5.8	3.4	1.8	1.3
Free oper. cash flow/total debt (%)	84.2	25.2	15	8.5	2.6	-3.2	-12.9
FFO/ total debt (%)	128.8	55.4	43.2	30.8	18.8	7.8	1.6
Return on capital (%)	34.9	21.7	19.4	13.6	11.6	6.6	1.0
Operating income/ sales (%)	27	22.1	18.6	15.4	15.9	11.9	11.9
Long-term debt/ capital (%)	13.3	28.2	33.9	42.5	57.2	69.7	68.8
Total debt/ capital (incl. STD) (%)	22.9	37.7	42.5	48.2	62.6	74.8	87.7
Companies	8	29	136	218	273	281	22

Data for earlier years and in greater detail are available by subscribing to Standard and Poor's CreditStats.
Source:www.standardandpoors.com

The general level and time structure of the interest rate can be determined by ascertaining the going market yield on various bonds currently traded. The market yield—which is the important quantity—is not the same as the coupon or stated interest rate of the bond. The true yield on a bond is obtained by considering (1) the contractual or stated rate on the bond, (2) the remaining life to maturity of the bond, and (3) the current market price of the bond. The purchasers or investors in a bond obtain two items of value: they are to receive the face or par value of the bond at maturity, and they are to receive the stated interest rate annually or semiannually over the life of the instrument. The price they pay is the present value of these two features; the yield they obtain is that rate which equates the present value of the two items with the going market price. This is represented by the following equation:

$$\text{Market Price} = \frac{CR}{(1+i)} + \frac{CR}{(1+i)^2} + \frac{CR}{(1+i)^3} \cdots + \frac{CR}{(1+i)^n} + \frac{FV}{(1+n)^n}$$

CR is the contractual or coupon rate, FV equals the face value of the bond to be paid at maturity, n is the number of years to maturity, and i is the unknown yield. Although the formula appears complex, it is not difficult to use. Furthermore, every investment house, knowledgeable investor, or broker has a computer program table that quickly reveals the yield if the market price is known or gives the price which should be paid if the buyer wishes to stipulate the yield.

Bond Prices on the Market   Since the maturity date and the coupon or stated interest rate are fixed at the time of the initial flotation, the market price of the bond moves in response to the changing level of the interest rate on the market. If the current market interest rate is higher than the fixed rate on a bond, the bond sells below par value or at a discount. If the going interest rate is below the coupon rate, the bond sells at a premium. The general bond market moves inversely to the interest rate; when the rate is falling, bond prices move upward, and when the yield rate is rising, bond prices drop. A "strong" market signifies a lowering of interest rates, and a "weak" market means the opposite.

The general level of interest rates is determined by the supply and demand for investment funds on the capital and money markets. The bond market, however, is not the model of a perfectly free market because the interest rate is influenced by government monetary and fiscal policy as implemented by the Federal Reserve System and the Treasury. The bond market is also a speculative market; the market interest rate is influenced by forecasts or expectations as to the future level of the interest rate.

Supplying the capital funds market are the net purchasers of bonds—insurance companies, investment companies, banks, trust funds, pension funds, and individual investors. The demand for funds comes from the net issuers of bonds—state and municipal governments, the federal government and its agencies, foreign governments, international agencies, and private corporations. The market can be analyzed in terms of the supply and demand for funds, or in terms of the supply and demand for securities. When the demand for funds increases at a faster rate than supply, the demanders of funds (borrowers of capital) have to pay higher interest rates. Or, conversely, if the supply of bonds on the market increases relative to the demand for debt securities, bond prices fall, that is, the interest rate rises. Of course, the opposite of these statements holds, and during a period of easy money, when the supply of funds seeking investment is relatively greater than the demand for capital, the interest rate falls, and bond prices rise.

9.6 Level and Structure of the Interest Rates

Table 9.1 presents average interest rates prevailing on high-grade securities for the 1960–2020 period. Interest rates, on even the best securities, show considerable movement. Treasury bonds are considered the safest of all investments, and therefore their interest rate is lower than other bonds of comparative maturity and tax status. The yield curve shows the relationship between the time to maturity and the yield to maturity, for a given issuer (or risk of issuer). That is, as time to maturity increases, what happens to the yield to maturity of debt? From the view of default risk, it is usually felt that short-term issues (of equal quality) are safer, from the investor's point of view, than long-term issues.

9.7 The Liquidity Preference Theory of the Term Structure

If any credit risk is involved at all, the short-term issue is safer because the borrower's position at repayment time is more easily forecast. Hicks (1937) referred to the additional yield on longer-term instruments as a "liquidity premium." Furthermore, short-term debt instruments are less subject to the interest rate risk. If the basic interest rate rises, the near maturity date of the short-term issue prevents any significant capital loss. In the same circumstance, an investor in long-term bonds

Table 9.1   US Interest Rates, 1960 to 2020, Selected Years

	US Treasury securities Bills		Constant maturities			Corporate bonds		High-grade muni. bonds	Home mortgage rates	Prime Rate Charged	Federal unds
	3-Month	6-Month	3-Year	10-Year	30-Year	Aaa	Baa				
1960	2.93	3.25	3.98	4.12	......	4.41	5.19	3.73	......	4.82	3.53
1965	3.95	4.06	4.22	4.28	......	4.49	4.87	3.27	5.81	4.54	4.04
1970	6.46	6.56	7.29	7.35	......	8.04	9.11	6.51	8.45	7.91	5.95
1975	5.84	6.12	7.49	7.99	......	8.83	10.61	6.89	9.00	7.86	6.25
1980	11.51	11.37	11.55	11.46	11.27	11.94	13.67	8.51	12.66	15.27	11.77
1985	7.48	7.66	9.64	10.62	10.79	11.37	12.72	9.18	11.55	9.93	7.69
1990	7.51	7.47	8.26	8.55	8.61	9.32	10.36	7.25	10.05	10.01	6.98
1995	5.51	5.59	6.25	6.57	6.88	7.59	8.20	5.95	7.87	8.83	5.21
2000	5.85	5.92	6.22	6.03	5.94	7.62	8.36	5.77	7.52	9.23	5.73
2001	3.45	3.39	4.09	5.02	5.49	7.08	7.95	5.19	7.00	6.91	3.40
2002	1.62	1.69	3.10	4.61	......	6.49	7.80	5.05	6.43	4.67	1.17
2003	1.02	1.06	2.10	4.01		5.67	6.77	4.73	5.80	4.12	2.12
2004	1.38	1.57	2.78	4.27	-	5.63	6.39	4.63	5.77	4.34	1.35
2005	3.16	3.40	3.93	4.29	-	5.24	6.06	4.29	5.94	6.19	3.22
2007	4.41	4.48	4.35	4.63	4.84	5.56	6.48	4.42	6.41	8.05	5.02
2008	1.48	1.71	2.24	3.66	4.28	5.63	7.45	4.80	6.05	5.09	1.92
2009	0.16	0.29	1.43	3.26	4.08	5.31	7.30	4.64	5.14	3.25	.16
2010	0.14	0.20	1.11	3.22	4.25	4.94	6.04	4.16	4.80	3.25	0.18
2012	0.09	0.13	0.38	1.80	2.92	3.67	4.94	3.14	3.69	3.25	0.14
2015	0.06	0.17	1.02	2.14	2.84	3.89	5.00	3.48	4.01	3.26	0.13
2017	0.94	1.05	1.58	2.33	2.89	3.74	4.44	3.36	3.97	4.10	1.00
2018	1.94	2.10	2.63	2.91	3.11	3.93	4.80	3.53	4.53	4.91	1.83
2020	2.08	2.07	1.94	2.14	2.58	3.39	4.38	3.38	3.94	5.28	2.16

[1]Rate on new issues within period; bank-discount basis.
[2]Yields on the more actively traded issues adjusted to constant maturities by the Department of the Treasury. In February 2002 , the Department of the Treasury discontinued publication of the 30-year series.
[3]Beginning December 7, 2001, data for corporate Aaa series are industrial bonds only.
[4]Effective rate (in the primary market) on conventional mortgages, reflecting fees and charges as well as contract rate and assuming, on the average, repayment at end of 10 years. Rates beginning January 1973 not strictly comparable with prior rates.
Source: Economic Report of the President, table B-42, March 2021 p.506.

may take a considerable drop in price if he wishes to sell them on the market. Because by and large, the investor risks less on short-term issues, in a normal market the short-term rate is generally below the long-term rate.

9.7.1   The Pure Expectations Theory of the Term Structure

The pure expectation theory of term structure of interest rates holds that the expected one-period return on investments is the same, regardless of the maturity of the security. The rate of interest, or yield, on a bond, is its current, or spot, rate. A forward rate is the interest rate that will prevail in the future. If the current (spot) two-year rate is R2 and the current three-year rate is R3, then there is a one-year interest rate, two years from now, denoted r_1, that equates the two-year and three-year bond yields.

$$(1 + R_3)^3 = (1 + R_2^2)^2 (1 + r_1)$$

$$1 + r_1 = \frac{(1 + R_3)^3}{(1 + R_2)^2}$$

$$r_1 = \frac{(1 + R_3)^3}{(1 + R_2)^2} - 1$$

If the current two-year bond rate is 4% and the current three-year bond rate is 5%, then the one-year bond rate, two years from now, should be

$$r_1 = \frac{(1.05)^3}{(1.04)^2} - 1$$

$$r_1 = \frac{1.1576}{1.0816} - 1 = 7.03$$

If the one-year bond rate two years from now exceeds 7.03%, then the investor should have purchased the two-year bond and reinvested the proceeds, rather than purchase the current three-year bond. It is obvious to the reader that an infinite number of expectations are incorporated into the yield curve. When the short-term rate rises above the long-term rate, it may be because of an inflationary condition which is expected to decline in the future.[9] A general forecast that the long-term rate is relatively high and likely to decline will lead borrowers to postpone long-term borrowing and finance short term for the time being.[10]

9.7.2 The Market Segmentation Theory of the Term Structure

Life insurance companies are renowned for being able to forecast when their clients will die and thus can manage their cash flow, leading them to be one of the larger purchasers of corporate bonds and being able to operate in the longer time to maturity instruments. There is a "market segmentation" theory of the term structure of interest rates which holds that particular financial intermediaries, matching their asset and liability lives, operate in different regions of the yield curve.

Knowledge of the factors affecting the general level of the rates and the time structure of the rates is important to both the investor and the corporate financial management. If the management desires to save on costs, it needs some insight into

---

[9] Although the short term rate went to 22.0% as against 13% on long-term treasuries, subsequently as the rate of inflation fell, the interest rates declined, and the long-term bonds proved to be the better investment.

[10] For a period in the 1970s, the short-term rate went considerably above the long-term rate during a period of inflation.

the structure and level of the interest rate and the factors likely to affect its future course. The timing of an issue is an important consideration. By anticipating long-run needs for funds, management can advance issue dates to take advantage of prevailing low rates; by financing short term, they can defer an issue if the long-term rates appear cyclically high.

In addition, since the maturity date of issue is a variable feature up to the time it is floated, management should try to ascertain the best maturity for an issue, considering both present and future costs.

When a firm decides whether it can make profitable use of long-term borrowed capital, the question arises of exactly how high an interest rate it has to pay. As the previous discussion indicated, the going interest rate and the length of the maturity of the issue have a great deal to do with the effective interest rate the firm has to offer in order to float its bonds. Also to be considered are the general credit standing of the firm, the type of issue, and the financial structure the firm already carries. If the firm is considered sound, with a good history of repayment on past obligations, having a good market position for its product and relatively stable demand, its interest cost tends to be low. The greater the equity the firm has relative to its debt, the lower is the interest rate it has to pay. A heavy amount of debt in proportion to the firm's ownership capital (residual risk-bearing capital) calls for a higher interest rate on any given issue.

9.7.3   Setting the Rate on a New Issue

A firm's investment bankers can evaluate a prospective issue and indicate the approximate interest rate it will have to bear. Many issues are already being traded on the secondary market, and some are likely to resemble the position of the proposed issue. A fairly precise estimate of the necessary interest rate can be achieved by comparing the new security to those with similar features already on the market.

The commission to be paid the investment bankers for underwriting and selling the issue is another cost entering the calculation. These commissions, or flotation costs, are relatively low on high-grade debt issues. Flotation costs are brought into perspective by comparing the effective yield which the investors will obtain (as determined by the market price they pay) and the effective interest rate the company pays (as determined by the net proceeds of the issue). The difference between these two rates represents the flotation costs.

Often a new issue is not floated at par but at a small discount or a small premium. The coupon rate on bonds is customarily set at round percentages or fractions, such as 4%, 4 ½%, or 5%, occasionally at quarters, for example, 4 ¾% or 5 ¼%; hardly ever are any fractions below ¼ percent used. Effective yields on the market, however, move over a far more continuous range than the coupon rates. Thus, if the effective rate on the market is 4.68% for a given class of 30-year bonds and a firm floats a new issue with a coupon rate of 4 ¾%, the new issue will be priced out at

101 ½, a premium of 1 ½ points, making the effective rate 4.68%. Similarly, if a new issue bore a 5% rate and the going market rate was 5.10%, it would be issued at a discount; it would sell for 99 5/8. On the issuing company's books, bond discounts are carried as deferred debits; bond premiums are carried halfway between the liability and equity section as a deferred credit. In each interest payment period, a proportionate amount of the discount or premium is amortized and applied against the current charge for interest expense. As the discount is amortized, it increases the current charge for interest expense, whereas the amortization of a bond premium reduces the amount charged for interest expense. Although bond premium and discount complicate the accounting, there is no particular advantage or disadvantage to issuing bonds at a discount or premium; in most issues the coupon rate is set as close to the market as possible in order to float the issue near par. If the firm sets the coupon rate above the market, it can obtain more funds than the face value of the issue, but it pays for these funds through the higher than necessary coupon rate. On the other hand, a firm can set the coupon rate below market; it saves on the rate but does not obtain the full face value of the issue. In any case, the effective rate the firm pays on the net proceeds it receives from an issue is the true explicit cost of these funds.

9.8   The Call Feature on Bonds

About 95% of corporate bond issues outstanding today carry a call feature. A provision in the bond indenture enables the company to call, that is, retire or pay back, some or all of the bonds before maturity at the company's option. The firm exercises its option by paying a penalty or call premium of a few points above the face value of the bonds. The general call premium is usually set somewhat higher than the sinking fund call and declines over the life of the issue so that it reaches zero near the maturity date.

The call option gives the borrowing firm an opportunity to repay its obligations before maturity, should any of the following develop: (1) The corporation obtains a heavy inflow of funds, and the best use of these funds at the moment appears to be a repayment of debt in order to save the interest charges. (2) New financing seems desirable (possibly in connection with an acquisition or merger), and the old debt contains some onerous or restrictive provisions interfering with the new plans. (3) Interest rates in the market drop, and the company can save on interest costs by calling the old issue and replacing it with a new one at a lower effective rate.

Replacing one issue with another, refunding before maturity, has many interesting facets. In favor of the operation is the lower effective interest rate obtainable; opposed to it is the call premium, other costs of retiring the old issue, and flotation costs on the new issue. However, the profitability of the maneuver can be ascertained by studying the market. The price the old issue would sell at, if there were no call, can be calculated by applying going market yields to the issue and obtaining the estimated free market price.[11] The problem can be illustrated by an example:

Amount of bond issue	$10,000,000
Coupon rate	5 ¼%
Years to maturity	25 years
Call price at this time	104
Present market interest rate on similar bonds	4 ½%
Additional expenses of calling old bonds and floating new issue	$250,000

If there were no call feature, the old bonds would sell on the market for approximately 111 ¼ giving a yield to maturity of 4.5%. The net present value of the interest savings on a refunding operation can be obtained by subtracting the call price and the additional flotation expenses from the implicit market price, thus:

Implicit market price (old bonds) of the whole issue		$11,125,000
Less: call price	$10,400,000	
Additional expenses	250,000	10,650,000
Net present value of annual interest savings over the life of the issue		$475,000

In the problem above, the annual difference between the interest charge on the old bonds at 5 ¼% and the interest rate obtainable on a new issue at 4 ½% comes to $75,000 ($525,000 minus $450,000) for 25 years. Against this saving must be set the call premium of $400,000 and various additional flotation and retirement expenses of $250,000. The solution illustrated essentially brings all the items to a present value basis and gives the immediate value of the savings obtainable on a refunding operation.

The call feature operates essentially for the benefit of the borrowing firms because they can retire the bonds when it is to their advantage. For the investors in the bonds, the call feature may deprive them of capital appreciation on their bonds or of receiving the benefit of the higher contractual interest rate if the market interest rate falls. On the other hand, when easy money prevails and the long-term interest rate is low, investors are not too concerned about giving the corporation a relatively flexible call feature. With the interest rate already at a secular low, it does not seem

---

[11] This cannot be obtained for the bond directly from the market because the effect of a drop in interest rates combined with a call price may force the bond investors into a locked-in position. They cannot sell at the call price or lower without losing on the interest rate; potential buyers cannot purchase above the call price because of the obvious danger of having the bonds called in by the company at the lower price. Any sales recorded during such a period are not likely to be normal market sales.

that any further drop would be sufficient to make a refunding worthwhile. When interest rates are at a "normal" level, bond investors are more wary of allowing the company too lax a call feature. The call penalty or premium may be set a bit higher, perhaps the initial coupon rate may be set a fraction of a point higher, and/or the bonds may be made noncallable or nonredeemable for a certain number of years after issue.

9.9   Convertible Bonds and Bonds with Warrants Attached

A convertible bond or debenture may at the option of the holder be turned into the common stock of the company at some agreed ratio or rate of exchange. Sometimes the convertible feature involves a fixed ratio; a $500 debenture may be turned into ten shares of stock. More commonly, the debentures may be turned into common shares for a fixed amount of face value plus an additional set amount of cash. For example, $50 of face value of the debentures plus $20 in cash can be turned in for one share of common stock. Often the cash premium increases over time to compensate for the likely rise in the book value of the common shares. The agreement on the conversion privilege allows for a compensating adjustment in the rate of conversion in case of any stock splits, stock dividends, etc.

A warrant differs from a conversion feature in that the warrant can be detached from the bond and exercised separately. A warrant entitles its holder to a share of stock upon surrender of the warrant plus, usually, a set amount of money.[12] Some warrants run indefinitely, whereas others expire after a given term. Again, as with the convertible, the cash requirement may be increased at a set rate over time. Since the warrant can be detached and exercised separately, the original holder can sell his warrant, and the subsequent holder can exercise the warrants, hold them, or trade them as he sees fit. A regular trading market may develop for the warrants.

Whereas the call feature operates for the benefit of the corporation, a conversion feature is to the advantage of the security holder. If the company's fortunes are fair, the owners of a convertible security can hold on to their bonds and obtain the contracted income and repayment of the principal at maturity. If the firm does well, the owner of a convertible security can choose to change the bond into common stock, sharing its increased dividends and appreciation. (Of course, this does not preclude the possibility of a secondary purchaser of a convertible security paying a relatively high premium for the conversion feature and suffering a decline in the value of the investment if the underlying stock drops on the market.)

---

[12]Warrants sometimes appear after reorganizations, when they are given to the stockholders allowing them the chance to get back into the company if the reorganized company should do better than its failed predecessor.

The market price of a convertible security is subject to forces in addition to those affecting the ordinary bond or debenture. If the conversion feature is not immediately valuable, the bonds will sell close to their basic investment value, that is, their value as a normal credit instrument. But if the market price of the underlying stock is high enough, the market price of the convertible bonds is determined basically by the price level of the shares. In no case will the bonds sell below their conversion value. As a matter of fact, in an optimistic atmosphere, market forces may push the price above the conversion value, so that a premium exists for the conversion privilege beyond its explicit value.

Some issues are both callable and convertible. If the conversion value is high enough, such bonds furnish an exception to the rule that a security cannot sell above its call price. Here the rule that an issue never sells below its conversion value has more force. If a purchaser buys a bond above its redemption price but at its conversion value into common, he has nothing to fear from a call. If the company calls the bond, he can escape loss by turning the bond into its equivalent in common stock. If he wants his investment back, he can sell the shares. Or he can find someone willing to buy the bonds and make the conversion; there exists a going market.

These relationships make possible an operation known as forcing conversion. This is done simply by calling an issue when the market price (due to the conversion value) is above the call price. Since no one takes a loss deliberately, no securities will be surrendered to the company; almost all will be turned into stock. The only rationale for forcing a conversion would seem to be a desire to get rid of an issue that in one way or another is troublesome to the arrangement of future financing.

Since a convertible issue may be eventually turned into common shares, the common stockholders are generally given the preemptive right on new convertible issues. This right is especially important where the price on new securities is below their immediate conversion value.[13]

The disadvantage of the convertible security rests on the potential dilution of the per share earnings of the common stock when and if the senior issue is converted. On the other hand, there are occasions when the convertible is the easiest and cheapest issue to float and is most readily accepted by the market. For some issues, the risk position is such that at the given interest rate the convertible feature is necessary to make the issue attractive to outside investors.[14] Convertible debentures often appeal to institutional investors, who are generally limited to a small amount of stock in their portfolio. A convertible debenture may be on the legally approved list for insurance company portfolios and still give the investment department a chance to try their hand at making a little capital appreciation.

---

[13] American Telephone and Telegraph Company for years raised equity funds by issuing rights to the stockholders for convertible debentures priced below their conversion value.

[14] This is probably true of the subordinated debentures—a popular security of the large finance companies—which is made more acceptable to the market by its convertible feature.

Sometimes the convertible security is initially floated at a price below its value in underlying shares; basically, the company is floating an equity issue. But instead of coming in all at once, the equity funds flow in more conveniently over time. The company gains on the lower financing costs and saves the tax on the interest for the period investors hold the bonds before deciding to convert.

The use of a convertible debt issue seems most logical in the financing of an untried company. The outside investor may want more safety than is provided by common shares and yet desire some gain, if the company is successful, as compensation for the extra risk of investing in a newer or smaller company. The use of convertible debentures, therefore, as the prime financial device of the Small Business Investment Companies in providing capital for smaller firms (falling under the special small business classifications) appears eminently appropriate.[15]

Some explanations given for the use of convertible securities are not always valid. Thus it is suggested that the convertible security is a good method of raising delayed equity funds when the current market for equities (common stocks) is depressed. The convertible security brings in immediate funds; when the stock market recovers sufficiently, the debentures are turned into shares. Thus the company avoids floating new shares at a very low price and obtains equity funds on better terms after the market rises. The issue of the convertible should be regarded simply as a delayed issue of common stock.

The fallacy in the foregoing explanation lies in the notion that business activities necessarily proceed in a smooth upward trend. But if earnings take a few sharp dips, the in-service chases on the debt issue—convertible or otherwise—may prove a strain on the financial resources of the firm. (Under such circumstances, obviously, the debt holders do not convert.) It is precisely the initial risk-bearing characteristic of equity financing, that is, the non-fixity of its return in good or bad times, that makes it an important part of the financial mix. The convertible issue (as long as it is outstanding) provides none of the cushion against business fluctuations which constitutes the uniquely valuable aspect of common stock financing. Thus the convertible issue will do the company no good if its estimate of future business developments is wrong. Moreover, if the company were certain that its shares were undervalued, it would do better to raise present funds with a callable debt issue. When the company's shares rise in price later, new stock can be floated, and the debt retired. There is no additional risk in this procedure over the flotation of a convertible, and if the company's fortunes improve, the present shareholders do not have to sacrifice part of their earnings gain or capital appreciation to the holders of the convertibles.

---

[15]The Small Business Investment Companies are the outgrowth of the Federal Small Business Act of 1958. A major objective of the Act is to help make more long-term capital available for smaller firms.

9.10   The Advantages and Disadvantages
of Long-Term Debt

To the conservative investor, a bond represents a relatively safe investment of his
money at a stated rate of return. True, there have been some defaults on bond issues
with subsequent losses to the bondholders, but by and large a diversified portfolio of
bonds in the past has earned enough to cover any loss in principal and still give its
owner a positive rate of return. (Substantial evidence indicates that a diversified mix
of secondary bonds over time has outperformed prime bonds.[16]) Because of their
relative stability, bonds are a desirable component of the investment portfolio of
insurance companies, pension funds, and other financial intermediaries, whose own
obligations to policyholders, savers, pensioners, and annuitants are generally on a
fixed or contractual basis. These financial intermediaries have very substantial funds
to invest and are an important part of the capital market.

The desirability of bonds as investment media for large segments of the financial
market has enabled corporations to obtain funds on debt issues at comparatively low
out-of-pocket costs. This and the fact that bond interest charges are deductible before
computing the corporate profit tax are the main advantage of using debt financing for
the firm. On the other hand, the use of debt financing binds the firm to the obligation
of periodically meeting fixed interest charges and to the repayment of the principal as
agreed. These obligations add certain rigidity to the financial operations and increase
the financial risk of the firm.

9.11   Malkiel's Bond Theorems

Let us look at several aspects of the pricing of long-term debt, as seen with ten-year
bonds. Let us assume that the current ten-year bond yield is 5%. Let us further
assume that our bond pays an annual (once a year) interest payment. A bond with a
coupon rate of 5% would sell for $1000, its par value. If we purchased a 5% bond
today and interest rates fell to 4% tomorrow, then we would own a bond with a
higher yield than the current market bond selling at par, and our bond would sell for a
premium, a price exceeding its face, or par, value! How can we price us a bond?
Remember that a bond pays interest, and its interest payment is constant during the
life of the bond. That is, the interest payment is an annuity, a fixed dollar payment for
a specified time period. We can price a bond by calculating the present value of the
bond's interest payments plus the present value of the return of the bond principal. In
the case of our ten-year 5% coupon bond, with a 4% yield, the bond price is:

---

[16] See W. Bradford Hickman, *Corporate Bond Quality and Investor Experience*, Princeton University Press, 1958.

$$Bp = \$50 \ (7.7217) + \$1000 \ (.6139) = \$1081.15$$

The bond price rises to \$1081.15, an 8.1% increase. If the interest rate rose to 6%, then our ten-year 5% bond would not yield a rate equal to the current yield of bonds and would sell for a discount, a price less than its par value.

$$Bp = \$50 \ (7.3601) + \$1000 \ (.5584) = \$926.41$$

Note that the value of the bond falls to \$926.41, a loss of 7.36% if the interest rate rises to 6%. The reader notes that if the interest rate rises 1%, then the bond price falls 7.36%, whereas a 1% decrease in the interest rate creates an 8.12% bond price increase. We can establish two of Malkiel's bond theorems:

(1) A rise in the interest rate causes bond price to fall, whereas a decrease in the interest rate causes bond prices to rise.
(2) The bond price appreciation when interest rates fall is greater than the bond price depreciation when interest rates rise.

Let us now create a 3×3 bond price matrix to establish several additional bond theorems:
Coupon rates

Coupon Rates

		4.0	5.0	6.0
Yields (Interest rates)	4.0	\$1000.00	\$50 (8.1109) + \$1000 (.6756) = \$1081.15	\$60 (8.1109) + \$1000 (.6756) = \$1162.25
	5.0	\$40 (7.7217) + \$1000 (-6139) = \$922.77	\$1000.00	\$60 (7.7217) + \$1000 (.6139) = \$1077.20
	6.0	\$40 (7.3601) + \$1000 (.5584) = \$852.80	\$50 (7.3601) + \$1000 (.5584) = \$926.41	\$1000.00

OR

	4.0	5.0	6.0
4.0	\$1000.00	\$1081.15	\$1162.25
5.0	\$922.77	\$1000.00	\$1077.20
6.0	\$852.80	\$926.41	\$1000.00

If the initial interest rate was 5% and interest rates fell to 4%, which can be caused by a Federal Reserve (the "Fed") Federal Funds rate cut, then a 4% coupon bond rises in value from \$922.77 to \$1000, an increase of 8.37%. The corresponding appreciations for the 5% and 6% coupon bonds are 8.12% (\$1000 to \$1081.15) and 7.90% (\$1077.20 to \$1162.25), respectively. When interest rates fall and bond prices rise, an investor's return is higher in lower coupon bonds. If the Fed raised interest rates from 5% to 6%, the 4% coupon bond price falls to $852.80, a decline of 7.58%. The corresponding depreciation for 5% and 6% bonds are 7.36% ($1000 to $926.41) and 7.17% ($1077.20 to $1000), respectively. An investor's loss is minimized in higher coupon bonds when interest rates rise. It appears that the lower coupon bonds experience the highest appreciation when interest rates fall and the highest depreciation when interest rates rise. A third bond theorem by Malkiel holds that the lower the coupon rate, the greater the change in bond prices; for a given change in the interest rate. The relationship between coupon rates and changes in interest rates is an elasticity concept.[17] The ultimate interest rate play is to purchase zero-coupon bonds if interest rates are expected to decline. The fourth bond theorem holds that the longer the time to maturity of the bond, the greater the bond price volatility for a given change in the interest rate.

If the initial interest rate was 5% and interest rates fell to 4%, then a 5% bond would sell for $922.77, a loss of 7.72%. A 4% coupon bond would fall in price from $1081.15 at an interest rate of 5% to $1000, or par. The 4% coupon bond declines 8.12% of its value. A corresponding 6% coupon bond falls from a price of $926.41, at an interest rate of 5%, to $852.80, at an interest rate of 4%. The 6% coupon bond falls 7.58% when interest rates rise. If interest rates rise from 5% to 6%, the 4% coupon bond rises to $1162.25, an increase of 7.50%, while a 5% coupon bond rises to $1077.20, a 7.72% increase, and the 6% coupon bond rises from $926.41 to $1000, an increase of 7.94%.

---

[17] The elasticity concept is proportional to the duration concept. Duration measures the sensitivity of bond prices to changes in the yield curve. The yield curve can slope and shift in different scenarios. The duration of a bond may be found by solving the equation:

$$\frac{dP}{P} = -D_i dr$$

where
P = bond price
dP = change in bond price
dr = change in interest rates
and
D = duration

$$D_i = \frac{-\frac{dP}{P}}{dr}.$$

9.12 Retirement of Debt

Corporations find it profitable at times to add to their debt financing. Other circumstances may call for the reduction of debt. An actual reduction of the aggregate amount of debt financing should, however, be distinguished from merely shifting its distribution or form. A firm may retire particular debt instruments or issues, yet the total amount of debt financing remains constant. Moreover, if ownership funds and debt financing are substitutable for each other within limits, the range in which various types of liability financing, that is, bank borrowing, open-book credit, and funded debt, may be substituted for each other is far wider. It is easy to detect the ordinary refunding operation either at or before maturity when one issue is directly replaced by another, but if a firm gradually accumulates funds from many credit sources to repay other debt instruments, what appears to be a retirement of debt is actually a mere shifting within the debt structure.

For purposes of this discussion, debt reduction is defined as the retirement of a debt issue with no significant increase in other debt forms to replace it. Discovering when a firm has retired debt under this definition is more difficult than it sounds. If the balance sheet of a firm is compared immediately before and after a legal or accounting retirement of a debt issue, then, of course, there always appears to be a reduction of total debt. To be useful, however, balance sheet comparisons and application of funds analysis should be made over a relevant time period or periods. This time period need not necessarily be the regular or fiscal period. The period selected for review can be less or more than a year, depending upon what sort of information one hopes to uncover. At any rate, in order to measure debt reduction, a sufficient time span should be allowed in order to observe if there has been a buildup of floating debt on the way to the absorption of a funded issue.

There are really only two reasons for a decrease in total debt, as distinguished from a redistribution of debt. Either the firm is moving toward a more conservative capital structure, or it is moving toward a smaller scale of operations, or perhaps both. The capital structure can be modified in a conservative direction (i.e., toward less risk and less potential equity income) by processes other than reducing debt. If a firm's expansion is financed with a larger proportion of equity funds than existed in the original capital structure, the resulting capital structure will be more conservative. If the analyst concerns himself only with funded debt and capital stock as constituting the capital structure, it is possible to overlook another method by which the firm's financial structure can be made more conservative. A firm that floats bonds to provide net working capital—that is, to reduce its dependence on short-term debt—generally lessens its short- or intermediate-term financial risk. The firm has either decreased its current liabilities—funded its current debt—or added to its current assets, in any case improving its current ratio or working capital position. That a firm reduces risk by this type of action is widely acknowledged, yet it might be missed in practice by financial analysts who concentrate on the bond-to-stock relationships. The reduction of debt is not a necessary condition for a less risky capital structure since this can be accomplished by other means. Nor is it a sufficient condition, since if the equity contribution decreases at a more rapid rate than debt, the result will be greater risk than before.

The equity funds may be reduced by losses or withdrawals even while the firm is repaying debt. In this case, of course, the scale of operations is necessarily decreased. If, however, debt is reduced while net worth is held intact or decreased by only a moderate amount in comparison to the decrease in debt, there is both a reduction in the scale of operations and a more conservative capital structure.

9.12.1   Toward a More Conservative Capital Structure

The motives underlying the shifts in capital structure entailed by a reduction in debt vary. A move toward a more conservative capital structure is probably a sign of the advancing maturity or solidity of the firm. Many newly established firms commence operations with a high financing risk. This may be the only way they can get started. If the firm begins to operate successfully, management acts to strengthen its capital structure, possibly by funding short-term liabilities (since in new business sources of short-term capital are relatively less scarce than long-term funds), perhaps by accumulating net worth at a greater rate than liabilities, or perhaps by actually retiring some debt. As the firm approaches maturity and its rate of growth slows down, the risk attendant upon the capital structure of the intermediate still moderately growing firm may seem too great for a conservative management, and debt may be scaled down still further. Perhaps in older firms, management might judge that the only major use of retained earnings is to repay liabilities. In this case, the reduction of debt may be carried to a point where the capital structure is pushed past its optimum position on the conservative side.[18]

9.12.2   Decreasing the Level of Operations

A decrease in the level of output takes place whenever the firm faces a seasonal, cyclical, or secular decline in demand, or possibly a decline in its competitive position. A decrease in production also occurs if the firm's marginal costs rise without a compensating increase in demand. The change in the firm's financial structure corresponding to the decline in output insofar as the management can control these changes on the downturn is largely affected by the situation the management thought it faced. A seasonal decline in demand is traditionally taken

---

[18] One may be tempted to draw an analogy on the financial side to Alfred Marshall's categories of the young and vigorous, the mature, and finally, the old or senile firm.

up by the repayment of current liabilities. This is a task of working capital manage-ment—the control and financial management of the current cash flow. A cyclical or temporary decline in the firm's position probably finds the management retiring debt while attempting to hold the ownership capital constant except as it might be impaired by unavoidable losses. At junctions such as this flexibility in the capital structure is important. The firm would like to be able to manage an orderly retirement of debt, enabling it to maintain its credit and capital base intact. Thus its survival chances would be high, and it could resume expansion on borrowed funds when the economic climate once again became propitious.

If the decline in the scale of the firm's operations is expected to be secular or permanent, the company will move toward a smaller capital structure. The mixture of debt and equity in the new financial structure would be appropriate for the new size of operations, given risks and costs, and both liabilities and net worth would be reduced in proportions resulting in a balance sheet tailored to the new situation. Management occasionally resists the movement toward a smaller balance sheet. The result is likely to be a redundant asset structure. The firm is over-conservatively financed, and it holds far too much cash and reserves in relation to its real output. This situation, however, is an aberration and not a rational development of mana-gerial responsibility. Though it may exist for a while, it will hardly last indefinitely. Such a firm is ripe for a merger offer or a proxy fight.

A reduction in the absolute amount of debt is, in general, more probable during a decline in the rate of expansion—possibly a contraction in output and an actual decrease in earnings—than during periods when the firm appears most prosperous. To the unsophisticated, it may seem an anomaly that debt can be reduced when earnings are declining. But obligations are not necessarily paid with earnings; they are paid with cash. These are related but are not identical. Thus there may be an automatic increase in available cash (even with a decline in earnings) if accounts receivable are collected and new credit accounts are not created at the same rate, or if inventory is sold off and not restored to its previous level. Quite simply, a reduction in gross working capital assets releases cash funds, and these may be applied against debt if the level of operations is declining.[19]

Similarly, the noncash charges against earnings, such as depreciation, depletion, or amortization, represent a cash flow which is independent of net reported account-ing earnings and even, in a sense, of current losses. As long as current operating losses do not exceed such noncash charges as depreciation and there is no buildup of other nonliquid assets, these charges represent a net source of funds to the firm. In periods of contracting activity, a logical use of these funds would be the reduction of outstanding debt.

---

[19] Of course, if the firm suffers steep losses in the liquidation of current assets, it may experience a stringency of cash.

9.12.3 Methods for Retiring Specific Issues

A whole system and complex of clauses can be written into the bond indenture for the purpose of setting debt retirement terms. There are four basic methods, however, by which specific debt issues can be repaid, and all debt retirement plans consist of one or a combination of these four methods:

1. repurchase on open market
2. repayment at maturity
3. repayment of the bond issue by periodic payments

   (a) sinking fund bonds
   (b) serial bonds

4. (a) callable bonds permitting retirement before maturity in whole or part
   (b) callable convertible bonds which may permit forced conversion

Most marketable bond issues today have become fairly standard. They call for a definite maturity date, averaging about 30 years before the bulk of the issue becomes due. They usually have a partial sinking fund provision requiring the company to retire about 2% of the original issue yearly, leaving about 40% of the debt to be met at maturity. The sinking fund trustees may purchase the yearly retirement quota at market or use a sinking fund call provision (paying a slight premium over par), whichever is cheaper. Most bonds contain a call provision enabling the issuer to call back the bonds in whole or part before maturity by paying a fixed call premium over the par value. This general call premium is higher than the one that prevails on bonds bought back under the sinking fund provision; however, both kinds of call premiums are highest near the date of issue and decrease in some regular manner over the life of the bonds. On occasion, bonds have been issued which are not redeemable under a general call for, say, the first five years. This enables the lenders to enjoy at least a few years of good interest rates if the bonds were floated in a period of relatively high rates.

9.12.4 Repurchase on the Open Market

A basic tool of the management used to retire debt, unless the issue has been privately placed, is repurchasing the debt on the market. If the company wishes to pay the going market price, there is no reason why it shouldn't buy back its own bonds or notes and thus effectively reduce its debt. This method is the only way a perpetual bond can be retired, but it could be used on all other issues. The disadvantage of market retirement arises if the company has made a bad bargain-sold bond at an interest rate that is now appreciably above the market rate. It would have to pay a heavy premium in the form of an increased market price to be released from its contract of debt.

The ability to repurchase on the open market can be valuable, and the loss of this ability is a chief drawback to private placement. Of course, if the firm presents an attractive proposition to the insurance companies or other financial institutions holding its bonds, these institutions would surely accept it. They would be able to use the funds released just as gainfully in an alternative investment. A repurchase of debt from private holders would have to be negotiated, however, and a price approximating the market price agreed upon. Purchase from the open market involves no such negotiation, and the amount to be bought at any particular time can be readily varied. A possible objection to a firm's repurchasing its debt on the open market is that the technique could be used to favor one set of creditors over the others. A shaky firm might use funds to buy back a junior issue, releasing these holders from risk while increasing the risk to the holders of the senior issue. That this is a possibility is attested by the attempt to guard against it by inserting minimum working capital maintenance requirements in most bond indentures. This abuse of purchasing on the open market would take place only if the management had questionable motives. In most cases, the management would not buy the junior issues if the firm's survival position were dubious, because the increased remaining risk to the senior securities would be even more intensified for the common stock.

9.12.5 Repayment or Refunding at Maturity

The due date, or date of maturity, is the stated time the corporation promises to pay a specific debt instrument. The maturity date permits the borrower and the suppliers of funds to reassess their positions, and by terminating the old debt contract, it allows them to negotiate a new loan under current market conditions. Thus the maturity date protects both the borrower and the lenders if later developments should make the initial debt terms disadvantageous to either party. If credit conditions have moved in its favor, the firm can get out of debt at par, without paying a premium. On the other hand, the maturity date assures the security owners of some positive action. The issue must be paid at maturity, as agreed, or else the bondholders begin action to enforce their claim. Thus the maturity date enables the bondholders, as well as the corporation, to limit the losses arising from an incorrect forecast of the general market interest rate or the specific risk rate for the firm.

Since a longer term to maturity on a bond lengthens the time during which the lenders cannot renegotiate, it increases the risk and interest charges. The due date is an important item upon which the firm and the investors must agree. The specific terms of an issue, including the length of maturity, are part of the bargain between the demands of the market as represented by the investment bankers and the concessions given by the borrower.

In most modern bond issues, a large part of the debt is retired before maturity by the sinking fund or by partial calls. Even so, the amount of debt to be repaid at maturity is usually too great to be paid back with the cash flow of a given year. If the company is intent on repaying the maturing debt, it must accumulate cash some time in advance. These funds can be invested in short-term liquid assets, but usually, the best return on the incoming funds is obtained if they are employed in buying back the issue. Sometimes the firm incurs other debt along the way to provide the funds to pay the bulk of the maturing issue. If so, what appears to be retirement of debt is actually a refunding of the issue once removed.

A financial operation known as refunding at maturity occurs when the firm either puts a new issue on the market to raise funds to repay the old or attempts an exchange of new bonds with the old bondholders. The latter, sometimes called refunding through exchange, generally indicates financial weakness in the firm. The old bondholders, however, often find it wiser to accept the new securities rather than call for repayment. Perhaps the firm has been meeting interest payments so far, but rather than force the firm to the market place to search for someone else to refinance the maturing debt, the present bondholders accept the offer of new securities, usually with some improvement in the terms. Accepting the exchange offer may be safer for the lenders than attempting to force payment and finding themselves in the midst of a reorganization.

9.12.6 *Gradual Reduction of Debt Issue by Sinking Fund Purchase*

Originally the sinking fund was thought of as assets set aside by the company for the repayment of the total bond issue at maturity. Today most sinking fund provisions retire only part of the issue, generally around 40–60%. Under modern sinking fund provisions, the trustees are directed to buy back a portion of the issue each year at the sinking fund call price or at the market price, whichever is lower. Instead of a formal device to retire the whole debt issue, the sinking fund effects a gradual reduction in the bonds outstanding and supports the market by giving the holders a probability that some of their holdings will be only short term.

In its earlier conception, the sinking fund consisted of assets, earmarked so that confusing them with other assets was impossible. The amount set aside annually was either fixed or contingent upon earnings. Eventually, it was to total the amount of the issue. A sinking fund reserve to balance the accumulation of assets in the fund was usually set up on the equity side of the balance sheet. This sinking fund reserve "appropriated" a portion of the surplus. Supposedly it served as a reminder to forestall too large a distribution of earnings by an overly liberal dividend policy. The sinking fund assets were far more important in repaying the debt than the so-called sinking fund reserve. A corporation showing a large sinking fund surplus reserve would find it difficult to redeem its bonds if it neglected to accumulate the sinking fund or suffered losses in the sinking fund assets. A surplus reserve cannot repay debt; only assets can do so.

Although the old-fashioned sinking fund is still used occasionally on municipal bond issues[20], it is rarely used for corporate issues. The basic disadvantage of an asset sinking fund is the difficulty of finding safe investments with a suitable yield. The sinking fund trustee had to minimize risk in investing his funds. Capital appreciation was not a prime consideration since the fund was to pay off a fixed sum of money. On the other hand, a loss in the sinking fund principal subjected the corporation to considerable hazard. If safety was the major requirement, obtaining a reasonable sinking fund income sometimes posed a dilemma. Suppose that safe investments for the sinking fund had a market yield of only 3%, whereas the market rate for the borrowing corporation was 5%. The real cost of the corporation's borrowed funds exceeded the nominal cost. Assume that the corporation had outstanding $1,000,000 in twenty-year sinking fund bonds, issued at par, with an interest rate of 5%. The yearly interest cost is $50,000. At the end of the tenth year, the sinking fund totals $500,000 and earns 3%. During the eleventh year, the firm's net interest cost is $35,000, that is, $50,000 less the $15,000 interest income earned on the sinking fund. The firm's net borrowed funds are only $500,000–$1,000,000 outstanding on the debt less the $500,000 tied up in sinking fund assets. Thus the real cost to this firm for the use of funds is $35,000 over $500,000 or 7%. Considering all the problems of sinking fund asset selection, the best investment for a bond sinking fund is the retirement of the bond issue itself. Such an investment eliminates all risks, including the risk of market fluctuations, present even in prime government bonds. Moreover, the interest cost saved by the sinking fund and the effective market rate on the bond issue are very close.

Thus the present type of sinking fund developed. The bond indentures carry provisions instructing the trustee to purchase a certain percentage of the issue (perhaps a minimum of 2–5%) annually. Actual retirement is accomplished by purchasing the bonds in the open market or by call.[21] The specific bonds to be called are determined by lot, so that no buyer knows at the time of his purchase the exact date he may be asked to surrender his bonds.

9.12.7   Reduction of Debt Through Serial Issues

An alternative to the modern sinking fund as a method of providing periodic retirement of bonded debt is the serial bond issue.

---

[20] Sinking funds may be advantageous on municipal issues if the yield on safe taxable Federal bonds is higher than the rate on the tax exempt local bonds. The sinking fund trustees should be careful, however, to choose the maturities of the sinking fund investments to coincide with the due dates of their own bonds if they wish to avoid the possibility of capital losses on the sinking fund assets.

[21] Bonds may also be retired by tender, where bondholders are invited to submit the price at which they are willing to sell their bonds back to the company. The company picks up the bonds tendered at the lowest prices

Under a serial bond issue, a definite schedule of maturity dates is established. In effect, there is a simultaneous issue of bonds with varying due dates. The corporation must pay each annual or semiannual fraction of the issue as it comes due. It usually has the choice of retiring a greater amount than the annual installment through the use of the call privilege. Given a plan of repayment so definite, the average interest charges on a serial issue are usually lower than under other systems of repayment.

Serial bonds, although prevalent as state or municipal issues, are not common corporation issues today. They are, however, the basic design of the equipment trust certificate. In general, the most common form for commercial bond issues is the partial sinking fund callable issue.

In order to give a serial bond issue some of the flexibility of the callable sinking fund bond issue, the serial bond issue also often contains a call feature. If the call privilege is exercised, the indenture frequently provides that the longest maturities be called first. This is in marked contrast to sinking fund bond indentures, which provide for call by lot.

An advantage of the sinking fund issue, for management, is that failing to make the annual sinking fund payments may be less serious than failing to pay a maturing serial bond installment. For one thing, the bondholders often may not know of defaults in sinking fund payments for some time after the event. Second, as long as interest payments are met, the bond trustees may not wish to take action other than registering a protest. Of course, the failure to make the regular sinking fund installments is a breach of contract, and the trustee may appeal to a court of equity to enforce payment of the installment. Such a course of action, however, could conceivably lead to the appointment of a receiver. The bondholders may hesitate to apply so drastic a remedy. The dilemma of the bondholders can be conceived as accruing to the benefit of the issuing corporation, giving it a flexibility not possible under a serial bond indenture. This argument can be overstated; obviously, no rational organization enters into a contract with the notion of possibly breaking it. By and large, the sinking fund requirement has to be considered as fixed an obligation as the serial bond maturities.

9.12.8   Debt Retirement and the Call Privilege

The call feature on a bond considerably increases managerial control over the issue; it facilitates a program of gradual debt reduction or perhaps refunding before maturity. Should there be a downward shift in market rates or a strengthening in the firm's credit standing, the call price places an effective ceiling on the premium that the company has to pay to retire its bonds.

If either the market interest rate or the specific risk rate applicable to the firm declines, then, other things being equal, a firm's securities rise in the market. However, were the call option an original term of the issue, the firm need not concern itself with what the market price would have otherwise been. It can pay the call price or the market price, whichever is lower. On the other hand, if there were no call provisions, a firm seeking to reduce its indebtedness would have no choice but to pay the market rate.

9.12.9 *Forced Conversion*

If a bond is both convertible and callable, the management may elect to reduce the risk of its capital structure by forcing conversion. If the market price of the common stock rises above the conversion rate, a convertible bond may command a market price higher than the call price. The management is then in a position to force the conversion from bonds to common stock. If the corporation issues a call for redemption, the security holders must convert or lose the difference between the market price of the bond and its call price. Under these circumstances, the firm can reduce its debt obligation and increase its equity capital at the same time.[22] The position of the original shareholders, however, will in a sense deteriorate, since the new stockholders pay for their shares with the surrender of debt whose value, as such, is below the going market price of the company's shares. On the other hand, the original stockholders are presumed to have gained on the initial issue of the convertible securities at a low-interest rate or have been recompensed with salable rights. Furthermore, the management's call only hastens the conversion which, under the assumed circumstances, is bound eventually to take place.

9.13 Summary

Bond issues are very important in corporate finance. Bond issues substantially exceed stock issuance. Bonds are refunded as interest rates hit historic (50-year lows) in 2020. The bond refunding decision has never been more relevant than in the past 3–5 years. An investment in long-term bonds must be aware of Malkiel's bond theorems.

References

Bierman, H. (1972). The bond refunding decision. *Financial Management*, 22–29.
Bradley, I. F. (1959). *Fundamentals of corporation finance*, Rev. ed.. Halt, Rinehart and Winston, Ch. 17.

---

[22] In rare cases the conversion must be accompanied by a payment of money. The company not only retires the debt issue, it increases equity capital and strengthens its working capital position.

Brealey, R. A., & Myers, S. C. (2003). *Principles of Corporate Finance* (7th ed.). McGraw-Hill/ Irwin, Chapter 26.

Fisher, L. (1959). Determinants of risk premiums on corporate bonds. *Journal of Political Economy, 47,* 217–237.

Gerstenberg, C. W. (1959). *Financial organization and management of business* (4th ed.). Prentice-Hall, Ch. 8.

Guerard, J. B., Jr. (2010). The corporate sector as a net exporter of funds. In J. R. Aronson, H. L. Parmet, & R. J. Thorton (Eds.), *Variations in economic analysis: Essays in Honor of Eli Schwartz*. Springer.

Hicks, J. R. (1937). *Value and capital*. Oxford University Press.

Macaulay, F. R. (1938). *Some theoretical problems suggested by the movements of interest rates*. Columbia University Press.

Malkiel, B. G. (1962a). Expectations, bond prices, and the term structure of interest rates. *Quarterly Journal of Economics, 76,* 197–218.

Malkiel, B. G. (1962b). The term structure of interest rates. *American Economic Review, 54,* 532–543.

Myers, S. (1977). Determinants of corporate borrowing. *Journal of Financial Economics, 5,* 147–176.

Schwartz, E. (1962). *Corporate finance*. St. Martin's Press. Chapter 9.

Schwartz, E. (1967). The refunding decision. *Journal of Business, 40,* 448–449.

Van Horne, J. C. (2002a). *Financial management & policy* (12th ed.). Prentice Hall, Chapter 18.

Van Horne, J. C. (2002b). *Financial market rates and flows* (6th ed.). Prentice-Hall, Inc.

Chapter 10
Debt, Equity, the Optimal Financial Structure, and the Cost of Funds

Traditionally the capital structure of a firm has been defined as the book value of its common stock, its preferred stock, and its bonds, or fixed liabilities. These items are considered to be the "permanent" financing of the firm. The special importance is given to them, however, which may lead to an error in financial analysis. Thus, a company which only has common shares in its capital structure is often described as conservatively or safely financed. But if, for example, the firm has considerable trade debt outstanding, owes on a bank loan, or is tied up with long-run rental contracts, it may not be "safely" financed.

Although distinguishing between current liabilities and longer-term financing is convenient in some analyses, the degree of difference between current and funded debt is often grossly exaggerated. The so-called permanent financing is not unalterable; bonds can be retired, reduced, or increased, so can preferred stock, and the book value of the total common stock equity may also be varied. On the other hand, no operating firm is likely to function without some amount of current liabilities; thus some current debt is permanent to the financial structure. Thus it would be better to consider a firm's capital or financial structure as consisting of all the items on the credit side of the balance sheet representing the equity and all the liability accounts.

An important general tool of financial structure analysis is the ratio of total debt to total assets. Of course, in a detailed financial analysis, the relationships and ratios among the items on the credit side of the balance sheet and among liability groupings and certain assets are significant and useful, but the usual financing analysis may be misleading when only the fixed debt is employed in depicting the capital structure of the firm.

Perhaps the most misleading ratio used as index of credit quality is the ratio of debt to the market value of the shares. Too often the market value of the stock is an ephemeral number based on a notion of future growth and hope for the future. If another recommended index is used, that is, debt service into cash flow, the potential operating coverage of the debt has already been captured. Because cash flow and the market value of the shares are related, the relation between debt and the market value of the stock is not an independent variable.

© The Author(s), under exclusive license to Springer Nature Switzerland AG 2022   215
J. B. Guerard Jr. et al., *Quantitative Corporate Finance*,
https://doi.org/10.1007/978-3-030-87269-4_10

10.1   Definition of Leverage: Profits and Financial Risk

An important concept in understanding the relationships in the financial structure of the firm is the ancient idea of "trading on the equity" now long going under the current term "leverage."[1] Leverage is the amount of outside funds (debt) the owners use in proportion to their own contributions to the financing of the firm. The use of debt is called leverage because these funds, acquired at a priority of repayment and given a priority of return, widen the potential swing of both gains and losses to the ownership shares. Any earnings on the assets acquired by borrowing in excess of the rate that has to be paid to the creditors belong to the owners and increase their net rate of return; however, if the earnings on the assets acquired with borrowed funds fall below the contracted rate, or if there are overall losses, the negative difference sharply reduces the rate of return or increases the loss on the equity. But as long as the marginal assets employed in the firm earn more than the cost of the borrowed funds, it will be profitable to use leverage, with the proviso that the financial risk of the firm is not thereby inordinately increased. The degree of leverage in a firm's capital structure is measured by noting how much the rate of return on equity would change with any change in the average rate of return on the total assets. The greater the proportion of outside funds to ownership capital, the more emphatic is the leverage effect.

Some financial analysts apparently recognize leverage only if the outside funds are acquired under a definite contract and the suppliers of these funds are paid a fixed positive rate of return. Leverage is thus limited to the use of bonds, preferred stock, or long-term bank loans. Under this concept, many banks, for example, are not considered as leveraged, since they often have no bonds or preferred stock outstanding in their capital structure. Nevertheless, authorities in the field of money and banking note the "highly leveraged aspect" of the typical bank's capital structure, the small percentage of equity in comparison to the total deposits or liabilities carried.

A broad definition of leverage covers the relationship between all the prior claim securities and obligations to the ownership capital. Trade accounts and other current liabilities are included in this concept of leverage. These obligations have priority over the ownership shares; they must be paid at least a zero rate of return. This seems a paradox until we remember that ours is a profit and loss economy. Shareholders may earn a negative rate of return, and the owners may absorb losses, but liability claims are not written down unless there is a failure or reorganization. The zero return placed on current liabilities is thus, in a sense, a fixed return, and it accordingly widens the possibilities for gains and losses on the ownership investment just as does any other fund borrowed from outside sources.

---

[1]Called "gearing" in England.

10.2   Illustrations of Leverage: Return and Risk

The effect of "zero" cost liabilities on the possible dispersion of return for the owners is demonstrated by the simplified example presented in Table 10.1. The Woodrow firm is financed solely with current liabilities and ownership capital. Whatever the rate of return on total assets, which by accounting definition is equal to the total liabilities and capital, the return on the ownership capital is two and a half times as large. Thus if the Woodrow Corporation earns 10% of its total assets, it would earn a before-taxes return of 25% on its owners' investment. If, however, the firm should experience even a small overall loss, the rate of loss to be absorbed by the equity capital is again two and a half times as large as the overall rate. Thus if the Woodrow Corporation lost 3% of its total assets, the loss on the ownership funds would be 7½%.

The use of current debt is a cheap method of financing; carried too far, it may become quite risky. Current liabilities constrict the firm's net working capital position. Although current liabilities carry a minimum interest (charge) if any, the principal amount is continually coming due. From this point of view, fixed debt, when it can be obtained on favorable terms, is a safer component of leverage than current debt. The interest charges on long-term debt reduce the profits derived from successful leverage and increase the possibilities of loss in case of a downturn, but at least the repayment of the principal of the debt is delayed into the future. Thus the firm has a chance to recover its financial position before the due date.

Table 10.1   Effect of leverage using current debt
Woodrow Corporation

Total Liabilities and Capital		
Current Liabilities		$300,000
Surplus	$100,000	
Capital Stock	100,000	
Total Common Stock Equity		200,000
Total Capital & Liabilities		$500,000

Rates of Return at Different Levels of Earnings on Total Assets

Rate of Return (EBIT)[a] on Total Assets	Total Profit (EBIT) or (Loss)	Rate of Return (EBT)[b] on Equity (Ownership Capital)
10%	$50,000	25%
5%	25,000	12 ½%
2%	10,000	5%
-0-	-0-	-0-
(3%) Loss	(15,000) Loss	(7½%) Loss
(5%) Loss	(25,000) Loss	(12½%) Loss

[a]EBIT, earnings before interest and taxes
[b]EBT earnings before taxes

Table 10.2 Effect of leverage using considerable fixed debt
Meredith Company

Liabilities and Capital		
Current Liabilities		$10,000,000
First Mortgage Bonds (4's)		40,000,000
Debentures (4 ½ 's)		15,000,000
Capital Stock 00 Common (1,000,000's)	$20,000,000	200,000
Surplus	15,000,000	35,000,000
Total Capital & Liabilities		$100,000,000

Effect of Leverage on the Rate of Return under Variations in the Profitability of the Assets

Rate of Return (EBIT) on Total Assets	Total Earnings before Interest and Taxes	Interest Charges	Earnings before Taxes (loss)	Rate of Return on Common Stock Equity Before Taxes
10%	$10,000,000	$2,275,000	$7,725,000	22.1%
8	8,000,000	2,275,000	5,725,000	16.3
6	6,000,000	2,275,000	3,725,000	10.6
4	4,000,000	2,275,000	1,725,000	4.9
2	2,000,000	2,275,000	(275,000) Loss	(0.8%)
0	-0-	2,275,000	(2,275,000) Loss	(6.5%)

Table 10.2 illustrates the possible behavior of the rate of earnings on the stock-holders' investment as the rate of earnings on the total assets changes, in a firm which has a considerable component of long-term debt leverage. For the Meredith Company, the rate of return on the common stock equity changes by approximately 2.85 percentage points for each percentage point change in the earnings on the total assets. If the company earned 10% on its total assets, the earnings on the common shareholders' equity would be 22.1%; at a 2% return on total assets, however, the company shareholders suffer a loss. The Meredith Company must gross 2.21½% on its total assets to break even.

An interesting method to calculate the break-even point of debt and equity financing is to express earnings before interest and taxes, EBIT, as a function of sales. If one calculates the earnings per share (eps) of the debt option, incurring higher interest costs and lower earnings, and equates the eps debt option with the eps equity option, keeping interest expenses and earnings constant, but issuing more shares, one can find the break-even sales level for the debt and equity decisions. For example, let us assume that EBIT equals 15% of sales. Interest charges are currently $10 million; there are 5,000,000 shares of stock outstanding, and the firm seeks to raise $100 million of new funds. If the firm's stock price is $100, then the firm must issue 1 million new shares of stock. If the current cost of debt is 8%, then the firm must pay an additional $8 million of interest expense. The break-even equation for indifference between debt and equity financing is:

$$\text{eps Equity} = (.15 * \text{Sales} - \$10\text{ MM})/(5\text{ MM} + 1\text{MM}) =$$
$$\text{eps Debt} = (.15 * \text{Sales} - (10\text{MM} + 8\text{MM}))/5\text{ MM}$$
$$(.15 * \text{Sales} - 10\text{ MM})/6\text{MM} = (.15 * \text{Sales} - 23\text{MM})/5\text{MM}$$
$$\text{Sales, Break-even} = \$586.67\text{ MM}.$$

If sales are expected to be less than $586.67 million, then the firm should issue equity; if sales are expected to exceed $586.67 million, then the firm should issue debt. Management should seek to maximize the earnings per share, eps, of the firm to maximize stockholder wealth.

Leverage is profitable if the rate of earnings on total assets is higher than the going rate of interest on the debt. Of course, the risk to the stockholders of loss and failure in case of a downturn must always be considered. It is generally felt that to finance safely with leverage, the stability of the earnings, or better the cash flow, is more important than its level. The reader is invited to follow the Lerner-Carleton derivation of a return on equity and the issue of leverage. The operating return on assets, ROA, or R is the ratio of the firm's EBIT to total assets. The firm pays interest on its liabilities, L, with a coupon rate of r.

$$\text{EBIT} = R(\text{Total Assets})$$
$$\text{Operating Income} = \text{EBIT} = R\ (\text{Liabilities} + \text{Equity}) = R(L + E)$$
$$\text{Less Interest Paid} = -I = r(\text{Liabilities}) = rL$$
$$\text{Earnings before Taxes} = \text{EBT} = R(L + E) - rL \qquad (10.1)$$
$$\text{Taxes Paid} = -\text{Taxes} = t[R(L + E) - rL]$$
$$\text{Earnings after Taxes} = \text{EAT} = (1 - t)[R(L + E) - rL]$$

The return on equity is given by earnings after taxes divided by equity and is a positive function of the liabilities-to-equity ratio.

$$\text{ROE} = \frac{\text{EAT}}{E} = \frac{(1 - t)\ [R(L + E) - rL]}{E}$$

$$= \frac{(1 - t)\ [RL + RE - rL]}{E} \qquad (10.2)$$

$$= (1 - t)\left[R + (R - r)\frac{L}{E}\right]$$

Thus, as long as the return on asset exceeds the cost of debt, then the return on equity rises linearly with leverage. Leverage is extremely important to the firm's stockholders. The choice of capital structure must be made with management's perception of the return on assets and its expected cost of debt.

10.3   Surrogate Evidence on the Development of "Optimum" Financial Structure

The basic types of firms have tended to develop characteristic financial structures. These regularities in the typical financial structures imply that certain similar forces operate on all the firms in an identifiable class. Occasionally an individual firm, because of special characteristics in its financial history or the predilections of its management, may diverge noticeably from the typical structure for a firm in the given industry. Nevertheless, a financial specialist can almost invariably determine whether she is reading the balance sheet of an electric utility, a railroad, a bank, a mercantile, or an industrial firm. Although one utility may differ in its financial structure ratios from another, these differences are usually small compared to the differences from a typical industrial or rail. The persistence of these differences seems to indicate that there is an "optimum" firm financial structure for various industries.

An optimum financial structure maximizes the long-run market value of the firm's common stock. This is not the same as asserting that the optimum capital structure maximizes profit or earnings per share. For both the earnings per share and the risk-adjusted rate at which the market capitalizes these earnings must be considered. The amount of financial risk or leverage a firm carries helps set the capitalization rate of the shares. If a firm's financial structure carries too much debt, that is, borrower's risk, the market may set a lower price for the shares than it would give for similar shares with smaller earnings but a more "conservative" financial structure.

Under "normal" conditions a company can add to its rate of profit by adding more debt to its financial structure. However, leverage increases the potential loss and lowers the survival rate in case of an economic downturn. In setting the value of the shares, investors consider both the financial risk of the firm's capital structure and the intensity of the economic or operating hazards typically faced by the industry. The investing market will accept the sacrifice of some financial safety for increased earnings, as long as other operating hazards are not too high, but it will discount the returns of a firm whose total risk seems too large. The hypothesis that the corporation management attempts to adjust to both economic risk and financial risk to obtain the best possible position for their shareholders may explain why companies in different industries have tended to develop typical financial structures.

If the theory of the optimum capital structure is correct, every firm has a preferred combination of risk and return.[2] Suppose three firms, A, B, C, are engaged in similar operations:

---

[2] A fuller theoretical discussion of the optimum financial structure may be found in E. Schwartz, "Theory of the Capital Structure of the Firm," *Journal of Finance*, March 1959. The arithmetical example follows the general pattern of Graham and Dodd, *Security Analysis*, McGraw-Hill, 1951, Chapter 37.

Earnings per share	$ 2.00	$ 2.50	$ 3.00
Market price	24.00	25.00	24.00
Price/earnings ratio	12X	10X	8X

Firm A carries very little leverage or financial risk, Firm B has some financial risk, and Firm C is highly leveraged. The earnings per share are for a "normal year," and the differences in earnings are due solely to the different financial structures. If these conditions prevailed then according to the revealed preferences, of the investors on the market, Firm A is too conservatively financed, Firm B has the optimum capital structure, and Firm C is "overleveraged."

The optimum capital structure not only maximizes the value of the shares, but it also implies the minimization of capital costs. The explicit cost of debt is cheaper than equity. As a moderate amount of debt is added to the financial structure, the cost of the financial mix decreases. However, as the proportion of debt in the total financial structure rises, the investors' appraisal of the quality of the debt falls. The cost of borrowed funds rises. In addition, as the internal risk of the firm increases, the price/earnings ratio falls, and the cost of equity funds rises. In an overleveraged company, the cost of the total financial mix is higher than the norm. For the ideal financial structure, the cost of capital mix is at a minimum.

10.4   The Pure Theory of the Optimal Financial Structure[3]

The pure theory of the optimal capital structure, as expressed by Schwartz (1959), is based on the assumption that the firm is a semi-monopsonistic demander of funds from the capital market. By discriminating against the suppliers of funds through employing varying debt instruments and judiciously balancing the total of financial risk and external risks, the firm can achieve an optimum financial structure, reducing total financing costs and maximizing the value of its shares. The parameters constituting the environment in which the firm exists are as follows:

1. The individual firm is confronted by two types of risks. One type we might call the "external risk" and the other type the "internal" or "financial risk."[4]
2. The external risks are a composite of the stability of earnings, or cash flow of the firm, and the liquidity, safety, and marketability of the assets typically held by the firm. The level of external risk is in large part dictated by the nature of the industry in which the firm is engaged and is not subject to any great extent to the control of the financial decision-makers.

---

[3] E. Schwartz, Op. Cit., *Journal of Finance*, March, 1959.

[4] Financial risk was divided into borrower's risk and lender's risk by Keynes in *The General Theory*, pp. 144–145.

3. Internal risk is the financial risk of the firm's capital structure. It is set by the types of liabilities (short term or funded) that the firm carries and the total amounts of the liabilities in proportion to the firm's equity capital. The factors constituting the firm's capital or financial structure can be varied considerably by the financial management.
4. The two types of risks together are the sum of the hazards to which the owners and the creditors of the firm may be subjected. The external risks are a parameter given by the nature of the industry; these external risks are borne in mind by both borrowers and lenders and influence the optimum financial risk that different types of firms are likely to carry.

The optimum capital structure for any widely held company is one which maximizes the long-run market value per share of the common stock. This is not quite the same as asserting that the optimum capital structure is one which will maximize profit or earnings per share. For both the earnings per share of stock and the rate at which they are capitalized must be considered. The amount of financial risk that a firm carries helps set the capitalization rate. If a firm's financial structure carries too much borrowers' risk, the market may set a lower price for the shares that it would give for similar shares with perhaps somewhat smaller earnings but less financial risk.

The ability of the firm to set up an optimum capital structure implies the ability to discriminate against suppliers of funds, investors, individuals, or financial institutions, with different preferences for income and aversions to risk. Discrimination on one level leads to complex financial structures. It means that by raising funds through securities and contracts with varying return and security provisions, the firm could lower its total financial costs. On a broader macro level, varying preferences for return and risk implies that by a judicious mix of overall debt (financial risk) and equity, the firm could maximize the value of its shares (minimize the cost of capital), that is, achieve an optimum capital structure.

The theoretical trade-off for a given firm between the rate of return on ownership capital (equity), the degree of financial risk (debt), and the market preference yielding the maximum price for the shares is illustrated in Fig. 10.1. The financial risk factor is indicated indirectly in Fig. 10.1. It is shown on the horizontal axis, inversely related to the proportion of equity (capital stock) in the capital structure. Thus as the amount of equity capital increases in a particular firm's capital structure, the debt-equity ratio and the degree of financial risk decrease. The conventional rate of return on the equity is depicted on the vertical axis. Because of the Pro-forma profitability of leverage, the rate of return on the equity falls as the proportional amount of share financing increase although volatility and financial risk decrease.

The transformation curve, D, gives the average rate of return for shares and degree of risk for a financial structure containing varying amounts of ownership capital. Sur-imposed in the figure are investor's indifference curves showing the investor substitution rate between earnings and the degree of risk for a firm of this type. Each indifference curve represents a given constant stock price. The tangency point E indicates the financial structure, the trade-off between risk and return which

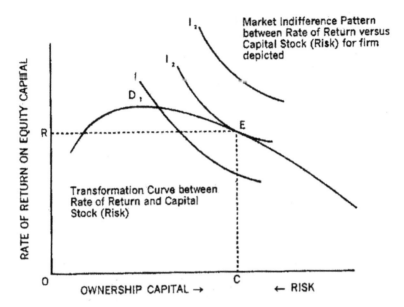

Fig. 10.1 Formal solution of a firm's optimum capita structure

will fetch the highest price for the shares on the market. It is the point where the earnings and the risk-adjusted discount rate yield the highest amount.

The tangency point E indicates the optimum amount of equity capital and rate of return on equity capital for the firm. In brief the conditions of the optimum are:

$$\frac{\text{Marginal sacrifice in earnings}}{\text{Marginal decrease in risk}} = \frac{\text{Marginal decrease in earnings}}{\text{Marginal increase in ownership}}$$

(investor's choice)   (decrease risk) (in the financial structure of the firm)

The letter R indicates the rate of earnings on equity investment and OC the optimum amount of equity capital for the firm, setting up the market capitalization rate and expected earnings that maximize the value of the shares.

The optimum capital structure varies for firms in different industries because the typical asset structures and the stability of earnings which determine inherent risks vary for different types of production. The theoretical solution of the optimum capital structure is made in a very formal manner since it must give consideration to many variables—increasing lender's risk, increasing borrower's risk, the interest rate structure, the forecasted earnings function, and the possibility of discriminating against the market supply of outside capital.

10.5   Modigliani and Miller: Constant Capital Costs

Contrary to traditional views of the relative costs of debt, equity, and the weighted cost of capital found in Schwartz (1959), Professors Franco Modigliani and Merton Miller (M&M) posited a model where in a non-tax world, for a firm of a given risk class, capital costs are constant regardless of the financial risk.[5] There is no optimum financial structure.

In the M&M model, the trade-off between financial risk and the cost of funds is unitary; if more debt is added to the financial mix, the cost of debt rises, and the desired rate of return on equity rises, so that the weighted cost of the financial mix remains constant. Let us briefly recount the three propositions of M&M in their seminal presentation of the cost of capital and valuation.[6] First, M&M hold for many firms in the same line of business, the cost of capital is a constant, ρ_o. The constant is determined by dividing the expected return per share by the stock price; that is, the cost of capital is determined by dividing net operating earnings of the firm by its total market value of the firm. Hence, the M&M hypothesis is often referred to as "The Net Operating Income" approach.[7] Thus, the average cost of capital is independent of capital structure. Second, the expected cost of equity rises linearly with the debt-to-equity ratio. The earlier Lerner-Carleton derivation is a variation of the M&M Proposition II. M&M argued that the firm must earn a return on investments exceeding ρ_o. M&M's Proposition III holds that if the firm earns at least ρ_o on its investments, the project(s) is acceptable regardless of the securities issued to finance the investment. M&M presented empirical evidence in their 1958 study, using the 40 firm electric utility study of Allen (1954) and the 42 firm oil company sample of Smith (1955). Both Allen and Smith provided data on the average values of debt and preferred stocks and market values of securities, such that M&M could calculate the debt-to-total value of securities ratio, d. M&M regressed the net returns, x, defined as the sum of interest, preferred dividends, and net income, as a function of ratio d. The Allen electric utility sample was covered from 1947 to 1948, and the Smith sample of oil companies was for year 1953. The M&M regressions were:

---

[5] Franco Modigliani and Merton H. Miller, "The Cost of Capital, Corporation Finance and the Theory of Investment," *American Economic Review*, June 1958.

[6] M&M discussed their initial proposition of capital structure irrelevance in terms of a class of homogeneous firms, firms that are perfect substitutes of each other, such as firms in the same industry.

[7] J.C. Van Horne, *Financial Management & Policy*, Prentice Hall, 2002, 12th edition, pp. 255–260.

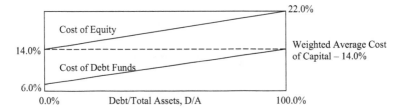

Fig. 10.2   A graphic depiction of the M&M hypothesis

$$\text{Electric Utilities} \quad x = 5.3 + .006d$$

$$\text{(s.e.)} \quad (.008)$$

$$\text{Oils} \qquad\qquad x = 8.5 + .006d$$

$$(.024)$$

M&M held that the regression results supported their Proposition I. The calculated t-statistics, found by the ratio of the regression slope, b, divided by its standard error (in parenthesis), should be 1.96 (or 1.645 at the 10% level), to be statistically significant. The calculated t-statistic on the electric utility sample is 0.75, far less than 1.645. The calculated t-statistic for the oil sample is 0.25. Thus, there is no statistically significance between net returns and the debt-to-asset ratio in the initial M&M study. We will take a detailed look at hypothesis testing in Chap. 11. M&M used the Allen and Smith samples to test their Proposition II. M&M regressed ROEs, defined by dividing net income by equity, as a function of the debt-to-equity ratio, b:

$$\text{Electric Utilities} = \text{ROE} = 6.6 + .017h$$

$$\text{(s.e.)} \quad (.004)$$

$$\text{Oil Companies} = \text{ROE} = 8.9 + .051h$$

$$(.012)$$

The estimated t-statistics of the electric utilities and oil companies sample of 4.25 and 4.35, respectively, rejected the null hypothesis of no association between ROE and the debt-to-equity ratio. Thus, support is found for Proposition M&M II that the cost of equity rises linearly with the debt-to-equity ratio.

A graphic depiction of the M&M model is shown in Fig. 10.2. The substitution rate between the risks and the costs of the elements of the financial mix is linear so that the average cost of capital is constant.

In an equation presenting the M&M hypothesis that is in a linear form, the function for the overall cost of capital appears as follows:

Table 10.3 Overall cost of funds in a simplified M&M model, where the unleveraged cost of equity is 14%, the initial cost of debt is 6%, and the financial risk premium is 8%

	A Total Assets	D Debt	E Equity	k_i Cost of Debt at the Given Financial Mix	K_e Cost of Equity at the Given Financial Mix	Total Cost of Debt Financing	Total Cost of Equity Financing	K_o Weighted Cost of Total Capital at the Given Financial Mix, k_i x D + k_e x E
1)	100	20	80	7.6%	15.6%	152	1248	14.0%
2)	100	50	50	10.0%	18.0%	500	900	14.0%
3)	100	90	10	13.2%	21.2%	1188	212	14.0%

$$k_o/A = D\left(i_j + D/A \cdot p\right) + E\left(e_j + D/A \cdot p\right).$$

where:

k_o/A = the overall cost of capital
D = the amount of debt the firm carries
i_j = the rate on the very first incremental amount of debt, the pure borrowing rate
$D/_A$ = equals the percentage of debt to total assets, the degree of leverage
A = total assets equals the sum of D and E
E = the amount of equity capital
e_j = the return required on a non-leveraged share of stock for an industry of this risk class
p = the sum of the borrower's and lender's risk premium, set equal to the difference between i_j and e_j
k_i = $(i_j + D/A \cdot p)$, it is the overall market interest rate on debt at the given degree of leverage
k_e = $(e_j + D/A \cdot p)$, it is the market's desired return on equity at the given level of leverage

Showing how this works in a simple arithmetical example, let us set:

i_j at 6.0%
ej at 14.0%
and p becomes 8.0%

The result is the same as that in the graph in Fig. 10.2. No matter what the degree of leverage, the non-tax cost of funds is constant at 14.0%. Moreover in Table 10.3, when the resultant change in the costs of the other components of the financial mix are considered, the marginal cost of capital is constant of 14.0%, again no matter what type of financing is employed.[8]

---

[8]Modern financial management uses the beta of a security to determine the firm's cost of equity. The reader will see the beta estimation in Chap. 14. The estimated beta is produced by regression a security monthly returns as a function of the corresponding market returns. Because the vast

The simplified version of the M&M model points out its difficulties. As the debt ratio reaches 90% of the financial structure, the least economic blip can bring about failure and bankruptcy. The costs of restructuring after bankruptcy are substantial. If the firm's capital structure is 100% equity, issuing debt increases the return per share without substantially increasing risk. Of course, Professors Modigliani and Miller did not present their hypothesis as a pure mathematical abstraction. They presented devices by which the outside investors could offset any anomalies in the financial structure of firms having marketable shares.

(a) Arbitrage: M&M argued that the external investors could engage in arbitrage to create an increased return on shares that were underleveraged. They could increase risk but raise the expected rate of return by employing debt to buy the shares.

The counterargument is that this is not a perfect offsetting tool. There are additional transaction costs, and external debt is more dangerous to maintain under varying returns and asset values and more dangerous in terms of legal liability to the external or individual investor than the internal debt of the firm.

(b) Hedging: External investors could reduce the risk on a firm which was over-leveraged by purchasing both the shares and the bonds (debt) of the firm. This might reduce the expected return, but it would provide a minimum fail-safe position (i.e., the final value of the bonds) in case of failure.

The counterargument involves the problem of transaction costs and the fact that the offset in holding the bonds is far from perfect in case of failure and bankruptcy. The legal and transaction costs of going through bankruptcy and reorganization is likely to erode a large part of the value of the firm.

The major contribution of the M&M model is to show that each type of financing, debt or equity, effects changes in the costs of the other. Nevertheless, when the costs of failure, bankruptcy, reorganization, and various transaction costs are considered, it is clear that the trade-off is not likely to be perfect. Perhaps we can sum up by noting that the central difficulty of the pure M&M model is the problem of the asymmetry of information. Should not a judicious financial management knowing the environmental conditions of their firm do a better job of setting up the financial structure than the outside investor? And if this is so, have we not arrived at the notion of an optimal capital structure?

Finally, the empirical evidence that financial structures are not random but appear to be significantly different for varying classes of firms points in the direction of the existence of optimal capital structure.[9]

---

majority of securities have issued long-term debt, the estimated betas involve leverage. The beta of a security rises with debt. M&M II is consistent with estimated betas and costs of equity.

[9] E. Schwartz and J.R. Aronson, "Some Surrogate Evidence in Support of the Concept of Optimal Financial Structures," *Journal of Finance*, March 1967.

Nevertheless, M&M made an important contribution. It is clear that the bounds of the optimal capital structure are much broader than theory would suggest. Secondly, the M&M hypothesis sharpens the argument or more clearly points out the tax advantage (the tax deductibility of interest) of debt under our current corporation income tax laws.

M&M recognized the cost of capital implications of interest deductibility in their original 1958 study. M&M held that the interest deductibility feature of corporate taxation leads to a decreasing cost of capital as the debt ratio rises. By 1963, M&M formulated the before-tax earnings yield, the ratio of expected earnings before interest and taxes, $\bar{x}$, to the market value of the firm, $\hat{v}$, as:

$$\frac{\bar{x}}{v} = \frac{\rho^T}{1-t}\left[1 - t \ \frac{\cdot D}{V}\right].$$

Due to the tax advantage, the cost of capital of the firm decreases with leverage, and the value of the firm will rise with the use of debt.

10.6   The Optimal Capital Structure and the M&M Hypothesis

The difference between optimal capital structure theory and the M&M hypothesis can be exaggerated. Both models emphasize the point that the use of one class of financing has rebound effects on the costs of the rest of the financial structure. In the optimal model, the overall cost of capital at any given time is constant within the range of the optimal capital structure. Debt or equity financing or some combination may be used for any particular project, as long as the financial mix is kept within an optimal range. Nevertheless, because every type of financing has interactions with the other sources of financing, the return on a project is not to be compared to the direct cost of its mode of financing but to the overall cost of the financial mix.

In the M&M model, the interaction between different types of financing is complete, so there is no optimal financial structure. Thus the firm's overall cost of capital at any point of time is constant at the proper financial mix, or it is constant regardless of the mix. Quite importantly, both of these views are in opposition to the sequential cost models, in which the cost of capital depends on the financing which is being used currently so that the cost is lowest when the firm uses retained earnings, rises for outside borrowing, and becomes still higher when borrowing capacity is strained and additional funds depend on the flotation of new shares. In short, in making real investment decisions, Schwartz and M&M agree that the appropriate discount variable is not the immediate financial source but the overall cost of capital.[10]

10.7 Empirical Factors Influencing Financial Structures

The two main external factors influencing the financial structure of a firm are the composition of its assets and the stability of its cash flow. Financial firms, such as banks and insurance companies, are prime examples of enterprises where the liquidity and marketability of their assets enable them to carry a high proportion of liabilities. Of course, in this instance the firm's selection of assets for safety, marketability, and liquidity may be predetermined by the heavy volume of the firms contingent or short-term liabilities, rather than the other way around. Nevertheless, a firm with safe marketable or short-term assets can finance these assets with a high proportion of debt with relatively matching maturities. Thus marketing firms carry short-term inventories, and creditable short-term accounts receivable can safely carry a relatively high proportion of short-term debt.

The stability of cash flow is influential in shaping the financial structure. The cash flow is the amount of free funds the firm can utilize over a short-run period. Cash flow and accounting profits or earnings may differ considerably. Cash flow is less than earnings, for example, by any increases in costs incurred on work in process; cash flow exceeds reported earnings by the extent of depreciation, depletion, and other book or noncash changes—that is, noncash charges representing the using up of assets acquired in the past. Although for internal control and budgeting purposes detailed analyses are made of the components of the cash flow, the rough rule of thumb for measuring the cash flow is reported earnings for the period plus depreciation, depletion, and any other noncash charges.

In calculating the leverage a firm might reasonably carry, the financial decision-makers must not only estimate the average level of the cash flow over time but the likelihood and extent of deviations from the norm. Where fluctuations from the average are not expected to be either deep or sustained, the firm may safely carry a high percentage of debt.

The inclusion of depreciation charges in the cash flow helps explain why firms with a good proportion of fixed assets may also carry more long-term debt. The fact that firms having a considerable fixed plant usually float bonds is not related to any physical attribute of the fixed plant; it is not dependent on any presumed safety that bricks and mortar bring to the bond mortgage. The affinity of fixed assets and long-term debt rests on the fact that the cash flow of firms holding considerable fixed assets must exceed their reported earnings. The depreciation charges taken against the fixed assets act as an extra cushion, which, added to the accounting net earnings,

---

[10] Schwartz has observed that the cost of capital is probably more invariant with respect to the debt-to-total assets ratio than he thought in 1959 and observed in 1966.

may help the firm meet its interest and principal obligations. A firm may show zero accounting profits after depreciation yet have a positive flow of cash. As long as reported losses do not exceed depreciation and depletion charges, some cash flow will be available to pay debt obligations. In other words, some cash is always generated as long as operating revenues are greater than out-of-pocket operating costs, no matter what the depreciation charges may be.

10.8   Measures for Approximating Financial Risk

There are no precise methods for measuring the degree of risk in a firm's financing. In practice, however, there are some rough measures of risk that can be computed and then compared to certain traditional "rules of thumb" or "benchmarks." These have been established from experience, and not surprisingly, econometric studies have shown that weakness in these measures to be quite predictive of financial failure or bankruptcy.

For current liabilities, the best measures of risk are the various working capital ratios. Chap. 5 focuses on credit analysis and risk using these ratios. For long-term debt, introduced in the previous chapter, the best measures of financial risk (or safety) are the various "times interest earned" ratios and the balance sheet financing structure percentages. The most insightful presentation of the financing structure shows the major financing categories as a percentage of total assets. A growing misleading financial structure ratio is the percentage of debt to the market value of the stock! In an overexuberant market, the price of the shares will balloon encouraging the incurring of debt levels which cannot be borne in more reasonable time.

Dividing the annual interest charges into the annual pre-tax earnings gives the interest earned coverage. Other things being equal, the larger the "times interest earned" ratio, the greater the safety of the funded debt. Of course, the coverage the financial analysts deem adequate is larger for an industrial, with its greater likelihood of fluctuating cash flow, than for the base service electric utility company. The rule of thumb minimum requirement for the highest grade industrial bonds is an interest coverage averaging eight times over the last 7 years, whereas only three times was required for the standard utility bonds.

The measure of earnings generally used in constructing the "times interest earned" ratio is EBIT, earnings before interest and taxes, or gross operating income. The earnings of a firm before interest and taxes represent the difference between costs and revenue; following economic theory, this is the variable the firm attempts to maximize. EBIT is the firm's overall economic return, given the best efforts of the marketing people and those in charge of production costs. Net profit, the earnings of the shareholders, is derived from EBIT, but it is also a function of the financial structure, how much goes to bondholders, and how much remains for the shareholder.

EBIT is the proper measure of earnings flow adequate to support a given debt structure. Interest is a deductible expense before income or profits taxes are computed. Net after-tax profit is an understatement of the firm's ability to pay its fixed charges. Thus, a firm with interest charges of, say, $200,000, reporting a zero income (no profits, no losses), will have covered its interest charges at least once. Assuming a 35% average tax rate, the same firm reporting a net after taxes income of $100,000 will have covered its fixed charges about 1.76 times.

The Nemo Corporation has the following financial structure:

Financial Structure

			Percent
Current Liabilities – Trade		$170,000	22
Bank Loan, 4 ½ %		50	
Long-term Debt (5% interest)		200,000	20
Preferred Stock (6% dividend)		30,000	3
Common Stock	$200,000		
Surplus	350,000	550,000	55
Total Capital & Liabilities		$1,000,000	100

and the following income statement:

Income Statement

Revenues	$1,500,000
Operating Expenses	1,300,000
	200,000
Depreciation	50,000
EBIT	150,000
Interest Expense	12,250
Earnings Before Taxes	137,750
Taxes (assume 50% approximate rate)	68,875
Earnings After Taxes	68,875
Preferred Stock Dividend	1,800
Earnings Available fro Common	67,075

($20,000 of fixed debt must be retired for sinking fund every year.)

Using the Nemo Company's figures, we can present examples of various measures of the times charges covered under various definitions and approaches. All of the methods are valid; they vary, but they show different useful relationships. The first example is the coverage ratio most commonly used, and the one which is usually indicated in the term "times interest or times fixed charges earned."

1. Times fixed charges earned (usual basis):

$$\frac{\text{EBIT}}{\text{Interest Charges}} = \frac{150,000}{12,250} = 12.3$$

All interest charges are divided into the EBIT. Preferred stock dividends are not included in the fixed charges, since they are not a legal liability of the company and do not have to be met in any given year. Although the sinking fund contribution is an annual obligation, it is not considered an expense or interest charge but a repayment of debt and thus is not normally included in the fixed charges. Nor is it deducted before computing normal accounting profits or income for tax purposes. In railroad financial analysis, contractual rental payments on leased lines are often included in the fixed charges. In analyzing other types of firms, rental obligations are generally not included in the fixed charges. A firm that rents a large part of its fixed assets may appear much more conservatively financed than a similar firm that has title to its fixed assets and has financed them by borrowing on mortgage bonds. Thus many financial analysts suggest that, at least for internal analysis, some proportion, perhaps a third, of annual rental payments be added to the fixed charges.

2. Times prior charges earned (includes preferred dividends):

$$\frac{\text{EBIT}}{\text{Interest} + \text{Preferred Dividends} + \text{Taxes on Preferred Dividends}}$$

$$= \frac{150,000}{12,250 + 1,800 + 1,800} = \frac{150,000}{15,850} = 9.5$$

This measure (times prior charges earned) includes the coverage on the preferred stock. Although it is a comprehensive measure, it is not commonly used. Since the preferred stock dividends are not deductible before income taxes are calculated, the taxes applicable to these dividends must be added back before the coverage is computed. In the example presented, $15,850 is the minimum the firm must earn, before interest and taxes, in order to cover interest, taxes, and preferred stock dividends. Nevertheless, the preferred dividends are not fixed charges since their payment is at the discretion of the directors, and a decision not to pay is not considered a failure to meet contractual obligations.

3. Times interest earned on a cash flow basis:

$$\frac{\text{EBIT} + \text{Depreciation}}{\text{Interest Charges}} = \frac{200,000}{12,250} = 16.3$$

The measurement of times interest earned on a cash flow basis is a gauge of the short-term ability of the firm to meet its interest charges. The annual depreciation allowance is not an actual expenditure in any given year, but a bookkeeping charge to approximate the firm's average annual capital consumption. Funds from both earnings and depreciation—as long as they haven't been shifted into nonliquid assets—are available to pay charges on the debt, at least on a short-term basis.

4. Times debt service charges earned on a cash flow basis:

$$\frac{\text{EBIT} + \text{Depreciation}}{\text{Interest Charges} + \text{Debt Retirement}} = \frac{200,000}{12,250 + 20,000} = \frac{200,000}{32,250} = 6.2$$

This ratio points out the significant relationship between the cash flow and all the service charges on the debt. Service charges here include interest and the annual obligation to repay part of the principal.[11] By relating debt service charges to the cash flow, we can achieve some estimate of the company's short-run ability to meet its debt obligations. In our example, no tax component is added to the sinking fund charge, although repayment of debt is not deductible before computing profit taxes. However, the funds available from depreciation, which are also nontaxable, more than cover the sinking fund requirement. If the sinking fund requirement exceeds the annual depreciation, it would be necessary to add income taxes to the excess sinking fund requirements before computing the coverage. If the ratio had been calculated as "times debt service charges covered on an earnings basis," the taxes applying to the sinking fund installment would have to be taken into account. Thus if a firm has to meet a $300,000 sinking fund installment and the income tax rate is approximately 50%, then the firm has to earn $600,000 before taxes if we are to consider the sinking fund as being paid out of earnings. However, relating the sinking fund payment and the interest charges to the estimated cash flow is the more relevant procedure.

It is apparent that various debt and prior charge coverages can be calculated in many ways. The stability of the coverage is generally more important than the absolute level of the coverage. Thus the ratios are often calculated for 5 or 10 years back; the average coverage for the period is determined and compared to the coverage of a few of the poorest years. A firm showing minimal fluctuations (i.e., one that is probably cyclically resistant) can be allowed a lower average coverage than one whose EBIT shows strong fluctuations and whose coverage of fixed charges in the poorer years shows a wide deviation from its average coverage.

---

[11] This is, of course, the concept generally used in local public finance. Since governments do not usually measure their success in terms of maximum money returns and minimum money costs, the commercial accounting that distinguishes between expenses and repayment of principal is not too important. A prime tool of municipal financial analysis is the level of debt service charges related to regular yearly revenues; this relation is some indication of the issuing area's economic ability to carry its debt.

10.9   Outside Financing Capacity

Until now we have discussed the theoretical foundations of optimal capital structure and seminal role of M&M theorems, these discussions are predicated on the notion that a firm that is willing to pay a sufficiently high-interest rate has theoretically unbounded borrowing capacity. In this section, we introduce the notion of credit rationing that questions the validity of this central assumption and discusses some practical ramifications of its failure. We refer the reader to Tirole [1] for more advanced analytical discussion on this topic and limit ourselves to key insights.

Bester and Hellwig (1987) argued that a would-be borrower is said to be *rationed* if he cannot obtain the loan that he wants even though he is willing to pay the interest that lenders are asking, perhaps even a higher interest. In practice such credit rationing seems to be commonplace. Some borrowers are constrained by fixed lines of credit which they must not exceed under any circumstances; others are refused loans altogether. As far as one can tell, these rationing phenomena are more than temporary dislocations in an otherwise efficient lending markets. Indeed, they seem to inhere in the very nature of the loan market.

Economists have studied credit rationing for several years and broadly attribute its existence to two key factors, namely, moral hazard and information asymmetry. Both of these factors pivot on a simple observation: an interest rate increase has no impact on the borrower in the event of bankruptcy as long as the borrower is protected by limited liability. Moral hazard explanation is that the reduced stake may demotivate the borrower and prompt him to engage in sub-optimal capital allocation by favoring projects with higher personal benefits as compared to projects with higher net present value (NPV) for the firm and, in extreme cases, engage in outright fraud. Adverse selection explanation is that, in a situation where the lenders cannot directly tell good and bad borrowers apart, higher interest rates tend to attract low-quality borrowers for low-quality borrowers are less affected by a rise in the interest rate than high-quality borrowers. Lenders may want to keep interest rates low in order to face a better sample of borrowers.

Credit rationing affects different firms differently depending on the strength of their balance sheet, nature of the underlying business, uncertainty of the future cash flows, and the ability of the lender to discern if the borrower used the funds for the overall benefit of the firm or for personal benefit. Next, we provide a simple example to illustrate circumstances that lead to credit rationing and effect of company fundamentals in alleviating or aggravating the resulting problems.

Consider an entrepreneur who has a project that requires fixed investment I with the following characteristics. If undertaken, the project either succeeds, that is, yields income $R > 0$, or fails and yields zero income. The probability of success is denoted by p. Note that p depends on both exogenous (macro factors, market conditions, competition, etc.) and endogenous (entrepreneur motivation) factors.

The entrepreneur initially has "assets" $A < I$. In order to undertake the said project, the entrepreneur needs to borrow I-A from lenders. Notably, once the entrepreneur has secured loan for his project, he can "behave" or "misbehave." Behaving in this context corresponds to efficient deployment of capital so as to maximize the chances of success for the project; we denote by *pH* the probability of success for the project when the entrepreneur "behaves." Misbehaving, on the other hand, corresponds to a scenario wherein the entrepreneur takes personal benefit B from the raised capital and operates in a manner that reduces the probability of success for the project; we denote by $pL(<pH)$ the probability of success for the project when the entrepreneur "misbehaves."

Among other things, the loan agreement also determines how the lender and entrepreneur will share the income from the project. Let RL denote the share of the lenders in the project profits, and let RB = R − RL denote the share of the entrepreneur. Note that expected profit for the lender in the "behave" scenario is given by pH*RL − (I − A). Assuming that the lending supply market is perfectly competitive, we can assume without loss of generality that the lender profits is equal to zero, that is, pH*RL = (I − A).

From the lender's perspective, it is imperative that the entrepreneur "behaves" and allocates the loaned capital in an efficient manner; in other words, the lender will enter into an agreement only if it can establish without doubt that the entrepreneur has sufficient monetary incentives to "behave." The entrepreneur's perspective is a little more complex. The entrepreneur has different incomes under the "behave" and "misbehave" scenarios as discussed below.

- "behave": entrepreneur income = pH * RB
- "misbehave": entrepreneur income = pL * RB + B

Consequently, a necessary and sufficient condition for the entrepreneur to behave is,

$$pH * RB >= pL * RB + B \, [\text{Viability Condition}]$$

Note that lenders will be unwilling to lend capital to the entrepreneur even for projects with positive NPV if the above condition is violated, resulting in credit rationing. We illustrate this statement through a numerical example.

Consider a hypothetical scenario when I = 100, A = 50, B = 20, pH = 70%, pL = 50% and R = 150. In this case, the entrepreneur needs to raise I − A = 50 units of capital from the lender for the project. The project produces income of R = 150 units when successful and zero units in case of failure. Furthermore, the probability of success for the project under the scenario that the entrepreneur behaves (misbehaves) is 70% (50%). Note that the project has a positive NPV under the "behave" scenario computed as pH*R − I = 5 > 0.

As discussed earlier, we can compute the lender's share of the income by solving the equation, pH*RL = (I − A) yielding RL = 71.43 and RB = R − RL = 78.57. Next, we evaluate the income of the entrepreneur under the two scenarios,

- "behave": entrepreneur income = pH * RB = 55
- "misbehave": entrepreneur income = pL * RB + B = 95.

Given that the entrepreneur has a higher income under the "misbehave" scenario, no lender will be willing to fund this positive NPV project resulting in credit rationing. Note that credit rationing in this particular instance results from a credibility gap. If the entrepreneur can convince the lender that he will "behave" once the funds have been received, the lender will be willing to fund the project. However, prima facie the high private benefits (B = 20) and high level of investment (I − A = 50) makes it difficult for the entrepreneur to formulate a credible story.

Lets revisit the viability condition which can be restated as follows:

$$pH * RB >= pL * RB + B$$
$$(D.p) * RB >= B\,(\text{where } D.p = pH - pL)$$
$$RB >= B/D.p$$
$$R - RL >= B/D.p\,(\text{since } RB + RL = R)$$
$$R - (I - A)/pH >= B/D.p$$
$$A >= I - pH * (R - B/D.p).$$

The final inequality has the following interpretation. An entrepreneur can obtain loan for a project only if the initial assets he has access to, namely, A, is greater than or equal to the threshold amount A(min) = I − pH* (R − B/D.p). Indeed, in the numerical example presented above A(min) = 65 < A = 50.

Factors that increase A(min)-A entrench the credit rationing phenomenon making it difficult for the entrepreneur to raise capital. On the other hand, factors that reduce A(min)-A mitigate credit rationing. We examine a few such factors and study their marginal effect on the credit rationing phenomenon.

Increasing the initial amount of fixed assets, namely, A, reduces A(min)-A and can go a long way in mitigating credit rationing problems. Stated differently, ceteris paribus firms with strong balance sheets that have adequate amount of liquid assets (cash and cash equivalents) are less likely to be exposed to credit rationing problems.

Reducing the private benefit, namely, B, that the entrepreneur can potentially derive by misbehaving reduces A(min)-A and can help avoid credit rationing problems. This can be arranged by carefully drafted loan agreement that earmarks the funds for project milestones and contractually disallow misuse of funds for entrepreneur personal use. Alternatively, the lender can provide fixed assets that have utility only in the context of the project under consideration thereby further reducing the private benefit that can accrue to the entrepreneur in the possibility he decides to misbehave.

Assuming that the differential, D.p = pH − pL, remains constant, increasing the probability of success pH reduces A(min)-A and can be helpful in avoid credit rationing problems. In other words, if the entrepreneur can demonstrate factors that have increased the likelihood of success for the project that can reduce the barriers to procuring credit. For example, successful completion of a critical R&D project or abnormal success of an initial limited launch of a product can provide evidence for higher probability of success for the project. Similarly, increasing values of R reduces A(min)-A and can be assist in alleviating credit rationing problems. In other words, reduced uncertainty (higher pH) and/or higher magnitude (high R) of future cash flows associated with a project reduce the likelihood of credit rationing.

References

Bester, H., & Hellwig, M. (1987). Moral hazard and equilibrium credit ratioing: An overview of the issues. In G. Bamberg & K. Spremenn (Eds.), *Agency theory: Information and incentives.* Springer.

Flannery, M. J., & Rangan, K. P. (2006). Partial adjustment toward target capital structures. *Journal of Financial Economics, 79,* 469–506.

Graham, B., & Dodd, D. (1951). *Security analysis.* McGraw-Hill. Ch. 37.

Guerard, J., Jr., Markowitz, H., & Xu, G. (2014). The role of effective corporate decisions in the creation of efficient portfolios. *IBM Journal of Research and Development, 58,* 6:1–6:11.

Harris, M., & Raviv, A. (1991). The theory of the optimal capital structure. *Journal of Finance, 48,* 297–356.

Hunt, P., Williams, C. M., & Donaldson, G. (1961). *Basic business finance* (Rev. ed.). Edited by R. D. Irwin, Chs. 15, 16.

Leary, M. T., & Roberts, M. R. (2005). Do firms rebalance their capital structures? *Journal of Finance, 60,* 2575–2619.

Lerner, E. M., & Carleton, W. T. (1966). *A theory of financial analysis.* Harcourt, Brace & World, Inc.

Miller, M. H. (1977). Debt and taxes. *Journal of Finance, 32,* 261–276.

Modigliani, F., & Miller, M. (1958). The cost of capital, corporate Finance, and the theory of investment. *The American Economic Review, 48,* 261–297.

Modigliani, F., & Miller, M. (1963). Corporate income taxes and the cost of capital: A correction. *American Economic Review, 53,* 433–443.

Modigliani, F., & Miller, M. H. (1966). Some estimates of the cost of capital to the electric utility industry. *American Economic Review, 56,* 333–339.

Myers, S. C., & Read, J.A., Jr. (2018). *Real options, taxes, and financial leverage.* MIT Working Paper.

Santos, J. C. (1997). Debt and equity as optimal contracts. *Journal of Corporate Finance, 3,* 355–366.

Schwartz, E. (1959, March). Theory of the capital structure of the firm. *Journal of Finance.*

Schwartz, E. (1962). *Corporate finance.* St. Martin's Press. Chapter 10.

Solomon, E. (1959). *The management of corporate capital.* The Free Press.

Solomon, E. (1963a). *Leverage and the cost of capital. Journal of Finance.*

Solomon, E. (1963b). *The theory of financial management.* Columbia University Press.

Strebulaev, I. A. (2007). What do tests of capital structure theory show? *Journal of Finance, 62,* 1747–1787.

Tirole, J. The Theory of Corporate Finance (Princeton:Princeton University Press)

Titman, S., & Wessels, R. (1988). The determinants of capital structure choice. *Journal of Finance, 43,* 1–19.

Van Horne, J. C. (2002). *Financial management and policy* (12th ed.). Prentice-Hall.

Weston, J. F. (1954, May). Norms for debt levels. *Journal of Finance.*

Chapter 11
Investing in Assets: Theory of Investment Decision-Making

Capital budgeting, or investment decision, depends heavily on forecasts of the cash and a correct calculation of the firm's cost of capital.[1] Given the cost of capital, that is, the appropriate discount rate and a reasonable forecast of the inflows, the determination of a worthwhile capital investment is straightforward. An investment is desirable when the present value of the estimated net inflow of benefits (or net cash inflow for pure financial investments) over time, discounted at the cost of capital, exceeds or equals the initial outlay on the project. If the project's present value of expected cash flow meets these criteria, it is potentially "profitable" or economically desirable; its yield equals or exceeds the appropriate discount rate. On a formal level, it does not appear too difficult to carry out the theoretical criteria. The stream of the forecasted net future cash flows must be quantified; each year's return must be discounted to obtain its present value. The sum of the present values is compared to the total investment outlay on the project; if the sum of the present values exceeds this outlay, the project should be accepted.[2]

The formula for obtaining the net present value of a project runs in this form:

$$PV = \frac{CF_1}{(1+i)} + \frac{CF_2}{(1+i)^2} + \frac{CF_n}{(1+i)^n} + \frac{S_n}{(1+i)^n} \tag{11.1}$$

$$\text{NPV} = \text{PV} - \text{I}.$$

PV is the present value of the net cash flow stream (CF_1, CF_2, etc.) over time to n years, S_n is the scrap value or the remaining value of the project at the end of its economic life at year n, and i is the applicable discount rate or cost of capital. NPV

---

[1] One need consider the case of Joel Dean's degree of necessity, in which projects must be undertaken regardless of their economic benefit (profit). Dean (1954) suggested that replacing a shop destroyed by fire could be an example of the degree of necessity, and he further suggested that you could not quantify the degree of necessity.

[2] The discounted cash flow approach has been widely accepted since Dean (1954).

J. B. Guerard Jr. et al., *Quantitative Corporate Finance*,
https://doi.org/10.1007/978-3-030-87269-4_11

equals the net present value, the present value of the benefit stream minus I, the full investment cost of the project.

If there is a cost for removing the project at the end of its economic life, then S (the scrap value) is negative. If the stream of returns is constant, their present value can be obtained by the summarization annuity formula: $PV = CF [1- (1 + i)^{-n}]/i$.

The table below provides an example of the mechanics of the capital evaluation problem. The project illustrated would be accepted because the present value of the estimated stream of net returns is $4,431,470.57, which is $431,470.57 above the project's initial cost of $4,000,000.

Net Present Value of Capital Project

YEARS	INVESTMENT COST OF PROJECT	ESTIMATED NET ANNUAL INFLOWS	DISCOUNT FACTOR (COST OF CAPITAL =12%) $\dfrac{1}{(1.12)^n}$	PRESENT VALUE OF INFLOWS
0	$4,000,000	0	0	0
1		$1,000,000	.8929	$892,857.14
2		1,500,000	.7972	1,195,800.00
3		2,000,000	.7118	1,423,600.00
4		1,500,000	.6355	635,500.00
5		1,000,000	.5674	283,713.43
6*		500,000	.5674	283,713.43
Total	$4,000,000			$4,431,470.57

*Period 6 includes a return of $250,000 and $250,000 scrap value.

Thus, the projected rate of return on the project is higher than the 12.0% discount rate, the estimated cost of capital.

11.1   Net Present Value and the Internal Rate of Return

There are two standard criteria for selecting desirable projects, but sometimes they may yield conflicting rankings.

1. Net present value. Net present value is obtained by subtracting the initial outlays from the gross present value of the benefits discounted at the given cost of capital. A project is acceptable if the NPV is positive.
2. The internal rate of return. The internal rate of return is the rate that brings the present value of the cash flows into equality with the initial outlay. The equation for the internal rate of return is formally similar to that for present value.

$$I = \frac{CF}{(1+r)} + \frac{CF_1}{(1+r)^1} + \ \cdots \ + \frac{CF_n}{(1+r)^n} \qquad (11.2)$$

The initial investment, I, is a given factor, and one solves for r, the internal rate of return, that is, the rate of discount, that brings the present value of the benefits equal to the outlay, I. If the internal rate of return (IRR) exceeds the cost of capital, the project is economically feasible.

In the vast majority of capital investment projects, both criteria give the proper signal as to whether a single project is acceptable. If a project's net present value is positive, it necessarily follows that its rate of return also exceeds the company's cost of capital. The project's positive net present value increases the firm's net income and cash flow and increases its stock price, ceteris paribus. However a selection conflict may arise when in comparing mutually exclusive projects, one project has a higher internal rate of return and one shows a higher net present value. If two projects are mutually exclusive, then one project can be accepted.

A sound capital budgeting takes into account increments to net working capital, additions to current assets in excess of current liabilities and annual incremental capital expenditures. Let us assume that we can introduce a new machine that costs $30 MM, which will produce a new product, producing sales in its initial year of production of $17.5 MM, and a 5% annual sales growth in sales during the 8-year life of the asset. These are several costs associated with the machine. Fixed costs, costs incurred regardless of production, will total $1.0 MM, and variable costs, such as materials, are estimated to be 65% of sales. Depreciation will be charged as allowed during a seven-year life by the accelerated cost recovery system, ACRS, reported in Chap. 3. Earnings before interest and taxes, EBIT, are found by sales-variable costs-fixed costs-depreciation. Taxes are paid at the incremental tax rate of 35%. The change in net working capital, dNWK, are $85,000 in year 1 and 10% of the change in sales, thereafter, and annual capital expenditures, CE, are $950,000 in year 1% and 15% of the change in sales, thereafter. The project operating cash flow is the machine's net income plus depreciation, and the machine's cash flow is its operating cash flow less its change in net working capital and its capital expenditures. The machine (project) discounted cash flow is discounted at a 10% cost of capital. The machine's present value, NPV, is $27.58 MM, which is less than its $30 MM cost (Table 11.1).

The machine's net present value, its PV less the cost, is −2.42 MM. The new machine should not be purchased because its NPV is negative.

11.2 Mutually Exclusive Projects

A problem of selecting projects apparently exists when more projects pass the economic test than a fixed capital budget will allow and/or when projects exist that are mutually exclusive alternatives (i.e., only one is to be taken from among several possible projects because they serve the same function). The NPV and the IRR can give conflicting signals in these cases where a selection must be made out of mutually exclusive projects.

Table 11.1 Finding the net present value of an investment

	Year							
	1	2	3	4	5	6	7	8
Sales	17,500,000	18,375,000	19,293,750	20,258,438	21,271,359	22,334,927	23,451,674	24,624,257
Variable Costs	11,375,000	11,943,750	12,540,938	13,167,984	13,826,384	14,517,703	15,243,588	16,005,767
Fixed Costs	1,000,000	1,000,000	1,000,000	1,000,000	1,000,000	1,000,000	1,000,000	1,000,000
Depreciation	4,287,000	7,347,000	5,247,000	3,747,000	2,679,000	2,679,000	2,679,000	1,335,000
EBIT	838,000	(1,915,750)	505,813	2,343,453	3,765,976	4,138,225	4,529,086	6,283,490
Taxes	293,300	(670,513)	177,034	820,209	1,318,092	1,448,379	1,585,180	2,199,222
Net Income	544,700	(1,245,238)	328,778	1,523,245	2,447,884	2,689,846	2,943,906	4,084,269
dNWK	85,000	87,500	91,875	96,469	101,292	106,357	111,675	117,258
CE	95,000	131,250	137,813	144,703	151,938	159,535	167,512	175,888
Project Op. CF	4,831,700	6,101,763	5,575,778	5,270,245	5,126,884	5,368,846	5,622,906	5,419,269
Project CF	4,651,700	5,883,013	5,346,091	5,029,073	4,873,654	5,102,954	5,343,719	5,126,123
Disct CF	4,228,860	4,855,839	4,016,518	3,434,857	3,026,052	2,880,618	2,742,397	2,391,336
PV(CF)	27,576,476							
Costs	30,000,000							
NPV (CF)	(2,423,524)							

As will be shown later, the constraints imposed by a prefixed capital budget is the result of a wrongful management decision. Nevertheless, in the general situation, supposedly a conflict in the rankings of mutually exclusive projects poses a problem. However, in every case, NPV indicates the optimum choice.

Projects may show up with conflicting rankings if the following three (not necessarily exclusive) conditions exist:

1. The shapes of the inflows over time differ.
2. The investment sizes of the projects differ.
3. The duration of the inflows differ.

The first condition, that is, the shape of the inflows, is illustrated in Table 11.2. Project A has a higher internal rate of return than B (11.3% as against 10.0%), but given the cost of funds rate of 6.0%, Project B's net present value is $113,500 as against $85,000 for A. When this type of conflict in ranking exists, the project with the higher internal rate of return has the higher earlier inflows: The rival project with the higher net present value (NPV) has a greater total inflow and relatively higher returns in later periods. If the decision between the projects is made on the basis of the rate of return, the earlier returns of A are acquired at the sacrifice of the greater returns from B later on, and this trade-off is made at a higher rate than the appropriate cost-of-funds warrants.

The effect of the differing shapes of the cash flows (the size of the investment constant) may be further explained with the aid of the graph shown in Fig. 11.1.

Project A has the advantage of a greater earlier cash inflow, but Project B promises higher later cash flow.

Table 11.2 Comparison of projects when the shapes of the cash flows differ

Project A

YEARS	OUTLAYS	INFLOWS	DISCOUNT RATE 6%	PV OF INFLOWS
1	$1,000,000	$700,000	.9434	$660,400
2	0	200,000	.8900	178,400
3	0	200,000	.8396	167,900
4	0	100,000	.7921	79,200
	$1,000,000	$1,200,000		$1,085,500

Project A has a net present value of $85,500. Its internal rate of return is calculated at 11.3%.

Project B

YEARS	OUTLAYS	INFLOWS	DISCOUNT RATE 6%	PV OF INFLOWS
1	$1,000,000	$100,000	.9434	$ 94,300
2	0	200,000	.8900	178,000
3	0	700,000	.8396	587,700
4	0	320,000	.7921	253,500
	$1,000,000	$1,320,000		$1,113,500

Project B has a net present value of $113,500. Its internal rate of return is calculated at 10.0%.
B offers $38,000 in additional net present value over Project A.

Fig. 11.1 Different shapes of cash flows of two investment options

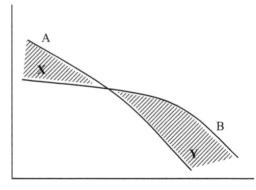

Cash Flows

Time

Indeed, project B offers higher total flow. The area X indicates the earlier differential inflow of A; the area Y depicts the later extra inflows of B. Absolute y is greater than absolute x. (If the absolute earlier returns were greater, there would be no problem; A would always win under either criteria.)

If the discount rate were zero, benefits would not be discounted, and B would always win. (B wins at lower rates.) On the other hand, at some high enough positive discount rate, A will win; however, if the going cost of capital is used as the discount rate and B still shows the higher present value, it follows that the sum of the present values of the funds depicted in y must be higher than the present value of x. Alternatively, if the funds depicted as X were to be invested to earn the cost of capital, they would not grow to equal the amount of Y.

In the case illustrated in Table 11.2, if the discount rate was 8.26%, the NPV of the two projects would come to the same amount—$47,661. However, if the discount rate goes past 8.26% (but is less than 11.7%), the present value of X will exceed Y. This involves no paradox; it simply means that when the cost of capital is high, earlier returns are preferred over later returns (because capital is relatively scarce and very productive), and the present value of Project A will exceed that of B.

Sometimes the argument is made that the comparison need not be made between projects A and B but that a whole investment strategy should be considered. Suppose the argument goes that X funds could be used in other projects, the returns for which show a higher present value than Y. Should not A be undertaken in preference to B? The answer is no. B is still preferable to A; X can be replaced by market supplied funds, which will be priced at the firm's cost of capital. The original argument is fallacious because it is based on a misuse of the concept of alternative or opportunity costs. In fact, the present value criterion is correct because it uses an objective measure in its calculations. The firm's cost of capital is a market-determined rate (modified by the firm's peculiarities), and it basically reflects the supply and demand for capital in the economy as a whole. Properly calculated, it is an opportunity cost; it is the rate the firm must pay to acquire additional new funds, and it is also the imputed value of the monies to stockholders should the firm elect to use the funds to repay some of its stockholders.

11.3   Difference in Project Size and Durations of Cash Flow

Condition 2, in which the sizes of the two projects differ, is illustrated in Table 11.3. Here, the smaller project, A, has a greater internal rate of return, but the larger project, B, has a higher net present value. This means that the size increment in Project B earns benefits at a rate higher than the cost of capital, even though there is an "averaging down" of the rate of return. A simple proof is to conceive of the bigger project in two parts, one of which (i.e., B-A) may be considered an increment to the smaller project. Since the incremental project has a positive NPV, it is a desirable or profitable addition. Here again, the criterion of net present value gives an unambiguous correct answer.

A simple intuitive example comparing rates of return to maximizing wealth is to consider two **exclusive** options, K and L. Option K requires a payment of $100 and will pay an assured return of $500 at the end of the year. Option L costs $1000 and returns.

$2000 at the end of year. The return on K is 500% and on L 100%. For any normal rate of interest and financing availability, L is far preferable. The extra return of $1500 for an additional outlay of $900 makes L a more valuable choice than K.

Table 11.3 Comparison of projects

Project	A	B
Cost	$1,000,000	$1,200,000
PV of Inflows at 6%	1,500,000	1,770,000
Internal Rate of Return	12.0%	11.5%
Benefits/Cost Ratio	1.5X	1.48X
Net Present Value	$500,000	$570,000

Consider the difference between A and B as an independent project, B-A

Project	B-A
Cost	$200,000
PV of Inflows at 6%	270,000
Internal Rate of Return	9.0%
Benefits/Cost Ratio	1.35
Net Present Value	$70,000

One rating criterion that has often been suggested is the benefit/cost ratio, that is, where the present value of the benefits divided by the cost of the project. However, as shown in Table 11.3, where the benefit/cost ratio favors Project A, this ratio is misleading when the projects differ in size. In fact, the optimal project size is reached when $dPV/dI = 1$, or when the discounted increase in inflows equals the marginal increase in the cost of the project. The proper criteria may not maximize the benefit/cost ratio; it does maximize net present value or total wealth.

The third condition, differing duration of the investment benefits, is a problem when the competing investment projects are renewable. If the projects are once-and-done activities, the one with the highest net present value is preferable. However, when the substitutable projects are renewable, not only may their internal rates of return differ but a net present value of, say, $100,000 for a stream of benefits lasting 5 years cannot be compared directly to the net present value of $120,000 for an alternate project of 7 years duration.

The problem involves a comparison of different strategies carried out over time, not merely a comparison of the return on the initial investments. The evaluation of strategies may not be very difficult if it can be assumed that each project could be renewed at the end of its life at similar costs and benefits as the present project. In this case, the net present values of each project can be annualized—reconverted into an equivalent flat annual amount over the life of each project. This is done by dividing the net present values by the present value annuity formula at the appropriate rate to yield the uniform annual equivalent return (UAER). The UAER amounts then can be compared directly. Table 11.4 illustrates such a problem. Here, although the NPV of Project B is larger, the equivalent annuity of Project A (at 10.0%) is $18,450 per annum; the equivalent annuity of B is $17,040 per annum. Thus, a series of A projects is preferred to one of B projects since it results in a higher stream of net benefits over time.

Table 11.4 Comparison of projects when durations differ (Using the method of annualization)

Project A

Year	Outlay	Returns	Disc.Factor (at 10.0%)	Present Value
1	80,000	60,000	.909	54,540
2		50,000	.826	41,300
3		40,000	.751	30,040
	80,000			125,880

NPV = $125,880 – 80,000 = $45,880
Equivalent annual stream of return for three years at 10% = $18,450.

Project B

Year	Outlay	Returns	Disc. Factor (at 10.0%)	Present Value
1	$170,000	80,000	.909	72,720
2		70,000	.826	57,820
3		60,000	.751	45,060
4		50,000	.683	34,150
5		40,000	.621	24,840
	$170,000			$234,590

NPV = $234,590 - $170,000 = $64,590
Equivalent annual stream of return for five years at 10% = $17,040

It might be noted at this point that, if the cost of capital (discount rate) should drop below 6.8%, Project B yields the greater annualized return and should be selected. This conforms to the basic rule that lower interest rates favor more capital intensive and longer-run projects.

In the case in which the future renewal costs and benefits of each project may not be exact replicas of the original projects, annualization will not work. The solution must be obtained by comparing the net present value of a series of linked, shorter-lived projects to the net present value of an alternate series of longer projects, both ending at a reasonably common time. Thus, if capital Project A lasts 4 years and rival Project B lasts 6 years, a comparison of the net present value of three A-type projects with forecasted costs and benefits renewed at the end of 4 years and 8 years, respectively, should be made with the net present value of two linked type B projects also lasting a total of 12 years. Forecasting costs and returns accurately is difficult, but the calculation itself is not so hard. Thus, if:

I = the project cost at each renewal period.
CF = benefits for each time period.
r = cost of capital (discount rate),
NPV = net present value.

Then:

$$1.\ \text{NPV}_A = -I_{A_1} + \frac{b_1}{(1+r)} + \cdots + \frac{b_4}{(1+r)^4} - \frac{I_{A4}}{(1+r)^4}$$

$$+ \frac{b_8}{(1+r)^8} + \cdots + \frac{b_8}{(1+r)^8} - \frac{I_{A8}}{(1+r)^8}$$

$$+ \frac{b_8}{(1+r)^8} + \cdots + \frac{b_{12}}{(1+r)^{12}} .$$

$$2.\ \text{NPV}_B = -I_{B_1} + \frac{CF_1}{(1+r)} + \cdots + \frac{CF_8}{(1+r)^8} - \frac{I_{B8}}{(1+r)^8}$$

$$+ \frac{CF_7}{(1+r)^7} + \cdots + \frac{CF_{12}}{(1+r)^{12}} .$$

The forecasted costs of renewing each project when the time comes, and estimates of the extended benefits must be made specific. Then NPV A may be compared with NPVB to see which is larger.

11.4   Lowest Annualized Total Costs

The annualization method comes into play in a variant problem where the benefits are a given and the question involves choosing the equipment or technique that shows the minimum cost. For example, suppose one piece of equipment has a lower initial cost but higher projected annual operating costs than an alternative which has a higher initial purchase price but lower projected ongoing operating and maintenance costs. The choice is made by taking the present value of the operating costs in each case and adding it to the purchase price. This total is then divided by the present value of the appropriate annuity factor to obtain the uniform annual equivalent cost (UAEC). The choice goes to the equipment showing the lowest UAEC.

11.5   The Irrational Fixed Capital Budget

As noted previously, in some circumstances, the decision-makers are forced to choose between projects because they are faced with a fixed capital investment budget. It should be noted before proceeding that an arbitrary fixed budget constraint is basically irrational. Presumably, an operating, viable corporation can acquire funds at an appropriate cost of capital, just as it can buy or hire any other productive factor. The proper capital budget is flexible and not fixed. It is set at the amount of profitable investment opportunities within the purview of the firm. Instead of equating the amount of funds to be made available to the opportunities for profitable real investment, a fixed capital budget arbitrarily rations the amount of capital investment. Such a budget can result in uneconomic behavior; it can force the substitution of less efficient projects or delay the implementation of worthwhile improvements. Any economically desirable project that cannot be undertaken because of a limited budget represents a loss of an opportunity to create net wealth represents a loss of wealth.[3] In short, the planned outlays for capital projects, the capital budget, should not be fixed a priori; the so-called budget should be constructed after all projects, the costs and returns for which have been properly counted, are evaluated and passed upon. Presumably, if all projects are evaluated at the ongoing and obtainable cost of funds, the ideal procedure is to enlarge the budget to accommodate all worthwhile noncompeting projects.

In operating practice, it may be well to show decision-makers both the results of the net present value criteria and the internal rate of return in order to demonstrate that the return on a feasible project having a positive net present value is above the cost of capital. Nevertheless, the optimum decision must go to the project showing the higher net present value. The internal rate of return is a ratio; a higher ratio is not the proper choice when it forecloses a project that has a higher absolute addition of wealth over cost. Essentially, the objective of capital choice is to maximize total wealth or value rather than the rate of return.

11.6 Real Investments and the Cost of Funds

The problem central to a firm's long-run financial policy is the selection of investment projects—that is, the choice of business projects based on the likely return compared to the cost of the funds needed to finance them. The theoretical concept for determining the desirability or profitability of a project does not entail any difficulties. In application, however, it may involve fairly complicated calculations.

The relation between the firm's cost of funds and the return from the earnings opportunities open to it is the major determinant of the success of the company. If the company can consistently apply funds to its operations at a rate of return above the cost of the funds used, it is considered a growth firm, and the owners of the common equity will experience financial happiness. In any case, as long as the management of the corporation does not regularly invest funds at a return lower than their costs—or consistently overlook profitable investment opportunities—the management must be deemed a success.

The application of investment theory requires the decision-makers to insert numerical estimates into the equations:

---

[3] In a case of capital rationing, the profitability index is calculated and used. The profitability index is the project's net present value divided by its initial cost [Brealey and Myers (2003).

$$PV = \frac{CF_1}{(1+i)} + \frac{CFn}{(1+i)^n} + \frac{Sn}{(1+i)^n} \qquad (11.3)$$

$$NPV = PV - I.$$

All the variables are subject to some degree of uncertainty in the calculation.

CFs the forecasted periodic estimated benefits. In the case of a pure business operation, the usual problem involves the calculation of the net after-tax cash flow of the project. Sometimes a project, for example, a recreation area for the employees, has no cash inflow but is estimated to yield benefits in terms of increased employee morale and employee improved retention rates. CF for each period is an estimate of some reasonable amount that the firm might pay for these beneficial improvements. If the project is a business operation, then CF is the periodic net after-tax cash flow the investment is forecasted to yield. This involves a forecast of the gross cash revenues minus the estimated out-of-pocket operating costs. These costs do not include economic depreciation which is implicit in the fact that n the life of the project is finite. The amount of depreciation allowable under the tax code is used to calculate the amount of cash flow subject to tax. The amount of tax, not the depreciation, is then subtracted to obtain the after-tax cash flow. The project should bear a share of managerial costs insofar as management costs increase with the volume of operations and presumably decrease with a cutback.

S is the forecasted scrap or withdrawal value of the project at the end of its economic life. S can include current or working capital no longer required when the operation ceases. S need not be a positive amount. It can be a negative amount if there are to be cleanup or demolition costs; on the other hand, there may be recoverable land value.

The estimated economic life of the investment in the project. The decision-maker needs to make a forecast as to when machinery or buildings will wear out and maintenance costs become excessive or when obsolesce could require the substitution of new technology. However, because of the effect of discounting mistakes in estimating n are inconsequential when the forecast for n is a long period.

The appropriate discount rate used to discount the inflows in order to obtain PV, the present value of the project, is depicted as i. For a project falling within the normal purview of the firm, the ideal discount rate is the weighted after-tax cost of the components making up the optimal financial structure for the firm. (An after-tax cost is employed because CF, the cash inflow, has been calculated on an after-tax basis.) Various adjustments are suggested where the level of risk differs from that of the usual operation.

I is the total investment outlay on the project. The investment in the project includes all the machinery, fixtures, and construction costs. It should include the imputed value of land costs. I should include working capital assets, cash needs and working inventory, that will be required by the project. If the project requires time to complete, a periodic charge at the cost of capital rate should be added to the investment cost of project. If projected initial operating returns are negative, show operating losses before the project gets "online," these start-up costs should be added to the total investment.

Table 11.5 Illustration of an investment decision, ABC, Inc. financial structure

	Normal Financial Structure Percent of Total Assets	Pre-tax Cost Current Market	After-tax Cost (Tax 35%)	Rate	Weighted After-tax Cost of K_0
Current Liability	30%	5.0%	3.25%		0.98%
L.T. Debt	30%	8.33%	5.42%		1.63%
Equity	40%	15.0% (desired pre-tax rate of return)	10.50%		4.20%
					Total 6.81%

The investment, I, is subtracted from PV, the present value of the inflows, to obtain NPV, the net present value of the project. If NPV is zero, the project is expected to generate a normal rate of return, one that will satisfy all the investors in the firm. If NPV is positive, the return is better than the norm; extra profits rebound to the shareholders. If investment, I exceeds PV, then the project does not cover the costs of capital (Table 11.5).

The project requires an outlay of $300,000 in fixed assets and $200,000 in gross working capital (i.e., current assets). That is, investment, I, equals $500,000.[4] $180,000 of the current assets will be available to the firm after the project ends in four years. The fixed assets will be worthless. The tax rate is 35%.

Year	Before Tax Cash Flow in Excess of Operating Costs	ACRS Depreciation Expenses	Effective Taxes (35%)	After Tax Cash Flow	Discount Factor (k=.0681)	P.V.
1	$120,000	$99,990	42,000	$177,990	.9363	$166,652
2	280,000	133,350	98,000	315,350	.8767	276,467
3	300,000	44,430	105,000	239,430	.8209	196,549
4	100,000	22,230	22,230	100,000	.7686	76,860
4a	180,000*			180,000	.7686	138,348
	Total PV					$854,876

*Return of working capital

NPV = $854,876 minus $500,000 or $354,876.

The project is worthwhile because its NPV exceeds zero.

Myers (1974) has suggested that the adjusted present-value model (APV) be considered relative to the weighted average cost of capital (WACC). In the APV approach, project cash flows are broken down into unleveraged, operating cash flow, and cash flow associated with financing the project. Different discount rates can be used to reflect the present value of the tax shield on debt used to finance the project. The reader is reminded that we discussed this tax shield on debt in Chap. 10.

---

[4]The ACRS depreciation charges for a three-year asset are 0.3363, 0.4445, 0.1481, and 0.0741, respectively.

11.7   The Adjusted Present Value (APV) Model

Modigliani and Miller (1958), in their seminal paper, introduced the paradigm of "law of one price" or "no arbitrage.[5]" No arbitrage concept is the foundation of modern finance as we know it. It is a very simple concept but powerful principle. It simply states that "two identical goods cannot sell at different prices." All the pricing and valuation models are based on this principle.

No arbitrage paradigm assumes that (a) markets are informationally and operationally efficient, (b) price is determined by market participants, and (c) market participants are rational. However, remember that time and place make the same good different! An identical product may sell at different prices at different places (distance) at the same time or at the same place at different times (time). Distance introduces transportation and transaction costs. In addition, distance and time introduces risk. No arbitrage paradigm requires that adjustment for distance, time, and risk to make the goods identical.

Here is a simple example to illustrate the no arbitrage concept in pricing and valuation—in finance we use value and price interchangeably.

You observe that in a market there are two goods, X and Y, traded. Good's payoff is determined by a random generator that generates a "G", (good) state, and "B" (bad) state of the economy. Both goods, X and Y, are trading at $1 each. You can participate at any level, that is, buy or sell any number of them. Payoff will scale accordingly. If the outcome is G, X pays $3, and Y pays $1. If the outcome is B, X pays $0, and Y pays $1. Payoff tree for someone buying, a long position, one of X or Y is given in Fig. 11.2. For the seller, a short position, signs will be reversed.

You may wonder the purpose of good Y? You pay $1 and receive $1 regardless of the outcome! In finance jargon it is the "risk-free" asset. It exists in all equilibrium pricing models (CAPM, CCAPM, and APM) developed so far. The reason is, there is one more unknown than number of equations in the setup. To be able to solve the problem, one of the variables must be set to a fix value. For example, if $a + b = 9$, there is no unique solution, that is, one cannot determine the value of a or b. The solution will be conditional. If $a = 0$, then, $b = 9$ etc. The fixed value in equilibrium models is known as the risk-free asset in finance. Hence, price of an asset is determined relative to the "risk-free" asset. In Black's version of the CAPM, the zero-beta asset is the shadow price of risk-free asset. That is, given the risk-return preference of participants in the market if there were a riskless asset, its price would be the same as zero-beta asset. In real world there is no risk-free asset. There are "default free" assets that are used as a proxy for risk-free asset.

Suppose another good, Z, is introduced to this market. Z's payoff is based on two random draws. It pays $9 if any of the draws is "G" and $0 if both draws are "B." Payoff diagram is shown in Fig. 11.3a. What will be the fair price of Z?

---

[5]F. Modigliani and M. H. Miller, "The Cost of Capital, Corporation Finance, and the Theory of Investment," American Economic Review, 48 (1958), pp. 261–297

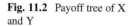
Fig. 11.2 Payoff tree of X
and Y

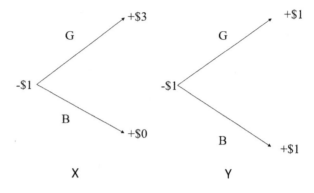

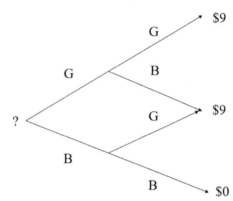
Fig. 11.3a Payoff tree of Z

Given the prices of X and Y, we can determine the fair price of Z in this market using the principle of law of one price! If the first draw is G, the payoff tree in Fig. 11.3a looks like nine times the payoff of Y. It is worth $9 to be there. If first draw is B, the tree looks like three times the payoff of X. It is worth $3 to be there. It is equivalent to tree having payoffs $9 and $3 (Fig, 11.3b).

Payoff of $9 and $3 can be broken into ($6 + $3) and ($0 + $3). The pair $6 and $0 is twice the payoff of X, and the pair $3 & $3 is three times the payoff of Y. Hence the fair price of Z is $2 + $3 = $5. If you observe that it is trading at $6, you can sell one of Z at $6, buy 2 of X for $2 and 3 of Z for $3. You will net $1 ($6–$2-$3) while completely hedging your obligation to pay $9 if the draw is G. In an efficient market prices will adjust and eliminate this arbitrage opportunity.

It is important to remember that price (i.e., value) is determined by the market participants and it is additive. By breaking the payoff of an asset into pieces and match with the payoff of assets traded in the market, we can price it. The payoff of an asset overtime is simply the cash flows it generates.

When cash flows are certain, that is, no risk, the fundamental valuation equation under discrete time assumption is:

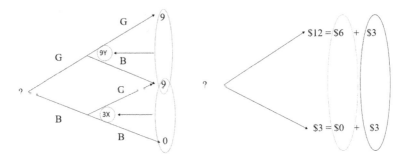

Fig. 11.3b Simplified payoff tree of Z

$$V_0 = \sum_{t=0}^{\infty} \frac{C_t}{(1+r_t)^t} = \sum_{t=0}^{T} \frac{C_t}{(1+r_t)^t} + \frac{V_T}{(1+r_T)^T} \qquad (11.4)$$

where C_t is the cash flow at time t, r_t is the discount rate, and V_T is the terminal value at time T. Remember that time $t = 0$ always represents "now." V_0 is the value at time $t = 0$. The only adjustment required is for the difference in time. Discount rate is the adjustment factor.

If cash flows are uncertain, the fundamental valuation equations is:

$$\tilde{V}_0 = \sum_{t=0}^{\infty} \frac{\tilde{C}_t}{(1+\tilde{r}_t)^t} = \sum_{t=0}^{T} \frac{\tilde{C}_t}{(1+\tilde{r}_t)^t} + \frac{\tilde{V}_T}{(1+\tilde{r}_T)^T} \qquad (11.5)$$

It requires estimation of cash flows and discount rates at each period. Except for some simple cases with finite paths and outcomes, it is impossible to do the calculations! One approach is to estimate the expected present value:

$$E\left[\tilde{V}_0\right] = E\left[\sum_{t=0}^{\infty} \frac{\tilde{C}_t}{(1+\tilde{r}_t)^t}\right] \neq \left[\sum_{t=0}^{\infty} \frac{E\left[\tilde{C}_t\right]}{(1+E[\tilde{r}]_t)^t}\right] \qquad (11.6)$$

However, mathematically it is not tractable and cannot be simplified to the form where expected value of cash flows and discount rates estimated separately. Instead, we use the following equation:

$$E[V_0] \cong \sum_{t=0}^{T} \frac{E\left[\tilde{C}_t\right]}{(1+r_{RADR})^t} \quad (C_T \text{ represents the terminal value } V_T) \qquad (11.7)$$

Notice that it is an approximation. Discount rate, r_{RADR}, is no longer time dependent. It is called "risk-adjusted discount rate." It captures the *risk of the cash flows* and *time value*. The problem is reduced to estimation of expected cash flows and corresponding risk-adjusted discount rates.

If outcomes and probabilities are *known* in each state of the world, expected value of cash flows can be estimated using the following equation:

$$E\left[\tilde{C}_t\right] = \sum_{s=1}^{S} p_{s,t} \times C_{s,t} \quad t = 0, \ldots, T \qquad (11.8)$$

Under uncertainty probabilities and outcomes are not known. They are estimated jointly. To simplify and make it more tractable, we make assumptions about distribution of outcomes. There is a rich literature about the decision-making under uncertainty, preferences, and distributional assumptions. It is shown that if participants' preferences are restricted to quadratic utility functions, the first two moments, mean and variance of the distribution are sufficient in decision-making. It is a very restrictive condition. Market participants' preferences vary. Rational market participant assumption imposes only two mathematical conditions for the choice of utility functions: First derivative must be positive, and the second derivative must be negative. Quadratic utility functions meet this requirement for a limited range of outcomes. However, if the distributions of outcomes are restricted to normal, or close to normal, mean-variance criteria hold, and market participants are not restricted to a single class of utility function.

Equation 11.7 is the fundamental valuation used in finance. It is important to remember the underlying assumptions baked into that formula: Cash flow distribution is restricted to normal, or close to normal distributions. And discount rate, r_{RADR}, captures the variance, that is, the risk, of the distribution as well as the time value of money.

Risk-adjusted discount rate is always have the following form:

$$r_{RADR} = r_{RF} + \text{Risk Premium} \qquad (11.9)$$

where r_{RF} is the risk-free rate and accounts for time value of money. Estimation of risk premium is another complex topic. Various models are used in estimation of the risk premium, including equilibrium asset pricing models like CAPM or APT, single, or multifactor models, and some ad hoc models. Decisions will be based on the evaluation of the distribution of values, and decision-makers preferred criteria.

11.7.1   Estimating Cash Flows

We utilize financial statements to estimate *expected* cash flows. It is crucial to understand and learn how to read financial statements. Accounting, "the language of business," is a standard set of rules for measuring firms' financial performance. Generally Accepted Accounting Principles (GAAP) varies across countries. There is convergence among various accounting standards, but there is still significant variance. In the US SEC authorizes the Financial Accounting Standards Board (FASB) to determine US GAAP accounting rules. All the European companies report under International Financial Reporting Standards (IFRS) since 2005. The International Accounting Standards Board (IASB) working with most of the countries around the world in adapting IFRS. The IASB and FASB have been working to converge US. GAAP and IFRS rules, but there are many differences. PricewaterhouseCoopers periodically publishes "Similarities and Differences —A comparison of IFRS and U.S. GAAP" reports. It is a must-read report if you are doing valuation and comparative analysis across different countries.

11.7.2 Gathering Information

In the United States, the SEC requires all public companies to file annual (10-K) and quarterly (10-Q) reports. Annual reports are filed within 60 days at the end of each fiscal year. Fiscal year varies across companies and may not be the same as calendar year. Quarterly reports are filed within 35 days at the end of each quarter (for the first three quarters) of the fiscal year.

We start with historical 10-K and 10-Qs. 10-Ks are more detailed, and it is audited by an independent authorized entity to make sure that company followed all the standards established by SEC and FASB. It contains company-specific and industry information, financial statements, (income statement, balance sheet, cash flow statements, and footnotes), details of stock options, debt schedules, management commentary, and forward-looking management discussion that may shed some insights.

In addition, companies are required to report previous two fiscal years' financial statements by applying the same accounting principles used in preparing the current fiscal year financial statements. It gives a good insight for the impact of accounting rule changes and, more importantly, facilitates correct financial and performance analysis. Quarterly reports shed light to seasonality of business and industry dynamics.

11.7.3 Financial Statement: Pitfalls

The three core financial statements are (1) income statement, (2) balance sheet, and (3) cash flow statement. These three statements are crucial in estimation of cash flows for valuation purposes. Timing of cash flows is extremely important in valuation. Financial statements may not reflect the actual timing of cash flows, and in some cases it may be misleading.[6] It is important to read the footnotes in financial statements carefully. Comparing previous years' financial statements posted at that

---

[6]"Tread Lightly Through These Accounting Minefields," by H. David Sharman and S. David Young Harvard Business Review on Point, (2002) highlights most common tricks companies use to adjust financial statement to fit their purpose.

time with the same statements reposted with the current accounting method the company is using is good starting point to understand the impact of switching methods. Most of common areas of abuse are:

- **Accrual Accounting**: How does the company measure and recognize revenues? If a car is shipped to a dealer, is it a sale?
- **Provisions for Future Costs**: It is usually uncertain and may be attributed many sources: Uncollectible debts, obsolete inventory, legal claims, restructuring and acquisition costs. . . .
- **Asset Valuation and Research and Development (R&D) Expense**: Assets are carried at book value, net of depreciation, and amortization. Companies can change depreciation methods or useful life of assets. US GAAP allows R&D costs expensed at will. Another source for the company to manipulate bottom line. . . .
- **Derivatives**: Derivatives are used for hedging purposes. They are very useful for risk management and help companies plan better for future. However, misuse or mistakes can be extremely costly. Details are not reported in financial statements to assess the future impact, and some may not be reported at all.
- **Related-Party Transactions**: These are the transactions made with other entities controlled by the company or management. It is rare case in the United States due to disclosure requirements, but in other countries with slack disclosure requirements, it could be extremely costly to shareholders!

11.7.4 Anatomy of Core Financial Statements

In Chapters 3 through 10, we covered the details of financial statements and line items covered in these statements as well as the performance analysis based on these statements. Companies may change the line items reported over time. There is no standard format to follow. In the United States, companies must electronically file all financial statements and required filings. These reports and filings are publicly available on EdgerOnline right after filing. Electronic filings use the tagging system called XBRL. They are like the HTML programing language used in website design. There are 5000+ unique codes corresponding to each accounting entry and explanation type. Data vendors like FactSet, Bloomberg, Value Line, and many more use electronic filings and standardize financial statements that they report. Each data vendor has their own standardization based on the type of company or industry for line items to report and aggregate the information accordingly. You should not be surprised if you see a different financial statement for a firm for the same year from various vendors. One advantage of standard financial statements is the ease of comparing financial statements analysis reports and creating composite financial statements for a group of firms or the whole industry.

Typical line items we need for financial modeling and use in estimating valuation cash are reported in following Tables 11.6, 11.7 and 11.8:

Table 11.6 Income statement

Net Revenues	Total payment for goods and services for the period.
Cost of Goods Sold (COGS)	Direct cost of manufacturing, procurement, and service company sells to generate revenue.
Gross Profit	Revenues – Cost of Goods Sold
Selling, General & Administrative Costs (SG&A)	Operating costs not directly related to COGS. Payroll, commissions, travel, advertising, marketing, office expenses, etc.
Research & Development (R&D)	Activities directed at development of new products or procedures.
Earnings Before Interest, Taxes, Depreciation & Amortization (**EBITDA**)	Gross Profit – SG&A – R&D
Depreciation & Amortization (**D&A**)	Allocation of costs over a fixed and internally developed software, customer data asset's useful life.
Other Operating Expenses/Income	Any operating expense not allocated in COGS, S&GA, R&D, and D&A or recognized in Net Revenue
Earnings Before Interest & Taxes (**EBIT**)	EBITDA – D&A – Other Operating Expense/Income
Interest Expense	Interest expense paid on the debt owed
Interest Income	Income from interest earned cash holdings, investments
Other Income/Expense	Gain/Loss on sale assets, disposal of business segment, write-offs, restructuring costs.
Income Tax Expense	Tax liability reported. (Not the actual taxes paid)
Net Income	EBIT – Interest Expense + Interest Income +/– Other Income/Expense

Note: In proforma financial statements it is a common practice to aggregate goodwill, intangible assets, and deferred taxes into other long-term assets. Similarly, deferred taxes and minority interest are aggregated into other long-term liabilities.

Balance sheet asset line item order is based on liquidity from most liquid to least. Liability line item order is based on when paid.

Unlike income statement and balance sheet, cash flow statement is hard to reconcile from one year to another in reported filings of a company. In addition, line items may change from one year to next.

In finance balance sheet is most important financial statement—it is the original financial statement when accounting invented! Income and cash flow statement derived from balance sheet. Financial economists view "finance balance sheet" and "accounting balance sheet" differently. Finance version is based on market values where accounting version is based on book values. Figure 11.4 reflects the finance perspective. Left side represents assets in place. It includes human capital of the firm. Right side shows how these assets are financed.

Table 11.7 Balance sheet

Assets	
Cash and Equivalents	Money held by the company and in banks accounts.
Marketable Securities	Debt or equity securities held by the company.
Accounts Receivable	Payments owed by customers for products and services already delivered.
Inventories	Any finished or unfinished goods waiting to be sold, and direct costs associated with the production of these goods.
Other Current Assets	Any other short-term assets not included in line items above – like prepaid expenses, short term investments.
Total Current Assets	Sum of the Line items above.
Property, Plant & Equipment ("Fixed Assets") (PP&E)	Land, buildings, and machinery used in manufacture of products and services.
Goodwill and Intangible Assets	Non-Physical assets; brands, patents, trademarks, and goodwill. Goodwill is difference between what the company paid and the book value from acquisition of a company or business unit.
Deferred Taxes	Future tax savings when taxes payable to tax authorities are higher than recorded taxes.
Other Long-Term Assets	Catch for all items that do not fit other categories.
Total Long-Term Assets	
Total Assets	Total Current Assets + Total Long-Term Assets
Liabilities	
Accounts Payable	Company's unpaid bills to suppliers.
Notes Payable	Debt or equity security held by the company.
Current Portion of Long-Term Debt	Portion of the long-term debt due within 12 months.
Total Current Liabilities	Sum of Line Items above.
Long-Term Debt	Portion of debt with maturity more than 12 months.
Deferred Taxes	Future tax obligation when taxes payable to tax authorities are lower than recorded taxes.
Minority Interest	
Other Long-Term Liabilities	Catch for all items that do not fit other categories.
Total Long-Term Liabilities	Long-Term Debt + Deferred Taxes + Minority Interest
Preferred Stock	Stock that has special rights and have priority over common stock.
Common Stock	Par value + capital received when shares sold above par value.
Treasury Stock	Common stock of the company bought back.
Retained Earnings	Total earnings of the company since inception minus dividends paid and losses,
Total Equity	Preferred Stock + Common Stock + Treasury Stock + Retained Earnings
Total Liabilities	Total Current Liabilities + Total Long-Term Assets + Total Equity

Table 11.8 Cash flow statement

Cash Flow from Operating Activities (CFfOA)
Net Income
Depreciation & Amortization
Change in Networking Capital (Increase)/decrease in Accounts Receivable
(Increase)/decrease in inventories
(Increase)/decrease in other operating expenses
Increase/(decrease) in accounts payable
Increase/(decrease) taxes payable

Cash Flow from Investing Activities (CFfIA)
Capital Expenditures
Proceeds from sales of equipment
Proceeds from sales of investments
Investment in subsidiary

Cash Flow from Financing Activities (CFfFA)
Proceeds/(payments) of long-term debt
Proceeds/(payments) of short-term debt; Line of Credit etc.
Proceeds/(purchase) of equity Proceeds/(purchase) of Common stock
Proceeds/(purchase) of Preferred stock
Dividends paid

Increase/(decrease) in Cash   CFfOA + CFfIA + CFfFA

11.7.5 Definition of Cash Flows

Unlevered free cash flows (UFCF) is defined as the cash flows generated by the assets of a firm that does not have any debt in its capital structure. Assets of the firm is always financed with equity. Components and the formulas are given in Table 11.10. Subscript t indicates the time, T is the tax rate, and Δ represents the change (Table 11.9).

Cash flows to invested capital (CFIC) is defined as the cash flows to equity and debt holders. It is derived from financial statements. Components and the formulas are given in Table 11.10.

Value of the firm, V, is simply the present value of unlevered free cash flows discounted at risk-adjusted discount rate that captures the risk of these cash flows. For firms with identical assets operating in the same industry, this discount rate will be the same. We will use r_A to represent the risk-adjusted discount rate that captures the *exogenous* business risk. For firms operating in the same industry, that is, business, this discount rate will be identical. It is independent of human capital, another asset of the firm. We will assume that human capital will impact the magnitude of cash flows.

We will use r_E to represent the risk-adjusted discount rate that captures the risk of free cash flows to equity. Firms in the same business with no debt in their capital structure r_E will be same the same as r_A. If there are different classes of equity with different rights to cash flows r_E will be different for each class.

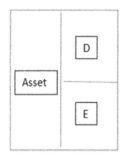

Firm's assets generate Cash Flows. Cash flows are claimed by:

1. Tax authorities
2. Debt holders (D)
3. Equity holders (E)

Value of the firm, V = D + E

D = PV (Expected Cash Flows to Debt Holders)

E = PV (Expected Cash Flows to Equity Holders)

Fig. 11.4 Finance view of balance sheet

Table 11.9 Unlevered free cash flows

Sales
  -   Cost of Goods Sold
  -   Depreciation
  - Other Fixed Costs	

= Earnings Before Interest and Taxes (Net Operating Income)	**EBIT$_t$**
- Taxes	- **EBIT$_t$ x T**
= Net Operating Income After Taxes	= **EBIT$_t$ x (1 -T)**
+ Depreciation	+ **Dep$_t$**

-    Investments into:
        Net Working Capital
        Capital Expenditures
	- **ΔNWC$_t$**
	  - **CapX$_t$**

= Unlevered Free Cash Flows (UFCF)	= **(EBIT$_t$ -I$_t$) x (1 – T) + Dep$_t$ – ΔNWC$_t$ – CapX$_t$**

We will use r_D to represent the risk-adjusted discount rate that captures the risk of cash flows to debt holders. Similarly, if there are debt with different rights and classes r_D be different for each class.

Value of a firm is then:

$$V = f(UFCF_t, r_A) = D + E$$
$$D = f(CFD_t, r_D)$$
$$E = f(FCFE_t, r_E)$$

Table 11.10   Cash flows to invested capital

Earnings Before Interest and Taxes	$EBIT_t$	
- Interest	-	I_t
= Earnings Before Taxes	EBT_t	
- Taxes	-	~~$EBT_t \times T$~~
= Net Income	$= NI_t$	
+ Depreciation	$+ Dep_t$	
- Investments into: Net Working Capital Capital Expenditures	- -	ΔNWC_t $CapX_t$
- Principal Payments of Debt	-	ΔD_t
= Free Cash Flows to Equity (FCFE)	$= NI_t + Dep_t - \Delta NWC_t - CapX_t - \Delta D_t$ $= (EBIT_t - I_t) \times (1 - T) + Dep_t - \Delta NWC_t - CapX_t - DD_t$ $= EBIT_t \times (1 - T) - I_t \times (1 - T) + Dep_t - \Delta NWC_t - CapX_t - \Delta D_t$ $= UFCF_t - [I_t \times (1 - T) + \Delta D_t\}$	
+ Cash Flows to Debt Holders (CFD)	$\Delta D_t + I_t$	
= Cash Flows to Invested Capital (CFIC)	$= EBIT_t \times (1 - T) + Dep_t - \Delta NWC_t - CapX_t + I_t \times T$ $= UFCF_t + I_t \times T$	

It is easier to estimate UFCF, r_A, CFD, and, r_D than FCFE and r_E. To estimate the value of equity, E, top-down (E = V-D) method is most common and widely used in the industry.

11.8   CFO Practice

In their survey of 392 Chief Financial Officers (CFOs), Graham and Harvey (2001) found that approximately 76% of the CFOs almost always used internal rate of return and net present value calculations in project or acquisition decisions. These decisions are well grounded in the financial decisions previously discussed. Graham and Harvey (2001) also found that approximately 57% of the CFOs surveyed used payback period and hurdle rate calculations. The payback period and hurdle rate calculations were the third and fourth methods, respectively, of analyzing projects. Adjusted present value and the profitability index, the latter being optimal for capital rationing, or a fixed capital budget, were the least used techniques. Let us discuss the payback period. The payback period is the ratio of the cost of capital investment divided by its annual earnings before depreciation [Gordon (1955)]. The payback period gives equal weight to the cash flow before the payback period, or "cutoff date," and ignores all cash flow after the cutoff date [Brealey and Myers (2003)].[7] The hurdle rate is a minimum acceptable rate of return established by management. It is generally higher than the firm's cost of capital [Lanzilotti (1958) and Lerner and Carleton (1966)].

---

[7] Gordon (1955) revealed that the payback period is proportional to the reciprocal of the project's rate of profit.

I = Cost of the project
E = Earnings or savings before depreciation of the project
n = Number of years of the project
s = Scrap value at time t = n
k = Project's expected rate of profit

$$C = \frac{E_1}{1+k} + \frac{E_2}{(1+k)^2} + \cdots + \frac{E_k}{a+k} + \frac{E_n + S_n}{(1+k)^n} \tag{11.9}$$

Gordon found if one ignored the scrap value, that:

$$C = \sum_{t=1}^{n} \frac{S}{(a+k)^t} = \frac{S}{k} - \frac{S}{k}\left(1 + \frac{1}{(a+k)}\right)^n \tag{11.10}$$

where (11.5) is the present value of an annuity

$$k = \frac{E}{C} - \frac{E}{C}\left(\frac{1}{1+k}\right)^n \tag{11.11}$$

As the life of the project becomes very large, equation (11.6) becomes

$$k = \frac{E}{C}$$

and the rate of profit is the reciprocal of the payback period. Furthermore, if the project life is infinite, then the rate of profit is the reciprocal of the payback period. The reader immediately recognizes the Gordon payback period analysis is consistent with the Gordon valuation of stock model shown in Chap. 8. If we buy a stock, we purchase a stream of expected future dividends. The stock price, P_{CS}, or P_0, is determined by the current dividends per share, D_0, and the growth rate of future dividends, g.

$$P_0 = \frac{D_0}{d - g} \tag{11.12}$$

where $k = \frac{D_0}{P_0 + g}$

The rate of profits earned by purchasing stock is its current dividend yield, $\frac{D_0}{P_0}$, plus the rate at which dividends are expected to grow, as long as k > g [Gordon and Shapiro (1956)].

11.9   Current Costs of the "Optimum" Financial Mix

For normal projects falling within the regular operations of the firm, the basic discount rate uses the cost of the financial mix which appears optimum for the type of firm. This involves some qualitative judgments as to the long-run reaction of the financial and investment market to financial risk. The analyst must estimate the amount of debt of all types the firm can safely and profitably carry, given the typical nature of its operations and of its asset structure. An important feature of this approach is its attempt to take into account the "full cost" of capital, including costs which may not be immediately explicit. If, for example, the firm already has a large equity base, it may be able to borrow at a relatively low rate to finance new operations. To assume that the cost of capital for the new projects is merely the external interest rate is to fail to account properly for the implicit costs of the "extra" equity capital already existing in the firm. Although a new project might be financed by debt to bring the firm's financial structure into balance, the firm must consider the full imputed costs of its capital mix. A given project is not burdened or credited with just the cost of its particular financial source, but with the costs of the overall "optimum" capital costs for the firm.

A rather subtle but perhaps useful concept in getting at the cost of capital for a particular project is to consider it, if it is large enough, as a financial entity. Potential projects differ not only in their forecasted rate of return but also in the certainties of the forecast and the possibilities of asset losses. Thus risky projects implicitly require heavier admixtures of relatively expensive equity capital (in a non-M&M world), whereas less risky ventures can be financed more heavily with cheaper debt. This approach furnishes one method of dealing with risk; the more risky projects are discounted at a higher rate before they are accepted.

If in actuality as the firm expands, it could adjust its financing to the type of projects it undertakes, and the capital structure will change over time. Such a policy slowly but continually moves the firm's financial structure into line with the type of enterprise, product, operations, and assets it currently maintains.

A difficulty may arise in calculating the desirability of additional investment in projects or areas already in operation. In some circumstances, outlays for the purpose of preventing a fall in earnings may be profitable. The difference in the forecasted flow of earnings with the project and without the project is the implied earnings of the investment. If, however, the mistake is made of setting the investment at just the cost of the additional assets, the rate of return will be grossly exaggerated. Properly, the amount of the investment is the sum of the new assets plus the liquidation or withdrawal value of the assets presently used in the operation. For example, a firm may contemplate additional outlays in a sector of its operations, not with the hope of increased returns but to prevent deterioration in its competitive position. The value of the losses forestalled may appear high compared to the cost of the additional assets. Yet the firm has another alternative, to withdraw entirely from this area of its activities. The salvage value of the existing investment plus the cost of the new assets must be compared to the present value of the flow of funds forecasted from the operation.[8] Only if the return is large enough on this basis should the operation be continued.[9]

On the other hand, the recoverable value of assets already committed to a given line of production is usually quite low; this furnishes one explanation why firms often continue to invest in activities whose return is stable or declining. (The return on the sum of the new repair and maintenance assets plus the low liquidation value of the old assets may be relatively good.) Thus economically it may be difficult for a firm to withdraw from a going operation.

11.10   The Effect of Taxes on the Financial Structure

A corporate tax system that allows the deduction of interest on debt but taxes the return to the stockholders encourages the use of debt financing or leverage. An illustration of this effect, under the assumption of a defined optimal financial mix, is shown in Table 11.11. In a non-tax world, the optimal financial mix would be at line 3, (60% equity, 40% debt) with the lowest weighted capital costs, 12.15% to be applied to *nontaxable* profits. Under a 34.0% tax rate on equity profits, the optimum financial mix shifts to line 5, (60% debt and 40% equity), showing a cost of capital 10.45% to be applied however to the after-tax cash flow. The example presented illustrates the existence of an optimal financial mix. Presumably, in a theoretical frictionless, M&M world, the introduction of taxes would lead to a corner solution, with the lowest amount of equity possible as the optimum.

---

[8] If a firm does not make the proper cost-of-investment determination, it could develop an unprogressive policy in regard to equipment replacement. The firm could repair its equipment long past an economic optimum because the fairly large total earnings on a repaired piece of equipment is credited just to the costs of repair. The economic investment in a reconstruction job, however, is the repair costs plus the salvage value of the unrepaired equipment. Furthermore, the forecasted cash flow should subtract the estimate of maintenance costs that may still have to be made in the future. When the problem is thus properly formulated, it may be that a replacement will show a higher rate of return than a repair job.

[9] In a sense, the existence of the whole firm may be periodically measured against the same criterion: how does the withdrawal value of the firm's assets compare to the discounted value of its projected cash flow.

Table 11.11 Effect of the corporate income tax on the optimum financial structure of the firm

Overall Financial Mix	Percent of Each	Pre-tax Cost (Investors' Desired Rate-of-Return or Current Cost)	Adjusted After-Tax Cost (Assume 34% Tax Rate)	Weighted Cost of Capital (Assuming to Corp. Tax)	After-Tax Weighted Cost of Capital (Used to Discount After-Tax Cash Flows)
1) Total Debt	20.0%	9.00%	5.94%	1.80%	1.19%
Equity	80.0%	13.25%	13.25%	10.60%	10.60%
Total Weighted Cost				12.40%	11.79%
2) Total Debt	30.0%	9.25%	6.11%	2.78%	1.83%
Equity	70.0%	13.50%	13.50%	9.45%	9.45%
Total Weighted Cost				12.23%	11.28%
3) Total Debt	40.0%	9.75%	6.44%	3.90%	2.57%
Equity	60.0%	13.75%	13.75%	8.25%	8.25%
Total Weighted Cost				12.15%*	10.82%
4) Total Debt	50.0%	10.75%	7.10%	5.38%	3.55%
Equity	50.0%	14.00%	14.00%	7.00%	7.00%
Total Weighted Cost				12.38%	10.55%
5) Total Debt	60.0%	12.00%	7.92%	7.20%	4.75%
Equity	40.0%	14.25%	14.25%	5.70%	5.70%
Total Weighted Cost				12.90%	10.45%**
6) Total Debt	70.0%	13.75%	9.08%	9.63%	6.35%
Equity	30.0%	16.78%	16.78%	5.03%	5.03%
Total Weighted Cost				14.66%	11.39%

*Least Costly financial mix without taxes
** Least costly financial mix after accounting for tax effect

11.11   Costing the Components of the Financial Mix

In evaluating a project, the decision-maker will employ an overall cost of capital. But arriving at this figure requires an understanding and a casting of the components going into the financial mix. The financial includes all the sources used to acquire and support all the assets employed by the firm.

11.11.1   Cost of Trade Credit

If the firm pays its trade creditors within the cash discount period, the cost of the credit is zero to the firm which obtains it.[10] There is a cost if the accounts are not paid

---

[10]The cost is not zero to the economy as a whole, but the individual firm is not concerned with aggregate economic costs at this point.

within the allotted time, that is, the loss of the discount. The supply of trade or book credit is not unlimited; the sellers ration it. In addition, the use of excessive book credit strains the debtor firm's liquidity position.

11.11.2   Cost of Bank Credit

Bank credit may be needed only seasonally to help finance peak operations. Its cost can be charged in at its estimated average annual use multiplied by the going interest rate on bank loans. Since the rate of return on the project has been calculated after taxes and interest is tax-deductible, the cost of loan capital to the firm can be lowered by the reduction in the tax bill. If the tax rate is approximately 35%, then a bank loan at 5% costs the firm only 3.25% net.

11.11.3   Cost of Long-Term Debt

The cost of various long-term loans (bonds, debentures, subordinated debt, etc.) is averaged into the overall cost at their relative weights in the financial mix multiplied by their current market interest rates. The current rate of interest is the one that must be considered. The company may have bonds outstanding bearing a 5%% coupon rate, but if similar new bonds cannot be floated today at less than 7%, the cost of debt capital is 7%. The final after-tax cost of long-term debt capital is reduced by the equivalent amount of tax saving.

11.11.4   Cost of Preferred Stock

Given an ongoing corporate profits tax, preferred stock is an expensive form of equity capital. It is generally used only because its subordinate position helps secure the safety of debt and thus lowers its cost. If preferred stock is employed as part of the capital mix, its cost to the firm is its current dividend yield. There is no tax saving on preferred since its dividends are not considered a cost, but a division of profits and come out after taxes are calculated.

11.11.4.1   Cost of Common Stock

The various uncertainties involved in the market evaluation of common shares make the costs of common equity funds the hardest to measure. A few guidelines help narrow the approximate range of these costs. The cost of common stock financing is based on its earnings yield and not its current dividend yield.

For normal stocks in a moderate market, the cost of equity capital may be approximated by the ratio $\frac{\text{earnings per share}}{\text{market price per share}}$, or the inverse of the traditional price/earnings ratio. The earnings used in the numerator must be a "normal" run of earnings. Neither the earnings per share of a poor year nor a boom year are appropriate.

The cost of equity funds is calculated on the market price of the shares. The amount earned on the book or asset value of the stock is of minor importance, since with passing of time the book value of the equity may diverge further and further from current economic realities and values.[11]

The earnings yield ratio is only a starting approximation of the cost of common stock financing. The current yield of a share on the market is influenced by the security markets' expectations and evaluation of the firm's future prospects. To get a usable cost of equity financing, the earnings yield has to be corrected for normal growth, potential recovery, forecasted decline, and super normal growth. In the case of normal sustainable growth, if the earnings yield on market is 6% (i.e., a P/E ratio of 16.6 times) and the sustainable growth rate of earnings is 4.0% per annum, the after-tax desired return is 10.0%. When the market expects a potential recovery from a depressed period, the P/E ratio may appear quite high, for example, 40 times or an earnings yield of $2\frac{1}{2}\%$. Here, the current yield may be ignored. The cost of equity financing is imputed from the market, that is, the normal average return for the general run of stocks in the broad industry class. Chap. 14 further develops and estimates the return on equity by formally introducing the reader to the required rate of return on equity. The required rate of return on equity and the weighted average cost of capital are used to value a potential merger candidate in Chap. 18.

When the market expects a decline in firms earnings, the current P/E ratio will be rather low, perhaps showing an earnings yield of 25%. Here again, after allowing for variant risk (i.e., the beta function), the cost of equity capital has to be imputed from that prevailing on the market.

In the case of super normal growth, the firm may possess technological or marketing advantages which enable it to obtain earnings growth well over the norm, for example, 15%–20% per annum. The P/E may be as high (or higher) than 50 times earnings or 2% earnings yield. Of course, such a growth rate cannot be sustained indefinitely; as someone has remarked, over time a firm with such a growth rate would soon constitute the whole economy. The 2% earnings yield is not the true cost of equity; the investors would soon be extremely dissatisfied if the firm invested in projects that returned a mere 2%. Once again the cost of equity has to be imputed from the long-run norm on the market.

A countercheck to the earnings yield is to consider the earnings yield relative to the long-run interest rate in a reasonably noninflationary market. A risk premium of perhaps 4%–6% might be added to the rate on top-grade long-term debt to obtain the cost of equity capital.

---

[11] The amount made on the asset value of the equity—especially its growth or trend over time—may, however, be a fairly good index of the firm's economic success.

11.11.4.2   Internal Funds

The net internal funds available to the firm for additional investment constitute the money inflow for the period in excess of the costs of operations less interest, taxes, preferred stock dividends, and required repayment of debt. Cash flow is a broad undifferentiated concept; nevertheless, for purposes of exposition, we should divide the internal funds to those derived from earnings and those derived from true economic depreciation.

11.11.5   The Cost of Retained Earnings

Historically the cost of using retained earnings to help finance a firm has been the subject of considerable debate. Some writings in corporation finance seemed to imply that retained earnings were costless; the use of retained earnings in lieu of other sources of funds was sometimes given almost "moral" approval. Internal financing supposedly avoided the use of debt, prevented the firm from having to deal with bankers, or from having to appeal for new stockholders on the capital market. A careful reading of these arguments shows them to be mostly homiletical; they gave no guides as to the cost of using retained earnings in the firm, and no aid in deciding how much of a given flow of earnings may be reasonably kept and how much should be paid out in dividends or used for stock buybacks.

A more scientific approach to the costing retained earnings is to impute to the retained funds their alternative or opportunity costs—that is, how much they could earn if they were paid out to the stockholders in dividends or in stock buybacks and reinvested in the best opportunities (of equal risk) available elsewhere. In effect, the firm should retain earnings only to the point where it can make a return on these funds equal to the normal market rate on equity funds. The cost of retained earnings should be imputed at the same rate as the normal return on shares of the same industry class. The general rule that retained earnings be given the imputed costs of equity is useful in solving the problem of capital cost, and it provides an answer to the problem of the amount of cash dividends or cash buybacks the firm should disperse and how much earnings it might retain. All cash earnings which cannot be employed at the proper rate of profit should be returned to the shareholders.

For the average firm in the economy, there is no shortage of internal funds. Overall the corporate sector generates a greater cash inflow rate then the growth of the economy. The aggregate inflow of cash exceeds the amount of economically feasible real investment, and thus the corporate sector is a net exporter of cash to the rest of the economy.[12]

---

[12]"The Corporate Sector: A Net Exporter of Funds," J.R. Aronson and E. Schwartz, *Southern Economic Journal*, October 1966.

11.11.6   Other Internal Funds-Depreciation, Depletion, etc.

As a sort of "shorthand" method of expression, depreciation, depletion, amortization of intangible assets, etc. are commonly listed as sources of funds. But deeper analysis makes it clear that an excess of *revenues* over out-of-pocket *operating* costs is only source of funds. Annual charges for depreciation and other such noncash charges are accounted as funds because reported accounting earnings are less than the total current funds flow. Depreciation, estimated amount of past capital dollar investment in fixed assets used up in the appropriate accounting period, is a charge against earnings but not against funds.

True economic depreciation based on the declining replacement value of the fixed assets represents a reemployment in the investment in capital. Funds arising from true economic depreciation carry the cost of the financial mix used to fund capital assets. In a theoretical sense, these funds might be returned to the original investors in the capital assets. The cost of keeping them in the business is what could be saved by not paying off an equivalent amount of debt and equity. Insofar as new projects are financed with funds arising from true depreciation, these must be given the imputed costs of the suitable financial mix. Under the current tax code, the allowable depreciation deduction exceeds the amount of true economic depreciation. Funds arising from this source must be considered under a different rubric than true economic depreciation. These funds constitute a sort of deferred equity fund granted to the company by the taxing authority. The proper charge for these funds, when retained in the business, is the imputed cost of equity funds.

11.12   Investments Under Negative Interest Rates and Hyperinflation

For the better part of the past century, most developed markets have been characterized by a positively sloped yield curve, positive interest rates, and "normal" rate of inflation (1%–10%). However, there have been exceptions when the yield curve has gotten inverted, the interest rates have turned negative or economies have suffered from hyperinflation (30%–100% or more). While the inverted yield curve does not create any technical issues with the material presented in this chapter, negative interest rates and hyperinflation do require special treatment as discussed below.

11.12.1   Negative Interest Rates

It is important to note that the discount rate computation starts with the risk-free rates and then applies a series of usually positive adjustments such as beta adjustment for covariance with market returns, country/industry premiums, currency adjustment,

Table 11.12 Present value of $1 for various combinations of discount rates (−5%, 0%, and 5%) and durations (1, 2, ... 10 years)

Year 1	-5%	0%	5%
1	1.05	1.00	0.95
2	1.11	1.00	0.91
3	1.17	1.00	0.86
4	1.23	1.00	0.82
5	1.29	1.00	0.78
6	1.36	1.00	0.75
7	1.43	1.00	0.71
8	1.51	1.00	0.68
9	1.59	1.00	0.64
10	1.67	1.00	0.61

and private equity or illiquidity discount. In the presence of these adjustments, the discount rate can turn positive even when the underlying risk-free rates are negative. However, in a few cases, the discount rate can still be negative, adjustments notwithstanding. We illustrate some of the pitfalls of using the NPV approach directly in such cases and offer guidance on how to proceed.

Table 11.12 reports the present value of $1 for various combinations of discount rates (−5%, 0%, and 5%) and durations (1, 2, ... 10 years).

As it is evident from the table, negative discount rates have an amplifying effect on the present value computation—present value of $1 for any duration is greater than $1. This creates a unique problem with respect to the computation of the terminal value; specifically, in the presence of negative discount rates, constant or positively growing stream of perpetual cash flows has infinite present value. In order to circumvent this problem, we need to understand and question the assumption of constant or positively growing perpetual cash flow stream in the presence of negative interest rates. Usually, negative interest rates arise in peculiar economic regimes that are also accompanied with deflationary tendencies, transient or short-term fiscal/ monetary imbalances or unusual geopolitical situations.

In such cases, an investor needs to determine if it is reasonable to assume negative interest rates in perpetuity and if so characterize the economic regime that will be concomitant with such a scenario. Among other things, perpetual negative interest rates will be accompanied with deflationary pressures and anemic growth that contradicts the assumption of constant or positive growing perpetual cash flow streams. Alternatively, one can conclude that negative interest rates are a short-term phenomenon, and they will revert back to their "normal" (positive) behavior after a certain period of time thereby resolving the issues with terminal value computation.

11.12.2 Hyperinflationary Regime

In economics, hyperinflation refers to an excessive amount of high and accelerating inflation that quickly erodes the real value of local currency resulting in rapid and continuing rise in nominal prices of goods/services. Countries such as Brazil, Germany, Greece, Peru, the Philippines, etc. have experienced hyperinflation in the past century.

Performing NPV computation in a hyperinflationary suffers from three key challenges. First, it becomes extremely difficult to forecast future cash flows given the inherent uncertainty around supply and demand of goods/services. Even in situations where such a demand can be forecasted within reasonable economic precision in real terms, computing the nominal value of these cash flows becomes problematic due to the unstable value of the local currency. Second, it is questionable whether the value of real assets (land, building, etc.) and present values of future cash flow streams are fungible in a hyperinflationary environment. This problem gets magnified while valuing firms and enterprises, one of the key practical application of NPV methodology. Third, in multinational companies that do business in different parts of the world, hyperinflation presents a particularly unique conundrum, namely, how to combine the NPV of cash flows in a stable currency with those derived from a currency suffering from hyperinflation.

There are two ways of circumventing the hyperinflation problems in the context of NPV methods. First, one can move to performing all of the computations in real terms. While this approach does not relieve the investor from forecasting the supply and demands of goods/services, it does annul the need to forecast unstable inflation rates into the future. The second approach is to perform computations in a stable foreign currency that offers the best de facto stable proxy for the local currency.

11.13 Summary

A crucial problem in the management of a firm's capital is the selection of assets or projects whose estimated return will justify the cost of the funds invested in them. Although the details of the calculations vary, essentially projects are worthwhile when the cash inflow discounted at the appropriate cost of capital shows a present value equal or in excess of the original required investment. If the data underlying the calculation is correct, the project is estimated to have a return equal to or better than the cost of capital. Because the net cash flow on the project is a forecast and the required investment is an estimate, the actual profitability is also only a forecast; the project in operation may realize more or less.[13] Nevertheless, such an initial forecast and calculation of expected value is an essential step in real asset investment decision-making.

If the management of a firm does its best in selecting assets and weighing their potential return against the cost of funds, then within the context of the prevailing economic conditions, the best possible return will be maintained on the stockholders' investment without subjecting the firm to inordinate risk of failure.

Appendix A: Application of APV

In this appendix, we will develop a simple economic model of a firm like the one Modigliani and Miller (M&M) (1961) used in their seminal paper that developed the relationship of discount rates and capital structure of the firm[14]. The basic assumptions are:

- No revenue growth, and operating efficiency of the firm is stable.
- EBIT is perpetual.

    - $\Delta NWC = 0$ (corollary of no growth assumption).
    - Dep = CapX (firm is maintaining operating efficiency. Depleted assets and Human Capital Replaced).

- If there is debt in capital structure, i.e., levered, D is constant.

UnleveredFree Cash Flow under these assumptions will be:

$$\begin{aligned} UFCF_t &= EBIT_t \times (1 - T) + Dep_t - \Delta NWC_t - CapX_t \\ &= EBIT \times (1 - T) \quad t = 0, \cdots, \infty \end{aligned} \tag{11.10}$$

Value of an unlevered firm, V_U, under these assumptions will be:

$$V_U = \frac{UFCF}{r_A} = \frac{EBIT(1 - T)}{r_A} = E_U \tag{11.11}$$

Let us assume an unlevered firm, U, with 100 shares traded in the market. Firm U's assets are generating \$100 in EBIT every period. Firm is facing a tax rate, T, of 40% and risk-adjusted discount rate for this type of business, r_A, is 20%.

---

[13] It should also be recognized that the accounting results and the estimated rate of return may not necessarily coincide because accounting conventions on depreciation and calculation of profits may differ from the assumptions underlying the cash-flow rate of return analysis. This need not invalidate the usefulness of either the discounted cash flow techniques or of accounting procedures.

[14] M. H. Miller and F. Modigliani, "Dividend Policy, Growth, and Valuation of Shares," Journal of Business, 34 (1961), pp. 311–433.

Under these assumptions, cash flows to tax authorities will be $40, and cash flows to equity holders will be $60 every period. Value of the firm, V_U, and value of the equity E_U, will be PV[$60,20%] = 60/20 = $300. Shares of this firm will be trading at $3 per share.

Let us introduce a levered firm, L, with the same assets in place, in the same business, and with 100 shares traded in the market. Firm L has $100 perpetual debt with 10% market interest rate. Risk-adjusted discount rate, r_D, is the same. Firm L will be generating $100 in EBIT every period as well. Under these assumptions, cash flows will be:

- UFCF = EBIT $\times$ (1 -T) = 100 x 0.60 = $60
- Taxes = (EBIT – I) x T = (EBIT – r_D x D) = (100 – 0.10 $\times$ 100) $\times$ 0.40 = $36
- CFD = $r_D \times$ D = 0.10 $\times$ 100 = $10
- FCFE = UFCF – [I $\times$ (1-T) + ΔD] = 60 – [10 $\times$ 0.60 + 0] = $54
- CFIC = UFCF + I $\times$ T = 60 + 10 $\times$ 0.40 = 60 + 4 = $64

How do we value this firm, V_L? Value of D, and value of E_L? Value of D is $100 for sure since the interest is the same as the risk-adjusted required rate for future cash flows to debt holders. However, the risk-adjusted required rate of return, r_E, is not known. Cash flows to debt holders have seniority over cash flows to equity holders. So, r_E must be greater than r_A. We will follow M&M's proposition: Value cash flows to invested capital (CFIC) to value firm L by applying the value additivity concept.

$$V_L = PV[CFIC, r_?] = PV[UFCF + I \times T, r_?] = PV[UFCF, r_?] + PC[I \times T, r_?]$$
$$= PV[60, r_?] + PV[4, r_?]$$

First term, UFCF, is from operations. It is subject to business risk. Relevant risk-adjusted discount rate, $r_?$, is r_A. It should be the same as firm U, 20%. It is independent of capital structure, and both firms are in the same business. Second term, I x T, is from tax shield. Interest paid is tax deductible under US and many countries' tax regimes. M&M asserted that appropriate applicable risk-adjusted discount rate for the tax shields is r_D.[15]

$$V_L = PV[UFCF, r_A] + PC[I \times T, r_D] = PV[60, 0.20] + PV[4, 0.10] = 300 + 40 = \$340$$
$$E_L = V_L - D = \$300\text{-}\$100 = \$240$$

Shares of firm L will be trading at $2.40 per share in the market.

In essence: V_L = V_U + Present Value of Tax Shiels (PVTS). This approach is called the Adjusted Present Value method (APV). It is the most general valuation, and correct, method with minimal implied assumptions!

---

[15] After a long academic debate, consensus agreement is M&M's assumption is correct.

Firm L's total value is $40 more than firm U. The economic contribution of both firms is the same. Assets are generating the same cash flow. This additional value is not a value creation but a subsidy to firm L by tax authorities.[16]

If firm U announces that it is issuing $100 perpetual debt, with the same risk characteristics of firm L's debt, and repurchase shares, at what price will shares of U trade right after the announcement? At the same price, it was trading before, higher, or lower?

We know the value of firm L, $V_L = D + E_L$, is $340. Since both firms' assets are identical, right after the announcement, the market value of firm U will be $340 as well. Shares of firm U will be trading at $3.40. Firm U will repurchase ~29.41 shares (100/3.40). The remaining ~70.59 shares in the market will be trading at $3.40 share. Otherwise, there will be an arbitrage opportunity! In essence going forward, $54, cash flows due to shareholders will be going to 70.59 shares traded in the market.

Firm L's present value of perpetual cash total flow of $54 is $240. We can easily calculate the implied risk-adjusted discount rate; $E_L = 240 = 54/r_E \rightarrow r_E = 54/240 = 22.5\%$. It is 2.5% higher than firm U's r_E due to increased risk induced by leverage. Since both firm's unlevered free cash flows are the same, what would be the implied risk-adjusted discount rate for UFCF to estimate the levered value of the firm? This discount should capture both the business and leverage risk. Let us call it weighted average cost of capital, r_{WACC}.

$$V_L = \text{UFCF}/r_{WACC} = 340 = 60/r_{WACC} \rightarrow r_{WACC} = 60/340 = 17.65\%.$$

M&M derived the following equations using the no arbitrage concept they introduced:

$$r_E = r_A + (r_A - r_D)(1 - T)\frac{D}{E_L}$$
$$= 0.20 + (0.20 - 0.10)(1 - 0.40)\frac{100}{240} \qquad (11.12)$$
$$= 22.5\%$$

$$r_{WACC} = r_E\frac{E_L}{V_L} + r_D(1 - T)\frac{D}{V_L}$$
$$= 0.225\frac{240}{340} + 0.10(1 - 0.40)\frac{100}{340} \qquad (11.13)$$
$$= 17.65\%$$

This has been M&M's most important contribution and changed world of the finance! It is important to note that both debt and equity in Eqs. (11.12 and 11.13) are

---

[16]Majority of academics agree that tax rules create distortions in the economy. In an ideal world to avoid distortions and opaque subsidies, corporate taxes should be 0%. Recipients of cash flows should be taxed.

market values, and Eq. (11.13) holds if and only if market value of debt to market value of equity, D/E_L, is constant. Given the interdependency of variables, r_E, E_L, r_D, and D, in Eqs. (11.12 and 11.13), it is impossible to get a unique solution without setting some conditions. This condition is known as "stable or constant capital structure" in finance.

It is important to remember that when weighted average cost of capital is used in decision-making for capital budgeting or in general valuation, it imposes restrictive constraints. For capital budgeting, it is assumed that the project under consideration is "average risk," that is, it does not change the risk of assets already in place, i.e., r_A stays the same, and if external financing is required, it is financed with equity and debt such that the capital structure of the company does not change. Otherwise, the weighted average cost of capital and traditional Net Present Value approaches will be wrong and misleading.

Basic Extensions of Valuation Theory

If an all-equity financed firm has multiple business lines, we can easily extend the fundamental valuation equation as:

$$E[\widetilde{V}_{U,0}] = \sum_{j=1}^{K} \sum_{t=0}^{T} \frac{E[UF\widetilde{C}F_{j,t}]}{(1+r_{A,j})^t} = \sum_{j=1}^{K} E[V_{U,j,0}] \qquad (11.14)$$

where $j = 1, \ldots, K$ represent each business line, and $r_{A,j}$ is the risk-adjusted discount rate for business line j. It is a simple sum of the parts approach.

If a levered firm has multiple debt with different seniority, that is different risk, the basic extension is:

$$E[\widetilde{V}_{L,0}] = E[\widetilde{V}_{U,0}] + E[\text{Present Value of Tax Shiels (TS)}]$$

$$E[\widetilde{V}_{L,0}] = \sum_{t=0}^{T} \frac{E[UF\widetilde{C}F_t]}{(1+r_A)^t} + \sum_{l=1}^{L} \sum_{t=0}^{T} \frac{T\widetilde{S}_{t,l}}{(1+r_{D,l})^t} \qquad (11.15)$$

where $r_{D,l}$ is the risk-adjusted discount rate for debt type $l = 1,\ldots,L$.

These two basic extensions can be combined to address many different cases. Adjusted Present Value (APV) in most general form is:

$$V_L = V_U + PVTS + / - Other\,Benefits/Costs$$

Tax shields are not the only source benefits. For example, if a company is offered debt below market rate, it is a form of subsidy. In estimating tax shields, market rates must be used. In the case of below market rate, debt interest expense is based on actual interest rate, but resulting tax shields must be discounted at market rate. The resulting benefit, subsidy, is the difference between the amount of debt taken and the amount the company borrows against the cash flows to debt holders at the market rate. For example, if the company is offered a perpetual debt of $100 at 6% to entice it to take a project, while the market rate for similar debt is 10%, the company can borrow $60, ($6/0.10), against the $6 promised cash flows. Thus, the subsidy is $40 ($100 – $60). Tax shields resulting from $6 interest payments must be discounted at 10%, Subsidy is the benefit, and it is added separately.

When a company borrows at higher interest rates than available, for example, the manager in charge chooses to borrow from a source at 10% because the source is offering kickbacks in the form of a low personal mortgage, free travel, jobs for relatives, etc. (an unethical and illegal in many cases), rather than at 6% available at another source, it is a cost to the company. If the company borrows $100 or perpetuity at 10%, the tax shield is based on $10 interest payments, but it must be discounted at 6%, the market rate. Company could borrow $166.67 ($10/0.06) against the $10 cash flows to the debtholders it is promising. The cost to the company is $66.67. It must be deducted as cost in this case. There are many forms of opaque subsidies offered by politicians or executives in charge to avoid public scrutiny. The APV method is the best and most transparent valuation technique available in decision-making.[17]

Application of APV

For the application and the steps involved in using the APV method, we will use the following investment decision problem GGI Technologies is facing:

GGI Technologies is planning to develop an Internet-based company as a wholly owned new subsidiary, B&M.com. The new subsidiary requires an initial investment of $100 million for software development, plant, and equipment. There is also a need for $8 million working capital (10% of expected first-year sales of $80 million). Working capital levels are expected to be 10% of sales in future years. Sales are expected to rise 10% per year over the next two years and stabilize at a rate of 3% indefinitely thereafter, reflecting the expected rate of inflation. Annual cash expenses are estimated to be 50% of revenues. Depreciation expense and annual investment in fixed assets are expected to be 10% of the initial investment in plant and equipment in the first year of operation, then grow at the rate of sales growth. The required rate of return on the assets, r_A, of the subsidiary is 20% and the tax rate is 40%. Assume all additional financing needs, i.e., short fall, of the subsidiary will be covered by the parent company. Any excess funds will be returned to the parent company.

---

[17] See teaching note "Using APV: A Better Tool for Valuing Operations" by Timothy L. Luehrman, Harvard Business Review, (1997) for an extended discussion.

1. Should GGI Technologies establish the new subsidiary if the initial investment is financed with all equity from the parent? (GGI Technologies does not have any debt in its capital structure at present.)
2. What is the value of B&G.com if GGI Technologies capitalizes the new division with 50% debt? Assume that this is the target capital structure for B&G.com and the borrowing rate is 8% at this capital structure.
3. What is the value of B&G.com if the parent borrows the entire amount of $108 million required for the initial investment and working capital? The debt will be a non-recourse loan to the parent, that is, it will be guaranteed by the subsidiary by using its assets as collateral. The borrowing rate is 14% at this level of debt financing. B&G.com will repay the loan at $10 million per year and will bring its debt level from the initial level of $108 million to $78 million by the end of year 3. This debt level is around the target level of 50% debt to total capitalization. The firm will maintain this capital structure ratio from that point on.
4. The parent plans to take B&G.com public at the end of year 3 (or at the beginning of year 4). GGI Technologies will refinance the old debt at 8% with new issue of unsecured debt at the public offering.

   a. What will be the expected value of B&G.com at the time of the initial public offering?
   b. What is the rate of return on the equity investment of GGI Technologies in B&G.com?

Figure 11.5 shows the diagram of steps GGI management should follow in applying the APV method in their decision-making process. We will use a simple

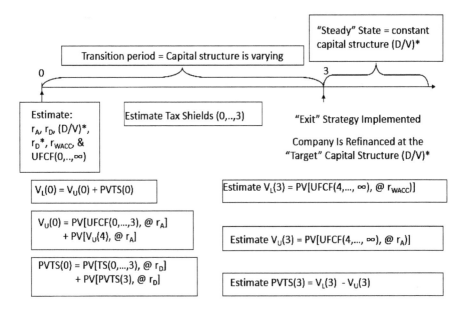

Fig. 11.5 Steps in applying APV method

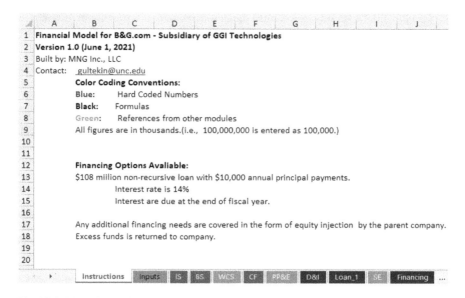

Fig. 11.6 Model instruction module

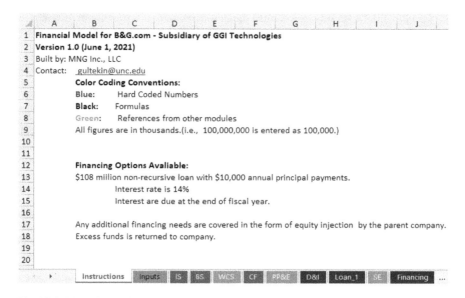

Fig. 11.7 Model parameters. (Note: Cell C15 on/off toggle switch activates the Loan.)

integrated financial Excel model to address these questions. Snapshots of the model used are shown in this chapter.

Note that time 0 represents "now" in finance jargon. In the Excel model, Year 0 represents the end of year 0, or the first day of year 1. This is important for timing of the cash flows and use of Excel's financial functions!

	A	B	C	D	E	F	G
1	B&G.com						
2	($ in thousands)						
3	December 31						
4	Income Statement						
5							
6	Net Sales			0	80,000	88,000	96,800
7	Cost of Goods Sold before D&A			0	40,000	44,000	48,400
8	Gross Profit/EBITDA			0	40,000	44,000	48,400
9							
10	Depreciation			0	10,000	11,000	12,100
11	EBIT			0	30,000	33,000	36,300
12							
13	Interest Expense			0	0	0	0
14	EBT			0	30,000	33,000	36,300
15							
16	Income Taxes			0	12,000	13,200	14,520
17	Net Income			0	18,000	19,800	21,780
18							
19							
20	Staging Area						
21	Sales & growth rates			$0	$80,000	10.00%	10.00%
22	COGS as % of sales			0.00%	50.00%	50.00%	50.00%
23	Effective Tax Rates			40.00%	40.00%	40.00%	40.00%
24							

Instructions | Inputs | IS | BS | WCS | CF | PP&E | D&I | Loan_1 | SE | Financing | ...

Fig. 11.8 Income statement

	A	B	C	D	E	F	G
1	B&G.com						
2	($ in thousands)						
3	December 31						
4	Balance Sheet						
5	Assets						
6	Cash			0	0	0	0
7	Accounts Receivable, net			0	8,000	8,800	9,680
8	Inventory			8,000	8,800	9,680	10,648
9	Total Current Assets			8,000	16,800	18,480	20,328
10							
11	PP&E - net			100,000	100,000	100,000	100,000
12	Total Assets			108,000	116,800	118,480	120,328
13							
14	Liabilities						
15	Accounts Payable			0	8,000	8,800	9,680
16	Total Current Liabilities			0	8,000	8,800	9,680
17							
18	Long-Term Debt - Current			0	0	0	0
19	Long-Term Debt			0	0	0	0
20	Total Liabilities			0	8,000	8,800	9,680
21							
22	Total Equity			108,000	108,800	109,680	110,648
23							
24	Total Liabilities and Equity			108,000	116,800	118,480	120,328
25	Check						
26							

Fig. 11.9 Balance sheet

The core financial statements are income statement, balance sheet, and cash flow statement. All other modules are supporting schedules. Balance sheet and cash flow statements are "dumb" statements. All the projections are done in other modules and

	A	B	C	D	E	F	G
1	B&G.com						
2	($ in thousands)						
3	December 31						
4	Cash Flow Statement						
5	**Operating Activities**						
6	Net Income			0	18,000	19,800	21,780
7	Depreciation			0	10,000	11,000	12,100
8	Change in Working Capital			(8,000)	(800)	(880)	(968)
9	**Cash Flow from Operating Activities:**			(8,000)	27,200	29,920	32,912
10							
11	**Investing Activities**						
12	Capital Expenditures			(100,000)	(10,000)	(11,000)	(12,100)
13	**Cash Flow from Investing Activities:**			(100,000)	(10,000)	(11,000)	(12,100)
14							
15	**Financing Activities**						
16	Proceeds from / (Repayment of) Long Term Debt			0	0	0	0
17	Proceeds from / (Repurchase of) Equity plug			108,000	(17,200)	(18,920)	(20,812)
18	**Cash Flow from Financing Activities:**			108,000	(17,200)	(18,920)	(20,812)
19	Net Change in Cash			0	0	0	0
20	Beginning Cash Balance			0	0	0	0
21	**Ending Cash Balance**			0	0	0	0
22							

Fig. 11.10 Cash flow statement

	A	B	C	D	E	F	G
1	B&G.com						
2	($ in thousands)						
3	December 31						
4	Working Capital Schedule						
5	**Working Capital Balances**						
6	Accounts Receivable, net			0	8,000	8,800	9,680
7	Inventory			8,000	8,800	9,680	10,648
8	**Total Non-cash Current Assets**			8,000	16,800	18,480	20,328
9							
10	Accounts Payable			0	8,000	8,800	9,680
11	**Total Non-Debt Current Liabilities**			0	8,000	8,800	9,680
12							
13	NET WORKING CAPITAL / (DEFICIT)			8,000	8,800	9,680	10,648
14							
15	**(Increase)/Decrease in Working Capital**			(8,000)	(800)	(880)	(968)
16							
17	**Ratios and Assumptions**						
18	Accounts Receivable (Collection period in days)			0	37	37	37
19	Inventory (Days outstanding)			0	80	80	80
20	Accounts Payable (Days payable)			0	73	73	73
21	Other Current Liabilities as % of COGS			0.00%	20.00%	20.00%	20.00%
22							
23	**Cash Flows by Individual Accounts**						
24	Accounts Receivables - net			0	(8,000)	(800)	(880)
25	Inventory			(8,000)	(800)	(880)	(968)
26	Accounts Payable			0	8,000	800	880
27	**(Increase)/Decrease in Working Capital**			(8,000)	(800)	(880)	(968)
28							
29	Staging Area						
30	Net Sales			0	80,000	88,000	96,800
31	Accounts Receivable as a % of Sales			0	10.00%	10.00%	10.00%
32	Accounts Payable as a % of sales			0	10.00%	10.00%	10.00%
33	Inventories as a % of Sales			$8,000	11.00%	11.00%	11.00%
34	Cost of Goods Sold before D&A			0	40,000	44,000	48,400

Fig. 11.11 Working capital – supporting module for balance sheet

	A	B	C	D	E	F	G
1	B&G.com						
2	($ in thousands)						
3	December 31						
4	Plant, Property, & Equipment						
5	Capital Expenditures						
6	Beginning Net PP&E			0	100,000	100,000	100,000
7	Capital Expenditures			100,000	10,000	11,000	12,100
8	(Depreciation Expense)			0	(10,000)	(11,000)	(12,100)
9	Ending Net PP&E			100,000	100,000	100,000	100,000
10							
11							
12	Staging Area						
13	Net Sales			0	80,000	88,000	96,800
14	CapEx, CapEx & Depreciation growth rate			$100,000	10.00%	10.00%	10.00%

Fig. 11.12 PP&E – supporting module for balance sheet

	A	B	C	D	E	F	G
1	B&G.com						
2	($ in thousands)						
3	December 31						
4	Equity Schedule						
5	Beginning Equity Balance			0	108,000	108,800	109,680
6	Net Income			0	18,000	19,800	21,780
7	Issuance/ (Repurchase of) Equity			108,000	(17,200)	(18,920)	(20,812)
8	Ending Equity Balance			108,000	108,800	109,680	110,648

Fig. 11.13 Shareholders equity – supporting module for balance sheet

reported to the balance sheet and cash flow statement. The financing module, shown in Fig. 11.14, is another important and necessary module used in financial models. It reconciles all the cash flows and determines the "plug" needed to balance the balance sheet. All the information reported in this module is linked from the cash flow statement. In this model, the plug is the equity injection from the parent company. Amounts shown in row 13 represent the shortfall needing to be financed. It is linked to the equity schedule. In general, a form of "Line of Credit" debt financing is used as plug. Since there is interest expense associated, it creates "circularity" in the model.

Next step is estimating valuation cash flows. Figure 11.15 shows valuation cash flows estimated in the model.

Terminal value at the end of year 3 is estimated using the Gordon growth model.

$$TV_U = \frac{UFCF_T \times (1+g)}{r_A - g} = \frac{20,812 \times 1.03}{0.20 - 0.03} = \$126,096$$

The NPV of the project is positive. It indicates that it is acceptable, but the final decision will depend on the risk tolerance of the managers of GGI Technologies.

	A	B	C	D	E	F	G
1	B&G.com						
2	*($ in thousands)*						
3	December 31						
4			Year	0	1	2	3
5	Cash Flow from Operating Activities:			(8,000)	27,200	29,920	32,912
6	Cash Flow from Investing Activities:			(100,000)	(10,000)	(11,000)	(12,100)
7	Cash flow Available for financing Activities			(108,000)	17,200	18,920	20,812
8	Beginning Cash Balance			0	0	0	0
9	Less: Minimum Cash Balance			0	0	0	0
10	Cash Available For Debt Repayment			(108,000)	17,200	18,920	20,812
11							
12	Proceeds from / (Repayment of) Long Term Debt			0	0	0	0
13	Cash Surplus/(Shortfall)			(108,000)	17,200	18,920	20,812
14	Proceeds from / (Repurchase of) Equity plug			108,000	(17,200)	(18,920)	(20,812)
15	Cash Surplus/(shortfall) after Equity Injection			0	0	0	0
16							
17	**Cash Balances**						
18	Estimated Minimum Cash Balance			0	0	0	0
19							
20	Staging Area						
21	Net Sales			0	80,000	88,000	96,800
22	Minimum Cash Balance as % sales			0	0.00%	0.00%	0.00%

Fig. 11.14 Financing – supporting model for balance sheet and cash flow statement

	A	B	C	D	E	F	G
1	B&G.com						
2	*($ in thousands)*						
3	December 31						
4	Valuation Cash Flows						
5	EBIT			0	30,000	33,000	36,300
6	Tax on EBIT			0	12,000	13,200	14,520
7	EBIAT			0	18,000	19,800	21,780
8	+ Depreciation			0	10,000	11,000	12,100
9	- CapEx			(100,000)	(10,000)	(11,000)	(12,100)
10	- Change in NWC			(8,000)	(800)	(880)	(968)
11	UFCF			(108,000)	17,200	18,920	20,812
12							
13	**Cash Flows to Debt Holders:**						
14	Principal			0	0	0	0
15	Interest			0	0	0	0
16	Total Cash Flow to Debt Holders			0	0	0	0
17							
18	**Cash Flow to Equity**			(108,000)	17,200	18,920	20,812
19	Cash Flow to Invested Capital			(108,000)	17,200	18,920	20,812

Fig. 11.15 Valuation cash flows

One advantage of building the integrated financial model of the firm is ability to see future financing needs of the firm. It is very easy to do sensitivity analysis and plan for contingencies.

Remember that figures shown for year 0 represent end of period for year 0, or the very first day of year 1. Balance sheet reported in Fig. 11.9 shows the book value of B&G.com at the end of year 0. Assets of the company are the cost of inventory for

	A	B	C
21	**Discount Rates & LT Growth Rate**		
22	Tax Rate		40.0%
23	Required Rate of Retun on Assets (r_A)		20.0%
24	Target Debt to Value (D/V)*		50.0%
25	Borrowing Rate at Target Capital Structure (r^*_D)		8.0%
26	Borrowing Rate withh All Debt Financing (r_D)		14.0%
27	WACC at Target Capital Structure (M&M) (r_{WACC})		16.0%
28	Long Term Growth Rate		3.0%

Fig. 11.16 Risk-adjusted discount rates and long-term growth rate (UFCF) assumption

1	**B&G.com**				
2	*($ in thousands)*				
3	**December 31**				
4	**Valuation with All Equity Financing**				
5	UFCF	(108,000)	17,200	18,920	20,812
6	Unlevered Terminal Value (UTV)				$126,096
7	PV of UFCF (years 1-3) @ r_A	39,516			
8	PV of UTV @ r_A	72,972			
9	Value from Operations	112,489			
10	Initial Investment	(108,000)			
11	NPV$_U$	4,489			

Fig. 11.17 NPV of the project if it is financed with equity investment from the parent company forever

$8 million and the cost of PP&E for $108 million. Liability of the subsidiary is the equity investment by the parent company of $108 million. It is the accounting view of the balance sheet. Market value of B&G.com is $112.489 million.

If GGI finances the subsidiary at target capital structure, that is, borrows or issues bonds at 8% for the amount $D such that D/V$_L$ is 50%, or D/E$_L$ = 1, what will be the NPV of this investment and what is the amount of debt that will be issued?

Figure 11.18 shows the valuation of the subsidiary with debt financing at target capital structure from the very beginning.

The NPV of the project is positive, and significantly higher than all-equity financing. The market value of the project is $147.863 million. Book value of the project is the same as before, $108 million, $8 million in inventories $100 million in PP&E in investment. GGI will borrow $73.931 million (50% of market value of B&G.com) and invest $34.069 million in the form of equity in the subsidiary! GGI invested only $34.069 million, but the market value of this investment is $73.931 million. The market and the book value of the debt, if interest rate is the market rate, is always the same at the time of issuance. However, if new information arrives after the issuance of debt, its market price will reflect it. For example, if interest rates in the market drop, bonds issued by GGI will be selling at premium if the risk of the bond does not change. One of the assumptions, and an extremely

	A B	C	D	E	F	G
1	B&G.com					
2	($ in thousands)					
3	December 31					
4	Valuation at Target Capital Structure					
5	UFCF		(108,000)	17,200	18,920	20,812
6	Unlevered Terminal Value (UTV)					164,895
7	PV of UFCF (years 1-3) @ r_WACC		42,222			
8	PV of UTV @ r_A		105,641			
9	Value from Operations		147,863			
10	Initial Investment		(108,000)			
11	NPV_L		39,863			

Fig. 11.18 Valuation at target capital structure

	A B	C	D	E	F	G
1	B&G.com					
2	($ in thousands)					
3	December 31					
4	Model Parameters & Assumptions	Year	0	1	2	3
5	Exhibit 1: New Project Investment and Financial Performance Estimates					
8	Minimum Cash Balance as % sales		$0	0.00%	0.00%	0.00%
9	Accounts Receivable as a % of Sales		$0	10.00%	10.00%	10.00%
10	Inventories as a % of Sales		$8,000	11.00%	11.00%	11.00%
11	CapEx, CapEx & Depreciation growth rate		$100,000	10.00%	10.00%	10.00%
12	Accounts Payable as a % of sales		$0	10.00%	10.00%	10.00%
13	Effective Tax Rates		40.00%	40.00%	40.00%	40.00%
14	Exhibit 2: Assumptions of financing options					
15	Loan 1	Yes				
16	Maximum	$108,000				
17	Annual Payment Amount	$10,000				
18	Interest Rate	14.00%				
19	Issuance/(Payment)		$108,000	($10,000)	($10,000)	($10,000)
20	Misc. Assumptions					
21	Days in a year	365				

Fig. 11.19 Model parameters and assumption

strong one, is that in future years the company will be issuing and buying back debt and equity such that market value of debt-to-equity ratio will be always 1. Market values will depend on the prevailing risk-adjusted discount rate in the market; it is not known at this time. Our estimates are based on the current information and market's risk-adjusted discount rates. Under no circumstance can time-varying WACC be estimated. **Any valuation done with time-varying WAAC is wrong!**

Valuation of the project with $108 million debt financing: The following figures will show the screenshots of the model. In the Excel model, toggle for Loan 1, as shown in Fig. 11.19, is turned on to activate the loan.

Discounted Value of Tax Shields in TV (DVTS), (164,895 – 126,096 = 38,799), in cell G8, is the "baked-in" present value of the tax shield in terminal levered value of the firm at the end of year 3. It is assumed that the company is recapitalizing at the target capital structure at the end of year 3 and staying at target capital structure forever.

	A B	C	D	E	F	G
1	B&G.com					
2	*($ in thousands)*					
3	December 31					
4	Income Statement					
5						
6	Net Sales		0	80,000	88,000	96,800
7	Cost of Goods Sold before D&A		0	40,000	44,000	48,400
8	Gross Profit/EBITDA		0	40,000	44,000	48,400
9						
10	Depreciation		0	10,000	11,000	12,100
11	EBIT		0	30,000	33,000	36,300
12						
13	Interest Expense		0	15,120	13,720	12,320
14	EBT		0	14,880	19,280	23,980
15						
16	Income Taxes		0	5,952	7,712	9,592
17	Net Income		0	8,928	11,568	14,388
18						
19						
20	Staging Area					
21	Sales & growth rates		$0	$80,000	10.00%	10.00%
22	COGS as % of sales		0.00%	50.00%	50.00%	50.00%
23	Effective Tax Rates		40.00%	40.00%	40.00%	40.00%

Fig. 11.20 Income statement – with debt financing

	A B	C	D	E	F	G
1	B&G.com					
2	*($ in thousands)*					
3	December 31					
4	Balance Sheet					
5	Assets					
6	Cash		0	0	0	0
7	Accounts Receivable, net		0	8,000	8,800	9,680
8	Inventory		8,000	8,800	9,680	10,648
9	Total Current Assets		8,000	16,800	18,480	20,328
10						
11	PP&E - net		100,000	100,000	100,000	100,000
12	Total Assets		108,000	116,800	118,480	120,328
13						
14	Liabilities					
15	Accounts Payable		0	8,000	8,800	9,680
16	Total Current Liabilities		0	8,000	8,800	9,680
17						
18	Long-Term Debt - Current		10,000	10,000	10,000	10,000
19	Long-Term Debt		98,000	88,000	78,000	68,000
20	Total Liabilities		108,000	106,000	96,800	87,680
21						
22	Total Equity		0	10,800	21,680	32,648
23						
24	Total Liabilities and Equity		108,000	116,800	118,480	120,328
25	Check					

Fig. 11.21 Balance sheet with debt financing

Value from Operations, cell D16, 112,489 is the V_U. It is independent of the financing. It will be same in all financing options.

Cell D18, 12,854, is the present value of interim tax shields (cells E12:G12) discounted at 14%, current market borrowing rate.

	A	B	C	D	E	F	G
1	**B&G.com**						
2	*($ in thousands)*						
3	December 31						
4	**Cash Flow Statement**						
5	**Operating Activities**						
6	Net Income			0	8,928	11,568	14,388
7	Depreciation			0	10,000	11,000	12,100
8	Change in Working Capital			(8,000)	(800)	(880)	(968)
9	**Cash Flow from Operating Activities:**			(8,000)	18,128	21,688	25,520
10							
11	**Investing Activities**						
12	Capital Expenditures			(100,000)	(10,000)	(11,000)	(12,100)
13	**Cash Flow from Investing Activities:**			(100,000)	(10,000)	(11,000)	(12,100)
14							
15	**Financing Activities**						
16	Proceeds from / (Repayment of) Long Term Debt			108,000	(10,000)	(10,000)	(10,000)
17	Proceeds from / (Repurchase of) Equity plug			0	1,872	(688)	(3,420)
18	**Cash Flow from Financing Activities:**			108,000	(8,128)	(10,688)	(13,420)
19	Net Change in Cash			0	0	0	0
20	Beginning Cash Balance			0	0	0	0
21	**Ending Cash Balance**			0	0	0	0
22							

Fig. 11.22 Cash flow statement with debt financing

	A	B	C	D	E	F	G
1	**B&G.com**						
2	*($ in thousands)*						
3	December 31						
4	**Debt & Interest**						
11	Long-Term Portion			98,000	88,000	78,000	68,000
12	Total			108,000	98,000	88,000	78,000
13			*Check*				
14							
15	**Total Long-Term Debt (Repayments) / Borrowings**			108,000	(10,000)	(10,000)	(10,000)
16	**Total Interest Expense - Long-Term Loans**			0	15,120	13,720	12,320
17							
18	**Principal Issuance/(Payments):**						
19	Loan 1			108,000	(10,000)	(10,000)	(10,000)
20							
21							
22	**Total Interest Expense:**						
23	Loan 1			0	15,120	13,720	12,320
24							
25							
26	**Current Portions:**						
27	Loan 1			10,000	10,000	10,000	10,000
28							
29							
30	**Long-Term Portions:**						
31	Loan 1			98,000	88,000	78,000	68,000
32							

Fig. 11.23 Debt and interest module with debt financing

	A	B	C	D	E	F	G
1	B&G.com						
2	($ in thousands)						
3	December 31						
4	Loan 1						
5	Beginning Balance			0	108,000	98,000	88,000
6	Issuance/(Payment)			108,000	(10,000)	(10,000)	(10,000)
7	Ending Balance			108,000	98,000	88,000	78,000
8							
9	Current			10,000	10,000	10,000	10,000
10	Long Term			98,000	88,000	78,000	68,000
11	Total			108,000	98,000	88,000	78,000
12							
13	Payments to Debt						
14	Interest			0	15,120	13,720	12,320
15	Principal			(108,000)	10,000	10,000	10,000
16	Total payment			(108,000)	25,120	23,720	22,320
17							
18	Staging Area						
19	Interest Rate			14.00%			

Fig. 11.24 Details of loan 1

	A	B	C	D	E	F	G
1	B&G.com						
2	($ in thousands)						
3	December 31						
4			Year	0	1	2	3
5	Cash Flow from Operating Activities:			(8,000)	18,128	21,688	25,520
6	Cash Flow from Investing Activities:			(100,000)	(10,000)	(11,000)	(12,100)
7	Cash flow Available for financing Activities			(108,000)	8,128	10,688	13,420
8	Beginning Cash Balance			0	0	0	0
9	Less: Minimum Cash Balance			0	0	0	0
10	Cash Available For Debt Repayment			(108,000)	8,128	10,688	13,420
11							
12	Proceeds from / (Repayment of) Long Term Debt			108,000	(10,000)	(10,000)	(10,000)
13	Cash Surplus/(Shortfall)			0	(1,872)	688	3,420
14	Proceeds from / (Repurchase of) Equity plug			0	1,872	(688)	(3,420)
15	Cash Surplus/(shortfall) after Equity Injection			0	0	0	0
16							
17	Cash Balances						
18	Estimated Minimum Cash Balance			0	0	0	0
19							
20	Staging Area						
21	Net Sales			0	80,000	88,000	96,800
22	Minimum Cash Balance as % sales			0	0.00%	0.00%	0.00%

Fig. 11.25 Financing module with debt financing

Cell D19, 26,188, is the present value of DVTS in cell G8, 38,799, discounted at 14%.

Total value with Debt Financing, cell D22, 151,531, is $V_L = V_U$ + PV of Tax Shield in interim stage + PV of Tax Shields in Terminal Value = 112,889 + 12,854 + 26,188.

	A	B	C	D	E	F	G
1	B&G.com						
2	($ in thousands)						
3	December 31						
4	Valuation Cash Flows		Year	0	1	2	3
5	EBIT			0	30,000	33,000	36,300
6	Tax on EBIT			0	12,000	13,200	14,520
7	**EBIAT**			0	18,000	19,800	21,780
8	+ Depreciation			0	10,000	11,000	12,100
9	- CapEx			(100,000)	(10,000)	(11,000)	(12,100)
10	- Change in NWC			(8,000)	(800)	(880)	(968)
11	**UFCF**			(108,000)	17,200	18,920	20,812
12							
13	**Cash Flows to Debt Holders:**						
14	Principal			(108,000)	10,000	10,000	10,000
15	Interest			0	15,120	13,720	12,320
16	**Total Cash Flow to Debt Holders**			(108,000)	25,120	23,720	22,320
17							
18	**Cash Flow to Equity**			0	(1,872)	688	3,420
19	**Cash Flow to Invested Capital**			(108,000)	23,248	24,408	25,740

Fig. 11.26 Valuation of cash flows with debt financing

	A	B	C	D	E	F	G
1	B&G.com						
2	($ in thousands)						
3	December 31						
4	Valuation with All Debt Financing		Year				
5	UFCF			(108,000)	17,200	18,920	20,812
6	Unlevered Terminal Value (UTV)						126,096
7	Levered Terminal Value (LV)						164,895
8	Discounted Value of Tax Shields in TV (DVTS)						38,799
9							
10	Taxes with no Debt			0	12,000	13,200	14,520
11	Taxes with Debt Financing			0	5,952	7,712	9,592
12	Tax Savings due to Interest (=Txr$_D$xD)			0	6,048	5,488	4,928
13							
14	PV of UFCF (years 1-3) @ r$_A$			39,516			
15	PV of UTV @ r$_A$			72,972			
16	Value from Operations			112,489			
17							
18	PV of Interest tax shield @ r$_D$			12,854			
19	PV of DVTS @ r$_D$			26,188			
20	PV of Tax Shields			39,042			
21							
22	Total Value with Debt Financing			151,531			
23	Initial Investment			(108,000)			
24	NPV$_L$			43,531			

Fig. 11.27 Valuation with debt financing

Net present value is positive. Again, it is an acceptable project. End of year 3 transactions will be: New debt in the amount of $82.448 million ($D_3 = V_{L,3} \times$ (D/V)$^* = \$164{,}895 \times (\frac{1}{2})$) will be issued. Old debt, balance of $78 million, is paid back.

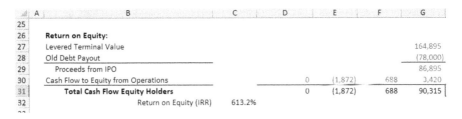

Fig. 11.28 Return on equity – IRR

Figure 11.28 shows return on equity investment of GGI. IRR on the investment is 613.2%! In this case, it represents all debt financing at the beginning and equity injection when needed or withdrawal when there is surplus. This represents "Venture Capital Perspective." Cash flows to equity from operations and proceeds from IPO are the relevant cash flows. Decision is based on the IRR of the project and VC's hurdle IRR. If the project's IRR is higher than the hurdle rate, it is acceptable. It is commonly used in practice, and in many cases, the terminal value is estimated by multiples etc., bypassing the estimation of discount rates.

References

Bierman, H., & Smidt, S. (1993). *The capital budgeting decision* (8th ed.). Macmillan.
Brealey, R. A., & Myers, S. C. (2003). *Principles of corporate finance* (7th ed.). McGraw-Hill/Irwin.
Dean, J. (1951). *Managerial economics*. Prentice-Hall. Chapters 4, 10.
Dean, J. (1954). Measuring the productivity of capital. *Harvard Business Review*.
Gordon, M. J. (1955). The payback period and the rate of profit. *Journal of Business*.
Gordon, M. J., & Shapiro, E. (1956). Capital equipment analysis: The required rate of profit. *Management Science*.
Graham, J. R., & Harvey, C. R. (2001). The theory and practice of corporate finance: Evidence from the field. *Journal of Finance Economics*, 187–243.
Hackbarth, D., & Johnson, T. C. (2015). Real options and risk dynamics. *Review of Economic Studies, 82*, 1449–1482.
Lanzilotti. (1958). Pricing objectives in large companies. *The American Economic Review*, 921–940.
Lerner, E. M., & Carleton, W. (1964). The integration of capital budgeting and stock valuation. *American Economic Review*, 683–702.
Lerner, E. M., & Carleton, W. T. (1966). *A theory of financial analysis*. Harcourt, Brace, and World, Inc.
Levy, H., & Sarnat, M. (1982). *Capital investment and financial decisions*. Prentice-Hall International. 2nd edition, Chapter 4.
Lewellen, W. G. (1969). *The cost of capital*. Wadsworth Publishing Company.
Mauer, D. C., & Sarkar, S. (2005). Real options, agency conflicts, and optimal capital structure. *Journal of Banking & Finance, 29*, 1405–1428.
Myers, S. C. (1974). Interactions of corporate financing and investment decisions: Implications for capital budgeting. *Journal of Finance, 29*, 1–25.
Myers, S. C., & Reade, J. A. (2018). Real options, taxes, and financial leverage. *Working Paper*.

Robichek, A. A., & Myers, S. C. (1965). *Optimal financial decisions*. Prentice-Hall, Inc.

Rubinstein, M. E. (1973). A mean-variance synthesis of corporate financial theory. *Journal of Finance*, 167–181.

Schall, L. D. (1972). Asset valuation, firm investment, and firm diversification. *Journal of Business, 45*, 13–21.

Schwartz, E. (1959). Theory of the capital structure of the firm. *Journal of Finance, 24*.

Schwartz, E. (1962). *Corporate finance*. St. Martin's Press. Chapters 10–11.

Schwartz, E. (1993). *Theory and application of the interest rate*. Westport, CT, Praeger.

Solomon, E. (Ed.). (1959). *The management of corporate capital*. University of Chicago Press.

Solomon, E. (1963). Leverage and the cost of capital. *Journal of Finance, 28*, 273–279.

Spencer, M. H., & Siegelman, L. (1959). *Managerial economics*. Richard D. Irwin. Chapters 10, 11.

Chapter 12
Regression Analysis and Estimating Regression Models

In February 2020, the US unemployment rate was at a 50-year low. One of the great economic problems of the COVID pandemic is the dramatic increase in unemployment claims. In this chapter, we specifically address the unemployment rate in 2020 as the COVID virus closed down the US economy in March 2020, with profound and highly significant impacts on US output, as measured by gross domestic product. Regression analysis is the primary technique discussed and estimated in this chapter. More specifically, regression analysis seeks to find the "line of best fit" through the data points. The regression line is drawn to best approximate the relationship between the two variables. Techniques for estimating the regression line (i.e., its intercept on the Y axis and its slope) are the subject of this chapter.

In simple regression analysis, one seeks to measure the statistical association between two variables, X and Y. Regression analysis is generally used to measure how changes in the independent variable, X, influence changes in the dependent variable, Y. Regression analysis shows a statistical association or correlation among variables, rather than a causal relationship among variables. In this chapter, we address the relationship between the US unemployment rate, the leading economic indicator (LEI), and weekly unemployment claims, a component of the LEI.

12.1   Estimating an Ordinary Least Squares Regression Line

The case of simple, linear, least squares regression may be written in the form:

$$Y = \alpha + \beta X + \varepsilon \tag{12.1}$$

where Y, the dependent variable, is a linear function of X, the independent variable. The parameters α and β characterize the population regression line, and ε is the randomly distributed error term. The regression estimates of α and β will be derived from the principle of least squares. In applying least squares, the sum of the squared regression errors will be minimized; our regression errors equal the actual dependent variable minus the estimated value from the regression line. If Y represents the actual value and $\widehat{Y}$ the estimated value, their difference is the error term, e. Least squares regression minimizes the sum of the squared error terms. The simple regression line will yield an estimated value of Y, $\widehat{Y}$ by the use of the sample regression:

$$\widehat{Y} = \widehat{a} + \widehat{b}X \tag{12.2}$$

In the estimation equation (12.2), a is the least squares estimate of α and b is the estimate of β. Thus, a and b are the regression constants that must be estimated. The least squares regression constants (or statistics) a and b are unbiased and efficient (smallest variance) estimators of α and β. The error term, $\widehat{e}_i$, is the difference between the actual and estimated dependent variable value for any given independent variable values, X_i.

$$\widehat{e}_i = Y_i - \widehat{Y}_i \tag{12.3}$$

The regression error term, $\widehat{e}_i$, is the least squares estimate of ε_i, the actual error term.[1]

To minimize the error terms, the least squares technique minimizes the sum of the squared error terms of the N observations,

$$\sum\nolimits_{i=1}^{N} \widehat{e}_i^2 \tag{12.4}$$

Thus, least squares regression minimizes:

$$\sum\nolimits_{i=1}^{N} \widehat{e}_i^2 = \sum\nolimits_{i=1}^{N}\left[Y_i - \widehat{Y}_i\right]^2 = \sum\nolimits_{i=1}^{N}\left[Y_i - \left(\widehat{a} + \widehat{b}X_i\right)\right]^2 \tag{12.5}$$

To assure that a minimum is reached, the partial derivatives of the squared error terms function

$$\sum\nolimits_{i=1}^{N} = \left[Y_i - \left(\widehat{a} + \widehat{b}X_i\right)\right]^2$$

will be taken with respect to $\widehat{a}$ and $\widehat{b}$.

---

[1]The reader is referred to an excellent statistical reference, such as Irwin Miller and J.E. Freund, *Probability and Statistics for Engineers*, (Englewood Cliffs, NJ: Prentice-Hall, 1965).

$$\frac{\partial \sum_{i=1}^{N}\widehat{e}_i^2}{\partial \widehat{a}} = 2\sum_{i=1}^{N}\left(Y_i - \widehat{a} - \widehat{b}X_i\right)(-1)$$

$$= -2\left(\sum_{i=1}^{N}Y_i - \sum_{i=1}^{N}\widehat{a} - \widehat{b}\sum_{i=1}^{N}X_i\right)$$

$$\frac{\partial \sum_{i=1}^{N}\sigma_i^2}{\partial \widehat{b}} = 2\sum_{i=1}^{N}\left(Y_i - \widehat{a} - \widehat{b}X_i\right)(-X_i)$$

$$= -2\left(\sum_{i=1}^{N}Y_iX_i - \widehat{a}\sum_{i=1}^{N}X_i - \widehat{b}\sum_{i=1}^{N}X_i^2\right)$$

The partial derivatives will then be set equal to zero.

$$\frac{\partial \sum_{i=1}^{N}\widehat{e}_i^2}{\partial \widehat{a}} = -2\left(\sum_{i=1}^{N}Y_i - \sum_{i=1}^{N}\widehat{a} - \widehat{b}\sum_{i=1}^{N}X_i\right) = 0$$

$$\frac{\partial \sum_{i=1}^{N}\widehat{e}_i^2}{\partial \widehat{b}} = -2\left(\sum_{i=1}^{N}YX_i - \widehat{a}\sum_{i=1}^{N}X_l - \widehat{b}\sum_{i=1}^{N}X_i^2\right) = 0$$

(12.6)

Rewriting these equations, one obtains the normal equations:

$$\sum_{i=1}^{N}Y_i = \sum_{i=1}^{N}\widehat{a} + \widehat{b}\sum_{i=1}^{N}X_i$$

$$\sum_{i=1}^{N}Y_iX_i = \widehat{a}\sum_{i=1}^{N}X_i + \widehat{b}\sum_{i=1}^{N}X_i^2$$

(12.7)

Solving the normal equations simultaneously for $\widehat{a}$ and $\widehat{b}$ yields the least squares regression estimates:

$$\widehat{a} = \frac{\left(\sum_{i=1}^{N}X_i^2\right)\left(\sum_{i=1}^{N}Y_i\right) - \left(\sum_{i=1}^{N}X_iY_i\right)}{N\left(\sum_{i=1}^{N}X_i^2\right) - \left(\sum_{i=1}^{N}X_i\right)^2}$$

$$\widehat{b} = \frac{N\left(\sum_{i=1}^{N}X_iY_i\right) - \left(\sum_{i=1}^{N}X_i\right)\left(\sum_{i=1}^{N}Y_i\right)}{N\left(\sum_{i=1}^{N}X_i^2\right) - \left(\sum_{i=1}^{N}X_i\right)^2}$$

(12.8)

An estimation of the regression line's coefficients and goodness of fit also can be found in terms of expressing the dependent and independent variables in terms of deviations from their means, that is, their sample moments. The sample moments will be denoted by M.

$$M_{XX} = \sum_{i=1}^{N} x_i^2 = \sum_{i=1}^{N} (x_i - \bar{x})^2$$

$$= N \sum_{i=1}^{N} X_i - \left(\sum_{i=1}^{N} X_i \right)^2$$

$$M_{XY} = \sum_{i=1}^{N} x_i y_i = \sum_{i=1}^{N} (X_i - \bar{X})(Y_i - \bar{Y})$$

$$= N \sum_{i=1}^{N} X_i Y_i - \left(\sum_{i=1}^{N} X_i \right) \left(\sum_{i=1}^{N} Y_i \right)$$

$$M_{YY} = \sum_{i=1}^{N} y_i^2 = \sum_{i=1}^{N} (Y - \bar{Y})^2$$

$$= N \left(\sum_{i=1}^{N} Y_i^2 \right) - \sum_{i=1}^{N} (Y_i)^2$$

The slope of the regression line, $\hat{b}$, can be found by:

$$\hat{b} = \frac{M_{XY}}{M_{XX}} \tag{12.9}$$

$$\hat{a} = \frac{\sum_{i=1}^{N} Y_i}{N} - \hat{b} \frac{\sum_{i=1}^{N} X_i}{N} = \bar{y} - \hat{b}\bar{X} \tag{12.10}$$

The standard error of the regression line can also be expressed in terms of the sample moments:

$$S_e^2 = \frac{M_{XX}(M_{YY}) - (M_{XY})^2}{N(N-2)M_{XX}}$$

$$S_e = \sqrt{S_e^2} \tag{12.11}$$

The major benefit in calculating the sample moments is that the correlation coefficient, r, and the coefficient of determination, R^2, can easily be found.

$$r = \frac{M_{XY}}{(M_{XX})(M_{YY})}$$

$$R^2 = (r)^2 \tag{12.12}$$

The coefficient of determination, R^2, is the percentage of the variance of the dependent variable explained by the independent variable. The coefficient of determination cannot exceed 1 nor be less than zero. In the case of $R^2 = 0$, the regression line is $Y = \bar{Y}$, and no variation in the dependent variable is explained. If the dependent variable pattern continues as in the past, the model with time as the independent variable should be of good use in forecasting.

The firm can test whether the a and b coefficients are statistically different from zero, the generally accepted null hypothesis. A t-test is used to test the two null hypotheses:

$$H_{o_1} : a = 0$$

$$H_{A_1} : a \text{ ne } 0$$

$$H_{o_2} : \beta = 0$$

$$H_{A_2} : \beta \text{ ne } 0$$

where ne denotes not equal.

The H_o represents the null hypothesis, while H_A represents the alternative hypothesis. To reject the null hypothesis, the calculated t-value must exceed the critical t-value given in the t-tables in the appendix. The calculated t-values for $\widehat{a}$ and $\widehat{b}$ are found by:

$$t_a = \frac{\widehat{a} - \alpha}{S_e} \sqrt{\frac{N(M_{XX})}{M_{XX} + (N\overline{X})^2}}$$

$$t_b = \frac{\widehat{b} - \beta}{S_e} \sqrt{\frac{(M_{XX})}{N}}$$

(12.13)

The critical t-value, t_c, for the .05 level of significance with $N - 2$ degrees of freedom, can be found in a t-table in any statistical econometric text. One has a statistically significant regression model if one can reject the null hypothesis that the estimated slope coefficient is equal to zero.

We can create 95% confidence intervals for $\widehat{a}$ and $\widehat{b}$, where the limits of $\widehat{a}$ and $\widehat{b}$ are:

$$\widehat{a} + t_{a/2} S_e^+ \sqrt{\frac{(N\overline{X})^2 + M_{XX}}{N(M_{XX})}}$$

$$\widehat{b} + t_{a/2} S_e \sqrt{\frac{N}{M_{XX}}}$$

(12.14)

To test whether the model is useful, an F-test is performed where:

$$H_o = \alpha = \beta = 0$$

$$H_A = \alpha \text{ ne } \beta \text{ ne } 0$$

$$F = \frac{\sum_{i=1}^{N} Y^2 \div 1 - \widehat{b}^2 \sum_{i=1}^{N} X_i^2}{\sum_{i=1}^{N} \widehat{e}^2 \div N - 2} \tag{12.15}$$

As the calculated F-value exceeds the critical F-value with (1, N- 2) degrees of freedom of 5.99 at the .05 level of significance, the null hypothesis must be rejected. The 95% confidence level limit of prediction can be found in terms of the dependent variable value:

$$\left(\widehat{a} + \widehat{b}X_0\right) + t_{a/2} S_e \sqrt{\frac{N(X_0 - \overline{X})^2}{1 + N + M_{XX}}} \tag{12.16}$$

12.2   Autocorrelation

An estimated regression equation is distorted by the first-order correlation of residuals. That is, the regression error terms are not white noise (random) as is assumed in the general linear model, but are serially correlated where

$$\widehat{\varepsilon}_t = \rho \, \widehat{\varepsilon}_{t-1} + u_t \qquad\qquad t = 1, 2, \ldots N \tag{12.17}$$

$\widehat{\varepsilon}_t$ = regression error term at time t
ρ = first-order correlation coefficient and
u_t = normally and independently distributed random variable

The serial correlation of error terms, known as autocorrelation, is a violation of a regression assumption and may be corrected by the application of the Cochrane-Orcutt procedure.[2] Autocorrelation produces unbiased, the expected value of parameter is the population parameter, but inefficient parameters. The variances of the parameters are biased (too low) among the set of linear unbiased estimators, and the sample t and F-statistics are too large. The Cochrane-Orcutt (CORC) procedure was developed to produce the best linear unbiased estimators (BLUE) given the autocorrelation of regression residuals. The CORC procedure uses the information implicit in the first-order correlative of residuals to produce unbiased and efficient estimators

$$Y_t = \alpha + \beta X_t + \varepsilon_t$$

---

[2]Cochrane D. and G.H. Orcutt. 1949. "Application of Least Squares Regression to Relationships Containing Autocorrelated Error Terms," *Journal of the Amencan Statistical Association* 44: 32-61.

$$\hat{\rho} = \frac{\sum \hat{e}_t \hat{e}_{t-1}}{\sum \hat{e}_t^2 - 1} \tag{12.18}$$

The dependent and independent variables are transformed by the estimated rho, $\hat{\rho}$, to obtain more efficient ordinary least squares estimates:

$$Y_t - \hat{\rho}\, Y_{t-1} = \alpha(1 - \hat{\rho}) + \beta(X_t - \rho X_{t-1}) + ut \tag{12.19}$$

The CORC procedure is an iterative procedure that can be repeated until the coefficients converge. One immediately recognizes that as ρ approaches unity the regression model approaches a first-difference model.

The Durbin-Watson, D-W, statistic was developed to test for the absence of autocorrelation:

$$H_o : \rho = 0$$

One generally tests for the presence of autocorrelation ($\rho = 0$) using the Durbin-Watson statistic:

$$D - W = d = \frac{\sum_{t=2}^{N}(\hat{e}_t - \hat{e}_{t-1})^2}{\sum_{t=2}^{N} \hat{e}_t^2} \tag{12.20}$$

The $\hat{e}$'s represent the ordinary least squares regression residuals, and a two-tailed tail test is employed to examine the randomness of residuals. One rejects the null hypothesis of no statistically significant autocorrelation if:

$$d < d_L \text{ or } d > 4 - d_u$$

where

d_L is the "lower" Durbin-Watson level

and

d_u is the "upper" Durbin-Watson level.

The upper and lower level of Durbin-Watson statistic levels are given in Johnston (1972). The Durbin-Watson statistic is used to test only for first-order correlation among residuals.

$$D = 2(1 - \rho) \tag{12.21}$$

If the first-order correlation of model residuals is zero, the Durbin-Watson statistic is 2. A very low value of the Durbin-Watson statistic, $d < d_L$, indicates positive autocorrelation between residuals and produces a regression model that is not statistically distorted by autocorrelation.

The inconclusive range for the estimated Durbin-Watson statistic is

$$d_L < d < d_u \text{ or } 4 - d_u < 4 - d_u.$$

One does not reject the null hypothesis of no autocorrelation of residuals if $d_u < d < 4 - d_u$.

One of the weaknesses of the Durbin-Watson test for serial correlation is that only first-order autocorrelation of residuals is examined, and one should plot the correlation of residual with various time lags

$$corr\ (\widehat{e}_t, \widehat{e}_{t-k})$$

to identify higher-order correlations among residuals.

In a nonlinear least squares (NLLS) model, one seeks to estimate an equation in which the dependent variable increases by a constant growth rate rather than a constant amount.[3] The nonlinear regression equation is:

$$Y = ab^x \tag{12.22}$$

$$\text{or } \log\ Y = \log\ a + \log\ b\ X$$

The normal equations are derived from minimizing the sum of the squared error terms (as in ordinary least squares) and may be written as:

$$\Sigma(\log\ Y) = N(\log\ a) + (\log\ b) \sum_{i=1}^{N} X$$

$$\Sigma(X \log\ Y) = (\log\ a) \sum_{i=1}^{N} X + (\log\ b) \sum_{i=1}^{N} X^2 \tag{12.23}$$

The solutions to the simplified NLLS estimation equation are:

$$\log \widehat{a} = \frac{\sum\limits_{i=1}^{N} (\log Y)}{N}$$

$$\log \widehat{b} = \frac{\sum\limits_{i=1}^{N} (X \log Y)}{\sum\limits_{i=1}^{N} X^2} \tag{12.24}$$

---

[3] The reader is referred to C.T. Clark and L.L. Schkade, Statistical Analysis for Administrative Decisions (Cincinnati: South-Western Publishing Company, 1979) for an excellent treatment of this topic.

12.3 Estimating Multiple Regression Lines

It may well be that several economic variables influence the variable that one is interested in forecasting. For example, the levels of the gross domestic product (GDP), personal disposable income, or price indices can assert influences on the firm. Multiple regression is an extremely easy statistical tool for researchers and management to employ due to the great proliferation of computer software. The general form of the two independent variable multiple regression is:

$$Y_t = \beta_1 + \beta_2 X_{2t} + \beta_3 X_{3t} + \varepsilon_t, t = 1, \ldots, N \tag{12.25}$$

In matrix notation multiple regression can be written:

$$Y = X\beta + \varepsilon \tag{12.26}$$

Multiple regression requires unbiasedness, the expected value of the error term is zero, and the X's are fixed and independent of the error term. The error term is an identically and independently distributed normal variable. Least squares estimation of the coefficients yields:

$$\widehat{\beta} = \left(\widehat{\beta}_1, \widehat{\beta}_2, \widehat{\beta}_3\right)$$
$$Y = X\widehat{\beta} + e \tag{12.27}$$

Multiple regression, using the least squared principle, minimizes the sum of the squared error terms:

$$\sum_{i=1}^{N} \widehat{e}_1^2 = \widehat{e}'\widehat{e}$$
$$\left(Y - X\widehat{\beta}\right)' \left(Y - X\widehat{\beta}\right) \tag{12.28}$$

To minimize the sum of the squared error terms, one takes the partial derivative of the squared errors with respect to $\widehat{\beta}$ and the partial derivative and set equal to zero.

$$\partial \frac{\left(\widehat{e}'\widehat{e}\right)}{\partial \beta} = -2X'Y + 2X'X\widehat{\beta} = 0$$
$$\widehat{\beta} = (X'X)^{-1}X'Y \tag{12.29}$$

Alternatively, one could solve the normal equations for the two variables to determine the regression coefficients.

$$\Sigma Y = \widehat{\beta}_1 N + \widehat{\beta}_2 \Sigma X_2 + \widehat{\beta}_3 \Sigma X_3$$

$$\Sigma X_2 Y = \widehat{\beta}_1 \Sigma X_2 + \widehat{\beta}_2 X_2{}^2 + \widehat{\beta}_3 \Sigma X_3{}^2 \qquad (12.30)$$

$$\Sigma X_3 Y = \widehat{\beta}_1 \Sigma X_3 + \widehat{\beta}_2 \Sigma X_2 X_3 + \widehat{\beta}_3 \Sigma X_3{}^2 \qquad (12.31)$$

When we solved the normal equations, (12.7), to find the a and b that minimized the sum of our squared error terms in simple linear regression and when we solved the two variable normal equations, equation (12.31) to find the multiple regression estimated parameters, we made several assumptions. First, we assumed that the error term is independently and identically distributed, that is, a random variable with an expected value, or mean of zero, and a finite and constant standard deviation. The error term should not be a function of time, as we discussed with the Durbin-Watson statistic, equation (12.21), nor should the error term be a function of the size of the independent variable(s), a condition known as heteroscedasticity. One may plot the residuals as a function of the independent variable(s) to be certain that the residuals are independent of the independent variables. The error term should be a normally distributed variable. That is, the error terms should have an expected value of zero, and 67.6% of the observed error terms should fall within the mean value plus or minus one standard deviation of the error terms (the so-called "bell curve" or normal distribution). Ninety-five percent of the observations should fall within the plus or minus two standard deviation levels, the so-called 95% confidence interval. The presence of extreme, or influential, observations may distort estimated regression lines and the corresponding estimated residuals. Another problem in regression analysis is the assumed independence of the independent variables in equation (12.31). Significant correlations may produce estimated regression coefficients that are "unstable" and have the "incorrect" signs, conditions that we will observe in Chap. 12. Let us spend some time discussing two problems discussed in this section, the problems of influential observations, commonly known as outliers, and the correlation among independent variables, known as multicollinearity.

There are several methods that one can use to identify influential observations or outliers. First, we can plot the residuals and 95% confidence intervals and examine how many observations have residuals falling outside these limits. One should expect no more than 5% of the observations to fall outside of these intervals. One may find that one or two observations may distort a regression estimate even if there are 100 observations in the database. The estimated residuals should be normally distributed, and the ratio of the residuals divided by their standard deviation, known as standardized residuals, should be a normal variable. We showed, in equation (12.29), that in multiple regression:

$$\widehat{\beta} = (X'X)X'Y$$

The residuals of the multiple regression line is given by:

$$\widehat{e} = Y' - \widehat{\beta}X$$

12.4   Influential Observations and Possible Outliers and the Application of Robust Regression

The standardized residual concept can be modified such that the reader can calculate a variation on that term to identify influential observations. If we delete observation i in a regression, we can measure the impact of observation i on the change in estimated regression coefficients and residuals. Belsley, Kuh, and Welsch (1980) showed that the estimated regression coefficients change by an amount, DFBETA, where:

$$DFBETA_i = \frac{(X'X)^{-1}X'\widehat{e}_i}{1 - h_i} \tag{12.32}$$

where

$$h_i = X_i(X'X)^{-1}X_i'$$

The h_i or "hat" term is calculated by deleting observation i. The corresponding residual is known as the studentized residual, sr, and defined as:

$$sr_i = \frac{\widehat{e}_i}{\widehat{\sigma}\sqrt{1 - h_i}} \tag{12.33}$$

where $\widehat{\sigma}$ is the estimated standard deviation of the residuals. A studentized residual that exceeds 2.0 indicates a potential influential observation [Belsley, et al., 1980]. Another distance measure has been suggested by Cook (1977), which modifies the studentized residual, to calculate a scaled residual known as the Cook distance measure, CookD. As the researcher or modeler deletes observations, one needs to compare the original matrix of the estimated residuals with the modified variance matrix. The COVRATIO calculation performs this calculation, where:

$$COVRATIO = \frac{1}{\left[\frac{n-p-1}{n-p} + \frac{\widehat{e}_{i*}}{(n-p)}\right]^p (1 - h_i)} \tag{12.34}$$

where

n = number of observations
p = number of independent variables
$\widehat{e}_{i*}$ = deleted observations

If the absolute value of the deleted observation is > 2, then the COVRATIO calculation approaches:

$$1 - \frac{3p}{n} \tag{12.35}$$

A calculated COVRATIO that is larger than $\frac{3p}{n}$ indicates an influential observation. The DFBETA, studentized residual, CookD, and COVRATIO calculations may be performed within SAS. See Appendix A for the Belsley, Kuh, and Welsch (1980) influential observations and outlier analysis. The identification of influential data is an important component of regression analysis. One may create variables for use in multiple regression that make use of the influential data, or outliers, to which they are commonly referred.

The modeler can identify outliers, or influential data, and rerun the ordinary least squares regressions on the re-weighted data, a process referred to as robust (ROB) regression. In ordinary least squares, OLS, all data is equally weighted. The weights are 1.0. In robust regression one weights the data universally with its OLS residual, that is, the larger the residual, the smaller the weight of the observation in the robust regression. In robust regression, several weights may be used. We will report the Huber (1973), Beaton-Tukey (1974), and Tukey MM estimate with 99% efficiency, discussed in Maronna, Martin, Yohai, and Salibian-Barrera (2019) weighting schemes in our analysis. In the Huber robust regression procedure, one uses the following calculation to weigh the data:

$$w_i = \left(1 - \left(\frac{|\widehat{e}_i|}{\sigma_i} \right)^2 \right)^2 \tag{12.36}$$

where

$\widehat{e}_i = $ residual i
$\sigma_i = $ standard deviation of residual
$w_i = $ weight of observation i

The intuition is that the larger the estimated residual, the smaller the weight. A second robust re-weighting scheme is calculated from the Beaton-Tukey bisquare criteria where:

$$w_i = \left(1 - \left(\frac{|\widehat{e}_i|}{\frac{\sigma_e}{4.685}} \right)^2 \right)^2, \quad if \ \frac{|\widehat{e}_i|}{\sigma_e} > 4.685;$$

$$1, \quad if \ \frac{|\widehat{e}_i|}{\sigma_e} < 4.685. \tag{12.37}$$

The second major problem is one of multicollinearity, the condition of correlation among the independent variables. If the independent variables are perfectly correlated in multiple regression, then the $(X'X)$ matrix of equation (12.30) cannot be inverted, and the multiple regression coefficients have multiple solutions. In reality, highly correlated independent variables can produce unstable regression coefficients due to an unstable $(X'X)^{-1}$ matrix. Belsley et al. advocate the calculation of a condition number, which is the ratio of the largest latent root of the correlation matrix relative to the smallest latent root of the correlation matrix. A condition number exceeding 30.0 indicates severe multicollinearity.

Efron, Hastie, Johnstone, and Tibshirayi (2004) introduce LAR to the reader by discussing automatic model-building algorithms, including forward selection, all subsets, and back elimination. One can measure the goodness of fit in terms of predictive accuracy, but one uses a different manner, how well model subsets perform in terms of portfolio geometric means using out-of-sample variable weights. LAR is a variation of forward selection; that is the technique selects the variable with the largest absolute correlation, x_{j1}, with the response variable, y, and performs simple linear regression of y on x_{j1}. The regression produces a residual vector orthogonal to x_{j1}, now considered to be the response or dependent variable. One projects the other predictor variables orthogonally to x_{j1} and repeats the selection process. The application of K steps produces a set of predictor variables, x_{j1}, x_{j2}, x_{j3}, ..., x_{jk}, to construct a K-parameter linear model. Hastie et al. (2013) state that forward selection is an aggressive fitting technique that can be overly greedy, eliminating at a second step useful prediction correlated with x_{j1}.

We will use robust regression and LAR techniques in Chap. 14 to model US stock returns.

12.5   The Conference Board Composite Index of Leading Economic Indicators and Real US GDP Growth: A Regression Example Including Much of the Pandemic Period

The composite indexes of leading economic indicator (LEI), coincident, and lagging economic indicators produced by The Conference Board reflect summary statistics for the US economy. Wesley Clair Mitchell of Columbia University constructed the indicators in 1913 to serve as a barometer of economic activity. The leading indicator series was developed to turn upward before aggregate economic activity increased and turn downward before aggregate economic activity diminished. Historically, the cyclical turning points in the leading index have occurred before those in aggregate economic activity, cyclical turning points in the coincident index have occurred at about the same time as those in aggregate economic activity, and cyclical turning points in the lagging index generally have occurred after those in aggregate economic activity.

The Conference Board's components of the composite leading indicators for the year 2018 reflect the work and variables shown in Levanon, Manini, Ozyildirim, and Schaitkin (2015) which continued work of Mitchell (1913, 1927, 1951), Burns and

Table 12.1 TCB LEI descriptions

Monthly Data (Click on series to chart data)

G0M910 - Composite index of 10 leading indicators (2016=100)
A0M001 - Average weekly hours, mfg. (hours)
A0M005 - Average weekly initial claims, unemploy. insurance (thous.)
A0M008 - Mfrs' new orders, consumer goods and materials (mil. 1982 $)
A0M033 - Mfrs' new orders, nondefense capital goods excl. aircraft (mil. chain 1982 $)
A0M029 - Building permits for new private housing units (thous.)
U0M019 - Index of stock prices, 500 common stocks, NSA (1941-43=10)
A0M106 - Money supply, M2 (bil. chain 2012 $) (LEI comp. until Apr '90)
U0M107 - Leading Credit Index™ (std. dev.)
U0M129 - Interest rate spread, 10-year Treasury bonds less federal funds
U0M083 - Consumer expectations, NSA (Copyright, Univ. of Michigan) (LEI comp. until Dec '77)
A0M125 - Avg. consumer expectations for business conditions (std. dev.)
G0M920 - Composite index of 4 coincident indicators (2016=100)
G0M930 - Composite index of 7 lagging indicators (2016=100)
A0M043 - Civilian unemployment rate (pct.)
A0M044 - Unemployment rate 15 weeks and over (pct.)
A0M052 - Personal income (AR bil. chain 2012 $)
A0M051 - Personal income less transfer payments (AR bil. chain 2012 $)
A0Q055 - Gross domestic product (AR bil. Chain 2012 $)

Mitchell (1946), Moore (1961a, b), and Zarnowitz (1992). The Conference Board index of leading indicators, in November 2019, is composed of:

Average weekly hours, manufacturing
Average weekly initial claims for unemployment insurance
Manufacturers' new orders, consumer goods, and materials
ISM® Index of New Orders
Manufacturers' new orders, nondefense capital goods excluding aircraft orders
Building permits, new private housing units
Stock prices, 500 common stocks
Leading Credit Index™
Interest rate spread, 10-year treasury bonds less federal funds[4] (Table 12.1)

The Conference Board composite index of leading economic indicator, LEI, is an equally weighted index in which its components are standardized to produce constant variances. Details of the LEI can be found on The Conference Board website, www.conference-board.org, and the reader is referred to Zarnowitz (1992) for his seminal development of underlying economic assumptions and the theory of the LEI and business cycles. See Appendix B for the current LEI and its components. We transform both real GDP, the output of the economy, and LEI series to stationarity

---

[4]The one change with the first edition of Guerard and Schwartz (2007) is the exclusion of the money supply.

and constant variance by taking the logarithm of each series and creating a first difference of the data, 1959Q1–2020Q3. The growth of real US GDP is correlated with the contemporaneous growth of the US LEI index. The regression coefficient on the LEI variable, 0.132, is statistically significant because the calculated t-value of 3.91 exceeds 1.96, the 5% critical level. One can reject the null hypothesis of no association between the growth rate of US GDP and the growth rate of the LEI. The reader notes, however, that we estimated the regression line with current or contemporaneous, values of the LEI series. There is no forecasting test of the regression relationship.

What we need to know is whether past growth rates in the LEI are statistically correlated with US real GDP growth.

The LEI series was developed to "forecast" future economic activity such that the current growth of the LEI series should be associated with future US GDP growth rates. Alternatively, one can examine the regression association of the current values of real US GDP growth and previous or lagged values of the LEI series. How many lags might be appropriate? Let us estimate regression lines using up to four lags of the US LEI series.

If one estimates a simple regression model using the SAS software, as shown in Table 12.2, the first lag of the DLLEI series produces the expected positive regression coefficient of 0.411, which is highly statistically significant at the 10% level,

Table 12.2 Regression analysis of real GDP

Time Period: 1959Q1 -2020Q3
Dependent: DLGDP
Ordinary Least Squares

constant	DLLEI	L1LEI	L2LEI	L3LEI	L4LEI	R-Squared	F-Statistic
0.007	0.132					0.055	15.26
(9.11)	(3.91)						
0.005		0.411				0.573	328.9
(10.45)		(18.13)					
0.005		0.454	-0.104			0.605	186.7
(11.42)		(18.95)	(-4.23)				
0.005		0.443	-0.135	0.065		0.609	126.7
(10.94)		(18.08)	(-4.66)	(1.97)			
0.005		0.441	-0.131	0.073	-0.016	0.608	94.3
(10.82)		(17.99)	(-4.58)	(1.76)	(-.44)		
(t-statistic)							

Robust Regression constant		L1LEI	L2LEI	L3LEI	L4LEI	Adjusted R-Squared	
0.006		0.29				0.266	
[.0004]		[.0221]					
0.006		0.257	0.040	-0.018	0.040	0.275	
[.0004]		[.0312]	[.0383]	[.0379]	[.0317]		
[standard error]							

having an estimated t-value of 18.13. Many statistical studies often use the 5% level as a minimum acceptable critical level.

The residual plots associated with Table 12.2 suggest that regression outliers can be a large problem, distorting the statistical association between US real GDP growth and the growth in the US LEI index. The largest RStudent observations occur in 2020, as one might expect, particularly in the announcements of Q2 and Q3. The regression diagnostics show influential data, where Studentardized Residuals, exceeding 2 and −2, with large Cook's D statistics, and where DFFITS exceed 2p/sqrt(n), reported in Appendix C, and possible outliers (Fig. 12.1).

If one includes two LEI lags in the estimated regression line, only the first lag in LEI is positive and statistically significant. The second lag of LEI is negative and statistically significant at the 15% level, having an estimated t-value of −4.23. In the regression analysis using three lags of the LEI series, the first and third quarter LEI lags are statistically significant. As in Guerard and Schwartz (2007), we report four

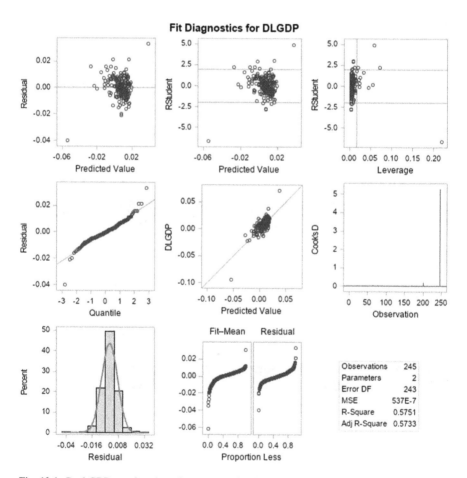

Fig. 12.1 Real GDP as a function of one quarter lag in LEI

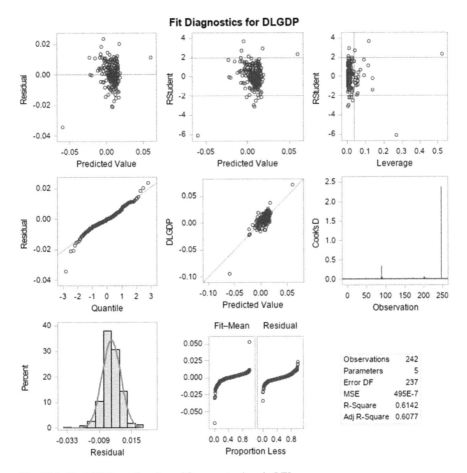

Fig. 12.2 Real GDP as a function of four quarter lags in LEI

quarters of LEI lags, based on Guerard (2001). The first and third quarter LEI lags are statistically significant in the GDP OLS regression analysis (Figs. 12.2, 12.3, 12.4, 12.5 and 12.6).

We use the SAS robust regression (PROC ROBUSTREG) procedure and the Huber and Tukey bisquare weighting scales.

Huber ROB US Q GDP with One Quarter Lag in LEI, 1959-2020

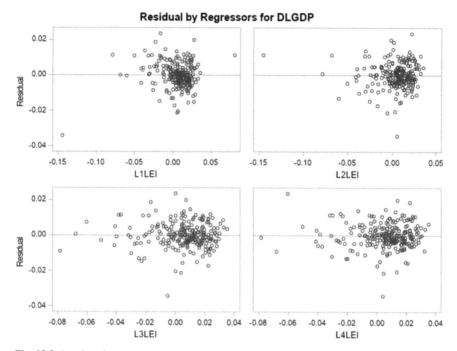

Fig. 12.2 (continued)

The ROBUSTREG Procedure

Model Information

Data Set	WORK.USLEI
Dependent Variable	DLGDP
Number of Independent Variables	1
Number of Observations	245
Missing Values	2
Method	M Estimation

Number of Observations Read	247
Number of Observations Used	245
Missing Values	2

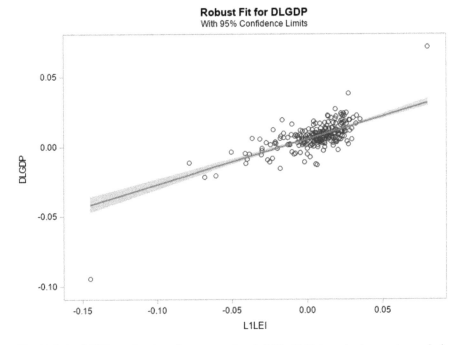

Fig. 12.3 Real GDP as a function of one quarter lags in LEI with Huber robust regression analysis

Parameter Information

Parameter	Effect
Intercept	Intercept
L1LEI	L1LEI

Parameter Estimates

Parameter	DF	Estimate	Standard Error	95% Confidence Limits		Chi-Square	Pr > ChiSq
Intercept	1	0.0055	0.0004	0.0047	0.0063	186.25	<.0001
L1LEI	1	0.3252	0.0188	0.2883	0.3621	298.23	<.0001
Scale	1	0.0056					

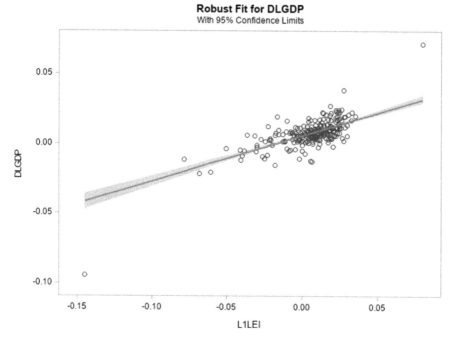

Fig. 12.4 Real GDP as a function of one quarter lags in LEI with the Beaton-Tukey bisquare robust regression analysis

Diagnostics

Obs	Standardized Robust Residual	Outlier
6	-3.6753	*
76	4.2124	*
87	3.4701	*
243	-3.5655	*
244	-9.3974	*
245	7.1275	*

The (largest) regression outliers are identified in 2020Q1, 2020Q2, and 2020Q3.

Diagnostics Summary

Observation Type	Proportion	Cutoff
Outlier	0.0245	3.0000

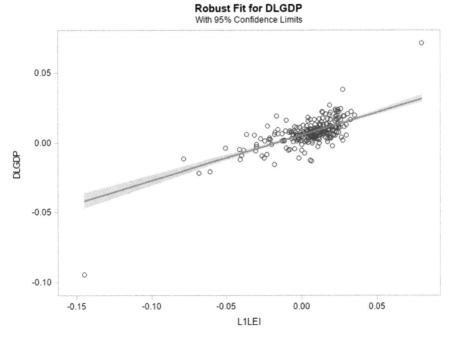

Fig. 12.5 Real GDP as a function of four quarter lags in LEI with the Beaton-Tukey bisquare M-estimation robust regression analysis

Goodness-of-Fit	
Statistic	**Value**
R-Square	0.3527
AICR	305.3211
BICR	312.6367
Deviance	0.0094

The Beaton-Tukey bisquare estimation is not statistically different from the Huber robust estimation.

The Beaton- Tukey Bisquare ROB US Q GDP with One Quarter Lag in LEI, 1959-2020

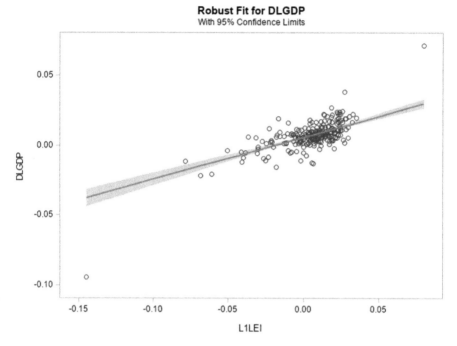

Fig. 12.6 Real GDP as a function of a quarter lag in LEI with the Beaton-Tukey bisquare S-estimation robust regression analysis

The ROBUSTREG Procedure

Model Information	
Data Set	WORK.USLEI
Dependent Variable	DLGDP
Number of Independent Variables 1	
Number of Observations	245
Missing Values	2
Method	M Estimation

Number of Observations Read	247
Number of Observations Used	242
Missing Values	5

Parameter Information

Parameter	Effect
Intercept	Intercept
L1LEI	L1LEI

Parameter Estimates

Parameter	DF	Estimate	Standard Error	95% Confidence Limits		Chi-Square	Pr > ChiSq
Intercept	1	0.0055	0.0004	0.0047	0.0063	186.25	<.0001
L1LEI	1	0.3252	0.0188	0.2883	0.3621	298.23	<.0001
Scale	1	0.0056					

Diagnostics

Obs	Standardized Robust Residual	Outlier
6	-3.6753	*
76	4.2124	*
87	3.4701	*
243	3.5655	*
244	9.3974	*
245	7.1275	*

The (largest) regression outliers are again identified in 2020Q1, 2020Q2, and 2020Q3 with the Beaton-Tukey bisquare regression procedure.

Diagnostics Summary

Observation Type	Proportion	Cutoff
Outlier	0.0245	3.0000

Goodness-of-Fit

Statistic	Value
R-Square	0.3527
AICR	305.3211
BICR	312.6367
Deviance	0.0094

We use the SAS robust regression (PROC ROBUSTREG) procedure and the Beaton-Tukey bisquare weighting scales with four quarter lags of LEI. The first quarter lag in LEI is highly statistically significant in US real GDP regression modeling.

The Beaton-Tukey Bisquare ROB US Q GDP with Four Quarter LEI Lags 1959-2020

The ROBUSTREG Procedure

Model Information

Data Set	WORK.USLEI
Dependent Variable	DLGDP
Number of Independent Variables	4
Number of Observations	242
Missing Values	5
Method	M Estimation

Number of Observations Read	247
Number of Observations Used	242
Missing Values	5

Parameter Estimates

Parameter	DF	Estimate	Standard Error	95% Confidence Limits		Chi-Square	Pr > ChiSq
Intercept	1	0.0055	0.0004	0.0047	0.0063	187.44	<.0001
L1LEI	1	0.2510	0.0206	0.2107	0.2912	149.10	<.0001
L2LEI	1	0.0514	0.0243	0.0039	0.0990	4.49	0.0340
L3LEI	1	-0.0044	0.0344	-0.0718	0.0630	0.02	0.8984
L4LEI	1	0.0207	0.0295	-0.0371	0.0785	0.49	0.4828
Scale	1	0.0058					

Diagnostics

Obs	Standardized Robust Residual	Outlier
3	-3.5436	*
73	4.3725	*
84	3.1382	*
240	-3.3421	*
241	-10.9850	*
242	9.1967	*

The (largest) regression outliers are identified in the 2019 time period.

Diagnostics Summary

Observation Type	Proportion	Cutoff
Outlier	0.0248	3.0000

Goodness-of-Fit

Statistic	Value
R-Square	0.3200
AICR	238.2955
BICR	258.2680
Deviance	0.0078

Beaton-Tukey S-Estimation Robust Regression EFF=0.99 US Q GDP LEI 1959-2020

The ROBUSTREG Procedure

Model Information

Data Set	WORK.USLEI
Dependent Variable	DLGDP
Number of Independent Variables	1
Number of Observations	245
Missing Values	2
Method	S Estimation

Number of Observations Read	247
Number of Observations Used	245
Missing Values	2

S Profile

Total Number of Observations	245
Number of Coefficients	2
Subset Size	2
Chi Function	Tukey
K0	7.0410
Breakdown Value	0.0570
Efficiency	0.9900

Parameter Estimates

Parameter	DF	Estimate	Standard Error	95% Confidence Limits		Chi-Square	Pr > ChiSq
Intercept	1	0.0057	0.0004	0.0049	0.0065	194.33	<.0001
L1LEI	1	0.2994	0.0220	0.2563	0.3424	185.62	<.0001
Scale	0	0.0067					

Diagnostics

Obs	Mahalanobis Distance	Robust MCD Distance	Leverage	Standardized Robust Residual	Outlier
6	0.0687	0.3656		-3.0799	*
43	1.2293	2.3623	*	-0.1653	
44	1.4513	2.7038	*	0.4737	
59	1.3779	2.5910	*	-1.1002	
60	1.3151	2.4944	*	0.4909	
61	1.5225	2.8134	*	-1.0938	
62	2.7033	4.6300	*	0.8397	
63	2.2261	3.8958	*	-0.8515	
76	1.0698	1.1745		3.6001	*
82	1.2546	2.4012	*	0.4531	
83	1.2758	2.4339	*	0.5682	
84	3.1835	5.3686	*	-1.2472	
89	1.3555	2.5566	*	1.9531	
90	1.4765	2.7427	*	-1.3541	
126	2.1795	3.8241	*	-0.4329	
127	2.2595	3.9472	*	0.3100	
167	1.7087	3.0999	*	0.0730	
168	2.0371	3.6051	*	1.6772	
170	1.5835	2.9073	*	0.7917	
195	1.7417	3.1507	*	-0.3320	
196	1.7891	3.2235	*	1.3455	
197	2.1092	3.7159	*	0.0587	
198	3.5533	5.9376	*	-1.0643	
199	4.0451	6.6942	*	0.9758	
200	1.7148	3.1092	*	0.2905	
244	7.2539	11.6305	*	-8.4535	*
245	3.6287	5.1110	*	6.2439	*

Huge outliers remain in Tukey S-Estimation in 2020Q2 and 2020Q3.

Diagnostics Summary

Observation Type	Proportion	Cutoff
Outlier	0.0163	3.0000
Leverage	0.1020	2.2414

Goodness-of-Fit

Statistic	Value
R-Square	0.4180
Deviance	0.0000

The Beaton-Tukey bisquare M and MM estimation models do not produce statistically different from the Huber robust estimation. The presence of influential observations and outliers requires a diligent researcher to employ robust regression modeling techniques. The choice of the weighting function, ten of which can be accessed in PROC REBOUSTREG in SAS produce virtually identical estimated coefficients and statistics of goodness-of-fit, a result reported in Guerard, Xu, and Wang (2019).

12.6 The Conference Board Composite Index of Leading Economic Indicators and the US Unemployment Rate: Another Regression Example Including Much of the Pandemic Period

The unemployment rate is one of the most important measures of an economy and has been extensively studied since the early twentieth century. The US unemployment rate is clearly a major political issue in the current pandemic. The unemployment rate has been linked to traditional business cycles (Mitchell, 1913; Keynes, 1936; Harberler, 1937; Burns and Mitchell, 1946; Samuelson, 1948; Klein, 1950; and Zarnowitz, 1992). It was generally agreed that during an expansion, employment, production, prices, money, wages, interest rates, and profits are usually rising, while the reverse occurs during a contraction.

Economic theory indicates that an appropriate level of unemployment is the key of economic development. A high unemployment rate leads to human suffering and many other negative societal consequences, while an extremely low unemployment rate leads to an increase in labor cost and subsequent inflation. One of the certain missions of government monetary policy is to control the unemployment rate at the ideal level while controlling inflation. Hence it is vital to be able to accurately forecast the unemployment rate in the near and long-term future, based on limited current information of the economy.

In this section, we revisit the approaches used in the seminal paper of Montgomery, Zarnowitz, Tsay, and Tiao (Montgomery et al., 1998), hereafter denoted as MZTT, on forecasting the US unemployment rate. Based on monthly observations in the period from 1959 to 1993, MZTT demonstrated that time series models are useful in predicting 1-month to 5-month ahead unemployment rates and they compared the out-of-sample prediction performance of a range of models. More than 20 years have passed and it is useful and necessary to re-exam these models post-publication, as well as to investigate the impact of new phenomena since, including the financial crisis of 2008. We pay special attention to the relationship between the US unemployment rate and the lagged US weekly unemployment claim, as well as the US leading economic indicator (LEI) series constructed by The Conference Board.

MZTT modeled the change in the US unemployment rate as a function of the weekly unemployment claims time series, 1948–1993. MZTT reported that nonlinear models, such as TAR reduced forecasting errors by as much as 28%, being more effective in periods of economic contraction and rising unemployment. One would expect that an increase in weekly unemployment claims should lead to an increase in the US unemployment rate. One would expect that an increase in the US leading economic indicator (LEI) time series should lead to a decrease in the US unemployment rate.

We replicate and extend the MZTT analysis. We report that the US LEI is negatively and statistically highly significantly associated with the change in the US unemployment rate during the 1959–2020 time period. This is particularly true for a one-month lag in the LEI. This is true during the COVID pandemic. See Table 12.3. We also report that the US weekly unemployment claims time series, reported on Thursday mornings, is positively and highly statistically significantly associated with the US unemployment rate during the 1959–2020 time period, particularly for lags one- and three-month variables. This is true during the COVID pandemic. See Tables 12.3a and 12.3b.

A simple OLS estimation of a model of the change in the unemployment with (differenced-logged) one-month LEI and WKUCL claims data is highly statistically significant, but the LEI variable has an incorrect (positive) sign of its coefficient.

$$DUE = -.010 + 3.817 * L1LEI + 3.364 \, L1WKUCL, \text{Adjusted R-squared}$$
$$= .657, F = 708.6$$

$$(t) \, (-1.05) \, (2.56) \, (31.00)$$

A robust (bisquare) estimation of a model of the change in the unemployment with (differenced-logged) one-month LEI and WKUCL claims data is highly statistically significant, and the LEI variable has the correct (negative) sign of its coefficient.

$$DUE = .011 - 8.894 * L1LEI + .264 \, L1WKUCL, \text{Adjusted R-squared}$$
$$= .118, AICR = 791.3$$

$$[s.e.] \, [.006] \, [1.053] \, [.137]$$

How can one optimally estimate the number of lags in a regression model? This topic is addressed in the next chapter

Table 12.3a Regression analysis of DUE

Time Period: 1959 -2020
Dependent: First-Differenced U.S. Unemployment Rate

Ordinary Least Squares

| constant | Independent Variable: U.S. Leading Economic Indicator Time Series | | | | | | Adjusted | |
	L1LEI	L2LEI	L3LEI	L4LEI	L5LEI	L6LEI	R-Squared	F-Statistic
0.046	-24.946						0.21	197.9
(3.16)	(-14.07)							
0.036	-31.992	12.684					0.245	121.6
(2.50)	(-15.34)	(2.09)						
0.034	-32.099	11.774	1.828				0.246	81.1
(2.40)	(-15.33)	(5.02)	(2.09)					
0.034	-32.283	11.322	1.321	1.085			0.245	60.8
(2.45)	(-15.26)	(5.03)	(0.56)	(0.51)				
0.033	-32.419	11.679	1.285	0.227	1.711		0.245	48.7
(2.26)	(-15.27)	(4.94)	(0.55)	(0.10)	(0.80)			
0.034	-32.322	11.729	1.452	0.211	2.447	-1.537	0.244	40.6
(2.31)	(-15.17)	(4.96)	(0.61)	(0.09)	(1.03)	(-.72)		

(t-statistic)

Robust Regression

| constant | Independent Variable: U.S. Leading Economic Indicator Time Series | | | | | | Adjusted | |
	L1LEI	L2LEI	L3LEI	L4LEI	L5LEI	L6LEI	R-Squared	AICR
-0.003	-9.975						0.115	791.4
(-.33)	[1.646]							
-0.006	-5.453	-7.356					0.155	803.8
[.006]	[1.057]	[1.052]						
-0.003	-3.823	-5.491	-4.098				0.16	823.1
[.006]	[1.102]	[1.126]	[1.076]					
-0.003	-3.278	-4.651	-3.64	-2.338			0.164	818.6
[.006]	[1.118]	[1.178]	[1.171]	[1.103]				
0.013	-3.277	-4.591	-3.553	-2.28	-0.384		0.163	814.5
[.006]	[1.120]	[1.182]	[1.194]	[1.148]	[0.944]			
0.013	-3.324	-4.689	-3.342	-2.016	0.254	-1.206	0.165	819.1
[.006]	[1.117]	[1.18]	[1.194]	[1.16]	[1.102]	[1.012]		

[standard error]

12.7   Summary and Conclusions

This chapter develops the theory of ordinary least squares, influential observations, and possible outlier issues. We examined the regressive association between real US GDP and the leading economic indicators during the 1959Q2–2020Q3 time period. We report that The Conference Board leading economic indicators are statistically significant in the regression analysis of real US GDP, in OLS and robust regression analysis. We report that The Conference Board leading economic indicators are statistically significant in the regression analysis of changes in the US unemployment rate, in OLS and robust regression analysis during the 1959–2020 time period. Yes, we included the pandemic period in our estimation period. We will use robust regression in Chap. 14, for estimating the determinants of cross-sectional stock returns in Japan and the United States and in Chap. 19, for non-US and global stock universes. Because of its importance, there is extensive literature on forecasting the unemployment rate, which we further address in Chap. 13.

Table 12.3b Regression analysis of DUE

Time Period: 1959 -2020
Dependent: First-Differenced U.S. Unemployment Rate

Ordinary Least Squares

constant	Independent Variable: U.S. Weekly Unemployment Claims Time Series						Adjusted R-Squared	F-Statistic
	L1WKUCL	L2WKUCL	L3WKUCL	L4WKUCL	L5WKUCL	L6WKUCL		
-0.003	3.192						0.654	1399.1
(-.33)	(37.40)							
-0.002	3.192	-0.554					0.673	761.5
(-.24)	(37.40)	(-6.61)						
-0.002	3.192	-0.588	0.216				0.675	511.9
(-.326)	(37.40)	(-6.95)	(2.05)					
-0.003	3.192	-0.584	0.213	0.020			0.675	382.9
(-.327)	(37.40)	(-6.76)	(2.47)	(0.23)				
-0.001	3.295	-0.592	0.153	0.061	-0.290		0.679	312.8
(-.318)	(38.83)	(-6.90)	(1.75)	(-0.71)	(-3.40)			
-0.002	3.316	-0.584	0.164	0.109	-0.320	0.220	0.682	263.3
(-.25)	(39.01)	(-6.83)	(1.88)	(1.24)	(-3.72)	(2.54)		

(t-statistic)

Robust Regression constant	Independent Variable: U.S. Weekly Unemployment Claims Time Series						Adjusted R-Squared	AICR
	L1WKUCL	L2WKUCL	L3WKUCL	L4WKUCL	L5WKUCL	L6WKUCL		
-0.008	0.912						0.05	839
[.006]	[.117]							
-0.006	0.892	0.911					0.117	803.6
[.006]	[.117]	].107]						
-0.003	0.89	1.02	0.642				0.144	903.9
[.006]	[.117]	[.105]	[.105]					
-0.003	0.794	0.895	0.615	0.32			0.135	816.5
[.006]	[.117]	[.114]	[.111]	[.105]				
0.013	0.792	0.864	0.601	0.352	0.414		0.15	803.7
[.006]	[.114]	[.114]	[.110]	[.1104]	[.100]			
0.013	0.819	0.88	0.654	0.422	0.38	0.102	0.153	807.2
[.006]	[.115]	[.114]	[.112]	[.109]	[.101]	[0.60]		

[standard error]

Appendices

Appendix A: Influential Observations and Outlier Detection

Influential observations are identified as observations, that when omitted, significantly influence estimated model coefficients and errors.

$$\widehat{y} = x\widehat{\beta} \qquad \widehat{y}(i) = x\widehat{\beta}_\varepsilon$$

$$z_i = (x_i, y_i) \text{ when } z_i \text{ is omitted}$$

$$r_i = r_i(\beta)$$

Belsley, Kuh, and Welsch (1980) reported a set of criteria to identify influential observations. SAS reports the Cook Distance, Rstudent, COVRATIO, DFBETA, and DFFITS criteria.

The Cook distance of z_i is

$$D_i = \frac{1}{p^* s^2} \| \widehat{y}(i) - \widehat{y} \|^2$$

$$p^* = rank\ (X), s^2 = \frac{1}{n - p^*} \sum_{i-1}^{n} r_i^2$$

H, the Hat Matrix, has a diagonal element, $h_1, \ldots h_n$ are the leverages of $x_1, \ldots x_n$

$$H = X(X'X)^{-1} x_i$$

$$h_i = X_i^i(X'X)^{-1} x_i$$

When observation i is omitted

$$D_i = \frac{r_i^2}{s^2} \frac{h_i}{p^*(1 - h_i)^2}$$

Student's version of $r_{(i)} =$

$$s_{(i)}^2 = s(i) = \frac{n - p - 1}{n - p} s^2 - \frac{e_i^2}{(n - p - 1)(1 - h_i)}$$

$$COVRATIO = \left[\left(\frac{n - p - 1}{n - p} \right) + \frac{e_i^{*2}}{n - p} \right)^P (1 - h_i) \right]$$

$DFBETA = change\ in\ regression\ coefficients\ iluminating\ observation\ i$

Compare to variance of b_j

$$\sigma^2 (X^1 X)^{-1} x_i$$

$$= b_j - b_{j(i)} = \frac{c_{ji} - e_i}{1 - h_i}$$

$$var(b_j) = \sigma^2 \sum_{k-1}^{n} c_j k^2$$

$$DFFITS = \left[\frac{h_i}{1 - h_i}\right]^{1/2} \frac{e_i}{s_i\left(\sqrt{1 - h_i}\right)}$$

RSTUDENT

$$e_i^* = \frac{e_i}{s(i)\sqrt{1 - h_i}}$$

$$COVRATIO = \frac{1}{\left[\frac{n-p-1}{n-p} + \frac{e_i^{*2}}{n-p}\right]^P (1 - h_i)}$$

Points with $COVRATIO - 1$ investigated for $3p/n$

$$\sqrt{1 - h_i}\, \frac{r(i)}{s(i)} = \frac{1}{\sqrt{1 - h_i}}\, \frac{v_i}{s(i)}$$

The US Leading Economic Indicators

Let us follow The Conference Board Components and their definitions, as of November 29, 2019:

BCI-01 Average Weekly Hours, Manufacturing

The average hours worked per week by production workers in manufacturing industries tend to lead the business cycle because employers usually adjust work hours before increasing or decreasing their workforce.

BCI-05 Average Weekly Initial Claims for Unemployment Insurance

The number of new claims filed for unemployment insurance is typically more sensitive than either total employment or unemployment to overall business conditions, and this series tends to lead the business cycle. It is inverted when included in the leading index; the signs of the month-to-month changes are reversed, because initial claims increase when employment conditions worsen (i.e., layoffs rise, and new hirings fall).

BCI-08 Manufacturers' New Orders, Consumer Goods and Materials (in 1982 $)

These goods are primarily used by consumers. The inflation-adjusted value of new orders leads actual production because new orders directly affect the level of both unfilled orders and inventories that firms monitor when making production decisions. The Conference Board deflates the current dollar orders data using price indexes constructed from various sources at the industry level and a chain-weighted aggregate price index formula.

BCI-130 ISM New Order Index

This index reflects the levels of new orders from customers. As a diffusion index, its value reflects the number of participants reporting increased orders during the previous month compared to the number reporting decreased orders, and this series tends to lead the business cycle. When the index has a reading of greater than 50, it is an indication that orders have increased during the past month. This index, therefore, tends to lead the business cycle. ISM new orders are based on a monthly survey conducted by Institution for Supply Management (formerly known as National Association of Purchasing Management). The Conference Board takes normalized value of this index as a measure of its contribution to LEI.

BCI-33 Manufacturers' New Orders, Nondefense Capital Goods Excl. Aircraft (in 1982 $)

This index, combing with orders from aircraft (in inflation-adjusted dollars), is the producers' counterpart to BCI-08.

BCI-29 Building Permits, New Private Housing Units

The number of residential building permits issued is an indicator of construction activity, which typically leads most other types of economic production.

BCI-19 Stock Prices, 500 Common Stocks

Standard and Poor's 500 stock index reflects the price movements of a broad selection of common stocks traded on the New York Stock Exchange. Increases (decreases) of the stock index can reflect both the general sentiments of investors and the movements of interest rates, which is usually another good indicator for future economic activity.

BCI-107 Leading Credit Index™

This index consists of six financial indicators: 2-years Swap Spread (real-time), LIBOR 3 month less 3-month Treasury Bill yield spread (real time), Debit balances at margin account at broker dealer (monthly), AAII Investors Sentiment Bullish (%) less Bearish (%) (weekly), Senior Loan Officers C&I loan survey–Bank tightening Credit to Large and Medium Firms (quarterly), and Security Repurchases (quarterly) from the Total Finance Liabilities section of Federal Reserve's flow of fund report. Because of these financial indicators' forward looking content, LCI leads economic activities.

BCI-129 Interest Rate Spread, 10-Year Treasury Bonds Less Federal Funds

The spread or difference between long and short rates is often called the yield curve. This series is constructed using the 10-year treasury bond rate and the federal funds rate, an overnight interbank borrowing rate. It is felt to be an indicator of the stance of monetary policy and general financial conditions because it rises (falls) when short rates are relatively low (high). When it becomes negative (i.e., short rates are higher than long rates and the yield curve inverts), its record as an indicator of recessions is particularly strong.

BCI-125 Avg. Consumer Expectations for Business and Economic Conditions

This index reflects changes in consumer attitudes concerning future economic conditions and, therefore, is the only indicator in the leading index that is completely expectations-based. It is an equally weighted average of consumer expectations of business and economic conditions using two questions, Consumer Expectations for Economic Conditions 12-months ahead from Surveys of Consumers conducted by Reuters/University of Michigan and Consumer Expectations for Business Conditions 6-months ahead from Consumer Confidence Survey by The Conference Board. Responses to the questions concerning various business and economic conditions are classified as positive, negative, or unchanged.

Appendix C: Identifying Influential Observations in a Regression

In the following regression estimate, is the differenced-log (DLOG) Leading Economic Indicator, LEI, lagged one period, L1LEI, and the differenced-log (DLOG) Weekly Unemployment Claims, lagged one period, L1WKUCL, a statistically significant determinant of the differenced unemployment rate (DUE)? Why?

The REG Procedure

Model: MODEL1
Dependent Variable: DUE
Analysis of Variance

Source	DF	Sum of Squares	Mean Square	F Value	Pr > F
Model	2	1.30406	0.65203	496.48	<.0001
Error	737	0.96790	0.00131		
Corrected Total	739	2.27196			

Root MSE	0.03624	R-Square	0.5740	
Dependent Mean	0.00021158	Adj R-Sq	0.5728	
Coeff Var	17128			

Parameter Estimates

Variable	DF	Parameter Estimate	Standard Error	t Value	Pr > |t|
Intercept	1	-0.00035475	0.00139	-0.26	0.7985
L1LEI	1	0.01676	0.21061	0.08	0.9366
L1WKUCL	1	0.37861	0.01530	24.74	<.0001

Note in the OLS-estimated regression only the one-period lagged Weekly Unemployment Claims variable is statistically significant.

Let us identify potentially influential observation on the basis of Studentized Residuals, the hat diagonal elements, DFFITS, and DFBETAS.

Model: MODEL1
Dependent Variable: DUE

Output Statistics

Obs	Residual	RStudent	Hat Diag H	Cov Ratio	DFFITS	DFBETAS Intercept	L1LEI	L1WKUCL
1	.	.	.	.	.	.	.	.
2	.	.	.	.	.	.	.	.
3	-0.0418	-1.1557	0.0055	1.0041	-0.0858	-0.0201	-0.0737	-0.0372
4	-0.0372	-1.0287	0.0044	1.0041	-0.0681	-0.0239	-0.0455	-0.0019
5	0.001244	0.0343	0.0017	1.0058	0.0014	0.0013	-0.0002	-0.0006
6	-0.0225	-0.6227	0.0039	1.0064	-0.0390	-0.0130	-0.0316	-0.0206
7	0.003085	0.0852	0.0020	1.0060	0.0038	0.0025	0.0018	0.0021
8	0.0109	0.3013	0.0016	1.0053	0.0122	0.0093	0.0045	0.0045
9	0.0185	0.5120	0.0024	10055	0.0253	0.0185	-0.0021	0.0118
10	0.0635	1.7568	0.0028	0.9944	0.0937	0.0774	-0.0525	-0.0663
11	-0.0341	-0.9421	0.0034	1.0039	-0.0550	-0.0313	-0.0051	-0.0364
12	-0.1357	**-3.7855**	0.0033	**0.9508**	**-0.2170**	-0.1517	0.0701	-0.0747
13	0.0731	**2.0322**	0.0098	**0.9971**	**0.2019**	0.0491	0.0880	-0.0750
14	-0.0870	**-2.4100**	0.0014	**0.9821**	-0.0909	-0.0888	0.0140	-0.0025
15	0.1318	**3.6705**	0.0019	**0.9527**	**0.1609**	0.1512	-0.0746	-0.0820
16	-0.0802	**-2.2226**	0.0027	0.9868	**-0.1161**	-0.0778	0.0013	-0.0637
17	-0.007407	-0.2044	0.0015	1.0054	-0.0078	-0.0072	-0.0001	0.0017
18	0.0303	0.8354	0.0020	1.0032	0.0370	0.0273	0.0067	0.0195
19	0.0119	0.3291	0.0016	1.0052	0.0130	0.0104	0.0044	0.0041
20	0.002460	0.0679	0.0015	1.0056	0.0027	0.0024	0.0000	0.0007
21	-0.0483	-1.3352	0.0029	0.9997	-0.0723	-0.0355	-0.0397	-0.0522
22	0.1168	**3.2478**	0.0019	**0.9639**	**0.1411**	0.1333	-0.0644	-0.0700
23	-0.0224	-0.6197	0.0023	1.0048	-0.0298	-0.0175	-0.0150	-0.0188
24	0.0674	1.8644	0.0019	0.9919	0.0820	0.0772	-0.0410	-0.0110
25	0.003986	0.1100	0.0015	1.0056	0.0043	0.0043	-0.0015	-0.0012
26	0.0329	0.9095	0.0045	1.0052	0.0609	0.0179	0.0501	0.0381
27	-0.0332	-0.9166	0.0032	1.0038	-0.0517	-0.0239	-0.0289	-0.0386
28	0.0621	1.7199	0.0041	0.9962	**0.1109**	0.0461	0.0550	-0.0227
29	0.0121	0.3343	0.0036	1.0072	0.0200	0.0074	0.0157	0.0102
30	-0.004667	-0.1289	0.0026	1.0066	-0.0066	-0.0035	-0.0037	-0.0001
31	0.0410	1.1325	0.0039	1.0027	0.0705	0.0261	0.0501	0.0102
32	-0.0748	**-2.0715**	0.0029	0.9896	**-0.1114**	-0.0512	-0.0761	-0.0696
33	0.0524	1.4511	0.0053	1.0008	0.1056	0.0300	0.0773	0.0104
34	-0.0460	-1.2718	0.0015	0.9990	-0.0499	-0.0446	-0.0004	-0.0138
35	-0.0332	-0.9180	0.0036	1.0043	-0.0552	-0.0225	-0.0360	-0.0030
36	-0.0169	-0.4684	0.0073	1.0105	-0.0401	-0.0063	-0.0361	-0.0224
37	-0.0233	-0.6422	0.0017	1.0041	-0.0266	-0.0197	-0.0106	-0.0018
38	-0.0583	-1.6129	0.0015	0.9950	-0.0620	-0.0522	-0.0165	-0.0158
39	0.0251	0.6942	0.0042	1.0063	0.0449	0.0142	0.0367	0.0192
40	0.0108	0.2970	0.0014	1.0052	0.0113	0.0106	-0.0003	-0.0024
41	-0.0124	-0.3411	0.0017	1.0054	-0.0143	-0.0102	-0.0066	-0.0027
42	-0.0223	-0.6161	0.0017	1.0043	-0.0258	-0.0207	-0.0030	-0.0112
...								
75	-0.0894	**-2.4768**	0.0019	**0.9812**	-0.1076	-0.0717	-0.0549	-0.0467
76	0.0381	1.0538	0.0026	1.0022	0.0539	0.0278	0.0338	0.0084
77	-0.0426	-1.1768	0.0016	1.0001	-0.0474	-0.0360	-0.0194	-0.0119
78	0.0231	0.6380	0.0018	1.0042	0.0269	0.0226	0.0002	-0.0101
79	-0.0455	-1.2575	0.0016	0.9993	-0.0509	-0.0383	-0.0213	-0.0139
80	-0.0103	-0.2842	0.0020	1.0057	-0.0126	-0.0081	-0.0066	-0.0061
81	-0.0499	-1.3798	0.0031	0.9994	-0.0768	-0.0348	-0.0478	-0.0549
82	0.0248	0.6859	0.0032	1.0054	0.0387	0.0255	-0.0031	-0.0248
83	-0.007031	-0.1942	0.0034	1.0074	-0.0114	-0.0045	-0.0084	-0.0029
84	-0.0311	-0.8593	0.0035	1.0045	-0.0506	-0.0192	-0.0393	-0.0280
85	0.0110	0.3030	0.0025	1.0063	0.0153	0.0079	0.0100	0.0039
86	-0.0781	**-2.1625**	0.0030	**0.9882**	**-0.1187**	-0.0552	-0.0724	-0.0843

Obs	Residual	RStudent	Hat Diag H	Cov Ratio	DFFITS	DFBETAS Intercept	L1LEI	L1WKUCL
87	0.005916	0.1633	0.0015	1.0055	0.0063	0.0053	0.0017	0.0004
88	0.0696	1.9304	0.0051	0.9941	**0.1384**	0.0674	0.0079	-0.0879
89	0.0331	0.9151	0.0015	1.0021	0.0353	0.0299	0.0086	0.0007
90	-0.0521	-1.4414	0.0028	0.9984	-0.0760	-0.0628	0.0433	0.0012
91	-0.004675	-0.1290	0.0018	1.0058	-0.0054	-0.0053	0.0026	0.0012
92	-0.008599	-0.2373	0.0017	1.0056	-0.0098	-0.0095	0.0041	0.0012
93	-0.0181	-0.4987	0.0023	1.0053	0.0237	-0.0218	0.0146	0.0120
94	-0.005170	-0.1427	0.0018	1.0058	-0.0060	-0.0058	0.0028	0.0013
95	-0.0366	-1.0106	0.0017	1.0016	-0.0417	-0.0404	0.0171	0.0044
96	0.0440	1.2151	0.0014	0.9995	0.0461	0.0440	-0.0044	0.0055
97	0.006667	0.1840	0.0017	1.0057	0.0077	0.0070	-0.0018	0.0013
98	-0.004221	-0.1165	0.0019	1.0059	-0.0050	-0.0037	-0.0014	0.0009
99	-0.0710	-1.9670	0.0054	0.9937	**-0.1444**	-0.0606	-0.0261	-0.1119
100	0.006951	0.1918	0.0015	1.0054	0.0074	0.0062	0.0019	0.0002
101	-0.0234	-0.6458	0.0021	1.0045	-0.0296	-0.0193	-0.0118	-0.0176
102	0.0702	1.9423	0.0029	0.9917	0.1056	0.0645	0.0175	-0.0484
103	-0.004190	-0.1157	0.0023	1.0063	-0.0056	-0.0033	-0.0028	0.0000
104	-0.0171	-0.4714	0.0022	1.0054	-0.0223	-0.0131	-0.0123	-0.0130
105	0.0352	0.9724	0.0030	1.0032	0.0535	0.0274	0.0257	-0.0077
106	0.0450	1.2437	0.0017	0.9995	0.0514	0.0374	0.0225	0.0191
107	-0.0273	-0.7530	0.0014	1.0031	-0.0278	-0.0259	-0.0022	-0.0023
108	0.005600	0.1546	0.0021	1.0061	0.0072	0.0053	0.0006	-0.0030
109	-0.0545	-1.5096	0.0066	1.0015	**-0.1234**	-0.0235	-0.1040	-0.0931
110	0.0256	0.7061	0.0014	1.0034	0.0260	0.0245	0.0015	0.0013
111	0.007985	0.2205	0.0024	1.0063	0.0108	0.0085	-0.0024	-0.0068
112	-0.0539	-1.4896	0.0014	0.9964	-0.0549	-0.0525	-0.0001	0.0018
113	-0.004857	-0.1340	0.0014	1.0054	-0.0050	-0.0045	-0.0006	-0.0008
114	0.0614	1.6976	0.0018	0.9942	0.0722	0.0501	0.0352	0.0147
115	0.008295	0.2289	0.0015	1.0053	0.0087	0.0076	0.0016	-0.0003
116	-0.0631	-1.7438	0.0014	0.9932	-0.0660	-0.0583	-0.0109	-0.0155
117	-0.0313	-0.8642	0.0018	1.0028	-0.0365	-0.0355	0.0179	0.0098
118	0.0141	0.3886	0.0029	1.0064	0.0211	0.0096	0.0147	0.0053
119	0.0117	0.3243	0.0022	1.0059	0.0153	0.0090	0.0089	0.0028
120	0.001589	0.0438	0.0015	1.0056	0.0017	0.0017	-0.0005	-0.0004
121	-0.0305	-0.8421	0.0022	1.0034	-0.0392	-0.0269	-0.0089	-0.0230
122	0.0103	0.2841	0.0015	1.0052	0.0109	0.0096	0.0015	-0.0011
123	-0.000652	-0.0180	0.0014	1.0054	-0.0007	-0.0006	-0.0000	-0.0000
124	0.0190	0.5251	0.0016	1.0046	0.0213	0.0188	-0.0005	-0.0073
125	0.006469	0.1785	0.0014	1.0054	0.0068	0.0059	0.0014	0.0000
126	0.0309	0.8539	0.0035	1.0046	0.0504	0.0413	-0.0394	-0.0256
127	-0.0189	-0.5214	0.0020	1.0050	-0.0235	-0.0213	0.0106	-0.0001
128	-0.0173	-0.4791	0.0027	1.0058	-0.0248	-0.0213	0.0158	0.0042
...								
570	-0.0304	-0.8388	0.0026	1.0038	-0.0428	-0.0349	0.0197	-0.0052
571	0.0513	1.4201	0.0034	0.9993	0.0830	0.0655	-0.0526	-0.0619
572	-0.0139	-0.3846	0.0024	1.0059	-0.0187	-0.0167	0.0113	0.0034
573	-0.0359	-0.9920	0.0025	1.0026	-0.0495	-0.0444	0.0328	0.0256
574	-0.0256	-0.7064	0.0016	1.0036	-0.0279	-0.0277	0.0100	0.0048
575	0.0239	0.6604	0.0017	1.0040	0.0269	0.0266	-0.0115	-0.0078

...

Obs	Residual	RStudent	Hat Diag H	Cov Ratio	DFFITS	DFBETAS Intercept	L1LEI	L1WKUCL
576	-0.0368	-1.0157	0.0018	1.0017	-0.0434	-0.0408	0.0182	0.0014
577	0.0504	1.3937	0.0014	0.9976	0.0531	0.0460	0.0114	0.0011
578	-0.0150	-0.4130	0.0025	1.0059	-0.0205	-0.0184	0.0135	0.0104
579	-0.0309	-0.8542	0.0015	1.0026	-0.0333	-0.0326	0.0090	0.0007
580	0.0447	1.2344	0.0017	0.9996	0.0516	0.0429	0.0033	-0.0170
581	-0.0385	-1.0628	0.0020	1.0015	-0.0481	-0.0441	0.0239	0.0032
582	0.0603	1.6688	0.0018	0.9946	0.0709	0.0667	-0.0265	-0.0349
583	0.0121	0.3331	0.0019	1.0055	0.0144	0.0137	-0.0071	-0.0023
584	-0.0182	-0.5032	0.0020	1.0050	-0.0223	-0.0213	0.0123	0.0088
585	0.0139	0.3852	0.0030	1.0065	0.0211	0.0179	-0.0154	-0.0076
586	0.008349	0.2307	0.0040	1.0079	0.0147	0.0115	-0.0118	-0.0087
587	-0.0175	-0.4828	0.0027	1.0059	-0.0252	-0.0215	0.0162	0.0043
588	0.0489	1.3534	0.0045	1.0011	0.0906	0.0686	-0.0741	-0.0340
589	-0.009354	-0.2586	0.0055	1.0093	-0.0192	-0.0138	0.0165	0.0086
590	-0.008997	-0.2492	0.0086	**1.0125**	-0.0232	-0.0148	0.0212	0.0151
591	0.0342	0.9474	0.0059	1.0064	0.0731	0.0516	-0.0641	-0.0360
592	-0.0432	-1.1972	0.0083	1.0066	-0.1093	-0.0695	0.0969	0.0409
593	0.0822	**2.2789**	0.0040	**0.9871**	0.1450	0.1140	-0.1178	-0.0814
594	0.0341	0.9431	0.0056	1.0061	0.0711	0.0509	-0.0619	-0.0370
595	0.0182	0.5035	0.0022	1.0052	0.0234	0.0211	-0.0122	-0.0018
596	0.0191	0.5318	0.0190	**1.0223**	0.0740	0.0385	-0.0698	-0.0317
597	-0.0105	-0.2914	0.0071	1.0109	-0.0246	-0.0165	0.0220	0.0115
598	0.0351	0.9752	0.0157	**1.0161**	**0.1231**	0.0669	-0.1151	-0.0522
,,,								
722	-0.0621	-1.7183	0.0014	0.9935	-0.0654	-0.0619	0.0053	-0.0093
723	0.001005	0.0277	0.0014	1.0054	0.0010	0.0010	-0.0000	-0.0000
724	-0.0443	-1.2228	0.0015	0.9995	-0.0470	-0.0457	0.0085	0.0137
725	0.002790	0.0770	0.0014	1.0055	0.0029	0.0029	-0.0006	-0.0005
726	0.0253	0.6985	0.0014	1.0035	0.0265	0.0264	-0.0064	-0.0031
727	-0.007587	-0.2094	0.0014	1.0053	-0.0079	-0.0077	0.0011	-0.0004
728	0.0142	0.3923	0.0015	1.0050	0.0153	0.0139	0.0001	-0.0038
729	-0.0606	-1.6768	0.0017	0.9943	-0.0687	-0.0674	0.0293	0.0126
730	0.0347	0.9590	0.0020	1.0024	0.0433	0.0409	-0.0245	-0.0196
731	-0.0308	-0.8503	0.0017	1.0029	-0.0353	-0.0346	0.0163	0.0087
732	-0.001433	-0.0395	0.0014	1.0054	-0.0015	-0.0014	-0.0000	-0.0001
733	0.0114	0.3150	0.0017	1.0053	0.0128	0.0119	-0.0030	0.0017
734	-0.000093	-0.002567	0.0020	1.0061	-0.0001	-0.0001	-0.0000	0.0000
735	0.2219	**6.2864**	0.0016	**0.8597**	0.2554	0.2501	-0.1017	-0.0339
736	0.2528	**15.7616**	0.7384	**1.6038**	**26.4777**	-0.9960	**6.2446**	**24.0222**
737	-0.2689	**-8.1582**	0.0994	**0.8599**	**-2.7108**	-0.9589	**2.3713**	**0.4742**
738	0.0472	1.3314	0.0413	**1.0398**	**0.2764**	0.0503	0.0033	-0.2109
739	0.0754	**2.1109**	0.0242	1.0105	**0.3324**	0.0451	**0.1196**	-0.1608
740	-0.1514	**-4.2429**	0.0087	0.9421	**-0.3975**	-0.0594	**-0.3278**	-0.0772
741	0.0521	1.4478	0.0113	1.0069	**0.1547**	0.0515	0.0057	-0.1101
742	-0.0714	-1.9775	0.0046	0.9929	**-0.1351**	-0.0778	0.0230	0.1016

Sum of Residuals	0
Sum of Squared Residuals	0.96790
Predicted Residual SS (PRESS)	1.86050

The **BOLD (RStudent) Studentized Residuals** exceed 2.00 or are less than −2.00, indicating influential observations.

The **BOLD DFFITS** exceed 0.109 indicating influential observations.

The **BOLD DFBETAS** are large and are paired with DFFITS exceed 0.109 indicating influential observations.

The bold RStudent, DFFITS, and DFBETAS statistics indicate that the 2020 time period is dominated by influential observations

References

Ball, L., & Mazumder, S. (2019). A Phillips curve with anchored expectations and short term unemployment. *Journal of Money, Credit, and Banking.*

Ball, L., Mankiw, N., & Romer, D. (1988). The new Keynesian economics and the output-inflation trade-off. *Brookings Papers on Economic Activity,* 1–65.

Beaton, A. E., & Tukey, J. W. (1974). The fitting of power series, meaning polynomials, illustrated on bank-spectroscopic data. *Technometrics, 16,* 147–185.

Belsley, D. A., Kuh, E., & Welsch, R. E. (1980b). *Regression diagnostics: Identifying influential data and sources of collinearity.* John Wiley & Sons. Chapter 2.

Brunner, K., & Meltzer, A. (1993). *Money and the economy: Issues in monetary analysis.* Cambridge University Press.

Burns, A. F., & Mitchell, W. C. (1946). *Measuring business cycles.* NBER.

Castle, J., & Shepard, N. (2009). *The methodology and practice of econometrics.* Oxford University Press.

Castle, J., Clements, M. P., & Hendry, D. F. (2015). Robust approaches to forecasting. *International Journal of Forecasting, 31,* 99–112.

Castle, J. L., & Hendry, D. F. (2019). *Modelling our changing world.* Palgrave.

Clements, M. P., & Hendry, D. F. (2005). Evaluating a model by forecast performance. *Oxford Bulletin of Economics and Statistics, 67,* 931–956.

Cochrane, D., & Orcutt, G. H. (1949). Application of least squares regression to relationships containing autocorrelated error terms. *Journal of the American Statistical Association, 44,* 32–61.

Dhrymes, P. (2017a). *Introductory econometrics* (Rev ed.). Springer.

Diebold, F., & Rudebusch, G. D. (1999). *Business cycles.* Princeton University Press.

Doornik, J. A., & Hendry, D. F. (2015). Statistical model selection with big data. *Cogent Economics & Finance, 3,* 1–15.

Efron, B., Hastie, T., Johnstone, J., & Tibshirani, R. (2004). Least angle regression. *The Annals of Statistics, 32,* 407–499.

Granger, C. W. J., & Newbold, P. (1977a). *Forecasting economic time series.* Academic Press, Inc..

Granger, C. W. J. (2001a). Essays in econometrics. In E. Ghysels, N. R. Swanson, & M. W. Watson (Eds.), *Cambridge university press.*

Guerard, J. B., Jr., Xu, G., & Wang, Z. (2019). *Portfolio and investment analysis with SAS: Financial modeling techniques for optimization.* SAS Press.

Guerard, J. B., Jr., Thomakos, D. D., & Kyrizai, F. S. (2020). Automatic time series modelling and forecasting: A replication case study of forecasting real GDP, the unemployment rate, and the impact of leading economic indicators. *Cogent Economics and Finance.*

Guerard, J. B., Jr. (2001). A note on the forecasting effectiveness of the U.S. leading economic indicators. *Indian Economic Review, 36,* 251–268.

Guerard, J. B., Jr. (2004). The forecasting effectiveness of the U.S. leading economic indicators: Further evidence and initial G7 results. In P. Dua (Ed.), *Business cycles and economic growth: An analysis using leading indicators* (pp. 174–187). Oxford University Press.

Gunst, R. F., & Mason, R. L. (1980). *Regression analysis and its application.* Marcel Dekker, Inc.

Hahn, F., & Solow, R. (1995). *A critical essay on modern macroeconomic theory.* MIT Press.

Harberler, G. (1937). *Prosperity and depression.* League of Nations, Geneva. Reprinted 1963. Atheneum.

Hastie, T., Tibshirani, R., & Friedman, J. (2016). *The elements of statistical learning: Data mining, inference, and prediction* (2nd ed.). 11th printing, Springer.

Hendry, D. F., & Nielsen, B. (2007). *Econometric modeling: A likelihood approach.* Princeton University Press.

Hendry, D. F., & Doornik, J. A. (2014a). *Empirical model discovery and theory evaluation.* MIT Press.

Hendry, D. F., & Krolzig, H. M. (2005a). The properties of automatic gets modelling. *The Economic Journal, 115*, c32–c61.

Huber, P. J. (1973). Robust regression: Asymptotics, conjectures, and Monte Carlo. *Annals of Statistics, 1*, 799–821.

Keynes, J. (1936). *The general theory of employment, interest, and money*. Macmillan.

Klein, L. (1950). *Economic fluctuations in the United States, 1941*. Wiley & Sons.

Krolzig, H. M., & Hendry, D. F. (2001b). Computer automation of general-to-specific model selection procedures. *Journal of Economic Dynamics and Control, 25*, 831–866.

Kyriazi, F. S., Thomakos, D. D., & Guerard, J. B. (2019a). Adaptive learning forecasting with applications in forecasting agricultural prices. *International Journal of Forecasting*. forthcoming.

Makridakis, S., Wheelwright, S. C., & Hyndman, R. J. (1998). *Forecasting: Methods and applications* (3rd ed.). John Wiley & Sons. Chapters 5, 6.

Mansfield, E. (1994). *Statistics for business and economics* (5th ed.). W.W. Norton & Company.

Maronna, R. A., Martin, R. D., & Yohai, V. J. (2006). *Robust statistics; Theory and methods with R*. Wiley.

Maronna, R. A., Martin, R. D., Yohai, V. J., & Salibian-Barrera, M. (2019). *Robust statistics; Theory and methods with R* (2nd ed.). Wiley.

Miller, I., & Freund, J. E. (1965). *Probability and statistics for engineers*. Prentice-Hall.

Mincer, J., & Zarnowitz, V. (1969). The Evaluation of Economic Forecasts. In J. Mincer (Ed.), *Economic forecasts and expectations*. Columbia University Press.

Mitchell, W. C. (1913). *Business cycles*. Burt Franklin reprint.

Mitchell, W. C. (1951). *What happens during business cycles: A progress report*. NBER.

Montgomery, A. L., Zarnowitz, V., Tsay, R., & Tiao, G. C. (1998). Forecasting the U.-S. unemployment rate. *Journal of the American Statistical Association, 93*, 478–493.

Moore, G. H. (1961a). *Business cycle indicators* (Vol. 1). Princeton University Press.

Moore, G. H. (1961b). *Business cycle indicators* (Vol. 2). Princeton.

Murphy, J. L. (1973). *Introductory Econometrics*. Richard D. Irwin, Inc.

Nelson, C. R., & Plosser, C. I. (1982). Trends and random walks in macroeconomic time series. *Journal of Monetary Economics, 10*, 139–162.

Nelson, C. R., & Plossner, C. I. (1982). Trends and random walks in macroeconomic time series: Some evidence and implications. *Journal of Monetary Economics, 10*, 139–162.

Nikolopoulos, K. I., & Thomakos, D. D. (2019). *Forecasting with the theta method*. Wiley.

Persons, W. (1931). *Forecasting business cycles*. Wiley & Sons, Inc.

Phillips, A. (1958). The relationship between unemployment and the rate of change of money wage rates in the United Kingdom, 1861–1957. *Economica, 25*, 283–299.

Romer, D. (2019). *Advanced macroeconomics* (5th ed.). McGraw-Hill.

Samuelson, P. (1948). *Economics*. McGraw-Hill.

Singer, A. C., & Feder, M. (1999). Universal linear prediction by model order weighting. *IEEE Transactions on Signal Processing, 47*, 2685–2699.

Taylor, J. (1999). *Monetary policy rules*. University of Chicago Press.

Tsay, R. S. (1988). Outliers, level shifts, and variance changes in time series. *Journal of Forecasting, 7*(1988), 1–20.

Tsay, R. S. (1989). Testing and modelling threshold autoregressive processes. *Journal of the American Statistical Association, 84*, 231–249.

Tsay, R. S. (2010). *Analysis of financial time series*. Wiley, Third Edition.

Tsay, R. S., & Chen, R. (2019). *Nonlinear time series analysis*. Wiley.

Vinod, H. D. (2008). *Hands-on intermediate econometrics using R: Templates for extending dozens of practical examples*. World Scientific. ISBN 10-981-281-885-5.

Zarnowitz, V. (1992). *Business cycles: Theory, history, indicators, and forecasting*. University of Chicago Press.

Chapter 13
Time Series Modeling and the Forecasting Effectiveness of the US Leading Economic Indicators

An important aspect of financial decision-making may depend on the forecasting effectiveness of the composite index of leading economic indicators, LEI. The leading indicators can be used as an input to a transfer function model of real gross domestic product, GDP. The previous chapter employed four quarterly lags of the LEI series to estimate regression models of association between current rates of growth of real US GDP and the composite index of leading economic indicators. This chapter examines whether changes in forecasted economic indexes help forecast changes in real economic growth. The transfer function model forecasts are compared to several naïve models in order to test which model produces the most accurate forecast of real GDP. No-change forecasts of real GDP and random walk with drift models may be useful when forecasting benchmarks (Mincer & Zarnowitz, 1969; Granger & Newbold, 1977). Economists have constructed leading economic indicator series to serve as a business barometer of the changing US economy since the time of Wesley C. Mitchell (1913). The purpose of this study is to examine the time series forecasts of composite economic indexes produced by The Conference Board (TCB) and test the hypothesis that the leading indicators are useful as an input to a time series model to forecast real output in the United States.

Economic indicators are descriptive and anticipatory time series data used to analyze and forecast changing business conditions. Cyclical indicators are comprehensive series that are systemically related to the business cycle. Business cycles are recurrent sequences of expansions and contractions in aggregate economic activity. Coincident indicators have cyclical movements that approximately correspond with the overall business cycle expansions and contractions. Leading indicators reach their turning points before the corresponding business cycle turns. The lagging indicators reach their turning points after the corresponding turns in the business cycle.

An example of business cycles can be found in the analysis of Irving Fisher (1911), who discussed how changes in the money supply led to rising prices and an initial fall in the rate of interest and how this results in raising profits, creating a boom. The interest rate later rises, reducing profits and ending the boom. A financial

J. B. Guerard Jr. et al., *Quantitative Corporate Finance*,
https://doi.org/10.1007/978-3-030-87269-4_13

crisis ensues when businessmen, whose loan collateral is falling as interest rates rise, run to cash and banks fail. The money supply is one series in The Conference Board index of leading economic indexes, LEI.

In this chapter, we introduce the reader to time series modeling tools. Section 13.2 of this chapter addresses time series modeling in practice. Section 13.3 presents the empirical evidence to support time series modeling of the US real GDP time series. In Sect. 13.4 we estimate a transfer function to test the forecasting effectiveness of the leading indicators in modeling real GDP. We use Autometrics, the automatic time series analysis and forecasting system developed by David Hendry and his colleagues to estimate structural breaks and outliers in US real GDP time series in Sect. 13.5 and changes in the US unemployment rate in Sect. 13.6. Outlier analysis and automatic time series modeling are highly statistically significant in the 1959–2018 time period and particularly relevant for root mean square error reductions in 2020.

13.1 Basic Statistical Properties of Economic Series

This chapter develops and forecasts models of economic time series in which we initially use only the history of the series. The chapter later explores explanatory variables in the forecast models. The time series modeling approach of Box and Jenkins involves the identification, estimation, and forecasting of stationary (or series transformed to stationarity) series through the analysis of the series autocorrelation and partial autocorrelation functions.[1] The autocorrelation function examines the correlations of the current value of the economic times series and its previous k-lags. That is, one can measure the correlation of a daily series of share prices, for example, by calculating:

$$p_{jt} = a + b \, p_{jt-1} \qquad (13.1)$$

where

p_{jt} = today's price of stock j
p_{jt-1} = yesterday's price of stock j and
b is the correlation coefficient

In a daily share price series, b is quite large, often approaching a value of 1.00. As the number of lags, or previous number of periods increase, the correlation tends to fall. The decrease is usually very gradual.

The partial autocorrelation function examines the correlation between p_{jt} and p_{jt-2}, holding constant the association between p_{jt} and p_{jt-1}. If a series follows a random walk, the correlation between p_{jt} and p_{jt-1} is one, and the correlation

---

[1] This section draws heavily from Box and Jenkins, *Time Series Analysis*, Chapters 2 and 3.

between p_j and p_{jt-2}, holding constant the correlation of p_{jt} and p_{jt-1}, is zero. Random walk series are characterized with decaying autocorrelation functions and a partial autocorrelation function by a "spike" at lag one and zeros thereafter.

Let Z_t be a time series. In the general case, stationarity implies that the joint probability [p(Z)] distribution $P(Z_{t1}, Z_{t2})$ is the same for all times t, t_1, and t_2 where the observations are separated by a constant time interval. The autocovariance of a time series at some lag or interval, k, is defined to be the covariance between Z_t and Z_{t+k}

$$\gamma_k = \text{cov}[Z_t, Z_{t+k}] = E[(Z_t - \mu)(Z_{t+k} - \mu)]. \tag{13.2}$$

One must standardize the autocovariance, as one standardizes the covariance in traditional regression analysis before one can quantify the statistically significant association between Z_t and Z_{t+k}. The autocorrelation of a time series is the standardization of the autocovariance of a time series relative to the variance of the time series, and the autocorrelation at lag k, ρ_k, is bounded between +1 and −1.

$$\rho_k = \frac{E[(Z_t - \mu)(Z_{t+k} - \mu)]}{\sqrt{E\left[(Z_t - \mu)^2\right] E\left[(Z_{t+k} - \mu)^2\right]}}$$

$$= \frac{E[(Z_t - \mu)(Z_{t+k} - \mu)]}{\sigma_Z^2} = \frac{r_k}{r_0} \tag{13.3}$$

The autocorrelation function of the process, $\{\rho k\}$, represents the plotting of r_k versus time, the lag of k. The autocorrelation function is symmetric and thus $\rho_k = \rho_{-k}$; thus, time series analysis normally examines only the positive segment of the autocorrelation function. One may also refer to the autocorrelation function as the correlogram. The statistical estimates of the autocorrelation function are calculated from a finite series of N observations, $Z_1, Z_2, Z_3, \ldots, Z_n$. The statistical estimate of the autocorrelation function at lag k, r_k, is found by:

$$r_k = \frac{C_k}{C_0}$$

where

$$C_k = \frac{1}{N} \sum_{t=1}^{N-k} (Z_t - \overline{Z})(Z_{t+k} - \overline{Z}), k = 0, 1, 2, \ldots, K.$$

C_k is, of course, the statistical estimate of the autocovariance function at lag k. In identifying and estimating parameters in a time series model, one seeks to identify orders (lags) of the time series that are statistically different from zero. The implication of testing whether an autocorrelation estimate is statistically different from zero leads one back to the t-tests used in regression analysis to examine the statistically significant association between variables. One must develop a standard error of the autocorrelation estimate such that a formal t-test can be performed to measure the statistical significance of the autocorrelation estimate. Such a standard error, S_e, estimate was found by Bartlett and, in large samples, is approximated by:

$$Var[r_k] \cong \frac{1}{N}, \text{ and}$$

$$S_e[r_k] \cong \frac{1}{\sqrt{N}}. \tag{13.4}$$

An autocorrelation estimate is considered statistically different from zero if it exceeds approximately twice its standard error.

A second statistical estimate useful in time series analysis is the partial autocorrelation estimate of coefficient j at lag k, ϕ_{kj}. The partial autocorrelations are found in the following manner:

$$\rho_j = \phi_{k1}\rho_{j-1} + \phi_{k2}\rho_{j-2} + \ldots + \phi_{k(k-1)}\rho_{jk-1} + \phi_{kk}\rho_{j-k} \qquad \text{or}$$

$$j = 1, 2, \ldots, k$$

$$\begin{bmatrix} 1 & \rho_1 & \rho_2 & \cdots & \rho_{k-1} \\ \rho_1 & 1 & \rho_1 & \cdots & \rho_{k-2} \\ \vdots & \vdots & \vdots & \cdots & \vdots \\ \rho_{k-1} & \rho_{k-2} & \rho_{k-3} & \cdots & 1 \end{bmatrix} \begin{bmatrix} \phi_{k1} \\ \phi_{k2} \\ \vdots \\ \phi_{kk} \end{bmatrix} \begin{bmatrix} \rho_1 \\ \rho_2 \\ \vdots \\ \rho_k \end{bmatrix}$$

The partial autocorrelation estimates may be found by solving the above equation systems for k = 1,2,3,... k.

$$\phi_{11} = \rho_1$$

$$\phi_{22} = \frac{\rho_2 - \rho_1^2}{1 - \rho_1^2} = \frac{\begin{vmatrix} 1 & \rho_1 \\ \rho_2 & \rho_2 \end{vmatrix}}{\begin{vmatrix} 1 & \rho_1 \\ \rho_1 & 1 \end{vmatrix}}$$

$$\phi_{33} = \frac{\begin{vmatrix} 1 & \rho_1 & \rho_1 \\ \rho_1 & 1 & \rho_2 \\ \rho_2 & \rho_1 & \rho_3 \end{vmatrix}}{\begin{vmatrix} 1 & \rho_1 & \rho_2 \\ \rho_1 & 1 & \rho_1 \\ \rho_2 & \rho_1 & 1 \end{vmatrix}}$$

The partial autocorrelation function is estimated by expressing the current autocorrelation function estimates as a linear combination of previous orders of autocorrelation estimates:

$$\widehat{r}_1 = \widehat{\phi}_{k1}r_{j-1} + \widehat{\phi}_{k2}^2r_{j-2} + \ldots + \widehat{\phi}_{k(k-1)}r_{j+k-1} + \widehat{\phi}_{kk^2}r_{j-k} \quad j = 1, 2, \ldots, k.$$

The standard error of the partial autocorrelation function is approximately:

$$Var\left[\widehat{\phi}_{kk}\right] \cong \frac{1}{N}, \quad \text{and}$$

$$S_e[\phi_{kk}] \cong \frac{1}{\sqrt{N}}.$$

13.1.1 The Autoregressive and Moving Average Processes

A stochastic process, or time series, can be repeated as the output resulting from a white noise input, α_t.[2]

$$\widetilde{Z}_t = \alpha_t + \psi_1\alpha_{t-1} + \psi_2\alpha_{t-2} + \ldots$$

$$= \alpha_t \sum_{j=1}^{\infty} \psi_j\alpha_{t-j}. \tag{13.5}$$

The filter weight, ψ_j, transforms the input into the output series. One normally expresses the output, $\widetilde{Z}_t$, as a deviation of the time series from its mean, μ, or origin

$$\widetilde{Z}_t = Z_t - \mu.$$

---

[2]Please see Box and Jenkins, *Time Series Analysis*, Chapter 3, for the most complete discussion of the ARMA (p,q) models.

The general linear process leads one to represent the output of a time series, $\tilde{Z}_t$, as a function of the current and previous value of the white noise process, α_t which may be represented as a series of shocks. The white noise process, α_t, is a series of random variables characterized by:

$$E[\alpha_t] \cong 0$$

$$Var[\alpha_t] = \sigma_\alpha^2$$

$$\gamma_k = E[\alpha_t \alpha_{t+k}] = \sigma_\alpha^2 \quad k \neq 0$$

$$0 \quad k = 0.$$

The autocorrelation function of a linear process may be given by:

$$\gamma_k = \sigma_\alpha^2 \sum_{j=0}^{\infty} \psi_j \psi_{j+k}.$$

The backward shift operator, B, is defined as $BZ_t = Z_{t-1}$ and $B_j Z_t = Z_{t-j}$. The autocorrelation generating function may be written as:

$$\gamma(B) = \sum_{k=-\infty}^{\infty} \gamma_k B^k$$

For stationarity, the ψ weights of a linear process must satisfy that $\psi(B)$ converges on or lies within the unit circle.

In an autoregressive, AR, model, the current value of the time series may be expressed as a linear combination of the previous values of the series and a random shock, α_t.

$$\tilde{Z}_t = \phi_1 \tilde{Z}_{t-1} + \phi_2 \tilde{Z}_{t-2} + \ldots + \phi_p \tilde{Z}_{t-p} + \alpha_t$$

The autoregressive operator of order P is given by:

$$\phi(B) = 1 - \phi_1 B^1 - \phi_2 B^2 - \ldots - \phi_p B^p$$

or

$$\phi(B)\tilde{Z}_t = \alpha_t \tag{13.6}$$

In an autoregressive model, the current value of the time series, $\widetilde{Z}_t$, is a function of previous values of the time series, $\widetilde{Z}_{t-1}$, $\widetilde{Z}_{t-2}$, ... and is similar to a multiple regression model. An autoregressive model of order p implies that only the first p order weights are nonzero. In many economic time series, the relevant autoregressive order is one, and the autoregressive process of order p, AR(p) is written as

$$\widetilde{Z}_t = \phi_1 \widetilde{Z}_{t-1} + \alpha_t$$

or

$$(1 - \phi_1 B)\widetilde{Z}_t = \alpha_t$$
$$\widetilde{Z}_t = \phi^{-1}(B)\alpha_t.$$

The relevant stationarity condition is $|B| < 1$ implying that $|\phi_1| < 1$. The autocorrelation function of a stationary autoregressive process,

$$\widetilde{Z}_t = \phi_1 \widetilde{Z}_{t-1} + \phi_2 \widetilde{Z}_{t-2} + \ldots + \phi_p \widetilde{Z}_{t-p} + \alpha_t$$

may be expressed by the difference equation:

$$P_k = \phi_1 P_{k-1} + \phi_2 P_{k-2} + \ldots + \phi_k P_{k-p} k > 0$$

or expressed in terms of the Yule-Walker equation as:

$$\rho_1 = \phi_1 + \phi_2 \rho_1 + \ldots + \phi_p \rho_{p-1}$$
$$\rho_2 = \phi_1 \rho_1 + \phi_2 + \ldots + \phi_p \rho_{p-2}$$
$$\widehat{\rho}_p = \widehat{\phi}_1 \rho_{p-1} + \widehat{\phi}_2 \rho_{p-2} + \ldots + \widehat{\phi}_p$$

For the first-order AR process, AR(1)

$$\rho_k = \phi_1 \rho_{k-1} = \widehat{\phi}_p.$$

The autocorrelation function decays exponentially to zero when ϕ_1 is positive and oscillates in sign and decays exponentially to zero when ϕ_1 is negative.

$$\rho_1 = \phi_1$$

and

$$\sigma_2 = \frac{\sigma_a^2}{1 - \phi_1^2}.$$

The partial autocorrelation function cuts off after lag one in an AR(1) process. For a second-order AR process, AR(2)

$$\tilde{Z}_t = \phi_1 \tilde{Z}_{t-1} + \phi_2 \tilde{Z}_{t-k} + \alpha_t$$

with roots:

$$\phi(B) = 1 - \phi_1 B - \phi_2 B^2 = 0$$

and, for stationarity, roots lying outside the unit circle, ϕ_1 and ϕ_2 must obey the following conditions:

$$\phi_2 + \phi_1 < 1$$
$$\phi_2 - \phi_1 < 1$$
$$-1 < \phi_2 < 1.$$

The autocorrelation function of an AR(2) model is:

$$\rho_k = \phi_1 \rho_{k-1} + \phi_2 \rho_{k-2} \tag{13.7}$$

The autocorrelation coefficients may be expressed in terms of the Yule-Walker equations as

$$\rho_1 = \phi_1 + \phi_2 \rho_2$$
$$\rho_2 = \phi_1 \rho_1 + \phi_2$$

which implies

$$\phi_1 = \frac{\rho_1(1 - \rho_2)}{1 - \rho_1^2}$$

$$\phi_2 = \frac{\rho_2(1 - \rho_1^2)}{1 - \rho_1^2}$$

and

$$\rho_1 = \frac{\phi_1}{1 - \phi_2} \text{ and } \rho_2 = \phi_2 + \frac{\phi_1^2}{1 - \phi_2}.$$

For a stationary AR(2) process,

$$-1 < \phi_1 < 1$$
$$-1 < \rho_2 < 1$$
$$\rho_1^2 < \frac{1}{2}(\rho_2 + 1).$$

In an AR(2) process, the autocorrelation coefficients tail off after order two, and the partial autocorrelation function cuts off after the second-order (lag).[3] The estimated autocorrelation function tails off after lag p in an AR(p) process, and the PACF cuts off after lag p.

In a q-order moving average (MA) model, the current value of the series can be expressed as a linear combination of the current and previous shock variables:

$$\tilde{Z}_t = \alpha_1 - \theta_1\alpha_{t-1} - \ldots - \alpha_q\theta_{t-q}$$
$$= (1 - \theta_1 B_1 - \ldots - \theta_q B_q)\alpha_t$$
$$= \theta(B)\alpha_t$$

The autocovariance function of a q-order moving average model is:

$$\gamma_k = E\left[(\alpha_t - \theta_1\alpha_{t-1} - \ldots - \theta_q\alpha_{t-q})(\alpha_{t-k} - \theta_1\alpha_{t-k-1} - \ldots - \theta_q\alpha_{t-k-q})\right]$$

The autocorrelation function, ρ_k, is:

$$\rho_k = \begin{array}{ll} \dfrac{-\theta_k + \theta_1\theta_{k+1} + \cdots + \theta_{q-k}\theta_q}{1 + \theta_1^2 + \ldots + \theta_q^2} & k = 1, 2, \ldots, q \\ 0 & k > q \end{array}$$

The autocorrelation function of a MA(q) model cuts off, to zero, after lag q and its partial autocorrelation function tail off to zero after lag q. There are no restrictions on the moving average model parameters for stationarity; however, moving average parameters must be invertible. Invertibility implies that the π weights of the linear filter transforming the input into the output series, the π weights, lie outside the unit circle.

---

[3] A stationary AR(p) process can be expressed as an infinite weighted sum of the previous shock variables

$$\tilde{Z}_1 = \phi^{-1}(B)\alpha_t.$$

In an invertible time series, the current shock variable may be expressed as an infinite weighted sum of the previous values of the series

$$\theta^{-1}(B)\tilde{Z}_t = \alpha_t.$$

$$\pi(B) = \psi^{-1}(B)$$

$$= \sum_{j=0}^{a} \phi^j B^j$$

In a first-order moving average model, MA(1)

$$\tilde{Z}_t = (1 - \theta_1 B)\alpha_t$$

and the invertibility condition is $|\theta_1| < 1$. The autocorrelation function of the MA (1) model is:

$$\rho_k = \frac{-\theta_1}{1 + \theta_1^2} \quad \begin{array}{l} k = 1 \\ k > 2. \end{array}$$

The partial autocorrelation function of an MA(1) process tails off after lag one and its autocorrelation function cut off after lag one.

In a second-order moving average model, MA(2)

$$\tilde{Z}_t = \alpha_t - \theta_1 \alpha_{t-1} - \theta_2 \alpha_{t-2}$$

the invertibility conditions require:

$$\theta_2 < \theta_1 < 1$$

$$\theta_2 - \theta_1 < 1$$

$$-1 < \theta_2 < 1$$

The autocorrelation function of the MA(2) is:

$$\rho_1 = \frac{-\theta_1(1 - \theta_2)}{1 + \theta_1^2 + \theta_1^2}$$

$$\rho_2 = \frac{-\theta_2}{1 + \theta_1^2 + \theta_1^2}$$

and

$$\rho_k = 0 \text{ for } k > 3.$$

The partial autocorrelation function of an MA(2) tails off after lag two.

In many economic time series, it is necessary to employ a mixed autoregressive-moving average (ARMA) model of the form:

$$\tilde{Z}_t = \phi_1 \tilde{Z}_{t-1} + \ldots + \phi_p \tilde{Z}_{t-p} + \alpha_t - \theta_1 \alpha_{t-1} - \ldots - \theta_q \alpha_{t-q} \qquad (13.8)$$

or

$$(1 - \phi_1 B - \phi_2 B^2 - \ldots - \theta_p B^p) \tilde{Z}_t = (1 - \theta_1 B - \theta_2 B^2 - \ldots - \theta_q B^q) \alpha_t$$

that may be more simply expressed as

$$\phi(B) \tilde{Z}_t = \theta(B) \alpha_t$$

The autocorrelation function of the ARMA model is

$$\rho_k = \phi_1 \rho_{k-1} + \phi_2 \rho_{k-2} + \ldots + \phi_p \rho_{k-p}$$

or

$$\phi(B) \rho_k = 0.$$

The first-order autoregressive moving average operator ARMA(1,1) process is written

$$\tilde{Z}_t - \phi_1 \tilde{Z}_{t-1} = \alpha_t - \theta_1 \alpha_{t-1}$$

or

$$(1 - \phi_1) \tilde{Z}_t = (1 - \theta 1 B) \alpha_t.$$

The stationary condition is $-1 < \phi_1 < 1$, and the invertibility condition is $-1 < \phi_1 < 1$. The first two autocorrelations of the ARMA (1,1) model are

$$\rho_1 = \frac{(1 - \phi_1 \theta_1)(\phi_1 - \theta_1)}{1 + \theta_1^2 - 2\phi_1 \theta_1}$$

and

$$\rho_2 = \phi_1 \rho_1.$$

The partial autocorrelation function consists only of $\phi_{11} = \rho_1$ and has a damped exponential.

An integrated stochastic progress generates a time series if the series is made stationary by differencing (applying a time-invariant filter) the data. In an integrated process, the general form of the time series model is

$$\phi(B)(1 - B)^d X_t = \theta(B)\varepsilon_t \qquad (13.9)$$

where $\phi(B)$ and $\theta(B)$ are the autoregressive and moving average polynominals in B of orders p and q, ε_t is a white noise error term, and d is an integer representing the order of the data differencing. In economic time series, a first difference of the data is normally performed.[4] The application of the differencing operator, d, produces a stationary ARMA(p,q) process. The autoregressive integrated moving average (ARMA) model is characterized by orders p, d, and q [ARIMA (p,d.q)]. Many economics series follow a random walk with drift, and an ARMA (1,1) may be written as:

$$\overline{V}^d X_t = X_t - X_{t-1} = \varepsilon_t + b\varepsilon_{t-1}. \qquad (13.10)$$

An examination of the autocorrelation function estimates may lead one to investigate using a first-difference model when the autocorrelation function estimates decay slowly. In an integrated process, the corr(X_t, $X_{t-\tau}$) is approximately unity for small values of time, τ.

13.2 ARMA Model Identification in Practice

Time series specialists use many statistical tools to identify models; however, the sample autocorrelation and partial autocorrelation function estimates are particularly useful in modeling. Univariate time series modeling normally requires larger data sets than regression and exponential smoothing models. It has been suggested that at least 40–50 observations be used to obtain reliable estimates.[5] One normally calculates the sample autocorrelation and partial autocorrelation estimates for the raw time series and its first (and possibly second) differences. The failure of the autocorrelation function estimates of the raw data series to die out as large lags implies that a first difference is necessary. The autocorrelation function (ACF) estimates of an AR (p) process tail off after p. The autocorrelation function estimates of an MA (q) process should cut off after q.[6] To test whether the autocorrelation estimates are statistically different from zero, one uses a t-test where the standard error of $\upsilon\tau$ is:[7]

---

[4] Box and Jenkins, *Time Series Analysis*. Chapter 6; C.W.J. Granger and Paul Newbold, *Forecasting Economic Time Series*. Second Edition (New York: Academic Press, 1986), pp. 109–110, 115–117, 206.

[5] Granger and Newbold, *Forecasting Economic Time Series*. pp. 185–186.

[6] See Box and Jenkins, *Time Series Analysis*, p. 79.

[7] Box and Jenkins, *Time Series Analysis*. pp. 173–179. Nelson (1973) is one of the texts used in the 1970s for forecasting and is still a very useful guide to applied time series modeling.

$$n^{-1/2}\left[1 + 2\left(\rho_1^2 + \rho_2^2 + \ldots + \rho_q^2\right)\right]^{1/2} \quad \text{for } \tau > q.$$

The partial autocorrelation function (PACF) estimates of an AR(p) process cut off after lag p and tail off after q for an MA(q) process. A t-test is used to statistically examine whether the partial autocorrelations are statistically different from zero. The standard error of the partial autocorrelation estimates is approximately:

$$\frac{1}{\sqrt{N}} \text{ for } K > p. \tag{13.11}$$

One can use the normality assumption of large samples in the t-tests of the autocorrelation and partial autocorrelation estimates. The identified parameters are generally considered statistically significant if the parameters exceed twice the standard errors.

The ARMA model parameters may be estimated using nonlinear least squares. Given the following ARMA framework generally pack-forecasts the initial parameter estimates and assumes that the shock terms are to be normally distributed.

$$\alpha_t = \widetilde{W}_t - \phi_1 \widetilde{W}_{t-1} - \phi_2 \widetilde{W}_{t-2} - \ldots - \phi_p \widetilde{W}_{t-p} + \theta_1 \alpha_{t-1} + \ldots + \theta_q \alpha_{t-q}$$

where

$$W_t = \overline{V}^d Z_t \text{ and } \widetilde{W}_t = W_t - \mu.$$

The minimization of the sum of squared errors with respect to the autoregressive and moving average parameter estimates produces starting values for the p order AR estimates and q order MA estimates.

$$\frac{\partial e_t}{-\partial \phi_j}\bigg|_{\beta_0} = \mu_{j,t} \text{ and } \frac{\partial e_t}{-\partial \theta_i}\bigg|_{\beta_0} = X_{j,t}$$

It may be appropriate to transform a series of data such that the residuals of a fitted model have a constant variance or are normally distributed. The log transformation is such a data transformation that is often used in modeling economic time series. Box and Cox (1964) put forth a series of power transformations useful in modeling time series.[8] The data is transformed by choosing a value of λ that is suggested by the relationship between the series amplitude (which may be approximated by the range of subsets) and mean.[9]

---

[8]G.E. Box and D.R. Cox, "An Analysis of Transformations," *Journal of the Royal Statistical Society*, B 26 (1964), 211–243.

[9]G.M. Jenkins, "Practical Experience with Modeling and Forecasting Time Series," *Forecasting* (Amsterdam: North-Holland Publishing Company, 1979).

$$X_t^\lambda = \frac{X_t^\lambda - 1}{\overline{X}^{\lambda-1}} \tag{13.12}$$

where X is the geometric mean of the series. One immediately recognizes that if $\lambda = 0$, the series is a logarithmic transformation. The log transformation is appropriate when there is a positive relationship between the amplitude and mean of the series. A $\lambda = 1$ implies that the raw data should be analyzed and there is no relationship between the series range and mean subsets. One generally selects the λ that minimizes the smallest residual sum of squares, although an unusual value of λ may make the model difficult to interpret. Some authors may suggest that only values of λ of -0.5, 0, 0.5, and 1.0 be considered to ease in the model building process.[10]

Many time series, involving quarterly or monthly data, may be characterized by rather large seasonal components. The ARIMA model may be supplemented with seasonal autoregressive and moving average terms:

$$\left(1 - \phi_1 B - \phi_2 B^2 - \ldots - \phi_p B^p\right)\left(1 - \phi_{1,s} B^s - \ldots - \phi_{p,s} B^p S^s\right)(1 - B)^d$$

$$(1 - B^s)^{ds} X_t = \left(1 - \theta_1 B - \ldots - \theta_q B^q\right)(1 - \theta_{1,s} B^s - \ldots$$

$$-\theta_{q,s} B^{q,s})\alpha_t \text{ or } \theta_p(B)\Phi_p(B^s)\overline{V}^d \overline{V}_x^D Z_t = \theta_q(B)\theta_Q(B^s)\alpha_t \tag{13.13}$$

One recognizes seasonal components by an examination of the autocorrelation and partial autocorrelation function estimates. That is, the autocorrelation and partial autocorrelation function estimates should have significantly large values at lags one and 12 as well as smaller (but statistically significant) values at lag 13 for monthly data.[11] One seasonally differences the data (a twelfth-order seasonal difference for monthly data and estimates the seasonal AR or MA parameters.) A random walk with drift model with a monthly component may be written as

$$\overline{V}\overline{V}_{12} Z_t = (1 - B)\left(1 - \theta B^{12}\right)\alpha_t \tag{13.14}$$

The multiplicative form of the (0,1,1) X (0,1,1) 12 model has a moving average operator that may be written as:

$$(1 - \theta B)\left(1 - \theta B^{12}\right) = 1 - \theta B - \theta B^{12} + \theta B^{13}$$

---

[10] Jenkins, *op. cit.*, pp. 135–138.

[11] Box and Jenkins, *Time Series Analysis*, pp. 305–308.

The random walk with drift with the monthly seasonal adjustments is the basis of the "airline model" in honor of the analysis by Professors Box and Jenkins of total airline passengers during the 1949–1960 period.[12] The airline passenger data analysis employed the natural logarithmic transformation.

There are several tests and procedures that are available for checking the adequacy of fitted time series models. The most widely used test is the Box-Pierce test, where one examines the autocorrelation among residuals, α_t:

$$\widehat{v}_k = \frac{t = \sum_{k+1}^{n} \alpha_t \alpha_{t-k}}{\sum_{t=1}^{n} \alpha_t^2}, k = 1, 2, \ldots$$

The test statistic, Q, should be X^2 distributed with (m-p-q) degrees of freedom.

$$Q = n \sum_{k=1}^{m} \widehat{v}_k^2.$$

The Ljung-Box statistic is a variation on the Box-Pierce statistic, and the Ljung-Box Q statistic tends to produce significant levels closer to the asymptotic levels than the Box-Pierce statistic for first-order moving average processes. The Ljung-Box statistic, the model adequacy check reported in the SAS system, can be written as

$$Q = n(n + 2) \sum_{k=1}^{m} (n = k)^{-1} \widehat{v}_k^2. \tag{13.15}$$

Residual plots are generally useful in examining model adequacy; such plots may identify outliers as we noted in the chapter. The normalized cumulative periodogram of residuals should be examined.

Granger and Newbold (1977) and McCracken (2000) use several criteria to evaluate the effectiveness of the forecasts with respect to the forecast errors. In this chapter, we use the root mean square error (RMSE) criteria. One seeks to minimize the square root of the sum of the absolute value of the forecast errors squared. That is, we calculate the absolute value of the forecast error, square the error, sum the squared errors, divided by the number of forecast periods, and take the square root of the resulting calculation. Intuitively, one seeks to minimize forecast errors. The absolute value of the forecast errors is important because if one calculated only a mean error, a 5% positive error could "cancel out" a 5% negative error. Thus, we minimize the out-of-sample forecast errors. We need a benchmark for

---

[12] Box and Jenkins, *op. cit.*

forecast error evaluation. An accepted benchmark [Mincer and Zarnowitz (1969)] for forecast evaluation is a no-change, NoCH. A forecasting model should produce a lower root mean square error (RMSE) than the no-change model. If several models are tested, the lowest RMSE model is preferred.

In the world of business and statistics, one often speaks of autoregressive, moving average, and random walk with drift models, or processes, as we have just introduced.

It is well known that the majority of economic series, including real gross domestic product (GDP) in the United States, follow a random walk with drift, RWD, and are represented with autoregressive integrated moving average (ARIMA) model with a first-order moving average operator applied to the first difference of the data. The data is differenced to produce stationary, where a process has a (finite) mean and variance that do not change over time and the covariance between data points of two series depends upon the distance between the data points, not on the time itself. The RWD process, estimated with an ARIMA (0,1,1) model, is approximately equal to a first-order exponential smoothing model (Cogger, 1974). The random walk with drift model has been supported by the work of Nelson and Plosser (1982). Let us return to The Conference Board data used in the previous chapter on regression and estimate an ARIMA RWD Model for US GDP for the 1959Q2–2020Q3 time period.

13.3   Estimating an ARIMA RWD for US GDP, 1959Q2–2020Q3

In this section we use SAS, proc. ARIMA, to estimate time series models. In SAS, one can obtain estimates of the autocorrelation, the partial autocorrelation function, and the autocorrelation check of the estimated time series model residuals. These estimations allow the reader to identify and estimate the model and test if the estimated residuals are consistent with the model being adequately fit. Let us estimate an ARIMA RWD model for US GDP for the 1959Q2–2020Q3 time period.

GDP RWD, 1959 -2020Q3

The ARIMA Procedure

Name of Variable = GDP_Known

Mean of Working Series 9993.123

Standard Deviation   4843.7

Number of Observations 247

Autocorrelation Check for White Noise

To Lag	Chi-Square	DF	Pr > ChiSq	Autocorrelations
6	1401.08	6	<.0001	0.989 0.980 0.968 0.956 0.944 0.932
12	2628.14	12	<.0001	0.919 0.907 0.895 0.882 0.870 0.858
18	3691.59	18	<.0001	0.846 0.834 0.823 0.811 0.799 0.787
24	4603.32	24	<.0001	0.776 0.764 0.752 0.740 0.729 0.717

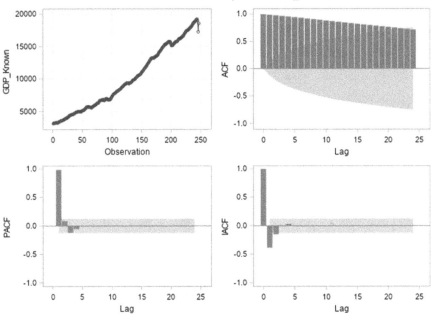

Trend and Correlation Analysis for GDP_Known

The reader notes the (gradual) tailing off of the ACF estimates. Hence the data should be the difference to stationarity. Once differenced, the data is not statistically different from white noise, see the chi-square value of 11.11 at lag 18, having an estimated probability of 88.95% of being white noise. There is a one-period spike in the estimated partial autocorrelation function. The "cutting off" of the PACF is consistent with an AR(1) process.

Name of Variable = GDP_Known	
Period(s) of Differencing	1
Mean of Working Series	62.85203
Standard Deviation	155.4497
Number of Observations	246
Observation(s) eliminated by differencing	1

Autocorrelation Check for White Noise

To Lag	Chi-Square	DF	Pr > ChiSq	Autocorrelations					
6	10.42	6	0.1081	-0.196	-0.015	0.033	0.033	-0.010	0.031
12	10.61	12	0.5626	-0.009	-0.024	-0.002	0.008	0.000	-0.005
18	11.11	18	0.8895	-0.020	-0.008	-0.006	0.017	-0.020	0.027
24	12.00	24	0.9799	0.010	-0.016	-0.016	-0.005	-0.051	0.004

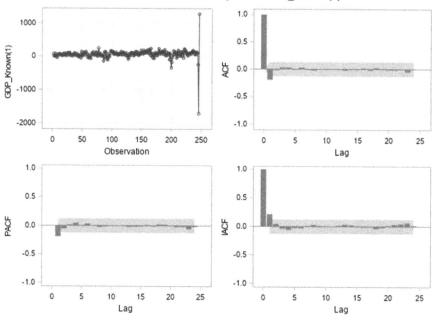

Trend and Correlation Analysis for GDP_Known(1)

Conditional Least Squares Estimation

Parameter	Estimate	Standard Error	t Value	Approx Pr > |t|	Lag
MU	62.13043	7.60279	8.17	<.0001	0
MA1,1	0.21864	0.06337	3.45	0.0007	1

Constant Estimate	62.13043
Variance Estimate	23241.25
Std Error Estimate	152.4508
AIC	3173.316
SBC	3180.326
Number of Residuals	246

The reader notices immediately the pronounced movements in GDP near the end of the GDP time series. The last two to three GDP observations of 2020 would appear to be outliers (and they will be shown to be so later in the chapter).

The MA (1) term is 0.218, with a t-statistics of 3.45, highly statistically significant.[13]

Correlations of Parameter Estimates

Parameter	MU	MA1,1
MU	1.000	-0.014
MA1,1	-0.014	1.000

Autocorrelation Check of Residuals

To Lag	Chi-Square	DF	Pr > ChiSq	Autocorrelations					
6	2.32	5	0.8032	-0.014	0.048	0.061	0.045	0.004	0.030
12	2.64	11	0.9947	-0.011	-0.028	-0.008	0.003	-0.006	-0.016
18	3.20	17	0.9999	-0.029	-0.016	-0.008	0.011	-0.013	0.027
24	4.57	23	1.0000	0.010	-0.018	-0.024	-0.024	-0.058	0.004
30	7.68	29	1.0000	0.063	-0.036	0.004	0.046	0.011	0.060
36	9.12	35	1.0000	0.037	0.002	-0.016	-0.021	0.012	-0.053
42	9.91	41	1.0000	0.000	-0.038	-0.032	-0.012	0.004	-0.005
48	22.12	47	0.9992	0.053	0.087	0.128	0.116	0.006	0.001

The residuals are random, having a chi-square statistic of 3.20 at lag 18. The estimated probability of the estimated residuals being random is 99.99%.

The ARIMA (0,1,1) model residuals indicated that the ARIMA RWD model is adequately fit for the 1959Q2–2020Q3 time period.

---

[13] The cutting off of the PACF suggests an AR(1) process. Had one estimated an ARIMA (1,1,0), the estimated AR (1) = −0.261 (t = − 3.63) and the estimated variance = 23116.6, slightly less than the ARIMA (0,1,1). The estimated time series models are essentially identical. The residuals are random, having a chi-square statistics of 2.93 at lag 18.

Model for variable GDP_Known

Estimated Mean   62.13043

Period(s) of Differencing 1

Moving Average Factors

Factor 1:   1 - 0.21864 B**(1)

246	19136.6161	152.4508	18837.8180	19435.4142	17302.5000	-1834.1161
247	17765.6406	152.4508	17466.8425	18064.4387	18583.5000	817.8594
248	18466.8141	152.4508	18168.0160	18765.6122	.	.
249	18528.9445	193.4699	18149.7505	18908.1385	.	.
250	18591.0749	227.2002	18145.7707	19036.3792	.	.
251	18653.2054	256.5332	18150.4095	19156.0012	.	.
252	18715.3358	282.8403	18160.9790	19269.6926	.	.
253	18777.4662	306.9006	18175.9521	19378.9804	.	.
254	18839.5967	329.2071	18194.3626	19484.8308	.	.
255	18901.7271	350.0952	18215.5531	19587.9012	.	.

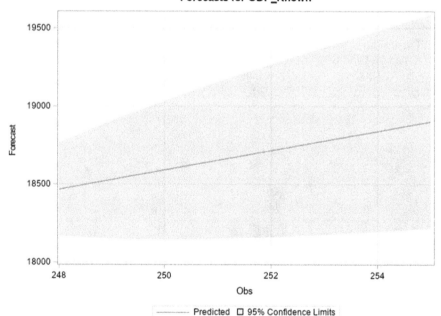

Forecasts for GDP_Known

---- Predicted □ 95% Confidence Limits

13.3.1   Estimating an ARIMA RWD for DLGDP, 1959Q2–2020Q3

Let us create the DLGDP time series, where we apply a logarithmic transformation to the GDP time series and first difference the logged GDP time series to stationarity.

DL GDP

The ARIMA Procedure

Name of Variable = DLGDP

Mean of Working Series 0.007251

Standard Deviation   0.01121

Number of Observations 246

Autocorrelation Check for White Noise

To Lag	Chi-Square	DF	Pr > ChiSq	Autocorrelations					
6	4.65	6	0.5899	0.008	0.102	0.057	0.061	-0.023	0.021
12	5.91	12	0.9204	-0.032	-0.029	0.013	0.039	0.010	-0.035
18	7.73	18	0.9824	-0.033	-0.011	-0.040	0.044	-0.020	0.042
24	9.06	24	0.9975	0.001	0.038	0.001	0.025	-0.046	0.026

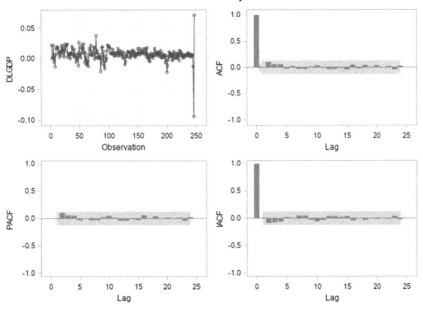

Trend and Correlation Analysis for DLGDP

Conditional Least Squares Estimation

Parameter	Estimate	Standard Error	t Value	Approx Pr > |t|	Lag
MU	0.0072538	0.0007236	10.02	<.0001	0
MA1,1	-0.0075659	0.06904	-0.11	0.9128	1

Constant Estimate	0.007254
Variance Estimate	0.000127
Std Error Estimate	0.011255
AIC	-1507.45
SBC	-1500.44
Number of Residuals	246

The ARIMA RWD model of DLGDP MA(1) term estimate is not statistically different from zer. Thus, the ARIMA RWD is actually an ARIMA random walk.

* AIC and SBC do not include log determinant.

Correlations of Parameter Estimates		
Parameter	MU	MA1,1
MU	1.000	-0.025
MA1,1	-0.025	1.000

Autocorrelation Check of Residuals

To Lag	Chi-Square	DF	Pr > ChiSq	Autocorrelations					
6	4.49	5	0.4817	0.001	0.100	0.055	0.061	-0.023	0.022
12	5.74	11	0.8899	-0.032	-0.029	0.013	0.039	0.010	-0.034
18	7.58	17	0.9747	-0.032	-0.011	-0.040	0.045	-0.020	0.042
24	8.91	23	0.9962	0.000	0.038	0.001	0.025	-0.046	0.026
30	11.42	29	0.9986	0.064	-0.021	0.007	0.059	0.022	0.021
36	13.87	35	0.9995	0.068	0.016	-0.021	-0.019	0.002	-0.054
42	15.47	41	0.9999	-0.017	-0.026	-0.034	0.003	0.024	0.052
48	21.73	47	0.9994	0.033	0.067	0.101	0.064	0.024	-0.015

The ARIMA RWD model is adequately fit.

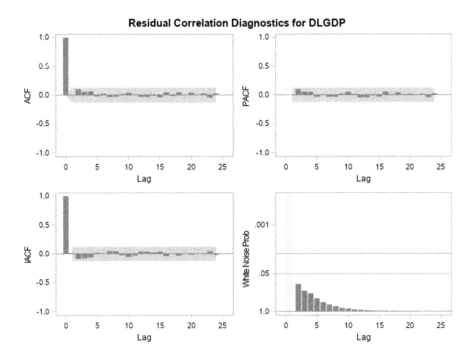

Residual Correlation Diagnostics for DLGDP

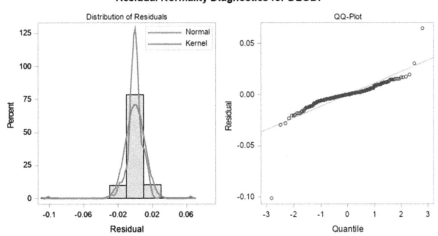

Residual Normality Diagnostics for DLGDP

Model for variable DLGDP
Estimated Mean   0.007254

Moving Average Factors
Factor 1: $1 + 0.00757\,B**(1)$

13.4   Estimating a Transfer Function Time Series Model of US GDP with the US Leading Economic Indicator Time Series as Its Input

Let us now estimate a transfer function in which the LEI time series is an input to the output time series, US GDP, the market value of goods and services produced in the economy.

TF DLGDP with L1LEI, 1959 – 2020Q3

The ARIMA Procedure

Name of Variable = DLGDP
Mean of Working Series 0.007251
Standard Deviation   0.01121
Number of Observations 246

Autocorrelation Check for White Noise

To Lag	Chi-Square	DF	Pr > ChiSq	Autocorrelations					
6	4.65	6	0.5899	0.008	0.102	0.057	0.061	-0.023	0.021
12	5.91	12	0.9204	-0.032	-0.029	0.013	0.039	0.010	-0.035
18	7.73	18	0.9824	-0.033	-0.011	-0.040	0.044	-0.020	0.042
24	9.06	24	0.9975	0.001	0.038	0.001	0.025	-0.046	0.026

Correlation of DLGDP and DLLEI
Variance of input =   0.000427
Number of Observations 246

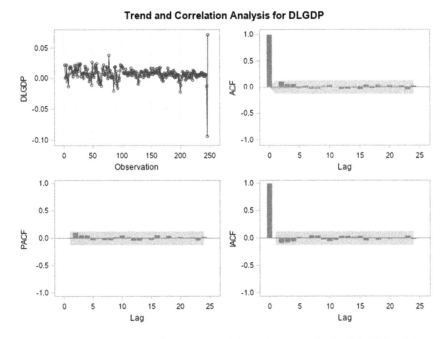

The cross-correlogram estimates reveal that DLLEI leads the DLGDP with one and possibly three-period lags. We estimate the transfer function and find that the DLLEI leads DLGDP by one quarter, having an estimated coefficient of 0.683 with a t-statistics of 12.46, which is highly statistically significant. The reader notices immediately the pronounced movements in DLGDP near the end of the DLGDP time series. The last two to three GDP observations of 2020 would appear to be outliers. They are shown to be outliers in this section.

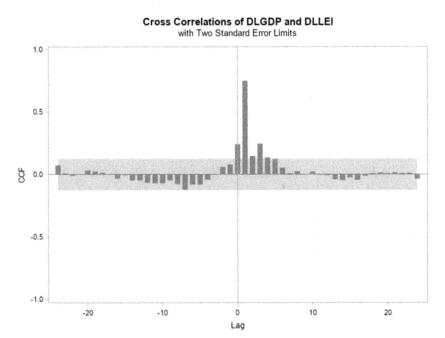

Maximum Likelihood Estimation

Parameter	Estimate	Standard Error	t Value	Approx Pr > |t|	Lag	Variable	Shift
MU	0.0049723	0.0004578	10.86	<.0001	0	DLGDP	0
MA1,1	0.31630	0.06242	5.07	<.0001	1	DLGDP	0
NUM1	0.14815	0.01807	8.20	<.0001	0	DLLEI	0
DEN1,1	0.63843	0.05125	12.46	<.0001	1	DLLEI	0

Both the MA(1) drift term and the contemporaneous and lagged DLLEI transfer function terms are highly statistically significant. One could estimate a three-period lag in the DLLEI time series, but the estimated transfer function variance estimates are not statistically different than the model reported.

Constant Estimate	0.004972
Variance Estimate	0.000086
Std Error Estimate	0.009292
AIC	-1593.15
SBC	-1579.14

Number of Residuals 245

Autocorrelation Check of Residuals

To Lag	Chi-Square	DF	Pr > ChiSq	Autocorrelations					
6	5.32	5	0.3781	-0.031	0.082	0.023	0.073	0.018	0.086
12	10.69	11	0.4695	0.036	0.019	0.069	0.103	0.063	-0.006
18	12.07	17	0.7961	-0.020	-0.004	-0.042	0.043	-0.030	0.017
24	13.66	23	0.9363	-0.026	0.012	-0.014	0.007	-0.067	0.017
30	16.33	29	0.9716	0.068	-0.013	0.015	0.067	0.008	0.011
36	20.84	35	0.9722	0.076	0.038	0.012	0.029	0.078	0.039
42	23.49	41	0.9872	0.069	0.036	-0.009	0.013	0.037	0.037
48	24.38	47	0.9974	-0.003	0.009	0.023	0.008	-0.018	-0.043

The transfer function model estimated residuals are not statistically different from zer; thus the model is adequately fit.

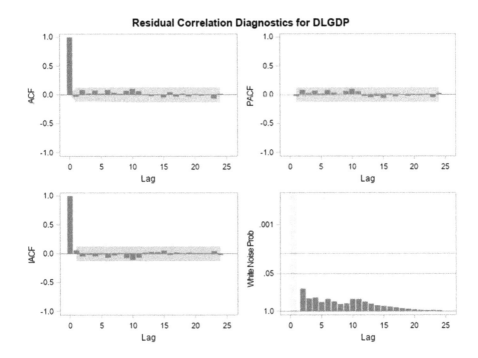

Residual Correlation Diagnostics for DLGDP

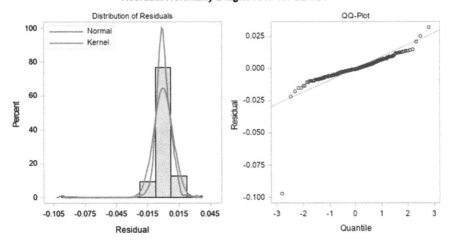

Residual Normality Diagnostics for DLGDP

Model for variable DLGDP

Estimated Intercept 0.004972

Moving Average Factors

Factor 1: 1 - 0.3163 B**(1)

Input Number 1

Input Variable   DLLEI

Overall Regression Factor 0.148147

Denominator Factors

Factor 1: 1 - 0.63843 B**(1)

Outlier Detection Summary

Maximum number searched 5

Number found   5

Significance used   0.05

Outlier Details

Obs	Type	Estimate	Chi-Square	Approx Prob>ChiSq
246	Additive	-0.07888	186.44	<.0001
247	Additive	0.05689	88.16	<.0001
78	Additive	0.02601	20.94	<.0001
205	Shift	-0.0028273	22.12	<.0001
8	Additive	-0.02259	22.78	<.0001

There are Tsay (1989) outliers present in the transfer function models, just as we reported outliers in the regression model. One notes the very large additive outliers in 2020Q2 and 2020Q3.

13.5 Forecasting Effectiveness of Time Series Modeling Using Autometrics to Estimate Outliers and Breaks: Studies of the US Real GDP, 1959–2020

In this section, we test and report on time series modeling and forecasting using the US leading economic indicator (LEI) as an input to forecasting real GDP and the unemployment rate. These time series have been addressed before, but our results are more statistically significant using more recently developed time series modeling techniques and software. We employ the automatic time series modeling, and forecasting of Hendry and Doornik (2014) and Doornik and Hendry (2015) with its emphasis on structural breaks is very relevant for modeling the MZTT unemployment rate data. Montgomery, Zarnowitz, Tsay, and Tiao (MZTT, 1998) modeled the US unemployment rate as a function of the weekly unemployment claims time series, 1948–1993. A similar conclusion is found for the impact of the LEI and weekly unemployment claims series leading the unemployment rate series. We report statistically significant breaks in these data, 1959–1993 and 1959 to the present.

As an introductory example, let us consider the US real GDP as can be represented by an autoregressive integrated moving average (ARIMA) model, The data is differenced to create a process that has a (finite) mean and variance that do not change over time, and the covariance between data points of two series depends upon the distance between the data points, not on the time itself—a transformation to stationarity. In economic time series, a first difference of the data is normally performed.[14] The application of the differencing operator, d, produces a stationary autoregressive moving average ARMA(p, q) model when all parameters are constant across time. Many economics series can be modeled with a simple subset of the class of ARIMA(p, d, q) models, particularly the random walk with drift and a moving average term. The random walk with drift economic time series behavior is not new and can be traced back to the works of Granger and Newbold (1977) and Nelson and Plossner (1982).

Automatic time series models have recently been discussed in Hendry (1986), Hendry and Krolzig (2001, 2005), Hendry and Nielsen (2007), Castle et al. (2013), Hendry and Doornik (2014), and Castle and Hendry (2019) and implemented in the Autometrics software.[15] Hendry sets the tone for automatic modeling by contrasting how statistically based his PC-Give and Autometrics work in contrast to the "data mining" and "garbage in, garbage out" routines, citing their forecasting efficiency and performance. If one starts with a large number of predictors, or candidate explanatory variables, say n, then the general model can be written:

$$y_t = \sum_{i=1}^{n} \gamma_i Z_{it} + u_t. \tag{13.16}$$

The (conditional) data generating processes is assumed to be given by:

$$y_t = \sum_{i=1}^{n} \beta_i Z_{(i),t} + \epsilon_t, \tag{13.17}$$

where

$$\epsilon_t \cong IN\left(0, \sigma_e^2\right) \text{ for any } n \leq N.$$

---

[14] Box and Jenkins, *Time Series Analysis*. Chapter 6; C.W.J. Granger and Paul Newbold, *Forecasting Economic Time Series*. Second Edition (New York: Academic Press, 1986), pp. 109–110, 115–117, 206.

[15] Automatic time series modelling has advocated since the early days of Box and Jenkins (1970). Reilly (1980), with the Autobox System, pioneered early automatic time series model implementation. Tsay (1988) identified outliers, level shifts, and variance change models that were implemented in PC-SCA. SCA was used in modelling time series in MZTT (1998).

One must select the relevant regressors where $\beta_j \neq 0$ in (13.5). Hendry and his colleagues refer to Eq. (13.16) as the most general, statistical model that can be postulated, given the availability of data and previous empirical and theoretical research as the general unrestricted model (GUM). The Hendry general-to-specific modeling process is referred to as *Gets*. One seeks to identify all relevant variables, the relevant lag structure and cointegrating relations, forming near orthogonal variables, Z. The general unrestricted model, GUM, with s lags of all variables can then be written:

$$y_t = \sum_{i=1}^{n} \sum_{j=0}^{s} \beta_{i,j} x_{i,t-j} + \sum_{i=1}^{n} \sum_{j=0}^{s} k_{i,j} z_{i,t-j} + \sum_{j=1}^{s} \theta_j y_{t-j} + \sum_{i=1}^{T} \delta_i 1_{\{i=t\}} + e_t,$$

(13.18)

where

$$\varepsilon_t \sim IN\left(0, \sigma_e^2\right).$$

Furthermore, outliers and shifts for T observations can be modeled with saturation variables, see Doornik and Hendry (2015) and Hendry and Doornik (2014, Chapters 7 and 14).

Automatic modeling seeks to eliminate irrelevant variables, variables with insignificant estimated coefficients, lag-length reductions, and reducing saturation variables (for each observation), the nonlinearity of the principal components, and combinations of "small effects" represented by principal components.[16] One can consider the orthogonal regressor case in which one ranks the variables by their t-statistics, highest to lowest, and defines m to be the smallest, but statistically significant t-statistic, t_m^2, and discards all variables with t-statistics below the m largest t-values. One must be reminded that every test statistic has a distribution occurring in different samples by the specification of Eqs. (13.5) and (13.6). We seek to select a model of the form:

$$y_t = \sum_{r=1}^{m} \delta_r Z_{\{r\},t} + n_t,$$

(13.19)

where $Z_{\{r\},t}$ is a subset of the initial N variables, and that model may differ from either the one postulated in (13.4), (13.5), (13.6), or (13.7) depending on which variables remain at the end of the selection process.[17] One progresses from the

---

[16] Doornik and Hendry (2013) remind the reader that the data generation process (DGP) is impossible to model, and the best solution that one can achieve is estimate the models to reflect the local DGP, through reduction, described above. The Automatic *Gets* algorithm reduces GUM to nest LGDP, the locally relevant variables. Congruency, in which the LGDP has the same shape and size as the GUM or models, reflects the local DGP.

[17] In the selection process, one tests the null hypothesis that the parameter in front of a variable is zero. The relevant t-statistic from a two-sided test is used.

general unrestricted model to the "final" model in (13.7) by establishing that model residuals are approximately normal, homoscedastic, and independent. Model reduction proceeds by tree searches of insignificant variables. The last, non-rejected model is referred to as the terminal equation. Selected model regressors have coefficients that are large relative to their estimated standard errors; since the estimators obtained by the initial model (13.5) are unbiased, the selected estimators are upward biased conditional on retaining $Z_{(j), t}$. The unselected variables will have downward biased estimators. By omitting irrelevant variables, the selection model does not "overfit" the model, and the relevant (retained) variables have estimated standard errors close to those from fitting Eq. (13.7).

The automatic time series modeling program (PCGets) or (Autometrics) is efficient, but Hendry and Nielsen (2007) state that the largest selection bias can arise from strongly correlated regressors. Autometrics deals with outliers and breaks in its automatic time series modeling. The regression sum of squares, RSS, rises as the outlier criteria shrink. Autometrics apply can impulse indicator saturation variables IIS, step indicator saturation variables (SIS), differenced IIS (DIIS), and trend saturation (TIS) to all marginal models. Where there are significant indicators. The step-indicator saturation (SIS) variables are generalized IIS variables with higher statistical power to detect location shifts. One can include outlier detection indicators (impulse-indicator saturation, I, and step-indicator saturation, S) in the Autometrics analysis of the LEI component effectiveness estimates.[18]

```
---- OxMetrics 8.10 started at 16:05:03 on 11-Apr-2021 ----

Ox Professional version 8.10 (Windows_64/U) (C) J.A. Doornik, 1994-2019

EQ(13.1) Modelling GDP Known by OLS

    The estimation sample is: 2 - 247
```

	Coefficient	Std.Error	t-value	t-prob	Part.R^2
GDP Known_1	0.968159	0.01396	69.4	0.0000	0.9519
Constant	92.3114	29.82	3.10	0.0022	0.0379
Trend	2.31018	0.9478	2.44	0.0155	0.0239

```
sigma                154.284  RSS                  5784263.02
R^2                  0.998994 F(2,243) =     120600 [0.000]**
Adj.R^2              0.998985 log-likelihood      -1587.09
no. of observations      246 no. of parameters          3
mean(GDP Known)      10021.1  se(GDP Known)          4843.5
```

---

[18] The use of saturation variables avoids the issue of forcing a unit root to capture the shifts, leading to an upward biased estimate of the lagged dependent variable coefficient. The authors are indebted to Jenny Castle for her comments on the application of saturation variables, which she observes addresses this very well.

```
When the log-likelihood constant is NOT included:
AIC                 10.0897    SC                  10.1325
HQ                  10.1069    FPE                 24093.8
When the log-likelihood constant is included:
AIC                 12.9276    SC                  12.9703
HQ                  12.9448    FPE                 411510.
```

```
AR 1-2 test:        F(2,241) =  6.6939 [0.0015]**
ARCH 1-1 test:      F(1,244) =  52.524 [0.0000]**
Normality test:     Chi^2(2) =  506.53 [0.0000]**
Hetero test:        F(4,241) =  6.5819 [0.0000]**
Hetero-X test:      F(5,240) =  5.2439 [0.0001]**
RESET23 test:       F(2,241) =  10.026 [0.0001]**
```

  The residuals of the AR(1) model are not random. The model fit in Eq. (13.1) is inadequate. The chi-squared score of the residuals exceeded the score consistent with residuals being normally distributed at the 10% level.

```
EQ(13.2) Modelling GDP Known by OLS
```

   The estimation sample is: 2 - 247

	Coefficient	Std.Error	t-value	t-prob	Part.R^2
GDP Known_1	1.00313	0.0009685	1036.	0.0000	0.9998
Constant	34.1261	10.64	3.21	0.0015	0.0408
I:246	-1801.99	73.06	-24.7	0.0000	0.7154
I:247	1192.66	72.88	16.4	0.0000	0.5253

```
sigma               72.3727    RSS              1267550.14
R^2                 0.999779   F(3,242) =   365700 [0.000]**
Adj.R^2             0.999777   log-likelihood    -1400.37
no. of observations  246 no. of parameters         4
mean(GDP Known      10021.1    se(GDP Known)       4843.5
When the log-likelihood constant is NOT included:
AIC                 8.57979    SC                  8.63678
HQ                  8.60274    FPE                 5322.98
When the log-likelihood constant is included:
AIC                 11.4177    SC                  11.4747
HQ                  11.4406    FPE                 90913.6
```

```
AR 1-2 test:        F(2,240) =  16.990 [0.0000]**
ARCH 1-1 test:      F(1,244) =  17.942 [0.0000]**
Normality test:     Chi^2(2) =  56.546 [0.0000]**
Hetero test:        F(2,241) =  4.1280 [0.0173]*
Hetero-X test:      F(2,241) =  4.1280 [0.0173]*
RESET23 test:       F(2,240) =  1.9166 [0.1493]
```

  The residuals of the AR(1) model are not random, violating normality and RESET, polynominal function parameters. The model fit in Eq. (13.2) is inadequate, despite identifying large outliers in 2020Q2 and 2020Q3. The chi-squared score of the Eq. (13.2) model residuals exceeded the score consistent with residuals being normally distributed at the 10% level.

```
Robust standard errors
          Coefficients        SE         HACSE         HCSE
GDP Known_1    1.0031   0.00096852   0.0013705    0.0010794
Constant      34.126      10.641      11.936       9.4322
I:246        -1802.0      73.057      16.533      12.711
I:247         1192.7      72.875      14.355      10.982

          Coefficients      t-SE       t-HACSE      t-HCSE
GDP Known_1    1.0031      1035.7       731.93       929.36
Constant      34.126       3.2070       2.8591       3.6180
I:246        -1802.0      -24.666     -108.99      -141.77
I:247         1192.7       16.366       83.085      108.60
Model saved to C:\JBG\JGResearch\MZTT 122020\QCF 3rdEd\Model AR1 Large
Residuals.pdf
Model saved to C:\JBG\JGResearch\MZTT 122020\QCF 3rdEd\Model AR1 Large
Residuals.pdf.gwg


EQ(13.3) Modelling GDP Known by OLS

        The estimation sample is: 13 - 247

          Coefficient   Std.Error   t-value   t-prob   Part.R^2
GDP Known_1   0.792120     0.03614     21.9    0.0000    0.6969
LEI_1        30.8807        2.307      13.4    0.0000    0.4617
LEI_2       -22.1552        2.995      -7.40   0.0000    0.2075
DI:90       -75.9363       26.05       -2.91   0.0039    0.0391
DI:165     -105.872        26.05       -4.06   0.0001    0.0733
DI:170       95.2226       26.27        3.62   0.0004    0.0591
DI:221     -111.122        26.05       -4.27   0.0000    0.0801
DI:245      1222.86       283.4         4.31   0.0000    0.0818
DI:246      2955.56       818.5         3.61   0.0004    0.0587
I:78        152.787        37.39        4.09   0.0001    0.0740
I:89        134.962        37.12        3.64   0.0003    0.0595
I:93       -100.984        37.29       -2.71   0.0073    0.0339
I:179        94.5089       37.46        2.52   0.0124    0.0296
I:200      -184.170        39.54       -4.66   0.0000    0.0941
I:209      -139.574        37.54       -3.72   0.0003    0.0620
I:211      -118.133        37.42       -3.16   0.0018    0.0455
I:212        92.6899       38.09        2.43   0.0158    0.0275
I:247      6737.77       1328.          5.07   0.0000    0.1096
T1:76        -4.10865       0.7842      -5.24   0.0000    0.1161
T1:147       -8.85875       2.374       -3.73   0.0002    0.0625
T1:158      -63.6550       25.70       -2.48   0.0140    0.0285
T1:159       67.6584       24.50        2.76   0.0063    0.0352
T1:197        4.89988       1.513       3.24   0.0014    0.0478
T1:220       -5.32882       1.884      -2.83   0.0051    0.0368
T1:243      1561.81       279.0         5.60   0.0000    0.1304
T1:245     -1545.91       276.1        -5.60   0.0000    0.1304
```

```
sigma                 36.6303    RSS                     280432.239
R^2                   0.999946   F(25,209) = 1.557e+05 [0.000]**
Adj.R^2               0.99994    log-likelihood          -1165.88
no. of observations     235      no. of parameters           26
mean(GDP Known)     10336.9      se(GDP Known)           4724.55
When the log-likelihood constant is NOT included:
AIC                   7.30578    SC                       7.68854
HQ                    7.46009    FPE                      1490.23
When the log-likelihood constant is included:
AIC                  10.1437     SC                       10.5264
HQ                   10.2980     FPE                      25452.4
```

```
AR 1-2 test:          F(2,207)  =    1.1393 [0.3220]
ARCH 1-1 test:        F(1,233)  =0.0033928 [0.9536]
Normality test:       Chi^2(2)  =    1.3203 [0.5168]
Hetero test:          F(27,196) =    1.3208 [0.1440]
RESET23 test:         F(2,207)  = 0.026778 [0.9736]
```

Robust standard errors

	Coefficients	SE	HACSE	HCSE
GDP Known_1	0.79212	0.036138	0.041027	0.039704
LEI_1	30.881	2.3067	2.5721	2.7371
LEI_2	-22.155	2.9953	3.4543	3.6726
DI:90	-75.936	26.054	7.6395	15.451
DI:165	-105.87	26.048	3.7185	5.3620
DI:170	95.223	26.269	6.0646	8.8012
DI:221	-111.12	26.053	9.6911	21.366
DI:245	1222.9	283.41	329.27	310.75
DI:246	2955.6	818.47	949.85	895.82
I:78	152.79	37.393	8.3392	7.7006
I:89	134.96	37.123	6.8288	6.2306
I:93	-100.98	37.287	7.6328	7.2250
I:179	94.509	37.461	7.9125	8.6738
I:200	-184.17	39.536	15.437	19.043
I:209	-139.57	37.540	7.3325	9.4750
I:211	-118.13	37.420	7.3451	8.6624
I:212	92.690	38.092	11.471	12.183
I:247	6737.8	1328.1	1551.0	1463.0
T1:76	-4.1086	0.78421	0.91995	0.86213
T1:147	-8.8587	2.3736	2.2065	2.3205
T1:158	-63.655	25.696	18.644	25.780
T1:159	67.658	24.503	18.409	25.219
T1:197	4.8999	1.5132	1.6751	1.7042
T1:220	-5.3288	1.8845	2.0622	2.1265
T1:243	1561.8	278.97	327.41	308.48
T1:245	-1545.9	276.09	323.93	305.30

	Coefficients	t-SE	t-HACSE	t-HCSE
GDP Known_1	0.79212	21.919	19.307	19.951
LEI_1	30.881	13.388	12.006	11.282

LEI_2	-22.155	-7.3967	-6.4137	-6.0326
DI:90	-75.936	-2.9146	-9.9399	-4.9146
DI:165	-105.87	-4.0646	-28.472	-19.745
DI:170	95.223	3.6248	15.701	10.819
DI:221	-111.12	-4.2653	-11.466	-5.2010
DI:245	1222.9	4.3147	3.7139	3.9352
DI:246	2955.6	3.6111	3.1116	3.2993
I:78	152.79	4.0860	18.322	19.841
I:89	134.96	3.6355	19.764	21.661
I:93	-100.98	-2.7083	-13.230	-13.977
I:179	94.509	2.5229	11.944	10.896
I:200	-184.17	-4.6583	-11.930	-9.6712
I:209	-139.57	-3.7180	-19.035	-14.731
I:211	-118.13	-3.1570	-16.083	-13.637
I:212	92.690	2.4333	8.0804	7.6081
I:247	6737.8	5.0732	4.3442	4.6054
T1:76	-4.1086	-5.2392	-4.4661	-4.7657
T1:147	-8.8587	-3.7321	-4.0149	-3.8177
T1:158	-63.655	-2.4772	-3.4143	-2.4692
T1:159	67.658	2.7613	3.6752	2.6829
T1:197	4.8999	3.2382	2.9252	2.8751
T1:220	-5.3288	-2.8277	-2.5841	-2.5060
T1:243	1561.8	5.5984	4.7702	5.0629
T1:245	-1545.9	-5.5993	-4.7723	-5.0636

Autometrics retains two lags of LEI, starting with four LEI lags, in the US GDP estimated time series models with several impulse indicator saturation variables (I, see observation 246, 2020Q2), and several differenced impulse indicator variables (DI, see observations 245 and 246, 2020Q1 and 2020Q2), respectively. Trend saturation variables (TI) are identified at observations 243 and 245 in 2019–2020. The RSS falls from the AR(1) estimate of 5784263.0 to the Autometrics estimated RSS of 280432.2 The estimated residual diagnostics of Eq. (13.3) reveal no statistically different differences from normality. That is, the chi-squared score of the residuals is less than the score consistent with residuals being normally distributed at the 10% level. Clearly, Autometrics is the most appropriate analysis software system for the pandemic period of the US GDP.

Model saved to C:\JBG\JGResearch\MZTT 122020\QCF 3rdEd\Model GDP AR1 LEIL4 IIS DIIS TIS.pdf.

13.6 Automatic Time Series Modeling of the Unemployment Rate Using Leading Economic Indicators (LEI)

Another widely studied time series is the (US) unemployment rate. We discussed the Montgomery, Zarnowitz, Tsay, and Tiao (MZTT, 1998) unemployment rate variables in the regression chapter. We address it now in its forecasting application.

MZTT modeled the quarterly unemployment rate for the 1948–1993 period and reached several very interesting conclusions. Among the conclusions, the unemployment rate contained no consistent trend, and in times of rising unemployment, the weekly unemployment insurance claims, WKUCL, were a useful input—however unemployment claims were not useful over the entire 1948–1993 time period. MZTT suggested that although future models could build upon asymmetric modeling analysis, long-run models had to forecast stable, slowing declining periods of unemployment. As we did in the previous section, let us use Autometrics to estimate monthly models of the US unemployment rate, UER, time series for the January 1959 to November 2020 time period.

EQ(13.4) Modelling UER by OLS, 1959 -2020
    The dataset is: C:\JBG\JGResearch\Univ Washington CFRM\LEI UER WKUCL.csv
    The estimation sample is: 13 - 741

	Coefficient	Std.Error	t-value	t-prob	Part.R^2
UER_1	0.966259	0.009651	100.	0.0000	0.9325
Trend	3.42129e-05	7.702e-05	0.444	0.6570	0.0003
Constant	0.192907	0.06476	2.98	0.0030	0.0121

sigma	0.436328	RSS	138.217505
R^2	0.932881	F(2,726) =	5045 [0.000]**
Adj.R^2	0.932697	log-likelihood	-428.299
no. of observations	729	no. of parameters	3
mean(UER)	5.99739	se(UER)	1.68188
AR 1-2 test:	F(2,724) =	3.1499 [0.0434]*	
ARCH 1-1 test:	F(1,727) =	0.20376 [0.6518]	
Normality test:	Chi^2(2) =	64199. [0.0000]**	
Hetero test:	F(4,724) =	2.2186 [0.0654]	
Chow test:	F(218,508)=	16.592 [0.0000]** for break after 523	

Summary of Autometrics search
initial search space	2^6	final search space	2^5
no. estimated models	5	no. terminal models	1
test form	LR-F	target size	Small:0.01
large residuals	0.005	presearch reduction	lags
backtesting	GUM0	tie-breaker	SC
diagnostics p-value	0.1	search effort	standard
time	0.08	Autometrics version	2.0a

The residuals of the AR(1) model are not random, violating normality and RESET, polynominal function parameters. The model fit of Eq. (13.4) is inadequate.

EQ(13.5) Modelling UER by OLS with Large Outliers, 1959 -2020
    The dataset is: C:\JBG\JGResearch\Univ Washington CFRM\LEI UER WKUCL.csv
    The estimation sample is: 13 - 741

	Coefficient	Std.Error	t-value	t-prob	Part.R^2
UER_1	0.985526	0.004253	232.	0.0000	0.9867
Constant	0.0814603	0.02638	3.09	0.0021	0.0130
I:736	10.2822	0.1896	54.2	0.0000	0.8025
I:738	-2.08896	0.1920	-10.9	0.0000	0.1405
I:740	-1.73383	0.1903	-9.11	0.0000	0.1028

sigma	0.189343	RSS	25.9558363
R^2	0.987396	F(4,724) =	1.418e+04 [0.000]**
Adj.R^2	0.987326	log-likelihood	181.302
no. of observations	729	no. of parameters	5
mean(UER)	5.99739	se(UER)	1.68188

When the log-likelihood constant is NOT included:

AIC	-3.32156	SC	-3.29007
HQ	-3.30941	FPE	0.0360965

When the log-likelihood constant is included:

AIC	-0.483683	SC	-0.452190
HQ	-0.471532	FPE	0.616509

AR 1-2 test:	$F(2,722)$ =	30.695 [0.0000]**
ARCH 1-1 test:	$F(1,727)$ =	6.7061 [0.0098]**
Normality test:	$Chi^2(2)$ =	286.76 [0.0000]**
Hetero test:	$F(2,723)$ =	97.572 [0.0000]**
Hetero-X test:	$F(2,723)$ =	97.572 [0.0000]**
RESET23 test:	$F(2,722)$ =	27.556 [0.0000]**

Robust standard errors

	Coefficients	SE	HACSE	HCSE
UER_1	0.98553	0.0042526	0.010447	0.0074034
Constant	0.081460	0.026384	0.058252	0.042096
I:736	10.282	0.18959	0.014645	0.011119
I:738	-2.0890	0.19201	0.081880	0.057247
I:740	-1.7338	0.19032	0.049813	0.034544

	Coefficients	t-SE	t-HACSE	t-HCSE
UER_1	0.98553	231.75	94.333	133.12
Constant	0.081460	3.0875	1.3984	1.9351
I:736	10.282	54.233	702.11	924.73
I:738	-2.0890	-10.879	-25.513	-36.490
I:740	-1.7338	-9.1100	-34.807	-50.191

Model saved to C:\JBG\JGResearch\Univ Washington CFRM\XCG 1959 2019
\Model UER AR1 Large Residuals N741.pdf
Model saved to C:\JBG\JGResearch\Univ Washington CFRM\XCG 1959 2019
\Model UER AR1 Large Residuals N741.pdf.gwg

The residuals of the AR(1) model are not random, violating normality and RESET, polynominal function parameters. The model fit of Eq. (13.5) is inadequate, despite identifying large outliers, fitted with impulse indicators saturation variables in 2020.

The estimation of impulse-indictor saturation variables differenced impulse-indictor saturation variables, and trend saturation variables within Autometrics are again needed to obtain an adequate model.

EQ(13.6) Modelling UER by OLS

sigma	0.105795	RSS	6.81631162
R^2	0.99669	F(119,609) =	1541 [0.000]**
Adj.R^2	0.996043	log-likelihood	668.667
no. of observations	729	no. of parameters	120
mean(UER)	5.99739	se(UER)	1.68188

When the log-likelihood constant is NOT included:

AIC	-4.34314	SC	-3.58731
HQ	-4.05151	FPE	0.0130350

When the log-likelihood constant is included:

AIC	-1.50526	SC	-0.749429
HQ	-1.21364	FPE	0.222632

AR 1-2 test:	F(2,607)	=	2.1504 [0.1173]
ARCH 1-1 test:	F(1,727)	=	0.41450 [0.5199]
Normality test:	Chi^2(2)	=	0.89062 [0.6406]
Hetero test:	F(157,541)	=	0.85068 [0.8884]
RESET23 test:	F(2,607)	=	1.0927 [0.3360]

Robust standard errors

	Coefficients	SE	HACSE	HCSE
UER_1	0.83339	0.0097388	0.012302	0.010887
Trend	-0.0033954	0.00039337	0.00043073	0.00038428
LEI_1	-0.0058106	0.0020007	0.0020786	0.0019870
WKUCL_1	0.0020021	3.2142e-05	1.5984e-05	1.5374e-05
DI:14	-0.47923	0.075131	0.024927	0.064519
DI:24	0.22765	0.075452	0.034070	0.089303
DI:26	0.15676	0.074847	0.0053966	0.013161
DI:31	0.25121	0.074833	0.015932	0.041349
DI:33	0.25816	0.086454	0.022827	0.046108
DI:34	0.21455	0.086452	0.022751	0.046102
DI:44	0.17222	0.074849	0.051053	0.13433
DI:47	0.24725	0.075378	0.0086139	0.0093184
DI:59	0.17637	0.074818	0.025027	0.066142
DI:66	0.18451	0.074813	0.010376	0.027467
DI:74	0.28768	0.074817	0.0045858	0.011454
DI:193	0.37563	0.075211	0.023671	0.065059
DI:208	0.20833	0.074859	0.014544	0.036061
DI:227	0.18586	0.074817	0.023767	0.062849
DI:255	-0.27405	0.074822	0.049975	0.13215
DI:257	0.27387	0.074849	0.0073981	0.018240

DI:260	-0.26330	0.10899	0.018886	0.024946
DI:264	-0.27349	0.074831	0.013976	0.036343
DI:461	-0.15232	0.074835	0.0036185	0.0086286
DI:465	0.11719	0.074841	0.0062291	0.016043
DI:471	0.23789	0.074810	0.022884	0.060558
DI:496	-0.17971	0.074816	0.0073385	0.019078
DI:520	0.14003	0.074820	0.0016735	0.0014605
DI:529	-0.14762	0.074814	0.015550	0.040980
DI:542	-0.14378	0.074811	0.035134	0.092906
DI:576	-0.14702	0.074810	0.0033324	0.0087384
DI:592	-0.23942	0.074818	0.015718	0.041159
DI:666	-0.13336	0.074812	0.0035371	0.0092171
DI:671	0.14903	0.074812	0.0013005	0.0026758
DI:677	0.22439	0.074820	0.0023466	0.0047022
DI:713	-0.18231	0.074814	0.0089661	0.023530
DI:729	-0.13724	0.074816	0.0023867	0.0053266
DI:736	6.1990	0.083398	0.042334	0.038838
DI:738	-2.7861	0.10303	0.035984	0.033168
DI:739	-2.3854	0.11852	0.043506	0.040246
DI:740	-2.0464	0.097737	0.029678	0.027214
I:22	0.50176	0.11573	0.031150	0.062681
I:71	-0.21380	0.10714	0.019261	0.018854
I:131	-0.26572	0.10793	0.019426	0.021389
I:132	-0.16743	0.10947	0.023360	0.028486
I:143	0.25706	0.10815	0.016709	0.023456
I:155	0.19082	0.10785	0.011319	0.021378
I:222	0.24597	0.10740	0.019134	0.022781
I:235	0.38962	0.10924	0.018610	0.030459
I:245	-0.32859	0.10744	0.013683	0.021757
I:261	-0.56575	0.15882	0.038413	0.050712
I:287	0.28965	0.10793	0.028640	0.025501
I:295	-0.39634	0.10795	0.031242	0.026755
I:296	0.35453	0.10725	0.024835	0.021795
I:307	0.33096	0.10671	0.012769	0.014860
I:326	0.51869	0.10680	0.011763	0.015480
I:436	0.42924	0.10798	0.0099799	0.019055
I:534	0.19425	0.10703	0.014605	0.014887
I:597	-0.23044	0.10712	0.018902	0.017864
I:658	0.18841	0.10781	0.022775	0.022889
I:664	-0.38406	0.10737	0.019883	0.020055
I:735	0.93986	0.10843	0.018978	0.023016
T1:20	-0.051115	0.030722	0.027685	0.037360
T1:24	0.055219	0.022099	0.023786	0.026679
T1:44	-0.066418	0.020313	0.018549	0.017685
T1:49	0.39117	0.13208	0.044005	0.038064
T1:50	-0.52664	0.17749	0.029776	0.026271
T1:52	0.45564	0.16342	0.027783	0.027050
T1:53	-0.26239	0.099709	0.028333	0.027571
T1:128	-0.053169	0.011672	0.0095514	0.011815
T1:132	0.052589	0.012602	0.0097616	0.012704
T1:165	0.080836	0.033007	0.012028	0.024756
T1:168	-0.40585	0.13288	0.036430	0.099014
T1:169	0.35506	0.11491	0.032063	0.087559

T1:177	-0.34485	0.11041	0.033260	0.068683
T1:178	0.39644	0.12170	0.033729	0.069782
T1:182	-0.13462	0.031867	0.0088305	0.021895
T1:192	0.12799	0.021618	0.010789	0.022591
T1:197	-0.087236	0.019213	0.012710	0.021263
T1:209	0.022358	0.0067520	0.0065482	0.0080879
T1:233	-0.013396	0.0034714	0.0028594	0.0037318
T1:267	-0.31620	0.10337	0.026228	0.037201
T1:268	0.62045	0.16545	0.025764	0.038900
T1:270	-0.79143	0.16413	0.022883	0.033334
T1:271	0.49340	0.10110	0.022178	0.030881
T1:350	0.21752	0.10856	0.016885	0.023949
T1:351	-0.65662	0.23775	0.015562	0.023148
T1:352	0.79941	0.25557	0.0064995	0.0068856
T1:353	-0.53591	0.19318	0.0047439	0.0070370
T1:355	0.47229	0.18099	0.018217	0.053610
T1:356	-0.35334	0.14603	0.023880	0.069188
T1:360	0.23024	0.088516	0.016131	0.033585
T1:362	-0.44296	0.16441	0.039502	0.080391
T1:363	0.27460	0.11411	0.040820	0.074722
T1:375	0.074283	0.061374	0.029993	0.054892
T1:377	-0.44622	0.16199	0.057264	0.096469
T1:378	0.38167	0.12240	0.045633	0.068069
T1:386	0.22883	0.13078	0.038214	0.049509
T1:387	-0.52931	0.22676	0.035768	0.062198
T1:388	0.27112	0.11864	0.026205	0.049322
T1:404	0.33524	0.11864	0.038938	0.065884
T1:405	-0.67703	0.22469	0.037140	0.076210
T1:406	0.35981	0.12424	0.030512	0.052891
T1:417	0.17789	0.11582	0.033026	0.064184
T1:418	-0.31920	0.16429	0.033493	0.067392
T1:420	0.19486	0.080840	0.018175	0.039247
T1:424	-0.059240	0.023931	0.0090347	0.017555
T1:448	0.22980	0.11433	0.016087	0.029246
T1:449	-0.68799	0.24011	0.015917	0.027792
T1:450	1.0124	0.25918	0.0054144	0.0049737
T1:451	-0.96222	0.24125	0.023132	0.040150
T1:452	0.41722	0.11738	0.024935	0.043246
T1:466	-0.011481	0.0039601	0.0024373	0.0035166
T1:581	-0.0087932	0.0018301	0.0015522	0.0017370
T1:621	-0.38757	0.11152	0.031932	0.036893
T1:622	0.87816	0.21784	0.077685	0.074007
T1:623	-0.48079	0.11418	0.070453	0.070288
T1:648	0.15317	0.049017	0.049219	0.063130
T1:649	-0.15080	0.047187	0.046387	0.059879
T1:734	0.65345	0.021103	0.015574	0.014810
T1:739	-0.65201	0.020635	0.014990	0.014202

	Coefficients	t-SE	t-HACSE	t-HCSE
UER_1	0.83339	85.573	67.743	76.550
Trend	-0.0033954	-8.6315	-7.8830	-8.8358
LEI_1	-0.0058106	-2.9043	-2.7954	-2.9243
WKUCL_1	0.0020021	62.290	125.26	130.23

DI:14	-0.47923	-6.3786	-19.225	-7.4278
DI:24	0.22765	3.0172	6.6820	2.5492
DI:26	0.15676	2.0944	29.048	11.911
DI:31	0.25121	3.3569	15.767	6.0752
DI:33	0.25816	2.9861	11.310	5.5991
DI:34	0.21455	2.4919	9.4307	4.6539
DI:44	0.17222	2.3009	3.3734	1.2821
DI:47	0.24725	3.2801	28.704	26.533
DI:59	0.17637	2.3573	7.0474	2.6666
DI:66	0.18451	2.4663	17.783	6.7176
DI:74	0.28768	3.8452	62.734	25.116
DI:193	0.37563	4.9944	15.869	5.7738
DI:208	0.20833	2.7829	14.324	5.7771
DI:227	0.18586	2.4842	7.8200	2.9572
DI:255	-0.27405	-3.6628	-5.4839	-2.0738
DI:257	0.27387	3.6590	37.019	15.015
DI:260	-0.26330	-2.4158	-13.942	-10.555
DI:264	-0.27349	-3.6548	-19.568	-7.5253
DI:461	-0.15232	-2.0354	-42.096	-17.653
DI:465	0.11719	1.5659	18.814	7.3047
DI:471	0.23789	3.1799	10.395	3.9283
DI:496	-0.17971	-2.4021	-24.489	-9.4197
DI:520	0.14003	1.8715	83.674	95.877
DI:529	-0.14762	-1.9732	-9.4936	-3.6023
DI:542	-0.14378	-1.9219	-4.0923	-1.5476
DI:576	-0.14702	-1.9653	-44.119	-16.825
DI:592	-0.23942	-3.2000	-15.232	-5.8169
DI:666	-0.13336	-1.7826	-37.702	-14.468
DI:671	0.14903	1.9921	114.60	55.696
DI:677	0.22439	2.9991	95.623	47.720
DI:713	-0.18231	-2.4369	-20.334	-7.7480
DI:729	-0.13724	-1.8343	-57.500	-25.765
DI:736	6.1990	74.331	146.43	159.61
DI:738	-2.7861	-27.041	-77.425	-83.999
DI:739	-2.3854	-20.127	-54.831	-59.272
DI:740	-2.0464	-20.937	-68.951	-75.194
I:22	0.50176	4.3357	16.108	8.0049
I:71	-0.21380	-1.9956	-11.101	-11.340
I:131	-0.26572	-2.4618	-13.678	-12.423
I:132	-0.16743	-1.5294	-7.1673	-5.8775
I:143	0.25706	2.3769	15.384	10.959
I:155	0.19082	1.7692	16.858	8.9259
I:222	0.24597	2.2902	12.855	10.797
I:235	0.38962	3.5667	20.936	12.792
I:245	-0.32859	-3.0583	-24.014	-15.103
I:261	-0.56575	-3.5622	-14.728	-11.156
I:287	0.28965	2.6837	10.113	11.358
I:295	-0.39634	-3.6714	-12.686	-14.814
I:296	0.35453	3.3055	14.275	16.267
I:307	0.33096	3.1016	25.918	22.272
I:326	0.51869	4.8566	44.097	33.508
I:436	0.42924	3.9753	43.011	22.527
I:534	0.19425	1.8148	13.300	13.048

I:597	-0.23044	-2.1512	-12.192	-12.900
I:658	0.18841	1.7475	8.2726	8.2315
I:664	-0.38406	-3.5771	-19.316	-19.150
I:735	0.93986	8.6678	49.523	40.836
T1:20	-0.051115	-1.6638	-1.8463	-1.3682
T1:24	0.055219	2.4987	2.3215	2.0698
T1:44	-0.066418	-3.2698	-3.5806	-3.7557
T1:49	0.39117	2.9617	8.8892	10.277
T1:50	-0.52664	-2.9672	-17.687	-20.046
T1:52	0.45564	2.7881	16.400	16.844
T1:53	-0.26239	-2.6316	-9.2610	-9.5170
T1:128	-0.053169	-4.5552	-5.5666	-4.5003
T1:132	0.052589	4.1730	5.3873	4.1396
T1:165	0.080836	2.4490	6.7205	3.2653
T1:168	-0.40585	-3.0542	-11.140	-4.0989
T1:169	0.35506	3.0900	11.074	4.0551
T1:177	-0.34485	-3.1233	-10.368	-5.0209
T1:178	0.39644	3.2574	11.754	5.6811
T1:182	-0.13462	-4.2244	-15.245	-6.1485
T1:192	0.12799	5.9203	11.863	5.6653
T1:197	-0.087236	-4.5404	-6.8634	-4.1028
T1:209	0.022358	3.3113	3.4144	2.7644
T1:233	-0.013396	-3.8590	-4.6850	-3.5897
T1:267	-0.31620	-3.0590	-12.056	-8.5000
T1:268	0.62045	3.7501	24.082	15.950
T1:270	-0.79143	-4.8220	-34.586	-23.743
T1:271	0.49340	4.8806	22.247	15.977
T1:350	0.21752	2.0037	12.882	9.0825
T1:351	-0.65662	-2.7618	-42.194	-28.366
T1:352	0.79941	3.1280	123.00	116.10
T1:353	-0.53591	-2.7742	-112.97	-76.157
T1:355	0.47229	2.6094	25.926	8.8098
T1:356	-0.35334	-2.4197	-14.796	-5.1070
T1:360	0.23024	2.6012	14.273	6.8555
T1:362	-0.44296	-2.6942	-11.214	-5.5100
T1:363	0.27460	2.4065	6.7271	3.6750
T1:375	0.074283	1.2103	2.4767	1.3533
T1:377	-0.44622	-2.7546	-7.7923	-4.6255
T1:378	0.38167	3.1181	8.3639	5.6071
T1:386	0.22883	1.7497	5.9882	4.6221
T1:387	-0.52931	-2.3342	-14.798	-8.5101
T1:388	0.27112	2.2852	10.346	5.4969
T1:404	0.33524	2.8257	8.6098	5.0884
T1:405	-0.67703	-3.0132	-18.229	-8.8838
T1:406	0.35981	2.8962	11.793	6.8029
T1:417	0.17789	1.5359	5.3865	2.7716
T1:418	-0.31920	-1.9430	-9.5303	-4.7364
T1:420	0.19486	2.4104	10.721	4.9649
T1:424	-0.059240	-2.4755	-6.5569	-3.3745
T1:448	0.22980	2.0100	14.285	7.8574
T1:449	-0.68799	-2.8653	-43.222	-24.755
T1:450	1.0124	3.9060	186.98	203.55
T1:451	-0.96222	-3.9885	-41.597	-23.965

T1:452	0.41722	3.5545	16.732	9.6477
T1:466	-0.011481	-2.8993	-4.7106	-3.2649
T1:581	-0.0087932	-4.8047	-5.6651	-5.0624
T1:621	-0.38757	-3.4754	-12.137	-10.505
T1:622	0.87816	4.0313	11.304	11.866
T1:623	-0.48079	-4.2107	-6.8243	-6.8403
T1:648	0.15317	3.1248	3.1120	2.4262
T1:649	-0.15080	-3.1957	-3.2508	-2.5183
T1:734	0.65345	30.965	41.958	44.123
T1:739	-0.65201	-31.597	-43.495	-45.911

Model saved to C:\JBG\JGResearch\Univ Washington CFRM\XCG 1959 2019
\Model UER AR1 IIS DIIS TIS LEI11 WKUCLL1 N741.pdf
Model saved to C:\JBG\JGResearch\Univ Washington CFRM\XCG 1959 2019
\Model UER AR1 IIS DIIS TIS LEI11 WKUCLL1 N741.pdf.gwg

We highlight, in bold, the statistically significant AR(1) and lagged input variables.

Autometrics retains one-month lags of LEI and WKUCL, starting with four LEI lags, in the US UER estimated time series models with several impulse indicator saturation variables, and several differenced impulse indicator variables (DI, see observations 736, 738, 739, and 740 in 2020), respectively. Trend saturation variables (TI) are identified at observations 734 and 735 in 2020. The RSS falls from the AR(1) estimate of 25.9 to the Autometrics estimated RSS of 6.8. The estimated residual diagnostics reveal no statistically different differences from normality.[19] Indeed, Autometrics is a most appropriate analysis software system for the pandemic period of the US unemployment rate. We allowed 12 lags of LEI and WKUCL to be modeled in the GUM.

EQ(13.7) Modelling UER by OLS with LEI and WKUCL, Lags to 12

The dataset is: C:\JBG\JGResearch\Univ Washington CFRM\LEI UER WKUCL.csv
The estimation sample is: 13 - 742

	Coefficient	Std.Error	t-value	t-prob	Part.R^2
sigma	0.0991424	RSS		5.88770531	
R^2	0.997142	F(130,599) =		1608 [0.000]**	
Adj.R^2	0.996522	log-likelihood		723.54	
no. of observations	730	no. of parameters		131	
mean(UER)	5.99863	se(UER)		1.68106	
When the log-likelihood constant is NOT included:					
AIC	-4.46127	SC		-3.63704	
HQ	-4.14328	FPE		0.0115931	
When the log-likelihood constant is included:					

---

[19]The authors estimated separate transfer function models using the WKUCL time series as input to the UER time series. The WKUCL input produced positive and statistically significant coefficients in the Autometrics estimations.

AIC	-1.62340	SC	-0.799166
HQ	-1.30540	FPE	0.198004

AR 1-2 test:	F(2,597)	= 0.94150 [0.3906]
ARCH 1-1 test:	F(1,728)	= 1.4837 [0.2236]
Normality test:	Chi^2(2)	= 0.46182 [0.7938]
Hetero test:	F(128,549)	= 0.91178 [0.7355]
RESET23 test:	F(2,597)	= 0.80208 [0.4489]

Robust standard errors

	Coefficients	SE	HACSE	HCSE
UER_1	0.51867	0.024475	0.024455	0.025611
Constant	1624.8	234.75	192.07	183.12
Trend	-2.1929	0.31728	0.25963	0.24757
LEI_2	-0.025952	0.0029114	0.0028707	0.0029763
WKUCL_1	0.0014813	9.5735e-05	4.7811e-05	5.9361e-05
WKUCL_2	0.00081458	0.00014364	7.7952e-05	8.9312e-05
WKUCL_6	0.0013461	0.00015188	0.00012782	0.00012307
WKUCL_9	0.00084001	0.00017490	0.00019229	0.00019345
DI:14	-0.41523	0.07034	0.017853	0.045818
DI:17	-0.22196	0.070344	0.0065730	0.013286
DI:22	0.53868	0.10267	0.011066	0.023087
DI:23	0.71694	0.14999	0.027905	0.049768
DI:24	-3.3857	0.40876	0.23278	0.31620
DI:25	-3.0634	0.37656	0.20880	0.28426
DI:26	-2.5182	0.34466	0.18651	0.25355
DI:27	-2.1786	0.31008	0.16007	0.21879
DI:28	-1.7370	0.27467	0.13483	0.18449
DI:29	-1.2441	0.23690	0.10737	0.14757
DI:30	-0.95275	0.19633	0.077059	0.10812
DI:31	-0.43501	0.15374	0.050556	0.071559
DI:32	-0.42544	0.10319	0.020810	0.032092
DI:57	0.21977	0.10409	0.042343	0.033288
DI:58	0.38396	0.15363	0.083567	0.065865
DI:59	0.78023	0.19543	0.12371	0.097768
DI:60	0.90969	0.23378	0.16259	0.12945
DI:61	1.2144	0.26968	0.20105	0.16040
DI:62	1.2741	0.30401	0.23876	0.19108
DI:63	1.4882	0.33677	0.27564	0.22079
DI:64	1.6166	0.36859	0.31170	0.25008
DI:65	1.6170	0.39924	0.34688	0.27842
DI:66	1.8138	0.42849	0.38145	0.30580
DI:67	1.7051	0.45742	0.41507	0.33302
DI:68	1.8847	0.48496	0.44774	0.35877
DI:69	2.1110	0.51215	0.47977	0.38437
DI:70	2.3367	0.53911	0.51078	0.40981
DI:71	2.2791	0.56551	0.54096	0.43480
DI:72	2.5603	0.59052	0.57034	0.45827
DI:73	2.6588	0.61545	0.59892	0.48185
DI:74	3.0625	0.63976	0.62657	0.50466
DI:75	-0.23155	0.10128	0.026429	0.021629

DI:129	0.25210	0.097413	0.030896	0.031037
DI:130	0.41558	0.13608	0.065053	0.062672
DI:131	0.37888	0.16494	0.10072	0.095075
DI:132	0.41646	0.18739	0.13502	0.12576
DI:133	0.81533	0.20622	0.16961	0.15678
DI:134	0.66623	0.18468	0.13290	0.12067
DI:135	0.50130	0.16155	0.098403	0.087517
DI:136	0.37854	0.13411	0.064989	0.056917
DI:137	0.21162	0.097090	0.032617	0.028898
DI:194	-0.19754	0.074098	0.0055254	0.0071029
DI:222	0.20061	0.070283	0.018627	0.045611
DI:263	0.19994	0.075850	0.015648	0.029002
DI:471	0.19953	0.070163	0.026472	0.069336
DI:677	0.17706	0.070198	0.016498	0.043164
DI:736	7.1216	0.23661	0.11225	0.13687
DI:737	3.4161	0.29882	0.14095	0.16825
I:24	4.7075	0.48623	0.26851	0.37973
I:44	0.28106	0.10534	0.019686	0.033150
I:48	-0.36255	0.11670	0.035817	0.057255
I:53	0.41201	0.10490	0.045326	0.035827
I:75	3.2074	0.68701	0.67967	0.54909
I:143	0.20275	0.10250	0.027915	0.031451
I:211	0.21543	0.10234	0.032391	0.035243
I:235	0.40345	0.10935	0.032494	0.036826
I:326	0.39668	0.10316	0.029204	0.034983
I:378	-0.29485	0.10173	0.014959	0.021517
I:406	-0.23651	0.10232	0.028534	0.025973
I:436	0.36233	0.10085	0.019062	0.017889
I:450	-0.25770	0.10015	0.014565	0.013359
I:452	-0.36916	0.10015	0.013821	0.013239
I:509	-0.30378	0.10105	0.016723	0.018167
I:554	0.21543	0.10093	0.025229	0.018282
I:588	0.27166	0.10245	0.028988	0.029506
I:597	-0.29523	0.10383	0.019071	0.019086
I:735	0.91553	0.10882	0.022755	0.039768
T1:45	-0.36001	0.084781	0.040458	0.078614
T1:46	0.42981	0.096570	0.048696	0.089371
T1:52	-0.063255	0.015090	0.012182	0.014081
T1:132	-0.57985	0.062505	0.076406	0.071978
T1:133	0.56982	0.063036	0.076586	0.072684
T1:165	0.13574	0.031579	0.012340	0.020543
T1:168	-0.35739	0.11764	0.038161	0.073238
T1:169	0.23350	0.093423	0.035973	0.062425
T1:187	-0.096620	0.048023	0.025992	0.049592
T1:189	-0.19421	0.081267	0.036430	0.083968
T1:192	0.66215	0.14739	0.041387	0.10645
T1:193	-0.42952	0.15191	0.029778	0.072416
T1:196	0.55490	0.13711	0.056639	0.064438
T1:197	-0.51505	0.10156	0.056695	0.065079
T1:217	0.13242	0.034612	0.035224	0.038835
T1:219	-0.091210	0.035671	0.036721	0.038321
T1:235	-0.17636	0.070478	0.065451	0.065351
T1:236	0.16193	0.069385	0.064289	0.065059

T1:254	-0.53444	0.080500	0.048084	0.098443
T1:255	0.61166	0.089737	0.054666	0.11133
T1:262	-0.27498	0.068954	0.038998	0.063221
T1:264	0.26338	0.092935	0.039486	0.077808
T1:267	-0.59319	0.15108	0.035903	0.055576
T1:268	0.68577	0.16639	0.031243	0.065938
T1:270	-0.30870	0.076401	0.028532	0.059845
T1:274	0.11919	0.024160	0.013146	0.022350
T1:293	0.52460	0.11136	0.036205	0.049531
T1:294	-0.85617	0.21259	0.081497	0.11768
T1:295	0.39464	0.11891	0.089394	0.11930
T1:305	-0.47272	0.10989	0.069117	0.088402
T1:306	0.56044	0.14899	0.065543	0.084698
T1:308	-0.11653	0.054545	0.029389	0.036452
T1:324	-0.075815	0.030594	0.039546	0.039021
T1:326	0.071211	0.027319	0.036201	0.034200
T1:375	-0.050068	0.0092539	0.0060220	0.0088572
T1:382	0.075829	0.019912	0.012095	0.020275
T1:388	-0.090938	0.021264	0.016275	0.024124
T1:395	0.034748	0.013175	0.012836	0.014881
T1:410	0.026714	0.010116	0.0087893	0.010083
T1:420	0.055726	0.016759	0.012421	0.017413
T1:425	-0.058222	0.011239	0.0082901	0.011574
T1:510	-0.020937	0.0018287	0.0020453	0.0019240
T1:545	0.020685	0.0022575	0.0027640	0.0024124
T1:591	-0.26654	0.057056	0.044529	0.048869
T1:592	0.22995	0.057791	0.044337	0.048808
T1:621	-0.36503	0.10816	0.030403	0.039146
T1:622	0.99649	0.21221	0.041388	0.052922
T1:623	-0.63938	0.12455	0.031656	0.045904
T1:629	0.059445	0.013584	0.0099907	0.011000
T1:662	0.41623	0.10506	0.028744	0.042591
T1:663	-0.70159	0.20330	0.038835	0.053292
T1:664	0.27426	0.10374	0.025593	0.034442
T1:712	-0.16673	0.051709	0.046040	0.060647
T1:713	0.16478	0.053745	0.047815	0.063326
T1:735	0.37004	0.039868	0.022120	0.026657
T1:739	1.8177	0.32622	0.26015	0.24718

	Coefficients	t-SE	t-HACSE	t-HCSE
UER_1	0.51867	21.192	21.210	20.252
Constant	1624.8	6.9215	8.4597	8.8731
Trend	-2.1929	-6.9115	-8.4461	-8.8578
LEI_2	-0.025952	-8.9140	-9.0404	-8.7197
WKUCL_1	0.0014813	15.473	30.982	24.954
WKUCL_2	0.00081458	5.6709	10.450	9.1206
WKUCL_6	0.0013461	8.8630	10.531	10.937
WKUCL_9	0.00084001	4.8027	4.3684	4.3422
DI:14	-0.41523	-5.9029	-23.259	-9.0627
DI:17	-0.22196	-3.1554	-33.769	-16.707
DI:22	0.53868	5.2465	48.678	23.332
DI:23	0.71694	4.7799	25.692	14.406

DI:24	-3.3857	-8.2830	-14.544	-10.708
DI:25	-3.0634	-8.1351	-14.671	-10.777
DI:26	-2.5182	-7.3062	-13.501	-9.9317
DI:27	-2.1786	-7.0257	-13.610	-9.9571
DI:28	-1.7370	-6.3240	-12.883	-9.4152
DI:29	-1.2441	-5.2517	-11.587	-8.4307
DI:30	-0.95275	-4.8527	-12.364	-8.8123
DI:31	-0.43501	-2.8295	-8.6045	-6.0791
DI:32	-0.42544	-4.1229	-20.444	-13.257
DI:57	0.21977	2.1114	5.1902	6.6021
DI:58	0.38396	2.4994	4.5947	5.8296
DI:59	0.78023	3.9923	6.3071	7.9803
DI:60	0.90969	3.8912	5.5949	7.0274
DI:61	1.2144	4.5033	6.0405	7.5711
DI:62	1.2741	4.1909	5.3362	6.6677
DI:63	1.4882	4.4190	5.3989	6.7404
DI:64	1.6166	4.3860	5.1866	6.4645
DI:65	1.6170	4.0503	4.6617	5.8079
DI:66	1.8138	4.2330	4.7550	5.9312
DI:67	1.7051	3.7277	4.1080	5.1202
DI:68	1.8847	3.8863	4.2094	5.2532
DI:69	2.1110	4.1220	4.4001	5.4923
DI:70	2.3367	4.3343	4.5747	5.7018
DI:71	2.2791	4.0301	4.2130	5.2416
DI:72	2.5603	4.3356	4.4890	5.5868
DI:73	2.6588	4.3202	4.4394	5.5181
DI:74	3.0625	4.7869	4.8877	6.0684
DI:75	-0.23155	-2.2862	-8.7613	-10.706
DI:129	0.25210	2.5879	8.1596	8.1225
DI:130	0.41558	3.0538	6.3883	6.6310
DI:131	0.37888	2.2971	3.7616	3.9851
DI:132	0.41646	2.2224	3.0844	3.3116
DI:133	0.81533	3.9537	4.8072	5.2004
DI:134	0.66623	3.6075	5.0129	5.5212
DI:135	0.50130	3.1030	5.0943	5.7280
DI:136	0.37854	2.8226	5.8247	6.6508
DI:137	0.21162	2.1796	6.4880	7.3229
DI:194	-0.19754	-2.6659	-35.751	-27.811
DI:222	0.20061	2.8542	10.770	4.3981
DI:263	0.19994	2.6360	12.778	6.8941
DI:471	0.19953	2.8437	7.5372	2.8777
DI:677	0.17706	2.5222	10.732	4.1020
DI:736	7.1216	30.099	63.445	52.033
DI:737	3.4161	11.432	24.236	20.303
I:24	4.7075	9.6816	17.532	12.397
I:44	0.28106	2.6681	14.277	8.4782
I:48	-0.36255	-3.1067	-10.122	-6.3322
I:53	0.41201	3.9277	9.0900	11.500
I:75	3.2074	4.6686	4.7191	5.8413
I:143	0.20275	1.9780	7.2632	6.4466
I:211	0.21543	2.1051	6.6511	6.1127
I:235	0.40345	3.6894	12.416	10.956
I:326	0.39668	3.8455	13.583	11.339

I:378	-0.29485	-2.8984	-19.710	-13.703
I:406	-0.23651	-2.3115	-8.2886	-9.1061
I:436	0.36233	3.5929	19.008	20.255
I:450	-0.25770	-2.5732	-17.693	-19.290
I:452	-0.36916	-3.6861	-26.710	-27.884
I:509	-0.30378	-3.0061	-18.165	-16.722
I:554	0.21543	2.1344	8.5388	11.784
I:588	0.27166	2.6516	9.3713	9.2069
I:597	-0.29523	-2.8433	-15.481	-15.468
I:735	0.91553	8.4134	40.234	23.022
T1:45	-0.36001	-4.2463	-8.8983	-4.5795
T1:46	0.42981	4.4508	8.8264	4.8093
T1:52	-0.063255	-4.1918	-5.1923	-4.4922
T1:132	-0.57985	-9.2768	-7.5890	-8.0559
T1:133	0.56982	9.0396	7.4402	7.8397
T1:165	0.13574	4.2983	11.000	6.6074
T1:168	-0.35739	-3.0379	-9.3654	-4.8799
T1:169	0.23350	2.4994	6.4910	3.7405
T1:187	-0.096620	-2.0119	-3.7173	-1.9483
T1:189	-0.19421	-2.3897	-5.3309	-2.3128
T1:192	0.66215	4.4926	15.999	6.2202
T1:193	-0.42952	-2.8274	-14.424	-5.9313
T1:196	0.55490	4.0472	9.7971	8.6114
T1:197	-0.51505	-5.0713	-9.0846	-7.9142
T1:217	0.13242	3.8257	3.7592	3.4097
T1:219	-0.091210	-2.5570	-2.4838	-2.3801
T1:235	-0.17636	-2.5023	-2.6945	-2.6986
T1:236	0.16193	2.3337	2.5187	2.4889
T1:254	-0.53444	-6.6390	-11.115	-5.4289
T1:255	0.61166	6.8161	11.189	5.4944
T1:262	-0.27498	-3.9878	-7.0511	-4.3494
T1:264	0.26338	2.8340	6.6700	3.3849
T1:267	-0.59319	-3.9264	-16.522	-10.673
T1:268	0.68577	4.1216	21.950	10.400
T1:270	-0.30870	-4.0405	-10.819	-5.1583
T1:274	0.11919	4.9333	9.0668	5.3328
T1:293	0.52460	4.7107	14.490	10.592
T1:294	-0.85617	-4.0274	-10.506	-7.2751
T1:295	0.39464	3.3187	4.4146	3.3079
T1:305	-0.47272	-4.3018	-6.8395	-5.3474
T1:306	0.56044	3.7617	8.5507	6.6169
T1:308	-0.11653	-2.1364	-3.9650	-3.1968
T1:324	-0.075815	-2.4781	-1.9171	-1.9429
T1:326	0.071211	2.6067	1.9671	2.0822
T1:375	-0.050068	-5.4104	-8.3141	-5.6528
T1:382	0.075829	3.8082	6.2695	3.7400
T1:388	-0.090938	-4.2766	-5.5877	-3.7696
T1:395	0.034748	2.6373	2.7070	2.3350
T1:410	0.026714	2.6406	3.0394	2.6493
T1:420	0.055726	3.3250	4.4863	3.2003
T1:425	-0.058222	-5.1802	-7.0231	-5.0305
T1:510	-0.020937	-11.449	-10.236	-10.882
T1:545	0.020685	9.1628	7.4836	8.5744
T1:591	-0.26654	-4.6715	-5.9857	-5.4542
T1:592	0.22995	3.9789	5.1864	4.7112

```
T1:621        -0.36503        -3.3751        -12.007        -9.3249
T1:622         0.99649         4.6958         24.077        18.829
T1:623        -0.63938        -5.1336        -20.198       -13.929
T1:629         0.059445        4.3762          5.9500        5.4039
T1:662         0.41623         3.9617         14.480         9.7728
T1:663        -0.70159        -3.4509        -18.066       -13.165
T1:664         0.27426         2.6438         10.716         7.9631
T1:712        -0.16673        -3.2243         -3.6213       -2.7491
T1:713         0.16478         3.0660          3.4462        2.6021
T1:735         0.37004         9.2816         16.728        13.882
T1:739         1.8177          5.5720          6.9872        7.3538
Model saved to C:\JBG\JGResearch\Univ Washington CFRM\XCG 1959 2019
\Model UER AR1 IIS DIIS TIS LEI112 WKUCLL12.pdf
Model saved to C:\JBG\JGResearch\Univ Washington CFRM\XCG 1959 2019
\Model UER AR1 IIS DIIS TIS LEI112 WKUCLL12.pdf.gwg
```

Autometrics retains only a two-month lag in the LEI and 1-, 2-, 6-, and 9-month lags in the WKUCL time series in the US unemployment model, a result reported in Dhrymes (2017). What is the real advantage of an automatic time series system such as Autometrics? The researcher can estimate naïve AR(1) models, with and without outliers. One can then apply Autometrics for model validation and/or obtain a better estimate, one with normal residuals. Second, the researcher can select a larger number of lags, and Autometrics will retain the statistically significant variables.[20]

We allowed 24 lags of LEI and WKUCL to be modeled in the GUM and tested various critical break levels.[21]

```
Equation (13.8) Autometrics Analysis of the U.S. Unemployment Rate
using the Tiny (0.001) Critical level
```

```
Summary of Autometrics search
initial search space    2^96  final search space      2^90
no. estimated models     2502  no. terminal models        20
test form                LR-F  target size       Tiny:0.001
large residuals            no  presearch reduction      lags
backtesting              GUM0  tie-breaker                SC
diagnostics p-value      0.01  search effort       standard
time                 23:40.84  Autometrics version      2.0a
```

```
                Coefficients        t-SE     t-HACSE      t-HCSE
A0M043_1             0.61372      32.395      28.622      31.099
Trend               -2.4100      -6.5184     -5.2055     -5.6524
```

---

[20] Guerard et al. (2020) reported that the LEI time series was a useful input in the real GDP time series and changes in the US unemployment rate models estimated with Autometrics during the 1959–2018 period.

[21] The authors are indebted to Professor Jennifer Caste of Oxford for her support in reading and helping the authors efficiently explore structural break levels. Any errors remaining are solely those of the authors.

Constant	1787.0	6.5314	5.2148	5.6631
AOM005_1	0.0016573	23.438	31.242	30.746
GOM910_2	-0.046634	-8.9797	-7.9418	-9.1175
GOM910_5	0.021737	3.2645	3.2037	3.4068
GOM910_13	-0.016418	-5.6157	-5.4011	-5.5176
AOM005_6	0.0014841	8.0052	6.5593	6.9641
AOM005_9	0.0010774	5.9688	5.6974	5.4746
AOM005_24	0.00047262	4.1642	3.8517	3.8300
DI:26	0.31377	3.0330	30.829	30.360
DI:27	0.41550	2.9508	21.519	20.778
DI:28	0.66233	4.0090	24.036	22.963
DI:29	0.91140	5.0226	27.964	26.527
DI:30	0.95480	4.9588	23.577	22.178
DI:31	1.2265	6.2026	28.130	26.446
DI:32	0.93837	4.7380	20.701	19.537
DI:33	1.1405	5.9029	25.889	24.429
DI:34	1.0360	5.6595	25.181	24.080
DI:35	0.66158	3.9961	21.464	21.779
DI:36	0.56817	4.0285	26.029	26.571
DI:37	0.33177	3.2001	27.576	27.595
DI:46	-0.19991	-2.6339	-24.552	-10.275
DI:53	0.20425	2.6908	12.129	4.6285
DI:74	0.25153	3.3139	14.127	5.3751
DI:193	0.42111	5.5193	65.868	54.265
DI:222	0.20734	2.7282	15.924	6.4959
DI:235	0.27080	3.5652	15.628	6.1665
DI:256	-0.49334	-4.4921	-21.443	-19.704
DI:736	4.9411	44.188	61.392	68.664
I:256	1.0456	6.6085	23.785	21.484
I:287	0.34061	3.0861	18.278	13.283
I:295	-0.43094	-3.7144	-9.0249	-7.6879
I:326	0.45045	4.1266	19.119	19.974
I:378	-0.31217	-2.8774	-17.784	-19.041
I:406	-0.28989	-2.6585	-12.802	-14.900
I:436	0.34941	3.2314	22.637	26.397
I:450	-0.26543	-2.4548	-17.458	-20.210
I:452	-0.39285	-3.6372	-28.119	-32.176
I:472	-0.27773	-2.5685	-17.622	-21.215
I:509	-0.33641	-3.0936	-16.702	-19.503
I:544	-0.27376	-2.4479	-6.9932	-9.8564
I:623	0.52251	4.6795	16.500	15.888
I:664	-0.44956	-4.1560	-29.449	-29.398
I:738	-2.7481	-19.672	-31.026	-33.975
T1:131	-0.39849	-9.6707	-8.7228	-10.230
T1:132	0.39825	9.6152	8.6886	10.154
T1:189	-0.15402	-9.4504	-12.732	-10.576
T1:194	0.36322	9.7215	14.329	12.216
T1:198	-0.20219	-6.5297	-10.650	-8.3256
T1:208	-0.38181	-4.2759	-6.5867	-3.6050
T1:209	0.40712	4.7087	7.3944	4.1318
T1:220	-0.023552	-3.4431	-4.6498	-3.9785
T1:267	-0.35707	-3.4400	-8.1883	-7.8357
T1:268	0.53366	3.2818	9.8809	7.4286

T1:270	-0.35115	-3.7679	-12.524	-5.3658
T1:273	0.16662	4.8187	11.660	5.8521
T1:294	0.031143	2.8355	2.9360	2.5386
T1:305	-0.30612	-4.4314	-5.0945	-3.8820
T1:306	0.27798	4.4405	5.1191	3.8613
T1:398	-0.28141	-7.9876	-6.8136	-7.7672
T1:399	0.27984	7.9984	6.8279	7.7874
T1:541	-0.37170	-9.1638	-7.4381	-9.6846
T1:542	0.36924	8.9999	7.2979	9.5206
T1:623	-0.041382	-4.6744	-4.5486	-4.3380
T1:628	0.050542	5.5844	5.4087	5.1797
T1:686	-0.012696	-7.7513	-6.9985	-7.7585
T1:733	-0.80692	-7.8667	-10.472	-10.546
T1:735	1.5497	10.638	14.375	14.603
T1:739	1.6749	4.4133	3.4949	3.7968

Both LEI and WKUCL variables are highly statistically associated with the US unemployment rate time series. We report in Table 13.1 that breaks are identified at all critical levels of significance and reduce the RSS.

First, note that a simple autoregression does not capture the cyclical variability of the level of unemployment as expected, with a low estimate of persistence and a relatively low R^2. The RSS is large, and there is autocorrelation, heteroscedasticity, and non-normality in the residuals. Adding indicators in this first model we find, as before in the case of the real GDP series, that the RSS is greatly reduced, autocorrelation disappears as does heteroscedasticity, but not non-normality—note that we

Table 13.1 US unemployment rate (UER) analysis
From 1959 to November 2020
Autometrics, AR(1), 24 lags of LEI, WKUCL with IIS, DIIS, TIS variables

Autometrics, AR(1), 24 Lags of LEI, WKUCL with IIS, DIIS, TIS variables

	ARI	OLS	Large	Huge	Minute	Tiny
Critical Levels of Breaks			0.1	0.0001	0.001	
AR(1) (t)	.965 (100.0)	.923 (54.4)	.944 (97.5)	.758 (72.6)	.861 (140.6)	.801 (45.2)
constant	.196 (3.07)	.472 (3.62)	.318 (3.61)			
trend	.001 (.40)	.001 (2.90)	.001 (3.59)			
LEI_1		.012 (0.51)	-.014 (-6.06)			
LEI_2				-.051 (-6.66)	-0.12 (-6.06)	-.012 (-10.0)
...						
WkUNCL_1		.004 (32.6)	.002 (10.5)	.001 (21.09)	.002 (96.1)	.002 (18.07)
...						
sigma	0.435	0.204	0.142	0.079	0.123	0.122
RSS	139.79	27.77	14.18	2.93	10.17	10.02
F-Stat	5082	973.6	7226			
AIC (C Not Included)	-1.66 *	-3.11	-3.88	-4.81	-4.14	-4.15
AR 1-2	2.597 *	125.48 *	17.699 *	2.178 *	3.95 *	4.565 *
ARCH1-1	0.202	603.44 *	9.57 *	38.143	0.312	0.004
Normality	61612 *	282.02 *	2.967	45.887	0.39	0.812
HERERO	2.229 *	173.77 *	2.495 *	0.56 *	1.232	1.045
RESET	10.989 *		1.354	0.026 *	0.575	1.706
Chow		3.661 *				

* = violates the null hypothesis at the 10* level

Violates the null hypothesis at the 10 level

have a total of 47 significant indicators. Adding to this last model the weekly claims, we can obtain even further RSS reduction, at the expense of about 20 more indicators, but the weekly claims add explanatory power (although at about the 10% partial R^2); thus, Autometrics does capture the empirical finding of weekly claims being a reasonable explanatory variable for the unemployment rate. Note, the normality test indicates now some improvement although we have a deterioration to the p-values of autocorrelation and heteroscedasticity that nevertheless remain into the close to 5–10% territory. Finally, the fourth model we examine contains the LEI lags in addition to the variables of the second model, but not the unemployment claims. The number of indicator variables increases to 65, all significant, and we can clearly see that the lags of the LEI variable not only are significant, but they have a relatively high contribution, by their partial R^2; the first lag of the LEI has a 30% contribution alone in explaining unemployment rate variability.

We report that the unemployment rate time series of the United States can be modeled with an AR(1) process, a one-period lag in weekly unemployment claims time series, with the expected positive coefficient, and a one-period lag in the LEI time series, with the expected negative coefficient. We report validation of the MZTT analysis. See the estimated Eq. (13.6). There is a substantial RSS reduction and normal and homoscedastic error terms. Causality testing with the SCA system, using the Box and Jenkins (1970) and Tsay (1988, 2010) methodologies produced statistical evidence that the DLLEI time series causes changes in the differenced US unemployment rate time series. See Appendix A.

13.7   Summary and Conclusions

We examined the regression and time series association between real US GDP and the leading economic indicators during the 1959–2020Q3 time period. We report that The Conference Board leading economic indicators are statistically significant once the regression line addresses outlier estimations in SAS (ETS). We applied the Hendry and Doornik automatic time series PC Give (OxMetrics) methodology to several well-studied macroeconomics series, real GDP, and the unemployment rate. We report that the OxMetrics and Autometrics system substantially reduces the regression sum of squares measures relative to a traditional variation on the random walk with drift model. The modeling process of including the leading economic indicator in forecasting real GDP has been addressed before, but our results are more statistically significant. The use of SAS-estimated transfer functions, Autometrics-estimated time series models with outliers, and SCA-estimated causality models leads researchers to one conclusion: Read GDP and the unemployment rate are significantly influenced, if not, caused, by the leading economic indicators. The automatic time series analysis and forecasting system developed by David Hendry and his colleagues to estimate structural breaks and outliers in time series data is relevant in almost all time periods, but the Autometrics software is particularly useful in modeling time series in 2020.

Appendix A: Granger Causality Modeling of the Change in the US Unemployment Rate with First-Differenced Changes in the LEI Time Series

One can use the SCA system, see Liu (1999), and tested for causality, as an application of the Chen and Lee (1990) test, one finds causality in the DUE, DLEI relationship with both 4-month and 18-month DLEI lags. Moreover, the use of SCA system produces univariate DUE time series results that the four lags of DLEI data are statistically significant, with an R-squared of 0.188. DLEI lags of 2 and 3 are statistically significant at the 5% level, lag one is significant at the 10% level, and the application of the Tsay (1988) outlier procedure reduces the RSS from 0.161 to 0.113 (Table 13.2).

Table 13.2 Summary for univariate time series model for the US unemployment rate

---

VARIABLE   TYPE OF   ORIGINAL   DIFFERENCING

          VARIABLE   OR CENTERED

DUE	RANDOM	ORIGINAL	NONE
DLEI	RANDOM	ORIGINAL	NONE

PARAMETER LABEL	VARIABLE NAME	NUM./ DENOM.	FACTOR	ORDER	CONS-TRAINT	VALUE	STD ERROR	T VALUE
1	DLEI	NUM.	1	1	NONE	-.0402	.0171	-2.35
2	DLEI	NUM.	1	2	NONE	-.0629	.0183	-3.45
3	DLEI	NUM.	1	3	NONE	-.0464	.0182	-2.55
4	DLEI	NUM.	1	4	NONE	-.0321	.0171	-1.88
5	DUE	D-AR	1	1	NONE	-.0807	.0376	-2.15

```
EFFECTIVE NUMBER OF OBSERVATIONS . .          710
R-SQUARE . . . . . . . . . . . . . .          0.188
RESIDUAL STANDARD ERROR. . . . . . .   0.161307E+00
(-2)*LOG LIKELIHOOD FUNCTION . . . .  -0.575818E+03
AIC. . . . . . . . . . . . . . . . .  -0.563818E+03
SIC. . . . . . . . . . . . . . . . .  -0.536427E+03
--
```

THE CRITICAL VALUE FOR SIGNIFICANCE TESTS OF ACF AND ESTIMATES IS 1.960

Table 13.3 Summary for univariate time series model with outliers—TFM1

VARIABLE	TYPE OF VARIABLE	ORIGINAL OR CENTERED	DIFFERENCING
DUE	RANDOM	ORIGINAL	NONE
DLEI	RANDOM	ORIGINAL	NONE

SUMMARY FOR UNIVARIATE TIME SERIES MODEL -- UTSMODEL

VARIABLE	TYPE OF VARIABLE	ORIGINAL OR CENTERED	DIFFERENCING
NS	RANDOM	ORIGINAL	NONE

PARAMETER LABEL	VARIABLE NAME	NUM./ DENOM.	FACTOR	ORDER	CONS-TRAINT	VALUE	STD ERROR	T VALUE
1		CNST	1	0	NONE	.0193	.0060	3.20

TOTAL NUMBER OF OBSERVATIONS   711
EFFECTIVE NUMBER OF OBSERVATIONS . .   711
RESIDUAL STANDARD ERROR. . . . . . . 0.160581E+00
--

PARAMETER LABEL	VARIABLE NAME	NUM./ DENOM.	FACTOR	ORDER	CONS-TRAINT	VALUE	STD ERROR	T VALUE
1	DLEI	NUM.	1	1	NONE	-.0233	.0148	-1.57
2	DLEI	NUM.	1	2	NONE	-.0628	.0154	-4.08
3	DLEI	NUM.	1	3	NONE	-.0490	.0154	-3.18
4	DLEI	NUM.	1	4	NONE	-.0154	.0149	-1.04

SUMMARY OF OUTLIER DETECTION AND ADJUSTMENT

TIME	ESTIMATE	T-VALUE	TYPE
11	-0.458	-5.47	TC
14	0.783	6.72	AO
17	0.268	3.21	TC
21	0.597	6.26	TC
22	-0.449	-3.44	AO
31	-0.380	-3.92	TC
32	0.396	3.00	AO
34	-0.238	-2.02	AO
43	0.306	2.71	AO
46	0.323	2.86	AO
66	-0.259	-2.29	AO
70	-0.260	-2.30	AO
74	-0.353	-3.13	AO
132	0.370	4.58	TC

140	0.338	4.12	TC
144	-0.277	-2.42	AO
177	-0.257	-2.28	AO
188	0.315	2.79	AO
190	0.566	6.20	TC
192	0.511	4.20	AO
193	-0.302	-2.58	AO
194	0.285	2.48	AO
197	-0.231	-2.85	TC
209	0.228	2.82	TC
216	-0.269	-2.38	AO
222	-0.257	-2.27	AO
227	-0.380	-3.36	AO
234	0.361	3.19	AO
235	-0.266	-2.35	AO
247	0.269	2.38	AO
255	0.587	6.35	TC
257	-0.358	-3.85	TC
263	-0.254	-2.68	TC
264	0.505	3.86	AO
268	0.335	2.93	AO
270	-0.295	-2.61	AO
273	0.340	4.18	TC
279	0.216	2.62	TC
284	0.325	3.92	TC
288	-0.455	-3.97	AO
294	-0.620	-5.49	AO
296	-0.359	-4.44	TC
306	0.356	3.15	AO
319	-0.259	-2.29	AO
324	-0.272	-2.41	AO
325	0.535	4.74	AO
335	-0.278	-2.46	AO
339	-0.260	-2.30	AO
378	0.263	3.26	TC
395	0.202	2.47	TC
400	0.202	2.45	TC
405	-0.298	-2.62	AO
435	0.414	3.67	AO
450	0.278	2.98	TC
451	-0.525	-4.03	AO
471	-0.337	-2.98	AO
508	-0.259	-2.30	AO
511	0.262	3.25	TC
542	0.330	2.92	AO
591	-0.273	-2.42	AO
592	0.183	2.27	TC
599	0.203	2.51	TC
608	0.278	3.44	TC
622	0.448	3.97	AO
623	-0.450	-3.98	AO
633	-0.242	-3.00	TC
644	-0.308	-2.72	AO
649	-0.269	-2.39	AO
663	-0.338	-2.99	AO
688	-0.298	-2.64	AO
694	-0.279	-2.47	AO
713	0.259	2.29	AO

MAXIMUM NUMBER OF OUTLIERS IS REACHED

** THE OUTLIER(S) AFTER TIME PERIOD 710 OCCURS WITHIN THE
   LAST FIVE OBSERVATIONS OF THE SERIES. THE IDENTIFIED TYPE
   ANS THE ESTIMATE OF THE OUTLIER(S) MAY NOT BE RELIABLE

TOTAL NUMBER OF OBSERVATIONS.   715
EFFECTIVE NUMBER OF OBSERVATIONS.   711
RESIDUAL STANDARD ERROR (WITHOUT OUTLIER ADJUSTMENT). . 0.162276E+00
RESIDUAL STANDARD ERROR (WITH OUTLIER ADJUSTMENT) . . . 0.112951E+00
--

Let us use SCA for causality testing, as an application of the Chen and Lee (1990) test, one finds

causality in the DUE, DLEI relationship with both 4-month DLEI lags.

VARIABLES ARE DUE, DLEI.

TIME PERIOD ANALYZED . . . . 1 TO 715
EFFECTIVE NUMBER OF OBSERVATIONS (NOBE). . .   715

SERIES NAME   MEAN   STD. ERROR

  1 DUE   -0.0029   0.1791
  2 DLEI   0.1175   0.4798

NOTE: THE APPROX. STD. ERROR FOR THE ESTIMATED CORRELATIONS BELOW
      IS (1/NOBE**.5) =   0.03740

SAMPLE CORRELATION MATRIX OF THE SERIES

  1.00
 -0.35 1.00

SUMMARIES OF CROSS CORRELATION MATRICES USING +,-,., WHERE
   + DENOTES A VALUE GREATER THAN 2/SQRT(NOBE)
   - DENOTES A VALUE LESS THAN -2/SQRT(NOBE)
   . DENOTES A NON-SIGNIFICANT VALUE BASED ON THE ABOVE CRITERION

BEHAVIOR OF VALUES IN (I,J)TH POSITION OF CROSS CORRELATION MATRIX OVER
ALL OUTPUTTED LAGS WHEN SERIES J LAGS SERIES I

        1   2

1 +++++++++.+- ------------
1 ....- ------.-....

2 ---........ ++++++++++++
2 ......++.+++ +..........-

CROSS CORRELATION MATRICES IN TERMS OF +,-,.

LAGS 1 THROUGH 6

```
+ -   + -   + -   + -   + -   + -
- +   - +   - +   - +   . +   . +
```

LAGS 7 THROUGH 12

```
+ -   + -   + -   . -   + -   - -
. +   . +   . +   . +   . +   . +
```

LAGS 13 THROUGH 18

```
. -   . -   . -   . -   . -   . -
. +   . .   . .   . .   . .   . .
```

LAGS 19 THROUGH 24

```
. .   . -   . .   . .   . .   - .
+ .   + .   . .   + .   + .   + -
- -
```

STEPAR VARIABLES ARE DUE,DLEI. @
ARFITS ARE 1 to 4. rccm 1 to 4.

TIME PERIOD ANALYZED . . . . 1 TO 715
EFFECTIVE NUMBER OF OBSERVATIONS (NOBE). . .   715

SERIES	NAME	MEAN	STD. ERROR
1	DUE	-0.0029	0.1791
2	DLEI	0.1175	0.4798

NOTE: THE APPROX. STD. ERROR FOR THE ESTIMATED CORRELATIONS BELOW
    IS (1/NOBE**.5) =   0.03740

SAMPLE CORRELATION MATRIX OF THE SERIES

```
 1.00
-0.35  1.00
```

SUMMARIES OF CROSS CORRELATION MATRICES USING +,-,., WHERE
    + DENOTES A VALUE GREATER THAN 2/SQRT(NOBE)
    - DENOTES A VALUE LESS THAN -2/SQRT(NOBE)
    . DENOTES A NON-SIGNIFICANT VALUE BASED ON THE ABOVE CRITERION

BEHAVIOR OF VALUES IN (I,J)TH POSITION OF CROSS CORRELATION MATRIX OVER
ALL OUTPUTTED LAGS WHEN SERIES J LAGS SERIES I

```
    1       2
```

```
1  ++++++++++.+-  ------------
1  ...........-  ------.-....


2  ----........  ++++++++++++
2  ......++.+++  +..........-
```

CROSS CORRELATION MATRICES IN TERMS OF +,-,.

LAGS 1 THROUGH 6

```
   + -    + -    + -    + -    + -    + -
   - +    - +    - +    - +    . +    . +
```

LAGS 7 THROUGH 12

```
   + -    + -    + -    . -    + -    - -
   . +    . +    . +    . +    . +    . +
```

LAGS 13 THROUGH 18

```
   . -    . -    . -    . -    . -    . -
   . +    . .    . .    . .    . .    . .
```

LAGS 19 THROUGH 24

```
   . .    . -    . .    . .    . .    - .
   + .    + .    . .    + .    + .    + -
```

DETERMINANT OF S(0) = 0.644754E-02

NOTE: S(0) IS THE SAMPLE COVARIANCE MATRIX OF W(MAXLAG+1),...,W(NOBE)

AUTOREGRESSIVE FITTING ON LAG(S) 1

SUMMARIES OF CROSS CORRELATION MATRICES USING +,-,., WHERE
   + DENOTES A VALUE GREATER THAN 2/SQRT(NOBE)
   - DENOTES A VALUE LESS THAN -2/SQRT(NOBE)
   . DENOTES A NON-SIGNIFICANT VALUE BASED ON THE ABOVE CRITERION

BEHAVIOR OF VALUES IN (I,J)TH POSITION OF CROSS CORRELATION MATRIX OVER
ALL OUTPUTTED LAGS WHEN SERIES J LAGS SERIES I


```
          1         2


1  .+++++....+-  +---.--.-.-.
1  .-.........-  --...-.-....


2  +......+.++.  -+++.+..+.+.
2  .........+..  ............
```

CROSS CORRELATION MATRICES IN TERMS OF +,-,.

LAGS 1 THROUGH 6

```
.+   +-   +-   +-   +.   +-
+-   .+   .+   .+   ..   .+
```

LAGS 7 THROUGH 12

```
.-   ..   .-   ..   +-   -.
..   +.   .+   +.   ++   ..
```

LAGS 13 THROUGH 18

```
.-   --   ..   ..   ..   .-
..   ..   ..   ..   ..   ..
```

LAGS 19 THROUGH 24

```
..   .-   ..   ..   ..   -.
..   ..   ..   +.   ..   ..
```

AUTOREGRESSIVE FITTING ON LAG(S) 1 2

SUMMARIES OF CROSS CORRELATION MATRICES USING +,-,., WHERE
 + DENOTES A VALUE GREATER THAN 2/SQRT(NOBE)
 - DENOTES A VALUE LESS THAN -2/SQRT(NOBE)
 . DENOTES A NON-SIGNIFICANT VALUE BASED ON THE ABOVE CRITERION

BEHAVIOR OF VALUES IN (I,J)TH POSITION OF CROSS CORRELATION MATRIX OVER
ALL OUTPUTTED LAGS WHEN SERIES J LAGS SERIES I

```
        1        2

1  ..++.......-  .+......-...
1  ...........-  .-..-..-....

2  ...+...+..+.  .-+.....+...
2  .........+..  ............
```

CROSS CORRELATION MATRICES IN TERMS OF +,-,.

LAGS 1 THROUGH 6

```
..   .+   +.   +.   ..   ..
..   .-   .+   +.   ..   ..
```

LAGS 7 THROUGH 12

```
..   ..   .-   ..   ..   -.
..   +.   .+   ..   +.   ..
```

LAGS 13 THROUGH 18

```
..   .⁻   ..   ..   .⁻   ..
..   ..   ..   ..   ..   ..
```

LAGS 19 THROUGH 24

```
..   .⁻   ..   ..   ..   ⁻.
..   ..   ..   +.   ..   ..
```

AUTOREGRESSIVE FITTING ON LAG(S) 1 2 3

SUMMARIES OF CROSS CORRELATION MATRICES USING +,-,., WHERE
  + DENOTES A VALUE GREATER THAN 2/SQRT(NOBE)
  - DENOTES A VALUE LESS THAN -2/SQRT(NOBE)
  . DENOTES A NON-SIGNIFICANT VALUE BASED ON THE ABOVE CRITERION

BEHAVIOR OF VALUES IN (I,J)TH POSITION OF CROSS CORRELATION MATRIX OVER ALL OUTPUTTED LAGS WHEN SERIES J LAGS SERIES I

```
        1          2

1   ...+.......-   ........-...
1   ............-   ............

2   ...+......+.   ........+...
2   ............   ............
```


CROSS CORRELATION MATRICES IN TERMS OF +,-,.

LAGS 1 THROUGH 6

```
..   ..   ..   +.   ..   ..
..   ..   ..   +.   ..   ..
```

LAGS 7 THROUGH 12

```
..   ..   .⁻   ..   ..   ⁻.
..   ..   .+   ..   +.   ..
```

LAGS 13 THROUGH 18

```
..   ..   ..   ..   ..   ..
..   ..   ..   ..   ..   ..
```

LAGS 19 THROUGH 24

```
..   ..   ..   ..   ..   ⁻.
..   ..   ..   ..   ..   ..
```


AUTOREGRESSIVE FITTING ON LAG(S) 1 2 3 4

SUMMARIES OF CROSS CORRELATION MATRICES USING +,-,., WHERE
  + DENOTES A VALUE GREATER THAN 2/SQRT(NOBE)
  - DENOTES A VALUE LESS THAN -2/SQRT(NOBE)

. DENOTES A NON-SIGNIFICANT VALUE BASED ON THE ABOVE CRITERION

BEHAVIOR OF VALUES IN (I,J)TH POSITION OF CROSS CORRELATION MATRIX OVER
ALL OUTPUTTED LAGS WHEN SERIES J LAGS SERIES I

            1   2

1 ....- ....
1 ....- ....

2 ....+.....+. ....+...
2 ....+.. ....

CROSS CORRELATION MATRICES IN TERMS OF +,-,.

LAGS 1 THROUGH 6

      ..   ..
      ..   ..   +.   ..

LAGS 7 THROUGH 12

      ..   ..   -.
      ..   ..   .+   ..   +.   ..

LAGS 13 THROUGH 18

      ..   ..
      ..   ..

LAGS 19 THROUGH 24

      ..   ..   -.
      ..   ..   ..   +.   ..   ..

Table 13.4 Stepwise autoregression summary

```
-----------------------------------------------------------------------
    I RESIDUAL I EIGENVAL.I CHI-SQ  I            I SIGNIFICANCE
LAG I VARIANCESI OF SIGMA I  TEST   I   AIC      I OF PARTIAL AR COEFF.
----+----------+----------+---------+----------+----------------------
  1 I .279E-01 I .267E-01 I 340.10 I  -5.514 I . -
    I .148E+00 I .149E+00 I         I          I . +
----+----------+----------+---------+----------+----------------------
  2 I .257E-01 I .253E-01 I 136.90 I  -5.696 I + -
    I .129E+00 I .130E+00 I         I          I . +
----+----------+----------+---------+----------+----------------------
  3 I .251E-01 I .248E-01 I  43.44 I  -5.747 I + -
    I .124E+00 I .124E+00 I         I          I . +
----+----------+----------+---------+----------+----------------------
  4 I .247E-01 I .243E-01 I  18.65 I  -5.762 I + .
    I .123E+00 I .124E+00 I         I          I + .
-----------------------------------------------------------------------
```

NOTE: CHI-SQUARED CRITICAL VALUES WITH   4 DEGREES OF FREEDOM ARE

            5 PERCENT: 9.5   1 PERCENT: 13.3

NOTE: THE PARTIAL AUTOREGRESSION COEFFICIENT MATRIX FOR LAG L IS THE
       ESTIMATED PHI(L) FROM THE FIT WHERE THE MAXIMUM LAG USED IS L
       (I.E. THE LAST COEFFICIENT MATRIX). THE ELEMENTS ARE
       STANDARDIZED BY DIVIDING EACH BY ITS STANDARD ERROR.
--

   MTSMODEL ARMA12. SERIES ARE DUE,DLEI. @
   MODEL IS (1-PHI*B)SERIES=C+(1-TH1*B)NOISE.

SUMMARY FOR MULTIVARIATE ARMA MODEL -- ARMA12

VARIABLE   DIFFERENCING

 DUE
 DLEI

 PARAMETER   FACTOR   ORDER   CONSTRAINT

 1   C   CONSTANT   0   CC
 2   PHI   REG AR   1   CPHI
 3   TH1   REG MA   1   CTH1
--

   CAUSALTEST MODEL ARMA12. OUTPUT PRINT(CORR). alpha .01

SUMMARY OF THE TIME SERIES

 SERIES   NAME   MEAN   STD DEV   DIFFERENCE ORDER(S)

   1   DUE   -0.0029   0.1791
   2   DLEI   0.1175   0.4798

ERROR COVARIANCE MATRIX
----

               1   2
   1   .032113
   2   -.030341   .230366

ITERATIONS TERMINATED DUE TO:
  MAXIMUM NUMBER OF ITERATIONS 10 REACHED
  TOTAL NUMBER OF ITERATIONS IS   14

MODEL SUMMARY WITH MAXIMUM LIKELIHOOD PARAMETER ESTIMATES

----- CONSTANT VECTOR (STD ERROR) -----

    0.052 (0.016)
   -0.138 (0.044)

----- PHI MATRICES -----

ESTIMATES OF   PHI(1) MATRIX AND SIGNIFICANCE

   -1.377   -.531   - -
    6.484   2.414   + +

```
STANDARD ERRORS

        .263        .064
        .782        .204

----- THETA MATRICES -----

ESTIMATES OF   THETA(  1 ) MATRIX AND SIGNIFICANCE


      -1.323      -.475         - -
       6.300      2.112         + +

STANDARD ERRORS

        .264        .066
        .780        .209


        ----------------------
        ERROR COVARIANCE MATRIX
        ----------------------


                    1               2
        1        .025766
        2       -.004718        .125043
```

Table 13.5 summary of final parameter estimates and their standard errors

PARAMETER NUMBER	PARAMETER DESCRIPTION	FINAL ESTIMATE	ESTIMATED STD. ERROR
1	CONSTANT (1)	0.052124	0.016013
2	CONSTANT (2)	-0.138469	0.044296
3	AUTOREGRESSIVE (1, 1, 1)	-1.377163	0.263322
4	AUTOREGRESSIVE (1, 1, 2)	-0.530786	0.064085
5	AUTOREGRESSIVE (1, 2, 1)	6.484085	0.782178
6	AUTOREGRESSIVE (1, 2, 2)	2.413705	0.204443
7	MOVING AVERAGE (1, 1, 1)	-1.322794	0.263844
8	MOVING AVERAGE (1, 1, 2)	-0.475195	0.065672
9	MOVING AVERAGE (1, 2, 1)	6.300207	0.780190
10	MOVING AVERAGE (1, 2, 2)	2.111769	0.209422

```
-------------------------------------
CORRELATION MATRIX OF THE PARAMETERS
-------------------------------------

         1     2     3     4     5     6     7     8     9    10
  1    1.00
  2   -.73  1.00
  3   -.33  -.36  1.00
  4   -.41  -.21   .85  1.00
  5   -.30  -.40   .77   .76  1.00
  6   -.18  -.48   .77   .49   .88  1.00
  7   -.33  -.37  1.00   .85   .79   .78  1.00
  8   -.40  -.24   .88   .98   .78   .55   .88  1.00
  9   -.30  -.40   .78   .76  1.00   .87   .79   .78  1.00
 10   -.18  -.47   .76   .49   .87   .99   .77   .52   .86  1.00
```

THE RESIDUAL COVARIANCE MATRIX IS SET TO FULL MATRIX
ALL ELEMENTS IN THE MATRIX PARAMETERS ARE ALLOWED TO BE ESTIMATED
-2*(LOG LIKELIHOOD AT FINAL ESTIMATES UNDER H5) IS -0.26704472E+04

THE RESIDUAL COVARIANCE MATRIX IS SET TO DIAGONAL MATRIX
ALL ELEMENTS IN THE MATRIX PARAMETERS ARE ALLOWED TO BE ESTIMATED
-2*(LOG LIKELIHOOD AT FINAL ESTIMATES UNDER H5*) IS -0.26876968E+04

THE RESIDUAL COVARIANCE MATRIX IS SET TO FULL MATRIX
THE (2,1)TH ELEMENTS IN THE MATRIX PARAMETERS ARE SET TO ZERO
-2*(LOG LIKELIHOOD AT FINAL ESTIMATES UNDER H4) IS -0.26703238E+04

THE RESIDUAL COVARIANCE MATRIX IS SET TO DIAGONAL MATRIX
THE (2,1)TH ELEMENTS IN THE MATRIX PARAMETERS ARE SET TO ZERO
-2*(LOG LIKELIHOOD AT FINAL ESTIMATES UNDER H4*) IS -0.26660243E+04

THE RESIDUAL COVARIANCE MATRIX IS SET TO FULL MATRIX
THE (1,2)TH ELEMENTS IN THE MATRIX PARAMETERS ARE SET TO ZERO
-2*(LOG LIKELIHOOD AT FINAL ESTIMATES UNDER H3) IS -0.26014689E+04

THE RESIDUAL COVARIANCE MATRIX IS SET TO DIAGONAL MATRIX
THE (1,2)TH ELEMENTS IN THE MATRIX PARAMETERS ARE SET TO ZERO
-2*(LOG LIKELIHOOD AT FINAL ESTIMATES UNDER H3*) IS -0.25922485E+04

THE RESIDUAL COVARIANCE MATRIX IS SET TO FULL MATRIX
THE (2,1)TH ELEMENTS IN THE MATRIX PARAMETERS ARE SET TO ZERO
THE (1,2)TH ELEMENTS IN THE MATRIX PARAMETERS ARE SET TO ZERO
-2*(LOG LIKELIHOOD AT FINAL ESTIMATES UNDER H2) IS -0.25870429E+04

THE RESIDUAL COVARIANCE MATRIX IS SET TO DIAGONAL MATRIX
THE (2,1)TH ELEMENTS IN THE MATRIX PARAMETERS ARE SET TO ZERO
THE (1,2)TH ELEMENTS IN THE MATRIX PARAMETERS ARE SET TO ZERO
-2*(LOG LIKELIHOOD AT FINAL ESTIMATES UNDER H1) IS -0.25733307E+04

CAUSALITY TEST BETWEEN VARIABLES   DUE   AND   DLEI

P, PP, Q, QQ, NP, NAR, NSAR, NMA, NSMA:
     1   0   1   0   0   4   0   4   0

-2*(LOG LIKELIHOOD) UNDER H5,H5*,H4,H4*,H3,H3*,H2,H1 ARE:
  1   -0.26704472E+04
  2   -0.26876968E+04
  3   -0.26703238E+04
  4   -0.26660243E+04
  5   -0.26014689E+04
  6   -0.25922485E+04
  7   -0.25870429E+04
  8   -0.25733307E+04

-2*(LOG LIKELIHOOD) UNDER H1,H2,H3,H3*,H4,H4*,H5,H5* ARE:
 -0.25733306E+04
 -0.25870430E+04
 -0.26014690E+04
 -0.25922485E+04
 -0.26703237E+04
 -0.26660242E+04
 -0.26704473E+04
 -0.26876968E+04

```
  LR1   TO   LR10:       68.97827148437500    0.1235351562500000    14.42602539062500
83.28076171875000       13.71240234375000    83.40429687500000      9.220458984375000
4.299560546875000 -17.24951171875000 97.11669921875000
  DDF1   TO   DDF10:      2.000000000000000    2.000000000000000     2.000000000000000
2.000000000000000    1.000000000000000       4.000000000000000      1.000000000000000
1.000000000000000  1.000000000000000  5.000000000000000
  CHI1   TO   CHI10:      9.210340416679674    9.210340416679674     9.210340416679674
9.210340416679674    6.634896640840623       13.27670418742523      6.634896640840623
6.634896640840623  6.634896640840623  15.00627252353956
  T1  TO  T10.    59.7679329  -9.08680534 5.21568489 74.0704193 7.07750559 70.1275940
2.58556223  -2.33533621  -23.8844090 82.0304260
  IT1 TO IT10:  1 -1 1 1 1 1 1 -1 -1 1
  I1 TO I6:  1 3 1 3 3 3

RESULT BASED ON THE BACKWARD PROCEDURE ( Y:DUE      , X: DLEI     )
      DUE <<= DLEI      (Y IS STRONGLY CAUSED BY X)

RESULT BASED ON THE FORWARD PROCEDURE  ( Y:DUE      , X: DLEI     )
      DUE <<= DLEI      (Y IS STRONGLY CAUSED BY X)
```

Thus, the application of the Chen and Lee (1990) test leads one finds causality in the DUE, DLEI relationship with both 4-month DLEI lags for the entire 1959-2018 time period.

Append B: Autometrics and Saturation Variables

When observations on a given phenomenon, such as real GDP or the unemployment rate, come from a process whose properties remain constant over time—for example, having the same mean and variance at all points in time—they are said to be stationary. That is, a time series is stationary when its first two moments, namely the mean and variance, are finite and constant over time and is said to be integrated of order zero, denoted I(0). Economies evolve and change over time in both real and nominal terms, as in major wars, the "Oil Crises" of the mid 1970s, or the more recent "Financial Crisis and Great Recession" over 2008–2012.

There are two important sources of non-stationarity often visible in time series: evolution and sudden shifts. The former reflects slower changes, such as enhancements in capital equipment, whereas the latter occurs from wars and policy regime changes. Developing a viable analysis of non-stationarity in economics really commenced with the discovery of the problem of "nonsense correlations." These high correlations are found between variables that should be unrelated: the correlation could be that the two variables might be related to a third variable, inducing a spuriously high correlation. Granger and Newbold (1977) re-emphasized that an apparently "significant relation" between variables, but where there remained substantial serial correlation in the residuals from that relation, was a symptom associated with nonsense regressions.

As Castle and Hendry (2019) discuss in their Palgrave monograph, on *Modelling our Changing World*, a structural break denotes a shift in the behavior of a variable over time, such as a jump in leading economic indicators, LEI, and unemployment, or real GDP. Many sudden changes, particularly when unanticipated, cause links between variables to shift. This is a problem that is especially prevalent in economics as many structural breaks are induced by events outside the purview of most economic analyses, such as COVID-19. Hendry and Doornik (2014) stated that the consequences of not taking breaks into account include poor models, large forecast errors after the break, and inappropriate tests of theories. The simplest to visualize is a shift in the mean of a variable. Forecasts based on the zero mean will be systematically badly wrong. Knowing of or having detected breaks, a common approach is to "model" them by adding appropriate indicator variables, namely artificial variables, that are zero for most of a sample period but unity over the time that needs to be indicated as having a shift. Indicators can be formulated to reflect any relevant aspect of a model, such as changing trends, or multiplied by variables to capture when parameters shift. It is possible to design model selection strategies that tackle structural breaks automatically as part of their algorithm, as advocated by Hendry and Doornik (2014). Even though such approaches, called indicator saturation methods, IIS, lead to more candidate explanatory variables than there are available observations, it is possible for a model selection algorithm to include large blocks of indicators for any number of outliers and location shifts, and even parameter changes. IIS creates a complete set of indicator variables. Each indicator takes the value 1 for a single observation, and 0 for all other observations. As many indicators as there are observations are created, each with a different observation corresponding to the value 1. So, for a sample of T observations, T indicators are then included in the set of candidate variables. However, all those indicators are most certainly not included together in the regression, as otherwise a perfect fit would always result and nothing would be learned. Although saturation creates T additional variables when there are T observations, Autometrics provides an expanding and contracting block search algorithm to undertake model selection when there are more variables than observations. To aid exposition, we shall outline the "split-half" approach analyzed in Hendry et al. (2008), which is just the simplest way to explain and analyze IIS, so bear in mind that such an approach can be generalized to a larger number of possibly unequal "splits," and that the software explores many paths. Including an impulse indicator for a particular observation in a static regression delivers the same estimate of the model's parameters as if that observation had been left out. Consequently, the coefficient of that indicator is equal to the residual of the associated observation when predicted from a model based on the other observations.

A step shift is just a block of contiguous impulses of the same signs and magnitudes. Although IIS is applicable to detecting these, then the retained indicators could be combined into one dummy variable taking the average value of the shift over the break period and 0 elsewhere, perhaps after conducting a joint F-test on the ex post equality of the retained IIS coefficients, there is a more efficient method for detecting step shifts. We can instead generate a saturating set of $T - 1$ step-shift indicators which take the value 1 from the beginning of the sample up to a given observation, and 0 thereafter, with each step switching from 1 to 0 at a different observation. Step indicators are the cumulation of impulse indicators up to each next observation. The "T"th step would just be the intercept. The $T - 1$ steps are included in the set of candidate regressors. The split-half algorithm is conducted in exactly the same way, but there are some differences. First, while impulse indicators are mutually orthogonal, step indicators overlap increasingly as their second index increases. Second, for a location shift that is not at either end, say from T1 to T2, two indicators are required to characterize it. Third, for a split-half analysis, the ease of detection is affected by whether or not T1 and T2 lie in the same split, and whether location shifts occur in both halves with similar signs and magnitudes. Castle et al. (2015) derive the null retention frequency of SIS and demonstrate the improved potency relative to IIS for longer location shifts.

Trend-Indicator Saturation: Thus, one way of capturing a trend break would be to saturate the model with a series of trend indicators, which generate a trend up to a given observation and 0 thereafter for every observation. However, trend breaks can be difficult to detect as small changes in trends can take time to accumulate, even if they eventually lead to very substantial differences.

Information criteria have a long history as a method of choosing between alternative models. Various information criteria have been proposed, all of which aim to choose between competing models by selecting the model with the smallest information loss. The trade-off between information loss and model "complexity" is captured by the penalty, which differs between information criteria. For example, the AIC sought to balance the costs when forecasting from a stationary infinite autoregression of estimation variance from retaining small effects against the squared bias of omitting them. Automated general-to-specific (Gets) approaches in Hendry and Krolzig (2001), Doornik (2009), and Hendry and Doornik (2014). This approach will be the one mainly used in this book when we need to explicitly select a model from a larger set of candidates, especially when there are more such candidates than the number of observations.

In a stationary world, many famous theorems about how to forecast optimally can be rigorously proved (summarized in Clements and Hendry, 1998):

1. Causal models will outperform non-causal (i.e., models without any relevant variables).
2. The conditional expectation of the future value delivers the minimum mean-square forecast error (MSFE).
3. Mis-specified models have higher forecast-error variances than correctly specified ones.
4. Long-run interval forecasts are bounded above by the unconditional variance of the process.
5. Neither parameter estimation uncertainty nor high correlations between variables greatly increase forecast-error variances.

Unfortunately, when the process to be forecast suffers from location shifts and stochastic trends, and the forecasting model is mis-specified, then:

1. Non-causal models can outperform correct in-sample causal relationships.
2. Conditional expectations of future values can be badly biased if later outcomes are drawn from different distributions.
3. The correct in-sample model need not outperform in forecasting, and can be worse than the average of several devices.

4. Long-run interval forecasts are unbounded.
5. Parameter estimation uncertainty can substantively increase interval forecasts; as can.
6. Changes in correlations between variables at or near the forecast origin. The problem for empirical econometrics is not a plethora of excellent forecasting models from which to choose, but to find any relationships that survive long enough to be useful: as we have emphasized, the stationarity assumption must be jettisoned for observable variables in economics. Location shifts and stochastic trend non-stationarities can have pernicious impacts on forecast accuracy and its measurement: Castle et al. (2019) provide a general introduction.

Castle et al. (2018) have found that selecting a model for forecasting from a general specification that embeds the DGP does not usually entail notable costs compared to using the estimated DGP—an infeasible comparator with non-stationary observational data. Indeed, when the exogenous variables need to be forecast, selection can even have smaller MSFEs than using a known DGP. That result matches an earlier finding in Castle et al. (2011) that a selected equation can have a smaller root mean square error (RMSE) for estimated parameters than those from estimating the DGP when the latter has several parameters that would not be significant on conventional criteria. The difficulty is when an irrelevant variable that happens to be highly significant by chance has a location shift, which by definition will not affect the DGP but will shift the forecasts from the model, so forecast failure results. Here rapid updating after the failure will drive that errant coefficient towards zero in methods that minimize squared errors, so will be a transient problem.

Castle et al. (2018) also conclude that some forecast combination can be a good strategy for reducing the riskiness of forecasts facing location shifts. Although no known method can protect against a shift after a forecast has been made, averaging forecasts from an econometric model, a robust device and a simple first-order autoregressive model frequently came near the minimum MSFE for a range of forecasting models on 1-step ahead forecasts in their simulation study. This result is consistent with many findings since the original analysis of pooling forecasts in Bates and Granger (1969), and probably reflects the benefits of "portfolio diversification" known from finance theory. Clements (2017) provides a careful analysis of forecast combination. A caveat emphasized by Hendry and Doornik (2014) is that some pre-selection is useful before averaging to eliminate very bad forecasting devices. For example, the GUM is rarely a good device as it usually contains a number of what transpire to be irrelevant variables, and location shifts in these will lead to poor forecasts. Granger and Jeon (2004) proposed "thick" modeling as a route to overcoming model uncertainty, where forecasts from all non-rejected specifications are combined. However, Castle (2017) showed that "thick" modeling by itself neither avoids the problems of model misspecification nor handles forecast origin location shifts. Although "thick" modeling is not formulated as a general-to-simple selection problem, it could be implemented by pooling across all congruent models selected by an approach like Autometrics.

References

Ashley, R. (1998). A new technique for postsample model selection and validation. *Journal of Economic Dynamics and Control, 22*, 647–665.

Ashley, R. A. (2003). Statistically significant forecasting improvements: How much out-of-sample data is likely necessary? *International Journal of Forecasting, 19*, 229–240.

Ashley, R., Granger, C. W. J., & Schmalensee, R. (1980). Advertising and aggregate consumption: An analysis of causality. *Econometrica, 48*, 149–167.

Box, G. E. P., & Jenkins, G. M. (1970). *Time series analysis: Forecasting and control.* Holden-Day.

Box, G. E. P., & Cox, D. R. (1964). An analysis of transformations. *Journal of the Royal Statistical Association Series B, 26*, 211–243.

Burns, A. F., & Mitchell, W. C. (1946). *Measuring business cycles.* NBER.

Castle, J., & Hendry, D. F. (2019). *Modelling our changing world.* Palgrave.

Castle, J., & Shepard, N. (2009). *The methodology and practice of econometrics.* Oxford University Press.

Castle, J., Doornik, J. A., & Hendry, D. F. (2013). Model selection in equations with many 'small' effects. *Oxford Bulletin of Economics and Statistics, 75*, 6–22.

Castle, J., Clements, M. P., & Hendry, D. F. (2015). Robust approaches to forecasting. *International Journal of Forecasting, 31*, 99–112.

Chen, C., & Lee, C. J. (1990). A VARMA test on the Gibson paradox. *Review of Economics and Statistics, 72*, 96–107.

Chen, C., & Liu, L.-M. (1993a). Joint estimation of model parameters and outliers in time series. *Journal of the American Statistical Association, 88*, 284–297.

Chen, C., & Liu, L.-M. (1993b). Forecasting time series with outliers. *Journal of Forecasting, 12*, 13–35.

Clemen, R. T., & Guerard, J. B. (1989). Econometric GNP forecasts: Incremental information relative to naive extrapolation. *International Journal of Forecasting, 5*, 417–426.

Clements, M. P., & Hendry, D. F. (1998). *Forecasting economic time series.* Cambridge University Press.

Dhrymes, P. J. (2017). *Introductory econometrics.* Springer.

Diebold, F. X., & Mariano, R. S. (1995). Comparing predictive accuracy. *Journal of Business and Economic Statistics, 13*, 253–263.

Diebold, F. X., & Rudebusch, G. D. (1999). *Business cycles: Durations, dynamics and forecasting.* Princeton University Press.

Fisher, I. (1911a). Recent changes in price levels and their causes. *The American Economic Review, 1*, 37–45.

Fisher, I. (1911b). *The purchasing power of money.* Macmillan.

Gordon, R. J. (1986). *The American business cycle.* University of Chicago Press.

Granger, C. W. J. (1969). Investigating casual relations by economic models and cross-spectral methods. *Econometrica, 37*, 424–438.

Granger, C. W. J. (1980). Testing for causality: A personal viewpoint. *Journal of Economic Dynamics and Control, 2*, 329–352.

Granger, C. W. J. (2001). In E. Ghysels, N. R. Swanson, & M. W. Watson (Eds.), *Essays in econometrics. Two volumes.* Cambridge University Press.

Granger, C. W. J., & Newbold, P. (1977). *Forecasting economic time series.* Academic Press.

Guerard, J. B. (1985). Mergers, stock prices, and industrial production: An empirical test of the Nelson hypothesis. In O. D. Anderson (Ed.), *Time series analysis: Theory and practice 7.* North-Holland Publishing Company.

Guerard, J. B., & McDonald, J. B. (1995). Mergers in the United States: A case study in robust time series. In C. F. Lee (Ed.), *Advances in financial planning and forecasting.* JAI Press.

Guerard, J., Thomakos, D., & Kyriazi, F. (2020). Automatic time series modeling and forecasting: A replication case study of forecasting real GDP, the unemployment rate and the impact of leading economic indicators. *Cogent Economics & Finance, 8*(1), 1759483.

Hamilton, J. R. (1994). *Time series analysis*. Princeton.

Hendry, D. F. (1986). Using PC-give in econometrics teaching. *Oxford Bulletin of Economics and Statistics, 48*, 87–98.

Hendry, D. F. (2000). *Econometrics: Alchemy or science?* Oxford University Press.

Hendry, D. F., & Doornik, J. A. (2014). *Empirical model discovery and theory evaluation*. MIT Press.

Hendry, D. F., & Krolzig, H. M. (2005). The properties of automatic gets modelling. *The Economic Journal, 115*, c32–c61.

Koopmans, T. C. (1947). Measurement without theory. *The Review of Economic Statistics*. Reprinted in R. Gordon & L. Klein, *Readings in Business Cycles*, Homewood, Illinois, Richard A. Irwin, Inc, 1965.

Koopmans, T. C. (1950). *When is an equation system complete for statistical purposes*. Technical report, Yale University. URL http://cowles.econ.yale.edu/P/cm/m10/m10-17.pdf

Krolzig, H.-M. (2001). Business cycle measurement in the presence of structural change: International evidence. *International Journal of Forecasting, 17*, 349–368.

Krolzig, H.-M., & Hendry, D. F. (2001). Computer automation of general-to-Specific model selection procedures. *Journal of Economic Dynamics & Control*, 831–866.

Maronna, R. A., Martin, R. D., Yohai, V. J., & Salibian-Barrera, M. (2019). *Robust statistics: Theory and methods (with R)*. Wiley.

McCracken, M. (2000). Robust out-of-sample inference. *Journal of Econometrics, 39*, 195–223.

Melicher, R., Ledolter, J., & D'Antonio, L. J. (1983). A time series analysis of aggregate merger activity. *Review of Economics and Statistics, 65*, 423–430.

Mincer, J., & Zarnowitz, V. (1969). The evaluation of economic forecasts. In J. Mincer (Ed.), *Economic forecasts and expectations*. Columbia University Press.

Mitchell, W. C. (1913). *Business cycles*. Burt Franklin reprint.

Mitchell, W. C. (1951). *What happens during business cycles: A Progress report*. NBER.

Montgomery, A. L., Zarnowitz, V., Tsay, R., & Tiao, G. C. (1998). Forecasting the U.-S. unemployment rate. *Journal of the American Statistical Association, 93*, 478–493.

Moore, G. H. (1961). *Business cycle indicators* (2 Vols). Princeton University Press.

Nelson, R. L. (1959). *Merger movements in American industry 1895–1954*. Princeton University Press.

Nelson, R. L. (1966). Business cycle factors in the choice between internal and external growth. In W. W. Alberts & J. E. Segall (Eds.), *The corporate merger*. The University of Chicago Press.

Nelson, C. R. (1973). *Applied time series analysis for managerial forecasting*. Holden-Day, Inc.

Nelson, C. R., & Plosser, C. I. (1982). Trends and random walks in macroeconomic time series. *Journal of Monetary Economics, 10*, 139–162.

Swanson, N. R. (1998). Money and output viewed through a rolling window. *Journal of Monetary Economics, 41*(1998), 455–473.

Theil, H. (1966). *Applied economic forecasting*. North-Holland.

Thomakos, D., & Guerard, J. (2004). Naïve, ARIMA, transfer function, and VAR models: A comparison of forecasting performance. *The International Journal of Forecasting, 20*, 53–67.

Thomakos, D., & Guerard, J.. (2022). *Causality testing and forecasting of the US unemployment rate*. Unpublished Working Paper. University of Athens.

Tsay, R. S. (1988). Outliers, level shifts, and variance changes in time series. *Journal of Forecasting, 7*, 1–20.

Tsay, R. S. (1989). Testing and modeling threshold autoregressive processes. *Journal of the American Statistical Association, 84*, 231–249.

Tsay, R. S. (2010). *Analysis of financial time series* (3rd ed.). Wiley.

Tsay, R. S., & Chen, R. (2019). *Nonlinear time series analysis*. Wiley.

Vining, R. (1949). Koopmans on the choice of variables to be studied and on methods of measurement. *The Review of Economics and Statistics*. Reprinted in R. Gorgon & L. Klein, *Readings in Business Cycles*, Homewood, Illinois: Richard A. Irwin, Inc, 1965.

Vinod, H. D. (2008). *Hands-on intermediate econometrics using R: Templates for extending dozens of practical examples*. World Scientific. ISBN 10-981-281-885-5. URL http://www.worldscibooks.com/economics/6895.html

Vinod, H. D. (2014). Chapter 4: Matrix algebra topics in statistics and economics using R. In M. B. Rao & C. R. Rao (Eds.), *Handbook of statistics: Computational statistics with R* (Vol. 34, pp. 143–176).

Vinod, H. D. (2017). Causal paths and exogeneity tests in generalCorr package for air pollution and monetary policy. A vignette accompanying the package "generalCorr" of R. URL: https://cloud.r-project.org/web/packages/generalCorr/vignettes/generalCorr-vignette3.pdf

Vinod, H. D. (2019). Chapter 2: New exogeneity tests and causal paths. In H. D. Vinod & C. R. Rao (Eds.), *Handbook of statistics: Conceptual econometrics using R* (Vol. 41, pp. 33–64). North Holland, Elsevier. https://doi.org/10.1016/bs.host.2018.11.011

West, K., & McCracken, M. (1998). Regression-based tests of predictive ability. *International Economic Review, 39*, 817–840.

Zarnowitz, V. (1992). *Business cycles: Theory, history, indicators, and forecasting*. University of Chicago Press.

Zarnowitz, V. (2001). *The old and the new in the U.S. economic expansion*. The Conference Board. EPWP #01-01.

Zarnowitz, V. (2004). The autonomy of recent US growth and business cycles. In P. Dua (Ed.), *Business cycles and economic growth: An analysis using leading indicators* (pp. 44–82). Oxford University Press.

Zarnowitz, V., & Ozyildirim, A. (2001). On the measurement of business cycles and growth cycles. *Indian Economic Review, 36*, 34–54.

Chapter 14
Risk and Return of Equity, the Capital Asset Pricing Model, and Stock Selection for Efficient Portfolio Construction

Individual investors must be compensated for bearing risk. It seems intuitive to the reader that there should be a direct linkage between the risk of a security and its rate of return. We are interested in securing the maximum return for a given level of risk, or the minimum risk for a given level of return. The concept of such risk-return analysis is the efficient frontier of Harry Markowitz (1952, 1959). If an investor can invest in a government security, which is backed by the taxing power of the federal government, then that government security is relatively risk-free. The 90-day Treasury bill rate is used as the basic risk-free rate. Supposedly, the taxing power of the federal government eliminates default risk of government debt issues. A liquidity premium is paid for longer-term maturities, due to the increasing level of interest rate risk. Investors are paid interest payments, as determined by the bond's coupon rate, and may earn market price appreciation on longer bonds if market rates fall or losses if market rates rise. During the period from 1928 to 2017, Treasury bills returned 3.44%, longer-term (10-year Treasury) government bonds earned 5.15%, and corporate stocks, as measured by the stock of the S&P 500 index, earned 11.53% annually, as measured by the mean annual return. The annualized standard deviations are 3.06%, 7.72%, and 19.66%, respectively, for Treasury bills, Treasury bonds, and S&P stocks.[1] The risk-return trade-off has been relevant for the 1928–2017 period. The correlation coefficient between annual returns for Treasury bills and the S&P 500 stock returns were −0.030 for the 1928–2017 time period. This was essentially no correlation between Treasury bills and large stocks, as measured by the S&P 500 stock. The correlation coefficient between annual returns for Treasury bonds and the S&P 500 stock returns was 0.30 for the 1928–2017 time period. Why do corporate stocks offer investors higher returns for stocks than bonds?

---

[1] Ibbottson and Sinquefield, *Stocks, Bonds, and Bills Yearbook*, 2018.

© The Author(s), under exclusive license to Springer Nature Switzerland AG 2022
J. B. Guerard Jr. et al., *Quantitative Corporate Finance*,
https://doi.org/10.1007/978-3-030-87269-4_14

First, as a stockholder, one owns a fraction, a very small fraction for many investors, of the firm. When one owns stocks, one is paid a dividend and hopes to earn stock price appreciation over time. That is, an investor buy a stock when he or she expects its stock price to increase enough to compensate the investor for bearing the risk of the stock's investment. Investors have become aware in recent years that not all price movements are in positive directions. In this chapter, we introduce the reader to (1) the Markowitz and Sharpe risk-return trade-off models, (2) the calculation of historic average returns and standard deviations of stocks, (3) beta estimations in the capital asset pricing model, and (4) a series of financial variables and regression-based models of those variables that affect stock returns. One can calculate expected returns on stocks with several different methods. This chapter and Chap. 15, on multifactor risk models, will be the two chapters to allow the reader to understand quantitative modeling and serve as an introduction to financial engineering in portfolio construction and management.

14.1   Calculating Holding Period Returns

A very simple concept is the holding period return (HPR) calculation, in which one assumes that the stock was purchased at last (period's) year's price and the investor earns a dividend per share for the current year and a price appreciation or depreciation.

$$HPR_t = \frac{D_t + P_t - P_{t-1}}{P_{t-1}}$$

where

D_t = current year's dividend
P_t = current year's stock price
P_{t-1} = last year's stock price and
HPR_t = current year's holding period return

The assumption of annual returns is common in finance. Markowitz (1959) used annual returns in Chap. 2 and Chap. 6 of *Portfolio Selection* to illustrate holding period returns and investment in the long run.[2] We calculate annual holding period stock returns from the Wharton Research Data Service (WRDS) Compustat database by calculating average price (AvgP) simply summing the high (PrcH) and low prices (PrcL) and dividing by two. See Table 14.1.

---

[2] Guerard and Schwartz (2007) examined three widely held stocks: DuPont, Dominion Resources, and IBM, for 1994–2003 period in our first edition. The pricing data was taken from the Standard & Poor's *Stock Guide*. The *S &P Stock Guide* presents high and low prices during the calendar year. An average price (AvgP) was be calculated by simply summing the high and low prices and dividing by two.

Table 14.1 Using high and low stock prices and dividends to calculate annual stock holding period returns

Year	IBM PrcH	PrcL	DPS	AvgPrc	HPR
2008	130.93	69.50	1.93	100.22	
2009	132.85	81.76	2.19	107.31	0.093
2010	147.53	116.00	2.59	131.77	0.252
2011	194.90	146.64	2.99	170.77	0.319
2012	211.79	177.35	3.38	194.57	0.159
2013	215.90	172.57	3.85	194.24	0.018
2014	199.21	150.50	4.31	174.86	-0.078
2015	176.30	131.65	5.07	153.98	-0.090
2016	169.95	118.90	5.57	144.43	-0.026
2017	182.79	139.13	5.97	160.96	0.156
2018	171.13	105.94	6.35	138.54	-0.100
				Mean	0.070
				STD	0.149

Year	BA PrcH	PrcL	DPS	AvgPrc	HPR
2008	88.29	36.17	1.70	62.23	
2009	56.56	29.05	1.70	42.81	-0.285
2010	76.00	54.80	1.70	65.40	0.568
2011	80.65	56.00	1.67	68.33	0.070
2012	77.83	66.82	1.75	72.33	0.084
2013	142.00	72.68	1.96	107.34	0.511
2014	144.57	116.32	2.99	130.45	0.243
2015	158.83	115.14	3.74	136.99	0.079
2016	160.07	102.10	4.47	131.09	-0.010
2017	299.33	155.21	5.78	227.27	0.778
2018	394.28	282.47	6.95	338.38	0.519
				Mean	0.256
				STD	0.327

Year	D PrcH	PrcL	DPS	AvgPrc	HPR
2008	48.5	31.26	1.57	39.88	
2009	39.79	27.15	1.76	33.47	-0.117
2010	45.12	36.12	1.88	40.62	0.270
2011	53.59	42.06	2.01	47.83	0.227
2012	55.62	48.87	2.12	52.25	0.137
2013	67.98	51.92	2.27	59.95	0.191
2014	80.89	63.14	2.41	72.02	0.241
2015	79.89	64.54	2.58	72.22	0.039
2016	78.97	66.25	2.75	72.61	0.044
2017	85.3	70.87	2.99	78.09	0.117
2018	81.67	61.53	3.21	71.60	-0.042
				Mean	0.111
				STD	0.128

Table 14.2 Annual holding period returns, 2009–2018

Year	IBM HPR	BA HPR	D HPR
2009	0.0926	-0.2848	-0.1166
2010	0.2521	0.5676	0.2698
2011	0.3187	0.0703	0.2269
2012	0.1592	0.0842	0.1367
2013	0.0181	0.5112	0.1909
2014	-0.0776	0.2431	0.2415
2015	-0.0904	0.0788	0.0386
2016	-0.0258	-0.0104	0.0436
2017	0.1558	0.7779	0.1166
2018	-0.0999	0.5194	-0.0419

As shown in Table 14.2, the IBM holding period returns (HPRs) were consistently positive and rather large during 2009–2013 and mostly negative during 2014–2018. The HPRs of stocks are often far more dependent upon stock price movements than the dividends received by investors. The HPRs of IBM range from −9.99% in 2018 to 31.87% in 2011. The HPRs of Boeing (BA) range from −28.48% in 2009 to 26.98% in 2010. The HPRs of Dominion Energy (D) range from −11.66% in 2009 to 77.79% in 2017. Boeing stock had much higher returns and standard deviations, a measure of total risk, than IBM or Dominion Energy. One can estimate an expected return for IBM, BA, and D by calculating the mean value of the annual HPRs during the 2009–2018 period. The expected return for IBM during the 2009–2018 period was 7.00%, with a standard deviation of 14.90%. If the annual HPRs of IBM are normally distributed, that is, returns are distributed as a "bell" curve, then 95% of the annual observations of IBM returns should fall within the −22.80% and 36.80% range (two standard deviations). The corresponding expected return for Boeing during the 2009–2018 period was 25.60%, with a standard deviation of 32.70%. If the annual HPRs of BA are normally distributed, then 95% of the annual observations of BA returns should fall within the −39.8% and 91.0% range (two standard deviations). The reader immediately sees how wide the two standard deviation range is for annual returns. It may be worthwhile to calculate a coefficient of variation (CV) in which the standard deviation is divided by the expected return. See Table 14.3.

These data indicate that IBM produces greater variation for a given level of expected returns than BA or D.

Let us assume that the student has access to the CRSP Database, the Center for Research in Security Prices database, developed by James Lorie and Lawrence Fisher at the University of Chicago during the 1960–1965 time period. The CRSP database contains many data, including monthly stock prices, returns, dividends, and

Table 14.3 Annual stock holding period return

Security	Ticker	E(R)	σ	CV
Boeing	BA	.256	.327	1.279
Dominion Resources	D	.111	.128	1.158
IBM	IBM	.070	.149	2.117

Source: SAS

Table 14.4 CRSP monthly returns January 2015 to December 2019

				Rho		
Stock	Mean	STD	IBM	D	BA	
IBM	0.0024	0.0649	1.0000	0.0831	0.3264	
D	0.0054	0.0391	0.0831	1.0000	0.0307	
BA	0.0206	0.0791	0.3264	0.0307	1.0000	

Source: CRSP database.

shares outstanding for publicly traded US stocks, from 1926 to present. Instead of tracking down stock prices and dividends, the student can access stock returns for virtually all US stocks traded. Let us continue to use IBM, Boeing, and Dominion Energy, the stocks used in our financial statement and valuation analysis. We can use 60 months of data, from January 2015 to December 2019, to calculate average returns, standard deviations, and correlation coefficients, to build risk-minimizing portfolios. See Table 14.4.

The portfolio expected return is a weighted combination of asset expected returns. In an equally weighted two-asset portfolio, $x_1 = x_2 = 0.50$. Let $x_1 =$ weight of IBM and $x_2 =$ weight of D.

$$
\begin{aligned}
E(R_p) &= x_1 E(R_1) + x_2 E(R_2) \\
&= .5(.0024) + .5(.0054) \\
&= .0012 + .0027 = .0039 \text{(monthly)}
\end{aligned}
\tag{14.1}
$$

The annual expected return of IBM and D equally weighted portfolio is $12*0.0039 = 0.0468$, or 4.68%.

The portfolio variance is given by the weighted asset variances and covariances.

$$\sigma_p^2 = x_1^2\sigma_1^2 + x_2^2\sigma_2^2 + 2x_1(1 - x_1)\sigma_{12}$$
$$\sigma_{12} = \rho_{12}\sigma_1\sigma_2 \tag{14.2}$$

ρ_{12} = correlation coefficients of assets 1, 2

σ_{12} = covariance of assets 1, 2

$$\begin{aligned}
\sigma_p^2 &= (.5)^2(.0649)^2 + (.5)^2(.0391)^2 + 2(.5)(.5)(0.0831)(.0649)(.0391) \\
&= .25(.0042) + .25(.0015) + (-.0000) \\
&= .0011 + .0004 + .0001 = 0.0016
\end{aligned} \tag{14.3}$$
$$\sigma_p = \sqrt{.0010} = 0.0400$$

The expected return on an equally weighted portfolio of IBM and D stock is 4.68%, and its corresponding standard deviation is 3.46%. The portfolio return should lie within the range of -3.32% and 12.568% approximately 95% of the time. This range corresponds to the expected return plus and minus two standard deviations of return.

To minimize risk, one seeks to allocate resources to assets such that the change in risk goes to zero with the change in the percentage invested in the asset. That is, risk minimization implies a first partial derivative of zero, with respect to the change in the asset weight.

Let $x_2 = 1 - x_1$

$$\begin{aligned}
\sigma_p^2 &= x_1^2\sigma_1^2 + (1 - x_1)^2\sigma_2^2 + 2x_1(1 - x_1)\sigma_{12} \\
&= x_1^2\sigma_1^2 + (1 - x_1)^2\sigma_2^2 + 2x_1(1 - x_1)\rho_{12}\sigma_1\sigma_2 \\
&= x_1^2\sigma_1^2 + (1 - x_1)(1 - x_1)\sigma_2^2 + 2x_1\rho_{12}\sigma_1\sigma_2 - 2x_1^2\rho_{12}\sigma_1\sigma_2 \\
\sigma_p^2 &= x_1^2\sigma_1^2 + (1 - 2x_1 + x_1^2)\sigma_2^2 + 2x_1\rho_{12}\sigma_1\sigma_2 - 2x_1^2\rho_{12}\sigma_1\sigma_2
\end{aligned}$$

$$\frac{\partial\sigma p^2}{\partial x_1} = 2x_1\sigma_1^2 - 2\sigma_2^2 + 2x_1\sigma_2^2 + 2\rho_{12}\sigma_1\sigma_2 - 4x_1\rho_{12}\sigma_1\sigma_2 = 0$$

$$2\sigma_2^2 - 2\rho_{12}\sigma_1\sigma_2 = 2x_1\sigma_1^2 + 2x_1\sigma_2^2 - 4x_1\rho_{12}\sigma_1\sigma_2$$

$$(\sigma_1^2 + \sigma_2^2 - 2\rho_{12}\sigma_1\sigma_2)x_1 = \sigma_2^2 - \rho_{12}\sigma_1\sigma_2$$

$$x_1 = \frac{\sigma_2(\sigma_2 - \rho_{12}\sigma_2)}{\sigma_1^2 + \sigma_2^2 - 2\rho_{12}\sigma_1\sigma_2} \tag{14.4}$$

Equation (14.4) shows the risk-minimizing weight (percentage invested) of asset one in the portfolio. The reader notes immediately the importance of minimizing the correlation coefficient between stocks to minimize risk.

In an optimally weighted portfolio, $x_1 = 0.245$ and $x_2 = 0.755$. Let x_1 = weight of IBM and x_2 = weight of D. The portfolio expected return is a weighted combination of asset expected returns.

$$E(R_p) = x_1 E(R_1) + x_2 E(R_2)$$
$$= .245(.0024) + .755(.0054)$$
$$= .0006 + .0041 = .0047 (\text{monthly})$$

The annual expected return of IBM and D equally weighted portfolio is $12*0.0047 = 0.0564$, or 5.64%.

$$\sigma_p^2 = (.245)^2(.0649)^2 + (.755)^2(.0391)^2 + 2(.245)(.755)(0.0831)(.0649)(.0391)$$
$$= .0600(.0042) + .5700\ (.0015) + (-.0000)$$
$$= .0003 + .0009 + (0000) = 0.0012$$
$$\sigma_p = \sqrt{.0010} = .0346$$

The expected return on an optimally weighted portfolio of IBM and D stock is 5.64%, and its corresponding standard deviation is 3.46%.

Markowitz analysis seeks to minimize the risk of a portfolio for a given level of return. In chapters 7 and 8 of Markowitz (1959), we are taught to estimate portfolios that minimize risk for a given level of return. Markowitz portfolio theory requires several inputs, the expected returns and standard deviations of assets, turnover constraints, and constraints on asset weights, such as maximum and minimum weights. Markowitz created Fig. 14.1 for Bloch et al. (1993) to summarize the investment process.

The goal in investment research is to operate as closely to the efficient frontier as possible. The efficient frontier, shown in Fig. 14.2, depicts the maximum return for a given level of risk.

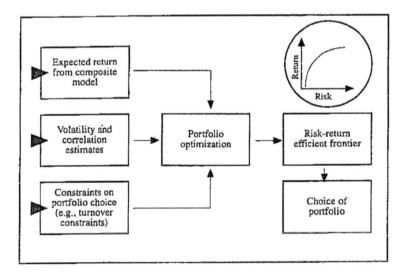

Fig. 14.1 Investment process

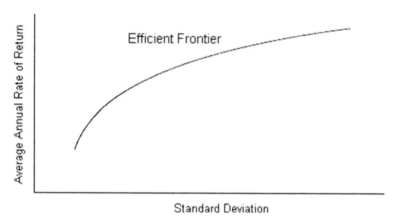

Fig. 14.2 The efficient frontier

Markowitz analysis sought to minimize risk for a given level of return. Thus, one could construct an infinite number of portfolios, by varying security weights, but the efficient frontier contained securities with weights that maximized return for a given level of risk.

The capital market line (CML) was developed to describe the return-risk trade-off assuming that investors could borrow and lend at the risk-free rate, RF, and that investors must be compensated for bearing risk. Investors seek to hold mean-variance efficient portfolios, invest for a one-period horizon, pay no taxes or transaction costs (we wish), and have homogeneous beliefs. All investors have identical probabilities of the distribution of future returns of securities.

$$E\left(\widetilde{R}_p\right) = R_F + \left[\frac{E\left(\widetilde{R}_M\right) - R_F}{\sigma_M}\right]\sigma_P \qquad (14.5)$$

$E(R_p)$ = Expected return on the portfolio
$E(R_M)$ = Expected return on the market portfolio, where all securities are held
     relative to their market value
σ_M = Standard deviation of the market portfolio
and σ_p = standard deviation of the portfolio.

The reader notes that as the standard deviation of the portfolio rises, its expected return must rise.

14.2 An Introduction to Modern Portfolio Theory and Betas

Markowitz created a portfolio construction theory in which investors should be compensated with higher returns for bearing higher risk. The Markowitz framework measured risk as the portfolio standard deviation, its measure of dispersion, or total risk. The Sharpe (1964), Lintner (1965), and Mossin (1966) development of the capital asset pricing model (CAPM) held that investors are compensated for bearing not total risk, but rather market risk, or systematic risk, as measured by the stock beta. An investor is not compensated for bearing risk that may be diversified away from the portfolio. The beta is the slope of the market model, in which the stock return is regressed as a function of the market return.

The CAPM holds that the return to a security is a function of the security beta.

$$R_{jt} = R_F + \beta_j[E(R_{Mt}) - R_F] + e_j \tag{14.6}$$

where

R_{jt} = expected security return at time t
$E(R_{Mt})$ = expected return on the market at time t
R_F = risk-free rate
β_j = security beta and
e_j = randomly distributed error term.

Let us examine the capital asset pricing model beta, its measure of systematic risk, from the capital market line equilibrium condition.

$$\beta_j = \frac{\mathrm{Cov}(R_j, R_M)}{\mathrm{Var}(R_M)} \tag{14.7}$$

$$E(R_j) = R_F + \left[\frac{E(R_M) - R_F}{\sigma_M^2}\right]\mathrm{Cov}(R_j, R_M)$$

$$= R_F + [E(R_M) - R_F]\frac{\mathrm{Cov}(R_j, R_M)}{\mathrm{Var}(R_M)} \tag{14.8}$$

$$E(R_j) = R_F + [E(R_M) - R_F]\beta_j$$

The security market line, SML, shown in Eq. (14.8), is the linear relationship between return and systematic risk, as measured by beta.

Let us estimate beta coefficients to be used in the capital asset pricing model (CAPM), to determine the rate of return on equity. One can regress monthly HPRs against a value-weighted Center for Research in Security Prices (CRSP) index, an index of all publicly traded securities. Most security betas are estimated using 5 years of monthly data, from January 2015 to December 2019, 60 observations, although one can use almost any number of observations.[3] We use proc reg in SAS to estimate ordinary least squares (OKS) betas and SAS PROC ROBUSTREG to estimate robust regression stock betas. One can use the Standard and Poor's 500 Index, or the CRSP value-weighted stock index as the market index in the beta estimation. The IBM beta is 1.331 when estimated using the S&P 500 Index and 1.372 when estimated using value-weighted CRSP index. There were significant outliers present in the OLS beta estimations. We estimate a robust regression beta using the MMEFF99 regression methodology as discussed in Maronna et al. (2019) and in Chap. 12 and implemented in SAS.

IBM OLS Beta

The REG Procedure
Model: MODEL1
Dependent Variable: ret

Number of Observations Read 60

Number of Observations Used 60

Analysis of Variance

Source	DF	Sum of Squares	Mean Square	F Value	Pr > F
Model	1	0.12461	0.12461	58.26	<.0001
Error	58	0.12405	0.00214		
Corrected Total	59	0.24866			

Root MSE	0.04625	**R-Square**	0.5011
Dependent Mean	0.00243	**Adj R-Sq**	0.4925
Coeff Var	1903.19419		

---

[3]One generally needs 30 observations for normality of residuals to occur, from the central limit theorem of statistics.

Parameter Estimates

Variable	DF	Parameter Estimate	Standard Error	t Value	Pr > |t|
Intercept	1	-0.00838	0.00614	-1.37	0.1773
sprtrn	1	1.33013	0.17426	7.63	<.0001

The estimated IBM beta of 1.331 is statistically significantly different from zero with a t-statistics of 7.63. The beta is statistically different from 1.0 at the 10% level, having a t-statistics of 1.90 when we use the null hypothesis of one, the beta of a stock that moves proportionally to the market.

IBM OLS Beta

The REG Procedure
Model: MODEL1
Dependent Variable: ret

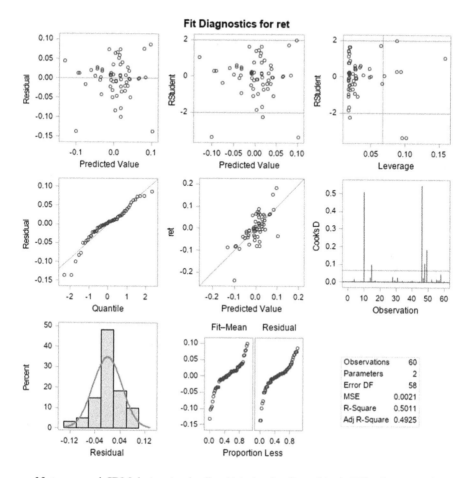

Note several IBM beta standardized(studentized) residual, RStudent, are less than −2.

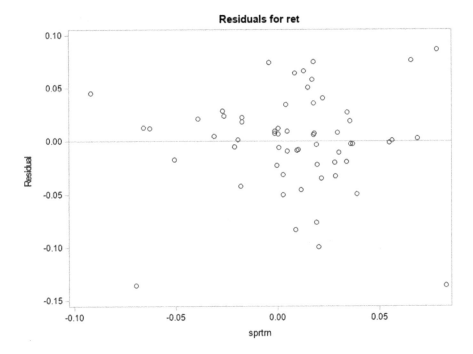

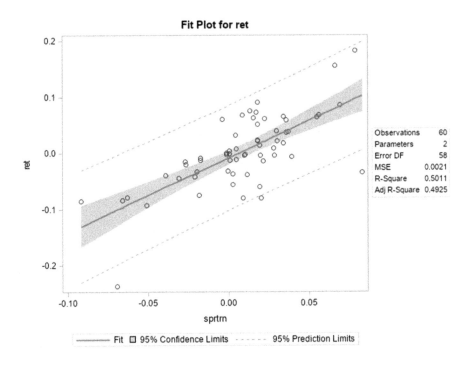

The Beaton-Tukey (1974) bisquare-estimated beta is 1.253 and is statistically different from zero, but not one.

IBM Robust M Bisquare Beta

The ROBUSTREG Procedure

Model Information

Data Set	WORK.IBM
Dependent Variable	ret
Number of Independent Variables	1
Number of Observations	60
Method	M Estimation

Parameter Estimates

Parameter	DF	Estimate	Standard Error	95% Limits	Confidence	Chi-Square	Pr > ChiSq
Intercept	1	-0.0027	0.0053	-0.0131	0.0077	0.26	0.6090
sprtrn	1	1.2534	0.1506	0.9581	1.5486	69.22	<.0001
Scale	1	0.0322					

Diagnostics Summary

Observation Type	Proportion	Cutoff
Outlier	0.0500	3.0000

The OLS and robust-estimated betas exceed one and are statistically significant from zero.

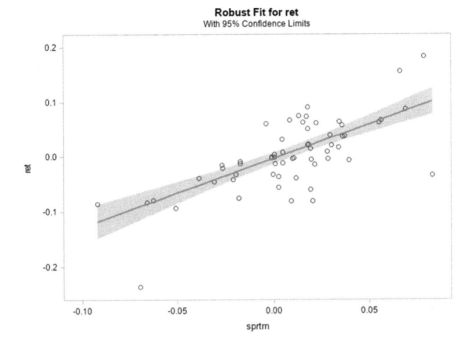

Goodness-of-Fit

Statistic	Value
R-Square	0.3557
AICR	84.8310
BICR	89.3105
Deviance	0.0841

The Tukey MM bisquare-estimated beta is 1.31 and is statistically different from zero, but and not one, at the 10% level.

IBM Robust MM EFF=0.99 Beta

The ROBUSTREG Procedure

Model Information

Data Set	WORK.IBM
Dependent Variable	ret
Number of Independent Variables	1
Number of Observations	60
Method	MM Estimation

Profile for the Initial LTS Estimate

Total Number of Observations	60
Number of Squares Minimized	45
Number of Coefficients	2
Highest Possible Breakdown Value	0.2667

MM Profile

Chi Function	Tukey
K1	7.0410
Efficiency	0.9900

Parameter Estimates

Parameter	DF	Estimate	Standard Error	95% Confidence Limits		Chi-Square	Pr > ChiSq
Intercept	1	-0.0065	0.0058	-0.0178	0.0048	1.26	0.2612
sprtrn	1	1.3100	0.1687	0.9794	1.6406	60.33	<.0001
Scale	0	0.0416					

Diagnostics Summary

Observation Type	Proportion	Cutoff
Outlier	0.0333	3.0000

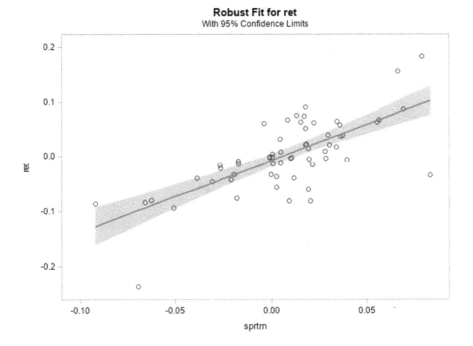

Robust Fit for ret
With 95% Confidence Limits

Goodness-of-Fit

Statistic	Value
R-Square	0.4257
AICR	68.1364
BICR	72.6551
Deviance	0.1117

One should estimate the IBM beta using robust regression, as we discussed in Chap. 12.

We perform similar beta estimations for BA and D and report the estimated betas in Table 14.5. BA and IBM betas exceed 1.0, whereas Dominion Energy is approximately 0.22, far below 1.0, and the excess return on D, reported in Chap. 8, is magnified by its lower beta. D has been a great stock since 1982, outperforming the S&P 500 with a beta of less than 0.60 for most of the 38 years.

If an investor estimates that the security's expected return exceeds the required rate of return from the CAPM and its beta, then the security should be purchased. The incremental purchase of the security drives up its price and lowers its expected return.

The difficulty of measuring beta and its corresponding security market line gave rise to extra-market measures of risk, found in the work of King (1966), Farrell

Table 14.5 Estimated stock betas

Regression	Stock	S&P 500	V-L CRSP
OLS	IBM	1.330	1.372
(t)		(7.63)	(8.16)
MMEFF99	IBM	1.310	
[se]		[.169]	
OLS	D	.224	.165
		(1.54)	(1.12)
MMEFF99	D	.231	
		[.149]	
OLS	BA	1.191	1.207
		(4.64)	(4.73)
MMEFF99	BA	1.190	
		[.248]	

(1973), Rosenberg (1973, 1976, 1979), Stone (1974, 2002), Ross (1976), and Ross and Roll (1980). The BARRA risk model was developed in the series of studies by Rosenberg and completely discussed in Grinold and Kahn (1999), which is discussed in the next chapter.

The total excess return for a multiple factor model, referred to as the MFM, in the Rosenberg methodology for security j, at time t, dropping the subscript t for time, may be written:

$$E(R_j) = \sum_{k=1}^{K} \beta_{jk}\tilde{f}_k + \tilde{e}_j \qquad (14.9)$$

The nonfactor, or asset-specific, return on security j, is the residual risk of the security, after removing the estimated impacts of the K factors. The term, f, is the rate of return on factor k. A single factor model, in which the market return is the only estimated factor, is obviously the basis of the capital asset pricing model. We discuss multifactor risk models in the next chapter.

14.3 Expected Returns Versus Historic Mean Returns

The expected returns on assets are not completely explained by using only historic means of the securities. One may estimate models of expected return using expectation data and reported financial data. There are several approaches to security valuation and the creation of expected returns. Graham and Dodd (1934) recommended that stocks be purchased on the basis of the price-earnings ratio. Graham and Dodd suggested that no stock should be purchased if its price-earnings ratio exceeded 1.5 times the price-earnings multiple of the market. Thus, the "low price-earnings" (PE) criteria were established. It is interesting that Graham and Dodd put forth the low PE model at the height of the Great Depression. Basu (1977) reported evidence supporting the low PE model. Academicians often prefer to test the low P/E approach by testing its reciprocal, the "high EP" approach. The high EP approach specifically addresses the issue of negative earnings per share, which can confuse the low P/E test. Hawawini and Kreim (1995) found statistical support for the high EP variable of NYSE and AMEX stocks from April 1962 to December 1989. At a minimum, Graham and Dodd advocated the calculation of a security's net current asset value, NCAV, defined as its current assets less all liabilities. A security should be purchased if its net current value exceeded its current stock price. The price-to-book (PB) ratio should be calculated, but not used as a measure for stock selection, according to Graham and Dodd (1962).

14.4   Fundamental Analysis and Stock Selection

Fundamental variables such as cash flow and sales have put used in composite valuation models for security selection [Ziemba (1990, 1992) and Guerard (1992, 1993)]. Livnat (1994) advocated the calculation of free cash flow, which subtracts capital expenditures from the operating cash flow. In addition to the income statement indicators of value, such as earnings, cash flow, and sales, many value-focused analysts also consider balance sheet variables, especially the book-to-market ratio. The income statement measures are dividends, earnings, cash flow, and sales, and the key balance sheet measure is common equity per share outstanding, or book value. Expected returns modeling has been analyzed with a regression model in which security returns are functions of fundamental stock data, such as earnings, book value, cash flow, and sales, relative to stock prices, and forecast earnings per share [Fama and French (1992, 1995), Bloch et al. (1993), Guerard et al. (1993), Ziemba (1992), Guerard (1997), and Stone et al. (1997, 2002)]. Hawawini and Keim (1995) found statistical support for the high cash flow-to-price (CP) and low price-to-book (P/B) variables of NYSE and AMEX stocks from April 1962 to December 1989.

Fama and French (1995) argue that the contrarian investment strategy is inconsistent with market data. Fama and French report results of a three-factor model that incorporates the market factor, size factor, and book-to-market ratio. In the Fama and French analysis of the CRSP database for the 1963–1994 period, the value-weighted market return exceeded the risk-free rate by approximately 5.2%, annually, whereas smaller stocks outperformed larger stocks by 3.2% annually, and higher book-to-market (price) outperformed smaller book-to-market stocks by 5.4%, annually. The Fama and French model may be written as:

$$R_j - R_F = \beta_M R_M + \beta_{size} R_{size} + \beta_{BM} R_{BM} \qquad (14.10)$$

where

R_j = Stock return
β_M = Market beta
R_M = Return on market index less the risk-free rate
β_{size} = Size beta
R_{size} = Return on size variable, defined as the return on small stocks less the return on large stocks
β_{BM} = Book-to-market beta

and

R_{BM} = Returns on high book-to-market stocks less the returns on low book-to-market stocks

The authors report the Fama and French three-factor model because of its importance in the literature. The results of the stock selection model introduced and estimated in the following section does not place a higher weight on the book-to-market (price) ratio than on the cash flow, relative book-to-price (and relative earnings, cash flow, and sales-to-price ratios), and analysts' forecast variables.

In 1975, a database of earnings per share (eps) forecasts was created by Lynch, Jones, and Ryan, a New York brokerage firm, by collecting and publishing the consensus statistics of one-year-ahead and two-year-ahead eps forecasts [Brown (1999)]. The database has evolved to be known as the Institutional Brokers' Estimate System (I/B/E/S) database. There is extensive literature regarding the effectiveness of analysts' earnings forecasts, earnings revisions, earnings forecast variability, and breadth of earnings forecast revisions, summarized in Bruce and Epstein (1994) and Brown (1999). The vast majority of the earnings forecasting literature in the Bruce and Brown references find that the use earnings forecasts do not increase stockholder wealth, as specifically tested in Elton et al. (1981). Reported earnings follow a random walk with drift process, and analysts are rarely more accurate than a no-change model in forecasting earnings per share (Cragg & Malkiel, 1968; Guerard & Stone, 1992). Analysts become more accurate as time passes during the year, and quarterly data is reported. Analyst revisions are statistically correlated with stockholder returns during the year (Hawkins et al., 1984; Arnott, 1985). Wheeler (1994) developed and tested a strategy in which analyst forecast revision breadth, defined as the number of upward forecast revisions less the number of downward forecast revisions, divided by the total number of estimates, was the criteria for stock selection. Wheeler found statistically significant excess returns from the breadth strategy. A composite earnings variable, CTEF, is calculated using equally weighted revisions, forecasts, and breadth of FY1 and FY2 forecasts, a variable put forth in Guerard (1997).

Guerard (1992, 1993), Ziemba (1990, 1992), and et al. (1997) employed annual fundamental Compustat variables, such as earnings, book value, cash flow, and sales, in addition to the composite earnings forecasting model in a regression model to identify the determinants of quarterly equity returns. The regression models used in the Guerard et al. (1997) studies employed the Beaton-Tukey robust regression procedure and latent root regression techniques to address the issues of outliers and multicollinearity. The reader was introduced to the robust regression and multicollinearity techniques in Chap. 12. Ziemba used capitalization-weighted regressions. Both sets of studies found statistical significance with expectation and reported fundamental data.

14.5 Modern Portfolio Theory and GPRD: An Example of Markowitz Analysis

In 1990, Harry Markowitz became the Head of the Global Portfolio Research Department (GPRD) at Daiwa Securities Trust. His department used fundamental data to create models for Japanese and US securities. The basic models tested included the earnings-to-price strategy (EPR), which is the inverse of the low P/E strategy, the book value-to-price (BPR), the cash flow-to-price (CPR), and sales-to-price (SPR) ratios. These variables should be positively associated (correlated) with subsequent returns. Single variable and composite model strategies were tested in the Japan and the United States, from 1974 to 1990. The composite models could be created by combining the variables using ordinary least squares (OLS), outlier-adjusted or robust regression (ROB), or weighted latent root regression (WLRR) modeling techniques, in which outliers and the high correlations among the variables are used in the estimation procedures. The reader is referred to Chap. 12 for a discussion of ROB and WLRR techniques. The Markowitz group tested single variables and composite model strategies, using the outlier-adjustment and multicollinearity modeling techniques shown in Chap. 12. The proprietary models combined the outlier adjustments and multicollinearity techniques to substantially outperform the Japanese equity market, from 1990 to 1994. The Markowitz group found that the use of the more advanced statistical techniques produced higher relative out-of-sample portfolio geometric returns and Sharpe ratios. Statistic modeling is not just fun, but it is consistent with maximizing portfolio returns. The quarterly estimated models outperformed the semiannual estimated models, although the underlying data was semiannual in Japan. The dependent variable in the composite model is total security returns, and the independent variables are the EPR, BPR, CPR, and SPR variables and their respective relative ratios, the current ratio divided by the average of the past 60 months of the ratios. Guerard and Schwartz (2007) reprinted the original Table 1 of Bloch et al. (1993); Table 14.6 reports backtests underlying the GPRD Daiwa Portfolio Optimization System (DPOS). We continue the table was a benchmark for stock modeling.

In Bloch et al. (1993), Markowitz and his research department estimated efficient frontiers for Japanese stocks for the 1974–1990 period and US stocks for the 1975–1989 period. The Bloch et al. efficient frontier is shown in Fig. 14.3. The

Table 14.6 DPOS Japanese and US simulations

D-POS, Japan. Simulation results: sorted by geometric mean. Let UL = 2.0 TCR = 2.0
PM = −1 PPar = 0.90 Begin = 7412 End = 9012.

SID	OP	TOV	RP	Rtri	ERET	GM	Shrp
900869	3	10.00	1	25.0	PROPRIETARY	25.60	0.89
900867	3	10.00	0	0.0	PROPRIETARY	25.39	0.89
900868	3	10.00	1	20.0	PROPRIETARY	25.32	0.87
900860	3	15.00	1	25.0	PROPRIETARY	24.28	0.84
900853	3	15.00	0	0.0	PROPRIETARY	24.19	0.85
900858	3	10.00	1	25.0	PROPRIETARY	24.04	0.85
900852	3	12.50	0	0.0	PROPRIETARY	23.94	0.85
900855	3	10.00	1	20.0	PROPRIETARY	23.93	0.86
900856	3	12.50	1	20.0	PROPRIETARY	23.90	0.84
900854	3	17.50	0	0.0	PROPRIETARY	23.89	0.84
900859	3	12.50	1	25.0	PROPRIETARY	23.89	0.83
900857	3	15.00	1	20.0	PROPRIETARY	23.81	0.82
900819	3	10.00	0	0.0	REGR(WLRR,4Q,4)	22.74	0.83
900820	3	10.00	1	25.0	REGR(WLRR,4Q,4)	22.68	0.82
900944	3	10.00	0	0.0	BPR	22.43	0.78
900908	3	10.00	1	20.0	REGR(LRR,4Q,9.l)	22.23	0.75
900874	3	10.00	0	0.0	REGR(OLS,4Q,8)	22.16	0.79
900878	3	10.00	0	0.0	REGR(OLS,4Q,9.1)	22.16	0.79
900903	3	10.00	0	0.0	REGR(OLS,4Q,8)	22.16	0.79
900914	3	10.00	0	0.0	REGR(OLS,4Q,9.1)	22.16	0.79
900841	3	10.00	1	25.0	REGR(WLRR,1Q,4)	22.00	0.79
900817	3	10.00	0	0.0	REGR(LRR,4Q14)	21.99	0.76
900983	3	10.00	1	20.0	REGR(WLRR,4Q,9.1)	21.93	0.75
900984	3	10.00	1	20.0	REGR(WLRR,4Q,9.l)	21.86	0.75
900794	3	15.00	1	20.0	REGR(WLRR,1Q,4)	21.84	0.76
900818	3	10.00	1	25.0	REGR(LRR,4Q,4)	21.84	0.75
900877	3	10.00	0	0.0	REGR(WLRR,4Q,8)	21.84	0.78
900906	3	10.00	0	0.0	REGR(WLRR,4Q,8)	21.84	0.78
900985	3	12.50	1	20.0	REGR(WLRR,4Q,9.1)	21.84	0.75
900913	3	10.00	0	0.0	REGR(WLRR,4Q,9.2)	21.83	0.77
900793	3	12.50	1	20.0	REGR(WLRR,1Q,4)	21.78	0.78
900791	3	12.50	0	0.0	REGR(WLRR,1Q,4)	21.75	0.79
900792	3	15.00	0	0.0	REGR(WLRR,1Q,4)	21.68	0.77
900982	3	10.00	1	20.0	REGR(WLRR,4Q,9.1)	21.66	0.75
900842	3	10.00	1	25.0	REGR(WLRR,10,4)	21.55	0.79
900766	3	10.00	1	20.0	REGR(WLRR,1Q,4)	21.49	0.78
900810	3	15.00	0	0.0	REGR(WLRR,1Q,4)	21.47	0.76
900901	3	10.00	0	0.0	REGR(LRR,4Q,9.1)	21.45	0.72
900813	3	10.00	0	0.0	REGR(OLS,4Q,4)	21.42	0.78
900840	3	10.00	1	25.0	REGR(WLRR,1Q,4)	21.41	0.76
900838	3	10.00	1	25.0	REGR(WLRR,1Q,4)	21.40	0.76
900909	3	10.00	1	20.0	REGR(WLRR,4Q,9.1)	21.40	0.75
900910	3	10.00	0	0.0	REGR(LRR,4Q,9.2)	21.34	0.75
900816	3	10.00	1	25.0	REGR(ROB,4Q,4)	21.30	0.76
900839	3	10.00	1	25.0	REGR(WLRR,1Q,4)	21.30	0.75

(continued)

Table 14.6 (continued)

SID	OP	TOV	RP	Rtri	ERET	GM	Shrp
900912	3	10.00	0	0.0	REGR(LRR,4Q,9.2)	21.29	0.71
900765	3	10.00	0	0.0	REGR(WLRR,1Q,4)	21.24	0.76
900815	3	10.00	0	0.0	REGR(ROB,4Q,4)	21.23	0.76
900902	3	10.00	0	0.0	REGR(WLRR,4Q,9.1)	21.16	0.74
900986	3	15.00	1	20.0	REGR(WLRR,4Q,9.1)	21.09	0.72
900954	3	10.00	0	0.0	REGR(OLS,4Q,4)	20.91	0.72
900876	3	10.00	0	0.0	REGR(LRR,4Q,8)	20.90	0.74
900905	3	10.00	0	0.0	REGR(LRR,4Q,8)	20.90	0.74
900911	3	10.00	0	0.0	REGR(ROB,4Q,9.2)	20.66	0.72
900907	3	10.00	1	20.0	REGR(ROB,4Q,9.1)	20.36	0.74
900763	3	10.00	0	0.0	REGR(LRR,1Q,4)	20.21	0.71
900875	3	10.00	0	0.0	REGR(ROB,4Q,8)	20.15	0.71
900904	3	10.00	0	0.0	REGR(ROB,4Q,8)	20.15	0.71
900787	3	12.50	0	0.0	REGR(LRR,1Q,4)	20.08	0.71
900900	3	10.00	0	0.0	REGR(ROB,4Q,9.1)	20.07	0.72
900781	3	12.50	1	20.0	REGR(OLS,1Q,4)	19.96	0.71
900788	3	15.00	0	0.0	REGR(LRR,1Q,4)	19.92	0.70
900764	3	10.00	1	20.0	REGR(LRR,1Q,4)	19.88	0.70
900790	3	15.00	1	20.0	REGR(LRR,1Q,4)	19.81	0.70
900789	3	12.50	1	20.0	REGR(LRR,1Q,4)	19.78	0.70
900779	3	12.50	0	0.0	REGR(OLS,1Q,4)	19.77	0.67
900786	3	15.00	1	20.0	REGR(ROB,1Q,4)	19.76	0.71
900780	3	15.00	0	0.0	REGR(OLS,1Q,4)	19.72	0.69
900784	3	15.00	0	0.0	REGR(ROB,1Q,4)	19.67	0.71
900782	3	15.00	1	20.0	REGR(OLS,1Q,4)	19.41	0.69
900759	3	10.00	0	0.0	REGR(OLS,1Q,4)	19.40	0.67
900785	3	12.50	1	20.0	REGR(ROB,1Q,4)	19.33	0.69
900760	3	10.00	1	20.0	REGR(OLS,1Q,4)	19.31	0.66
900783	3	12.50	0	0.0	REGR(ROB,1Q,4)	19.10	0.69
900761	3	10.00	0	0.0	REGR(ROB,1Q,4)	19.03	0.68
900931	3	10.00	0	0.0	CPR	19.01	0.68
900762	3	10.00	1	20.0	REGR(ROB,1Q,4)	19.00	0.67
900932	3	10.00	0	0.0	SPR	18.63	0.61
900716	3	10.00	1	20.0	benchmark	17.25	0.60
900927	3	10.00	0	0.0	EPR	16.82	0.57
900826	6	20.00	3	25.0	PROPRIETARY	24.63	0.84
900709	6	20.00	3	20.0	PROPRIETARY	23.61	0.81
900710	6	20.00	3	25.0	PROPRIETARY	23.44	0.82
900733	6	25.00	1	20.0	PROPRIETARY	23.34	0.80
900773	6	17.50	3	20.0	PROPRIETARY	23.26	0.78
900707	6	20.00	3	20.0	PROPRIETARY	23.08	0.79
900847	6	20.00	0	0.0	PROPRIETARY	22.62	0.81
901030	6	20.00	3	20.0	BPR	22.42	0.78
900796	6	20.00	3	20.0	REGR(OLS,2S,4)	22.33	0.79
901047	6	20.00	3	20.0	BPR	22.20	0.77
900770	6	22.50	0	0.0	REGR(OLS,1S,4)	22.17	0.77
900795	6	20.00	0	0.0	REGR(OLS,2S,4)	22.14	0.79
900749	6	20.00	3	25.0	REGR(OLS,1S,4)	22.03	0.78
900800	6	20.00	3	20.0	REGR(LRR,2S,4)	21.98	0.78
900849	6	20.00	0	0.0	REGR(LRR,3S,4)	21.98	0.77

(continued)

Table 14.6 (continued)

SID	OP	TOV	RP	Rtri	ERET	GM	Shrp
900748	6	20.00	3	20.0	REGR(OLS,1S,4)	21.80	0.77
900754	6	20.00	3	20.0	REGR(LRR,12,4)	21.68	0.74
900747	6	20.00	0	0.0	REGR(OLS,1S,4)	21.65	0.77
900802	6	20.00	3	20.0	REGR(WLRR,2S,4)	21.60	0.79
901029	6	20.00	0	0.0	BPR	21.59	0.76
900755	6	20.00	3	25.0	REGR(LRR,1S,4)	21.52	0.74
900799	6	20.00	0	0.0	REGR(LRR,2S,4)	21.51	0.77
901046	6	20.00	0	0.0	BPR	21.49	0.76
900801	6	20.00	0	0.0	REGR(WLRR,2S,4)	21.40	0.79
900769	6	17.50	0	0.0	REGR(OLS,1S,4)	21.34	0.76
900778	6	22.50	3	20.0	REGR(WLRR,1S,4)	21.30	0.79
900772	6	22.50	0	0.0	REGR(LRR,1S,4)	21.26	0.72
900753	6	20.00	0	0.0	REGR(LRR,1S,4)	21.20	0.72
900756	6	20.00	0	0.0	REGR(WLRR,1S,4)	21.10	0.77
900757	6	20.00	3	20.0	REGR(WLRR,1S,4)	21.06	0.78
900758	6	20.00	3	25.0	REGR(WLRR,1S,4)	21.02	0.78
900777	6	17.50	3	20.0	REGR(WLRR,1S,4)	20.95	0.78
900771	6	17.50	0	0.0	REGR(LRR,1S,4)	20.93	0.71
900848	6	20.00	0	0.0	REGR(ROB,3S,4)	20.74	0.76
900776	6	22.50	3	20.0	REGR(ROB,1S,4)	20.37	0.76
900797	6	20.00	0	0.0	REGR(ROB,22,4)	20.24	0.75
900798	6	20.00	3	20.0	REGR(ROB,2S,4)	20.12	0.76
900752	6	20.00	3	25.0	REGR(ROB,1S,4)	19.56	0.73
900751	6	20.00	3	20.0	REGR(ROB,1S,4)	19.35	0.73
900750	6	20.00	0	0.0	REGR(ROB,1S,4)	19.29	0.72
900775	6	17.50	3	20.0	REGR(ROB,1S,4)	19.26	0.72
901049	6	20.00	3	20.0	CPR	18.99	0.68
901051	6	20.00	3	20.0	SPR	18.69	0.61
901048	6	20.00	0	0.0	CPR	18.65	0.67
901050	6	20.00	0	0.0	SPR	17.87	0.59
901045	6	20.00	3	20.0	EPR	17.55	0.59
901044	6	20.00	0	0.0	EPR	17.34	0.59

SID = simulation ID; OP = period of re-optimization; TOV = turnover constraint; Rtri = rebalancing trigger; ERET = model description; GM = geometric mean; Shrp = Sharpe ratio.

REGR (technique, period, equation)

Technique = OLS for ordinary least-squares regression analysis,
LRR for latent root regression,
ROB for robust regression,
WLRR for weighted latent root regression;

Period = 1S for one-period semi-annual analysis,
2S for two-period semi-annual analysis,
3S for three-period semi-annual analysis,
1Q for one-period quarterly analysis,
4Q for four-period quarterly analysis;

Equation = 4; $TRR = a_0 + a_1 EPR + a_2 BPR + a_3 CPR + a_4 SPR + a_5 REPR + a_6 RPBR + a_7 RCPR + a_8 RSPR + e_t$

8; $TRR = a_0 + a_1 EPR + a_2 BPR + a_3 CPR + a_4 SPR + e_t$

9.1; $TRR = a_0 + a_1 EPR + a_2 BPR + a_3 CPR + a_4 SPR + a_5 REPR(2) + a_6 RPBR(2) + a_7 RCPR(2) + a_8 RSPR(2) + e_t$,

where (2) denotes 2-year averages of relative variables.

9.2; $TRR = a_0 + a_1 EPR + a_2 BPR + a_3 CPR + a_4 SPR + a_5 REPR(3) + a_6 RPBR(3) + a_7 RCPR(3) + a_8 RSPR(3) + e_t$,

where (3) denotes 3-year averages of relative variables.

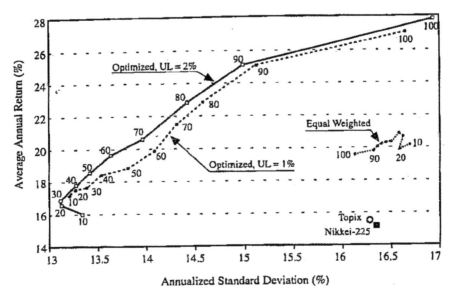

Fig. 14.3 DPOS public model risk-return trade-off curve

upper bounds on security weights are a decision variable in which an increase in the upper bound allows the investment manager to exercise more asset selection power and possibly shift out the efficient frontier, as shown in Fig. 14.3, from Bloch et al. (1993), in which the upper bound on security weights increases from 1% to2%.

The "public" model trade-off curve, or efficient frontier, is shown in Fig. 14.3, and the corresponding "proprietary" model is shown in Fig. 14.4. The trade-off curve contains portfolios of varying risk-return characteristics, with Japanese portfolios producing annualized returns from 17% to over 26%, annualized.

14.6 Further Estimations of a Composite Equity Valuation Model: The Roles of Analyst Forecasts and Momentum in Stock Selection

This section exposes the reader to issues of databases and the inclusion of variables in composite models to identify undervalued securities. The database for this analysis is created by the use of all securities listed on the Compustat database, the I/B/E/S database, and the Center for Research in Security Prices (CRSP) database during the 1976–2019 period. The annual Compustat file contains some 399 data items from the company income statement, balance sheet, and cash flow statement during the 1950–2018 period. The I/B/E/S database contains all earnings forecasts

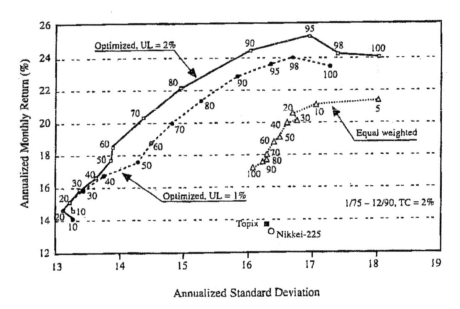

Fig. 14.4 DPOS proprietary model risk-return trade-off curve

made during the 1976–2019 period. The CRSP file contains monthly stock prices, shares outstanding, trading volumes, and returns for all traded securities from 1926 to 2019.

There are a seemingly infinite number of financial variables that may be tested for statistical association with monthly security returns. Bloch et al. (1993) tested a set of fundamental variables in the United States during the 1975–1990 period. Guerard (1997) tested a set of I/B/E/S variables for the 1982–1994 period. In this study, we test the variables of these two studies using both fundamental and expectation data. We initially test the effectiveness of the individual variables using the information coefficients (ICs) rather than the upper quintile excess returns or the excess returns of individual variable portfolio optimizations. The information coefficient is the slope of the regression estimation in which ranked subsequent security returns are a function of the ranked financial strategy. The advantage of the IC approach is that the slope has a corresponding t-statistics that allows one to test the null hypothesis that the strategy is uncorrelated with subsequent returns. In developing a composite model, one seeks to combine variables that are statistically associated with subsequent returns. Let us define the variables tested in this chapter.

EP = earnings per share /price per share
BP = book per share/price per share
CP = cash flow per share/price per share
SP = sales per share/price per share

DY = dividend yield, dividend per share/price per share
NCAV = net current asset value, net current assets per share/price per share
FEP1 = one-year-ahead forecast earnings per share /price per share
FEP2 = two-year-ahead forecast earnings per share/price per share
RV1 = one-year-ahead forecast earnings per share monthly revision/price per share
RV2 = two-year-ahead forecast earnings per share monthly revision/price per share
BR1 = one-year-ahead forecast earnings per share monthly breadth/price per share
BR2 = two-year-ahead forecast earnings per share monthly breadth/price per share

A consensus earnings-per-share I/B/E/S forecast, revisions and breadth variable, CTEF, has been created and tested using I/B/E/S data since January 1976.

The monthly ICs for all traded securities from January 1976 to November 2020 period for these variables are shown in Table 14.7. The majority of the variables are statistically associated with stockholder returns, a result consistent with Bloch et al. and Guerard studies.

We estimate a similar monthly model for the January 1996 to November 2020 period.

$$TR_{t+1} = a_0 + a_1 EP_t + a_2 BP_t + a_3 CP_t + a_4 SP_t + a_5 REP_t + a_6 RBP_t + a_7 RCP_t$$
$$+ a_8 RSP_t + a_9 CTEF_t + e_t$$

$$(14.13)$$

where:

EP = [earnings per share]/[price per share] = earnings-price ratio
BP = [book value per share]/[price per share] = book-price ratio
CP = [cash flow per share]/[price per share] = cash flow-price ratio
SP = [net sales per share]/[price per share] = sales-price ratio
REP = [current EP ratio]/[average EP ratio over the past 5 years]
RBP = [current BP ratio]/[average BP ratio over the past 5 years]
RCP = [current CP ratio]/[average CP ratio over the past 5 years]
RSP = [current SP ratio]/[average SP ratio over the past 5 years]
CTEF = consensus earnings-per-share I/B/E/S forecast, revisions and breadth
e = randomly distributed error term.

The monthly ordinary least squares (OLS) regression is plagued with approximately twice the number observations outside the 95% confidence interval as one might expect given a normal distribution of residuals. These aberrant observations, or outliers, lead us to re-estimate the monthly regression lines using a Beaton-Tukey bisquare (or robust, ROB) regression technique, in which each observation is weighted as the inverse function of its OLS residual. The application of the Beaton-Tukey ROB procedure addresses the issue of outliers. The weighted data is plagued with multicollinearity, the correlation among the independent variables, which may lead to statistically inefficient estimates of the regression coefficients. Bloch et al. (1993) and Guerard, Takano, and Yamane (1993) applied latent root regression (LRR) to the ROB-weighted data, referred to as weighted latent root

Table 14.7 Individual Variables Effectiveness in Stock Selection within Various Universes

Factors Testing, 12/1995 - 12/2020
Universe: Russell 3000

Factor	Quintile Universe Return					Excess vs. Benchmark				
	Spread	1	2	3	4	5	1	2	3	4	5
REG10	12.3905	14.7170	12.6186	10.6680	7.7934	-1.7863	4.5956	2.5448	0.7357	-1.6079	-9.9880
REG9	10.0218	14.2093	11.4590	9.9681	6.6946	1.4738	4.3621	1.4955	0.0230	-2.7300	-7.0484
RCP	10.0080	11.6632	12.5415	12.0564	9.2110	-0.3110	2.2490	2.3682	1.6588	-0.7682	-8.3664
CTEF	9.9940	15.3081	11.5383	8.7462	6.1485	2.9381	4.8942	1.5409	-0.8091	-2.9241	-5.8431
CP	9.2968	12.5079	12.2061	11.3907	7.3861	-0.0293	3.0222	1.9292	1.1492	-2.2398	-8.1121
REG8	8.6485	13.1164	11.6810	9.8030	7.4133	1.6611	3.4282	1.6334	-0.1316	-2.1365	-6.8220
EP	7.9756	12.8990	12.1365	10.4817	7.4389	0.3791	3.0689	1.7087	0.2172	-2.0289	-7.3651
SP	6.7802	10.6193	11.6481	11.1763	8.3071	1.9181	1.5503	1.7125	1.0914	-1.5482	-6.8284
REP	6.6344	11.5862	13.0724	11.1610	8.6679	1.4810	1.8831	2.5099	0.6969	-1.0216	-5.8845
PM	5.6018	12.3399	10.5971	10.9244	9.7091	-0.4320	2.4441	0.4778	0.7975	-0.0101	-8.3324
BP	2.2700	8.9046	11.2232	9.1154	8.6114	5.8685	-0.1321	1.1500	-0.8336	-1.0531	-3.1362
RBP	-1.5918	4.3298	11.2151	10.9576	10.5689	6.7110	-3.9797	1.2908	0.6879	0.3543	-2.3734
RSP	-1.6910	3.1071	11.6590	12.2451	10.8562	5.8415	-5.0168	1.7042	1.9373	0.6118	-3.3350

Factor	IC				IC T-Stat				Sharpe Ratio		
	1 Month	3 Month	6 Month	12 Month	1 Month	3 Month	6 Month	12 Month	1	2	3
REG10	0.0405	0.0568	0.0786	0.0945	5.8965	7.0267	8.1355	8.7843	0.5599	0.5296	0.4401
REG9	0.0344	0.0483	0.0662	0.0848	5.3906	6.5108	7.5806	8.4465	0.4849	0.4552	0.4083
RCP	0.0320	0.0482	0.0648	0.0830	4.7213	5.8537	6.8321	7.6704	0.3563	0.5437	0.5968
CTEF	0.0375	0.0463	0.0593	0.0745	5.4845	6.2657	7.0573	7.8563	0.6732	0.4792	0.3115
CP	0.0322	0.0509	0.0722	0.0924	5.2209	6.5624	7.7788	8.6822	0.3940	0.5313	0.5276
REG8	0.0292	0.0433	0.0612	0.0791	5.0337	6.2285	7.3818	8.2850	0.4263	0.4694	0.4043
EP	0.0348	0.0531	0.0761	0.1000	5.4301	6.7179	8.0165	9.1301	0.4485	0.5615	0.4932
SP	0.0212	0.0329	0.0456	0.0578	4.1949	5.2885	6.1774	6.8366	0.3028	0.4527	0.4800
REP	0.0329	0.0487	0.0672	0.0872	5.2486	6.3996	7.4583	8.3765	0.3952	0.6251	0.5636
PM	0.0234	0.0339	0.0486	0.0484	4.6109	5.4017	6.3663	6.3719	0.4599	0.4843	0.4839
BP	0.0053	0.0100	0.0174	0.0255	2.1263	2.9347	3.8535	4.5574	0.2353	0.4453	0.3703
RBP	-0.0042	-0.0096	-0.0143	-0.0139	-1.9445	-2.8012	-3.3255	-3.2370	0.0694	0.4295	0.5004
RSP	-0.0060	-0.0114	-0.0157	-0.0152	-2.2718	-3.0677	-3.4041	-3.3111	0.0304	0.4428	0.5632

Factor	Sharpe Ratio		Information % > Ratio	% > Up Benchmark	% > Down Benchmark	%Total Turnover				
	4	5	1	Benchmark	Benchmark	Benchmark	1	2	3	4	5
REG10	0.2673	-0.1322	0.3715	52.6490	57.4359	46.5347	50.9043	96.9393	101.8584	88.9520	43.4945
REG9	0.2188	-0.0233	0.3017	52.6490	60.5128	40.5941	38.4420	76.7027	80.7419	68.5232	32.6468
RCP	0.3889	-0.0821	0.1468	46.6887	55.3846	32.6733	24.9513	45.6086	45.1207	27.3393	10.0504
CTEF	0.1717	0.0336	0.5336	54.9669	59.4872	49.5050	78.4135	124.7229	136.0295	130.2927	76.4909
CP	0.2582	-0.0712	0.1908	50.6623	57.4359	40.5941	18.5972	34.9233	36.0569	24.6952	12.5309
REG8	0.2555	-0.0162	0.2225	50.3311	59.4872	35.6436	28.7298	58.0209	62.4276	52.5607	25.0815
EP	0.2463	-0.0531	0.2222	51.3245	58.4615	40.5941	22.5253	38.5140	34.4733	21.8211	12.0456
SP	0.3333	-0.0069	0.0910	48.0132	57.9487	31.6832	13.4408	26.6190	30.1520	26.4149	13.4195
REP	0.3381	-0.0211	0.1464	49.0066	55.3846	39.6040	23.0863	40.3828	34.5108	36.7983	8.2370
PM	0.3474	-0.0741	0.1921	48.6755	55.3846	38.6139	65.8848	113.0214	120.3136	110.7814	60.1641
BP	0.3287	0.1574	-0.0072	44.7020	52.8205	31.6832	20.6760	39.8696	41.2851	32.7192	15.8767
RBP	0.4956	0.1978	-0.1921	43.0464	56.4103	19.8020	26.8288	54.7448	60.6891	46.3582	20.6307
RSP	0.5163	0.1703	-0.2296	41.7219	53.8462	20.7921	27.3939	56.5918	63.3837	52.1516	23.6480

(continued)

regression, WLRR, and produced models with higher in-sample F-statistics and higher out-of-sample geometric means using WLRR than ROB and OLS techniques. We create a composite model weight using the average weight of the positive coefficients of the preceding 12 monthly regressions, a monthly equivalent to the four quarter averaging techniques used in Guerard et al. (1997). One notes the large weighting of the CTEF variable and the relatively small weighting of the BP variable, a result consistent with Guerard et al. (1997). In terms of information coefficients, ICs, the use of the WLRR procedure produces the highest IC for the models during the 1990–2001 period. The WLRR technique produces the largest

Table 14.7 (continued)

Factors Testing, 12/1995 - 12/2020
Universe: MSCI_ACWXUS

Factor	Quintile Universe Return						Excess vs. Benchmark				
	Spread	1	2	3	4	5	1	2	3	4	5
REG10	11.0270	13.5030	10.1821	7.6401	4.5619	2.1243	8.4394	4.9122	2.3509	-0.5450	-2.5810
CTEF	9.7430	14.3318	9.4565	6.3768	3.9768	4.0640	9.0739	4.1046	1.2420	-0.9057	-0.7580
REG9	9.2534	12.9607	9.2638	6.6775	5.1973	3.6680	8.0831	4.0779	1.5171	0.0076	-1.3404
CP	8.8767	11.6778	8.6766	7.4342	6.1499	2.5637	6.8774	3.4433	2.0829	0.7989	-2.0450
REG8	7.8132	12.1050	9.0334	7.0470	5.1272	4.2412	7.2929	3.8457	1.8863	-0.0841	-0.8007
SP	6.5435	11.0910	8.0797	7.5342	6.1713	4.5237	6.2392	2.9927	2.3818	0.9688	-0.6127
EP	5.2108	10.8041	7.6051	6.6812	6.5261	5.6284	6.1635	2.5029	1.3720	1.0806	0.7294
RCP	5.0303	8.4342	9.0103	8.1263	8.0956	2.9522	3.7218	3.6802	2.6139	2.7297	-1.4936
PM	3.9561	12.3587	8.6497	6.6161	5.4169	3.8142	6.8396	3.1366	1.2937	0.4475	-0.3228
BP	3.1302	10.0108	7.0644	6.7851	6.1766	7.0320	5.3083	1.9250	1.5951	0.9993	1.8353
REP	0.7837	7.0273	7.6027	8.1020	8.4844	6.1461	2.3874	2.3670	2.6764	3.1063	1.3753
RSP	0.3470	6.9845	7.3167	7.4626	8.1021	6.9594	2.6401	2.1712	2.0865	2.6041	1.8864
RBP	-3.1562	5.7702	6.6520	6.7165	8.2379	9.5668	1.4644	1.4982	1.3807	2.7999	4.3710

Factor	IC				IC T-Stat				Sharpe Ratio		
	1 Month	3 Month	6 Month	12 Month	1 Month	3 Month	6 Month	12 Month	1	2	3
REG10	0.0309	0.0400	0.0535	0.0530	4.5406	5.2205	5.9723	5.9386	0.6078	0.5370	0.4284
CTEF	0.0339	0.0376	0.0465	0.0479	4.6752	4.9905	5.4438	5.4839	0.7095	0.5327	0.3478
REG9	0.0247	0.0330	0.0420	0.0507	4.0469	4.7203	5.2284	5.6836	0.5442	0.4802	0.3683
CP	0.0206	0.0333	0.0446	0.0582	3.4288	4.3410	4.8961	5.5532	0.4872	0.4709	0.4367
REG8	0.0190	0.0280	0.0364	0.0463	3.4660	4.2082	4.6777	5.2112	0.4989	0.4670	0.3852
SP	0.0081	0.0165	0.0220	0.0314	1.9955	3.0238	3.3693	4.0376	0.4562	0.4217	0.4090
EP	0.0180	0.0266	0.0350	0.0502	3.2926	3.9046	4.3245	5.2359	0.4314	0.4005	0.3930
RCP	0.0148	0.0203	0.0272	0.0374	2.6152	3.0197	3.4805	4.0539	0.3585	0.5020	0.5029
PM	0.0216	0.0293	0.0449	0.0256	3.8421	4.5718	5.5325	4.4259	0.6404	0.5219	0.3878
BP	-0.0015	0.0027	0.0069	0.0179	-1.5498	-0.9161	0.4166	2.3623	0.3852	0.3731	0.3757
REP	0.0090	0.0113	0.0132	0.0229	2.2910	2.6276	2.8891	3.8162	0.2976	0.4171	0.4803
RSP	-0.0073	-0.0086	-0.0101	-0.0050	-2.2274	-2.4204	-2.6748	-2.4917	0.2507	0.3825	0.4401
RBP	-0.0124	-0.0185	-0.0238	-0.0216	-2.9010	-3.4966	-3.9751	-3.9504	0.2107	0.3515	0.3941

Factor	Information Ratio		% > Benchmark	% > Up Benchmark	% > Down Benchmark	% Total Turnover				
	4	5	1	1	1	1	2	3	4	5
REG10	0.2552	0.1058	0.7307	59.6552	64.5349	52.5424	49.0993	94.8198	102.2669	92.4564	47.0119
CTEF	0.2026	0.2016	0.9851	62.0690	65.6977	56.7797	83.0200	129.1228	135.6552	124.2430	66.0295
REG9	0.2974	0.1962	0.6096	58.6207	65.6977	48.3051	36.4511	74.2492	81.0278	69.8360	33.1509
CP	0.3683	0.1164	0.5143	54.8276	63.9535	41.5254	18.0741	33.8594	36.1680	28.1636	16.1825
REG8	0.2962	0.2289	0.5292	56.8966	63.3721	47.4576	31.0797	63.3403	70.0873	60.8550	29.4325
SP	0.3462	0.2484	0.4357	51.7241	59.3023	40.6780	12.8697	26.5662	29.6818	25.7459	12.5924
EP	0.4032	0.2711	0.4259	56.2069	63.9535	44.9153	21.3955	40.2006	42.5836	32.3160	17.2848
RCP	0.4734	0.1260	0.2890	52.7586	58.7209	44.0678	25.7654	48.0950	51.2264	34.7463	15.7543
PM	0.2766	0.1375	0.6529	57.9310	56.3953	60.1695	64.3660	115.3055	124.0192	115.0397	63.1612
BP	0.3500	0.3882	0.3221	52.4138	58.7209	43.2203	18.4094	36.8363	38.6550	31.4005	14.5249
REP	0.4887	0.2847	0.1848	55.5172	61.0465	47.4576	25.6506	47.9863	50.3129	33.6827	14.2877
RSP	0.4946	0.3641	0.1467	47.2414	55.8140	34.7458	28.3547	59.1573	65.9391	54.2392	23.7891
RBP	0.4961	0.4983	0.0842	49.3103	57.5581	37.2881	29.3718	60.7605	66.6907	54.1097	23.9083

and most statistically significant IC, a result consistent with the previously noted studies and the GPRD example. The t-statistics on the composite model exceed the t-statistics of its components. The purpose of a composite security valuation model is to identify the determinants of security returns and produce a statistically significant out-of-sample ranking metric of total returns.

14.6.1   REG8 Model

We estimate a monthly model, the original Markowitz model of GPRD DPOS (1993) which we refer to as REG8, for the January 1996 to November 2020 period.

$$
\begin{aligned}
TR_{t+1} = a_0 &+ a_1 EP_t + a_2 BP_t + a_3 CP_t + a_4 SP_t + a_5 REP_t + a_6 RBP_t \\
&+ a_7 RCP_t + a_8 RSP_t + e_{t+1}
\end{aligned}
\tag{14.14}
$$

where *EP, BP, CP, SP, REP, RBP, RCP, RSP* are above defined variables at time t and e is randomly distributed error term.

14.6.2   REG9 Model

In Guerard and Schwartz (2007), the authors added CTEF as a ninth factor in the regression, which we do in this chapter. In testing the REG9 model, we use a robust (MM-Tukey, EFF $= 0.99$) analysis robust regression on Eq. (14.15) to identify variables statistically significant at the 10% level. The REG9 model is written as:

$$
\begin{aligned}
TR_{t+1} = a_0 &+ a_1 EP_t + a_2 BP_t + a_3 CP_t + a_4 SP_t + a_5 REP_t + a_6 RBP_t \\
&+ a_7 RCP_t + a_8 RSP_t + a_9 CTEF_t + e_{t+1}
\end{aligned}
\tag{14.15}
$$

14.6.3   REG10 Model

The third regression model is a ten-factor model, REG10, publishes as USER with U.S. stocks and GLER with Global stocks. The GLER model is estimated monthly using a robust (MM-Tukey, EFF $= 0.99$) analysis regression on Eq. (14.16) to identify variables statistically significant at the 10% level. REG10 is written as:

$$
\begin{aligned}
TR_{t+1} = a_0 &+ a_1 EP_t + a_2 BP_t + a_3 CP_t + a_4 SP_t + a_5 REP_t + a_6 RBP_t \\
&+ a_7 RCP_t + a_8 RSP_t + a_9 CTEF_t + a_{10} PM71_t + e_{t+1}
\end{aligned}
\tag{14.16}
$$

where $\alpha = (a_1, \ldots, a_{10})$ is the rolling average of normalized coefficients of $\alpha_{t-1}, \ldots, \alpha_{t-12}$.

$$
PM71 = \text{Price Momentum},\ \ P(t-1)/T(t-7).
$$

The normalization of α_t takes two steps. First, we set nonpositive and nonstatistical significant components to be zero. Second, the remaining positive significant components are rescaled so that they add up to 1. The 12-month smoothing is consistent with the four quarter smoothing in Bloch et al. (1993).

While EP and BP variables are significant in explaining returns, the majority of the forecast performance is attributable to other model variables, namely, the relative earnings-to-price, relative cash-to-price, relative sales-to-price, price momentum, and earnings forecast variables. The weighting results are extremely consistent with McKinley Capital Management being a Global Growth specialist. The CTEF and PM variables accounted 40% of the weights in the GLER Model. We refer to using WLRR on the first eight variables, the Markowitz Model, as REG8. The model produced out-of-sample statistically significant excess returns in the portfolios. We refer to using WLRR on the ten variables, referred to as GLER in our published articles, as REG10. The REG9 and REG10 models produced out-of-sample statistically significant excess returns in the portfolios, and the excess returns most often exceeded the excess returns of REG8. The GLER model is very similar to the work by Bob Haugen. In a recent update of Guerard and Mark (2003) study of CTEF and the REG9 model, Guerard and Mark (2020) report that the CTEF variable continues to be a positive and statistically significant variable for stock selection in the US CTEF, REG9 and REG10, USER, have statistically significant ICs, Level I test, and Active Returns, Level II tests, but Active Returns and Specific Returns (stock selection) falls as one expands the models relative to CTEF in the U.S. Factor Contributions increase because of increased Earnings Yield, Value, and Industry exposures and returns rise, as we will discuss in Chap. 15.

We report the Guerard et al. (2021) FactSet Alpha Tester results for the 1996–2020 period in Table 14.8. The results of Table 14.8 support the estimation of the composite security valuation model reported in Guerard et al. (1997) and Guerard and Mark (2003, 2020). We report an equally weighted Alpha Tester analysis of US stocks and non-US in Table 14.8, respectively. The CTEF variable and REG8, REG9, and REG10 robust regression-based models produced the highest and statistically significant ICs for the 12/1995–12/2020 period for 1-, 3-, 6-, and

Table 14.8 Alpha testing

Security	Security	Mean	STD
IBM	S1	0.17%	5.03%
D	S2	1.00%	3.82%
BA	S3	10.64%	5.74%

Correlation Matrix	S1	S2	S3
S1	100%	0%	18%
S2	0%	100%	6%
S3	18%	6%	100%

12-month holding periods in the US, non-US stock, and global stock universes. Alpha Tester analysis is a similar (alternative) test to the information coefficients, ICs, reported in Guerard et al. (2015) in Level I testing.

That is, CTEF and WLRR lead a set of 20+ models in having statistically significant IC in the US Russell 3000 and non-US stock universes. The model incorporates reported earnings, book value, cash flow, and sales; the corresponding relative variables; and an equally weighted composite model of earnings forecasts, revisions, and breadth. One notes that the statistical significance of the CTEF variable is quite similar to the one-year-ahead EPS breadth.

14.7 Summary and Conclusions

We report that a stock selection model and an earnings forecasting model reported in Guerard and Schwartz (2007), built on data from 1975 to 2002, continues to be highly statistically significant from December 1996 to November 2020 period. We show how forecasted earnings acceleration produces highly statistically significant stock selection in global and US stock universes. CTEF, REG8, REG9, and REG10 models optimized portfolios produce higher active and specific returns in non-US stocks, whereas only CTEF works in US CTEF and PM complement the original eight-factor Markowitz model in non-US stocks. Have markets and stock selection models changed since Guerard and Mark (2003)? CTEF, REG8, REG9, and, REG10 still dominate most other models, including the 36 models tested in Guerard et al. (2018), including the post-global financial crisis. The most statistically significant variable in identifying security returns is the composite earnings forecast variable. We will use the nine-factor and ten-factor composite models in Chap. 21 to rank international securities.

Appendices

Appendix A: The Three-Asset Case

Let us now examine a three-asset portfolio construction process using IBM, Dominion Resources, and BA securities for the 2012–2016 time period.

$$E(Rp) = \sum_{i=1}^{N} x_i \, E(R_i) \qquad (14.17)$$

$$\sigma_p^2 \sum_{i=1}^{N} \sum_{N}^{j=1} x_i x_j \sigma_{ij} \tag{14.18}$$

$$E(Rp) = x_1 E(R_1) + x_2 E(R_2) + x_3 E(R_3)$$

let $x_3 = 1 - x_1 - x_2$

$$E(Rp) = x_1 E(R_1) + x_2 E(R_2) + (1 - x_1 - x_2) E(R_3)$$

$\sigma_p^2 = x_1^2 \sigma_1^2 + x_2^2 \sigma_2^2 + \sigma_3^2 x_3^2 + 2x_1 x_2 \sigma_{12} + 2x_1 x_3 \sigma_{23} + 2x_2 x_3 \sigma_{23}$

$\quad = x_1^2 \sigma_1^2 + x_2^2 \sigma_2^2 + (1 - x_1 - x_2)^2 \sigma_3^2 + 2x_1 x_2 \sigma_{12} + 2x_1 (1 - x_1 - x_2) \sigma_{13}$

$\quad + 2x_2 (1 - x_1 - x_2) \sigma_{23}$

$\quad = x_1^2 \sigma_1^2 + x_2^2 \sigma_2^2 + (1 - x_1 - x_2)(1 - x_1 - x_2) \sigma_3^2 + 2x_1 x_2 \sigma_{12} + 2x_1 \sigma_{13}$

$\quad - 2x_1^2 \sigma_{13} - 2x_1 x_2 \sigma_{13} + 2x_2 \sigma_{23} - 2x_1 x_2 \sigma_{23} - 2x_2^2 \sigma_{23}$

$\quad = x_1^2 \sigma_1^2 + x_2^2 \sigma_2^2 + \left(1 - 2x_1 - 2x_2 + 2x_1 x_2 + x_1^2 + x_2^2\right) \sigma_3^2 + 2x_1 x_2 \sigma_{12}$

$\quad + 2x_1 \sigma_{13} - 2x_1^2 \sigma_{13} - 2x_1 x_2 \sigma_{13} + 2x_2 \sigma_{23} - 2x_1 x_2 \sigma_{23} - 2x_2^2 \sigma_{23}$

$$\tag{14.19}$$

$$\frac{\partial \sigma_p^2}{\partial x_1} = 2x_1 \left(\sigma_1^2 + \sigma_3^2 - 2\sigma_{13}\right) + x_2 \left(2\sigma_3^2 + 2\sigma_{12} - 2\sigma_{13} - 2\sigma_{23}\right) - 2\sigma_3^2 + 2\sigma_{13}$$

$$= 0$$

$$\frac{\partial \sigma_p^2}{\partial x_2} = 2x_2 \left(\sigma_2^2 + \sigma_3^2 - 2\sigma_{23}\right) + x_1 \left(2\sigma_3^2 + 2\sigma_{12} - 2\sigma_{13} - 2\sigma_{23}\right) - 2\sigma_3^2 + 2\sigma_{23}$$

$$= 0$$

Let's assume

Asset	Security
1	IBM
2	D
3	BA

The optimal portfolio weights involve selling IBM short and going long on DuPont and Dominion. How does the portfolio using optimal weights compare to an equally weighted portfolio of the three assets? (Table 14.9)

The equally weighted portfolio has an expected return of 14.9% and a 13.42% standard deviation.

Table 14.9 Appendix. CRP Stock data

Security	Security	Mean	STD
IBM	S1	0.17%	5.03%
D	S2	1.00%	3.82%
BA	S3	10.64%	5.74%

Correlation Matrix	S1	S2	S3
S1	100%	0%	18%
S2	0%	100%	6%
S3	18%	6%	100%

$E(R_p) = .333(.110 + .095 + .242) = .149$

$\sigma_p^2 = (.333)^2(.0167)^2 + (.333)^2(.0407) + (.333)^2(.0799) + 2(.333)(.333)(-.0103) + 2$
$(.333)(.333)(-.0041) + 2(.333)(.333)(.0303)$

$$\sigma_p^2 = .0019 + .0045 + .0081 + (-.0023) + (-.0009) + .0067 = .0180$$

$$\sigma_p = .1342$$

The optimally weighted portfolio has an expected return of

$$E(R_p) = .795(.110) + .2517(.095) + (-.0467)(.242)$$
$$= .0875 + .0244 + (-.0113)$$
$$= .1006$$

$\sigma_p^2 = (.795)^2(.0167) + (.2517)^2(.0407) + (-.0467)^2(.0799) + 2\,(.795)(.2517)(-.0103) +$
$2(.795)(-.0467)(-.0041) + 2(.2517)(-.0467)(.0303)$
$= .0106 + .0026 + .0002 + (-.0041) + .0003 + (-.0007)$
$\sigma_p^2 = .0089$
$\sigma_p = .0943$

The optimally weighted portfolio has an expected return of 10.06% and a standard deviation of 9.43%. Portfolio risk can be significantly reduced through diversification.

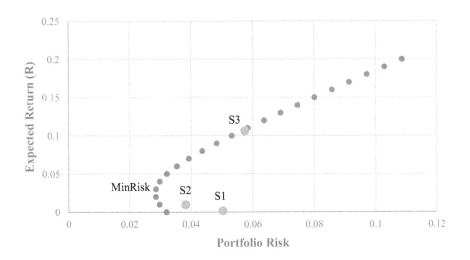

Appendix B: ICs

Time Period: 1/31/01 to 5/31/2021
Various U.S., non-US, and global universes

Universe	Factor	Mean	Std	T-stat
ACW	CTEF	0.040	0.080	7.67
ACW	FCF_YLD	0.024	0.058	6.44
ACW	BR2	0.029	0.080	5.63
ACW	BR1	0.026	0.074	5.28

ACW	FY2RV3	0.033	0.100	5.00
ACW	FY1RV3	0.025	0.091	4.28
ACW	ROIC	0.022	0.083	4.13
ACW	ROE	0.022	0.080	4.11
ACW	AX_Profitability	0.019	0.071	3.94
ACW	ROA	0.020	0.090	3.32
ACW	IBES_EPS_5Y_GTOS	0.015	0.069	3.25
ACW	CP	0.018	0.087	3.14
ACW	DP	0.019	0.094	3.07
ACW	FEP1	0.023	0.114	3.01
ACW	IBES_EXP_DY	0.019	0.099	3.00
ACW	OCFYLD	0.016	0.081	2.85
ACW	GROSS_MARGIN	0.015	0.078	2.84
ACW	AX_MidTermMomentum	0.029	0.155	2.84
ACW	EP	0.017	0.093	2.82
ACW	RCP	0.013	0.071	2.73
ACW	FEP2	0.024	0.133	2.69
ACW	IBES_FY1_EPS_G	0.018	0.103	2.68
ACW	IBES_EPS_5Y_GRO	0.013	0.076	2.66
ACW	IBES_FY1_DPS_G	0.015	0.089	2.63
ACW	EBITDA_EV	0.016	0.095	2.59
ACW	Sales_Assets	0.010	0.067	2.31
ACW	PM121	0.023	0.164	2.10
ACW	YOY_EPS_G	0.008	0.069	1.78
ACW	PM91	0.017	0.158	1.66
ACW	REP	0.007	0.066	1.59
ACW	SALES_EV	0.008	0.080	1.50
ACW	AX_Growth	0.007	0.075	1.48
ACW	ES	0.008	0.080	1.45
ACW	STDEV	0.015	0.186	1.25
ACW	ALTMANZ	0.007	0.088	1.22
ACW	PM61	0.011	0.145	1.15
ACW	PM31	0.009	0.119	1.11
ACW	CURRENT-RATIO	0.003	0.064	0.78
ACW	AX_Value	0.005	0.114	0.74
ACW	SP	0.005	0.105	0.72
ACW	IBES_PTG_RET	0.005	0.136	0.61
ACW	AX_Size	0.001	0.089	0.12
ACW	AX_Liquidity	0.000	0.118	−0.05
ACW	YOY_SALES_G	−0.001	0.086	−0.19
ACW	IBES_EPS_LTG	−0.002	0.079	−0.35
ACW	AX_DividendYield	−0.006	0.103	−0.36
ACW	AX_EarningsYield	−0.009	0.137	−0.39
ACW	BP	−0.004	0.118	−0.48

ACW	RSP	−0.008	0.118	−1.01
ACW	Debt_Equity	−0.005	0.065	−1.16
ACW	AX_ShortTermMomentum	−0.010	0.110	−1.33
ACW	RBP	−0.013	0.120	−1.66
ACW	AX_ExRateSensitivity	−0.006	0.051	−1.79
ACW	AX_Volatility	−0.022	0.173	−1.91
ACW	IBES_FY1_EPS_DISP	−0.018	0.136	−2.04
ACW	Percent_Accrual	−0.007	0.051	−2.16
ACW	CHG_DEBT	−0.006	0.044	−2.21
ACW	AX_Leverage	−0.008	0.048	−2.42
ACW	CAPEX_DEP	−0.014	0.082	−2.57
ACW	CHG_SHARES	−0.015	0.073	−3.25
ACW	IBES_REC_MEAN	−0.016	0.071	−3.41
ACW	IBES_REC_MEAN_3M	−0.015	0.043	−5.13
ACWXUS	CTEF	0.043	0.080	8.14
ACWXUS	FCF_YLD	0.024	0.052	6.92
ACWXUS	BR2	0.032	0.078	6.33
ACWXUS	BR1	0.028	0.073	5.94
ACWXUS	FY2RV3	0.033	0.100	5.09
ACWXUS	IBES_EXP_DY	0.030	0.095	4.78
ACWXUS	DP	0.027	0.088	4.59
ACWXUS	FY1RV3	0.027	0.091	4.48
ACWXUS	ROIC	0.024	0.089	4.17
ACWXUS	ROE	0.023	0.087	4.05
ACWXUS	AX_Profitability	0.021	0.077	3.98
ACWXUS	CP	0.020	0.081	3.72
ACWXUS	EP	0.020	0.093	3.36
ACWXUS	FEP1	0.025	0.117	3.33
ACWXUS	EBITDA_EV	0.019	0.089	3.30
ACWXUS	ROA	0.020	0.094	3.26
ACWXUS	AX_MidTermMomentum	0.031	0.151	3.10
ACWXUS	RCP	0.014	0.067	3.08
ACWXUS	FEP2	0.027	0.137	2.97
ACWXUS	IBES_EPS_5Y_GTOS	0.013	0.069	2.95
ACWXUS	GROSS_MARGIN	0.013	0.076	2.65
ACWXUS	IBES_FY1_EPS_G	0.017	0.102	2.60
ACWXUS	IBES_EPS_5Y_GRO	0.012	0.077	2.47
ACWXUS	PM121	0.024	0.159	2.31
ACWXUS	IBES_FY1_DPS_G	0.014	0.090	2.30
ACWXUS	YOY_EPS_G	0.010	0.072	2.14
ACWXUS	REP	0.009	0.069	1.99
ACWXUS	AX_Growth	0.010	0.077	1.90
ACWXUS	Sales_Assets	0.008	0.069	1.82
ACWXUS	PM91	0.018	0.155	1.74

ACWXUS	SALES_EV	0.008	0.078	1.58
ACWXUS	STDEV	0.017	0.180	1.42
ACWXUS	ES	0.008	0.083	1.40
ACWXUS	ALTMANZ	0.007	0.088	1.26
ACWXUS	PM61	0.011	0.146	1.14
ACWXUS	PM31	0.008	0.120	0.96
ACWXUS	AX_Value	0.007	0.118	0.94
ACWXUS	SP	0.005	0.103	0.71
ACWXUS	IBES_PTG_RET	0.005	0.130	0.63
ACWXUS	CURRENT_RATIO	0.002	0.060	0.47
ACWXUS	AX_Size	0.002	0.081	0.31
ACWXUS	YOY_SALES_G	0.000	0.089	0.07
ACWXUS	BP	0.000	0.115	−0.05
ACWXUS	IBES_EPS_LTG	−0.001	0.074	−0.12
ACWXUS	AX_DividendYield	−0.005	0.093	−0.28
ACWXUS	AX_EarningsYield	−0.016	0.134	−0.68
ACWXUS	AX_ExRateSensitivity	−0.003	0.055	−0.74
ACWXUS	AX_ShortTermMomentum	−0.008	0.112	−1.06
ACWXUS	RSP	−0.009	0.117	−1.13
ACWXUS	AX_Liquidity	−0.010	0.107	−1.36
ACWXUS	Debt_Equity	−0.007	0.073	−1.39
ACWXUS	RBP	−0.012	0.123	−1.44
ACWXUS	AX_Volatility	−0.017	0.162	−1.61
ACWXUS	CAPEX_DEP	−0.010	0.075	−2.00
ACWXUS	Percent_Accrual	−0.007	0.056	−2.03
ACWXUS	IBES_FY1_EPS_DISP	−0.017	0.113	−2.26
ACWXUS	CHG_SHARES	−0.010	0.070	−2.28
ACWXUS	AX_Leverage	−0.009	0.050	−2.64
ACWXUS	CHG_DEBT	−0.010	0.052	−2.89
ACWXUS	IBES_REC_MEAN	−0.016	0.073	−3.40
ACWXUS	IBES_REC_MEAN_3M	−0.018	0.045	−6.09
EAFE	CTEF	0.035	0.110	4.82
EAFE	FCF_YLD	0.017	0.061	4.17
EAFE	BR2	0.025	0.102	3.75
EAFE	FY2RV3	0.031	0.142	3.36
EAFE	RCP	0.016	0.080	3.04
EAFE	AX_Profitability	0.021	0.106	2.90
EAFE	BR1	0.018	0.097	2.82
EAFE	IBES_EXP_DY	0.027	0.147	2.79
EAFE	DP	0.024	0.133	2.73
EAFE	ROIC	0.021	0.116	2.73
EAFE	FY1RV3	0.021	0.126	2.60
EAFE	ROE	0.019	0.121	2.44
EAFE	ROA	0.019	0.124	2.34

EAFE	CP	0.014	0.097	2.22
EAFE	FEP1	0.021	0.150	2.16
EAFE	IBES_EPS_5Y_GTOS	0.013	0.090	2.12
EAFE	EP	0.016	0.118	2.01
EAFE	FEP2	0.022	0.176	1.92
EAFE	GROSS_MARGIN	0.011	0.097	1.73
EAFE	IBES_EPS_5Y_GRO	0.010	0.090	1.65
EAFE	AX_MidTermMomentum	0.020	0.181	1.65
EAFE	EBITDA_EV	0.011	0.111	1.56
EAFE	AX_Growth	0.009	0.095	1.42
EAFE	ES	0.010	0.108	1.41
EAFE	IBES_FY1_DPS_G	0.009	0.114	1.14
EAFE	PM121	0.013	0.189	1.07
EAFE	AX_Value	0.010	0.140	1.06
EAFE	PM31	0.011	0.157	1.03
EAFE	IBES_FY1_EPS_G	0.008	0.122	1.00
EAFE	PM91	0.012	0.181	0.99
EAFE	Sales_Assets	0.006	0.092	0.96
EAFE	STDEV	0.012	0.204	0.93
EAFE	REP	0.005	0.083	0.89
EAFE	YOY_EPS_G	0.004	0.087	0.77
EAFE	SALES_EV	0.005	0.106	0.77
EAFE	PM61	0.007	0.175	0.63
EAFE	IBES_PTG_RET	0.005	0.151	0.48
EAFE	SP	0.002	0.130	0.23
EAFE	CURRENT_R	0.001	0.083	0.20
EAFE	BP	0.002	0.142	0.17
EAFE	ALTMANZ	0.001	0.099	0.12
EAFE	RSP	0.000	0.130	0.03
EAFE	Percent_Accrual	0.000	0.070	−0.10
EAFE	RBP	−0.002	0.134	−0.23
EAFE	AX_EarningsYield	−0.008	0.168	−0.28
EAFE	YOY_SALES_G	−0.002	0.095	−0.36
EAFE	AX_DividendYield	−0.012	0.140	−0.47
EAFE	AX_ExRateSensitivity	−0.002	0.072	−0.47
EAFE	IBES_FY1_EPS_DISP	−0.005	0.123	−0.64
EAFE	CHG_SHARES	−0.005	0.092	−0.81
EAFE	AX_Size	−0.005	0.101	−0.83
EAFE	IBES_EPS_LTG	−0.005	0.081	−0.92
EAFE	Debt_Equity	−0.007	0.094	−1.14
EAFE	AX_Liquidity	−0.010	0.126	−1.19
EAFE	AX_ShortTermMomentum	−0.012	0.136	−1.22
EAFE	CHG_DEBT	−0.004	0.054	−1.23
EAFE	AX_Volatility	−0.017	0.192	−1.36

EAFE	AX_Leverage	−0.006	0.064	−1.41
EAFE	CAPEX_DEP	−0.010	0.070	−2.09
EAFE	IBES_REC_MEAN	−0.011	0.078	−2.19
EAFE	IBES_REC_MEAN_3M	−0.011	0.058	−3.00
Europe	CTEF	0.037	0.123	4.60
Europe	BR2	0.035	0.120	4.38
Europe	BR1	0.029	0.112	3.94
Europe	FY2RV3	0.034	0.151	3.43
Europe	FY1RV3	0.028	0.133	3.27
Europe	ROIC	0.028	0.151	2.85
Europe	AX_Profitability	0.027	0.148	2.61
Europe	ROE	0.022	0.143	2.36
Europe	ROA	0.026	0.167	2.36
Europe	AX_MidTermMomentum	0.033	0.215	2.34
Europe	PM121	0.033	0.215	2.33
Europe	ES	0.024	0.149	2.30
Europe	FCF_YLD	0.011	0.073	2.28
Europe	IBES_EPS_5Y_GRO	0.018	0.121	2.23
Europe	IBES_FY1_DPS_G	0.021	0.145	2.21
Europe	PM91	0.027	0.207	1.99
Europe	IBES_EPS_5Y_GTOS	0.016	0.127	1.87
Europe	Sales_Assets	0.016	0.134	1.83
Europe	AX_Growth	0.014	0.119	1.83
Europe	SALES_EV	0.013	0.111	1.81
Europe	IBES_FY1_EPS_G	0.016	0.156	1.57
Europe	YOY_EPS_G	0.010	0.102	1.57
Europe	Percent_Accrual	0.006	0.063	1.41
Europe	EP	0.011	0.123	1.36
Europe	CURRENT_R	0.009	0.106	1.36
Europe	GROSS_MARGIN	0.010	0.115	1.29
Europe	PM31	0.013	0.166	1.22
Europe	PM61	0.014	0.185	1.18
Europe	STDEV	0.017	0.234	1.13
Europe	ALTMANZ	0.008	0.129	0.97
Europe	EBITDA_EV	0.008	0.135	0.94
Europe	FEP1	0.010	0.167	0.94
Europe	REP	0.006	0.097	0.87
Europe	CP	0.005	0.124	0.64
Europe	IBES_EXP_DY	0.004	0.136	0.50
Europe	DP	0.004	0.132	0.46
Europe	FEP2	0.005	0.194	0.42
Europe	RCP	0.001	0.096	0.17
Europe	IBES_PTG_RET	0.002	0.206	0.14
Europe	AX_Value	0.001	0.188	0.05

Europe	SP	−0.001	0.161	−0.07
Europe	YOY_SALES_G	−0.002	0.104	−0.24
Europe	AX_EarningsYield	−0.017	0.208	−0.46
Europe	BP	−0.007	0.194	−0.58
Europe	IBES_EPS_LTG	−0.004	0.098	−0.69
Europe	AX_ExRateSensitivity	−0.006	0.117	−0.81
Europe	AX_ShortTermMomentum	−0.008	0.138	−0.82
Europe	AX_Size	−0.008	0.104	−1.22
Europe	AX_Liquidity	−0.011	0.130	−1.28
Europe	AX_Volatility	−0.019	0.216	−1.36
Europe	AX_DividendYield	−0.044	0.170	−1.47
Europe	IBES_FY1_EPS_DISP	−0.017	0.171	−1.54
Europe	RBP	−0.020	0.156	−1.94
Europe	RSP	−0.021	0.159	−2.02
Europe	IBES_REC_MEAN_3M	−0.010	0.071	−2.14
Europe	IBES_REC_MEAN	−0.013	0.091	−2.18
Europe	AX_Leverage	−0.011	0.078	−2.19
Europe	CHG_DEBT	−0.010	0.066	−2.32
Europe	CAPEX_DEP	−0.013	0.078	−2.60
Europe	Debt_Equity	−0.019	0.113	−2.62
Europe	CHG_SHARES	−0.014	0.066	−3.18
EM	CTEF	0.050	0.081	9.36
EM	BR1	0.039	0.075	7.90
EM	MQ	0.058	0.121	7.29
EM	FCF_YLD	0.030	0.064	7.14
EM	BR2	0.037	0.080	7.10
EM	FY2RV3	0.036	0.090	6.16
EM	FY1RV3	0.033	0.085	5.85
EM	ROE	0.030	0.092	4.95
EM	ROIC	0.029	0.093	4.71
EM	IBES_EXP_DY	0.030	0.101	4.59
EM	DP	0.029	0.097	4.57
EM	EP	0.027	0.093	4.44
EM	CP	0.025	0.085	4.38
EM	RCP	0.018	0.068	4.05
EM	FEP1	0.030	0.114	4.03
EM	IBES_FY1_EPS_G	0.026	0.101	3.98
EM	FEP2	0.033	0.128	3.97
EM	AX_MidTermMomentum	0.035	0.143	3.73
EM	AX_Profitability	0.022	0.085	3.68
EM	YOY_EPS_G	0.018	0.075	3.60
EM	REP	0.018	0.079	3.52
EM	IBES_EPS_5Y_GTOS	0.017	0.076	3.43
EM	ROA	0.020	0.093	3.37

EM	IBES_FY1_DPS_G	0.018	0.084	3.20
EM	EBITDA_EV	0.021	0.100	3.14
EM	PM121	0.030	0.154	2.96
EM	IBES_EPS_5Y_GRO	0.014	0.076	2.85
EM	GROSS_MARGIN	0.015	0.095	2.46
EM	ALTMANZ	0.017	0.113	2.33
EM	PM91	0.022	0.152	2.20
EM	AX_Growth	0.011	0.089	1.90
EM	Sales_Assets	0.009	0.071	1.85
EM	AX_Size	0.010	0.086	1.84
EM	STDEV	0.017	0.170	1.53
EM	PM61	0.013	0.139	1.48
EM	SALES_EV	0.007	0.085	1.32
EM	IBES_PTG_RET	0.011	0.130	1.25
EM	ES	0.007	0.084	1.20
EM	SP	0.009	0.112	1.19
EM	IBES_EPS_LTG	0.005	0.085	0.87
EM	YOY_SALES_G	0.004	0.086	0.78
EM	AX_Value	0.005	0.124	0.61
EM	CURRENT_RATIO	0.002	0.070	0.51
EM	AX_DividendYield	0.000	0.108	−0.01
EM	AX_ExRateSensitivity	−0.001	0.068	−0.13
EM	PM31	−0.002	0.120	−0.24
EM	BP	−0.003	0.128	−0.32
EM	RSP	−0.003	0.114	−0.38
EM	RBP	−0.007	0.119	−0.88
EM	AX_EarningsYield	−0.021	0.140	−0.88
EM	Debt_Equity	−0.005	0.077	−1.00
EM	AX_Liquidity	−0.010	0.145	−1.02
EM	AX_ShortTermMomentum	−0.010	0.110	−1.33
EM	AX_Volatility	−0.015	0.154	−1.50
EM	CAPEX_DEP	−0.007	0.072	−1.55
EM	AX_Leverage	−0.012	0.063	−2.84
EM	CHG_DEBT	−0.012	0.057	−3.15
EM	CHG_SHARES	−0.014	0.062	−3.37
EM	IBES_FY1_EPS_DISP	−0.026	0.111	−3.63
EM	Percent_Accrual	−0.013	0.056	−3.64
EM	IBES_REC_MEAN	−0.028	0.080	−5.30
EM	IBES_REC_MEAN_3M	−0.027	0.055	−7.58
Japan	RCP	0.028	0.106	4.05
Japan	FCF_YLD	0.021	0.088	3.71
Japan	IBES_EXP_DY	0.034	0.143	3.63
Japan	MQ	0.035	0.155	3.47
Japan	DP	0.030	0.146	3.13

Japan	CP	0.021	0.123	2.62
Japan	SALES_EV	0.022	0.129	2.56
Japan	BP	0.029	0.179	2.45
Japan	IBES_PTG_RET	0.027	0.179	2.27
Japan	SP	0.023	0.159	2.17
Japan	AX_Value	0.025	0.188	2.02
Japan	CTEF	0.019	0.142	2.01
Japan	BR2	0.017	0.131	1.95
Japan	RBP	0.022	0.177	1.89
Japan	RSP	0.021	0.172	1.83
Japan	FEP2	0.023	0.196	1.83
Japan	EBITDA_EV	0.018	0.153	1.77
Japan	AX_Profitability	0.015	0.136	1.63
Japan	FEP1	0.017	0.177	1.45
Japan	FY2RV3	0.015	0.161	1.38
Japan	CURRENT_R	0.010	0.120	1.30
Japan	Sales_Assets	0.010	0.120	1.29
Japan	ES	0.011	0.140	1.13
Japan	BR1	0.009	0.129	1.03
Japan	EP	0.008	0.164	0.78
Japan	ROA	0.008	0.162	0.75
Japan	AX_ExRateSensitivity	0.005	0.121	0.67
Japan	IBES_EPS_5Y_GRO	0.003	0.108	0.46
Japan	IBES_EPS_5Y_GTOS	0.003	0.108	0.40
Japan	IBES_FY1_EPS_DISP	0.004	0.148	0.37
Japan	ALTMANZ	0.003	0.148	0.35
Japan	FY1RV3	0.003	0.155	0.33
Japan	GROSS_MARGIN	0.003	0.144	0.33
Japan	STDEV	0.005	0.235	0.33
Japan	IBES_FY1_EPS_G	0.003	0.153	0.30
Japan	ROIC	0.003	0.151	0.28
Japan	REP	0.002	0.135	0.28
Japan	IBES_EPS_LTG	0.002	0.113	0.20
Japan	AX_MidTermMomentum	0.001	0.212	0.08
Japan	AX_DividendYield	0.002	0.207	0.05
Japan	AX_Growth	−0.001	0.130	−0.08
Japan	PM31	−0.003	0.167	−0.25
Japan	AX_EarningsYield	−0.011	0.196	−0.30
Japan	PM121	−0.005	0.211	−0.35
Japan	ROE	−0.003	0.137	−0.36
Japan	CHG_DEBT	−0.002	0.075	−0.36
Japan	AX_Volatility	−0.007	0.231	−0.44
Japan	AX_Liquidity	−0.006	0.200	−0.44
Japan	PM91	−0.008	0.200	−0.60

Japan	PM61	−0.008	0.188	−0.62
Japan	IBES_FY1_DPS_G	−0.006	0.138	−0.67
Japan	IBES_REC_MEAN	−0.005	0.115	−0.69
Japan	IBES_REC_MEAN_3M	−0.005	0.083	−0.83
Japan	CHG_SHARES	−0.005	0.084	−0.98
Japan	YOY_EPS_G	−0.008	0.118	−0.99
Japan	Debt_Equity	−0.013	0.143	−1.34
Japan	AX_Leverage	−0.013	0.125	−1.55
Japan	AX_Size	−0.011	0.112	−1.55
Japan	YOY_SALES_G	−0.015	0.129	−1.78
Japan	Percent_Accrual	−0.011	0.095	−1.79
Japan	CAPEX_DEP	−0.010	0.077	−2.02
Japan	AX_ShortTermMomentum	−0.033	0.175	−2.63
R1000	CTEF	0.030	0.117	3.88
R1000	FY2RV3	0.025	0.125	2.99
R1000	FCF_YLD	0.018	0.096	2.83
R1000	BR2	0.018	0.105	2.67
R1000	FY1RV3	0.019	0.107	2.64
R1000	Sales_Assets	0.018	0.106	2.53
R1000	FEP1	0.021	0.136	2.33
R1000	FEP2	0.022	0.146	2.28
R1000	BR1	0.014	0.097	2.25
R1000	DJ_SENT	0.011	0.082	2.00
R1000	ROIC_QTR	0.014	0.108	1.94
R1000	SALES_EV	0.015	0.120	1.91
R1000	IBES_EPS_5Y_GRO	0.013	0:108	1.90
R1000	ROE	0.012	0.105	1.79
R1000	IBES_EPS_5Y_GTOS	0.012	0.106	1.78
R1000	ROIC	0.013	0.110	1.76
R1000	AX_MidTermMomentum	0.021	0.192	1.67
R1000	PR_SENT	0.008	0.070	1.65
R1000	IBES_FY1_EPS_G	0.013	0.124	1.65
R1000	ROA	0.012	0.114	1.60
R1000	IBES_FY1_DPS_G	0.010	0.101	1.43
R1000	CP	0.012	0.131	1.40
R1000	SP	0.012	0.140	1.31
R1000	GROSS_MARGIN	0.009	0.104	1.26
R1000	PM91	0.013	0.184	1.11
R1000	RCP	0.008	0.106	1.09
R1000	OCFYLD	0.008	0.113	1.08
R1000	EP	0.008	0.122	1.06
R1000	EBITDA_EV	0.009	0.141	1.02
R1000	PM121	0.013	0.191	1.00
R1000	AX_Growth	0.007	0.104	0.99

R1000	ES	0.007	0.101	0.97
R1000	CHG_DEBT	0.003	0.051	0.81
R1000	REP	0.004	0.090	0.74
R1000	PM61	0.008	0.168	0.74
R1000	IBES_PTG_RET	0.008	0.177	0.69
R1000	PM31	0.007	0.146	0.69
R1000	ALTMANZ	0.005	0.115	0.67
R1000	STDEV	0.009	0.225	0.58
R1000	PMUS	0.008	0.188	0.58
R1000	DP	0.004	0.150	0.38
R1000	IBES_EXP_DY	0.001	0.149	0.12
R1000	Debt_Equity	0.001	0.091	0.10
R1000	AX_Size	0.001	0.112	0.08
R1000	AX_Leverage	0.000	0.093	0.06
R1000	IBES_REC_MEAN_3M	0.000	0.056	−0.01
R1000	CURRENT_RATIO	−0.001	0.110	−0.08
R1000	IBES_EPS_LTG	−0.002	0.135	−0.19
R1000	YOY_EPS_G	−0.001	0.080	−0.20
R1000	AX_DividendYield	−0.009	0.179	−0.28
R1000	AX_EarningsYield	−0.013	0.185	−0.39
R1000	RSP	−0.005	0.156	−0.49
R1000	AX_Value	−0.005	0.138	−0.58
R1000	BP	−0.005	0.133	−0.60
R1000	YOY_SALES_G	−0.005	0.107	−0.67
R1000	AX_Volatility	−0.015	0.215	−1.03
R1000	AX_MidCap	−0.018	0.080	−1.09
R1000	RBP	−0.011	0.135	−1.20
R1000	AX_Liquidity	−0.012	0.150	−1.20
R1000	CAPEX_DEP	−0.007	0.082	−1.26
R1000	AX_ShortTermMomentum	−0.013	0.142	−1.33
R1000	Percent_Accrual	−0.007	0.072	−1.46
R1000	IBES_REC_MEAN	−0.009	0.096	−1.48
R1000	IBES_FY1_EPS_DISP	−0.017	0.134	−1.90
R1000	AX_ExRateSensitivity	−0.016	0.116	−2.06
R1000	CHG_SHARES	−0.016	0.090	−2.78
R2000	CTEF	0.042	0.086	7.51
R2000	OCFYLD	0.033	0.072	7.02
R2000	FCF_YLD	0.033	0.080	6.37
R2000	FEP2	0.045	0.117	5.84
R2000	FEP1	0.046	0.121	5.82
R2000	DJ_SENT	0.022	0.056	5.57
R2000	AX_Size	0.031	0.091	5.21
R2000	ROA	0.035	0.104	5.12
R2000	BR2	0.022	0.066	5.04

R2000	ROIC_QTR	0.034	0.098	4.99
R2000	RCP	0.028	0.087	4.96
R2000	FY2RV3	0.025	0.077	4.94
R2000	BR1	0.021	0.065	4.86
R2000	ROIC	0.034	0.109	4.82
R2000	REP	0.032	0.104	4.77
R2000	ROE	0.034	0.111	4.66
R2000	Sales_Assets	0.025	0.083	4.58
R2000	EP	0.034	0.117	4.50
R2000	AX_Growth	0.021	0.074	4.38
R2000	CP	0.031	0.112	4.20
R2000	FY1RV3	0.019	0.068	4.19
R2000	STDEV	0.046	0.169	4.12
R2000	IBES_FY1_EPS_G	0.020	0.074	4.07
R2000	SALES_EV	0.024	0.096	3.87
R2000	EBITDA_EV	0.032	0.127	3.83
R2000	PR_SENT	0.011	0.042	3.75
R2000	IBES_EPS_5Y_GRO	0.017	0.068	3.73
R2000	IBES_EPS_5Y_GTOS	0.018	0.074	3.67
R2000	AX_MidTermMomentum	0.027	0.133	3.05
R2000	SP	0.024	0.120	3.00
R2000	IBES_FY1_DPS_G	0.020	0.104	2.81
R2000	PM91	0.021	0.124	2.56
R2000	PM61	0.018	0.113	2.41
R2000	GROSS_MARGIN	0.012	0.076	2.38
R2000	DP	0.019	0.122	2.31
R2000	PM121	0.018	0.132	2.10
R2000	PM31	0.011	0.097	1.72
R2000	ALTMANZ	0.008	0.076	1.70
R2000	PMUS	0.014	0.124	1.67
R2000	AX_MidCap	0.023	0.068	1.65
R2000	YOY_EPS_G	0.005	0.048	1.56
R2000	AX_EarningsYield	0.037	0.134	1.47
R2000	Debt_Equity	0.008	0.090	1.41
R2000	BP	0.009	0.105	1.31
R2000	ES	0.005	0.058	1.24
R2000	AX_Leverage	0.006	0.080	1.21
R2000	IBES_EXP_DY	0.008	0.140	0.86
R2000	AX_DividendYield	0.021	0.136	0.84
R2000	AX_Value	0.006	0.106	0.84
R2000	RBP	0.002	0.086	0.38
R2000	CHG_DEBT	0.001	0.040	0.23
R2000	YOY_SALES_G	0.000	0.069	0.04
R2000	RSP	−0.003	0.102	−0.52

R2000	IBES_PTG_RET	−0.006	0.133	−0.74
R2000	AX_ExRateSensitivity	−0.004	0.068	−0.89
R2000	IBES_EPS_LTG	−0.007	0.113	−0.95
R2000	AX_Liquidity	−0.010	0.112	−1.39
R2000	CAPEX_DEP	−0.006	0.052	−1.63
R2000	AX_ShortTermMomentum	−0.012	0.094	−1.77
R2000	IBES_REC_MEAN	−0.010	0.070	−2.26
R2000	CURRENT_RATIO	−0.014	0.088	−2.48
R2000	AX_Volatility	−0.038	0.169	−3.39
R2000	Percent_Accrual	−0.017	0.072	−3.61
R2000	LowPrice	−0.035	0.126	−4.26
R2000	CHG_SHARES	−0.023	0.079	−4.42
R2000	IBES_REC_MEAN_3M	−0.011	0.038	−4.58
R2000	IBES_FY1_EPS_DISP	−0.032	0.096	−5.14
WORLD	CTEF	0.031	0.094	4.76
WORLD	FCF_YLD	0.020	0.072	4.00
WORLD	BR2	0.025	0.094	3.77
WORLD	BR1	0.021	0.087	3.43
WORLD	ROIC	0.022	0.097	3.35
WORLD	FY2RV3	0.029	0.125	3.33
WORLD	IBES_EPS_5Y_GTOS	0.021	0.091	3.29
WORLD	ROE	0.021	0.095	3.25
WORLD	AX_Profitability	0.018	0.084	3.06
WORLD	ROA	0.020	0.106	2.75
WORLD	IBES_EPS_5Y_GRO	0.017	0.092	2.71
WORLD	FY1RV3	0.020	0.111	2.57
WORLD	GROSS_MARGIN	0.015	0.089	2.39
WORLD	Sales_Assets	0.012	0.083	1.99
WORLD	AX_MidTermMomentum	0.023	0.170	1.96
WORLD	IBES_FY1_DPS_G	0.013	0.101	1.86
WORLD	RCP	0.010	0.086	1.75
WORLD	ES	0.012	0.096	1.74
WORLD	OCFYLD	0.011	0.090	1.73
WORLD	FEP1	0.015	0.127	1.69
WORLD	IBES_FY1_EPS_G	0.013	0.120	1.62
WORLD	AX_Volatility	0.020	0.187	1.56
WORLD	AX_Liquidity	0.012	0.122	1.37
WORLD	FEP2	0.014	0.150	1.37
WORLD	ALTMANZ	0.009	0.098	1.36
WORLD	PM31	0.013	0.134	1.35
WORLD	AX_EarningsYield	0.010	0.107	1.32
WORLD	PM91	0.014	0.168	1.22
WORLD	EP	0.008	0.103	1.17
WORLD	PM121	0.013	0.175	1.09

WORLD	CURRENT_RATIO	0.006	0.076	1.06
WORLD	AX_DividendYield	0.007	0.107	0.98
WORLD	PM61	0.010	0.154	0.96
WORLD	IBES_EXP_DY	0.008	0.121	0.92
WORLD	DP	0.007	0.120	0.91
WORLD	CP	0.006	0.101	0.83
WORLD	STDEV	0.009	0.196	0.68
WORLD	AX_Growth	0.004	0.097	0.65
WORLD	SALES_EV	0.004	0.098	0.62
WORLD	AX_Size	0.003	0.086	0.57
WORLD	REP	0.003	0.073	0.56
WORLD	YOY_EPS_G	0.003	0.076	0.55
WORLD	EBITDA_EV	0.004	0.108	0.52
WORLD	IBES_PTG_RET	0.005	0.150	0.49
WORLD	CHG_DEBT	0.001	0.045	0.31
WORLD	IBES_EPS_LTG	0.001	0.094	0.14
WORLD	Percent_Accrual	−0.001	0.056	−0.14
WORLD	SP	−0.002	0.123	−0.23
WORLD	YOY_SALES_G	−0.003	0.091	−0.46
WORLD	RSP	−0.005	0.133	−0.60
WORLD	Debt_Equity	−0.004	0.074	−0.86
WORLD	AX_Value	−0.008	0.124	−0.90
WORLD	AX_ExRateSensitivity	−0.004	0.057	−0.92
WORLD	RBP	−0.010	0.128	−1.10
WORLD	BP	−0.013	0.134	−1.43
WORLD	CAPEX_DEP	−0.008	0.077	−1.51
WORLD	IBES_FY1_EPS_DISP	−0.018	0.142	−1.86
WORLD	IBES_REC_MEAN_3M	−0.006	0.048	−1.88
WORLD	AX_Leverage	−0.008	0.061	−1.91
WORLD	IBES_REC_MEAN	−0.014	0.079	−2.50
WORLD	CHG_SHARES	−0.016	0.081	−2.77

Variable	Definition
FY1RV3	Three-month revisions of 1-year-ahead I/B/E/S earnings per share revisions
FY2RV3	Three-month revisions of 2-year-ahead I/B/E/S earnings per share revisions
BR2	Two-tear-ahead I/B/E/S forecast breadth
BR1	One-tear-ahead I/B/E/S forecast breadth
FCF_YLD	Forecasted free cash flow yield
CTEFOCFROIC	Proprietary forecasted free cash flow
IBES_EPS_5Y_GTOS	I/B/E/S 3–5 year forecasted growth rate
MQ	McKinley capital management prioritary quant score
CURRENT_R	Current ratio
AX_Profitability	Axioma profitability factor return

ROIC	Return on invested capital
IBES_EPS_5Y_GRO	Five-year-ahead I/B/E/S forecast EPS growth rate
IBES_FY1_DPS_G	One-year-ahead I/B/E/S forecast dividend growth rate
STDEV	One year annualized standard deviation
ALTMANZ	Altman Z-score
AX_Size	Axioma size factor return
ES	Corporate exports
IBES_FY1_EPS_G	One-year-ahead I/B/E/S forecast EPS growth rate
REP	Relative earnings to price ratio
AX_MidTermMomentum	Axioma medium-term momentum factor return
ROA	Return on assets ratio
ROE	Return on equity ratio
Sales_Assets	Sales-to-assets ratio
GROSS_MARGIN	Profits-sales ratio
YOY_EPS_G	Year-to-year EPS growth
PM61	Price momentum 6-month momentum
AX_Growth	Axioma growth factor return
IBES_EPS_LTG	I/B/E/S 3–5 year forecasted growth rate
PM91	Price momentum 9-month (net reversion) momentum
PM121	Price momentum 12-month (net reversion) momentum
AX_Liquidity	Axioma liquidity factor return
SALES_EV	Sales-to-enterprise value ratio
DP	Dividends-to-price ratio
IBES_EXP_DY	I/B/E/S forecasted dividend yield to price ratio
PM31	Price momentum 3-month (net reversion) momentum
AX_Volatility	Axioma volatility factor return
AX_DividendYield	Axioma dividend yield factor return
AX_ShortTermMomentum	Axioma short-term momentum factor return
RCP	Relative earnings to price ratio
IBES_REC_MEAN	I/B/E/S recommendation mean ratio
Percent_Accrual	(Net income − Operating cash flow)/Net income ratio
SP	Sales to price ratio
FEP1	I/B/E/S 1-year-ahead forecasted earnings to price ratio
EP	Earnings to price ratio
CP	Cash flow to price ratio
FEP2	I/B/E/S 2-year-ahead forecasted earnings to price ratio
IBES_PTG_RET	I/B/E/S targeted price to current price ratio
EBITDA_EV	EBITDA-to-enterprise value
BP	Book value to price ratio
AX_Value	Axioma value factor return
AX_EarningsYield	Axioma earnings yield factor return

RSP	Relative sales to price ratio
RBP	Relative book value to price ratio
YOY_SALES_G	Year-to-year sales growth
AX_ExRateSensitivity	Axioma exchange ratio factor return
Debt_Equity	Debt-to-equity ratio
CHG_SHARES	Percentage change in stock shares
CHG_DEBT	Percentage change in debt
IBES_REC_MEAN_3M	I/B/E/S 1-year-ahead forecasted earnings to price ratio
IBES_FY1_EPS_DISP	I/B/E/S 1-year-ahead forecasted standard deviation earnings to price ratio
AX_Leverage	Axioma leverage factor return
CAPEX_DEP	Capital expenditures-to-depreciation ratio

Appendix C: Matrix Algebra

In finance matrix, algebra is very useful. Many programing languages, including Excel, have built in functions to do calculations easily with matrix operations. Matrices are tables of numbers with finite number of rows and columns. A matrix is described by giving the number of rows first followed by the number of columns, its dimensions. We define scalars (single numbers) as simply 1x1 matrices, vectors as 1xN, (a row vector with N elements), or as Mx1 (a column vector with M elements).

Let **A** denote an M x N and **B** denote an N x L matrix:

$$A_{M \times N} = \begin{pmatrix} a_{11} & a_{12} & \cdots & a_{1N} \\ a_{21} & a_{22} & \cdots & a_{2N} \\ \vdots & \vdots & \ddots & \vdots \\ a_{M1} & a_{M2} & \cdots & a_{MN} \end{pmatrix}, \quad B_{N \times L} = \begin{pmatrix} b_{11} & b_{12} & \cdots & b_{1L} \\ b_{21} & b_{22} & \cdots & b_{2L} \\ \vdots & \vdots & \ddots & \vdots \\ b_{N1} & b_{N2} & \cdots & b_{NL} \end{pmatrix}$$

Let **F** denote the product of **A** and **B**, then **F** is an M x L matrix:

$$F_{M \times L} = A_{M \times N} \times B_{N \times L}$$

Note that first matrix's number of columns and the second matrix's number of rows must be the same in the multiplication operation! If necessary, transpose one of the matrices to obtain this property. In Excel, multiplication operation is done by using the function MMULT(range 1, range 2).

Transpose of matrix A_{MxN} denoted by A^T, or A', is an N x M matrix:

$$A^T_{N \times M} = \begin{pmatrix} a_{11} & a_{21} & \cdots & a_{M1} \\ a_{12} & a_{22} & \cdots & a_{M2} \\ \vdots & \vdots & \ddots & \vdots \\ a_{1N} & a_{2N} & \cdots & a_{MN} \end{pmatrix}$$

In Excel, the function is TRANSPOSE(Range).

Let X and Y denote two N x 1 vectors, then the difference $X - Y$ is given by an N x 1 vector D:

$$D_{N \times 1} = X_{N \times 1} - Y_{N \times 1} = \begin{pmatrix} x_1 \\ x_2 \\ \vdots \\ x_N \end{pmatrix} - \begin{pmatrix} y_1 \\ y_2 \\ \vdots \\ y_N \end{pmatrix} = \begin{pmatrix} x_1 - y_1 \\ x_2 - y_2 \\ \vdots \\ x_N - y_N \end{pmatrix}$$

Matrix algebra in Excel

Let numbers in cells A1:C1 be the matrix A (1x3) and numbers in E1:G3 be the matrix B (3x3). Product of A and B will be matrix with 1 row and 3 columns. Select 3 columns, say A3:C3, and enter formula =MMULT(A1:C1,E1:G3). You must use Ctrl+Shift+Enter! It will show as { =MMULT(A1:C1,E1:G3)}. The {} indicates that it is an array function.

To transpose matrix A, select 3 rows, say A5:A7, and enter the formula =TRANSPOSE(A1:C1). Then Ctrl+Shift+Enter. It will show as {=TRANSPOSE (A1:C1)}.

Here is a simple example:

$$A_{1 \times 3} \times B_{3 \times 3} = Z_{1 \times 3} = \begin{bmatrix} 2 & 3 & 4 \end{bmatrix} \begin{bmatrix} 13 & -8 & -3 \\ -8 & 10 & -1 \\ -3 & -1 & 11 \end{bmatrix} = \begin{bmatrix} -10 & 10 & 35 \end{bmatrix}$$

$$A^T_{3 \times 1} = \begin{bmatrix} 2 \\ 3 \\ 4 \end{bmatrix}$$

Portfolio Statistics with Matrix Algebra

Expected return and variance of a portfolio of N securities is calculated using the following two equations:

$$E[r_P] = \sum_{i=1}^{N} x_i E[r_i]$$

$$\sigma_P^2 = \sum_{i=1}^{N} \sum_{j=1}^{N} x_i x_j \sigma_{ij}$$

where x_i is the weight of security i in the portfolio, σ_{ij} is the covariance of security i with the security j. Sum of weights is always 1.

Let X be Nx1 matrix, column vector, representing the weights of the securities in the portfolio, R be Nx1 matrix representing expected returns of securities, and Ω be NxN covariance matrix of securities. Then, expected return and the variance of portfolio in matrix algebra will be:

$$E[R_P] = X_{1xN}^T \times R_{Nx1} = \begin{pmatrix} x_1 & x_2 & \cdots & x_N \end{pmatrix} \begin{pmatrix} r_1 \\ r_2 \\ \vdots \\ r_N \end{pmatrix}$$

$$\sigma_P^2 = X_{1xN}^T \times \Omega_{NxN} \times X_{Nx1} = \begin{pmatrix} x_1 & x_2 & \cdots & x_N \end{pmatrix} \begin{pmatrix} \sigma_{11} & \sigma_{12} & \cdots & \sigma_{1N} \\ \sigma_{21} & \sigma_{22} & \cdots & \sigma_{2N} \\ \vdots & \vdots & \ddots & \vdots \\ \sigma_{N1} & \sigma_{N2} & \cdots & \sigma_{NN} \end{pmatrix} \begin{pmatrix} x_1 \\ x_2 \\ \vdots \\ x_N \end{pmatrix}$$

and,

$$X_{1xN}^T \times 1_{Nx1} = \begin{pmatrix} x_1 & x_2 & \cdots & x_N \end{pmatrix} \begin{pmatrix} 1 \\ 1 \\ \vdots \\ 1 \end{pmatrix} = 1$$

Unit vector, **1**, is used to make sure that sum of weights is 1!

To demonstrate it in Excel, we will use the example of two security portfolio of IBM and D presented in chapter 14. An equally weighted portfolio of IBM and D has the following statistics:

$$E[r_P] = \begin{pmatrix} 0.5 & 0.5 \end{pmatrix} \begin{pmatrix} 0.0024 \\ 0.0054 \end{pmatrix} = 0.0039$$

$$\sigma_P^2 = \begin{pmatrix} 0.5 & 0.5 \end{pmatrix} \begin{pmatrix} 0.004212 & 0.000211 \\ 0.000211 & 0.001529 \end{pmatrix} \begin{pmatrix} 0.5 \\ 0.5 \end{pmatrix} = 0.001541$$

$$\sigma_P = \sqrt[2]{0.001541} = 0.039251$$

▲	A	B	C	D	E	F	G	H	I
				Correlation Matrix		Covariance Matrix			
1									
2		Mean	STD	IBM	D	IBM	D	Weight	Unit Vec
3	IBM	0.0024	0.0649	1	0.0831	0.004212	0.000211	0.5	1
4	D	0.0054	0.0391	0.0831	1	0.000211	0.001529	0.5	1
5									
6	Porfolio	0.0039	0.039251					1	

Formulas in cells **B6**, **C6**, and **H6** are:

B6: =MMULT(TRANSPOSE(H3:H4),B3:B4)
C6: =SQRT(MMULT(MMULT(TRANSPOSE(H3:H4),F3:G4),H3:H4))
H6: =MMULT(TRANSPOSE(H3:H4),I3:I4)

If we change the weights in cells H3 and H4 to 0.245 and 0.755, respectively, portfolio stats will be:

▲	A	B	C	D	E	F	G	H	I
				Correlation Matrix		Covariance Matrix			
1									
2		Mean	STD	IBM	D	IBM	D	Weight	Unit Vec
3	IBM	0.0024	0.0649	1	0.0831	0.004212	0.000211	0.245	1
4	D	0.0054	0.0391	0.0831	1	0.000211	0.001529	0.755	1
5									
6	Porfolio	0.0047	0.034674					1	

References

Asness, C., Moskowitz, T., & Pedersen, L. (2013). Value and momentum everywhere. *Journal of Finance, 68*, 929–985.

Barillas, F., & Shanken, J. (2018). Comparing asset pricing models. *Journal of Finance, 73*, 715–755.

Basu, S. (1977). Investment performance of common stocks in relation to their price earnings ratios: A test of market efficiency. *Journal of Finance, 32*, 663–682.

Baumol, W. J. (1963). An expected gain-confidence limit criteria for portfolio selection. *Management Science, 10*, 174–182.

Beheshti, B. (2015). A note on the integration of the alpha alignment factor and earnings forecasting models in producing more efficient Markowitz frontiers. *International Journal of Forecasting, 31*, 582–585.

Beaton, A. E., & Tukey, J. W. (1974). The fitting of power series, meaning polynomials, illustrated on bank-spectroscopic data. *Technometrics, 16*, 147–185.

Belsley, D. A., Kuh, E., & Welsch, R. E. (1980a). *Regression diagnostics: Identifying influential data and sources of collinearity*. Wiley. Chapter 2.

Blin, J, S. Bender, & Guerard, J.B., Jr. 1997. Earnings forecasts, revisions and momentum in the Estimation of efficient market-neutral Japanese and U.S. portfolios. In A. Chen (Ed.), *Research in finance* (Vol. 15).

Black, F., Jensen, M. C., & Scholes, M. (1972). The capital asset pricing model: some empirical results. In M. C. Jensen (Ed.), *Studies in the theory of capital markets*. Praeger Publishers.

Bloch, M., Guerard, J. B., Jr., Markowitz, H. M., Todd, P., & Xu, G.-L. (1993). A comparison of some aspects of the U.S. and Japanese equity markets. *Japan and the World Economy, 5*, 3–26.

Borkovec, M., Domowitz, I., Kiernan, B., & Serbin, V. (2010). Portfolio optimization and the cost of trading. *Journal of Investing, 19*, 63–76.

Brennan, T. J., & Lo, A. (2010). Impossible frontiers. *Management Science, 56*, 905–923.

Bruce, B., & Epstein, C. B. (1994). *The handbook of corporate earnings analysis*. Probus Publishing Company.

Brush. J. (2001). Price momentum: A twenty-year research effort. *Columbine Newsletter*.

Brush, J., & Boles, K. E. (1983). The predictive power of relative strength and CAPM. *Journal of Portfolio Management*.

Chan, L. K. C., Hamao, Y., & Lakonishok, J. (1991). Fundamentals and stock returns in Japan. *Journal of Finance, 46*, 1739–1764.

Chu, Y., Hirschleifer, D., & Ma, L. (2017). *The casual effect of limits on arbitrage on asset pricing models*. NBER Working Paper 24144. http://www.nber.org/papers/w24144

Connor, G., Goldberg, L., & Korajczyk, R. A. (2010). *Portfolio risk analysis*. Princeton University Press.

Conrad, J., & Kaul, G. (1989). Mean reversion in short-horizon expected returns. *Review of Financial Studies, 2*, 225–240.

Conrad, J., & Kaul, G. (1991). Components of short-horizon individual security returns. *Journal of Financial Economics, 29*, 365–384.

Conrad, J., & Kaul, G. (1993). Long-term market overreaction or biases in compound returns. *Journal of Finance, 59*, 39–63.

Conrad, J., & Kaul, G. (1998). An anatomy of trading strategies. *Review of Financial Studies, 11*, 489–519.

Connor, G., Goldberg, L., & Korajczyk, R. A. (2010). *Portfolio risk analysis*. Princeton University Press.

Cootner, P. (1964). *The random character of stock market prices*. MIT Press.

Dhrymes, P. J. (2017). *Introductory econometrics* (2nd ed.). Springer.

Dimson, E. (1988). *Stock market anomalies*. Cambridge University Press.

Domowitz, I., & Moghe, A. (2018). Donuts: A picture of optimization applied to fundamental portfolios. *Journal of Portfolio Management, 44*, 103–113.

Dremen, D. (1979). *Contrarian investment strategy*. Random House.

Dremen, D. (1998). *Contrarian investment strategies: The next generation*. Simon and Schuster.

Elton, E. J., & Gruber, M. J. (1972). *Security analysis and portfolio analysis*. Prentice-Hall, Inc.

Elton, E. J., & Gruber, M. J. (1980). *Modern portfolio theory and investment analysis*. Wiley.

Elton, E. J., Gruber, M. J., & Gultekin, M. (1981). Expectations and share prices. *Management Science, 27*, 975–987.

Elton, E. J., Gruber, M. J., & Gultekin, M. (1983). Professional expectations: Accuracy and diagnosis of errors. *Journal of Financial and Quantitative Analysis*.

Elton, E. J., Gruber, M. J., Brown, S. J., & Goetzman, W. N. (2007). *Modern portfolio theory and investment analysis* (7th ed.). Wiley.

Fama, E. F. (1965). Portfolio analysis in a stable Paretian market. *Management Science, 11*, 404–419.

Fama, E. F. (1976). *Foundations of finance*. Basis Books.

Fama, E. F. (1991). Efficient capital markets II. *Journal of Finance, 46*, 1575–1617.

Fama, E. F., & French, K. R. (1992). Cross-sectional variation in expected stock returns. *Journal of Finance, 47*, 427–465.

Fama, E. F., & French, K. R. (1995). Size and the book-to-market factors in earnings and returns. *Journal of Finance, 50*, 131–155.

Fama, E. F., & French, K. R. (2008). Dissecting anomalies. *Journal of Finance, 63*, 1653–1678.

Fama, E. F., & MacBeth, J. D. (1973). Risk, return, and equilibrium: Empirical tests. *Journal of Political Economy, 81*, 607–636.

Graham, B. (1973). *The intelligent investor*. Harper and Row.

Graham, B., & Dodd, D. (1934). *Security analysis: Principles and technique* (1st ed.). McGraw-Hill Book Company.

Graham, B., Dodd, D., & Cottle, S. (1962). *Security analysis: Principles and technique* (4th ed.). McGraw-Hill Book Company.

Grinold, R., & Kahn, R. (1999). *Active portfolio management*. McGraw-Hill/Irwin.

Guerard, J.B., Jr. & Stone, B.K. 1992. Composite forecasting of annual corporate earnings. In A. Chen (Ed.), *Research in finance* (Vol. 10)

Guerard, J. B., Jr., Gultekin, M., & Stone, B. K. (1997). The role of fundamental data and analysts' earnings breadth, forecasts, and revisions in the creation of efficient portfolios. In A. Chen (Ed.), *Research in finance* (Vol. 15).

Guerard, J. B. & Mark, A. (2003). The optimization of efficient portfolios: The case for an R&D quadratic term. In A. Chen (Ed.), *Research in finance* (Vol. 20).

Guerard, J. B., Jr., Gillam, R. A., Markowitz, H., Xu, G., Deng, S., & Wang, E. (2018). Data mining corrections testing in Chinese stocks. *Interfaces, 48*, 108–120.

Guerard, J. B., Jr. (2012). Global earnings forecasting efficiency. In J. Kensinger (Ed.), *Research in finance 28*. JAI Press.

Guerard, J. B., Jr. (2016). Investing in global markets: Big data and applications of robust regression. *Frontiers in Applied Mathematics and Statistics, 1*, 1–16.

Guerard, J. B., Jr., & Mark, A. (2020). Earnings forecasts and revisions, price momentum, and fundamental data: Further explorations in financial anomalies. In C. F. Lee (Ed.), *Handbook of financial econometrics*. World Scientific Publishers.

Guerard, J. B., Jr., & Stone, B. K. (1992). Composite forecasting of annual corporate earnings. In A. Chen (Ed.), *Research in finance 10*. JAI Press.

Guerard, J. B., Jr., Gultekin, M., & Stone, B. K. (1997). The role of fundamental data and analysts' earnings breadth, forecasts, and revisions in the creation of efficient portfolios. In A. Chen (Ed.), *Research in finance 15*. JAI Press.

Guerard, J. B., Jr., Xu, G., & Gultekin, M. N. (2012). Investing with momentum: The past, present, and future. *Journal of Investing, 21*, 68–80.

Guerard, J. B., Jr., Rachev, R. T., & Shao, B. (2013). Efficient global portfolios: Big data and investment universes. *IBM Journal of Research and Development, 57*(5), Paper 11.

Guerard, J. B., Jr., Markowitz, H. M., & Xu, G. (2014). The role of effective corporate decisions in the creation of efficient portfolios. *IBM Journal of Research and Development, 58*(4) Paper 11.

Guerard, J. B., Jr., Markowitz, H. M., & Xu, G. (2015). Earnings forecasting in a global stock selection model and efficient portfolio construction and management. *International Journal of Forecasting, 31*, 550–560.

Gunst, R. F., Webster, J. T., & Mason, R. L. (1976). A comparison of least squares and latent root regression estimators. *Technometrics, 18*, 75–83.

Guerard, J. B., Jr., & Markowitz, H. M. (2018). The existence and persistence of financial anomalies: What have you done for me lately? *Financial Planning Review*. https://doi.org/10.1002/cfp2.1022

Harvey, C. R. (2017). Presidential address: The scientific outlook in financial economics. *Journal of Finance, 72*, 1399–1440.

Harvey, C. R., Lin, Y., & Zhu, H. (2016). The cross-section of expected returns. *Review of Financial Studies, 291*, 5–69.

Hastie, T., Tibshirani, R., & Friedman, J. (2016). *The elements of statistical learning: Data mining, inference, and prediction* (2nd ed., 11th printing). Springer.

Haugen, R. A., & Baker, N. (1996). Communality in the determinants of expected results. *Journal of Financial Economics, 41*, 401–440.

Kuhn, M., & Johnson, K. (2013). *Applied predictive Modeling*. Springer.

Haugen, R., & Baker, N. (2010). Case closed. In J. B. Guerard (Ed.), *The handbook of portfolio construction: Contemporary applications of Markowitz techniques*. Springer.

Hawkins, E. H., Chamberlain, S. C., & Daniel, W. E. (1984). Earnings expectations and security prices. *Financial Analysts Journal, 405*, 24–38.

Hirshleifer, D. (2001). Investor psychology and asset pricing. *Journal of Finance, 64*, 1533–1597.

Hirshleifer, D., Hon, K., & Teoh, S. H. (2012). The accrual anomaly: Risk or mispricing? *Management Science, 57*, 1–16.

Hirschleifer, D. (2014). Behavioral finance. *Annual Review of Economics, 7*. Submitted. https://doi.org/10.1146/annurev-financial-092214-043752.

Hochbaum, D. S. 2018. Machine learning and data mining with combinatorial optimization algorithms. *INFORMS Tut Orials in Operations Research*. Published online: 19 October 2018, pp. 109–129.

Hong, H., & Kubik, J. D. (2003). Analyzing the analysts: Career concerns and biased earnings forecasts. *Journal of Finance, 58*(2003), 313–351.

Hong, H., Kubik, J. D., & Solomon, A. (2000). Security analysts' career concerns and the herding of earnings forecasts. *RAND Journal of Economics, 31*, 121–144.

ITG. (2007). *ITG ACE – Agency cost estimator: A model description.*

Jacobs, B. I., & Levy, K. (1988). Disentangling equity return regularities: New insights and investment opportunities. *Financial Analysts Journal, 44*, 18–43.

Jacobs, B. I., & Levy, K. (2017). *Equity management: The art and science of modern quantitative investing* (2nd ed.). McGraw-Hill.

Jagadeesh, N., & Titman, S. (1993). Returns to buying winners and selling losers: Implications for stock market efficiency. *Journal of Finance, 48*, 65–91.

Kim, H. K., & Swanson, N. R. (2018). Mining big data using parsimonious factor, machine learning, variable selection, and shrinkage methods. *International Journal of Forecasting, 34*, 339–334.

Keane, M. P., & Runkle., D.E. (1998). Are financial analysts' forecasts of corporate profits rational? *The Journal of Political Economy, 106*, 768–805.

Konno, H., & Yamazaki, H. (1991). Mean – Absolute deviation portfolio optimization and its application to the Toyko stock market. *Management Science, 37*, 519–537.

Korajczyk, R. A., & Sadka, R. (2004). Are momentum profits robust to trading costs? *Journal of Finance, 59*, 1039–1082.

Kuhn, M., & Johnson, K. (2013). *Applied predictive modeling.* Springer.

Lakonishok, J., Shleifer, A., & Vishny, R. W. (1994). Contrarian investment, extrapolation and risk. *Journal of Finance, 49*, 1541–1578.

Lakonishok, J., Shleifer, A., & Vishny, R. W. (1994). Contrarian investment, extrapolation and risk. *Journal of Finance, 49*, 1541–1578.

Latane, H. A. (1959). Criteria for choice among risky ventures. *Journal of Political Economy, 67*, 144–155.

Leamer, E. E. (1972). A class of informative priors and distributed lag analysis. *Econometrica, 40*, 1059–1081.

Leamer, E. E. (1973). Multicollinearity: A Bayesian interpretation. *Review of Economics and Statistics, 55*, 371–380.

Leamer, E. E. (1978). *Specification searches: Ad hoc inference with nonexperimental data.* Wiley.

Leamer, E. E. (2012). The context matters: Comment on Jerome H. Friedman, fast sparse regression and classification. *International Symposium of Forecasting, 28*, 741–748.

Lee, J. H., & Stefek, D. (2008). Do risk factors eat alphas? *Journal of Portfolio Management, 344*, 12–25.

Lesmond, D. A., Schill, M. J., & Zhou, C. (2004). The illusory nature of trading profits. *Journal of Financial Economics, 71*, 349–380.

Levy, H. (1999). *Introduction to investments* (2nd ed.). South-Western College Publishing.

Levy, H., & Duchin, R. (2010). Markowitz's mean-variance rule and the Talmudic diversification recommendation'. In J. B. Guerard (Ed.), *The handbook of portfolio construction: Contemporary applications of Markowitz techniques.* Springer.

Levy, M. (2012). *The capital asset pricing model in the 21st century.* Cambridge University Press.

Lin, D., Foster, D. P., & Ungar, L. H. (2011). VIF regression: A fast regression algorithm for large data. *Journal of the American Statistical Association, 106*, 232–247.

Lintner, J. (1965). The valuation of risk assets and the selection of risky investments in stock portfolios and capital budgets. *Review of Economics and Statistics, 47*, 51–68.

Lo, A. W., Mamaysky, H., & Wang, J. (2000). Foundations of technical analysis: Computational algorithms, statistical inference, and empirical implementation. *Journal of Finance, 60*(2000), 1705–1764.

Lo, A. (2017). *Adaptive markets*. Princeton University Press.

Malkiel, B. (1996). *A random walk down wall street* (6th ed.). W.W. Norton and Company.

Markowitz, H. M. (1952). Portfolio selection. *Journal of Finance, 7*, 77–91.

Markowitz, H. M. (1959). *Portfolio selection: Efficient diversification of investment* (Cowles Foundation monograph no. 16). Wiley.

Markowitz, H. M. (1976). Investment in the long run: New evidence for an old rule. *Journal of Finance, 31*, 1273–1286.

Markowitz, H. (1987). *Mean-variance analysis in portfolio choice and capital markets*. Basil Blackwell.

Markowitz, H. M. (2013). *Risk-Return Analysis*. McGraw-Hill.

Markowitz, H. M., & Xu, G. L. (1994). Data mining corrections. *Journal of Portfolio Management, 21*, 60–69.

Markowitz, H. M., Guerard, Jr., J. B., Xu, G., & Beheshti, B. (2021, forthcoming). Financial anomalies in portfolio construction and management. *The Journal of Portfolio Management*.

Maronna, R. A., Martin, R. D., & Yohai, V. J. (2006). *Robust statistics: Theory and methods*. Wiley.

Maronna, R. A., Martin, R. D., Yohai, V. J., & Salibian-Barrera, M. (2019a). *Robust statistics; theory and methods with R*. Springer.

Menchero, J., Morozov, A., & Shepard, P. (2010). Global equity modeling. In J. B. Guerard (Ed.), *The handbook of portfolio construction: Contemporary applications of Markowitz techniques*. Springer.

Miller, W., Xu, G., & Guerard, J. B., Jr. (2014). Portfolio construction and management in the BARRA Aegis system: A case study using the USER data. *Journal of Investing, 23*, 111–120.

Mossin, J. (1966). Equilibrium in a capital asset market. *Econometrica, 34*, 768–783.

Mossin, J. (1973). *Theory of financial markets*. Prentice-Hall.

Mueller, P. (1993). Empirical tests of biases in equity portfolio optimization. In S. A. Zenios (Ed.), *Financial Optimization*. Cambridge University Press.

Ramnath, S., Rock, S., & Shane, P. (2008). The financial analyst forecasting literature: A taxonomy with suggestions for further research. *International Journal of Forecasting, 24*, 34–75.

Roll, R. (1979). Testing a portfolio for ex ante mean/variance efficiency. In E. Elton & M. J. Gruber (Eds.), *Portfolio theory: 25 years after*. North-Holland Publishing Company.

Rosenberg, B. (1974). Extra-market components of covariance in security returns. *Journal of Financial and Quantitative Analysis, 9*, 263–274.

Rosenberg, B., & Marathe, V. (1979). In H. Levy (Ed.), *Tests of capital asset pricing hypotheses* (Vol. 1). Research in finance, JAI Press.

Rosenberg, B., & McKibben. (1973). The reduction of systematic and specific risk in common stocks. *Journal of Financial and Quantitative Analysis, 8*, 317–333.

Ross, S. A. (1976). The arbitrage theory of capital asset pricing. *Journal of Economic Theory, 13*, 341–360.

Ross, S. A., & Roll, R. (1980). An empirical investigation of the arbitrage pricing theory. *Journal of Finance, 352*, 1071–1103.

Rudd, A., & Rosenberg, B. (1979). Realistic portfolio optimization. In E. Elton & M. J. Gruber (Eds.), *Portfolio theory: 25 years after*. North-Holland Publishing Company.

Rudd, A., & Clasing, H. K. (1982). *Modern portfolio theory: The principles of investment management*. Dow-Jones Irwin.

Sharpe, W. F. (1963). A simplified model for portfolio analysis. *Management Science, 9*, 277–293.

Sharpe, W. F. (1964). Capital asset prices: a theory of market equilibrium under conditions of risk. *Journal of Finance, 19*, 425–442.

Sharpe, W. F. (1971). Mean-absolute deviation characteristic lines for securities and portfolios. A simplified model for portfolio analysis. *Management Science, 18*, B1–B13.

Sharpe, W. F. (2012). *William F. Sharpe: Selected works*. World Scientific Publishing Company.

Stone, B. K., & Guerard, J. B., Jr. (2010). Methodologies for isolating and assessing the portfolio performance potential of stock market return forecast models with an illustration. In J. B. Guerard (Ed.), *The handbook of portfolio construction: Contemporary applications of Markowitz techniques*. Springer.

Subramanian, S., Suzuki, D., Makedon, A., & Carey, J. (2013, March). *A PM's guide to stock picking*. Bank of America Merrill Lynch.

Wheeler, L. B. (1994). Changes in consensus earnings estimates and their impact on stock returns. In B. Bruce & C. B. Epstein (Eds.), *The handbook of corporate earnings analysis*. Probus.

Williams, J. B. (1938). *The theory of investment value*. Harvard University Press.

Chapter 15
Multifactor Risk Models and Portfolio Construction and Management

The previous chapter introduced the reader to Markowitz mean-variance analysis and the Capital Asset Pricing Model. The cost of capital calculated in Chap. 10 assumes that the cost of equity is derived from the Capital Asset Pricing Model and its corresponding beta or measure of systematic risk. The Gordon Model, used for equity valuation in Chap. 8, assumes that the stock price will fluctuate randomly about its fair market value. The cost of equity is dependent upon the security beta. In this chapter, we address the issues inherent in a multi-beta or multiple factor risk model. The purpose of this chapter is to introduce the reader to multifactor risk models. There are academic multifactor risk models, such as those of Cohen and Pogue (1967), Farrell (1974), Stone (1974), Ross (1976), Roll and Ross (1980), Dhrymes et al. (1984, 1985), and Fama and French (1992, 1995, 2008). There are practitioner multifactor risk models, such as Barra, created during the 1973–1979 time period, Advanced Portfolio Technologies (APT), created in 1987, and Axioma, created in the late 1990s, which gained practitioner acceptance in the 2000–2019 time period. The former academicians who created these practitioner models are Barr Rosenberg, Andrew Rudd, John Blin, Steve Bender, and Sebastian Ceria. We will introduce the reader to the practitioner models and their academician creators in this chapter. Which models are best? We, at McKinley Capital Management, MCM, have tested these models. None of the models are perfect, but the models are generally statistically significant when the statistically significant tilt variables of Chap. 14 are used for portfolio construction. In this chapter, we discuss the MCM Horse Races of the 2010–2019 time period to test stock selection within the commercially available multifactor risk models.[1] We trace the development of the Barra, APT, and Axioma commercially available risk models. We conclude with an update of US and non-US portfolios for the 1996–2020 time period. Long-term portfolio strategies have worked for the past 24 years, not just out-of-sample, but post publication of Bloch et al. (1993) with the Axioma Statistical Risk Model.

---

[1] This chapter reflects our knowledge on November 30, 2019.

© The Author(s), under exclusive license to Springer Nature Switzerland AG 2022
J. B. Guerard Jr. et al., *Quantitative Corporate Finance*,
https://doi.org/10.1007/978-3-030-87269-4_15

15.1   The Barra System

Multifactor risk models were developed in the early 1970s. Barr Rosenberg and Walt McKibben (1973), Rosenberg (1974), and Rosenberg and Marathe (1979) created the academic support for the creation of the Barra risk model, the primary institutional risk model of the 1975–2005 time period. Accurate characterization of portfolio risk requires an accurate estimate of the covariance matrix of security returns. A relatively simple way to estimate this covariance matrix is to use the history of security returns to compute each variance, covariance, and security beta. The use of beta, the covariance of security and market index returns, is one method of estimating a reasonable cost of equity funds for firms. However, the approximation obtained from simple models may not yield the best possible cost of equity. The simple, single index beta estimation approach suffers from two major drawbacks:

- Estimating a covariance matrix for the Russell 3000 stocks requires a great deal of data.
- It is subject to estimation error. For example, in the previous chapter, the reader saw that the estimated correlation between two stocks such as Boeing and IBM had a higher correlation than that between Boeing and Dominion Resources. However, one might expect a higher correlation between DuPont and Dow than between DuPont and IBM, given that DuPont and Dow are both chemical firms.

Taking this further, one can argue that firms with similar characteristics, such as firms in their line of business, should have returns that behave similarly. For example, IBM, Boeing, and Dominion Resources will all have a common component in their returns because they would all be affected by news that affects the stock market, measured by their respective betas. The degree to which each of the three stocks responds to this stock market component depends on the sensitivity of each stock to the stock market component.

Additionally, one would expect DuPont and Dow to respond to news affecting the chemical industry, whereas IBM and Dell would respond to news affecting the computer industry. The effects of such news may be captured by the average returns of stocks in the chemical industry and the petroleum industry. One can account for industry effects in the following representation for returns:

$$
\begin{aligned}
\widetilde{r}_{DD} = E[\widetilde{r}_{DD}] + \beta \cdot [\widetilde{r}_M - E[\widetilde{r}_M]] \\
+ 1 \cdot [\widetilde{r}_{CHEMICAL} - E[\widetilde{r}_{CHEMICAL}]] + 0 \cdot [\widetilde{r}_P - E[\widetilde{r}_{DD}]] + \mu_{DD}
\end{aligned}
\tag{15.1}
$$

where:

$\widetilde{r}_{DD}$ = DD's realized return
$\widetilde{r}_M$ = the realized average stock market return
$\widetilde{r}_C$ = realized average return to chemical stocks
$\widetilde{r}_P$ = the realized average return to petroleum stocks
$E[.]$ = expectations

β_{DD} = DD's sensitivity to stock market returns.
μ_{DD} = the effect of DD specific news on DD returns

This equation simply states that DD's realized return consists of an expected component and an unexpected component. The unexpected component depends on any unexpected events that affect stock returns in general $[\tilde{r}_M - E[\tilde{r}_M]]$, any unexpected events that affect the chemical industry $[\tilde{r}_C - E[\tilde{r}_C]]$, and any unexpected events that affect DD alone (μ_{DD}). Similar equations may be written for IBM and Dominion Resources.

The sources of variation in DD's stock returns, thus, are variations in stock returns in general, variations in chemical industry returns, and any variations that are specific to DD. Moreover, DD and Dow returns are likely to move together because both are exposed to stock market risk and chemical industry risk. DD, IBM, and D, on the other hand, are likely to move together to a lesser degree because the only common component in their returns is the market return.

By beginning with our intuition about the sources of comovement in security returns, Rosenberg (1974) made substantial progress in estimating the covariance matrix of security returns. Rosenberg (1974) is the covariance matrix of common sources in security returns, the variances of security-specific returns, and estimates of the sensitivity of security returns to the common sources of variation in their returns, creating the Barra risk model. Because the common sources of risk are likely to be much fewer than the number of securities, we need to estimate a much smaller covariance matrix, and hence, a smaller history of returns is required. Moreover, because similar stocks are going to have larger sensitivities to similar common sources of risk, similar stocks will be more highly correlated than dissimilar stocks.

15.2 Barra Model Mathematics

The Barra risk model is a multiple factor model (MFM). MFMs build on single-factor models by including and describing the interrelationships among factors.[2]

For single-factor models, the equation that describes the excess rate of return is:

$$\tilde{r}_j = X_j \tilde{f} + \tilde{u}_j \tag{15.2}$$

where:

$\tilde{r}_j$ = total excess return over the risk-free rate
X_j = sensitivity of security j to the factor

---

[2]The reader is referred to the Barra (US-E3) United States Equity, Version 3, Risk Model Handbook. Guerard and Mark (2003, 2018) and Miller et al. (2014) used the US-E3 model in their analysis. The reader is also referred to Rosenberg and Marathe (1979), Rudd and Rosenberg (1980), Rudd and Clasing (1982), and Grinhold and Kahn (1999).

$\tilde{f}_j$ = rate of return on the factor

$\tilde{u}_j$ = nonfactor (specific) return on security j

We can expand this model to include K factors. The total excess return equation for a multiplier-factor model becomes:

$$\tilde{r}_j = \sum_{k=1}^{K} X_{jk}\tilde{f}_k + \tilde{u}_j \tag{15.3}$$

where:

X_{jk} = risk exposure of security j to factor k

$\tilde{f}_k$ = rate of return on factor k

Note that when $K = 1$, the MFM equation reduces to the earlier single-factor version; the CAPM was addressed in the previous chapter. When a portfolio consists of only one security, Eq. 15.3 describes its excess return. But most portfolios comprise many securities, each representing a proportion, or weight, of the total portfolio. When weights h_{p1}, h_{p2}, ..., h_{pN} reflect the proportions of N securities in portfolio P, we express the excess return in the following equation:

$$\tilde{r}_p = \sum_{k=1}^{K} X_{Pk}\tilde{f}_k + \sum_{j=1}^{N} h_{Pj}\tilde{u}_j \tag{15.4}$$

where:

$$X_{Pk} = \sum_{j=1}^{N} h_{Pj}X_{jk}$$

This equation includes the risk from all sources and lays the groundwork for further MFM analysis.

Investors look at the variance of their total portfolios to provide a comprehensive assessment of risk. To calculate the variance of a portfolio, one needs to calculate the covariances of all the constituent components. Without the framework of a multiple-factor model, estimating the covariance of each asset with every other asset is computationally burdensome and subject to significant estimation errors. Let us examine the risk structure of the Barra MFM (Fig. 15.1).

The Barra MFM simplifies these calculations dramatically, replacing individual company profiles with categories defined by common characteristics (factors). The specific risk is assumed to be uncorrelated among the assets, and only the factor variances and covariances are calculated during model estimation. Let us briefly review how Barr Rosenberg initially estimated the Barra factor structure (Fig. 15.2).

The Rosenberg MFM framework was developed and estimated in Rosenberg's study with McKibben (1973) and the Rosenberg extramarket component study (1974), in which security-specific risk could be modeled as a function of financial descriptors or financial characteristics of the firm. The financial characteristics that were statistically associated with beta during the 1954–1970 period were:

Fig. 15.1 Stock risk in the Barra system

$$V(i,j) = \text{Covariance}[r(\tilde{i}), r(\tilde{j})]$$

where $V(i,j)$ = asset covariance matrix, and

i,j = individual stocks.

$$V = \begin{bmatrix} V(1,1) & V(1,2) & \cdots & V(1,N) \\ V(2,1) & V(2,2) & \cdots & V(2,N) \\ V(3,1) & V(3,2) & \cdots & V(3,N) \\ \vdots & \vdots & & \vdots \\ V(N,1) & V(N,2) & \cdots & V(N,N) \end{bmatrix}$$

$$\tilde{r} = X\tilde{f} + \tilde{u}$$

where $\tilde{r}$ = vector of excess returns,

X = exposure matrix,

$\tilde{f}$ = vector of factor returns, and

$\tilde{u}$ = vector of specific returns.

$$\begin{bmatrix} \tilde{r}(1) \\ \tilde{r}(2) \\ \vdots \\ \tilde{r}(N) \end{bmatrix} = \begin{bmatrix} X(1,1) & X(1,2) & \cdots & X(1,K) \\ X(2,1) & X(2,2) & \cdots & X(2,K) \\ \vdots & \vdots & & \vdots \\ X(N,1) & X(N,2) & \cdots & X(N,K) \end{bmatrix} \begin{bmatrix} \tilde{f}(1) \\ \tilde{f}(2) \\ \vdots \\ \tilde{f}(K) \end{bmatrix} + \begin{bmatrix} \tilde{u}(1) \\ \tilde{u}(2) \\ \vdots \\ \tilde{u}(N) \end{bmatrix}$$

Fig. 15.2 Barra system factor loadings

1. Latest annual proportional change in earnings per share
2. Liquidity, as measured by die quick ratio
3. Leverage, as measured by die senior debt-to-total assets ratio
4. Growth measure of earnings per share
5. Book-to-price ratio
6. Historic beta
7. Logarithm of stock price
8. Standard deviation of earnings per share growth
9. Gross plant per dollar of total assets
10. Share turnover

Rosenberg et al. (1975), Rosenberg and Marathe (1979), and Rudd and Rosenberg (1979, 1980) expanded upon the initial Rosenberg MFM framework.

The statistically significant determinants of the security systematic risk became the basis of the Barra E1 Model risk indices. The domestic Barra E3 (US-E3) model uses 13 sources of factor, or systematic, exposures. The sources of extramarket factor exposures are volatility, momentum, size, size nonlinearity, trading activity, growth, earnings yield, value, earnings variation, leverage, currency sensitivity, dividend yield, and nonestimation universe.

The multiple-factor model significantly reduces the number of calculations. For example, in the US Equity Model (US-E3), 65 factors capture the risk characteristics of equities. This reduces the number of covariance and variance calculations; moreover, since there are fewer parameters to determine, they can be estimated with greater precision. The Barra US-E3 factor index definitions are shown in Appendix A. The Barra risk management system begins with the MFM equation:

$$\tilde{r} = X\tilde{f} + \tilde{u} \tag{15.5}$$

where:

$\tilde{r}$ = excess return on the asset
X = exposure coefficient on the factor
$\tilde{f}$ = factor return
$\tilde{u}$ = specific return

Substituting this relation in the basic equation, we find that:

$$\text{Risk} = \text{Var}(\tilde{r}) \tag{15.6}$$

$$= \text{Var}\left(X\tilde{f} + \tilde{u}\right) \tag{15.7}$$

Using the matrix algebra formula for variance, the risk equation becomes:

$$\text{Risk} = XFX^{T} + \Delta \tag{15.8}$$

where:

X = exposure matrix of companies upon factors
F = covariance matrix of factors
X^{T} = transpose of X matrix
Δ = diagonal matrix of specific risk variances

This is the basic equation that defines the matrix calculations used in risk analysis in the Barra equity models.

15.3   The Barra Multifactor Model and Analysts' Forecasts, Revisions, and Breadth

Let us address the estimated earnings forecasting components of the CTEF model discussed in the previous chapter for the Russell 3000 universe during the 1990–2001 period. The CTEF model produced not only higher ICs than its components (the reader is referred to Chap. 14) but also higher and more statistically significant asset selection than its components in the Russell 3000 universe. We retain the Guerard and Mark (2003) analysis of the CTEF variable into the Barra multifactor risk model, as presented in Guerard and Schwartz (2007). See Table 15.1 for Russell 3000 earnings component results in which portfolios of approximately 100 stocks are produced by tilting on the individual and component CTEF factors. The forecast earnings per share for the one-year-ahead and two-year-ahead periods, FEP1 and FEP2, offer negative, but statistically insignificant, asset selection. The total active returns are positive and not statistically significant. The asset selection is negative because the FEP variables have positive and statistically significant loadings on the risk indices, particularly the earnings yield index. The factor loading of the FEP variables on the earnings yield risk index is not unexpected, given that the earnings yield factor index in the US-E3 includes the forecast earnings-to-price variable. Thus, there is no multiple-factor model benefit to the FEP variables.

   The monthly revision variables, the RV variables, offer no statistically significant total active returns or asset selection abilities. The breadth variables, BR, produce statistically significant total active returns and asset selection, despite statistically significant risk index loadings. The breadth variable load on the earnings yield and growth risk indexes. Let us take a closer look at the BR1 factor risk index loading. The BR1 variable leads a portfolio manager to have a positive average active exposure to the earnings yield index, which incorporates the analyst predicted earnings-to-price and historic earnings-to-price measures. The BR1 tilt has a negative and statistically significant average exposure to size, nonlinearity, the cube of normalized market capitalization. This result is consistent with analyst revisions being more effective in smaller capitalized securities. The BR1 variable tilt leads the

Table 15.1 Components of composite earnings forecasting variable, 1990–2000, Russell 3000 universe

R3000 Analysis	Earnings Total Active	T-stat	Asset Selection	T-stat	Risk Index	T-stat	Sectors	T-stat
FEP1	2.14	1.61	-1.18	-1.17	**4.20**	**4.42**	-0.86	-1.34
FEP2	1.21	0.91	-1.43	-1.35	**3.33**	**3.35**	-0.78	-1.15
RV1	0.76	0.69	0.34	0.42	0.92	1.46	-0.34	-0.89
RV2	1.40	1.37	1.09	1.31	0.81	1.42	-0.39	-1.08
BR1	**2.59**	**2.83**	**1.85**	**2.43**	**1.08**	**2.15**	-0.20	-0.51
BR2	**2.43**	**2.36**	**1.51**	**1.75**	**1.09**	**2.04**	-0.01	-0.02
CTEF	**2.87**	**2.81**	**2.07**	**2.66**	**1.19**	**1.70**	-0.26	-0.66

Bold figures denote statistically significant coefficients at the 10% level

Table 15.2 Risk and return of mean-variance efficient portfolios, 1990–2001

Universe	Total Active	t-value	Asset Selection	t-value	Risk Index	t-value	Sectors	t-value
RMC	1.98	1.37	0.99	0.86	0.97	1.45	-0.88	-0.97
R1000	2.47	2.52	1.85	2.12	0.82	2.13	-.11	-.23
R2500	7.76	4.37	6.48	3.96	1.61	2.85	-0.33	-0.62
R2000	9.68	5.83	8.81	5.57	0.90	2.36	-.02	-.07

where *RMC* Frank Russell Mid-Cap Universe, *R1000* Frank Russell Largest 1000 Stock Universe, *R2000* Frank Russell Small Cap Universe, *R2500* Frank Russell Small- and Mid-Cap Universe

portfolio manager to have a positive and statistically significant exposure to the growth factor index, composed of the growth in the dividend payout ratio, the growth rates in total assets and earnings per share during the past 5 years, recent one-year earnings growth, and the variability in capital structure. Furthermore, the one-year-ahead BR is slightly better than the two-year-ahead BR, a result consistent with Stone et al. (2002). The CTEF variable produces statistically significant total active returns and asset selection. The CTEF variable loading on the risk index is statistically significant at the 10% level because of its loading on the earnings yield and nonlinear size indexes, as was the case with its breadth components. There are no statistically significant sector exposures in the CTEF variable. The CTEF model offers statistically significant asset selection in a multiple-factor model framework. See Guerard and Mark (2003) for a more complete discussion of the earnings forecasting alpha work within the Barra system.

The Frank Russell large market capitalization universe (the Russell 1000), middle-market capitalization (mid-cap), small-capitalization (Russell 2000), and small- and middle-market capitalization (Russell 2500) universes are used in the tests in this chapter (see Table 15.2). One can test the equally weighted composite model, CTEF, of I/B/E/S earnings forecasts, revisions, and breadth, described in the previous section. The portfolio optimization algorithm seeks to maximize the ranking of the CTEF variable while minimizing risk. The underlying CTEF variable is statistically significant, having a monthly information coefficient of 0.049 over the 491,119 observations. The CTEF variable is used as the portfolio tilt variable in the ITG optimization system using the Barra risk model, and statistically significant total excess returns are found in the Frank Russell universes. One can create 100 stock portfolios monthly during the 1990–2001 period. A lambda tilt value of one is initially used in producing efficient portfolios. Active returns rise as the average stock size diminishes a result consistent with the inefficient markets literature summarized in Dimson (1977) and Ziemba (1992). The highest total active returns are found in the Russell 2000 stocks, the smallest stocks in the largest 3000 stocks in the Frank Russell universes, each year (see Table 15.2). The CTEF tilt variable does produce statistically significant sector exposures in the Russell 2000 stocks, as reported in Table 15.2, as we previously noted in the Russell 3000 universe.

Table 15.3   CTEF variable, Russell 1000 universe

Attribution Report
Annualized Contributions To Total Return

Source of Return	Contribution (% Return)	Risk (% Std Dev)	Info Ratio	T-Stat
Risk-Free	4.93	N/A	N/A	N/A
Total Benchmark	13.58	14.52		
Market Timing	-0.02	0.25	-0.18	-0.61
Risk Indices	0.82	1.16	0.62	2.13
Sectors	-0.11	1.01	-0.07	-0.23
Asset Selection	1.85	2.76	0.61	2.12
Total Exceptional Active	2.54	3.07	0.75	2.59
Total Active	2.47	3.07	0.73	2.52
Total Managed	16.04	14.96		

Table 15.4   CTEF variable factor exposures, Russell 1000 universe

Attribution Analysis
Annualized Contributions To Risk Index Return

Source of Return	Average Active Exposure	Contribution (% Return) Average [1]	Variation [2]	Total [1+2]	Total Risk (% Std Dev)	Info Ratio	T-Stat
Volatility	-0.01	0.01	-0.07	-0.06	0.17	-0.32	-1.12
Momentum	0.12	-0.07	0.08	0.01	0.60	0.03	0.11
Size	-0.20	0.36	-0.09	0.27	0.93	0.24	0.83
Size Non-Linearity	-0.02	0.02	0.03	0.05	0.10	0.44	1.52
Trading Activity	0.00	0.00	0.01	0.01	0.11	0.11	0.37
Growth	-0.05	0.05	0.03	0.08	0.14	0.48	1.65
Earnings Yield	0.13	0.66	-0.12	0.55	0.40	1.20	4.13
Value	0.06	0.03	0.02	0.06	0.17	0.30	1.03
Earnings Variation	0.02	-0.02	0.00	-0.03	0.10	-0.21	-0.73
Leverage	0.06	-0.01	-0.04	-0.04	0.17	-0.23	-0.80
Currency Sensitivity	-0.02	0.01	-0.05	-0.04	0.11	-0.32	-1.11
Yield	0.04	0.01	-0.04	-0.04	0.14	-0.24	-0.81
Non-Est Universe	0.00	0.00	0.01	0.01	0.04	0.20	0.68
Total				**0.82**	**1.16**	**0.62**	**2.13**

The CTEF variable produces statistically significant asset selection in the Russell 1000 Universe, during the 1990–2001 period. The reader is referred to Table 15.3. The risk index exposure is statistically significant in the Russell 1000 Universe. The factor exposures of the CTEF variable in the Russell 1000 Universe are shown in Table 15.4. The CTEF variable has statistically significant factor loadings on earnings yield and growth, as was the case in the Russell 3000 universe.

Table 15.5 CTEF variable, Russell 2000 universe

Attribution Analysis
Annualized Contributions To Total Return

Source of Return	Contribution (% Return)	Risk (% Std Dev)	Info Ratio	T-Stat
Risk-Free	4.93	N/A	N/A	N/A
Total Benchmark	11.73	18.42		
Expected Active	-0.11	N/A	N/A	N/A
Market Timing	0.09	0.45	0.05	0.17
Risk Indices	0.90	1.11	0.68	2.36
Sectors	-0.02	0.95	0.02	0.07
Asset Selection	8.81	4.67	1.61	5.57
Total Exceptional Active	9.79	4.88	1.71	5.90
Total Active	9.68	4.88	1.69	5.83
Total Managed	21.41	17.67		

The total active return of CTEF variable for the Russell 2000 universe is shown in Table 15.5. Active returns to the CTEF variable are much larger in the Russell 2000 stocks than in the Russell 1000 stocks. The CTEF variable produces statistically significant asset selection and factor exposures, primarily due to the earnings yield exposure. The earnings forecasting variable produces over 600 basis points annually of greater asset selection in the Russell 2000 universe than it does in the Russell 1000 Universe. Earnings forecasts, revisions, and breadth generate greater asset selection in small stock universes than in larger stock universes. Moreover, as the firm size decreases, the CTEF variable is more statistically associated with risk index returns, such as earnings yield. The factor exposures increase as the size of firms decreases. The earning yield variable loading is statistically significant in the Russell 2000 universe. Miller et al. (2014) updated Guerard and Mark (2003) using US-E3 for the 1979–2009 time period. Miller et al. (2014) reported a "beautiful trade-off curve" and highly statistically significant positive Management Returns (Active Returns) and Asset Selection (Stock Selection), and a slightly edited form is in Guerard et al. (2019). As one increases the portfolio lambda, decreasing the risk aversion level, the number of assets in the portfolio falls, the Managed Returns rise, and the Information Ratio rises versus either the Russell 3000 Growth Index or the S&P 500 Index. Asset selection is statistically significant and substantial, exceeding 350 basis points for the 30-year backtest, and produced statistically significant portfolio returns passing the Markowitz-Xu's (1994) Data Mining Corrections test.

The BARRA risk model is an extramarket covariance model to describe the risk behavior of equity securities. Barr Rosenberg and his coauthors developed the Barra system in the mid-to-late 1970s, and the system has been a financial success, went public and was listed on NASDAQ (BARZ), and was acquired by Morgan Stanley in

2004 for over $800 million. The Barra system was by no means the only multifactor or extramarket covariance model.[3]

15.4   Early Alternative Multi-Beta Risk Models

Bernell Stone (1974) developed a two-factor index model which molded equity returns as a function of an equity index and debt returns.[4] In recent years, Stone et al. (2002) developed a response surface portfolio algorithm to generate portfolios that have similar stock betas (systematic risk), market capitalizations, dividend yield, and sales growth cross-sections, such that one can access the excess returns of the analysts' forecasts, forecast revisions, and breadth model, as one moves from low (least preferred) to high (most preferred) securities with regard to your portfolio construction variable (i.e., CTEF). Stone et al.'s (2002) response surface risk model identified several risk models of 3, 6, and 8 factors that created portfolios producing large excess returns. Excess returns similar to the excess returns to asset selection shown in Table 15.2 can be produced with Stone algorithm during the 1982–1998 period. Farrell (1974, 1997) estimated a four "factor" model extramarket covariance model.[5] Farrell took an initial universe of 100 stocks in 1973 (due to computer limitations) and ran market models to estimate betas and residuals from the market model:

$$R_{j_t} = a_j + b_j R_{M_t} + e_j \tag{15.9}$$

$$e_{j_t} = R_{j_t} - \widehat{a}_j - \widehat{b}_j R_{M_T} \tag{15.10}$$

The residuals of Eq. (15.10) should be independent variables. That is, after removing the market impact by estimating a beta, the residual of IBM should be independent of Dow, Merck, or Dominion Resources. The residuals should be independent, of course, in theory. Farrell (1973) examined the correlations among the security residuals of Eq. (15.10) and found that the residuals of IBM and Merck were highly correlated, but the residuals of IBM and D (then Virginia Electric and Power) were not correlated. Farrell used a statistical technique known as Cluster Analysis to create clusters, or groups, of securities, having highly correlated market model

---

[3] Rudd and Rosenberg (1979) published one of the first academic/commercial portfolio optimization analyses in the Elton and Gruber monograph to honor Harry Markowitz. Miller et al. (2014) used the BARRA optimizer.

[4] Bernell Stone is a dear friend, a professor, and coauthor with Guerard. His dissertation *Risk, Return, and Equilibrium: A General Single-Period Theory of Asset Selection and Capital Market Equilibrium* was published by MIT Press. Stone et al. (2002) received an Honorable Mention for the Mosokowitz Prize for research in socially responsible investing (please see Chap. 22).

[5] Jim Farrell was President of the Institute of Quantitative Research in Finance, "The Q-Group," for over 30 years. The Q-Group sponsors academic research and hosts outstanding seminars.

residuals. Farrell found four clusters of securities based on his extramarket covariance. The clusters contained securities with highly correlated residuals that were uncorrelated with residuals of securities in the other clusters. Farrell referred to his clusters as "Growth Stocks" (electronics, office equipment, drug, hospital supply firms, and firms with above-average earnings growth), "Cyclical Stocks" (metals, machinery, building supplies, general industrial firms, and other companies with above-average exposure to the business cycle), "Stable Stocks" (banks, utilities, retailers, and firms with below-average exposure to the business cycle), and "Energy Stocks" (coal, crude oil, and domestic and international oil firms). In 1976, Ross published his "Arbitrage Theory of Capital Asset Pricing," which held that security returns were a function of several (4–5) economic factors. In 1986, Chen, Ross, and Roll developed an estimated multifactor security return model based on:

$$R = a + b_{MP}MP + b_{DEI}DEI + b_{UI}UI + b_{UPR}UPR + b_{UTS}UTS + e_t \qquad (15.11)$$

where:

MP = monthly growth rate of industrial production
DEI = change in expected inflation
UI = unexpected inflation
UPR = risk premium and
UTS = term structure of interest rates

Chen, Ross, and Roll (CRR) defined unexpected inflation as the monthly (first) differences of the Consumer Price Index (CPI) less the expected inflation rate. The risk premia variable is the "Baa and under" bond return at time and less the risk-free rate. The term structure variable is the long-term government bond return less the Treasury bill rates, known at time $t - 1$, and applied to time t. When CRR applied their five-factor model in conjunction with the value-weighted index betas, during the 1958–1984 period, the index betas are not statistically significant, whereas the economic variables are statistically significant. The Stone, Farrell, Dhrymes et al. (1984) and Chen, Ross, and Roll multifactor models used 4 to 5 factors to describe equity security risk. The models used different statistical approaches and economic models to control for risk. The reader may now ask a simple question: If 4 or 5 betas are appropriate, why not estimate 20 betas? The important point is that the betas should be estimated on variables that are independent, or orthogonal, of the other variables. The estimation of 20 betas on orthogonal variables, 10–15 of which are not statistically significant, can produce the quite similar expected returns for securities as four or five betas estimated on economic variables or prespecified variables. Blin and Bender (1987–1997) estimated a 20 (factor) beta model of covariances based on 3.5 years of weekly stock returns data. The Blin and Bender Arbitrage Pricing Theory (APT) model followed the Ross factor modeling theory, but Blin and Bender estimated betas from 20 orthogonal factors. Blin and Bender never sought to identify their factors with economic variables.

This chapter addresses many aspects of estimating and using multifactor models of risk. What are portfolio managers and investors concerned about these risk models? In Chap. 8, the reader was introduced to equity valuation models, which held that stock prices are functions of discounted expected dividends. The equity discount rate can be determined by the multifactor risk model (see Roll & Ross, 1984).

15.5 APT Approach

John Blin et al. (1997) and Guerard (2012) demonstrated the effectiveness of the APT, Sungard APT, and FIS APT systems in portfolio construction and management. Let us review the APT approach to portfolio construction. The estimation of security weights, w, in a portfolio is the primary calculation of Markowitz's portfolio management approach. The issue of security weights will be now considered from a different perspective. The security weight is the proportion of the portfolio's market value invested in the individual security.

$$w_s = \frac{MV_s}{MV_p} \tag{15.12}$$

where w_s = portfolio weight in security s, MV_s = value of security s within the portfolio, and MV_p = the total market value of portfolio.

The active weight of the security is calculated by subtracting the security weight in the (index) benchmark, b, from the security weight in the portfolio, p

$$w_{sp} - w_{sb} \tag{15.13}$$

Blin and Bender created APT, Advanced Portfolio Technologies, Analytics Guide (2005), which built upon the mathematical foundations of their APT system, published in Blin et al. (1997). The following analysis draws upon APT analytics. Volatility can be broken down into systematic and specific risk:

$$\sigma_p^2 = \sigma_{\beta p}^2 + \sigma_{\varepsilon p}^2 \tag{15.14}$$

where σ_p = Total Portfolio Volatility, $\sigma_{\beta p}$ = Systematic Portfolio Volatility, and $\sigma_{\varepsilon p}$ =Specific Portfolio Volatility. Blin and Bender created a multifactor risk model within their APT risk model based on forecast volatility.

$$\sigma_p = \sqrt{52 * \left(\sum_{c=1}^{C}\sum_{i=1}^{S} w_i \beta_{i,c}\right)^2 + 52 * \sum_{i=1}^{S} w_i^2 \varepsilon_i^2} \tag{15.15}$$

where:

$\sigma_p =$ forecast volatility of annual portfolio return
$C =$ number of statistical components in the risk model
$w_i =$ portfolio weight in security i
$\beta_{i,c} =$ loading (beta) of security i on risk component c
$\varepsilon_{i,\,week} =$ weekly specific volatility of security i

The multiplier 52 is the annualization factor since there are 52 weeks per year.

Tracking error is a measure of volatility applied to the active return of funds (or portfolio strategies) against a benchmark, which is often an index. Portfolio tracking error is defined as the standard deviation of the portfolio return less the benchmark return over 1 year.

$$\sigma_{te} = \sqrt{E\big((r_p - r_b) - E(r_p - r_b)\big)^2} \qquad (15.16)$$

where:

$\sigma_{te} =$ annualized tracking error
$r_p =$ actual portfolio annual return
$r_b =$ actual benchmark annual return

Systematic tracking error of a portfolio is a forecast of the portfolio's active annual return as a function of the securities' returns associated with APT risk model components. Portfolio-specific tracking error can be written as a forecast of the annual portfolio active return associated with each security's specific behavior.

$$\sigma_{ete} = \sqrt{52\sum\nolimits_{i=1}^{N_s}\big(w_{i,p} - w_{i,b}\big)^2\varepsilon_{i,week}^2 R_p} \qquad (15.17)$$

The APT calculated portfolio tracking error versus a benchmark as:

$$\sigma_{p-b} = \sqrt{52\big(w_p - w_b\big)^T\big(b^T b + \varepsilon^T\varepsilon\big)\big(w_p - w_b\big)}, \qquad (15.18)$$

where:

$\sigma_{p-b} =$ forecast tracking error
$b = A(N_c \times N_s)$ matrix of component loadings; N_c components in the model and N_s securities in the portfolio
$\varepsilon =$ a diagonal matrix $(N_c \times N_s)$ of the specific loadings
$w_p - w_b =$ the (N_s-dimensional) vector security active weights

$$\sigma_{p-b} = \sqrt{52\left[\Big(\sum\nolimits_{c=1}^{N_c}\sum\nolimits_{s=1}^{N_s}\big(w_{sp} - w_{sb}\big)b_{sc}\Big)^2 + \sum\nolimits_{s=1}^{N_s}\big(w_{sp} - w_{sb}\big)^2\varepsilon_s^2\right]} \qquad (15.19)$$

where:

$w_{sp} - w_{sb}$ = portfolio active weights in security s
b_{sc} = the loading of security s on component c
ε_s = weekly specific volatility of security s

The marginal tracking error is a measure of the sensitivity of the tracking error of an active portfolio to changes in the active weight of a specific security.

$$\partial_s \left[\sigma_{p-b} \right] = \frac{\partial \sigma_{p-b}}{\partial w_{s,p-b}} \tag{15.20}$$

where: $w_{s, p-b}$ = active portfolio weights in security s.
    The APT calculated contribution to risk of a security is:

$$\Delta \left[\sigma_p \right] = \frac{52 \left(b^T b + s^T s \right) w}{\sqrt{52 \left(b^T b + s^T s \right) w}} \tag{15.21}$$

where $\Delta [\sigma_p] = $ A (N_s − *dimensional*) vector of contribution to volatility for securities in the portfolio with N_s securities.
    The portfolio Value at Risk (VaR) is a measure of the distribution of expected outcomes. If one is concerned with a 95% confidence level, α, and a 30-day time horizon, then the 95%, 30-day VaR of the portfolio is the minimum we would expect to lose, which is characterized by

$$P[V_T < V_0 - v(\alpha, T)] = 1 - \alpha \tag{15.22}$$

where:

$v(\alpha, T)$ = VaR at the confidence interval α for time T
V_T = portfolio value at time T
α = required confidence level

If the portfolio returns are normally distributed,

$$v_s(\alpha, T) = V_0 - E(V_T) - V_0 \phi^{-1}(\alpha) \sigma_p^T \tag{15.23}$$

where:

$v_s(\alpha, T)$ = Gaussian VaR
σ_p^T = forecast portfolio volatility
$\phi^{-1}(\alpha)$ = inverse cumulative normal distributed function

The total VaR is

$$v_s(\alpha, T) = V_0 \sqrt{\frac{\alpha}{1-\alpha}\left(\sigma_{\beta p}^T\left(\phi^{-1}(\alpha)\sigma_{sp}^T\right)^2\right)} \qquad (15.24)$$

The Tracking Error at Risk (TaR) is a measure of portfolio risk estimating the magnitude that the portfolio return may deviate from the benchmark return over time. The maximum deviation of a portfolio return over the time horizon T at given confidence level α is

$$v_{p-b}(\alpha, T) = V_0 \sqrt{\left(\frac{1}{\sqrt{1-\alpha}}\sigma_{\beta,p-b}^T\right)^2 + \phi^{-1}\left(\frac{1+\alpha}{2}\right)\sigma_{s,p-b}^2} \qquad (15.25)$$

How relevant is APT and Markowitz mean-variance analysis? We shall see in the next section. Spoiler alert: Harry Marowitz (1952, 1959) produces statistically significant portfolio Active returns and Specific Returns (stock selection) when used with a statistically significant tilt variable. A great recent book on portfolio risk is Connor et al. (2010). Korajcyzk has a direct link to an outstanding bibliography on APT and multifactor risk models (https://www.kellogg.northwestern.edu/faculty/korajczy/htm/aptlist.htm).

15.6   Applying the Blin and Blender APT Model

Blin et al. (1997) used an estimated 20-factor beta model of covariances based on 3.5 years of weekly stock returns data. The Blin and Bender Arbitrage Pricing Theory (APT) model followed the Ross factor modeling theory, but Blin and Bender estimated betas from at least 20 orthogonal factors. Empirical support is reported in Guerard et al. (2010, 2012) and Guerard (2012) for the application of mean-variance, enhanced index tracking, EIT, and tracking error at risk optimization techniques.

It is well known that as one raises the portfolio lambda, the expected return of a portfolio rises and the number of securities in the optimal portfolios falls (see Blin et al., 1997). Lambda, a measure of risk aversion, the inverse of the risk-aversion acceptance level of the Barra system, is a decision variable to determine the optimal number of securities in a portfolio. As lambda increases, the number of stocks in portfolios falls, and Sharpe Ratios rise. A more aggressive, growth-maximizing asset manager uses a larger lambda.

In spite of the Markowitz Mean Variance (M59) portfolio construction and management analysis being six decades old, it does very well in maximizing the Sharpe Ratio, geometric mean, and information ratio relative to newer approaches. See Table 15.6 for an adapted version of the Markowitz Mean-Variance and Mean-Variance Tracking Error at Risk (MVTaR) simulations of Guerard et al. (2015). A lambda of 200, an aggressive risk-tolerance parameter, is sufficient in US and global markets to maximize the Geometric Mean, Sharpe Ratio, and Information Ratio, IR,

Table 15.6 Efficient frontier of the global stock selection model, GLER, with various portfolio optimization techniques universe: (1) global I/B/E/S stocks with at least two analysts, 1999–2011, and (2) largest 7500 global stocks (in $USD)

APT Risk Model and Optimization System

Earnings Model Component	Mean-orVariance Methodology	Lambda	Annualized Return	Sharpe Ratio	Information Ratio	Tracking Error
GLER	M59	1000	15.84	0.590	0.78	13.11
		500	16.34	0.590	0.82	12.08
		200	16.37	0.610	0.85	12.68
		100	15.90	0.580	0.81	12.66
		5	10.11	0.440	0.51	8.81
Benchmark			5.59	0.240		
GLER	MVTaR	1000	16.10	0.660	0.94	11.18
		500	15.91	0.651	0.90	11.44
		200	16.09	0.691	0.97	10.83
		100	14.18	0.591	0.77	11.23
		5	8.51	0.344	0.33	8.75

Constraints:
1. Turnover is 8% (buys) monthly
2. Long-only holdings (a 4% maximum stock weight)
3. Threshold positions of 35 basis points

for the MCM variables. The lambda of 200 MVTaR GLER simulation outperforms the M59 (based on Markowitz's seminal Portfolio Selection, one of the two sleeping partners of the authors) on only the IR basis (0.97 versus 0.85). MVTaR does not "Beat the Tar" out of the traditional Markowitz Mean-Variance model, but it does produce lower tracking errors and higher IRs. A lambda of 5 is reported to illustrate an index-enhanced portfolio strategy.[6] The purpose of Guerard et al. (2015) was to update Guerard and Mark (2003) and report additional evidence on the continued statistical significance of earnings forecasting analysis and the 9-factor composite

---

[6] A very important result of APT portfolio modeling is reported in Guerard et al. (2012) for the application of mean-variance, enhanced index tracking, and tracking error at risk optimization techniques for the 1997–2009 time period. One can optimize US stocks to create portfolios in the Russell 3000 universe for the 12-year backtest period, 1997–2009, where the absolute value of the stock weight cannot deviate by more than 2%, which Guerard, Krauklis, and Kumar referenced to as an EIT portfolio, or equal active weight of 2% (EAW2). One needs 99 stocks in the mean-variance portfolios and 108 stocks in the EAW2 portfolios for a lambda of 200. One needs 159 stocks in the mean-variance portfolios and 158 stocks in the EAW2 portfolios for a lambda of 10. A lambda of 10 is used by an index-hugging investor who is afraid of taking active bets in the portfolio. An even less aggressive manager (a horrid benchmark-hugger) needs 161 stocks of an EAW1 portfolio for a lambda of 10, as opposed to 131 stocks for a Sharpe Ratio-maximizing asset manager using a lambda of 100.

stock selection model introduced in Chap. 14. In the world of applied investment research, the authors believe that all three levels of portfolio testing must be passed for statistical significance![7]

15.7   Applying the Blin and Blender APT Model, BARRA, and Axioma: The McKinley Capital Management (MCM) Horse Race Tests

The Axioma Robust Risk Model is a multifactor risk model, in the tradition of the Barra model and Eq. (15.20).[8] Axioma offers both US and World Fundamental and Statistical Risk Models. The Axioma Risk Models use several statistical techniques to efficiently estimate factors. The ordinary least squares residuals (OLS) are not homoscedastic; that is, when one minimizes the sum of the squared residuals to estimate factors using OLS, one finds that large assets exhibit lower volatility than smaller assets. A constant variance of returns is not found. Axioma uses a weighted least squares (WLS) regression, which scales the asset residual by the square root of the asset market capitalization (to serve as a proxy for the inverse of the residual variance). Robust regression, using the Huber M Estimator, addresses the issue and problem of outliers. (Asymptotic) principal components analysis (PCA) is used to estimate the statistical risk factors. A subset of assets is used to estimate the factors and the exposures, and factor returns are applied to other assets. In 2011, McKinley Capital Management, LLC (MCM) initiated a "Horse Race" testing procedure to test if all optimizers were created equal. The Spring 2012 *Journal of Investing* featured a Quantitative Risk Management Special edition, reporting the MCM US Horse Race and featuring articles by Markowitz; Tsuchida, Zhou, and Rachev (the Cognity system of FinAnalytica); Wormald and van Der Merwe (APT); Saxena and Stubbs (Axioma); and Guerard, Xu, and Gultekin (APT, BARRA, and Cognity). They are not. In 2011, APT and Axioma were the winners among several (4–5) optimization systems using the MCM US "Public Models" CTEF and USER. Both APT and Axioma optimization systems produced highly statistically significant asset selection (the BARRA attribution system was used as a judge). See Table 15.7.

What factor contributions of the Axioma portfolio were rewarded by the market, 1998–2009? Small size (the exposure of -1.17 produced 329 basis points of return

---

[7]Guerard brought the Blin and Bender APT system to MCM in August 2005. The APT system produced real-time portfolios that generated statistically significant Active Returns and Specific Returns (see Guerard et al. 2019). The real-time Specific Returns exceeded 300 basis points, very consistent with its backtests. Guerard is highly biased in his assessment of risk models; Blin and Bender offered MCM a great discount, and the modeling worked. Long live APT and the "Blender Boys," as we called them at Drexel, Burnham, Lambert, where Guerard met them in 1988. Blin and Bender sold APT to Sungard, as Rudd sold BARRA to Morgan Stanley. Guerard has great respect for entrepreneurs who build great models and sell them to become great commercial successes.

[8]*Axioma Robust Risk Model Handbook*, January 2010.

Table 15.7   APT and axioma statistical risk models, US stocks, 1998–2009

Assumptions and Constraints
Universe: All US stocks in the WRDS Database
Time Period: Jan1998 to Dec2009

Position Constraint :
      Minimum Position =0.0
      Maximum Position=4.0
Position Threshold Holding = 0.35%,
Turnover =8% Monthly; and
Transaction Cost: 1.25% each way

Risk Model :
      Axioma - Axioma Statistical
      Model
      APT - APT Statistical Model

Risk Setting:

      Axioma -   Tracking error target of 8%

      APT -   Lambda of 200

Model :   Regression Weighted EP BP CP SP REP RBP RCP RSP
      CTEF PM

	SR	IR	TE	SD	AR	Asset Selection	T-Stat
Axioma	0.177	0.69	7.78	21.92	7.06	6.00	5.09
APT	0.117	0.40	10.29	23.02	5.87	6.77	3.92
R3000G	-0.074			19.73	1.73		

with a corresponding t-statistic of 3.16) and earnings yield (the exposure of 0.09 produced 41 basis points of return with a corresponding t-statistic of 1.85, statistically significant at the 10% level). Growth, leverage, and yield produced negative statistically significant factor returns in the BARRA USER attributions for 1998–2009. APT and Axioma both had statistically based risk models; Axioma and BARRA had fundamentally based risk models. Which is better for CTEF and USER analyses?

15.8   Why Use the Axioma Statistical Model?

In 2012, MCM ran a second Horse Race competition using a two-I/B/E/S analyst, 7500-largest market-capitalized global universe. The Axioma Statistical Risk Model produced higher Geometric Means, Sharpe Ratios, and Information Ratios using GLER data than its Fundamental (WW21AxiomaMH) Model.

Table 15.8　Axioma analysis of the FactSet Global 7500-stock data universe

Axioma Ranked Global Backtest
FSGLER Model using All Country World Growth Index Constituents
Universe: ACWG
Simulation Period: Jan. 1999 - Dec. 2011
Transactions Costs: 150 basis points each way, respectively.

Return Model	Risk Model	Tracking Error	Sharpe Ratio	Information Ratio	Ann. Active Return	Ann. Active Risk	N
GLER	STAT	4	0.554	1.475	9.99	6.78	144
		5	0.602	1.385	11.38	8.24	110
		6	0.656	1.409	13.25	9.40	87
		7	0.715	1.454	14.94	10.28	70
		8	0.748	1.451	16.20	11.16	58
	FUND	4	0.382	1.091	6.08	5.57	163
		5	0.460	1.151	7.73	6.72	129
		6	0.521	1.158	9.33	8.06	104
		7	0.582	1.217	11.02	9.06	83
		8	0.647	1.281	12.75	9.95	71

Guerard et al. (2013, 2015) reported efficient frontiers using both of the Axioma Risk Models, and found that the statistically based Axioma Risk Model, the authors denoted as "STAT," produced higher geometric means, Sharpe ratios, and information ratios than the Axioma fundamental Risk Model denoted as "FUND." The geometric means and Sharpe ratios increase with the targeted tracking errors; however, the information ratios are higher in the lower tracking error range of 3–6%, with at least 200 stocks, on average, in the optimal portfolios. See Table 15.8. The Guerard et al. studies assumed 150 basis points, each way, of transaction costs. The use of ITG cost curves produced about 115–125 basis points of transaction costs, well under the assumed costs. The Guerard et al. studies also used the Sungard APT statistical model which produced statistically significant asset selection in US and global portfolios. The dominance of the statistical risk model has been a prevailing result of the MCM Horse Races, 2010–2017.

In 2017, MCM ran the third Horse Race using Axioma with integrated ITG transactions costs, in the MSCI Non-US (XUS), Global (GL), and Emerging Markets (EM) universes. The results reported by XUS, GL, and EM show much higher Active and Specific Returns in non-US markets than in US stocks. International markets are more inefficient than US markets.

15.9   Alpha Alignment Factor

Axioma has pioneered several techniques to address the so-called underestimation of realized tracking errors, particularly during the 2008 Financial Crisis. This chapter and Appendix A report the Alpha Alignment Factor, AAF, which recognizes the possibility of missing systematic risk factors and makes amends to the greatest extent that is possible without a complete recalibration of the risk model that accounts for the latent systematic risk in alpha factors explicitly. In the process of doing so, AAF approach not only improves the accuracy of risk prediction but also makes up for the lack of efficiency in the optimal portfolios.

The empirical results in a test-bed of real-life active portfolios based on client data show clearly that the abovementioned unknown systematic risk is a significant portion of the overall systematic risk, and should be addressed accordingly. Saxena and Stubbs (2012) reported that the earning-to-price (E/P) and book-to-price (B/P) ratios used in USER Model and Axioma Risk Model have average misalignment coefficients of 72% and 68%, respectively. While expected-return and risk models are indispensable components of any active strategy, there is also a third component, namely the set of constraints that is used to build a portfolio. Saxena and Stubbs (2012) proposed that the risk variance–covariance matrix C be augmented with additional auxiliary factors in order to complete the risk model. The augmented risk model has the form of

$$C_{new} = C + \sigma_{\underline{\alpha}}^2 \underline{\alpha} \bullet \underline{\alpha}' + \sigma_{\underline{\gamma}}^2 \underline{\gamma} \bullet \underline{\gamma}' \qquad (15.26)$$

where $\underline{\alpha}$ is the alpha alignment factor (AAF), σ_α is the estimated systematic risk of $\underline{\alpha}$, $\underline{\gamma}$ is the auxiliary factor for constrains, and σ_γ is the estimated systematic risk of $\underline{\gamma}$. The alpha alignment factor $\underline{\alpha}$ is the unitized portion of the uncorrelated expected-return model, i.e., the orthogonal component, with risk model factors. Saxena and Stubbs (2012) reported that the AAF process pushed out the traditional risk model-estimated efficient frontier. Saxena and Stubbs (2015) refer to as alpha in the augmented regression model as the implied alpha. Saxena and Stubbs (2015) report that there is a small increment to specific risk compared to its true systematic risk.

Saxena and Stubbs (2012) applied their AAF methodology to the USER model, running a monthly backtest based on the above strategy over the time period 2001–2009 for various tracking error values of σ chosen from {4%, 5%... 8%}. For each value of σ, the backtests were run on two setups, which were identical in all respects except one, namely that only the second setup used the AAF methodology ($\sigma_\alpha = 20\%$). Axioma's fundamental medium-horizon risk model (US2AxiomaMH) is used to model the active risk constraints.

Saxena and Stubbs (2012) analyzed the time series of misalignment coefficients of alpha, implied alpha, and the optimal portfolio, and found that almost 40–60% of the alpha is not aligned with the risk factors. The alignment characteristics of the implied alpha are much better than those of the alpha. Among other things, this implies that the constraints of the above strategy, especially the long-only constraints, play a proactive role in containing the misalignment issue. In addition, not only do the orthogonal components of both the alpha and the implied alpha have systematic risk, but the magnitude of the systematic risk is comparable to that of the systematic risk associated with a median risk factor in US2AxiomMH.

Saxena and Stubbs (2012) showed the predicted and realized active risks for various risk target levels, and noted the significant downward bias in risk prediction when the AAF methodology is not employed.[9] The realized risk-return frontier demonstrates that not only does using the AAF methodology improve the accuracy of the risk prediction, it also moves the ex-post frontier upward, thereby giving ex-post performance improvements. In the process of doing so, AAF approach not only improves the accuracy of risk prediction but also makes up for the lack of efficiency in the optimal portfolios.[10] Saxena and Stubbs (2015) extended their 2012 *Journal of Investing* research and reported positive frontier spreads. The reader is referred to Appendix B for the technical development of AAF.

Saxena and Stubbs (2012) analyzed the time series of misalignment coefficients of alpha, implied alpha, and the optimal portfolio, and found that almost 40–60% of the alpha is not aligned with the risk factors. Additionally, not only do the orthogonal components of both the alpha and the implied alpha have systematic risk, but the magnitude of the systematic risk is comparable to that of the systematic risk associated with a median risk factor in US2AxiomMH. Guerard and Chettiappan (2017) reported stock selection in non-US, global, and EM markets during 2003–2016. We reported the Guerard and Chettiappan (2017) results in Table 15.9, that CTEF and GLER are highly statistically significant in producing statistically significant active and specific returns, which rose with higher targeted tracking errors. EM portfolios dominated XUS and GL portfolios in terms of producing higher geometric means, active returns, and specific returns.

Let us take a deep dive into the CTEF and GLER variables and their effectiveness. Let us examine the use of AAF in two analyses. First, Guerard (2012) used

---

[9]The bias statistic shown is a statistical metric that is used to measure the accuracy of risk prediction; if the ex-ante risk prediction is unbiased, then the bias statistic should be close to 1.0. Clearly, the bias statistics obtained without the aid of the AAF methodology are significantly above the 95% confidence interval, which shows that the downward bias in the risk prediction of optimized portfolios is statistically significant. The AAF methodology recognizes the possibility of inadequate systematic risk estimation, and guides the optimizer to avoid taking excessive unintended bets.

[10]Guerard et al. (2013, 2015) created efficient frontiers using both of the Axioma Risk Models, and found that the statistically based Axioma Risk Model, the authors denoted as "STAT," produced higher geometric means, Sharpe ratios, and information ratios than the Axioma fundamental Risk Model, denoted as "FUND." The AAF technique was particularly useful with composite models of stock selection using fundamental data, momentum, and earnings expectations data. Furthermore, the geometric means and Sharpe ratios increase with the targeted tracking errors; however, the information ratios are higher in the lower tracking error range of 3–6%, with at least 200 stocks, on average, in the optimal portfolios. The Guerard et al. studies assumed 150 basis points, each way, of transactions costs. The use of ITG cost curves produced about 115–125 basis points of transactions costs, well under the assumed costs. The Guerard et al. studies also used the Sungard APT statistical model which produced statistically significant asset selection in US and global portfolios.

Table 15.9 Stock selection modeling in global, non-US, and EM universes

MSCI Index Constituents-only
Axioma Statistical Risk Model and Optimizer
Mean-Variance Analysis

Model: Global CTEF GLER TE

Period: 2003-01-31 to 2016-12-30 (Monthly)

Benchmark: ACWG

	Portfolio Return	Active Return	T-Stat	Specific Return	T-Stat
GLOBAL_CTEF-TE4	12.44%	3.89%	2.28	1.36%	1.28
GLOBAL_CTEF-TE6	15.87%	7.32%	3.32	3.89%	2.89
GLOBAL_CTEF-TE8	16.31%	7.77%	3.13	4.13%	2.92
GLOBAL_GLER-TE4	13.28%	4.73%	2.95	1.54%	1.42
GLOBAL_GLER-TE6	16.03%	7.49%	3.61	3.26%	2.44
GLOBAL_GLER-TE8	16.87%	8.32%	3.45	4.27%	2.98

Model: XUS CTEF GLER TE

Period: 2003-01-31 to 2016-12-30 (Monthly)   Benchmark: ACWXUSG

	Portfolio Return	Active Return	T-Stat	Specific Return	T-Stat
XUS_CTEF_TE4	14.11%	6.17%	3.59	3.55%	3.10
XUS_CTEF_TE6	16.27%	8.32%	3.89	3.87%	2.85
XUS_CTEF_TE8	18.05%	10.10%	4.28	4.78%	3.15
XUS_GLER_TE4	13.78%	5.83%	3.67	1.93%	1.81
XUS_GLER_TE6	15.11%	7.17%	3.52	3.10%	2.27
XUS_GLER_TE8	16.22%	8.28%	3.82	4.36%	2.94

Model: EM CTEF GLER TE

Period: 2003-01-31 to 2016-12-30 (Monthly)

Benchmark: EMG

	Portfolio Return	Active Return	T-Stat	Specific Return	T-Stat
EM_CTEF_TE4	16.15%	5.43%	3.37	4.08%	3.27
EM_CTEF_TE6	17.34%	6.62%	3.00	3.76%	2.27
EM_CTEF_TE8	19.76%	9.04%	3.67	5.17%	2.97
EM_GLER_TE4	17.72%	7.00%	4.90	4.69%	3.97
EM_GLER_TE6	19.81%	9.09%	4.59	6.21%	3.91
EM_GLER_TE8	20.92%	10.20%	4.43	6.55%	3.81

Table 15.10   A deep dive into EM portfolio construction

January 2003 - December 2016
Axioma Risk Model and Optimizer

	FUND Risk Model Tracking Error			
Model: EM MQ	5.00	6.00	7.00	10.00
Ann. Port Return	16.76	17.34	18.36	20.62
Ann STD	22.06	22.19	22.26	22.40
Ann. Active Return	6.45	7.24	8.05	10.31
Ann. Active Risk	5.04	5.95	6.65	8.35
N	68.20	63.40	58.50	55.40
ShR	0.682	0.714	0.748	0.844
IR	1.281	1.217	1.208	1.235

	FUND AAF20% Risk Model Tracking Error			
Model: EM MQ	5.00	6.00	7.00	10.00
Ann. Port Return	16.53	17.19	18.87	10.44
Ann STD	22.11	22.09	22.22	22.35
Ann. Active Return	6.29	6.89	8.57	10.12
Ann. Active Risk	4.31	5.24	6.94	8.15
N	77.80	69.70	59.80	56.00
ShR	0.671	0.701	0.773	0.838
IR	1.447	1.309	1.234	1.242

AAF in creating global portfolios using GLER for the 1999–2009 time period. The Axioma Statistical Risk Model with an AAF of 20% produced realized tracking errors closer to targeted tracking errors and higher Information Ratios using GLER data than its Statistical Risk Model without AAF Model. See Table 5.7 for supporting Saxena and Stubbs' (2012) position on AAF pushing out the frontiers. Moreover, AAF helps enhance IRs at targeted tracking errors of 5%, 6%, and 7%, respectively, with the Axioma Fundamental Risk Model in the EM universe and targeted tracking errors of 5% and 7% with the Axioma Statistical Risk Model.

The application of the Axioma Fundamental Risk Model to the EM universe produced a real-time portfolio having Geometric Means, Sharpe Ratios, and IRs in the top decile of EM portfolios (see Table 15.10).

The application of the Axioma Fundamental Risk Model and AAF to the EM universe produced a real-time portfolio having Geometric Means, Sharpe Ratios, and IRs in the EM portfolio.

15.10   Financial Anomalies in Global Portfolio Management: Evidence Through COVID-19

In our final section of this chapter, we model financial anomalies through the pandemic of 2020.[11] The initial data used in this analysis are the Morgan Stanley Capital International (MSCI) Index-constituents' data for the All Country World, denoted Global, the Morgan Stanley Capital International (MSCI) Index-constituents' data for the All Country World ex USA Index, denoted non-US, and Russell 3000 (R3) Index, denoted US, for the December 1995 to December 2020 time period. We report an equally weighted Alpha Tester analysis of Global, non-US stock results, and US stocks in Table 5.11, respectively. The CTEF variable and REG8, REG9, and REG10 robust-regression-based models produced the highest and statistically significant ICs for the December 1995 to December 2020 period for 1-, 3-, 6-, and 12-month holding periods in the US, non-US stock, and Global stock universes. Alpha Tester analysis is a similar (alternative) test to the Information Coefficients, ICs, reported in Guerard et al. (2015) in Level I testing.

Our simulation conditions assume 20% quarterly turnover, 35 basis point threshold positions, an upper bound in mean-variance optimization of 4% on security weights, and fixed transactions costs of 125 basis points in US stock portfolios and 150 basis points in non-US stock portfolios..[12] Country maximum weights were the maximum country benchmark weight for 5% or 1.5 times the country benchmark weight. Country minimum weights were the maximum country benchmark weight or zero. We use the mean-variance (MV) optimization techniques. Our portfolios' simulations reported in Table 15.12 use the Axioma Statistical Risk Models (Axioma Medium-Term US Model, version 4, Axioma World-Wide Model, version 4) for portfolio construction. Financial anomalies, as published in 1993 and 2003, continue to outperform, producing over 300–415 basis points, annually, in Risk Total Effect (Active Returns) in most (CTEF, REG9, and REG10) US and non-US stocks during the 1996–2020 time period. Corresponding Global portfolios returns, see Table 15.2, were 150–225 basis points, annually. Is that enough for investors? Should that be enough for investors? The excess returns reported in Table 15.2, produced tracking errors between 6% and 8%, should satisfy investors who seek to maximize the Geometric Mean and the utility of terminal wealth.[13] The Sharpe ratios and information ratios for US and non-US portfolios with tracking errors of 6–8%.

---

[11] This section draws heavily from Markowitz et al. (2021).

[12] ITG estimates McKinley Capital Management transactions costs to be about 60 basis points, each way, for 2011–2015.

[13] Fama (1991) hypothesized that anomalies could not be effectively tested because of changing asset pricing models and the number of factors in multifactor models. Barillas and Shanken (2018) addressed the issue of changing risk models and factors. We are using Statistical Risk Models to construct portfolios. Fundamental risk models, which evolve over time, are used only for attribution analysis. The Active Returns from statistical risk models of 15 factors are invariant to changing risk models for attributions.

The important result of the empirical update is that CTEF, REG8, REG9, and REG10 mean-variance portfolios produce statistically significant portfolio active returns and significant stock-specific returns for the December 1996 to December 2020 time period in non-US stocks. The non-US stock Active Returns, denoted Risk Total Effects, are produced by positive stock selection, denoted Risk Stock Specific Effect, and Value, Size, Medium-Term Momentum and Earnings Yield factor returns. The CTEF and REG8 non-US portfolios produced historic tracking errors of approximately 6%. The corresponding US stock Active Returns, denoted Total Effects, are produced by positive stock selection, denoted Risk Stock Specific Effect, and Value, Size, Medium-Term Momentum and Earnings Yield factor returns. The CTEF and REG8 US portfolios produced historic tracking errors of approximately 6.3–6.9%. We believe that no US portfolio should be run with a tracking error of less than 6% if one wants to outperform the market. In the real world of asset management, Domowitz and Moghe (2018) reported with their Donut strategy that it may well be better to be near-optimal in portfolio construction and its implementation than optimal in portfolio construction and avoid needless debate regarding the relative risk of large benchmark weight stocks.[14] Beheshti et al. (2020) tested the Domowitz and Moghe (2018) Donut strategy and substantiated its results. We continue to explore new modeling techniques in robust regression, test new multifactor risk models, and revise our heuristics in response to changing natural environments, such as the present pandemic, and its effect on analysts' forecasts and price momentum. We concur with the Adaptive Markets Hypothesis of Lo (2004, 2017) and its fifth key point that "survival is the ultimate force driving competition, innovation, and adaptation."

15.11   Summary and Conclusions

Multifactor risk models, MFM, may be used to analyze a stock selection or evaluating variable. Blin et al. (1997) reported that properly constructed risk models created portfolios in which tilt variables that have statistically significantly ICs produced statistically significant Active Returns. One must remove the market effect and extramarket covariances to properly estimate the contribution of a variable to the creation of efficient portfolios. It is important to see how quantitative analysis was developed in investment research and analysis. Harry Markowitz, Bill Sharpe, Jan Mossin, and Jack Treynor pioneered capital market equilibrium and the creation and estimation of the Capital Asset Pricing Model. In the 1970s, Barr Rosenberg and his colleagues at Barra; John Blin and Steve Bender at APT; and Sebastian Ceria at Axioma developed and estimated multifactor risk models. Recent research and

---

[14] Guerard et al. (2015) and Markowitz et al. (2019) reported at the Q-Group Meeting in October 2019 that the robust-regression based models passed the Markowitz-Xu (1994) Data Mining Corrections test.

commercialization by APT and Axioma have furthered portfolio optimization. MCM developed its Horse Race risk model philosophy to assess the most effective risk models to maximize the Geometric Means, Sharpe Ratios, and Information Ratios. APT and Axioma have reigned supreme in our tests. The integration of Axioma and AAF and ITG transactions costs on the FactSet Investment has led to an Axioma portfolio implementation. Recent portfolio modeling supports long-term multifactor risk modeling of portfolios using a regression-based stock selection model.

Appendices

Appendix A: US-E3 Descriptor Definitions

This Appendix gives the detailed definitions of the descriptors which underlie the risk indices in US-E3. The method of combining these descriptors into risk indices is proprietary to BARRA.

1. Volatility

(i) *BTSG: Beta times sigma*

This is computed as $\sqrt{\beta}\sigma_\varepsilon$, where β is the historical beta and σ_ε is the historical residual standard deviation. If β is negative, then the descriptor is set equal to zero.

(ii) *DASTD: Daily standard deviation*

This is computed as:

$$\sqrt{N_{days}\left[\sum\nolimits_{t=1}^{T} w_t r_t^2\right]}$$

where r_t is the return over day t, w_t is the weight for day t, T is the number of days of historical returns data used to compute this descriptor (we set this to 65 days), and N_{days} is the number of trading days in a month (we set this to 23).

(iii) *HILO: Ratio of high price to low price over the last month*

This is calculated as:

$$\log\left(\frac{P_H}{P_L}\right)$$

where P_H and P_L are the maximum price and minimum price attained over the last 1 month.

(iv) *LPRI: Log of stock price*

This is the log of the stock price at the end of last month.

(v) *CMRA: Cumulative range*

Let Z_t be defined as follows:

$$Z_t = \sum_{s=1}^{t} \log\left(1 + r_{i,s}\right) - \sum_{s=1}^{t} \log\left(1 + r_{f,s}\right)$$

where $r_{i,s}$ is the return on stock I in month s, and $r_{f,s}$ is the risk-free rate for month s. In other words, Z_t is the cumulative return of the stock over the risk-free rate at the end of month t. Define Z_{max} and Z_{min} as the maximum and minimum values of Z_t over the last 12 months. CMRA is computed as:

$$\log\left(\frac{1 + Z_{max}}{a + Z_{min}}\right)$$

(vi) *VOLBT: Sensitivity of changes in trading volume to changes in aggregate trading volume*

This may be estimated by the following regression:

$$\frac{\Delta V_{i,t}}{N_{i,t}} = a + b\, \frac{\Delta V_{M,t}}{N_{m,t}} + \xi_{i,t}$$

where $\Delta V_{i,t}$ is the change in share volume of stock I from week t − 1 to week t, $N_{i,t}$ is the average number of shares outstanding for stock I at the beginning of week t − 1 and week t, $\Delta V_{M,t}$ is the change in volume on the aggregate market from week t − 1 to week t, and $N_{M,t}$ is the average number of shares outstanding for the aggregate market at the beginning of week t − 1 and week t.

(vii) *SERDP: Serial dependence*

This measure is designed to capture serial dependence in residuals from the market model regressions. It is computed as follows:

$$SERDP = \frac{\frac{1}{T-2}\sum_{t=3}^{T}\left(e_t + e_{t-1} + e_{t-2}\right)^2}{\frac{1}{T-2}\sum_{t=3}^{T}\left(e^2_t + e^2_{t-1} + e^2_{t-2}\right)}$$

where e_t is the residual from the market model regression in month t, and T is the number of months over which this regression is run (typically, T = 60 months).

(viii) *OPSTD: Option-implied standard deviation*

This descriptor is computed as the implied standard deviation from the Black-Scholes option pricing formula using the price on the closest to at-the-money call option that trades on the underlying stock.

2. Momentum

(i) *RSTR: Relative strength*

This is computed as the cumulative excess return (using continuously compounded monthly returns) over the last 12 months – i.e.,

$$\text{RSTR} = \sum_{t=1}^{T} \log\left(1 + r_{i,t}\right) - \sum_{t=1}^{T} \log\left(1 + r_{f,t}\right)$$

where $r_{i,t}$ is the arithmetic return of the stock in month I, and $r_{f,t}$ is the arithmetic risk-free rate for month i. This measure is usually computed over the last 1 year—i.e., T is set equal to 12 months.

(ii) *HALPHA: Historical alpha*

This descriptor is equal to the alpha term (i.e., the intercept term) from a 60-month regression of the stock's excess returns on the SandP 500 excess returns.

3. Size

(i) *LNCAP: Log of Market capitalization*

This descriptor is computed as the log of the market capitalization of equity (price times number of shares outstanding) for the company.

4. Size Nonlinearity

(i) *LCAPCB: Cube of the log of market capitalization*

This risk index is computed as the cube of the normalized log of market capitalization.

5. Trading Activity

(i) *STOA: Share turnover over the last year*

STOA is the annualized share turnover rate using data from the last 12 months—i.e., it is equal to $V_{\text{ann}}/\overline{N}_{\text{out}}$, where V_{ann} is the total trading volume (in number of shares) over the last 12 months and $\overline{N}_{\text{out}}$ is the average number of shares outstanding over the previous 12 months (i.e., it is equal to the average value of the number of shares outstanding at the beginning of each month over the previous 12 months).

(ii) *STOQ: Share turnover over the last quarter*

This is computed as the annualized share turnover rate using data from the most recent quarter. Let V_q be the total trading volume (in number of shares) over the most recent quarter and let $\overline{N}_{\text{out}}$ be the average number of shares outstanding over the period (i.e., $\overline{N}_{\text{out}}$ is equal to the average value of the number of shares outstanding at the beginning of each month over the previous 3 months). Then, STOQ is computed as $4V_q/\overline{N}_{\text{out}}$.

(iii) *STOM: Share turnover over the last month*

This is computed as the share turnover rate using data from the most recent month —i.e., it is equal to the number of shares traded last month divided by the number of shares outstanding at the beginning of the month.

(iv) *STO5: Share turnover over the last 5 years*

This is equal to the annualized share turnover rate using data from the last 60 months. In symbols, STO5 is given by:

$$\text{STO5} = \frac{12 \left[\frac{1}{T} \sum_{s=1}^{T} V_s\right]}{\sigma_\varepsilon}$$

where V_s is equal to the total trading volume in month s, and $\overline{N}_{\text{out}}$ is the average number of shares outstanding over the last 60 months.

(v) *FSPLIT: Indicator for forward split*

This descriptor is a 0–1 indicator variable to capture the occurrence of forward splits in the company's stock over the last 2 years.

(vi) *VLVR: Volume to variance.*

This measure is calculated as follows:

$$\text{VLVR} = \log \frac{\frac{12}{T} \left[\sum_{s=1}^{T} V_s P_s\right]}{\sigma_\varepsilon}$$

where V_s equals the number of shares traded in month s, P_s is the closing price of the stock at the end of month s, and σ_ε is the estimated residual standard deviation. The sum in the numerator is computed over the last 12 months.

6. Growth

(i) *PAYO: Payout ratio over 5 years*

This measure is computed as follows:

$$\text{PAYO} = \frac{\frac{1}{T}\sum_{t=1}^{T} D_t}{\frac{1}{T}\sum_{t=1}^{T} E_t}$$

where D_t is the aggregate dividend paid out in year t, and E_t is the total earnings available for common shareholders in year t. This descriptor is computed using the last 5 years of data on dividends and earnings.

(ii) *VCAP: Variability in capital structure*

This descriptor is measured as follows:

$$\text{VCAP} = \frac{\frac{1}{T-1}\sum_{t=2}^{T}(|N_{t-1} - N_t|P_{t-1} + |LD_{t-1} - LD_t| + |PE_t - PE_{t-1}|)}{CE_T + LD_T + PE_T}$$

where N_{t-1} is the number of shares outstanding at the end of time $t-1$; P_{t-1} is the price per share at the end of time $t-1$; LD_{t-1} is the book value of long-term debt at the end of time period $t-1$; PE_{t-1} is the book value of preferred equity at the end of time period $t-1$; and $CE_T + LD_T + PE_T$ are the book values of common equity, long-term debt, and preferred equity as of the most recent fiscal year.

(iii) *AGRO: Growth rate in total assets*

To compute this descriptor, the following regression is run:

$$TA_{it} = a + bt + \xi_{it}$$

where TA_{it} is the total assets of the company as of the end of year t, and the regression is run for the period $= 1, \ldots, 5$. AGRO is computed as follows:

$$\text{AGRO} = \frac{b}{\frac{1}{T}\sum_{t=1}^{T} TA_{it}}$$

where the denominator average is computed over all the data used in the regression.

(iv) *EGRO: Earnings growth rate over last 5 years*

First, the following regression is run:

$$EPS_t = a + bt + \xi_t$$

where EPS_t is the earnings per share for year t. This regression is run for the period $t = 1, \ldots, 5$. EGRO is computed as follows:

$$EGRO = \frac{b}{\frac{1}{T}\sum_{t=1}^{T}EPS_t}$$

(v) *EGIBS: Analyst-predicted earnings growth*

This is computed as follows:

$$EGIBS = \frac{(EARN - EPS)}{(EARN + EPS)/2}$$

where EARN is a weighted average of the median earnings predictions by analysts for the current year and next year, and EPS is the sum of the four most recent quarterly earnings per share.

(vi) *DELE: Recent earnings change*

This is a measure of recent earnings growth and is measured as follows:

$$DELE = \frac{(EPS_t - EPS_{t-1})}{(EPS_t + EPS_{t-1})/2}$$

where EPS_t is the earnings per share for the most recent year, and EPS_{t-1} is the earnings per share for the previous year. We set this to missing if the denominator is nonpositive.

7. Earnings Yield

(i) *EPIBS: Analyst-predicted earnings-to-price*

This is computed as the weighted average of analysts' median predicted earnings for the current fiscal year and next fiscal year divided by the most recent price.

(ii) *ETOP: Trailing annual earnings-to-price*

This is computed as the sum of the four most recent quarterly earnings per share divided by the most recent price.

(iii) *ETP5: Historical earnings-to-price*

This is computed as follows:

$$\text{ETP5} = \frac{\frac{1}{T}\sum_{t=1}^{T}\text{EPS}_t}{\frac{1}{T}\sum_{t=1}^{T}P_t}$$

where EPS_t is equal to the earnings per share over year t, and P_t is equal to the closing price per share at the end of year t.

8. Value

(i) *BTOP: Book-to-price ratio*

This is the book value of common equity as of the most recent fiscal year end divided by the most recent value of the market capitalization of the equity.

9. Earnings Variability

(i) *VERN: Variability in earnings*

This measure is computed as follows:

$$\text{VERN} = \frac{\left(\frac{1}{T-1}\sum_{t=1}^{T}(E_t - \bar{E})^2\right)^{\frac{1}{2}}}{\frac{1}{T}\sum_{t=1}^{T}E_t}$$

where E_t is the earnings at time t($t = 1,\ldots,5$), and $\bar{E}$ is the average earnings over the last 5 years. VERN is the coefficient of variation of earnings.

(ii) *VFLO: Variability in cash flows*

This measure is computed as the coefficient of variation of cash flow using data over the last 5 years—i.e., it is computed in an identical manner to VERN, with cash flow being used in pace of earnings. Cash flow is computed as earnings plus depreciation plus deferred taxes.

(iii) *EXTE: Extraordinary items in earnings*

This is computed as follows:

$$\text{EXTE} = \frac{\frac{1}{T}\sum_{t=1}^{T}|\,\text{EX}_t + \text{NRI}_t\,|}{\frac{1}{T}\sum_{t=1}^{T}E_t}$$

where EX_t is the value of extraordinary items and discontinued operations, NRI_t is the value of nonoperating income, and E_t is the earnings available to common before extraordinary items. The descriptor uses data over the last 5 years.

(iv) *SPIBS: Standard deviation of analysts' prediction to price*

This is computed as the weighted average of the standard deviation of IBES analysts' forecasts of the firm's earnings per share for the current fiscal year and next fiscal year divided by the most recent price.

10. Leverage

(i) *MLEV: Market leverage*

This measure is computed as follows:

$$\text{MLEV} = \frac{\text{ME}_t + \text{PE}_t + \text{LD}_t}{\text{ME}_t}$$

where ME_t is the market value of common equity, PE_t is the book value of preferred equity, and LD_t is the book value of long-term debt. The value of preferred equity and long-term debt are as of the end of the most recent fiscal year. The market value of equity is computed using the most recent month's closing price of the stock.

(ii) *BLEV: Book leverage*

This measure is computed as follows:

$$\text{BLEV} = \frac{\text{CEQ}_t + \text{PE}_t + \text{LD}_t}{\text{CEQ}_t}$$

where CEQ_t is the book value of common equity, PE_t is the book value of preferred equity, and LD_t is the book value of the long-term debt. All values are as of the end of the most recent fiscal year.

(iii) *DTOA: Debt-to-assets ratio*

This ratio is computed as follows:

$$\text{DTOA} = \frac{\text{LD}_t + \text{DCL}_t}{\text{TA}_t}$$

where LD_t is the book value of long-term debt, DCL_t is the value of debt in current liabilities, and TA_t is the book value of total assets. All values are as of the end of the most recent fiscal year.

(iv) *SNRRT: Senior debt rating*

This descriptor is constructed as a multilevel indicator variable of the debt rating of a company.

11. Currency Sensitivity

(i) *CURSENS: Exposure to foreign currencies*

To construct this descriptor, the following regression is run:

$$r_{it} = \alpha_I + \beta_i r_{mt} + \varepsilon_{it}$$

where r_{it} is the excess return on the stock, and r_{mt} is the excess return on the SandP 500 Index. Let ε_{it} denote the residual returns from this regression. These residual returns are in turn regressed against the contemporaneous and lagged returns on a basket of foreign currencies, as follows:

$$\varepsilon_{it} = c_i + \gamma_{i1}(FX)_t + \gamma_{i2}(FX)_{t-1} + \gamma_{13}(FX)_{t-2} + \mu_{it}$$

where ε_{it} is the residual return on stock I, $(FX)_t$ is the return on an index of foreign currencies over month t, $(FX)_{t-1}$ is the return on the same index of foreign currencies over month $t - 1$, and $(FX)_{t-2}$ is the return on the same index over month $t - 2$. The risk index is computed as the sum of the slope coefficients γ_{i1}, γ_{i2}, and γ_{i3}—i.e., CURSENS $= \gamma_{i1} + \gamma_{i2} + \gamma_{i3}$.

12. Dividend Yield

(i) *P_DYLD: Predicted dividend yield*

This descriptor uses the last four quarterly dividends paid out by the company along with the returns on the company's stock and future dividend announcements made by the company to come up with a BARRA-predicted dividend yield.

13. Nonestimation Universe Indicator

(i) *NONESTU: Indicator for firms outside US-E3 estimation universe*

This is a 0–1 indicator variable: It is equal to 0 if the company is in the BARRA estimation universe and equal to 1 if the company is outside the BARRA estimation universe.

Appendix B: Factor Alignment Problems and Quantitative Portfolio Management

Until now we have discussed Markowitz mean-variance optimization (MVO) framework, evolution of factor models in modeling risk and return, and their application in building reliable and robust expected returns models. Next, we move our focus to the application of factor models within the broader aegis of quantitative portfolio managements (QPM).

Factor models play an integral role in quantitative equity portfolio management. Their applications extend to almost every aspect of quantitative investment methodology including construction of alpha models, risk models, portfolio construction, risk decomposition, and performance attribution. Given their pervasive presence in the field and the natural trend toward specialization, it comes as no surprise that different groups of researchers are often involved in developing factor models for each one of the aforementioned applications.

For instance, a team of quantitative portfolio managers (PM) can develop an in-house model for alpha generation and procure a factor model for purposes of risk management from a third-party risk model vendor. Subsequently, they can combine the two models within the framework of Markowitz mean-variance optimization (MVO) framework to construct optimal portfolios. A completely different factor model can then be used for purposes of performance attribution to identify the key drivers and detractors of performance. Notably, the choices of factors in each one of these factor models need not be identical, thereby introducing incongruity in the portfolio management process. To further complicate the matters, the constraints in the quantitative strategy can introduce additional systematic risk exposures that are not captured by the risk model. The problems that arise due to the interaction between the alpha model, the risk model, and constraints in MVO framework are collectively referred to as Factor Alignment Problems (FAP). Detailed theoretical investigation of FAP and subsequent evolution of augmented risk models has become an interesting line of research in recent years. We refer the readers to Saxena and Stubbs (2012, 2013, 2015), Ceria et al. (2012), and Martin et al. (2013) for detailed analysis on these topics. A brief summary of the key themes is presented next.

We initiate the discussion by focusing on the unconstrained MVO problem, namely,

$$\text{Maximize } \alpha * h - (1/2) * h^T * Q * h,$$

where α, h, and Q denote the expected returns (referred to as alpha), portfolio holdings, and covariance matrix of returns based on a factor risk model, respectively. Furthermore, let X denote the matrix of risk factors that are used in building the covariance matrix Q. It is instructive to examine the relationship between α and X. Specifically, consider the projection of α on the vector space spanned by the columns of X; the said projection, α_X, is referred to as the spanned component of α, whereas the residual $\alpha_O = \alpha - \alpha_X$ is referred to as the orthogonal component. Even though both the spanned and orthogonal components are derived from a common source, namely α, they have widely different risk characteristics especially when examined through the lens of the covariance matrix Q.

By virtue of being spanned by the risk factors X, the spanned component is deemed to have systematic risk when examined by the risk model. Among other things, this implies that any portfolio exposure to the spanned component is penalized in the MVO optimization framework for both systematic and specific risks. The orthogonal component, on the other hand, is assumed to have no systematic risk since $X^T * \alpha_X = 0$. This implies that any exposure to the orthogonal component gets a free pass since it is penalized only for the specific risk. Furthermore, as the number of holdings in the portfolio increases, specific risk can asymptotically go down to zero due to diversification benefits. This disparity between the risk treatment of the spanned and orthogonal component creates incentives for the optimizer to the load on the orthogonal component at the cost of exposure to the spanned component.

Such an overloading on the orthogonal component is in itself not particularly troublesome. In fact, if the risk model captures "all" possible sources of systematic risk, such an overloading is completely justified by the portfolio construction philosophy of MVO. Unfortunately, risk models are not designed to capture all sources of risk. Instead, they are designed to capture salient sources of systematic risk, and often omit systematic risk factors that are deemed to be insignificant. These design choices are motivated by a variety of reasons including but not limited to statistical power concerns, preference for parsimony, aversion to multicollinearity, etc. Consequently, choices made during the design of a risk model can result in risk "blind spots," namely, systematic risk factors that are missing from the risk model. If the alpha model happens to bet on these risk blind spots, the resulting systematic risk exposures will be left untreated during the portfolio construction phase resulting in latent systematic risk exposures. Next, we discuss some of the practical consequences of such latent risk exposures.

Most QPM employ an active or total risk constraint. These constraints are meant to limit the volatility of the portfolio and avoid wild swings in portfolio performance. For instance, a portfolio manager (PM) can include an active risk constraint as part of the portfolio construction process to limit the annualized volatility of active returns to be less than or equal to a predetermined limit, say 3%. The expression of this constraint during portfolio optimization necessarily requires a risk model. If the risk and alpha model suffer from FAP, the optimizer will load up on the orthogonal component of alpha, and the realized risk of the portfolio can be significantly higher than 3%. Saxena and Stubbs (2013) studied this phenomenon in detail and found that the divergences in risk prediction that result from FAP tend to be statistically significant and increase during periods of market stress. In other words, presence of FAP can lead of "volatility" surprises in the realized performance of portfolio, and these surprises tend to aggravate during periods of market turmoil. Additionally, they were able to quantify the amount of latent systematic risk in the orthogonal component of alpha and showed that α_O tends to be more volatile than almost half of the risk factors included in a typical commercial risk model. Ceria et al. examined potential sources of latent systematic risk factors and argued that proprietary definitions of well-known factors (momentum, value, growth, etc.) can introduce latent systematic risk factors. In other words, a portfolio manager's propensity to curate an alpha factor not only leads to better performance (aka alpha) but also introduces latent systematic risk exposures that often go undiagnosed during the portfolio construction process. Martin et al. (2013) examined the role of constraints in FAP and concluded certain kinds of portfolio constraints can also create unintended systematic risk factor exposures which can expose the portfolio to the vagaries of FAP.

Risk underestimation, as discussed here, is merely a symptom of a bigger issue, namely, compromise of the mean-variance optimality of the resulting portfolio holdings. Recall that the primary objective of portfolio optimization is to create a portfolio having an optial risk-return tradeoff. If a portion of systematic risk exposure of the portfolio is inadequately captured by the risk model, then the resulting portfolio cannot be expected to be optimal ex-post, its ex-ante optimality notwithstanding. Factor Alignment Problems (FAP) symbolize the difficulty that a portfolio manager (PM) faces in ensuring consistency between the ex-post characteristics of an optimized portfolio, and her ex-ante expectations. Lack of alignment between the alpha and risk factors creates risk "blind-spots" that get exaggerated by virtue of employing an optimizer. The resulting optimized portfolios take excessive exposure to systematic factors missing from the risk model, thereby compromising the overarching goal of optimal risk-return tradeoff, as originally envisaged by Harry Markowitz. Saxena and Stubbs (2015) demonstrate the usefulness of augmented risk models in addressing this issue. Their results indicate that augmenting the risk model with an appropriate augmenting factor not only remedies the risk underestimation problem but also improves risk adjusted returns, thereby restoring the notion of Markowitz efficiency.

Among other things, their results suggest strong synergistic advantages of integrating alpha and risk research processes. In other words, we need to abandon the "one-size-fits-all" approach to risk management and take a more nuanced approach that is sensitive to the specific requirements of a PM. Ultimately, the primary responsibility of a risk model is to capture all undiversifiable (i.e., systematic) sources of risk that are relevant to a given investment process. A risk model that is constructed in a manner which is agnostic to the very factors that the PM is betting on cannot be expected to accomplish that goal. Augmented risk models partly accomplish this goal and should act as precursor to fully customized risk models that shed the artificial barrier between alpha and risk research and take a holistic view of the investment process.

Bibliography

Martin, C., Saxena, A., & Stubbs, R. A. (2013). Constraints in quantitative strategies: An alignment perspective. *Journal of Asset Management, 14*(5).

Ceria, S., Saxena, A., & Stubbs, R. A. (2012). Factor alignment problems and quantitative portfolio management. *The Journal of Portfolio Management, 38*(2), 29–43.

Guerard, J. B. Jr., Krauklis, E., & Kumar, M. (2012). Further analysis of efficient portfolios with the user data. *Journal of Investing, 21*(1), 81–88.

Saxena, A., & Stubbs, R. A. (2012). An empirical case study of factor alignment problems using the user model. *Journal of Investing, 21*(1):25–43.

Saxena, A., & Stubbs, R. A. (2013). Alpha alignment factor: A solution to the underestimation of risk for optimized active portfolios. *The Journal of Risk 15*(3), 3–37.

Saxena, A., & Stubbs, R. (2015). Augmented risk models to mitigate factor alignment problems. *Journal of Investment Management, 13*(3).

References

Arnott, R. (1985). The use and misuse of consensus earnings. *Journal of Portfolio Management,* 18–27.

Barillas, F., & Shanken, J. (2018). Comparing asset pricing models. *Journal of Finance, 73,* 715–755.

Basu, S. (1977). Investment performance of common stocks in relations to their price earnings ratios: A test of market efficiency. *Journal of Finance, 32,* 663–682.

Baumol, W. J. (1963). An expected gain-confidence limit criteria for portfolio selection. *Management Science, 10,* 174–182.

Beaton, A. E., & Tukey, J. W. (1974). The fitting of power series, meaning polynomials, illustrated on bank-spectroscopic data. *Technometrics, 16,* 147–185.

Black, F., Jensen, M. C., & Scholes, M. (1972). The capital asset pricing model: Some empirical tests. In M. Jensen (Ed.), *Studies in the theory of capital markets.* Praeger.

Blin, J. M., Bender, S., & Guerard, J. B., Jr. (1997). Earnings forecasts, revisions and momentum in the estimation of efficient market-neutral Japanese and U.S. portfolios. In A. Chen (Ed.), *Research in finance* (Vol. 15). JAI Press.

Bloch, M., Guerard, J. B., Jr., Markowitz, H. M., Todd, P., & Xu, G.-L. (1993). A comparison of some aspects of the U.S. and Japanese equity markets. *Japan and the World Economy, 5,* 3–26.

Borkovec, M., Domowitz, I., Kiernan, B., & Serbin, V. (2010). Portfolio optimization and the cost of trading. *Journal of Investing, 19,* 63–76.

Chen, N.-F., Roll, R., & Ross, S. (1986). Economic forces and the stock market. *Journal of Business, 59,* 383–403.

Chopra, V. K., & Ziemba, W. T. (1993). The effect of errors in mean-variance and covariance on optimal portfolio choice. *Journal of Portfolio Management, 19,* 6–11.

Christopherson, J. A., Carino, D. R., & Ferson, W. E. (2009). *Portfolio performance measurement and benchmarking.* McGraw-Hill Companies.

Cohen, K. J., & Pogue, J. A. (1967). An empirical evaluation of alternative portfolio-selection models. *Journal of Business, 40,* 166–193.

Conner, G., & Korajczyk, R. A. (1995). The arbitrary theory and multifactor models of asset returns. In R. A. Jarrow, V. Malsimovic, & W. T. Ziemba (Eds.), *Handbooks in Operations Research and Management Science: Finance* (Vol. 9, pp. 87–144). Elsevier.

Connor, G., Goldberg, L., & Korajczyk, R. A. (2010). *Portfolio risk analysis*. Princeton University Press.

Dhrymes, P. J., Friend, I., & Gultekin, N. B. (1984). A critical re-examination of the empirical evidence on the arbitrage pricing theory. *Journal of Finance, 39*, 323–346.

Dhrymes, P. J., Friend, I., Gültekin, M. N., & Gültekin, N. B. (1985). New tests of the APT and their implications. *Journal of Finance, 40*, 659–674.

Dimson, E. (1988). *Stock market anomalies*. Cambridge University Press.

Domowitz, I., & Moghe, A. (2018). Donuts: A picture of optimization applied to fundamental portfolios. *Journal of Portfolio Management, 44*, 103–113.

Elton, E. J., & Gruber, M. J. (1970). Homogeneous groups and the testing of economic hypothesis. *Journal of Financial and Quantitative Analysis, 5*, 581–602.

Elton, E. J., & Gruber, M. J. (1973). Estimating the dependence structure of share prices—Implications for portfolio selection. *Journal of Finance, 28*, 1203–1232.

Elton, E. J., Gruber, M. J., & Padberg, M. (1978). Simple criteria for optimal portfolio selection: Tracing out the efficient frontier. *Journal of Finance, 33*, 296–302.

Elton, E. J., Gruber, M. J., & Padberg, M. W. (1979). Simple criteria for optimal portfolio selection: The multi-index case. In E. J. Elton & M. J. Gruber (Eds.), *Portfolio theory, 25 years after: Essays in honor of Harry Markowitz*. North-Holland.

Elton, E. J., Gruber, M. J., Brown, S. J., & Goetzman, W. N. (2007). *Modern portfolio theory and investment analysis* (7th ed.). Wiley.

Fama, E. F. (1970). Efficient capital markets: A review of theory and empirical work. *Journal of Finance, 26*, 383–417.

Fama, E. F. (1976). *Foundations of finance*. Basic Books.

Fama, E. F., & MacBeth, J. D. (1973). Risk, return, and equilibrium: Empirical tests. *Journal of Political Economy, 81*, 607–636.

Fama, E. F. (1991b). Efficient capital markets II. *Journal of Finance, 46*, 1575–1617.

Fama, E. F., & French, K. R. (1992b). The cross-section of expected stock returns. *Journal of Finance, 47*, 427–465.

Fama, E. F., & French, K. R. (1993). Common risk factors in the returns on stocks and bonds. *Journal of Finance, 50*, 131–155.

Fama, E. F., & French, K. R. (2008b). Dissecting anomalies. *Journal of Finance, 63*, 1653–1678.

Fama, E. F., & French, K. R. (2016). A five-factor asset pricing model. *Journal of Financial Economics, 118*(1), 1–22.

Farrell, J. L., Jr. (1974). Analyzing covariance of returns to determine homogeneous stock groupings. *Journal of Business, 47*, 186–207.

Farrell, J. L., Jr. (1997). *Portfolio management: Theory and applications*. McGraw-Hill/Irwin.

Ferson, W. E., & Harvey, C. R. (1995). Explaining the predictability of asset returns. In A. Chen (Ed.), *Research in finance* (Vol. 11). JAI Press.

Ferson, W., & Schadt, R. Measuring fund strategy and performance in changing economic conditions. *Journal of Finance, 51*, 425–462.

Fogler, R., John, K., & Tipton, J. (1981). Three factors, interest rate differentials, and stock groups. *Journal of Finance, 36*, 323–335.

Geczy, C., Guerard, J. B., Jr., & Samanov, M. W. (2020). SRI need not kill your Sharpe and information ratios—Forecasting of earnings and efficient SRI and ESG portfolios. *Journal of Investing, 29*, 110–127.

Gibbons, M. R., Ross, S. A., & Shanken, J. (1989). A test of efficiency of a given portfolio. *Econometrica, 57*, 1121–1152.

Graham, B. (1973b). *The intelligent investor*. Harper & Row.

Grinhold, R., & Kahn, R. (1999). *Active portfolio management*. McGraw-Hill/Irwin.

Guerard, J. B., Jr., & Stone, B. K. (1992b). Composite forecasting of annual corporate earnings. In A. Chen (Ed.), *Research in finance* (Vol. 10).

Guerard, J. B., Jr. (2012). Global earnings forecasting efficiency. In J. Kensinger (Ed.), *Research in finance* (Vol. 28). JAI Press.

Guerard, J. B., Jr., & Mark, A. (2003). The optimization of efficient portfolios: The case for a quadratic R&D term. In *Research in finance* (Vol. 20, pp. 213–247). JAI Press.

Guerard, J. B., Jr., & Mark, A. (2018, forthcoming). Earnings forecasts and revisions, price momentum, and fundamental data: Further explorations of financial anomalies. In C. F. Lee (Ed.), *Handbook of financial econometrics*. World Scientific.

Guerard, J. B., Jr., Gultekin, M., & Stone, B. K. (1997). The role of fundamental data and analysts' earnings breadth, forecasts, and revisions in the creation of efficient portfolios. In A. Chen (Ed.), *Research in finance* (Vol. 15). JAI Press.

Guerard, J. B., Jr., Xu, G., & Gultekin, M. N. (2012). Investing with momentum: The past, present, and future. *Journal of Investing, 21*, 68–80.

Guerard, J. B., Jr., Rachev, R. T., & Shao, B. (2013). Efficient global portfolios: Big data and investment universes. *IBM Journal of Research and Development, 57*, 11:1–11:11.

Guerard, J. B., Jr., Markowitz, H. M., & Xu, G. (2014). The role of effective corporate decisions in the creation of efficient portfolios. *IBM Journal of Research and Development, 586*, 1–11.

Guerard, J. B., Jr., Markowitz, H. M., & Xu, G. (2015). Earnings forecasting in a global stock selection model and efficient portfolio construction and management. *International Journal of Forecasting, 31*, 550–560.

Guerard, J. B., Jr., Gillam, R. A., Markowitz, H. M., Xu, G., Deng, S., & Wang, Z. (2018a). Data mining corrections testing in Chinese stocks. *Interfaces, 48*, 108–120.

Guerard, J. B., Jr., Markowitz, H. M., Xu, G., & Wang, E. (2018b). Global portfolio construction with emphasis on conflicting corporate strategies to maximize stockholder wealth. *Annals of Operations Research, 267*, 203–219.

Guerard, J. B., Jr., & Mark, A. (2020b). Earnings forecasts and revisions, Price momentum, and fundamental data: Further explorations of financial anomalies. In C. F. Lee (Ed.), *Handbook of financial econometrics*. World Scientific Handbook in Financial Economics. forthcoming.

Gunst, R. F., Webster, J. T., & Mason, R. L. (1976b). A comparison of least squares and latent root regression estimators. *Technometrics, 18*, 75–83.

Hirschleifer, D. (2014). Behavioral finance. *Annual Review of Financial Economics, 7*, Submitted. https://doi.org/10.1146/annurev-financial-092214-043752

Hirschleifer, D., Hon, K., & Teoh, S. H. (2012). The accrual anomaly: Risk or mispricing? *Management Science, 57*, 1–16.

Hirshleifer, D. (2001). Investor psychology and asset pricing. *Journal of Finance, 64*, 1533–1597.

Hong, H., & Kubik, J. D. (2003). Analyzing the analysts: Career concerns and biased earnings forecasts. *Journal of Finance, 58*(2003), 313–351.

Hong, H., Kubik, J. D., & Solomon, A. (2000). Security analysts' career concerns and the herding of earnings forecasts. *RAND Journal of Economics, 31*, 121–144.

Jacobs, B. I., & Levy, K. (1988). Disentangling equity return regularities: New insights and investment opportunities. *Financial Analysts Journal, 44*, 18–43.

Jacobs, B. I., & Levy, K. N. (2010). Reflections on portfolio insurance, portfolio theory, and market simulation with Harry Markowitz. In J. B. Guerard Jr. (Ed.), *The handbook of portfolio construction: Contemporary applications of Markowitz techniques*. Springer.

Jacobs, B. I., & Levy, K. (2017). *Equity management: The art and science of modern quantitative investing* (2nd ed.). McGraw-Hill.

Jagannathan, R., & Korajczyk, R. A. (1986). Assessing the market timing performance of managed portfolios. *Journal of Business, 59*, 217–235.

James, W., & Stein, C. (1961). Estimation with quadratic loss. *Proceedings of the Fourth Berkeley Symposium on Mathematical Statistics and Probability, 1*, 361–379.

Jensen, M. C. (1968). The performance of mutual funds in the period 1945–1964. *Journal of Finance, 23*, 389–416.

Jensen, M. C. (1972). Optimal utilization of market forecasts and the evaluation of investment performance. In G. P. Szego & K. Shell (Eds.), *Mathematical methods in investment and finance*. Elsevier.

King, B. F. (1966). Market and industry factors in stock Price behavior. *Journal of Business, 39*, 139–191.

Levy, H. (2012). *The capital asset pricing model in the 21st century*. Cambridge University Press.

Lo, A. W. (2004). The adaptive markets hypothesis: Market efficiency from an evolutionary perspective. *The Journal of Portfolio Management, 30*, 15–29.

Lo, A. (2017). *Adaptive markets*. Princeton University Press.

Lo, A. W., Mamaysky, H., & Wang, J. (2000). Foundations of technical analysis: Computational algorithms, statistical inference, and empirical implementation. *Journal of Finance, 60*, 1705–1764.

Malkiel, B. (1996). *A random walk down Wall Street* (6th ed.). W.W. Norton and Company.

Markowitz, H. M. (1959). *Portfolio selection: Efficient diversification of investment* (Cowles Foundation monograph no. 16). Wiley.

Markowitz, H. M. (2000). *Mean-variance analysis in portfolio choice and capital markets*. Frank J. Fabozzi Associates.

Markowitz, H. M., Guerard, J. B., Jr., Xu, G., & Behesti, B. (2021, May). Financial anomalies in portfolio construction and management. *Journal of Portfolio Management., 47*(6), jpm.2021.1.242.

Miller, W., Xu, G., & Guerard, J. B., Jr. (2014). Portfolio construction and management in the BARRA aegis system: A case study using the USER data. *Journal of Investing, 23*, 111–120.

Mossin, J. (1966). Equilibrium in a capital asset market. *Econometrica, 34*, 768–783.

Mossin, J. (1973). *Theory of financial markets*. Prentice-Hall, Inc.

Reinganum, M. (1981). The arbitrage pricing theory: Some empirical tests. *Journal of Finance, 36*, 313–321.

Roll, R., & Ross, S. A. The arbitrage pricing theory approach to strategic portfolio planning. *Financial Analysts Journal, 51*, 14–26.

Rosenberg, B. (1974). Extra-market components of covariance in security returns. *Journal of Financial and Quantitative Analysis, 9*, 263–274.

Rosenberg, B. (1976). Security appraisal and unsystematic risk in institutional investment. *Proceedings of the seminar for research in security prices*. University of Chicago.

Rosenberg, B., & Marathe, V. (1979). Tests of capital asset pricing hypotheses. In H. Levy (Ed.), *Research in finance* (Vol. 1). JAI Press.

Rosenberg, B., & McKibben, W. (1973). The prediction of systematic and unsystematic risk in common stocks. *Journal of Financial and Quantitative Analysis, 8*, 317–333.

Rosenberg, B., & Rudd, A. (1977). *The yield/beta/residual risk tradeoff*. Working paper no. 66. Research Program in Finance. Institute of Business and Economic Research. University of California, Berkeley.

Rosenberg, B., Hoaglet, M., Marathe, V., & McKibben, W. (1975). *Components of covariance in security returns*. Working paper no. 13, Research Program in Finance. Institute of Business and Economic Research. University of California. Berkeley.

Ross, S. A. (1976). The arbitrage theory of capital asset pricing. *Journal of Economic Theory, 13*, 341–360.

Ross, S. A., & Roll, R. (1980). An empirical investigation of the arbitrage pricing theory. *Journal of Finance, 35*, 1071–1103.

Rudd, A., & Clasing, H. K. (1982). *Modern portfolio theory: The principles of investment management*. Dow Jones-Irwin.

Rudd, A., & Rosenberg, B. (1979). Realistic portfolio optimization. In E. Elton & M. Gruber (Eds.), *Portfolio theory, 25 years after*. North-Holland.

Rudd, A., & Rosenberg, B. (1980). The 'market model' in investment management. *Journal of Finance, 35*, 597–607.

Saxena, A., & Stubbs, R. A. (2012). An empirical case study of factor alignment using the USER model. *Journal of Investing, 21*, 25–44.

Sharpe, W. F. (1963). A simplified model for portfolio analysis. *Management Science, 9*, 277–293.

Sharpe, W. F. (1964). Capital asset prices: A theory of market equilibrium under conditions of risk. *Journal of Finance, 19*, 425–442.

Sharpe, W. F. (1971). A linear programming approximation for the general portfolio analysis problem. *Journal of Financial and Quantitative Analysis, 6*, 263–1275.

Sharpe, W. F. (2012). *William F. Sharpe: Selected works*. World Scientific Publishing Company.

Stone, B. K. (1970). *Risk, return, and equilibrium: A general single-period theory of asset selection and capital market equilibrium*. MIT Press.

Stone, B. K. (1973). A linear programming formulation of the general portfolio selection problem. *Journal of Financial and Quantitative Analysis, 8*, 621–636.

Stone, B. K. (1974). Systematic interest-rate risk in a two-index model of returns. *Journal of Financial and Quantitative Analysis, 9*, 709–721.

Stone, B. K., Guerard, Jr., J. B., Gultekin, M., & Adams, G. (2002). *Socially responsible investment screening*. Working Paper. Marriott School of Management. Brigham Young University.

Von Hohenbalken, B. (1975). A finite algorithm to maximize certain pseudoconcave functions on polytopes. *Mathematical Programming, 9*, 189–206.

Ziemba, W. T. (1992). *Invest Japan*. Probus Publishing Co.

Chapter 16
Options

Options can be generalized as contracts that can be bought at a given price, enabling one to buy or sell an asset or security at a possible future profit. If the profitable opportunity does not arise, the price paid for the option is foregone. An understanding of options theory and analysis is useful to financial managers as it enables them to estimate trends and may be employed to temporarily secure assets until a decision is made whether to buy or not and to hold on to new projects or innovations until a final decision.

Options are financial contracts that allow one to purchase or sell assets at a predetermined price up to an expiration date. Perhaps the most recognized traded options appear in the financial markets as put and call options on common stocks. Speculators use options because of their low initial cost and possible high returns. Common stock options have been traded on the Chicago Board Options Exchange (CBOE) since 1973.

A call option gives its owner the right, not the obligation, to buy a specific number of shares at a specific price, known as the exercise price, up to the expiration date. A premium paid is the cost of purchasing the option. A buyer of the call option expects the stock price to rise, while a seller, or writer, expects the stock price to decline or remain the same. The seller retains the premium whether or not the option is exercised.

Let us look at an example of a common stock purchase and an option valuation. The common stock price of IBM was $134.04 on December 31, 2019. Let us assume that IBM is expected to reach $155 by mid-June 2020. If a trader believed that IBM would appreciate during the coming months, he could purchase 100 shares of stock for $1340.40 plus the brokerage commission. If IBM increases in value to $155 per share, then the investor has secured profits of $20.96 per share, or $209.60 on the 100 shares. The profit on such an investment is 15.645%.

J. B. Guerard Jr. et al., *Quantitative Corporate Finance*,
https://doi.org/10.1007/978-3-030-87269-4_16

$$R_t = \frac{P_t - P_{t-1}}{P_{t-1}} = \frac{\$155 - \$134.04}{\$134.04} = .1564$$

On the other hand, the price of IBM may fall or remain constant as well as it might rise. If the price of IBM fell to $90.00 per share, then the loss on the stock is −32.86%.

$$R_t = \frac{\$90.00 - 134.04}{134.04} = -.3286$$

A trader could purchase an IBM call option on December 31, 2019, with various exercise or strike prices, shown in Table 16.1. For call options expiring in June 2020, one notes that the price one must pay for the call option is a function of the strike price. A call option is the right to purchase stock, and that right has value only if one expects the stock price to exceed the exercise price. The intrinsic value of a call option is its common stock price less its exercise price. Let us look at the premia on IBM call options in Table 16.1.

Table 16.1 IBM June 2020 Call Options
Priced on December 31, 2019
June 19, 2020 (171d); CSize 100; IDiv 3.09 USD; R 1.90; IFwd 131.90

Strike	Ticker	BSze	Bid	Ask	ASze	Last	IVM	DM	Volm	OInt
90	IBM 6/19/20 C90	0	41.850	45.300	0	48.5200	0.0000	1.0000	0	0
95	IBM 6/19/20 C95	0	37.000	40.450	0	0.0000	0.0000	1.0000	0	0
100	IBM 6/19/20 C100	0	32.000	35.700	0	33.7000	0.0000	1.0000	0	11
105	IBM 6/19/20 C105	0	27.150	30.600	0	33.2500	0.0000	1.0000	0	0
110	IBM 6/19/20 C110	0	22.950	25.250	0	26.0000	21.5999	0.9645	0	5
115	IBM 6/19/20 C115	0	19.600	20.650	0	19.9200	25.2699	0.8527	1	160
120	IBM 6/19/20 C120	0	15.500	15.950	0	15.1000	22.8308	0.7924	21	510
125	IBM 6/19/20 C125	0	11.800	12.050	0	13.4100	21.6535	0.7001	0	176
130	IBM 6/19/20 C130	0	8.650	8.750	0	8.2500	20.8212	0.5920	10	1074
135	IBM 6/19/20 C135	0	5.950	6.100	0	5.8000	20.0183	0.4785	17	1023
140	**IBM 6/19/20 C140**	**0**	**3.850**	**3.950**	**0**	**3.7100**	**19.1620**	**0.3626**	**34**	**1789**
145	IBM 6/19/20 C145	0	2.370	2.450	0	2.2400	18.6163	0.2566	11	1342
150	**IBM 6/19/20 C150**	**0**	**1.390**	**1.440**	**0**	**1.3200**	**18.2172**	**0.1720**	**352**	**5074**
155	IBM 6/19/20 C155	0	0.790	0.820	0	0.7500	17.9622	0.1084	20	807
160	IBM 6/19/20 C160	0	0.420	0.470	0	0.4400	17.8109	0.0664	13	900
165	IBM 6/19/20 C165	0	0.210	0.290	0	0.3300	17.8518	0.0395	0	390
170	IBM 6/19/20 C170	0	0.120	0.190	0	0.1900	18.2436	0.0256	0	203
175	IBM 6/19/20 C175	0	0.070	0.130	0	0.1200	18.7634	0.0170	0	478
180	IBM 6/19/20 C180	0	0.040	0.110	0	0.0900	19.5545	0.0126	0	46

Source: Bloomberg, L.P

The premium that one must pay for an IBM call, an option to purchase stock, declines with the exercise price. As the exercise price rises, there is less of a chance that the stock price will exceed the exercise price and have intrinsic value.

Exercise Price	Premium
$125	$12.05
130	8.75
135	6.10
150	1.44
155	0.82
160	0.47

Note that if the common stock price rose to $157 by June 2020 and one purchased a call option to purchase IBM at a $150 exercise price, for $1.44, then one could earn a larger return than buying the stock. If the price of IBM rises to $157, then the call option must be worth at least its intrinsic value, or $7.00. The rate of return for the call option is at least 386%.

$$R_t \geq \frac{\$157 - \$150 - 1.44}{1.44} = 3.86$$

or 386%!

The common stockholder return was 17.19% for the similar price movement—options offer investors the advantage of leverage. One may deal in the movement of a $134.04 stock with lower costs by purchasing a call option. However, one may also lose a large portion of the amount at risk by dealing in options. If the price of IBM falls to $90, then the call option has zero value, and the rate of return on the call option is −100%, whereas the common stockholder lost only 32.86%. However, there is a limited loss on a call option; one may lose all of your premium, but no more. A call option on a stock whose stock price exceeds its exercise price is referred to as an "in the money" (ITM) option. The intrinsic value of an ITM option is positive, and one profits dollar for dollar on the call option as the stock price rises.

Note that if an investor was "bullish" on IBM, then one might prefer to purchase a call option whose exercise price is $135, approximately the same price as the stock price. One can purchase this option, known as an "at the money" (ATM) call option, for a premium of $6.10. If the price of IBM rises to $150, then the ATM call option return is at least 284.62%.

$$R = \frac{\$150 - \$135 - \$6.10}{6.10} = \frac{8.90}{6.10} = 145.90\%$$

There is a time component to options in which there is time for the stock price to rise and exceed the exercise price, producing a future positive intrinsic value. If IBM rises to $92, then the $90 exercise price call option buyer realizes a rate of return of at least 1900.00% for the month.

$$R_t = \frac{92 - 90 - .10}{.10} = 19.0000$$

Obviously, such a rapid rise of IBM is very unlikely. A call option whose stock price is less than the exercise price is referred to as an "out of the money" (OTM) option. The call option premium for an ITM option exceeds the premium of corresponding ATM or OTM options. For call option buyers, OTM offers the greatest leverage. The reader will note that the rate of return for a $90 exercise price exceeds that of the $85 exercise price, if the stock price rises.

A bullish investor who purchases OTM options also runs the chance that the options will expire useless. Thus, one sees a return and risks trade-off of call options in terms of deciding upon the exercise price. One can draw a graph to illustrate call option buying strategies.

One profits by purchasing call options as stock prices rise to exceed the exercise price and the premium paid for the option (Fig. 16.1).

Why would one write, sell, a call option? If an investor owns the stock and expects no significant appreciation in the near future, then one can write call options to provide additional income or returns. That is, the investor would receive the premiums. In the case of the previously examined IBM call options, one can write a September 2004 option and receive $1.50. If IBM's stock price is still about $85.00 at the expiration of the November option, then the call writer keeps the premium, and the call buyer's option expires without value. The call writer receives a $0.15 premium for writing the $90 exercise or strike price option, and all stock price appreciation above $90 is "traded away" for a $0.15 premium. In essence, the investor who owns the underlying stock and writes calls on the stock is known as a covered call writer, and all gains above the exercise price are called away for the premium. As the call writer writes OTM options with an exercise price that is rising, then the covered call writer returns approximate owning (long) the stock returns. The portfolio manager or writer of this covered call clearly expects the price of IBM to be

Fig. 16.1 Profiting on buying a call option

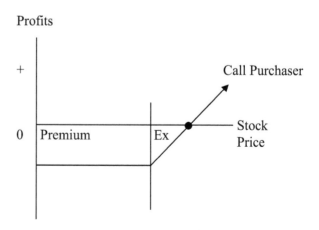

at or below $90 by the last week of September. The writing of covered calls by portfolio holders can be very profitable.

An investor who buys a put option buys the right to sell the stock. The put option has a specified strike or exercise price for a stated time period, just as does the call option. The put option is of interest to an investor who expects the stock price to decline. For example, if an investor buys an $85 put option on IBM, then the investor is paying $1.50 to sell IBM at $85. If the price of IBM falls $70, then the intrinsic value of the put is $15, and there will be an additional premium for time. If the price of IBM falls to $50, then the put option has at least $35 of intrinsic value. An investor who expects the common stock of a firm to fall may either buy a put or sell the stock short. An investor who writes or sells a put expects the stock not to fall and receives the premium.

Options may be used to protect stock gains. Suppose an investor bought Lockheed stock, LMT, on July 21, 2016, in anticipation of the run-up to the Presidential election of November 2016. The investor paid $253.60 a share for the stock in July 2016. Lockheed stock was priced at $393.14 a share on May 10, 2021. LMT had outperformed the S&P 500, leading to a very happy investor. The investor might be concerned about LMT falling, and the investor could buy a "protective" put, with an exercise price of $375, for approximately $13.00, the average of the bid-ask September 2021 $375 Put, see Table 16.2, or $19.15, the average of the bid-ask December $375 Put spread, see Table 16.3. The longer time to maturity costs the investor for the additional protection. Any loss of LMT below $375 would be hedged by the put for a cost of only $19.15 for the December $375 Put. If LMT fell back to $253.60 by mid-December, the investor would maintain a gain of

Table 16.2 Lockheed Martin September 2021 Put Prices

LMT US Equity	95)Actions	96)Export	97)Settings				Option Monitor		
↑393.1401		2.80	0.01	392.99	393.14	Hi	396.99	Lo	391.29
Center	393.14 Strikes	19 Exp 17-Sep-21		Exch US Composite			92)07/21/21 E | ERN »		
Strike	**Puts**								
	Ticker	BSze	Bid	Ask	Last	IVM	DM	Volm	OInt
	17-Sep-21 (130d); CSize 100; IDiv 4.28 USD; R .18; IFwd 390.05								
350	LMT 9/17/21 P350		6.20	6.50	6.40	22.85	-0.2	5	589
355	LMT 9/17/21 P355		7.20	7.50	7.00	22.39	-0.23	13	76
360	LMT 9/17/21 P360		8.40	8.60	8.30	22.01	-0.26	10	501
365	LMT 9/17/21 P365		9.80	10.30	9.50	21.95	-0.29	4	118
370	LMT 9/17/21 P370		11.30	11.50	11.00	21.43	-0.33	16	80
375	LMT 9/17/21 P375		12.70	13.30	12.71	21.08	-0.36	13	213
380	LMT 9/17/21 P380		14.60	15.20	14.77	20.92	-0.4	14	438
385	LMT 9/17/21 P385		16.80	17.30	16.45	20.51	-0.44	5	75
390	LMT 9/17/21 P390		19.00	19.60	18.40	20.26	-0.48	22	37

Source: Bloomberg, L.P

Table 16.3 Lockheed Martin December 2021 Put Prices

LMT US Equity	95)Actions	96)Export	97)Settings				Option Monitor		
↑393.1401		0.72%	392.99	/	393.14	Hi	396.99	Lo	391.29
Center	393.14 Strikes	19 Exp 17-Dec-21		Exch US Composite			92)07/21/21 E	ERN »	
Strike	**Puts**								
	Ticker	Bid	Ask	ASze	Last	IVM	DM	Volm	OInt
	17-Dec-21 (221d); CSize 100; IDiv 6.86 USD; R .20; IFwd 387.65								
350	LMT 12/17/21 P350	11.4	12.2		11.25	22.76	-0.26	1	10
355	LMT 12/17/21 P355	11.3	13.9		15.50y	21.94	-0.28		3
360	LMT 12/17/21 P360	12.2	14.9		16.48y	21.21	-0.3		14
365	LMT 12/17/21 P365	15	17.1		21.00y	21.69	-0.34		24
370	LMT 12/17/21 P370	17.6	18.7		18.70y	21.67	-0.37		2
375	LMT 12/17/21 P375	17.8	20.5		26.40y	20.63	-0.39		25
380	LMT 12/17/21 P380	21.6	22.9		26.04y	21.37	-0.42		85
385	LMT 12/17/21 P385	22.2	24.9		24	20.23	-0.46	30	96
390	LMT 12/17/21 P390	26.3	26.9		25.7	20.53	-0.49	1	101

Source: Bloomberg, L.P

($393.14–253.60 – $19.15, or $120.39), whereas the stock investor profits would fall to zero! The return on buying LMT and the subsequent protective put would be 44.20% ($120.39 / ($253.60 + $19.15)).

Options operate like financial leverage on stock prices.

Let us introduce the notation of Malkiel and Quandt that is very useful to address combinations of put and call strategies. The Malkiel–Quandt notation is a bit dated but, in the opinion of the authors, a very useful means to describe the basic premises of buying and writing puts, calls, and straddles, combinations of a call and a put. Later in the chapter, we will introduce the reader to the binomial options pricing model, the current means of illustrating the pricing structure of calls and puts.

16.1 The Malkiel–Quandt Notation

The authors have long admired the options notation put forth by Malkiel and Quandt (1969) in their very enjoyable and stimulating monograph on options markets. One profits from a stock price rise when one purchases a call option and loses only the premium from a stock price decline. As Malkiel and Quandt write, the option profits lose in terms of stock price advances and declines in a 2×1 matrix.

$$\begin{bmatrix} \dfrac{Pcs \uparrow}{Pcs \downarrow} \end{bmatrix}$$

If one purchases a call option, one profits as the stock price advances, and the Malkiel–Quandt notation would be:

$$\begin{bmatrix} +1 \\ 0 \end{bmatrix}$$

Note that the premiums are not included in the analysis, and one must be careful in examining option portfolio strategies. There may be a problem with naked call writing where the writer does not own the underlying stock, and where one loses in the event of a stock price advance in terms of unlimited stock advances and retains only the call premium in the event of a price decline:

$$\begin{bmatrix} -1 \\ 0 \end{bmatrix}$$

However, if one is a covered call writer, the Malkiel–Quandt notation is:

$$\begin{matrix} \text{Covered} \\ \text{Call} \end{matrix} = \begin{matrix} \text{long in} \\ \text{stock} \end{matrix} + \begin{matrix} \text{call} \\ \text{writer} \end{matrix}$$

$$\text{writer} = \begin{bmatrix} +1 \\ -1 \end{bmatrix} + \begin{bmatrix} -1 \\ 0 \end{bmatrix} = \begin{bmatrix} 0 \\ -1 \end{bmatrix}$$

It would appear that the covered call writer never wins; however, if one writes a call option with an exercise price in excess of the stock price (OTM), one can substantially profit on the stock price advance and receive a call premium to partially offset the stock price decline.

A put buyer return pattern can be written as:

$$\begin{bmatrix} 0 \\ +1 \end{bmatrix}$$

and a put writer return pattern is:

$$\begin{bmatrix} 0 \\ -1 \end{bmatrix}$$

The Malkiel–Quandt notation allows the reader to understand the basic concepts of buying and writing calls and puts.[1]

It should be obvious to the reader that an investor in IBM stock must have some expectations of future appreciation. Let us assume that IBM stock may rise 20% in

---

[1] Straddles, straps, and strips, combinations of calls and puts, might pose problems for the Malkiel–Quandt notation. For example, if one purchases one call and one put at the same exercise price on the identical stock, it would appear that the investor will never lose.

$$\underset{\text{buyer}}{\text{straddle}} = \underset{\text{buyer}}{\text{call}} + \underset{\text{buyer}}{\text{put}}$$

$$\begin{bmatrix} +1 \\ 0 \end{bmatrix} + \begin{bmatrix} 0 \\ +1 \end{bmatrix} = \begin{bmatrix} +1 \\ +1 \end{bmatrix}$$

One could purchase an 80 September call on IBM for $5.10 and an 80 IBM September put for $0.35. If the investor buys a straddle, one invests $5.45 (or $545 in the 100 share position) in the position. The investor purchases volatility in buying a straddle on profits only if the price rises above $59.38 or falls below $50.63. A straddle buyer does not care which event (advance or decline) occurs as long as the volatility is very large. A strike might lead one to buy a straddle on the firm's stock; one could profit on the put if the strike is long and violent and the call if the strike is settled quickly and cheaply. The Malkiel–Quandt positions do not offer the investor an insight into the volatility problems in purchasing a straddle. A straddle writer appears to consistently lose in the Malkiel–Quandt analysis:

$$\begin{bmatrix} -1 \\ 0 \end{bmatrix} + \begin{bmatrix} 0 \\ -1 \end{bmatrix} = \begin{bmatrix} -1 \\ -1 \end{bmatrix}$$

However, in reality, the writer wins as long as the price fluctuates between $50.63 and $59.38. The straddle writer wins as long as the stock volatility is low. Anything that reduces the stock volatility enhances the profitability of the straddle writer; for example a merger, particularly a conglomerate merger reduces systematic risk and total risk. A merger announcement could lend one to write straddles on the acquiring firm's stock.

A strap is when one has two calls and one put in an option portfolio. A strap buyer is not completely certain about the course of a stock's movement; however, the buyer is leaning toward an upward stock movement. One can easily see this in the Malkiel–Quandt notation:

$$\underset{\text{buyer}}{\text{strip}} \quad 2 \begin{bmatrix} +1 \\ 0 \end{bmatrix} + \begin{bmatrix} 0 \\ +1 \end{bmatrix} = \begin{bmatrix} +2 \\ +1 \end{bmatrix}$$

One must be aware of the numerous (3) transaction costs and premiums associated with straps. In our Alcoa example, the 55 October strap would cost $737.50 [2($3) + $1.375 = $7.375]. Therefore, the break-even price range is $47.63 and $58.69 [$55 + $7.28/2]. One profits at a 2:1 ratio as the stock price advances above $55. A strip is two puts and one call. A strip purchaser is somewhat confused but more pessimistic than optimistic about the stock price movement. The Malkiel–Quandt notation for a strip purchaser is:

$$\underset{\text{purchaser}}{\text{strip}} \quad \begin{bmatrix} +1 \\ 0 \end{bmatrix} + 2 \begin{bmatrix} 0 \\ +1 \end{bmatrix} = \begin{bmatrix} +1 \\ +2 \end{bmatrix}$$

Straddles, straps, and strips offer investors and portfolio managers opportunities to alter portfolio return distributions.

Table 16.4 IBM stock price example

t=0 (current)	t=1	t=2	t=3
			146.38
		121.98	109.78
			109.78
	101.65	91.49	
			82.34
			109.78
		91.49	
$84.71			82.34
			82.34
	76.24	68.62	
			61.76

The expected IBM prices are shown in Table 16-4 for our three-period analysis.

the next period from its current price of $55.00, or fall 10%. Let us consider a three-period analysis of possible prices. The possible prices can be determined by the binomial probability distribution (Table 16.4).

16.2   The Binominal Option Pricing Model

The binomial option pricing model (OPM) evolved from Sharpe (1978), Rendleman and Bartter (1979), and Cox, Ross, and Rubinstein (1979). Much easier than the Black and Scholes OPM derivation (the limiting case of the binominal OPM is its continuous-time partner, the traditional Black and Scholes OPM), the binomial OPM may be developed from a very practical example. Let us assume that the firm's current stock price is P_{cs} and either goes up u in the next period or falls by d. A call option is equal to the maximum of zero or $P_{cs} - Ex$, where Ex is the exercise price. In the King's English, a call option cannot be negative. In a three-period model, the value of the call option, V_c, may be developed:

t = 0	t = 1	t = 2	t = 3
			$P_{cs}(u)^3 - Ex$
		$P_{cs}(u)^2 - Ex$	$P_{cs}(u^2d) - Ex$
	$P_{cs}(u) - Ex$		$P_{cs}(udu) - Ex$
$P_{cs} - Ex$		$P_{cs}(ud) - Ex$	$P_{cs}(udd) - Ex$
			$P_{cs}(duu) - Ex$
		$P_{cs}(du) - Ex$	$P_{cs}(duu) - Ex$
	$P_{cs}(d) - Ex$		$P_{cs}(dud) - Ex$
		$P_{cs}(d)^2 - Ex$	$P_{cs}(ddu) - Ex$
			$P_{cs}(d)^3 - Ex$

The value of the call option can be found by multiplying the discounted value of the call in each of the eight possible states of nature by the respective probabilities of these states. The probability of an upward price movement, u, is p:

$$p = \frac{r - d}{u - d},$$

where r is one plus the risk-free rate. The probability of a downward movement, d, is q, where $q = 1 = p$. Thus, the value of the call option:

$$V_c = \frac{p^3[P_{cs}(u^3) - Ex] + 3p^2q[P_{cs}(u^2d) - Ex] + 3pq^2[P_{cs}(ud^2) - Ex] + q^3[P_{cs}(d^3) - Ex]}{(1 + r)^3}$$

$$(16.1)$$

This equation may seem rather formidable until one remembers that the reader solved this in sophomore statistics class. If k represents the number of upward movements in n periods:

$$V_c = \frac{1}{r^n} \sum_{k=0}^{n} \frac{k!}{(n-k)!} p^k q^{(n-k)} \left[\max \left(0, P_{cs} u^k p^{(n-k)} - Ex \right) \right] \qquad (16.2)$$

Let us work a very small numerical example that will illustrate the analysis. Assume the following:

$$P_{cs} = \$84.71$$

$$Ex = \$85.00$$

$$r = 1.2$$

$$d = .9$$

$$r = 1.02$$

$$p = \frac{1.02 - .9}{1.2 - .9} \quad \frac{.12}{.30} = .40$$

$$V_c = \frac{1}{(1.02)^3} \{(.4)^3[\max(0, 84.71(1.2)^3 - 85]$$
$$+ 3(.4)^2(.6)[\max(0, 84.71(1.2)^2(.9) - 85]$$
$$+ 3(.4)(.6)^2[\max(0, 84.71(1.2)(.9^2) - 85]$$
$$+ (.4)^3[\max(0, 84.71(.9^3) - 85]\}$$

$$V_c = .942[.064(61.38) + .288(24.79) + .432(0) + .216(0)]$$

$$V_c = .942(3.93 + 7.14)$$

$V_c = \$10.42$

The value of the three-period call is $10.42. The call option, if only two-period, would be:

$$V_c = (1.07)^2\{[(.4)^2(84.71(1.2)^2 - 85) + 2(.4)(.6)[\max(0, 84.71(1.2)(.9) - 85)]$$
$$+(.6)^2[\max(0, 84.71(.9)^2 - 85]\}$$

$$V_c = (.961)[.16(36.98) + .48(6.49) + (.36)(0.00)] = .961(5.92 + 3.12) = \$8.68$$

Note, as in the general case of options, as the time to maturity rises, the value of the option increases. The binomial option pricing model convergences to the traditional Black and Scholes Option Pricing Model as the number of periods goes to infinity (continuous-time), and the binomial approximation of the stock price distribution becomes normally distributed (for price relatives).

$$\log(P^*_{cs}/P_{cs}) = k \log(u/d) + n \log d, \qquad (16.3)$$

$$E[\log(P^*_{cs}/P_{cs})] = mn,$$

where * denotes the maturity date, and mn is the expected mean of the series.

$$\text{Prob}\left[\frac{\log(P^*_{cs}/P_{cs}) - mn}{\sigma\sqrt{n}} < Z\right] = N(Z)$$

The traditional Black and Scholes OPM can be written from the binominal option pricing model as:

$$V_c = P_{cs}N(X) - Exr^{-t}N(X - \sigma\sqrt{t}) \qquad (16.4)$$

where

$$X = \frac{\log(P_{cs}/Exr^{-t})}{\sigma\sqrt{t}} + \frac{1}{2}\sigma\sqrt{t}$$

The reader may well remember that the binomial distribution converges to the normal distribution with approximately 30 periods (the Central Limit Theorem).

16.3   The More Traditional Black and Scholes Option Pricing Model Derivation

The rudiments of the Black and Scholes OPM were known in the 1960s as shown in the works of Sprenkle (1962), Boness (1964), and Kassouf (1969); however, Black and Scholes (1973) developed their model with a hedge ratio that allowed not only a theoretical value but a risk-free hedging strategy.[2] The Black and Scholes OPM is characterized by rather restrictive assumptions: (1) the stock price series follows a lognormal distribution; that is, the distribution of price relatives is normally distributed; (2) the risk-free is constant; (3) the variance of the price series is constant and known; (4) the stock pays no dividend; and (5) the call option is a European call option that cannot be exercised prior to expiration (maturity).

The traditional Black and Scholes (1973) Option Pricing Model (OPM) finds a theoretical value of the call option, V_c, by creating a risk-free portfolio composed of a call option and shares of the underlying stock. The value of the portfolio, V_p, is found by the sum of the market values of equity and options.

$$V_p = hnV_c + mP_{cs} \qquad (16.5)$$

where

h = hedge ratio,
n = number of call options,
m = number of shares of stock,
and P_{cs} = price of common stock.

If there are no transactions in the stock and options, n and m are assumed to be constant, and the value of the portfolio may change as a result of fluctuating stock prices and call option prices. The reader is aware that changes in the stock price affect the price of a call; however, the call price is affected by changes in the time to maturity of the option. Although the price of the underlying stock may be unchanged for a trading day, the value of the call will fall as a result of time decay. Without time decay, the change in the portfolio is:

$$dV_p = hndP_{cs} + mdV_c \qquad (16.6)$$

If n and m are selected in this way, the risk-free hedge is created:

---

[2]Wall Street quickly adopted the Black and Scholes OPM to the extent that even practitioners into their 50s and 60s who cannot take a derivative, total or partial, have hired "computer jocks" to program the OPM into their computers or use their H–P calculators.

$$dV_p = hndP_{cs} + mdV_c \qquad (16.7)$$

One notices that if the call option is written ($m < 0$), then the stock is purchased ($n > 0$) to create the hedged portfolios. The stock price and option price changes are:

$$dP_{cs} = P_{cs_{t+dt}} - P_{cs_t},$$
$$dV_c = V_{c_{t+dt}} - V_{c_t}.$$

The change in the stock price series follows a random walk, that is, the stock price series follows a lognormal distribution and the distribution of price relatives. In $(P_{cs_t} / - P_{cs_{t-1}})$ is normally distributed. Black and Scholes (1973) and Smith (1976) write the change in the value of the call option may be written in terms of changes in time and the stock price.

$$dV_c = \frac{\partial V_c}{\partial t} dt + \frac{\partial V_c}{\partial P_{cs}} dP_{cs} + \frac{1}{2} \frac{\partial^2 V_c}{\partial P_{cs^2}} P_{cs}{}^2 \sigma^2 \qquad (16.8)$$

where

σ^2 = variance in stock price relatives.

The change in the value of the hedged portfolio is:

$$dV_p = hdP_{cs} - \left(\frac{\partial V_c}{\partial t} + \frac{1}{2} \frac{\partial^2 V_{c2}}{\partial P_{cs}} \sigma^2 P_{cs}^2 \right) dt \qquad (16.9)$$

The change in the value of the hedged portfolio should equal the risk-free rate, R_F.

$$R_F V_p = R_F P_{cs} \frac{\partial V_c}{\partial P_{cs}} + \frac{\partial V_c}{\partial^t} + \frac{1}{2} \frac{\partial^2 V_c}{\partial P_{cs^2}} \sigma^2 P_{cs}^2$$

$$0 = R_F P_{cs} \frac{\partial V_c}{\partial P_{cs}} + \frac{\partial V_c}{\partial^t} + \frac{1}{2} \frac{\partial^2 V_{c2}}{\partial P_{cs}} \sigma^2 P_{cs}^2 - R_F V_P$$

The value of the V_c is found by solving the above partial differential equation subject to:

$$V_c = \max \left(P_{cs} - Ex, 0 \right), 0 < t < T(\text{maturity date}), 0 < P_{cs}.$$

The Black and Scholes OPM value of the call option is found to be:

$$V_c = P_{cs}N(d_1) - \frac{Ex}{e^{rt}}N(d_2) \qquad (16.10)$$

where

$$d_1 = \frac{\ln\left(P_{cs}/Ex\right) + \left(r + \frac{\sigma^2}{2}\right)t}{\sigma\sqrt{t}},$$

$$d_2 = \frac{\ln\left(P_{cs}/Ex\right) + \left(r - \frac{\sigma^2}{2}\right)t}{\sigma\sqrt{t}},$$

$N(.)$ = cumulative normal distribution, and $e = 2.71828$.

The optimal hedge ratio, h, to create the risk-free hedge is $N(d_1)$. If the call option is purchased, $N(d_1)$ shares of stock are sold short. If the call is overvalued and written, the stock is purchased in the ratio of $N(d_1)$. The traditional Black and Scholes OPM assumes that no dividends are paid. The dividends may be handled in several manners: (1) the present value of the escrowed dividend may be subtracted from the stock price, and (2) the time to ex-dividend is substituted for the same variable, and the relevant stock price is the price less the present value of the escrowed dividends

$$0 = \frac{1}{2}\frac{\partial^2 V_{c_2}}{\partial P_{cs}}\sigma^2 P_{cs}^2 + \frac{\partial V_c}{\partial t} + (R_F - D)P_{cs}\frac{\partial V_c}{\partial P_{cs}}R_F V_c \qquad (16.11)$$

The value of the call option is clearly a function of six variables.

$$\begin{array}{ccccccc} & + & + & + & + & - & - \\ V_c = f(P_{cs}, & t, & R_F, & \sigma^2, & D, & Ex) \end{array}$$

As the stock price rises, the value of the call increases. This is immediately clear for an "in the money" call in which the stock price exceeds the exercise price. An increase in the exercise price reduces the value of the call. An increase in the time to maturity creates more time for the stock price to exceed the exercise price and thus raise the value of the option. The risk-free rate is used to discount the exercise price in determining the value of the call; an increase in the risk-free rate reduces the present value of the exercise price and increases the value of the call option. The dividend reduces the ex-dividend price and reduces the call. The variance of the price relative series positively affects the value of the call because an increase in the variability serves the probability that stock price will exceed the exercise price and raise the value of the call option.

The reader should be aware of the options tab of the www.pcquote.com website. An investor can access real-time option prices and calculated Black and Scholes implied option volatilities. Let us show an example of a Lockheed Martin call option with an exercise price of $350, priced on March 9, 2020, with data from Bloomberg, L.P. The option expires in over 3 months or 102 days. The stock price of LMT was $354.00, and the 3-month Treasury bill, the risk-free rate, was 0.90%. The historic volatility was 17.90% (the standard deviation). The last traded call was priced at an ask price of $17.60. Is this a fair value of the call option?

16.4 Black and Scholes Model Calculation

The calculated Black and Scholes Option Pricing Model value of the LMT June 2020 call with an exercise price of $350 is:

$$V_c = N(d_1)P_{cs} - \frac{Ex}{e^{rt}}N(d_2)$$

where

$$d_1 = \frac{\ln \frac{\$354}{\$350} + \left(.009 + \frac{(.179)^2}{2}\right)\frac{102}{252}}{.179\sqrt{102/252}}$$

$$= \frac{.0215}{.1139} = .189$$

$$d_2 = \frac{\ln \left(\frac{\$354}{\$350}\right) + \left(.009 - \frac{(.179)^2}{2}\right)\frac{102}{252}}{.179\sqrt{102/252}}$$

$$= \frac{.0085}{.1139} = .075$$

The cumulative normal distribution, found in the standard "area under the normal curve" statistical tables, produces cumulative areas of 57.5 and 53.19%, respectively, for d_1 and d_2.

$$V_c = \$354(.575) - \frac{350}{2.71828^{(.009)\left(\frac{102}{252}\right)}}(.5319)$$

$$= \$203.55 - 185.48 = \$18.07$$

The calculated Black and Scholes OPM value of $18.07 is less than the ask price of $34.80. One should (sell) write the LMT call option. One could write the call option at $34.80, buy 57.5 shares of IBM stock to hedge [$N(d_1)$ is the hedge ratio], and earn an excess return of 38.6%. The Black and Scholes LMT call option should sell for $18.07 because its 17.9% is substantially less than the 41.4% implied volatility used by the market to price the option.

The put option price is developed from the put–call parity relationship in which the purchase of a put is equal to the price of the call plus the present value of the strike price less the stock price. That is, the purchase of a put is analogous to the purchase of a call, shorting the stock, and borrowing the discounted exercise price.

$$V_c - P_{cs} - V_P + \frac{Ex}{1 + R_F} = 0$$

$$V_P - V_c + \frac{Ex}{1 + R_F} - P_{cs}$$

The empirical results of Klemkosky and Resnick (1979) and Gould and Galai (1974) substantiate the put–call parity theory and options market efficiency [Galai (1977) and Phillips and Smith (1980)]. One may wonder why options trading firms have emerged in options analysis. One can easily find and agree upon the stock price, exercise price, risk-free rate, time to maturity, and dividends. The relevant variance to be used in pricing the option is of great concern and interest as is the role that options and stocks play in a diversified portfolio. Latane and Rendleman (1976) found that if one assumes that the call option is correctly priced and solves the Black and Scholes OPM for the implied variance, the implied variances were superior to historic variances if forecasting future variances. Schmalensee and Trippi (1978) found that variances are not constant but are first-order negatively serially correlated, i.e., mean-reverting. In one of the best studies of options market efficiency, Whaley (1982) found the American call and European call options less the escrowed present values of dividends produced values within 3–5 cents of observed values (far within transactions costs, bid-ask spread) with the goodness of fit measures exceeding 0.98. Furthermore, the option pricing errors are due to variance errors and not to dividends, time to maturity (particularly the American call option), or the "in the moneyness" of the call option. Whaley noted the conclusion about the "in the money" nature of option problems. Whaley's option regression results indicated a tendency for "out of the money" ($P_{cs} < Ex$) to be underpriced by the traditional Black and Scholes models at the 10% level (although the regression coefficient $= 0.04$–0.10). Black (1976) noted that "out of the money" options tended to be underpriced. Most options theorists agree that implied variances are relevant variance for options investing, and these variances are not constant but rather mean-reverting.

The two major topics of option analysis concern the area of option portfolios and portfolio insurance. Merton, Scholes, and Gladstein (1978, 1982) developed portfolio models to analyze option portfolios. In their earlier price, Merton, Scholes, and Gladstein found using an option simulation period from June 1963 to December 1975, that "in the money" covered calls ($P_{cs} > Ex$) earned less return than merely owning the stock. This is hardly surprising because further stock price appreciation above the exercise price (above the current option premium) is called away by the option buyer. The "out of the money" covered calls earned higher returns than the "in the money" covered calls, but were still less than stock returns. The right tail of the stock price distribution is eliminated by writing calls on the held stock. Although covered call writing reduced returns relative to holding stock, the portfolio standard deviations also were lower involving call writers. The principal contribution of the Merton, Scholes, and Gladstein (1978) call option study was that portfolios composed of 90% commercial paper, and 10% purchased "out of the money" calls outperformed stock returns, also producing much higher portfolio variances. The "out of the money" calls produced tremendous returns, despite the fact that 70% of the options expired unused, earning approximately 400%. One notes that "out of the money" option premiums when the exercise price is 1.2 times the stock price (i.e., Ex = 96, P_{cs} = 80), it should be very small. Therefore, if the stock price rises slightly over the exercise price, the corresponding percentage option returns would be very large.

The Merton, Scholes, and Gladstein (1982) put study found that "in the money" puts (Ex > P_{cs}) provided portfolio insurance that was too costly for investors (reducing portfolio returns), whereas "out of the money" puts (Ex < P_{cs}) offered downside protection (protective puts) at a very reasonable (0.4%) cost to the portfolio manager. Writing uncovered puts was far less variable than fully covered call writing, although both strategies had identical Malkiel–Quandt notation; the stock variance led to the higher variability in covered call writing. If one combines protective put purchases with "out of the money" call writing, one could effectively shape the desired portfolio distribution. Bookstaber and Clarke (1983) developed a portfolio algorithm to set exercise price levels for puts and calls to manage an options portfolio (with as few as five options). The existence of the options markets should not influence the firm's stock price and cost of capital. Stock prices are determined by discounted expected earnings. Option prices are determined by stock prices and stock volatilities.

16.5   The OPM and Corporate Liabilities

Black and Scholes (1973) derived an economic value of the European call option, assuming no taxes or transactions costs, log-normally distributed stock prices, and a known and constant risk-free rate:

$$W = XN(d_1) - Ce^{-r}N(d_2),$$

where

W = Price of the call
X = Current value of the underlying asset, the stock price
C = Exercise price of the call option
t = Time to maturity

r_f = Instantaneous risk-free rate
N(...) = Standardized normal cumulative probability density function

$$d_1 = \frac{ln\left(\frac{x}{c}\right) + \left(r_f + \frac{\sigma^2}{2}\right)t}{\sigma\sqrt{t}};$$

$$d_2 = d_1 - \sigma\sqrt{t} = \text{Instantaneous variance of V's returns.}$$

The firm's equity can be viewed as a European call option in which stockholders sell the firm to the bondholders with an option to buy back the firm at time period t. The exercise price is the face value of the debt. The equity will have a positive value only if the terminal value of the firm exceeds the face value of the debt.[3] The face value of the debt can be thought of as the "limited liability" of the equity; it is the protection against the depreciation of the firm's value below the face value of the debt.[4]

Merton (1974) used the Black and Scholes options model to price corporate liabilities where the value of equity, f, may be expressed.

---

[3] The authors believe few stockholders view the call option feature as truly meaningful. If the market value of the firm falls below the value of debt, the stockholders' value is zero.

[4] The stockholders own the firm and agree to pay the bondholders the value of the firm until the value of the assets exceeds the face value of the debt. If the value of assets is less than the face value of debt, then the equity holders must declare bankruptcy, and the bondholders can take possession of the firm. When a firm uses leverage, then the risk of the firm increases and the return on equity rises (linearly). The beta of the firm rises with leverage as a linear function:

$$\beta_{jL} = \beta_{ju}\left[1 + \frac{D}{E}(1-t)\right]$$

where
β_{jL} = firm j beta with leverage

β_{ju} = firm j beta with leverage
D = debt
E = equity
T = corporate tax rate

The cost of equity is a linear function of the firm beta as the reader saw in Chap. 14 with the Capital Asset Pricing Model (CAPM):

$$Ke = R_F + \{E(R_M) - R_F\}b$$

The beta in the CAPM is a levered beta for the vast majority of firms in our economy. The concept of options and leverage is important because equity is like a call option on the firm. Equity holders can undertake risky projects that may increase the volatility of the firm. The value of the call is a function of volatility, and higher risk increases the value of equity at the expense of debt.

$$f = h \overset{+\;-\;+\;+\;+}{(V,\ B,\ r_f,\ \sigma^2,\ t)}.$$

The value of equity increases with the value of the firm, the risk-free rate, the variance, and the time to maturity of the debt. The equation for the value of equity can be written as:

$$\text{dist.}1 = \frac{\ln\left(\frac{V}{B}\right) + \left(r_f + \frac{\sigma^2}{2}\right)t}{\sigma\sqrt{t}}$$

$$\text{dist.}2 = d_1 - \sigma;$$

$$f = V[N(d_1)] - Be^{-r_f t}[N(d_2)].$$

Assume that a firm has a current market value of $5 million and that the face value of its debt is $2 million. The firm's variance of return is 10% and the risk-free rate is 5%. The firm's debt will mature in 20 years. The market values of its debt and equity can be found:

$$\text{dist.}1 = \frac{\ln\left(\frac{5,000,000}{2,000,000}\right) + (.05 + \frac{.10}{2})20}{\sqrt{.10}(\sqrt{20})}$$

$$= \frac{.9163 + 2.000}{.3162(4.4721)}$$

$$= 2.0623$$

From the table of areas under the normal curve,

$$N(d_2) = .2422 + .5000 = .7422.$$

The 0.5000 representing the area under the left-hand side of the normal curve is added as the cumulative normal function is employed. To find the value of equity,

$$S = \$5,000,000(.9803) - \$2,000,000(.7422)e^{-(.05)20}$$

$$= \$4,901,500 - \$546,080.24$$

$$= \$4,355,419.76$$

The market value of the firm's equity, from the options pricing model, is $4,355,419.76. Since its current market value is $5,000,000, the current market value of its debt must be $644,580.24. The market value of debt is the minimum value of the firm or the face value of debt.

If the firm engages in a merger that reduces its variance to 8%, the value of its debt will rise and the market value of its equity will decrease:

$$\text{dist.1} = \frac{\ln\left(\frac{5,000,000}{2,000,000}\right) + \left(.05 + \frac{.08}{2}\right)20}{\sqrt{.08}(\sqrt{20})}$$

$$= \frac{2.7163}{1.2647}$$

$$= 2.1478$$

$$N(\text{dist.1}) = .9842$$

$$\text{dist. } 2 = 2.1478 - \sqrt{.08}\left(\sqrt{20}\right)$$

$$N(\text{dist. } 2) = .8106$$

The value of equity is found:

$$S = \$5,000,000(.9842) - \$2,000,000\,(.8106)e^{-r/t}$$
$$= \$4,324,560.52$$

The value of equity has fallen from $4,355,419.76 to $4,324,560.52 entirely because of the reduction in the variance resulting from the merger. The value of the debt has risen:

$$\$5,000,000 - \$4,324,560.52 = \$675,439.48$$

Debt's value rises from $644,580.24 to $675,439.48 because of the reduction in the firm's variance.

If investors hold the market portfolio (equal amounts of the firm's debt and equity), a change in the market values of corporate liabilities will not affect the market values of their total holdings. The loss on equity holdings will be offset by the gain on debt holdings. If investors do not hold the market portfolio, wealth transfers from bondholders to stockholders must be made to leave stockholders indifferent to mergers. The merged firm will have a higher debt-to-equity ratio. Because interest on debt is deductible and the value of the firm rises with debt, the higher debt-to-equity ratio increases the value of the firm. The important aspect of this appendix is the stockholder–management relationships are discussed in Chap. 22.

References

Black, F. (1975). Fact and fantasy in the use of options. *Financial Analysts Journal*, 36–72.
Black, F., & Scholes, M. (1973). The pricing of options and corporate liabilities. *Journal of Political Economy, 81*, 637–654.
Boness, A. J. (1964). Elements of a theory of stock option value. *Journal of Political Economy, 72*, 163–175.

Bookstaber, R. (1987). *Options pricing and investing strategies*. Probus Publishing.

Bookstaber, R. (1981). *Option pricing and strategies in investing*. Addison-Wesley Publishing.

Bookstaber, R., & Clarke, R. (1983). An algorithm to calculate the return distribution of portfolios with option positions. *Management Science, 29*, 419–429.

Cootner, P. (Ed.). (1964). *The random character of stock market prices*. MIT Press.

Cox, J. C., & Rubinstein, M. (1985). *Options market*. Prentice-Hall.

Cox, J. C., Ross, S., & Rubinstein, M. (1979). Option pricing: A simplified approach. *Journal of Financial Economics, 6*, 229–263.

Galai, D. (1977). Tests of market efficiency of the Chicago Board of Options Exchange. *Journal of Business*, 167–197.

Gould, J. P., & Galai, D. (1974). Transactions costs and the relationship between put and call prices. *Journal of Financial Economics, 1*, 105–130.

Jarrow, R. A., & Rudd, A. (1983). *Option pricing*. Dow Jones-Irwin.

Jarrow, R. A., & Turnbull, S. (2000). *Derivatives securities* (2nd ed.). South-Western College Publishing.

Kassouf, S. T. (1969). An econometric model for option Price with implications for investors' expectations and audacity. *Econometrica, 37*, 685–694.

Klemkosky, R. C., & Resnick, B. G. (1979). Put-call parity and market efficiency. *Journal of Finance, 34*, 1141–1157.

Latane, H., & Rendleman, R. (1976). Standard deviations of stock Price ratios implied in option prices. *Journal of Finance, 31*, 369–381.

McDonald, R. L. (2003). *Derivatives Markets*. Addison Wesley.

Malkiel, B. G., & Quandt, E. (1969). *Strategies and rational decisions in the securities options market*. MIT Press.

Merton, R. (1974). On the pricing of corporate debt: The risk structure of interest rates. *Journal of Finance, 29*, 449–470.

Merton, R., Scholes, M., & Gladstein, M. (1982). The returns and risks of alternative put-option portfolio investment strategies. *Journal of Business, 55*, 1–55.

Merton, R., Scholes, M., & Gladstein, M. (1978). The returns and risks of alternative call option portfolio investment strategies. *Journal of Business, 51*, 183–242.

Merton, R. C. (1973). Theory of rational option pricing. *Bell Journal of Economics and Management Science, 4*, 141–183.

Phillips, S., & Smith, C. (1980). Trading costs for listed options: Implications for market efficiency. *Journal of Financial Economics, 8*, 179–201.

Schmalensee, R., & Trippi, R. (1978). Common stock volatility expectations implied by option Premia. *Journal of Finance, 33*, 129–148.

Sharpe, W. F. (1978). *Investments*. Prentice-Hall.

Smith, C. W. (1976). Option pricing. *Journal of Financial Economics, 3*, 3–51.

Sprenkle, C. M. (1961). Warrant prices and indicators of expectations and preferences. *Yale Economic Essays, 1*, 179–231.

Stroll, H. R. (1969). The relationship between put and call option prices. *Journal of Finance, 24*, 1–24.

Tsay, R. S. (2002). *Analysis of financial time series*. John Wiley & Sons. Chapter 6.

Whaley, R. (1982). Valuation of American call options on dividend-paying stocks. *Journal of Financial Economics, 10*, 29–58.

Chapter 17
Real Options

Chapter 11 dealt with the capital budgeting process in which a financial manager accepts a project only if the discounted cash flow of that project exceeds the initial costs of the project. The discount rate is the cost of capital. The difference between the discounted cash flow and the initial cash outlay is the net present value, NPV, which should be positive to accept a project. This chapter discusses another application of cash flow and valuation, the application of real option theory.

The application of real option analysis can take several forms. First, one can examine the possible complications of the strict application of the NPV rule to an R&D investment decision, and how the stockholder wealth may be enhanced by the use of real options analysis. An investment project may be viewed as the right to pay an initial cash outlay, or cost, to receive the present value of the project's future cash flow. The second application of real options strategies is the case of abandonment valuation. When one calculates the value of a real option, one equates the investment cost of the project with the exercise price of the real option. The present value of the project is equivalent to the price of the underlying asset. Real options are an application of options theory to the operation and valuation of real investment projects (McDonald, 2003).

Research and development expenditures are capital expenditures involving discounting cash flow such that the net present value is positive. The research and development expenditure leading to the implementation of new technology is the call premium, with the present value of the final project being the value of the call option. The R&D costs are the premium paid to acquire the future investment cash flow of the project resulting from the R&D activities. A pharmaceutical firm that engages in R&D expenditures may need to consider abandonment decisions and values. Additionally, current R&D projects uncover options in future and expansive R&D projects. The R&D option is an option to expend the firm when it may not be obvious to management, at the current time, that primary financial decision is profitable. A current or static, negative net present value need not lead management to eliminate the R&D project from its consideration. It is possible to reconsider the

J. B. Guerard Jr. et al., *Quantitative Corporate Finance*,
https://doi.org/10.1007/978-3-030-87269-4_17

project at a later date when initial cash outlays of projects change, costs of capital change, and higher estimates of future cash flows are different.

The traditional NPV capital budgeting criteria measures the profitability and acceptability of undertaking a project at the present time and ignores the potential profitability of delaying the project until a later date. McDonald (2003) makes the point that the traditional, or static, NPV criteria assumes that the project will not be undertaken if it is rejected in the current period. The project cost can be thought of as an exercise price, and the present value of the project is the value of the underlying asset. If we delay the investment, we lose the cash flow of the project, very similar to a stock dividend not received. If the project cash flows are less than the interest saved by deferring the payment of the project investment, then it is efficient to wait to invest in the project.

17.1   The Option to Delay a Project

Projects are traditionally analyzed using the expected cash flows and discount rates at the time of the analysis; the net present value computed on that basis is a measure of its value at that time. Expected cash flows and discount rates may change over time, however, and then so will the net present value. Information may change over time, and the option to wait, or delay a project, may have value. Thus, if expected cash flow rises or the discount rate falls, a project that has a negative net present value now may have a positive net present value in the future. In a competitive market, in which individual firms have no special advantages over their competitors in undertaking projects, this may not seem significant. However, in an environment in which only one firm, such as a firm with a patent, can take a project or barriers to entry extensive advertising, or other frictions prevail to create a change in the project's value over time to give it the characteristics of a call option.

In the abstract, assume that a project requires an initial investment, as the R&D program, C. The present value of expected cash inflows computed right now is PVCF. The net present value of this project is the difference between the two:

$$NPV = PVCF - C$$

Now assume that the firm has exclusive rights to this project for the next n years and that the present value of the cash inflows may change over that time, because of changes in either the cash flows or the discount rate. Thus, the project may have a negative net present value right now, but it may still be a good project if the firm waits. Defining PVCF again as the present value of the cash flows, the firm's decision rule to accept the project should be if PVCF > C. If the firm does not take the project, it incurs no additional cash flows, though it will lose what it originally invested in the project. The price of the project, such as an R&D project, is the price of the call option. The exercise price of the call option is the cost of future investments needed when an initial investment is made. The project's expected net present value is analogous to the price of stock in the Black and Scholes formulation. The underlying asset is the project, the strike price of the option is the investment needed to take the project, and the life of the option is the period for which the firm has rights to the project. The variance in this present value represents the variance of the underlying asset. The value of the option is largely derived from the variance in cash flows as the higher the variance, the higher the value of the project delay option. Thus, the value of an option to do a project in a stable business will be less than the value of one in a changing environment. Mitchell and Hamilton (1988) emphasize that management needs to identify strategic objectives, review the impact of strategic options, such as R&D projects directed toward strategic planning, and identify the strategic planning targets of future R&D projects.

17.2   Implications of Viewing the Right to Delay a Project as an Option

Several interesting implications emerge from the analysis of the ability to delay a project as an option. First, a project may have an initial negative net present value based upon current expected cash flows, but it may still be a "valuable" project because of the option characteristics. Thus, while a negative net present value should encourage a firm to reject a project, it should not lead it to conclude that the rights to this project are worthless. Second, a project may have a positive net present value but still not be accepted right away. This is because the firm may gain by waiting and accepting the project in a future period, particularly for risky projects. In static analysis, increasing uncertainty increases the riskiness of the project and may make it less attractive. When the project is viewed as an option, an increase in uncertainty may actually make the option more valuable.

17.3   Abandonment Value

A project does not always produce its expected cash flow, and the net present value of a project, initially calculated to be positive, does not always produce value for the shareholders. How does management come to grips with cash flow forecasts that turn out to be incorrect or were based on assumptions that are not substantiated? What can the firm's financial managers do to minimize stockholder's losses? A possible solution is abandonment of the project. The option to abandon a project may consist of selling the project's assets and not realizing future cash flow. The abandonment option is a put option, similar in logic to the put options on stocks discussed in Chap. 16.

   Let us develop an investment scenario where abandonment value enhances the decision-making process. An R&D project requires the construction of a new building near to, but off, the main corporate grounds. The new building will cost

Table 17.1 Economic scenarios of the economy and project cash flow

| | Year 1 | | Year 2 | |
State of the economy	Probabilities	($mm) Cash flow	Probabilities	Cash flow
Recession	.30	$20.0	.30	10
			.40	20
			.30	30
Normal	.30	30.0	.15	20
			.50	30
			.35	50
Boom	.40	50.0	.10	30
			.40	50
			.50	75

Table 17.2 Expected net present value ($MM)

| | Cash flow | | PVIF | | Present value | | | | |
State of nature	Year 1	Year 2	Year 1	Year 2	Year 1	Year 2	Present value	Joint probability	Expected present value
Recession	20	10	0.909	0.826	18.18	8.26	26.44	0.090	2.38
	20	20	0.909	0.826	18.18	16.52	34.70	0.120	4.16
	20	30	0.909	0.826	18.18	24.78	42.96	0.090	3.87
Normal	30	20	0.909	0.826	27.27	16.52	43.79	0.045	1.97
	30	30	0.909	0.826	27.27	24.78	52.05	0.150	7.81
	30	50	0.909	0.826	27.27	41.30	68.57	0.105	7.20
Boom	50	30	0.909	0.826	45.45	24.78	70.23	0.040	2.81
	50	50	0.909	0.826	45.45	41.30	86.75	0.160	13.88
	50	75	0.909	0.826	45.45	61.95	107.40	0.200	21.48
Expected present value									65.56
Expected net present value									20.56

$45,000,000 and can house a small production facility for 3 years even if management decides to forego or postpone the R&D project. Sales of the production facility are dependent upon the state of the economy. The corporate economists have prepared a set of three-year cash flow forecasts that are first-year probabilities and second-year conditional probabilities. That is, the cash flow forecasts are dependent upon particular states of nature occurring in years one and two. See Table 17.1. One should calculate the expected net present value of the projected cash flow, assuming a 10% cost of capital. The calculations of the expected net present value and internal rate of return are shown in Table 17.2. One multiplies the cash flow under the various economic scenario, depression, recession, normal, and boom, by the cash flow occurring in that state of the economy. The three scenarios, the two-period analysis, produce 9 possible states of the economy. The key to the analysis is to calculate the

Table 17.3 Expected internal rate of return

State of nature	Internal rate of return	Joint probability	E(IRR)
Recession	-25.66%	0.090	-2.31%
	-7.50	0.120	-.90
	6.84	0.090	0.62
Normal	7.87	0.045	0.35
	21.53	0.150	3.23
	43.89	0.105	4.61
Boom	33.33	0.040	1.33
	54.31	0.160	8.69
	96.10	0.200	19.22
Expected internal rate of return			34.84%

joint probabilities of each possible state. Each state of the economy is conditional upon the previous period's state of the economy. See Table 17.2 for the calculation of the joint probabilities, the expected present value, and the expected net present value.

The expected present value of cash flow is $65,560,000. Given the cost of the building of $45,000,000, the expected net present value of the new building is $20,560,000. The expected net present value exceeds zero, and the new building can be justified at a cost of capital of 10%. The expected internal rates of return (IRR) of the project for the nine scenarios are shown in Table 17.3. The expected internal rate of return is 34.84%, far exceeding the cost of capital, and the project is acceptable. The reader is referred to Chap. 11 in which we stated that in the vast majority of cases, a positive net present value implies that the internal rate of return exceeds the cost of capital.

The expected IRR for the project is 34.84%. The expected IRR exceeds the cost of capital, and hence the expected net present value is positive. The expected variance of the project is 14.94%; the calculations are shown in Table 17.4.

What is the economic benefit of being able to abandon the new building project after year one at an abandonment value of $24,000,000? If the abandonment value of $ 24.0 million exceeded the expected present value of cash flow of year two in any scenario or state of the economy, then the expected net present value calculation of the new building should be recalculated. The present value of the abandonment value of $ 24.0 million exceeds the expected present value of cash flow for year two in the recession ($16.52 million) mode. The abandonment value option increases the project net present value (NPV) to $67.15 MM (see Table 17.5), and the internal rate of return to 37.77%. The variance of the expected internal rate of return falls to 12.88% with abandonment value.

The recalculated expected net present value of the new building is shown in Table 17.6. The expected net present value increases by $1.59 million by the presence of the abandonment option. The project should be abandoned after year one.

Table 17.4 Variance of expected internal rate of return

State of nature	Joint prob	IRR	E(IRR)	Std (IRR)
Recession	0.090	−0.2566	−0.0231	0.0329
	0.120	−0.0750	−0.0090	0.0215
	0.090	0.0684	0.0062	0.0071
Normal	0.045	0.0787	0.0035	0.0033
	0.150	0.2153	0.0323	0.0027
	0.105	0.4389	0.0461	0.0009
Boom	0.040	0.3333	0.0133	0.0000
	0.160	0.5431	0.0869	0.0061
	0.200	0.9610	0.1922	0.0751
			0.3484	0.1494

Table 17.5 Expected net present value of year two states of nature

	Cash flow				Joint	Present value of
State of nature	Year 1	Year 2	Year 2	Year 2	probability	year 2
Recession	20	10	0.826	8.26	0.300	2.48
	20	20	0.826	16.52	0.400	6.61
	20	30	0.826	24.78	0.300	7.43
Recession expected (PV)						16.52
Normal	30	20	0.826	16.52	0.100	1.65
	30	30	0.826	24.78	0.500	12.39
	30	50	0.826	41.30	0.350	14.46
Normal expected (PV)						28.50
Boom	50	30	0.826	24.78	0.100	2.48
	50	50	0.826	41.30	0.400	16.52
	50	75	0.826	61.95	0.500	30.98
Boom expected (PV)						49.97

    The presence of an abandonment value of $24.0 million enhances the net present value of the project by $1.59 million because the abandonment value exceeds the expected present value of year two cash flows in the recession scenario. The abandonment analysis may not be complete until one calculates the present value of the cash flow foregone by abandoning the project. The present value of the abandoned cash flow is shown in Table 17.7 and is $33.74 million.

    This chapter follows Copeland and Weston (1992) and calculates the abandonment put value. One uses the present value of the abandoned cash flow as the equivalent of the stock price, the abandonment value as the exercise price, and a two-year period for the option. If the risk-free rate is 5%, the value of the put option is calculated to be $0.036 million.

Table 17.6 Expected net present value with abandonment value

	Cash flow		PVIF		Present value				
	Year	Year	Year	Year	Year	Year	Present	Joint	Expected
State of nature	1	2	1	2	1	2	value	probability	present value
Recession	44	0	0.909	0.826	40.00	0.00	40.00	0.090	3.60
	44	0	0.909	0.826	40.00	0.00	40.00	0.120	4.80
	44	0	0.909	0.826	40.00	0.00	40.00	0.090	3.60
Normal	30	20	0.909	0.826	27.27	16.52	43.79	0.045	1.97
	30	30	0.909	0.826	27.27	24.78	52.05	0.150	7.81
	30	50	0.909	0.826	27.27	41.30	68.57	0.105	7.20
Boom	50	30	0.909	0.826	45.45	24.78	70.23	0.040	2.81
	50	50	0.909	0.826	45.45	41.30	86.75	0.160	13.88
	50	75	0.909	0.826	45.45	61.95	107.40	0.200	21.48
Expected present value with abandonment value									67.15
Expected net present value with abandonment value									22.15

Table 17.7 Expected net present value of year two cash flow

State of nature	Year 2	Year 2	PV(year 2)	Joint probability	Present value of year 2
Recession	10	0.826	8.26	0.090	0.74
	20	0.826	16.52	0.120	1.98
	30	0.826	24.78	0.090	2.23
Normal	20	0.826	16.52	0.045	0.74
	30	0.826	24.78	0.150	3.72
	50	0.826	41.30	0.105	4.34
Boom	30	0.826	24.78	0.040	0.99
	50	0.826	41.30	0.160	6.61
	75	0.826	61.95	0.200	12.39
Expected (PV) of year 2 cash flow					33.74

$$v_c = P_{cs}N(d_1) - \frac{Ex}{e^{rt}}N(d_2)$$

$$d_1 = \frac{\ln\left(\frac{P_{cs}}{Ex}\right) + rt}{\sigma\sqrt{t}} + \frac{1}{2}\sigma\sqrt{t}$$

$$d_2 = d_1 = \sigma\sqrt{t}$$

$$d_1 = \frac{\ln\left(\frac{33.74}{24}\right) + .05 \ (1)}{.1288 \ \sqrt{1}}$$

$$+ \frac{1}{2}(.1288)$$

$$= \frac{0.3406 + .05}{.1288(1)} + \frac{1}{2}(.1288) \ \sqrt{1}$$

$$= \frac{.3906}{.1288} + .0644 = 3.097$$

$$d_2 = 3.907 - .1288\sqrt{1} = 3.907 - .1288 = 2.968$$
$$N(d_1) = 1.000$$
$$N(d_2) = 0.9986$$

$$v_c = 33.74(1.000) - \frac{24}{2.71828 - .05(1)} \ (.9986)$$

$$= 33.74 - \frac{24}{1.051} \ (.9986)$$

$$= 33.74 - 22.80 = 10.94$$

$$v_{c_0} - P_0 = P_{cs} + \frac{Ex}{e^{rt}}$$

$$P_0 = 10.94 - 33.74 + \frac{24}{1.051}$$

$$= \$10.94 - 33.74 + 22.84 = \$ \ .036$$

The value of the put option to abandon the project is worth \$0.036 MM.

17.4   Options in Investment Analysis/Capital Budgeting

In traditional investment analysis, a project or new investment should be accepted only if the returns on the project exceed the hurdle rate, the cost of capital, which leads to a positive net present value. Several additional aspects of real options are embedded in capital budgeting projects. The first is the option to delay a project, especially when the firm has exclusive rights to the project. The second is the option to expand a project to cover new products or markets sometime in the future.

References

Chance, D. M., & Peterson, P. P. (2002). *Real options and investment valuation*. The Research Foundation of AIMR.

Copeland, T., & Weston, J. F. (1992). *Financial policy and corporate policy* (3rd ed.). Addison-Wesley Publishing Company. Chapter 12.

Hackbarth, D., & Johnson, T. C. (2015). Real options and risk dynamics. *Review of Economic Studies, 82*, 1449–1482.

Hamilton, W. F., & Mitchell, G. R. (1990). R&D in perspective: What is R&D worth? *The McKinsey Quarterly*, 150–160.

Mauer, D. C., & Sarkar, S. (2005). Real options, agency conflicts, and optimal capital structure. *Journal of Banking & Finance, 29*, 1405–1428.

McDonald, R. L. (2003). *Derivatives Markets*. Addison Wesley.

McDonald, R. L. (2006). The role of real options in capital budgeting: Theory and practice. *Journal of Applied Corporate Finance, 18*, 28–39.

Mitchell, G. R. (1985). New approaches to the strategic Management of Technology. *Technology in Society*, 227–239.

Mitchell, G. R., & Hamilton, W. F. (1988). Managing R&D as a strategic option. *Research-Technology Management*, 15–22.

Trigeorgis, L. (1996). *Real options*. MIT Press.

Tserlukevich, Y. (2008). Can real options explain financing behavior? *Journal of Financial Economics, 89*, 232–252.

Van Horne, J. C. (2002). *Financial management and policy* (12th ed.). Prentice-Hall, Chapter 7.

Chapter 18
Mergers and Acquisitions

A company can grow by taking over the assets or facilities of another. The various methods by which one firm obtains or "marries into" the business, assets, or facilities of another company are mergers, combinations, or acquisitions.[1] These terms are not used rigidly. In general, however, a merger signifies that one firm obtains another by issuing its stock in exchange for the shares belonging to owners of the acquired firm, or buys another firm with cash. Company X gives some of its shares to Company Y shareholders for the outstanding Y stock. When the transaction is complete, Company X owns Company Y because it has all (or almost all) of the Y stock. Company Y's former stockholders are now stockholders in Company X. In a combination, a new corporation is formed from two or more companies who wish to combine. The shares of the new company are exchanged for those of the original companies. The difference between a combination and a merger lies more in legal distinctions than in any discernible differences in the economic or financial result. In practice, the terms merger and combination are often used interchangeably. The study of merger profitability is as old as corporate finance itself. Arthur S. Dewing (1921, 1953) reported on the relative unsuccessfulness of mergers for the 1893–1902 time period; and Livermore (1935) reported very mixed merger results for the 1901–1932 time period. Mandelker (1974) put forth the Perfectly Competitive Acquisitions Market (PCAM) hypothesis in which competition equates returns on assets of similar risk, such that acquiring firms should pay premiums to the extent that no excess returns are realized to their stockholders. The PCAM holds that only the acquired firms'

---

[1] Although mergers, combinations, and acquisitions are exciting events of great interest to the financial community, they do not represent the most common method through which even individual companies grow. Most firm growth is direct growth, which takes place quietly and is financed either by internal sources or by floating new securities. A study of 74 large companies showed that of their growth from 1900 to 1948, 75% was direct growth and only 25% was accounted for by mergers. See, J. F. Weston (1953).

© The Author(s), under exclusive license to Springer Nature Switzerland AG 2022
J. B. Guerard Jr. et al., *Quantitative Corporate Finance*,
https://doi.org/10.1007/978-3-030-87269-4_18

stockholders earn excess returns. Jensen and Ruback (1983) reported that acquired firms profited handsomely, while acquiring firms lost little money such that wealth was enhanced. Recent evidence by Jarrell et al. (1988), Ravenscraft and Scherer (1989), and Alberts and Varaiya (1989) finds little gain to the acquiring firm and significant merger premiums paid for the acquired firms.

Acquisition usually refers to a transaction in which one firm buys the major assets or the controlling shares of another company. On occasion, one corporation has purchased another corporation's subsidiaries. An acquisition differs from a merger in that generally (but not always) cash is used rather than an exchange of securities.[2]

No new net financial holdings are created in the economy by any of the forms of merger or acquisition [Mossin (1973)]. If the transaction involves an exchange of stock, the supply of shares of one company's stock is eliminated and is replaced by the shares of the surviving company. If the transaction is financed by cash, cash holdings by individuals go up, but cash held by corporations goes down; the supply of outstanding securities in the hands of the public goes down, but the amount held by corporations rises. If a corporation floats new securities to obtain funds to finance its acquisition, then the process is slightly roundabout, but the net results are the same. One section of the public surrenders its cash for new corporate securities; another group gives up a different issue of securities for cash.

In this chapter, we introduce the reader to the economic and noneconomic theories of mergers. We report the time-series history of mergers in the United States and estimate a time-series model to assess the statistical association between mergers and stock prices, and mergers and the LEI, long put forth by Nelson (1959, 1966). We introduce the reader to the Rappaport (1979) framework for the valuation of a merger candidate. We test for merger profit and synergy. We conclude by answering the question: Do mergers enhance stockholder wealth?

18.1   Noneconomic Motives for Combinations

Noneconomic motives may dominate in a number of mergers; sociological or psychological motives, possibly involving the personal well-being of the management, may lie behind some combinations. Management may feel that its prestige, status, or salary will rise if it is in charge of a larger business (even though the *rate* of profit may be no higher than that of the present firm). Management may favor a merger because the acquiring firm has better bonus or retirement provisions than the

---

[2]A firm which either by the exchange of securities or purchase of shares controls subsidiary companies is called a *holding* company. The holding company differs from a *parent* company in that the parent company has production functions of its own to perform, whereas a holding company exists *mainly* to control or coordinate its subsidiaries.

present company. A merger or combination may take place because the management of one company is getting old and a merger may appear to be the simplest method of obtaining new managerial personnel. A wave of mergers or acquisitions may take place to satisfy the desires for risk of speculative management.

Where an industry is characterized by oligopolistic competition, company growth, whether achieved by combination or direct investment, may be announced as being undertaken to "maintain the company's relative share of the market" or to "prevent deterioration of the company's competitive position." These announced reasons may appear to reflect desires to maintain corporate prestige or status. Actually, they may have considerable economic validity as rule-of-thumb guides to prevent erosion of the earning power of the firm. As noted in Chap. 11, sometimes the return on a new investment may be computed, not on the basis of potential additional earnings, but on the basis of losses averted or on a prevented decrease in earnings. Thus, growth can be defensive, a rational move to forestall a downward slide in the company's market position which, once started, may be difficult to halt.

By their nature, the psychological and sociological motives underlying some instances of corporation growth are obscure and complex and do not allow for easy analysis or any reliable quantification. Since economic analysis rests on the supposition of economic rationality on the part of decision-makers, it cannot furnish much light on growth undertaken mainly to satisfy desires for status or power.

18.2   Holding Companies

The use of the term "holding company" is not always clear in financial literature. In general, a holding company is a firm that owns controlling shares in other corporations and whose main function is to coordinate and direct the affairs of these subsidiaries. In contrast, many operating companies run subsidiaries and yet are not usually referred to as holding companies, because they have extensive operations of their own.[3] The holding company must also be distinguished from the investment company, which holds shares in other corporations mainly for the investment return and does not guide their internal management.

Among the advantages of the holding company is the flexibility of administration. It enables top management to centralize those functions which can be performed at lower costs in large volumes and to decentralize responsibility for programs and functions better performed or supervised at local levels. Of course, like other forms

---

[3]Thus, Standard Oil of New Jersey, the US Steel Corporation, and the original Bethlehem Steel Corporation, although they owned many subsidiaries, were not usually thought of as holding companies.

of organization, the holding company can become too large and, thus, inefficient in trying to coordinate its various enterprises.[4]

18.3   A Merger History of the United States

The height of the merger movement was reached in 1901 when 785 plants combined to form America's first billion-dollar firm, the United States Steel Corporation. The series of mergers creating US Steel allowed it to control 65% of the domestic blast furnace and finished steel output. This growth in concentration was typical of the first merger movement. The early mergers saw 78 of 92 large consolidations gain control of 50% of their total industry output, and 26 secure 80% or more.

The beginning of the major merger movement occurred during a period of rapid economic growth. The economic rationale for the large merger movement was the development of the modern corporation, with its limited liability, and the modem capital markets, which facilitated the consolidations through the absorption of the large security issues necessary to purchase firms. Nelson (1959) found the mergers were highly correlated with the period's stock prices (.613) and industrial production (.259) during the 1895–1904 period. The expansion of security issues allowed financiers the financial power necessary to induce independent firms to enter large consolidations. The rationale for the first merger movement was not that of trying to preserve profits despite slackening demand and greater competitive pressures.

---

[4]In the 1930s, investors in holding companies suffered heavy losses, and this fact, plus the exposure of legislative and operating scandals involving some promoters of holding companies, led to the passage by Congress of the Public Utility Company Act of 1935. Some of the major provisions of this legislation are:

1. Holding companies were prohibited over the third level-one operating company and two holding companies. That is, a holding company structure could go up to the grandfather level but not beyond. This "great-grandfather death clause" was designed to restrict the excessive leverage possible in the holding company structure.
2. Holding company systems were to be consolidated and simplified to develop some aspect of geographical contiguity. If the holding company is to serve an economic, function, the location of its operating companies should have some geographical logic.
3. "Upstream" loans were made, illegal. An upstream loan is made from a lower level company to a higher one. A holding company that serves a function should help to finance its operating companies, and not the other way around. The request for an upstream loan could not be resisted by a captive operating company even though the purpose of the loan was to bail the holding company out of a financial difficulty, and its net effect might be to deplete the assets of the operating company without necessarily rescuing the holding company.
4. Service and construction companies organized by the holding company system to work for operating companies were required to be nonprofit. In the usual case, there is no felt need to regulate intercompany prices because the resulting profits are only a shift of earnings among the various companies. However, in the public utility fields excessive service charges or construction costs could affect the rate base and thus may be borne eventually by the consumers.
5. These rules and other provisions of the Public Utility Act of 1935 aroused considerable antagonism in the business community.

Nor was the merger movement the result of the development of the national railroad system, which reduced geographic isolation and transportation costs. The first merger movement ended the 1904 depression, with the *Northern Securities* case. Here, it was held, for the first time, that antitrust laws could be used to attack mergers leading to market dominance.

A second major merger movement stirred the country from 1916 until the Depression of 1929. This merger movement was only briefly interrupted by the First World War and the recession of 1921 and 1922. Approximately 12,000 mergers of the period coincided with the stock market boom of the 1920s. Although mergers greatly affected the electric and gas utility industry, the market structure was not as severely concentrated by the second movement as it had been by the first. Stigler (1950) concluded that mergers during this period created oligopolies, such as Bethlehem Steel and Continental Can. Mergers, primarily vertical and conglomerate in nature as opposed to the essentially horizontal mergers of the first movement, did affect competition adversely. The conglomerate product-line extensions of the 1920s were enhanced by the high cross-elasticities of demand for the merging companies' products (Lintner, 1971). Antitrust laws, though not seriously enforced, prevented mergers from creating a single dominant firm. Merger activity diminished with the Depression of 1929 and continued to decline until the 1940s.

The third merger movement began in 1940; mergers reached a significant proportion of firms in 1946 and 1947. The merger action from 1940 to 1947, although involving 7.5% of all manufacturing and mining corporations and controlling 5% of the total assets of the firms in those industries, was quite small compared to the merger activities of the 1920s. The mergers of the 1940s included only one between companies with assets exceeding $50 million and none between firms with assets surpassing $100 million. The corresponding figures for the mergers of the 1920s were 14 and 8, respectively. Eleven firms acquired larger firms during the mergers of the 1920s than the largest firm acquired during the 1940s.

The mergers of the 1940s affected competition far less than did the two previous merger movements, with the exception of the food and textile industries. The acquisitions by the large firms during the 1940s rarely amounted to more than 7% of the acquiring firms' 1939 assets or as much as a quarter of the acquiring firm's growth from 1940 to 1947. Approximately $5 billion of assets were held by acquired or merged firms over the period 1940–1947. Smaller firms were generally acquired by larger firms. Companies with assets exceeding $100 million acquired, on average, firms with assets of less than $2 million. The larger firms tended to engage in a greater number of acquisitions than smaller firms. The acquisitions of the larger acquiring firms tended to involve more firms than did those of smaller acquiring firms. Mergers added relatively less to the existing size of the larger acquiring firms. The relatively smaller asset growth of the larger acquiring firms is in accordance with the third merger movement's generally small effects on competition and concentration.

The current merger movement is an extension of the 1940s' conglomerate movement beginning in 1951 and continuing to the present. Conglomerate mergers involve two firms in different, uncorrelated businesses. Of the nine mergers occurring in 1951 involved acquired firms with assets exceeding $10 million, four were conglomerate mergers, of which three were product-line extension combinations.

Table 18.1 Quarterly net merger announcements 1999–2018

	First	Second	Third	Fourth	Annual
1999	2,413	2,833	2,826	2,628	10,700
2000	3,634	3,440	3,125	2,937	13,136
2001	2,807	2,502	2,169	2,209	9,777
2002	2,281	2,278	2,074	2,110	8,743
2003	2,195	2,298	2,241	2,511	9,243
2004	2,756	2,827	2,708	2,624	10,915
2005	2,774	3,076	2,988	2,846	11,684
2006	2,972	3,162	3,011	2,964	12,109
2007	3,068	2,950	2,663	2,587	11,268
2008	2,339	2,145	2,206	1,705	8,395
2009	1,597	1,746	1,851	1,978	7,172
2010	2,294	2,306	2,310	2,191	9,431
2011	2,517	2,585	2,371	2,392	9,865
2012	2,345	2,377	2,423	2,890	10,035
2013	2,288	2,211	2,593	2,621	9,713
2014	2,880	3,020	3,060	3,140	12,100
2015	3,150	3,186	3,319	3,280	12,935
2016	3,073	2,929	3,101	2,943	12,049
2017	3,041	2,790	2,608	2,763	11,202
2018	2,848	3,172	3,099	2,889	12,008

Source: Mergerstat Review, December 2019

The growth of the large conglomerate mergers continued throughout the forecast period. In 1954, 21 of the 37 mergers involving acquired firms with assets exceeding $10 million were conglomerate in nature; 14 of the 21 conglomerate mergers were product-line extensions, while only two of the mergers were market-extension combinations. The mergers of the 1960s, 1970s, and 1980s were almost purely conglomerate in nature (Table 18.1).

Merger activity continues to be very high, but not increasing as dramatically as in much of the postwar conglomerate merger movement.

The time series of US mergers follows a near-random walk (Shugart and Tollison 1984). We test the work of Nelson (1959, 1966) that mergers are significantly associated with lags in stock prices or the Leading Economic Indicator (LEI) time series modeling the data with Autometrics, as we did in Chap. 13. Mergers follow an AR(1) process with a large first-order autoregressive parameter then is close to one; see Eq. (18.1). We report that mergers during the 1998–2018 time series are statistically associated with a one-period lag in the LEI variable, see Eq. (18.2).[5]

---

[5]The AR(1) model in least squares produced an RSS of 3563507.64, which is reduced to 1627855.60, with Autometrics. The Autometrics-estimated time-series model residuals were randomly distributed, as was the case of real GDP and the change in the US unemployment rate, reported in Chap. 13. Guerard (1985) and Guerard and McDonlad (1991) reported that the merger time was significantly associated with lagged stocks prices. The authors did not test a transfer function model using the LEI variable.

EQ(1) Modelling Mergers by OLS

	Coefficient	Std.Error	t-value	t-prob	Part.R^2
Mergers_1	0.855866	0.06048	14.2	0.0000	0.7329
Constant	353.503	160.6	2.20	0.0309	0.0622
Trend	0.755644	1.177	0.642	0.5229	0.0056

sigma	220.942	RSS	3563507.64
R^2	0.743791	F(2,73) =	106 [0.000]**
Adj.R^2	0.736772	log-likelihood	-516.549
no. of observations	76	no. of parameters	3
mean(Mergers)	2655.03	se(Mergers)	430.637

When the log-likelihood constant is NOT included:

AIC	10.8345	SC	10.9265
HQ	10.8712	FPE	50742.1

When the log-likelihood constant is included:

AIC	13.6723	SC	13.7643
HQ	13.7091	FPE	866648.

AR 1-2 test:	F(2,71)	=	0.48027 [0.6206]
ARCH 1-1 test:	F(1,74)	=0.0012872 [0.9715]	
Normality test:	Chi^2(2)	=	36.851 [0.0000]**
Hetero test:	F(4,71)	=	1.9242 [0.1158]
Hetero-X test:	F(5,70)	=	1.5648 [0.1814]
RESET23 test:	F(2,71)	=	0.57252 [0.5667]

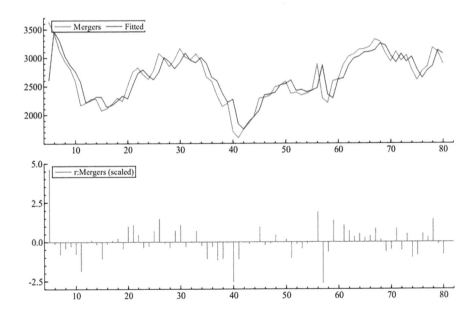

EQ(2) Modelling Mergers by OLS
        The estimation sample is: 5 - 80

	Coefficient	Std.Error	t-value	t-prob	Part.R^2
Mergers_1	0.750698	0.06129	12.2	0.0000	0.6788
LEI_1	7.07670	1.763	4.01	0.0001	0.1850
DI:56	477.350	108.0	4.42	0.0000	0.2158
I:5	1006.57	152.4	6.60	0.0000	0.3805
I:40	-562.467	153.1	-3.67	0.0005	0.1596

sigma	151.418	RSS	1627855.55
log-likelihood	-486.777		
no. of observations	76	no. of parameters	5
mean(Mergers)	2655.03	se(Mergers)	430.637

When the log-likelihood constant is NOT included:

AIC	10.1036	SC	10.2570
HQ	10.1649	FPE	24435.9

When the log-likelihood constant is included:

AIC	12.9415	SC	13.0948
HQ	13.0028	FPE	417353.

AR 1-2 test: $F(2,69)$ $=$ 2.9466 [0.0592]
ARCH 1-1 test: $F(1,74)$ $=$ 0.85078 [0.3593]
Normality test: $Chi^2(2)$ $=$ 0.20014 [0.9048]
Hetero test: $F(6,67)$ $=$ 0.78130 [0.5875]
Hetero-X test: $F(7,66)$ $=$ 0.70715 [0.6659]
RESET23 test: $F(2,69)$ $=$ 0.047961 [0.9532]

Robust standard errors

	Coefficients	t-SE	t-HACSE	t-HCSE	t-JHCSE
Mergers_1	0.75070	12.249	11.536	13.892	13.796
LEI_1	7.0767	4.0148	3.7241	4.4361	4.4133
DI:56	477.35	4.4204	26.376	18.399	10.200
I:5	1006.6	6.6040	46.262	55.697	0.023716
I:40	-562.47	-3.6727	-20.154	-23.382	-0.046384

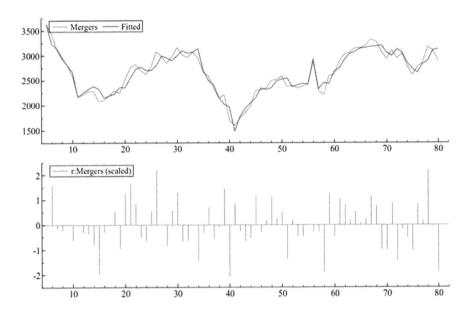

```
Model saved to C:\JBG\JGResearch\QCF3Ed\Model Mergers AR1 LEIL1 SPL1
IIS DIIS TIS.pdf
Model saved to C:\JBG\JGResearch\QCF3Ed\Model Mergers AR1 LEIL1 SPL1
IIS DIIS TIS.pdf.gwg
```

18.4 Using an Accounting Basis

If one company purchases another company and only the acquiring company survives, the combination is a merger. Consolidation involves the combination of at least two firms in which a new firm is created.

A company acquires another firm by the *purchase* or accounting method when the acquired firm is valued as the amount actually purchased (Weston et al., 2004). The parent firm records the acquired assets at their current market value. If the current value of the acquired firm exceeds its book value, the excess is goodwill that will be amortized over a period of no more than 40 years. The purchase method is the sole acceptable method of accounting for mergers and acquisitions since the Financial Accounting Standards Board (FASB) voted in January 2001 to eliminate the pooling-of-interests method of accounting for mergers and acquisitions.

A second accounting practice, the *pooling-of-interests* method, which is no longer allowed since the FASB released its "Business Combinations" statement (No. 141) in June 2001, had allowed acquisitions in which interests of both firms' stockholders are merged. No assets are revalued, and no goodwill was recognized as assets in pooling of interests mergers, and liabilities of the new firms were recorded as such on the unmerged firms' books. The pooling-of-interests method may be used only when one firm issues voting stock for the acquisition; no cash transactions are allowed. The stock issuance for the acquisition must be a single transaction, and the acquiring firm must acquire at least 90% of the acquired firm's stock. The FASB No. 141 Statement superseded the Accounting Principles Board (APB) Opinion 16, "Accounting for Business Combinations" of August 1970, which allowed both purchase and pooling-of-interests methods of accounting for mergers and acquisitions.

In the purchase method of accounting for a merger or acquisition, the purchase price is the basis for recording the assets and liabilities of the acquired firm. The earnings of the acquired firm are reported by the acquiring firm from the date of the acquisition forward. In the pooling-of-interests method of accounting, the original historic cost basis of the assets and liabilities of the acquired firm are carried forward, and the earnings of the combined entity are combined for any reporting periods. Prior year earnings would be the earnings of the pooling-of-interest combination. Assets acquired in a pooling-of-interest M&A could be sold for a higher price than was recorded in the original transaction, and gains in such a transaction would be earned income. The pooling-of-interests method had the potential to manipulate earnings of the firm (Weston et al., 2004).

18.5   The Economic Basis for Acquisitions

There are three economic reasons for seeking to acquire other firms, to pursue profits through monopoly power, to generate economies of scale in operations, and to contribute managerial abilities lacking in the acquired firm's management. Expectations of the three economic benefits of mergers can explain horizontal mergers, in which a firm acquires a firm in its own industry. But the economic reasons cannot explain the postwar trend of conglomerate mergers, in which a firm acquires a firm in an unrelated industry. Economists have held that mergers do not generally produce synergism, the "2 + 2 = 5" effect. When synergism is absent, the value of the mergers adheres to the additive property of value, in which the value of the merged film is the sum of the market values of the unmerged firms (Mossin, 1973):

$$MV_{ab} = MV_a + MV_b \qquad\qquad (18.1)$$

where

MV_{ab} = market value of the merged firm, a and b;
MV_a = market value of firm a;
MV_b = market value of firm b.

If no synergism exists, there would not seem to be any economic rationale for conglomerate mergers.

18.6   Theories of Conglomerate Mergers

Conglomerate mergers were explained by Mueller (1969) as resulting from firms' attempts to maximize growth. If a larger, more mature firm wants to consider the more profitable investment opportunities and the higher rates of return on projects of a smaller, younger firm, the larger firm might purchase the smaller firm. By purchasing the smaller firm, the larger firm has internalized the investment opportunities, and the more profitable marginal investment contributes to the larger firm's growth. Synergism, in which the merged firm's market value exceeds the sum of the unmerged firms' market values, would rarely be found in a conglomerate merger. Mueller hypothesized that the conglomerate merger could be justified only if the managers of the acquiring firm could find investment opportunities currently overlooked by the acquired firm's management. Mueller's theory of mergers has been questioned because its premise is that firms maximize growth. Corporate benefits tend to be distributed on the basis of profits, not growth of sales. Management would maximize profits, in a rational environment, if profits served solely as the basis for benefits.

Gort (1969) put forth a theory of mergers based on valuation discrepancies or disturbances. Acquisitions of other firms occur when a higher value is placed on the firm's assets by the potential purchaser than by the current owners. Valuation discrepancies tend to occur in times of high stock prices, as purchasers tend to believe the rapid changes in stock prices represent an increased economic disturbance. The higher disturbance increases the distribution of the firm's assets' valuations, raising the probability that an investor will place a higher value on the firm than the current owners. The empirical evidence has revealed that mergers occur during periods of high stock prices; thus, there could be a basis for Gort's disturbance theory.

One of the principal theories of conglomerate mergers is that mergers reduce the probability of bankruptcy for the merged firm. Lewellen (1971) advanced the theory that mergers reduce the variance of the merged firm's cash flow. Mergers may lessen the possibility of default and increase the firm's debt capacity. Debt capacity is increased as lenders are willing to establish a higher corporate lending limit to the merged entity. However, the merger must involve firms with less than perfectly correlated income streams. The merging firms must not have cash flows such that defaults on their borrowings occur simultaneously.

Assume that firms A and B are considering a merger. Given the following probability estimates of the possible states of the world, would the risk of default be lessened by the merger?

Cash Flow

Probability	State	Firm A	Firm B
0.2	Depression	$ 100	$920
0.3	Recession	300	700
0.4	Normal	700	240
0.1	Boom	1,000	0

It is obvious that the cash flows of firms A and B are negatively correlated. Firm A could be a venture capital firm that profits from a rising economy, while firm B could be a gold mining company that profits when gold prices are high, the economy falling.

Firms A and B have the same expected values E(CF) and essentially the same standard deviations of cash flows, σ_{CF}. The expected cash flow is of the form:

$$\sum_{i=1}^{4} p_i CF_i,$$

where p_i = probability of the occurrence of state of nature, I, and the variance is of the form

$$\sum_{i=1}^{4} [CF_i - E(CF)]^2 p_i. \tag{18.2}$$

We may calculate the expected values of the two cash flows, CF_A and CF_B.

.2($80)	=	$16	.2(920)	=	$184
.3($280)	=	84	.3(700)	=	210
.4($700)	=	280	.4(240)	=	96
.1($1,000)	=	100	.1(0)	=	0
E(CF$_A$)	=	490	E(CF$_B$)	=	490

And the variances:

.2($ 80 - $490)	=	$33,620
.3($ 280 - $490)	=	13,230
.4($ 700 - $490)	=	17,640
.1($1,100 - $490)	=	37,210
Var(CF$_A$)	=	$101,700
σ(CF$_A$)	=	$319

.2($920 - $490)	=	$36,980
.3($700 - $490)	=	13,230
.4($240 - $490)	=	25,000
.1($ 0 - $490)	=	24,010
Var(CF$_B$)	=	$99,200
σ(CF$_B$)	=	$315

The covariance of A's and B's cash flows, the covariance being:

$$\sum_{i=1}^{4} p_i[CF_{A_i} - E(CF_A)][E(CF_{B_i} - E(CF_B)],$$ (18.3)

is calculated:

$$
\begin{array}{lll}
.2(\$80 - \$490) & (\$920 - \$490) & = \$-35{,}260 \\
.3(\$\ 280 - \$490) & (\$700 - \$490) & = -13{,}230 \\
.4(\$\ 700 - \$490) & (\$240 - \$490) & = -21{,}800 \\
.1(\$1{,}100 - \$490) & (\$0 - \$490) & = \underline{-29{,}890} \\
& Cov(CF_A, CF_B) & = \$-99{,}380
\end{array}
$$

The correlation coefficient, r, of the cash flows of firms A and B will establish that these are imperfectly correlated firms.

$$r = \frac{Cov(CF_A, CF_B)}{\sqrt{Var(CF_A \times Var(CF_B)}}$$

$$\frac{\$ - 99{,}380}{\sqrt{\$101{,}700 \times \$99{,}200}} = \frac{\$ - 99{,}380}{\$100{,}452} = -.989$$

The correlation coefficient of -0.99 shows that the cash flows of firms A and B are almost perfectly negatively correlated.

If firms A and B are assumed to have assets of $10,000 each, a debt of $6000 each, with each firm's debt having a 4% cost, the unmerged firms should have essentially the same default probability. The interest costs for each firm are $240, that is, $6000 (0.04), and the probabilities of default, given by the area under the normal curve, are essentially equal. The general form is:

$$Z = \frac{X - \mu}{\sigma}$$ (18.4)

where

$Z =$ Standardized normal variable;
$X =$ Interest cost, $240;
$\mu =$ Expected value, $490 of cash flow;
$\sigma =$ Standard deviation of cash flow, $319 and $315 for firms A and B, respectively.

$$P_r(\text{Firm A's cash flow} > \$240)$$

$$Z_A = \frac{\$240 - \$490}{\$319} = -.784$$

$$Z_B = \frac{\$240 - \$490}{\$315} = -.794$$

The Z values of -0.784 and -0.794 correspond to bankruptcy probabilities of 0.2187 and 0.2175 for firms A and B, respectively. The Z value of -0.784 indicates an area under the normal curve such that there is a 0.2923 probability that the outcome cash flow will fall between $240 and $490. There is a probability of 0.5000 that the cash flow will exceed $490. Summing up these probabilities gives 0.7823, the probability that firm A's cash flow will exceed $240. Thus $(1 - 0.7823) = 0.2187$, the probability of firm A's cash flow falling below its interest obligations and bankruptcy ensuing.

The merger of firms A and B yields a cash flow of:

Probability	State	(M) Merged Firm's Cash Flow
.2	Depression	$1,020
.3	Recession	1,000
.4	Normal	940
.1	Boom	1,000

The expected value and the variance of the cash flow may be calculated:

.2($1,020)	=	$204
.3($1,000)	=	300
.4($490)	=	376
.1($1,000)	=	100
$E(CF_M)$	=	$980
.2($1,020 - $980)	=	$320
.3($1,000 - $980)	=	120
.4($940 - $980)	=	640
.1($1,000 - $980)	=	40
$Var(CF_M)$	=	$1,120

The variance of the merged firm's cash flow is only $1120. Its standard deviation is $33.47, and the probability of bankruptcy goes to zero:

$$Z = \frac{\$480 - \$980}{\$33.47} = -14.94 > -3.0$$

Even though the merged firm's interest cost is $480, the sum of the unmerged firms' interest expenses, the standard deviation is so reduced that the probability of bankruptcy is zero. The Z value far exceeds -3.0, the minimum value of the normal curve. With less than perfectly correlated cash flows, mergers reduce the merged firm's variance and risk of default.

Another way of examining the reduction in variance is to look at the merger in a portfolio approach. Assume that firms A and B have cash flow variances of 30% and a correlation coefficient of 0.40. If the merged firm is made up of 50% of firm A and 50% of firm B, the merged firm's variance and standard deviation will be

$$
\begin{aligned}
\sigma(CF_M) &= \sqrt{Var(CF_M)} \\
&= \sqrt{W_A^2\sigma_A^2 + W_B^2\sigma_B^2 + 2W_AW_B\sigma_{AB}\sigma_A\sigma_B} \\
&= \sqrt{(.50)^2(.30) + (.50)^2(.30) + 2(.50)(.50)(.40)(\sqrt{.30})(\sqrt{.30})} \\
&= \sqrt{.075 + .075 + .06} \\
&= \sqrt{.21} \\
&= .4583.
\end{aligned}
$$

The unmerged firms A and B had variances of 30%, whereas the merged firm has a variance of 21%. Firms A and B had standard deviations of 0.5477, that is, $\sqrt{.30}$, and the merged firm's standard deviation is only 0.4583. The portfolio approach reveals that mergers reduce the risk of the merged firm relative to the risks of the unmerged firms.

Since the variance of the merged firm falls, the merged firm could sustain a higher debt level than the sum of the unmerged firms. The existence of possibly additional debt capacity creates more highly leveraged firms. In the business world of taxation, leverage increases the stock prices and the total market value of the firm.

In contrast to those combinations that exhibit some organic consistency, there are those whose avowed purpose is *diversification*. Diversification has, at times, been promoted as a cure to almost all managerial difficulties. If one branch of the company runs into difficulties, it is expected that some other operation will show enough profits to assuage the economic hurt. But diversification may also make it possible for a losing division to drain away the returns from the profitable operation. On closer inspection, however, some diversified combinations reveal distinct economic or technical connections. The historic classic case is the ice and fuel company where men and equipment used for delivering ice in the summer were diverted to delivering fuel in the winter. A whiskey distillery may expand into petrochemicals because, among other things, some of the engineering problems are similar. Some combinations have been motivated by historical relationships that may no longer exist. Where there is no logical relationship in the structure of a widely diversified combination, the reasons for the merger may rest on temporary tax advantages or perhaps on speculative manipulations. This type of combination may develop managerial or administrative problems and often does not survive intact for long.

18.7 Combinations Correcting Economic or Financial Imbalances

Using the criterion that the economic or financial success of a combination is basically demonstrated by a rise in core earnings, potentially worthwhile merger situations have the following characteristics:

1. One or both of the firms have some imbalance in their financial or economic structure that can be offset by the merger.
2. The combination creates or increases market dominance.
3. The merger enables the firm to take certain tax savings.
4. The merger reduces unused duplicate facilities.

A merger that attempts to correct financial or economic imbalances is perhaps the one type that may lead to socially desirable results. For example, the new firm could lower costs by eliminating duplicate facilities. The merger releases productive factors for more valuable uses elsewhere. Presumably, these factors were previously underutilized. In other words, the independent firms' output structures were unbalanced—i.e., they did not have an optimum mix of the productive factors.

The following is a list of merger possibilities arising from various imbalances in the companies' production or financial structure. In every case, the firms are not operating with the optimum production or capital mix.

1. A firm with aggressive management and a relatively small asset base might combine with a firm with a large asset base but no new developing managerial talent.
2. A firm with fixed assets and a shortage of net working capital might merge with a firm with abundant working capital.
3. A firm with a heavily leveraged financial structure could combine with one that had an ultraconservative equity position.
4. A firm with a strong marketing position might unite with a firm, with a good production position.
5. A firm with solidly established products could combine with another company that had a top-notch research program.
6. A firm with heavy expansion plans could combine with a company having a strong liquidity position.
7. A firm faced with seasonal demand for its product could combine with another firm whose product showed a reverse seasonal pattern.

A merger is not the only cure possible for a firm with an imbalance. A company can absorb its redundant productive factors by self-expansion (or contraction) in the proper direction. A combination, nevertheless, often is the fastest or cheapest way of making the adjustment.

18.8   Combinations Increasing Market Dominance

A merger that increases the company's degree of market dominance can be potentially profitable to the stockholders. Since the merger removes some competition,
selling prices may be more easily maintained in the market, or the larger combined
company can bargain better with its suppliers.[6] If the combined company can raise
revenues or reduce costs, the net earnings available for all the stockholders rise.

Historically, financial promoters seem to have profited by putting together combinations involving increased market power. On the basis of its combined earning
power, the new company may be worth more than its parts and can carry a larger
capital structure without loss to the investors. This possibility has, on occasion, been
recognized by promoters who have organized the consolidation of two or more
companies and as their reward has taken some of the securities in the combination. If
the promoters were reasonable, the investors would still obtain increased profits even
after allowing for the return on the additional securities floated and given out as a
reward or an inducement to the promoters, lawyers, investment bankers, management people, and major stockholders involved. However, sometimes the organizers
were too acquisitive or were overly optimistic in forecasting the earnings of the
merger and, as a consequence, rewarded themselves excessively with new issues of
stocks or bonds. The result was an *overcapitalized* company, a company whose
earnings were inadequate to support a market price for its capital structure equal to
its *stated* or *par* value.

Although combining to increase market power can profit promoters and stockholders, its overall effect probably reduces the amount of workable competition in
the economy. In general, the American tradition has been against private attempts to
control market forces. This tradition has been expressed (although not always
strongly enforced) in a series of legislative acts including the Sherman Anti-Trust
Act of 1899, the Clayton Act of 1914, and the Celler Anti-Merger Amendment of
1950 to the Clayton Act. As the law now stands, mergers (or acquisitions of the
assets of other firms) that tend substantially to lessen competition are prohibited.
What constitutes a lessening of competition is, of course, subject to judicial interpretation. Nevertheless, the rule was construed widely enough in 1958 to prevent the
merger of Bethlehem Steel Corporation and Youngstown Sheet and Tube, even
though competition between these two geographically separated companies existed
in only a small part of their output.[7] The court held the view that both companies
were essentially strong and that their merger would eliminate a wide area of *potential*
competition. White (1982) found little evidence that the increase in the share of value
added in the United States was due to an increase in aggregate concentration. White
found that the increase occurred before the large conglomerate merger activity of the
late-1960s.

---

[6] A firm which is able to reduce the prices paid to its suppliers shows no true example of external
economies. The result is merely a backward shifting of costs.

[7] Bethlehem Steel is now in bankruptcy and no longer an independent company.

18.9   Combinations for Tax Advantages

The present corporation profits tax law allows for a tax loss carryback of 3 years and a tax loss carry-forward of 5 years from the year the loss occurred. For example, suppose a corporation shows a deficit on the current year's operations of $1,000,000. It can apply this loss against the profits of the last 3 years, and if it has earned *at least* $1,000,000 taxable profits in these years, the Treasury refunds the company approximately $350,000. If, however, past profits are insufficient to cover the loss, the excess loss may be used to reduce the taxable profits of the next 5 years. Of course, in any case, the total reduction in tax liability cannot exceed $350,000.

A *merger* for tax savings may take place when a firm with a tax loss carry-forward cannot earn enough itself to use its tax credit. Since the law allows the merged companies to consolidate their earnings, the tax loss carry-forward enables the new firm to reduce its tax liabilities regardless of which division originated the losses. Thus, a firm with a tax loss credit of $1,000,000 can offer to a potential merger partner the possibility of saving $350,000 in taxes on future income. In a sense, the loss firm brings an extra asset to the union which has to be considered in setting the merger terms.

The potential tax credit never equals the actual loss. In spite of tax credits, it is very difficult to get rich by taking losses. If a company's projected revenues do not cover out-of-pocket costs, its past losses have value in a merger, but merger or not, there is no reason to continue its operations. The company's operations will be closed down, and unless its assets can be turned to something else, it will be liquidated either *before* or after the merger. If it is liquidated before the merger, it becomes what is known as a corporate shell. As a matter of fact, there is occasional trading in these corporation shells-firms which have no assets but do have a charter and tax losses that may be used by other profitable companies.

Under the old tax laws, established stable earning firms could well undertake some risk projects since losses, if any, could be offset for tax purposes against solid current earnings. Some losses would be recovered through reduced taxes. On the other hand, a new firm was at a disadvantage because it would have no established earnings upon which the taxes might be reduced. Under the present tax law, however, a company experiencing losses has two alternatives. It may continue to operate, hoping eventually to make profits against which its loss carry-forward will be useful, or, if the situation is hopeless, it will close down and look for a tax merger which will return part of the losses it has suffered. The opportunity of obtaining some return of losses through a merger for tax purposes is thus the sole value of the loss carry-forward provision to an unsuccessful new enterprise which discontinues operations. The possibility of a new company obtaining some return of possible losses through a merger helps redress the odds on a risky venture, as opposed to the more favorable conditions for taking chances of older and larger companies. And, if the conditions were thus unaltered in favor of established firms, the force of dynamic new competition would be yet weaker, and the forces which encourage the growth of oligopoly and business concentration even stronger.

To sum up, the loss carry-forward tax credit was designed (1) to equalize the tax burden for firms having a given rate of return but whose income pattern was erratic with firms having the same rate of return at a more stable level; and (2) to improve the expected return for new risk ventures. The latter function is actually attained by allowing a loss company to bring its tax credit into a merger. Harris et al. (1982) did not find that the tax loss carry-forward variable is useful in predicting acquired firms in the 1974–1977 period.

18.10   The Larson–Gonedes Exchange Ratio Model

It has been held that mergers often involve firms with differing price-earnings (P/E) multiples. Larson and Gonedes (1969) developed an exchange ratio model that is both theoretical and highly testable. The model can be developed from an assumption that the acquiring firm wants to project a growth image so that the market will capitalize its earnings at a high price-earnings multiple. As the firm acquires smaller firms (in terms of the firms; price-earnings multiples), the earnings of the acquired firms are capitalized at the acquiring firm's higher price-earnings multiple. The merger results in a higher market price for the acquiring firm.

Assume two firms, A and B, can be represented by the following data:

	Firm A	Firm B
Earnings (E)	$20 million	$5 million
Shares Outstanding (S)	2 million	1 million
Earnings Per Share (EPS)	$10	$5
Stock Price Per Share (P)	$160	$50
Price-Earnings Multiple (P/E)	16	10

Firm A is interested in acquiring firm B and is willing to pay firm B's stockholders $60 per share for their stock. Firm B's stockholders are delighted because they have immediately profited $10 per share. If firm A is willing to pay $60 per share for firm B, the exchange ratio, ER, is

$$ER = \frac{P_B}{P_A} = \frac{\$60}{\$160} = .375.$$

To pay firm B's stockholders, firm A can issue firm B's stockholders (0.375) 1,000,000 shares or 375,000 shares of firm A's stock. If a stockholder in firm B before the merger owned 100 shares of stock, worth $5000, he now owns 37.5 shares of firm A's stock. Following the merger, the former stockholder of firm A has holdings worth $6000, that is, 37.5x ($160). His gain is $1000 because of the merger.

We must now examine whether firm A's stockholders will profit. If firm A pays firm B's stockholders $60 per share and issues 375,000 shares, the merger is very profitable for firm A's stockholders as long as the market allows firm A's price-earnings multiple to remain at 16. Following the merger, firm AB is represented as follows:

Firm AB		
Earnings	=	$20 million + $5 million = $25 million
Shares Outstanding	=	2,000,000 + 375,000 = 2,375,000
EPS	=	$10.53
Price	=	P/E (EPS) = 16 ($10.53) = $168.48

Firm A's stockholders have profited $8.48 per share because of the merger. Notice, however, that firm A's stockholders profited only because the market did not adjust the firm's price-earnings multiple after the merger. A decrease in the price-earnings multiple of the merged firm could well wipe out the acquiring firm's gains.

In the absence of synergy, Larson and Gonedes held that firm A's price-earnings multiple should fall as a result of the merger. The price-earnings multiple of the merged firm, θ, should be the weighted average of the merging firms' price-earnings multiples in the absence of synergy:

$$\theta = \frac{P_A(S_A) + P_B(S_B)}{E_A + E_B}$$
$$= \frac{\$160(2,000,000) + \$60(1,000,000)}{\$20,000,000 + \$5,000,000}$$
$$= 15.2.$$

If the market does not adjust for the decline in aggregate real growth after the merger, a fallacious increase in market price can take place. The whole phenomenon may be classed as illusory growth.[8] If the price-earnings multiple of the merged firm is equal to θ, there are no profits for the acquiring firm's stockholders; the price of firm A's stock remains at $160:

$$P_A = \$10.53(15.2) = \$160.06$$

The presence of synergy would be shown in a price-earnings multiple of the merged firm exceeding θ. If the price-earnings multiple of the merged firm is less than θ, the acquiring firm's stockholders suffer a loss and would oppose the merger.

Firm A's shareholders should be willing to pay firm B's stockholders a price dictated by a maximum acceptable exchange ratio, ER_A. Larson and Gonedes derived this to be

---

[8]This problem is described in more detail in Chap. 22.

$$ER_A = \frac{\theta(E_A + E_B) - (P/E_A(E_A))}{(P/E)_A E_A\left(\frac{1}{S_A}\right)(S_B)} \tag{18.6}$$

For example, if the market were to allow the merged firm's price-earnings multiple to remain at 16, firm A's stockholders could afford to pay firm B's stockholders a price far exceeding $60 per share and still reap merger profits:

$$ER_A = \frac{16(\$25,000,000) - 16(20,000,000)}{16(\$20,000,000)\left(\dfrac{1}{2,000,000}\right)(1,000,000)}$$

$$= .500.$$

An exchange ratio of 0.500 implies a maximum acceptable price of $80 per share for firm B's stock:

$$P_B = ER_A(P_A) = .500(\$160) = \$80.$$

If firm A pays firm B's stockholders a price of exactly $80 per share, firm A's stockholders are indifferent to the merger because the price of firm A's stock remains at $160 per share. Firm A must issue 500,000 shares, that is, 0.5 (1,000,000), to acquire firm B. Following the merger, firm A is represented as:

			Firm A
Earnings			$25,000,000
Shares Outstanding			2,500,000
EPS			$10
P/E			16
Price	=	$160 =	$10(16).

Firm A's stockholders profit as long as the exchange ratio is below 0.500. If the market allows a price-earnings multiple of the merged firm to remain at 16, firm A cannot pay a higher price than $80 per share for firm B and still profit from the merger. Of course, firm A would prefer to offer firm B's stockholders as small a price as possible for their stock to consummate the merger. If firm A paid firm B's stockholders a price of $60 per share, the merger premium, MP, would be given in terms of the actual exchange ratio, AER:

$$MP = \left(\frac{AER(P_A) - P_B}{P_B}\right)(100) \tag{18.7}$$

where

$$\text{AER} = \frac{\$60}{\$160} = .375$$

$$\text{MP} = \left(\frac{.375(\$160) - \$50}{\$50}\right)(100) = 20$$

A 20% merger premium would have been paid.[9]

Synergism will result only if θ exceeds the weighted average of the merging firm's price-earnings multiples. If

$$\theta > \frac{P_A(S_A) + P_B(S_B)}{E_A + E_B},$$

the market value of the merged firm exceeds the market values of the unmerged firms. If $\text{ER} > P_B/P_A$, the acquired firm profits more proportionally than the acquiring firm. The reverse occurs if $\text{ER} < P_B/P_A$. A firm should engage in a merger only if it benefits its stockholders. The relative bargaining strengths of the acquiring and acquired firms determine the distribution of the merger gains. Note that high price-earnings multiples, generally resulting from stock price, allow more mergers to be profitable. The history of mergers in the United States substantiates the positive association between stock prices and number of mergers.

---

[9] The minimum θ consistent with the profitability constraint, that the merged firm's price must be at least as great as the acquiring firm's price, is given by

$$\theta = \frac{P_A[\text{AER}(S_B) + S_A]}{E_A + E_B}$$
$$= \frac{\$160[.375(1,000,000) + 2,000,000]}{\$20,000,000 + 5,000,000}$$
$$= 15.2.$$

Notice the minimum θ consistent with the acquiring firm's profitability constraint equals the weighted average price-earnings multiple for the merging firms.

If firm A paid firm B's stockholders $70 per share for their stock, the actual exchange ratio would be .4375 and the merger premium would be 40%. The minimum θ consistent with merger profitability would be

$$\theta = \frac{\$160[(.4375)1,000,000 + 2,000,000]}{\$25,000,000}$$
$$= 15.6.$$

If firm A is willing to pay firm B's stockholders a 40% merger premium, the market must allow the merged firm to maintain at least a 15.6 price-earnings multiple for the merger to profit the acquiring firm's stockholders.

18.11 Valuation of a Merger Candidate

The use of discounted cash flows can value a potential merger candidate. Rappaport (1979, 1981) developed a logical and consistent framework for evaluating the expected cash flow of a merger candidate. The capital budgeting techniques of Chap. 11, the risk adjustments of a beta estimation illustrated in Chap. 14, and the cost of capital calculations of Chap. 10 are used in this section. The Rappaport (1979) framework for merger valuation addresses several issues:

1. What is the maximum pay to pay for the target company?
2. What are the primary areas of risk?
3. What are the earnings, cash flow, and balance sheet implications of the acquisition?
4. How should the acquisition be financed?

Let us analyze a potential merger candidate for an acquiring firm with the following capital structure:

Long-Term Debt	$5,647
Equity	9,063
Total Capital	$14,710

The firm has a bond rating of AA3 (Moody's AA) and a beta of 0.84 versus the S&P 500 Index.

The cost of capital, k_c, can be calculated by using an acceptable market risk premium and the current AA bond yield of 5.6%. The Ibbotson and Sinquefield market risk premium of 8.8% was based on the 1926–1993 period. If we use the Ibbotson and Sinquefield data for the 1951–2002 period, found on WRDS, we find an average annual rate of return on equities of 12.53%, and a corresponding average Treasury bill yield of 5.15%, implying a 7.38% market risk premium. The cost of equity capital for our acquiring firm is:

$$k_e = .0515 + (.0738).84 = .1135.$$

The cost of equity capital, via the Capital Asset Pricing Model, is 11.35%. The weighted average cost of capital may be calculated as:

$$k_c = k_e \left(\frac{E}{D+E}\right) + k_d \left(\frac{D}{D+E}\right)(1-t)$$

$$k_c = .1135 \left(\frac{9063}{5647+9063}\right) + .056 \left(\frac{5647}{5647+9063}\right)(1-.35)$$

$$= .084.$$

The acquiring firm's weighted average cost of capital is the appropriate discount rate for valuing merger candidates.

Table 18.2 Forecasted sales, cash flow growth rates

Years	1-5	6-10
Growth in sales	10%	15%
Profit margin	30%	32%
Tax rate	35%	35%
f = capital expenditures as a % of sales growth	25%	25%
w = working capital increase as a % of sales growth	10%	10%

Table 18.3 Forecasted sales, cash flow ($MM)

$Sales_0 = 100$	Year of Acquisition									
	1	2	3	4	5	6	7	8	9	10
Sales	110	121	133.1	146.4	161.1	185.3	213.1	245.1	281.8	324.0
Operating Expenses	77	84.7	93.2	102.5	112.8	126	144.9	166.7	191.6	220.3
EBIT	33	36.3	39.9	43.9	48.3	59.3	68.2	78.4	90.2	103.7
$(t=.35) = $ Taxes	11.6	12.7	14.0	15.4	16.9	20.8	23.9	27.4	31.6	36.3
Op.Earnings after Taxes	21.4	23.6	25.9	28.5	31.4	38.5	44.3	51.0	58.6	67.4
Depreciation	6.0	7.2	8.6	10.4	12.4	14.9	17.9	21.5	25.8	31.0
Δ Sales	10	11	12.1	13.0	14.7	24.2	27.8	32.0	36.7	42.2
$-$ CE (f = .25)	-2.5	-2.8	-3.0	-3.3	-3.7	-6.1	-7.0	-8.0	-9.2	-10.6
$-\Delta$ NWK (w = .10)	-1.0	-1.1	-1.2	-1.3	-1.5	-2.4	-2.8	-3.2	-3.7	-4.2
Cash Flow	23.9	26.9	30.3	34.3	38.6	44.9	52.4	61.3	71.5	83.6

A potential merger partner has a current sales level of $100 million, depreciation of $5MM, debt value of $50 million, and 10 million shares outstanding. Depreciation is assumed to increase by 20% annually. One can further assume the following scenarios over a 10-year valuation period:

The forecasted sales and cash flow levels are shown in Table 18.2, and the discounted cash flows are shown in Table 18.3. The discounted cash flow of the acquired firm is $281.52 million during the 10-year holding period. One capitalizes the 10-year earnings on an infinite period and discounts the result from its 10-year discounted cash flow.

Discounted Cash Flow

Year	CF	Discount Rate	PV(CF)
1	23.9	.9225	22.05
2	26.9	.8571	23.06
3	30.3	.7849	23.78
4	34.3	.7241	24.84
5	38.6	.6680	26.25
6	44.9	.6165	27.68
7	52.4	.5685	29.79
8	61.3	.5247	32.16
9	71.5	.4838	34.59
10	83.6	.4464	37.32
			$281.52

Present value of discounted 10-year earnings, $earn_{t+10}$.

$$PV(Earn_{t+10}) = .4464 \left(\frac{67.4}{.084} \right)$$

$$= \$358.18$$

The total present value is the present value cash flow in the 10-year period, $281.52 MM, plus the capitalized 10-year earnings, $358.18, or $639.70 million. One should subtract the assumed debt, $50 MM, from the present value of the firm's cash flow, to derive its estimated value of equity. The market value of equity is $589.70, and given the acquired firm's 10 million shares, the fair value of the acquired firm's equity should be $58.97 per share. The Rappaport merger valuation framework assumes that the acquiring firm can produce a reasonably accurate forecast of projected earnings and cash flow. One pays in an all-cash manner if consistent with the acquiring firm's current liquidity and targeted debt-to-equity ratio.

18.12 Testing for Synergism: Do Mergers Enhance Stockholder Wealth?

Research has established that the market anticipates mergers at least 7 months before they occur. Mergers generally occur during periods of rising stock prices, and thus tests for synergy must consider market movements. Adjusting for systematic risk, Halpern (1973) found that the merger gains before the announcement are divided equally between the acquiring and acquired firms. Conn and Nielsen (1977) tested evidence of merger synergism, as defined by the Larson and Gonedes model, at the month of the merger announcement, the month of the merger consummation, and the month following the merger consummation; they found support for the Larson and Gonedes model. Harris et al. (1982) provide evidence that during mergers in the 1970s, the price-earnings multiples of the acquired firms were significantly less than that of nonacquired firms.

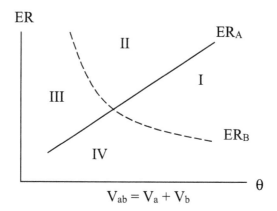

$$V_{ab} = V_a + V_b$$

In terms of the Larson and Gonedes exchange ratio model and the ER-0 diagram, mergers in quadrant 1 profit both firms' stockholders. Above the ER_B rectangular hyperbola, the acquired firm's stockholders win since the merger premium is relatively large. To the right of ER_A, the acquiring firm's stockholders win as the market maintains

$$\theta > \frac{P_A(S_A) + (S_B)}{E_A + E_B}. \tag{18.8}$$

Conn and Nielsen found that 78 of 131 mergers between 1960 and 1969 occurred in quadrant 1 in the month of announcement, supporting Larson and Gonedes' model.

However, by the time of the merger consummation and the month following the merger consummation, only 72 and 67 of the 131 mergers were profitable to both firms. In the month following consummations, only 51% of the mergers were profitable to both firms' stockholders. The acquired firms' stockholders lost in 24 of the 131 mergers in the month following consummation, whereas the acquiring firms' stockholders lost in 58 cases.[10]

The reader probably expects the stockholders of acquired firms to earn positive, and highly significant excess returns. After all, merger premiums rose to 25–30% during the 1958–1978 period (Dodd & Ruback, 1977). What about the acquiring firms? If the acquired firms' stockholders profit handsomely from a merger, should not the acquiring firm's stockholders lose? Are mergers zero-sum events? Is wealth created by mergers? Let us examine much of the empirical evidence. Mandelker (1974) put forth the Perfectly Competitive Acquisitions Market (PCAM) hypothesis in which competition equates returns on assets of similar risk, such that acquiring firms should pay premiums to the extent that no excess returns are realized to their stockholders. The PCAM holds that only the acquired firms' stockholders earn

---

[10]Harris et al. (1982) found that acquired firms had significantly lower price-earnings multiplier than acquiring firms.

excess returns. However, Mandelker studied the mergers of 241 acquiring firms during the 1948–1967 period and found that acquiring firms' stockholders earned 5.1% during the 40 months prior to the mergers, but excess returns decreased by 1.7% in the 40 months following the merger. Positive net excess returns (3.7%) were earned by the acquiring firms in the Mandelker study. Thus, Mandelker found no evidence that acquiring firms paid too much for the acquired firms. Moreover, the acquired firms' stockholders realized excess returns of 12% for the 40-month period prior to the merger, and 14% for the 7-month period prior to the merger.

The Mandelker results have been substantiated by much of the empirical literature. Dodd and Ruback (1977) found that successful acquiring firms' stockholders gained 2.8% in the month before the merger announcement during the 1958–1978 period, whereas the successful acquired firms' stockholders gained 20.9% excess returns. Dodd and Ruback found that the acquired firms' stockholders gained 19.0% even if the merger was unsuccessful, whereas the acquiring firms' stockholders gained less than 1%. The empirical evidence for the 1973–1998 period is consistent, from 20 months prior to the merger to its close, the combined firms' stockholders gain approximately 1.9% (Andrade et al., 2001). Moeller et al. (2003) analyzed 12,023 mergers during the 1980–2001 period and found a 1.1% gain to acquiring firm's shareholders.

Moeller et al. (2003) used a sample of public acquiring firms making acquisitions exceeding $1 million. Private acquired firms accounted for 5583 of the 12,023 acquisitions involving subsidiaries. The private firm acquisitions tended to be more cash-financed (50.56%) than public firm acquisitions (29.57% cash, 55.32% in equity), and the subsidiary acquisitions were predominantly cash (75.92%). The 3-day cumulative average returns (CARS) of the acquiring firms' deals were on average 1.1%; however, the acquisition of private firms produced higher CARS (1.50%) than the public firm acquisitions (−1.02%), but less than the subsidiary acquisitions (2.00%). Smaller-capitalized firms' acquisitions private firms financed by equity produced the highest CARS, whereas larger-capitalized, equity-financed acquirers produced the lowest CARS (−2.45%). The equity financing of subsidiaries produced excess returns of 5.40%. In general, Moeller et al. found that cash financing produced higher excess returns than equity deals, led by cash deals for private firms and subsidiaries. The cash acquisitions by small firms produced statistically significant excess returns. Large firms make poorer acquisitions. Rapport and Sirower (1999) noted that the large 1990s' mergers involved more stock and less cash than the large 1980s' mergers. Rappaport and Sirower state that the acquiring firm's management asks if its stock is undervalued; issuing new (more undervalued) shares would penalize current stockholders. Acquiring stocks should finance with stock if uncertain of synergy. Rappaport and Sirower (1997) calculate a stockholder value at risk (SVAR) which measures the merger premium divided by the market value of the acquiring firm before the merger announcement is made. The greater the merger premium and the greater the market value of the seller relative to the acquiring firm, then the higher the SVAR (if no postacquisition) synergies are realized.

Mergers may enhance stockholder wealth; however, whereas Andrade, Mitchell, and Stafford further found that the target or acquired stockholders gained about 23.8% for the 20-month period, consistent across the decades of the 1973–1998 period, the acquiring firms' stockholders lost about 3.8%, during the corresponding 20-month period. For the largest merger in US history prior to 1983, Ruback (1982) found that DuPont lost 9.89% ($789 million of stockholder wealth) in the month prior to the merger announcement, whereas Conoco stockholders gained 71.2% ($3201.2 million) for the 2-month period prior to the successful DuPont merger announcement.

Do mergers affect the firms' operations? Hall (1993) found that research and development (R&D) activities were not impacted significantly by mergers. Hall found no lessening of R&D spending. Healy et al. (1992) reported that mergers seeking strategic takeovers outperformed financially motivated takeovers. Strategic takeovers generally involved friendly takeovers financed with stock, whereas financial takeovers were hostile takeovers involving cash payments. During the 1979–1982 period, for the 50 largest mergers, Healy, Palepu, and Ruback found that strategic takeovers made money for the acquiring firms, whereas financial takeovers broke even. Acquiring stockholders of strategic acquisitions made 4.4% for 5 years post merger, assuming no premiums paid, whereas financial takeovers earned the acquiring stockholders 1.1%. The premiums paid in financial takeovers were higher (45%) than in strategic takeovers (35%), and the synergies were lower in financial takeovers. Trimbath (2002, p. 137) found "no significant merger effect on net profit, operating profit, or market value" when analyzing firms purchased by Fortune 500 firms during the 1981–1995 period. Mergers generate a net gain for stockholders in the US economy, but one prefers to be a stockholder in the acquired, rather than the acquiring, firm.

18.13   Divestment and Spinoff

A divestment or spinoff may be considered as a contrast to a merger. Spinoffs reduce the scale of operations. A company may wish to dispose of a subsidiary; such an action is called a divestment.[11] A firm may want to dispose of a subsidiary company for economic or financial reasons, or it may have to divest itself of a given operation as the result of antitrust action. A firm may voluntarily get rid of its interests in a subsidiary if the subsidiary operations do not mesh properly with the parent, or if the parent company needs funds for working capital or to finance other acquisitions or expansion.

---

[11] Sometimes the subsidiary interest is a branch, a by-product operation, or plant, and is not independently incorporated. If so, an initial step of incorporation would have to be taken before the subsidiary can be relinquished.

A divestment because of the antitrust laws takes place in compliance with court orders or a consent decree where the decision has been made that the ownership of the subsidiary gives the parent company too great an actual or potential dominance in the market. Divestments have also occurred in compliance with actions taken to simplify holding company systems.

Basically, a company can dispose of an unwanted subsidiary in three ways: (1) by the sale of the subsidiary's shares on the security markets, (2) by selling the subsidiary to another company for cash or securities, or (3) by a spinoff. When subsidiaries are sold to another company, payment is often made in the notes, bonds, or the common shares of the purchasing company. A company that has technological know-how but not the necessary development capacity may find it profitable to sell a subsidiary with an experimental product to a market-oriented firm for shares of stock.

A spinoff consists of distributing the ownership shares in the subsidiary on a pro rata basis to the stockholders of the major company. Whereas the sale of the subsidiary (whether on the market so that it becomes an independent company or to another major company) brings the parent company either cash or new security investments, a spinoff decreases the parent company's stated assets with no offsetting increase. On the other hand, a spinoff may be the least upsetting to the management of the subsidiary (a new owning company may begin by shifting personnel around). The spinoff also saves the management of the parent company the effort and difficulty of negotiating a sale or organizing a security flotation. A spinoff spares the parent company worry about whether it best served its stockholders in the sale price obtained for the subsidiary. Spinoffs have been criticized as an inefficient antimonopoly remedy since they leave the same ownership in charge of both companies. However, history seems to indicate that, in time, the sale and trading of stocks tend to change the ownership complexion in the separate firms. After a while, the management of the formerly related firms develops their own desires to excel, and competition between mother and daughter firm may be as intense as any other in the industry.

One factor for divestment is the shedding of an unsatisfactory acquisition or merger. Kaplan and Weisbach (1992) studied such divestitures. Kaplan and Weisbach found that 44% of the large acquisitions completed in the 1970s and early 1980s had been divested in the 1980s. Fifty-six percent of these divestments reported gains or no losses, whereas 44% reported losses on the sales. Kaplan and Weisbach found that the average sale price, deflated by the S&P 500, was 90% of the purchase price. Conglomerate mergers were most often divested.

18.14   Summary and Conclusions

Mergers and acquisitions have been a major source of corporate growth and economic concentration during the past 125 years. The empirical evidence is mixed; most acquired firms' stockholders profit handsomely with excess returns exceeding

25%, whereas acquiring firms' shareholders earn excess returns of only about 1–1.50%. We illustrate the Rappaport discounted cash flow model for assessing the market value of the target firm. The financial evidence is consistent with the Perfectly Competitive Acquisitions Market (PCAM) hypothesis in which competition equates returns on assets of similar risk, such that acquiring firms should pay premiums to the extent that no excess returns are realized to their stockholders. The PCAM holds that only the acquired firms' stockholders earn excess returns.

References

Alberts, W. W., & Segall, J. E. (1966). *The corporate merger*. The University of Chicago Press.

Alberts, W. W., & Varaiya, N. P. (1989). Assessing the profitability of growth by Acquistion. *International Journal of Industrial Organization, 7*, 133–149.

American Institute of Certified Public Accountants. (1970, August). *Opinions of the accounting review board* (No. 16). American Institute of Certified Public Accountants.

Andrade, G., Mitchell, M., & Stafford, E. (2001). New evidence and perspectives on mergers. *Journal of Economic Perspectives, 15*, 103–120.

Butters, J. K., Lintner, J., & Cary, W. L. (1951). *Corporate mergers*. Graduate School of Business, Harvard University.

Caves, R. E. (1989). Mergers, takeovers, and economic efficiency. *International Journal of Industrial Organization, 7*, 151–174.

Chandler, A. D., Jr. (1977). *The visible hand: The managerial revolution in American business*. The Belknap Press of Harvard University Press. Chapters 4, 9, and 10.

Conn, R., & Nielsen, J. (1977). An empirical test of the Larson-Gonedes exchange ratio determination model. *Journal of Finance, 32*, 749–760.

Dewing, A. S. (1921). A statistical test of the success of consolidations. *Quarterly Journal of Economics, 36*, 84–101.

Dewing, A. S. (1953). *The financial policy of corporations* (Vol. 2, 5th ed.). New York, Ronald Press. Chapters 30, 31, 32.

Dodd, P., & Ruback, R. (1977). Tender offers and stockholders returns: An empirical test. *Journal of Financial Economics, 4*, 351–374.

Gort, M. (1969). An economic disturbance theory of mergers. *Quarterly Journal of Economics, 83*, 724–742.

Guerard, J. B. (1985). Mergers, stock prices, and industrial production: An empirical test of the Nelson hypothesis. In O. D. Anderson (Ed.), *Time series analysis: Theory and practice 7*. Amsterdam North-Holland Publishing Company.

Guthman, H. G., & Dougall, H. E. (1955). *Corporate financial policy* (3rd ed.). Prentice-Hall. Chapters 25, 26, 27.

Hall, B. H. (1993). The effect of takeover activity on corporate research and development. In A. J. Auerback (Ed.), *Corporate takeovers: Causes and consequences*. The University of Chicago Press.

Halpern, P. (1973). Empirical estimates of the amount and distribution of gains to companies in mergers. *Journal of Business, 46*, 554–575.

Harris, R. S., Stewart, J. F., & Carleton, W. T. (1982). Financial characteristics of acquired firms. In Keenan and White (Ed.), *Mergers and acquisitions*. Lexington Books.

Haugen, R., & Langetieg, T. (1975). An empirical test for synergism in mergers. *Journal of Finance, 30*, 1003–1013.

Healy, P. M., Palepu, K. G., & Ruback, R. S. (1992). Does corporate performance improve after mergers. *Journal of Financial Economics, 21*, 135–175.

Healy, P. M., Palepu, K. G., & Ruback, R. S. (1997). Which takeovers are profitable? Strategic or financial. *Sloan Management Review, 37*, 45–57.

Jarrell, G. A., Brickley, J. A., & Netter, J. M. (1988). The market for corporate control: The empirical evidence since 1980. *Journal of Economic Perspectives, 2*, 49–68.

Jensen, M. C. (1983). Symposium on the market for corporate control: The scientific evidence. *Journal of Financial Economics, 11*, entire issue.

Jensen, M. C. (1986). Agency costs of free cash flow, corporate finance, and takeovers. *American Economic Review, 76*(1986), 323–329.

Jensen, M. C., & Ruback, R. S. (1983). The market for corporate control: The scientific evidence. *Journal of Financial Economics, 11*, 5–50.

Kaplan, S. N., & Weisbach, M. S. (1992). The success of acquisitions: Evidence from divestments. *Journal of Finance, 47*, 107–137.

Larson, K., & Gonedes, N. (1969). Business combinations: An exchange ratio determination model. *Accounting Review, 44*, 720–728.

Lewellen, W. G. (1971). A pure financial rationale for the conglomerate mergers. *Journal of Finance, 26*, 521–537.

Lintner, J. (1971). Expectations, mergers, and equilibrium in purely competitive markets. *American Economic Review, 61*, 101–112.

Livermore, S. (1935). The success of industrial mergers. *Quarterly Journal of Economics, 50*, 68–96.

Mandelker, G. (1974). Risk and return: The case of merging firms. *Journal of Financial Economics, 1*, 303–335.

Manne, H. G. (1965). Mergers and the market for corporate control. *Journal of Political Economy, 73*, 110–120.

Melicher, R., Ledolter, J., & D'Antonio, L. J. (1983). A time series analysis of aggregate merger activity. *Review of Economics and Statistics, 65*, 423–430.

Moeller, S. B., Schlingemann, F. P., & Stulz, R. M. (2003). *Do shareholders of acquiring firms gain from acquisitions?* National Bureau of Economic Research, Working Paper 9523.

Mossin, J. (1973). *The theory of financial markets*. Prentice-Hall.

Mueller, D. C. (1969). A theory of conglomerate mergers. *Quarterly Journal of Economics, 83*, 743–759.

Nelson, R. L. (1959). *Merger movements in American industry 1895–1954*. Princeton University Press.

Rappaport, A. (1979). Strategic analysis for more profitable acquisitions. *Harvard Business Review* (July/August), pp. 99–110.

Rappaport, A. (1981). Selecting strategies that create stockholder value. *Harvard Business Review* (May/June), pp. 139–149.

Rappaport, A., & Sirower, M. (1999). Stock or cash?, *Harvard Business Review* (November–December), pp. 147–158.

Ravencraft, D. J., & Scherer, F. M. (1989). The profitability of mergers. *International Journal of Industrial Organization, 7*, 101–116.

Ruback, R. S. (1982). The Conoco takeover and stockholder returns. *Sloan Management Review, 23*, 13–34.

Ruback, R. S. (1993). Do target stockholders lose in unsuccessful control contests? In A. J. Auerback (Ed.), *Corporate takeovers: Causes and consequences*. The University of Chicago Press.

Shugert, W. F., & Tollison, R. D. (1984). The random character of merger activity. *RAND Journal of Economics, 12*, 500–509.

Stigler, G. J. (1950). Monopoly and oligopoly by merger. *American Economic Review, 40*, 23–34.

Trimbath, S. (2002). *Mergers and efficiency: Changes across time*. Kluwer Academic Publishers.

Weston, J. F. (1953). *The role of mergers in the growth of large firms*. University of California Press.

Weston, J. F., Mitchell, M. L., & Mulhern, J. H. (2004). *Takeovers, restructuring, and corporate governance* (4th ed.). Pearson/Prentice-Hall.

Chapter 19
Liquidation, Failure, Bankruptcy, and Reorganization

Not every company justifies the confidence of its original investors placed in it. A free enterprise system is one of profit and loss. There is no guarantee that all capital will earn the "normal" rate of return. In a world of change, where sure knowledge of the future is lacking and decisions are made under conditions of more or less uncertainty, the operation of any business is a calculated risk. The data in Fig. 19.1 depict the failure rate per 10,000 firms in the United States during the 1930–1998 period.[1] The reader obviously notes the high level of business failures

---

[1] The Dunn & Bradstreet (D&B) data are used in this chapter. However, the D&B data are not included in the economic report of the President after 1998. A second source of data is the American Bankruptcy Institute, ABI, which carries a 1980–2002 series on its website. The business bankruptcy filings on the ABI database for the 1998–2002 period are:

Year	Business filings
1980	43,694
1987	82,446
1990	64,853
1992	70,643
2000	35,472
2006	19,695
2007	28,332
2008	43,546
2009	60,837
2010	56,282
2015	24,745
2019	22,780
2020	21,655

The decline in filings may be surprising given the stock market performance of the period.
Source: https://abi-org.s3.amazonaws.com/Newsroom/Bankruptcy_Statistics/Total-Business-Consumer1980-Present.pdf

© The Author(s), under exclusive license to Springer Nature Switzerland AG 2022
J. B. Guerard Jr. et al., *Quantitative Corporate Finance*,
https://doi.org/10.1007/978-3-030-87269-4_19

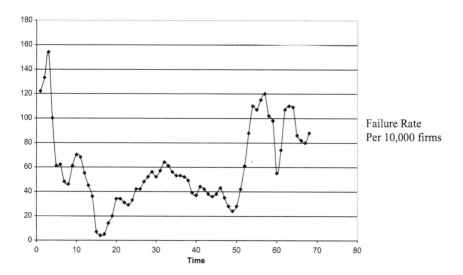

Fig. 19.1 US failure rate, 1920–1998

during the Great Depression. The reader may be surprised by the high level of failures during the 1980s, given the excellent stock market performance. Failures have risen substantially during the past 40 years and were very high during the 1990–1992 period. As noted in Chap. 5, there was a deterioration in the Altman Z score, the bankruptcy prediction model, during the 1963–2003 period as a result of falling profit margins and sales efficiency. The large debt issuance of the 1975–2003 period, combined with high interest rates during several years of the period, was no doubt an important factor in the rising failure rate. Buell and Schwartz (1981) modeled the failure rate in the United States during the 1950–1978 period as a function of the total debt to total assets ratio and the variation about the time trend of the level of employment. Buell and Schwartz reported the increases in the debt ratio were (highly) statistically associated with the failure rate, whereas the failure rate is negatively associated with employment. An economic downturn is associated with decreasing cash flows and lower operating cash flows of firms. Under an increase in financial leverage, a favorable turn in economic events can bring a firm high earnings; however, on the other hand, a recession in demand or a miscalculation of costs may entail substantial losses.

The broad question explored in this chapter concerns the course of action open to the various classes of investors (suppliers of funds to the firm) in the event the company does not perform at a minimum level.

Bankruptcies fell in the United States during the 1980–2019 time period as the US economy soared. Bankruptcies spiked in 1987, with the "Crash" and again in 2008–2009, the "Global Financial Crisis" (GFC). Overall, bankruptcies fell during the 1980–2019 time period.

The Bankruptcy Reform Act of 1978 provides the formal procedures for the resolution of claims against the firm. The bankruptcy act provides for liquidation (Chap. 7) or reorganization (Chap. 11). A bankruptcy filing may be voluntary or involuntary.

19.1   Voluntary Liquidation

Liquidation is the sale of the operating assets of a firm and their conversion into cash or other liquid assets. In most cases, the cash is returned to the investors according to their legal priorities. The trustee appointed by the court shuts down the firm and sells its assets. Assets are distributed by the absolute priority rule (APR), where the court establishes the seniority, or hierarchy, of claims. No junior claims are paid until senior claims are fully paid.

Liquidations may be voluntary or involuntary. If it cannot properly meet its obligations, an involuntary liquidation may be forced upon the company by its creditors. Involuntary liquidation is one of the remedies in bankruptcy, whereas a voluntary liquidation is undertaken as a decision of the management passed upon by the shareholders. Whether voluntary or involuntary, liquidation is recommended whenever the capitalized value of the reasonably projected cash flow of the firm, i.e., its value as a "going concern," is below the appraised liquidation value of the company's assets.[2] In the case of voluntary liquidation, funds may be obtained from the sale of the assets that could be used to buy market stock or bonds (in which case the liquidated firm will turn into an investment company[3]), or an entirely different type of business may be purchased. A partial liquidation means that some portion of the claims against the company is paid off and the other investors remain with the scaled-down firm.

The level of the cash flow of the firm determines the decision to liquidate. A firm may show accounting losses, and yet the best course of action may be to continue operations. On the other hand, it may be eminently reasonable to withdraw funds from a firm that shows no explicit accounting loss.

---

[2]Sometimes a firm's shares sell on the market consistently below their liquidation value. This indicates that the net liquidating asset value of the firm is greater than its value as a going concern. Of course, the inside holders or the management may strongly resist any proposal to liquidate.

[3]For example, the Adams Express Company and Berkshire Hathaway are companies which have been turned from operating firms into investment companies.

19.2   A Liquidation Example

Table 19.1 shows an example of a firm that might consider dissolution even though its accounting profits are positive.

1. Appraised liquidation value for the sale of the total assets = $13,000,000.
2. Going concern value of the firm on a cash flow basis assuming a necessary overall rate of return of 10% and a time horizon of 25 years, $1,000,000 $a_{25|10\%} = \$9,077,000$.
3. Rate of return on the liquid value of the equity on a net basis $\frac{300,000}{6,000,000} = 5.0\%$.

If the stockholders should decide to dissolve the firm, they would receive $6,000,000 (after paying off the creditors) for the $9,000,000 book value of their investment. On the other hand, the gross going concern value of their firm is only $9,077,000, leaving the shareholders with only $2,077,000 as the net ongoing value of their equity.

Viewing the situation from the net basis, let us assume that the going rate of earnings on equity capital for Company A's type of risk and potential return is 10%; then the $6,000,000 obtained from liquidation could be invested to bring in at least $600,000 net instead of the present $300,000 net. Basing the decision on the overall approach (i.e., on the use of EBID) is preferable because it is always possible that an inadequate return on the equity is due to a suboptimal capital structure or an especially high allowance for depreciation. One should emphasize that the book value of the assets (or equity) really is irrelevant to the decision except as the book value of certain assets may give a rough indication of their liquidation value. The book value of fixed assets may diverge widely from liquidation, but the liquidation value and the book value of current assets may run fairly close. The argument for liquidation, in this case, is the assumption that there is no strong reason for forecasting any appreciable improvement in the earnings of the firm for 25 years.

Table 19.1 Company A
Condensed operating statement

EBITD	$1,000,000	
Less depreciation	200,000	
EBIT	800,000	
Interest	200,000	
Earnings before taxes	600,000	
Less taxes	300,000	
Income available for common	300,000	

Position statement

Assets		Liabilities	
Current assets	$11,000,000	Current liabilities	3,000,000
Fixed assets	5,000,000	Fixed liabilities	4,000,000
		Common stock equity	9,000,000
	16,000,000		16,000,000

Table 19.2 Company B
Condensed operating statement

EBITD	$1,000,000		
Less depreciation	900,000		
EBIT	100,000		
Interest	200,000		
	(100,000)		
Profit taxes	-0-		
Deficit on the common equity	(100,000)		

Position Statement

Assets		Liabilities	
Current assets	$ 3,000,000	Current liabilities	1,000,000
Fixed assets	13,000,000	Fixed liabilities	4,000,000
		Common equity	11,000,000
	16,000,000		16,000,000

19.3 Remaining in Business

Table 19.2 illustrates the position of a firm that might as well stay in business even though it shows an accounting deficit of $100,000 annually. It can realize only $5,500,000 from the sale of its assets leaving only $500,000 for the equity holders after paying the creditors. On the other hand, it has a gross going concern value of $7,600,000 leaving a net value of $2,600,000 for the equity after allowance is made for liabilities. It may be difficult to understand how this firm can have any value when the operating statement shows a net loss. However, after considering noncash depreciation charges, Company B presently returns $800,000 cash flows annually in available net funds after the payment of interest.

1. Appraised liquidation value for sale of total assets = $5,500,000.
2. Going concern value of the firm on a cash flow basis, assuming a necessary overall rate of return of 10% and a 15-year projection basis; $1,000,000 $a_{15|}$ $_{10\%}$ = $7,600,000.

These funds could be reinvested as risk capital to bring a return of perhaps 10% ($80,000 additional per year) or used to repay the long-term debt, reducing the principal and interest costs. It follows that, in a comparatively brief period, the accounting statements would show a profit.[4]

In addition to the return on the investment of the net new funds of $800,000, the original investment in the firm will, under our assumptions, continue to produce a

---

[4]Furthermore, under the present tax loss carry-forward provision, Company B would probably not have to pay any profits tax for some time.

gross cash flow of $1,000,000 annually for at least 15 more years. The owners of the firm will retrieve more of their investment if they continue to operate than if they liquidate.

To sum up, the decision to liquidate should not be based on the accounting net profits or losses, but on whether the company earns an "adequate" rate of return on the "withdrawable funds" within the firm. A firm can show an accounting profit because all the implicit costs of capital are not recognized. If the firm were explicitly charged for all its funds (including equity) at the going rate, it would show a loss. It might be preferable to liquidate such an enterprise. On the other hand, a company may show an accounting loss because in the past it has misdirected some of its investment. However, the firm may show a good return on the net withdrawable funds. Neither one of these enterprises can be considered "successful," but in the second case, a smaller net loss results if operations continue.

19.4   Failure

Failure is the inability of a firm to meet its obligations as they come due.[5] Failure may result from physical losses or disasters, robbery, or embezzlement; usually, however, it occurs because of a more or less prolonged period of operating losses, or because funds have been invested, at too heavy a rate, in assets that are not efficient or easily marketable. The two main reasons for failure are often given as the lack of adequate working capital or poor management. But inadequate working capital may be the proximate cause of a firm's failure and not the basic cause. Every business that fails has inadequate working capital at the time of failure. A firm that shows earnings or potential earnings should be able to keep or acquire working capital. Perhaps inadequate working capital is a cause of failure in new small businesses when the organizers fail to provide enough working capital to survive a period of initial losses. Even here, we may doubt whether more working capital would have insured success or merely prolonged the period of duration.

Poor management may be a basic cause for the firm's financial disaster. Mismanagement shows itself in many ways: in a failure to control costs, in the inability to foresee and provide funds for obvious contingencies, or in wasting the firm's resources in unproductive programs. The ability of the management is a qualitative matter; it cannot be adduced from a single mishap, but the management may be suspect if the firm seems subject to a succession of economic errors.

The most important reason for failure is simply bad business conditions. Bad business conditions may be general or centered in a particular industry, or in a particular geographic area. High interest rates and high unemployment are often associated with bad economic conditions. One could alternatively examine the relationship between failures and the composite index of leading economic

---

[5]Failure, of course, should not be confused with an occasional slow payment on a bill or statement.

indicators discussed and analyzed in Chaps. 12 and 13. If demand falls off sharply, even competent management may be unable to keep its company afloat. Of course, some will argue that really good management would have foreseen the economic storm far enough in advance to have a program for saving the firm. But if a firm attempts to cover all risks, it is likely to earn little or no profits. Moreover, the management of a company whose assets are specialized for the requirements of a particular industry may not be able to do much if there is a fall in demand.

19.5   Informal Remedies

The failure of a firm puts its future into the hands of its creditors. Since the firm cannot meet its obligations, i.e., cannot pay its debts, the law removes the major control from the owners and gives it to creditors so that they may salvage whatever they can to reduce their losses. The creditors (liability holders) can press their claims against the firm through various devices of differing degrees of legal formality. The debt holders may, without going through the courts, form a creditors' committee to guide and supervise the management of the company through a period of hard times, and by pressing for operational economies, holding sales, collecting accounts, and minimizing purchases, attempt to bring the firm back to a current basis.

Other procedures for dealing with failure, without necessary recourse to the courts, are composition and quasi-reorganization. The firm may be wholly or partially liquidated by a composition of creditors. After the creditors are paid a proportion of their claims on a pro-rata basis out of the surrendered assets of the company, a composition relieves the owners of any further liability. In some cases, various sorts of quasi-reorganization may be attempted. For example, the creditors may all take an agreed-upon cut in their claims if the stockholders put additional funds into the businesses; if the plan goes through, the stockholders may continue to run the company.

The device likely to be used in more complicated cases (or where perhaps there is distrust among the creditors or between the creditors and the owners) is bankruptcy. Bankruptcy places the affairs of the company under the custody and protection of the courts until the disposition of the various claims can be decided.

No matter what legal device the creditors use to handle the situation of a failed company, there are only two basic functional remedies, reorganization or liquidation. The next section on bankruptcy discusses the remedies in more detail.

19.6   Bankruptcy

Bankruptcy is a process by which a financially distressed individual or firm can purge itself of debt by surrendering all its assets to the court. These assets are used, as far as they extend, to satisfy the creditors' claims. Once an individual passes through

Table 19.3 The largest bankruptcies, 1980–2005

Company	Bankruptcy date	Total Assets pre-bankruptcy
WorldCom, Inc.	7/21/02	$103,914,000,000
Enron Corporation*	12/2/01	63,392,000,000
Conseco, Inc.	12/18/02	61,392,000,000
Texaco, Inc.	4/12/87	35,892,000,000
Financial Corp. of America	9/9/88	33,864,000,000
Global Crossing Ltd.	1/28/02	30,185,000,000
UAL Corp.	12/9/02	25,197,000,000
Adelphia Communications	6/25/02	21,499,000,000
Pacific Gas and Electric Co.	4/6/01	21,470,000,000
MCorp	3/31/89	20,228,000,000
Mirant Corporation	7/14/03	19,415,000,000
First Executive Corp.	5/13/91	15,193,000,000
Gibraltar Financial Corp.	2/8/90	15,011,000,000
Kmart Corporation	1/22/02	14,600,000,000
FINOVA Group, Inc. (The)	3/7/01	14,050,000,000
HomeFed Corp.	10/22/92	13,885,000,000
Southeast Banking Corp.	9/20/91	13,390,000,000
NTL, Inc.	5/8/02	13,003,000,000
Reliance Group Holdings, Inc.	6/12/01	12,598,000,000
Imperial Corp. of America	2/28/90	12,263,000,000
Federal-Mogul Corp.	10/1/01	10,150,000,000
First City Bancorp. of Texas	10/31/92	9,943,000,000
First Capital Holdings	5/30/91	9,675,000,000
Baldwin-United	9/26/83	9,383,000,000

[a]The Enron assets were taken from the 10-Q filed on 11/19/02
Source: http://www.bankruptcydata.com/Research/15_Largest.htm

bankruptcy, he can start his economic life anew. He can acquire new holdings, and he need not pay his old debts from these new holdings unless he so volunteers. Bankruptcy is a humane innovation in the law when one considers that it was not so long ago that persons with unpaid debts might be plunged into debtor's prison. The concept of a legal release from hopeless debt was so important to our American forefathers that the right to establish uniform bankruptcy laws was given to Congress in Article I, Section 8, of the Constitution.

The WorldCom Case Looking at current developments, the largest bankruptcies of the 1980–2005 period are shown in Table 19.3. The largest bankruptcy was that of WorldCom in 2002. Worldcom, Inc. declared bankruptcy in July 2002. Using 2001 data to calculate the Altman Z, as was done in Chap. 5 for Dupont and IBM, one finds a value of 1.30 for WorldCom in 2001. The WorldCom value was less than the critical level of 2.000 and indicated potential bankruptcy problems. There can be several criticisms of the Altman Z calculation of this period. First, the end-of-year financials may not have been known much before May, and there may have been

Table 19.4 Largest bankruptcies of 2019[1]

Company	Assets[2] ($Million)	Filing date
PG&E	71,385	1/29/19
Ditech Holding	14,164	2/11/19
Windstream Holdings	13,126	2/25/19
Weatherford Intl	6,601	7/1/19
Thomas Cook Group	6,569	9/16/19
EP Energy Corp	4,181	10/3/19
Bristow Group	2,861	5/11/19
Dean Foods	2,322	11/12/19
Sanchez Energy	2,160	8/11/19
Hexion Holdings	2,097	4/1/19

1. Through 12/24/2019
2. Assets at fiscal year and prior to filing

Source: *BankrupcyData.com*

little time for persons to calculate the Altman Z score. Second, if one used data of the year 2000 to calculate the Altman Z, one obtains a 1.47 value, less than its critical score, but not a terrible value, such as we saw for Lucent in Chap. 5. A third problem is that WorldCom is a telecommunications firm, a utility firm, and was shown in Chap. 5, utility firms had lower Altman Z scores than manufacturing or industrial firms. In fact, the WorldCom Altman Z score exceeded the utility median (1.18) in 2001.

There were several large bankruptcies in 2019, particularly PG&E, the California public utility associated with enormous wildfires in the Los Angeles area. See Table 19.4.

19.6.1 Bankruptcy Procedures

Bankruptcy can be either voluntary or involuntary. Voluntary bankruptcy takes place when the owners, seeing the hopelessness of their position, petition for bankruptcy and place the firm's assets into the hands of the court. Involuntary bankruptcy may take place if creditors believe that the situation is deteriorating, that the owners may be attempting to withdraw funds to the eventual detriment of the creditors, or that the owners may be preferring (i.e., favoring) some creditors over the others, and they may "throw the firm into bankruptcy." The acts of bankruptcy consist of a list of actions, similar to the ones described above, which legally entitle the creditors to institute bankruptcy proceedings. The purpose of bankruptcy, in this case, is to conserve the firm's resources, such as they may be, for the benefit of all the creditors according to the legal strength of their claims.

Regardless of how the firm came into the hands of the law, the court's first task is to appoint an official entitled a "trustee in bankruptcy." The trustee runs the company until the final decision is made as to how best to dispose of the firm's affairs with the least loss to the creditors. The two major alternatives are liquidation or reorganization.

Just as in the case of voluntary liquidation, the rational economic basis for the choice rests on whether or not the going concern value of the projected funds inflow of the firm exceeds its liquidation value. However, where the firm (e.g., a railroad) is adjudged to perform a necessary public service, the remedy of liquidation may not be allowed; the only reorganization may be legally permissible. The creation of Amtrak is an example of a public enterprise created by the failures of many private railroads.

19.6.2   Priorities in Liquidation

If liquidation is the chosen course of action, the claims against the company (the liabilities) must be ranked in order of their legal strength, or priority. First comes the so-called preferred claims, established by law; these consist of such items as all accrued taxes, back salaries, and the actual expenses of the receiver (especially the lawyer) and the bankruptcy proceedings. Second in priority come the holders of liabilities with liens (or mortgages) against specific assets. Whatever funds these assets bring in are applied first to these "sheltered" claims; afterward any unsatisfied portion of the sheltered claim shares equally with the rest of the general creditors. Thus, having a specific lien or pledge or mortgage does not always protect the holder from loss. The protected claims, however, usually fare better than the general creditors, and they can never do worse. After the various protected claims come, the general creditors who share pro rata in the rest of the funds made available by liquidation. Following the debt, holders come the preferred stockholders, if any. If, as rarely happens, anything is left over, it goes to the common stockholders.

Table 19.5 illustrates the settlement of liabilities (after the preference claims have been paid) under different circumstances. Following the rules of allocation, it indicates how the final settlement for each class of liabilities was calculated. Note that the book value of the assets as such had no bearing on the settlements. The accrued interest takes on the same legal priority as the principal of the claim to which it is attached. Liquidations follow the doctrine of absolute priority; i.e., no junior claim is allocated any funds until all the claims senior to it have had their legal due.[6]

---

[6]A lawyer friend of one of the authors told us that in the world of business, it is not uncommon for a subordinated debt holder to threaten to hold up settlement. In such cases, it may be in the best interests to make a partial payment to the junior debtholders (claims) and allow settlement to proceed.

Table 19.5 Settling claims in liquidation under varying circumstances

Case 1

Book value of assets		Liquidation proceeds	
Fixed assets	$1,000,000	$ 424,000	
Current assets	2,000,000	1,500,000	
Total	$3,000,000	$1,924,000	

Liabilities	Principal	Accrued interest or dividends	Total claims presented	Settlement	Percent of claim
First mortgage bonds (on fixed assets)	$300,000	$6,000	$306,000	$306,000	100.0%
Second mortgage bonds (on fixed assets)	300,000	18,000	318,000	268,000	84.3
General creditors	1,800,000	0	1,800,000	1,350,000	75.0
Preferred stock	500,000	100,000	600,000	0	0.0
Common stock (book value)	1,000,000	0	1,000,000	0	0.0
			$4,024,000	$1,924,000	

Case 2

Liquidation value of assets	
Fixed assets	$1,500,000
Current assets	3,000,000
Available funds	$4,500,000

Liabilities	Claims presented	Settlement	Percent of claim
First mortgage bonds (on fixed assets)	$3,000,000	$2,250,000	75.0%
Second mortgage Bonds (on fixed assets)	500,000	250,000	50.0
General creditors	4,000,000	2,000,000	50.0
Common stock	5,000,000	0	0
	$12,500,000	$4,500,000	

Case 3

Liquidation value of assets	
Fixed assets	$2,000,000
Current assets	5,000,000
	$7,000,000

Liabilities	Claims presented	Settlement	Percent of claim
First mortgage bonds (on fixed assets)	$3,000,000	$3,000,000	100.0%
General creditors	3,500,000	3,500,000	100.0
Preferred stock (1,000,000 + 100,000 accumulated dividends)	1,100,000	500,000	45.5
Common stock	8,000,000	0	0.0
		$7,000,000	

19.6.3 Reorganization

If the reasonably capitalized value of the firm's cash flow is greater than the liquidation value, reorganization is the preferable remedy. The purpose of a reorganization is to trim down the existing claims so that the projected cash flow of the bankrupt firm can support its new financial structure. Sometimes, after passing through a reorganization, a firm's operations have taken a favorable turn, and it has moved on to economic success.

The reorganization plan is hammered out by the representatives of all the classes of claimants with the guidance of the SEC or the ICC. It must be passed by a vote of two-thirds of all classes of claims and approved by the court. When all this is accomplished, the reorganized firm is discharged from bankruptcy. Because of the legal intricacies involved, sometimes it may take years for a firm to be discharged from bankruptcy. Improving economic conditions may encourage a junior class of security holders to delay the final reorganization in hopes that the eventual settlement will be more favorable to them. In the meantime, the trustees run the firm, collecting revenues, paying currently incurred obligations, and perhaps servicing old debt that has a strong legal and financial position.

Bankruptcy implies losses for someone. How these losses are to be apportioned is the problem. The doctrine of absolute priority (APR) is followed in liquidation. This doctrine is also the basic rule, reiterated by the Supreme Court,[7] in reorganization. In other words, no value is to be passed on to any junior claim if any class of superior claims has suffered deterioration in the exchange of securities proposed in the reorganization.

In a recent study of 300 cases from the Arizona and New York federal bankruptcy courts from 1995 to 2001, Bris, Welch, and Zhu (2006) found that Chapter 7 liquidations offered few advantages relative to Chapter 11 reorganizations. The Chapter 7 liquidations took almost as long a time to resolve, cost almost as much in fees, and provided creditors with lower recovery rates than Chapter 11 liquidations. Bris, Welch, and Zhu found Chapter 7 firms had lower prebankruptcy assets than the corresponding Chapter 11 cases; they had less equity owned by managers and more unsecured debt than Chapter 11 firms. The mean number of days in bankruptcy exceeded 70 days for both Chapter 7 and 11 firms, with larger bankruptcies taking longer to resolve, and bankruptcy costs (expenses divided by prebankruptcy assets) were approximately 8% for Chapter 7 cases and 17% for Chapter 11 cases. The greater the number of creditors, then the lower is the unsecured recovery rate. Bris, Welch, and Zhu also found that APR tends to be violated when there are fewer secured creditors when secured creditors own a larger fraction of total debt, and the larger the debt.

A rival to the doctrine of absolute priority is that of relative priority. A reorganization following the lines of relative priority would be considered fair as long as the losses apportioned to each claim were smaller for those of higher senior positions. The first mortgage bonds might accept an adjustment in their claims

---

[7]Case v. Los Angeles Lumber Products Co., 308 U.S. 106 (1939).

representing an economic loss of 10%, the second mortgage bonds 30%, the general creditors 50%, and the equity holders 90%. (Under absolute priority, the claims of the first mortgage bonds would have been satisfied in full, and the stockholders would have received nothing as long as the general creditors suffered any loss. Whether or not the second mortgage holders would suffer any loss would depend on the estimate of the value of their specific lien.)

The theory of "joint venturers" has been promulgated in support of the doctrine of relative priority. The notion is that all the investors in a business are together in a risky enterprise. If the firm founders, some may lose more than others, then they should make sacrifices to get the company afloat again. If the initial venture is successful, the debt holders will receive a limited return, and all the extra rewards go to the stockholders. The advocacy of relative priority seems to rest on a certain sympathy to the shareholders—possibly deriving out of some image of the poor farm girl and the rich mortgage holder. In truth, in the modern corporation generally, the shareholders are risk-takers, "men-of-the-world," and the bondholders are mostly pension funds, insurance companies, or trustees, representing basically the holdings of the aged or funds designed for the support of widows and orphans.

An obvious objection to relative priority is that there are an infinite number of solutions within its framework. It becomes quite difficult for any court to decide whether the proposed plan calls for an excessive sacrifice on the part of the senior creditors.

Although absolute priority is, no doubt rightly, the basic rule of law, a small amount of relative priority tends to creep into most approved reorganizations. If a firm has a complicated financial structure, the protective liens may be quite entangled; for example, a railroad bond which has a first mortgage on one section of the track might have a second mortgage on another part of the roadbed and possibly even a third lien over another part of the road. It may be easier in such cases to make some compromises of conflicting claims than to try to make a sure determination of which has the highest priority. Another difficulty arises out of the politics of the reorganization proceeding itself; the junior security holders are often in a strategic position to delay the approval of the plan if they do not obtain some concessions. Furthermore, the trustees in bankruptcy are often recruited from the ranks of the old management, and their latent sympathies may lie with the stockholders. Last, reorganizations often take place during the lows of the business cycle. The courts during such a period may fear that the projected cash flows may be set at too pessimistic a level. Rather than cut off any claim completely, they prefer to give the junior claimants contingent values, such as warrants which can be exercised profitably if the company makes any recovery of earning power, and its new shares rise on the market.[8]

---

[8] Warrants are options to purchase shares at a fixed price. If a warrant carries an option price of $15, and the shares have a market value of $10, the warrant has little intrinsic value. If the shares should rise above $15, however, the warrants are worth at least the difference between $15 and the share price. Actually, because of their speculative leverage, warrants tend to sell somewhat above their intrinsic value.

Table 19.6 Analysis of a hypothetical reorganization plan of the hardnox corporation

1. If existing economic conditions were not to improve significantly, the earnings of theHardnox Corporation were conservatively expected to average about $1,100,000 per annum before depreciation (on the old base), interest, and taxes. Because of back losses and high depreciation, the company would not have to pay taxes on this section of thecash flow for 15 years. This level of cash flow on the existing plant should prevail for about 15 years with future returns running somewhat lower.

2. The liquidation value of the Hardnox Corporation was appraised at about $4,000,000.

3. If the cash flow of the firm is conservatively capitalized at 10 % for 15 years, the going concern value of Hardnox is a minimum of $8,367,000 ($1,100,000 $a_{15/10\%}$) against a liquidation value of $4,000,000. The reorganization is therefore clearly advantageous.

Through the issue of warrants, junior claimants might recoup part of their losses. It should be acknowledged, however, that the existence of the warrants constitutes a drag on the potential appreciation of the shares that may have been given out during the reorganization as part settlement of senior claims. The warrants, then, delay the rate at which the higher priority claimants recover their losses.

The following analysis, found in Table 19.6, illustrates the problems involved in setting up a reorganization plan. This plan could possibly be rejected by the court; it contains some amount of relative priority.

Financial Structure Before Reorganization*

Ranking claims	Principal	Accrued interest
1st mortgage bonds (30 years partial sink ing fund 1% annually) 4%	$10,000,000	$1,000,000
Debentures 5's	16,000,000	2,000,000
Common stock (200,000's at $100 par value)	20,000,000	0

* All the general creditors representing small or preferred claims were settled by the trustee in bankruptcy.

The proposed reorganization:

1. The senior bondholders to receive new $100 par value common stock for the accrued interest of $1,000,000, leaving the principal of the bonds intact (total distributed: $10,000,000 bonds and $1,000,000 common).

2. The debenture holders to receive $500 worth of 4% pfd. and $625 par value of common representing the rest of the principal and accrued interest (total distributed: $8,000,000 pfd. and $10,000,000 common).
3. The common stockholders to receive 1 share of common for every 10 they now held. (total distributed: $2,000,000 par value common).

Projected Reorganization
Capital Structure

1st mortgage bonds (unchanged) 4's	$10,000,000
Pfd. stock 4%	8,000,000
Common stock (130,000's)	13,000,000

Analysis of the Plan
A. *Feasibility*

Prior and Fixed Charges

	Before reorganization	After reorganization
Fixed (Interest)	1,200,000	400,000
Contingent	0	320,000
Total	1,200,000	720,000
Sinking fund installment	100,000	100,000
Cash requirements	1,300,000	820,000

If the cash flow runs slightly over $1,000,000 a year, the prior charges on the new financial structure will be met readily. The margin for the repayment of the principal on the bonds in the early years is not great, but if the worst should occur, dividends may be passed on the preferred. Of course, as the bonds are gradually repaid, the bond interest charges drop. If economic conditions improve, some funds may be available to improve the value of the common.

B. *Fairness*

Projected Operating Statement

EBITD	1,100,000
(Estimated true depreciation) *	120,000
	980,000
Interest	400,000
	580,000
Preferred dividends	320,000
Earnings on common	260,000
Earnings per common share	$2.00

19.6.4 Analysis of the Reorganization

1. The maximum likely market value of the new common shares would be $20, representing a price/earnings ratio of ten.
2. The bondholders would get back most of their principal value in the form of their present contract. However, for their accrued interest amounting to $100 a bond, they would receive a share of common having a value at best of $20. Therefore, for each $1100 value of claim, the bondholders would get back securities worth $1020 or 92.7% of their claim.
3. The preferred stock given to the debenture holders has contingent claim against dividends rather than a fixed claim; moreover, it carries only a 4% dividend against the previous 5% on the debentures. It is dubious if the preferred has a value of over $60 per share. Thus, the debenture holders would receive securities valued at $300 worth of preferred and $125 worth of common for each $1125 of principal and accrued interest or a settlement of 37.8%.
4. For each $1000 par value of old common, the shareholders are to get one share of common having a maximum value of $20, or a nominal settlement of 2%. From an aggregate point of view, however, they receive securities having an approximate value of $400,000. This would make up part of the loss suffered by the bondholders or debenture holders.
5. Conclusion: The proposed reorganization plan has a considerable element of relative priority. It is a moot question whether it would pass the scrutiny of the courts.

In practice, reorganization procedures are likely to be even more complex than those shown for the hypothetical Hardnox Corporation. Each class of claimants appoints a protective committee, and many proposals are argued back and forth. Nevertheless, the illustration of the Hardnox Corporation case demonstrates most of the essential steps of a corporate reorganization:

1. First, it is determined whether the positive cash How of the firm makes reorganization worthwhile.
2. All the claims are assembled and ranked in order of their priority.
3. New securities are proposed to be exchanged for the old claims.
4. Many of the new securities will be given a contingent or variable claim (income bonds, preferred stock, or common stock) in place of the fixed return securities they replace. (A strongly secured senior security may be left entirely untouched.)
5. The object of the reorganization is to reduce the fixed charges and debt of the firm so that it can meet its remaining obligations out of its projected cash flow. (This constitutes the overriding test of "feasibility.")
6. The values remaining in the firm must be divided according to the priority of the claimants. This constitutes the test of legal fairness.
7. If the reorganization is successful, the firm should not fail again, at least in the near future.

19.7 Summary

From the viewpoint of the various investors, any reorganization is a necessary unpleasant financial surgery. It distributes losses among the classes of claimants. But the reorganization in itself did not cause loss; it is only a recognition of the economic deterioration that has already occurred. The object of a reorganization is to tailor the financial structure of the firm to fit the going concern value of the remaining earnings.

References

Altman, E. (1984). A further empirical investigation of the bankruptcy cost question. *Journal of Finance, 39*, 1067–1089.

Bosland, C. C. (1949). *Corporate finance and regulation*. Ronald Press, Chapter.14.

Bris, A., Welch, I., & Zhu, N. (2006). The costs of bankruptcy: Chapter 7 liquidation versus Chapter 11 reorganization. *Journal of Finance, 61*, 1253–1303.

Buell, S. G., & Schwartz, E. (1981). Increasing leverage, potential failure rates, and possible impacts on the macro-economy. *Oxford Economic Papers, 33*, 442–458.

Buchanan, N. S. (1950). *The economics of business enterprise*. Holt, Rinehart and Winston. Chapters. 12, 13, 14.

Dewing, A. S. (1953). *The financial policy of corporations* (Vol. 2, 5th ed.). Ronald Press, Chapters. 37, 38, 39, 40, 41, 42, 43.

Guthman, H. G., & Dougall, H. E. (1955). *Corporate financial policy* (3rd ed.). Prentice-Hall. Chapters. 30, 31.

Haugen, R., & Senbet, L. (1979). The insignificance of bankruptcy costs in the theory of optimal capital structure. *Journal of Finance, 33*, 385–392.

Hunt, P., Williams, C. M., & Donaldson, G. (1961). *Basic business finance* (rev. ed.). Richard D. Irwin. Chapters. 27, 28.

Senbet, L. W., & Seward, J. K. (1995). Financial distress, bankruptcy, sand reorganization. In R. A. Jarrow, V. Maksimovic, & W. T. Ziemba (Eds.), *Handbooks in operations research and management science: Finance* (Vol. 9, pp. 921–961).

Titman, S. (1984). The effect of capital structure on a firm's liquidation decision. *Journal of Financial Economics, 11*, 137–152.

Warner, J. B. (1977). Bankruptcy cost: Some evidence. *Journal of Finance, 32*, 337–347.

Chapter 20
Corporation Growth and Economic Growth and Stability

Economic growth is usually defined as the rise in total measurable economic output over a given period, i.e., such as in the growth in real GDP modeled in Chaps. 12 and 13. This is a definition of gross growth. However, if the population increases significantly in the same period, output per capita may remain constant or even decline. The average citizen of the country is no better off than before, and thus in one very relevant sense, no economic growth has taken place.[1] A net concept of growth (closer to welfare criteria) can be developed by using the increase in average output per capita. Another problem of measuring growth arises if society shows a desire for an increase in leisure. If the average output does not decline while the workweek shortens and vacations lengthen, growth has taken place, although its fruits have been absorbed as an increase in leisure. Growth is difficult to discover between the dips and rises of the business cycle. The national per capita output at a peak period may appear to have grown very rapidly compared to that of a preceding recession in business activity. Actually, the potential output of the recession period, the output possible were the economy operating at capacity, might be relatively high. Growth should thus be measured from one equilibrium period to another, or, if possible, measured in terms of increases in potential capacity product per capita. If a nation's economy falls significantly from its capacity, the problem is essentially one of economic stability and not of economic growth.[2]

---

[1] Countries such as India or Egypt have increased their total output in the twentieth century and yet have had little or no increase in per capita real income because of the rapid rise in population.

[2] One suggested index of growth is the average product per man hour. This allows for variations in employment and working hours and does a fair job of reflecting growth in potential output.

© The Author(s), under exclusive license to Springer Nature Switzerland AG 2022 587
J. B. Guerard Jr. et al., *Quantitative Corporate Finance*,
https://doi.org/10.1007/978-3-030-87269-4_20

20.1 Factors in Economic Growth

Even if precise measurement is not entirely possible, it can be agreed that economic growth takes place when the potential product of the economy rises sufficiently to allow an increase in the real income per capita. Two factors contribute to a rise in average real income[3]—real capital formation and an increase in the skills of a society, that is an increase in human capital.

Capital formation is a net increase in the amount of tools, buildings, machinery, stocks of goods, soil drainage projects, port improvements, and other such facilities available to the community. The government may provide capital formation in the form of roads, dams, bridges, public buildings, schools, parks, and other similar items. Educational institutions and foundations may provide new capital in the form of classrooms, laboratories, and hospitals. The major contribution to capital formation by individuals is mostly in the form of house construction. In a modern economy, a large area of net capital formation derives from private businesses whose dominant form is the corporation. Whenever a company invests in new plant or equipment, increases the capacity or efficiency of old plants, builds up inventories, or in any way increases the productive potential of itself and the aggregate economy, it adds to the output capacity, the growth of the economy. Private capital formation or real investment is a dynamic element in our economic system. Most economists consider its behavior the major key to understanding the problems of economic stability and economic growth.

The other element making for economic growth—an increase in the skills of a society (human capital)—is less tangible than physical capital formation. An improvement in productive organization, in communicative ability, in manual or physical dexterity, and the development of new processes can add to the productive potential of a society, even if the total physical capital remains constant. Possibly, an increase in the skills and techniques of the population could be subsumed into the general category of capital formation. Resources can be estimated in physical capital and/or in human capital. The director of an economic program in an underdeveloped nation may have to decide whether to use some of his funds to send more of his people away for education and training or whether to build a fish-freezing plant.[4] City fathers may have to decide between a higher school budget and improved storm sewers, and the finance officer of a corporation may have to choose between a new milling machine and a bigger research program.[5]

---

[3] An increase in leisure may be considered a rise in real income or product.

[4] Of course, education is more than a capital investment in human skills. Insofar as it pleasures the spirit, it is also a consumer good.

[5] Although the important role of the private corporation in the formation of capital is clear, less obvious is the part that industry plays in developing the skills of society. Not only are there formal training programs and research, there is a continual "learning on the job," and most importantly inculcating industrial discipline. The crucial significance for economic progress of an understanding and disciplined labor force is evident to economists who have worked in underdeveloped areas.

20.2   Savings and Real Investment

Economic growth is made possible only if there is an increase in usable equipment in the society and/or an improvement of the skills and techniques of the populace. If the economy is operating near its capacity, the resources necessary for capital formation or for training programs must come either out of the country's own current production or from the current production of another country in the form of a loan or bridging grant. From a global view, the quantity of goods and services available to mankind is relatively scanty, and the amount of product that can be diverted from current consumption to capital formation of all kinds is not as abundant as one might wish. Yet capital can be made available to society only through voluntary savings or through compelling a decreased production of consumer goods. Thus, the classical economists, viewing humanity's material progress historically, did not err in emphasizing the importance of saving in the improvement of mankind's economic environment. The modern emphasis is, however, on the act of real investment itself. Capital formation may slacken, not from a lack of savings, but from a lack of profit-making opportunities or dynamic push on the part of the sectors of the economy that make investment decisions.

In summary, nations must rely on increased capital and improved technology for a sustained increase in income. Sacrifice of total possible current consumption can help provide savings to build new plants and machinery. Research may increase its soil fertility, and countries can build dams to irrigate dry areas and provide power. Governments can build schools and train their people, providing the seedbed for a more advanced economy. If the growth of capital exceeds the growth of the population and the decline of natural resources, the economy may continue to return increases in per capita income to its members. If, however, the rate of population growth exceeds the rate of provision of new capital or if the rate of capital growth and inventory is insufficient to make up the loss of natural resources, per capita real income will decline.

A high rate of savings is not a sufficient condition to ensure capital growth. Especially in a rich economy, entrepreneurs and investors may be unwilling to engage in capital formation during periods when the demand for finished products is slack. Thus, a growing rate of consumption and individual expenditures may be as necessary for the continuous formation of capital as the passive act of saving, which releases productive resources. The English economist, John Maynard Keynes, held that in normal times a rich economy is more than likely to provide enough savings (or nonconsumption) to enable capital growth to take place (Weintraub, 1978). The major problem may be to provide enough consumption expenditures and other incentives to ensure that sufficient active investment takes place to employ the amount that is saved.

20.3   Corporation Investment Spending and Economic Stability

Capital formation is necessary for sustaining economic progress. It is also needed to maintain full employment. According to national income theory, if the amount of investment undertaken is not large enough to offset the amount of money that would be saved at the full employment income level, output and income slide off and the aggregate economy reaches equilibrium at something less than full employment of labor and other resources.[6] The fiscal and monetary authorities in the government may then initiate various policies to attempt to offset the inadequacy of investment spending, but it is clear that the level of business investment spending in a free enterprise economy is a crucial variable in determining the level of national income and employment.[7]

The capital markets play a most important role in mobilizing consumer savings and allocating them to investment uses. The sale of new securities on the primary capital market taps personal savings and funnels them into capital formation. Even though a good deal of corporation investments are internally financed by business savings (retained earnings and current depreciation charges), the new funds provided by individuals (either directly, or indirectly through financial institutions) add a dynamic thrust to the economy. These funds add to the potential growth of existing companies and allow for the formation of new competitive companies. However, dealers and traders must operate fairly, and the management of corporations must be mindful of the interests of their security holders if the capital markets are to function properly. For if the ordinary investor doubts the fairness of the treatment he will receive if he puts his savings into corporation securities, the flow of savings into the capital markets will surely slow down, the cost of risk capital will rise, and there will be a brake on the rate of economic development.

In the United States, the corporation is a major catalyst in private capital formation. As shown in Table 20.1, in 2002 gross private investment in the United States totaled $1622.4 billion. Of this amount, $1185.3 billion or 73.1% was initiated by business enterprises. In 2019, 76.2% of GDP originated in business enterprises.[8]The factors influencing the amount of business investment are complex. The volume and trend of consumer spending, the trend of prices, the costs and risks of obtaining capital, and the optimism or pessimism of entrepreneurs and managers who forecast future demand affect investment decisions.

---

[6]The income theory given here is of course highly simplified.

[7]The responsibility of the government toward maintaining a reasonable level of employment in the economy is not clear-cut. However, passage of the Employment Act in 1946 and subsequent political developments indicate an implicit commitment to the use of stabilizing policies if the economy should veer too sharply from full employment.

[8]*The Economic Report of the President,* January 2021, Table 13-6, p. 464,

Table 20.1 Business contributions to gross private capital exp. ($1996 billion)

	Real GDP (in billions)	Real Gross Private Domestic Investment	Business Investment	Percent Private Business of Total Investment	Percent Private Business Investment/GDP
1947	1524.3	191.5	110.7	57.8%	12.6%
1950	1763.9	271.8	121.6	44.7	15.4
1960	2369.3	239.5	149.9	62.6	10.1
1970	3587.3	421.0	273.3	64.9	11.7
1980	4986.3	662.2	469.4	70.9	13.3
1990	6713.3	849.6	632.4	74.4	12.7
2000	9274.0	1755.2	1,329.9	75.8	18.9
2002	9512.1	1622.4	1,185.3	73.1	17.1

Source: US Department of Commerce, Bureau of Economic Analysis, http://reserch.stlouisfed.org/fed/data/gdp

In 2019, as reported in the 2021 *Economic Report of the President,* Gross Private Domestic Investment had returned to much of its 2000 level of GDP, reaching 17.5%, up from its low levels of 13.4 and 14.4%, of GDP, in 2009 and 2010, respectively. In Q3 of 2020, in the midst of the COVID pandemic, Gross Private Domestic Investment was 17.4% of GDP.[9]

The forecast of anticipated future returns has more effect in determining new investment than the volume of current profits. Sometimes it is assumed that if firms have high current earnings, they necessarily invest because they have the funds to do so. This assumption confuses the total amount of retained earnings with business investment spending. Retained earnings are more fruitfully regarded as business savings rather than investment. When funds generated internally are used to purchase new productive assets, the firm's savings are offset by capital spending. But if corporate managements generally are pessimistic as to future earnings, they may develop a high "liquidity preference." That is, they may keep a larger proportion of the firm's assets in cash, building up liquid holdings from their current funds flow. If increased spending in other sectors of the economy does not offset this development, the level of economic activity will decline.

On the other hand, if business pessimism can lead to a downturn in economic activity, an aggressive investment policy can contribute to inflation. When the aggregate demand for goods and services is strong, businessmen can create additional effective demand by borrowing funds or drawing down liquid balances and spending on equipment or inventory. This extra demand (financed by an increase in credit, or an increase in the velocity of money circulation, or both) pushes the price level upward. However, seldom does this type of demand inflation become severe or

---

[9] *The Economic Report of the President,* January 2021, Table 13-4, p. 460.

dangerous. It can be curbed by effective monetary measures, or eventually comes to a halt itself if the central monetary authorities create no net new bank reserves.[10] Runaway or hyperinflations seem to occur only when the fiscal responsibility or financial power of the central government breaks down, and it can find no source of funds other than to print money or create monetary credit. At some point in this process, the public loses confidence that the official money will ever have any scarcity value. The runaway inflation begins.

20.4 Monetary Policy, the Cost of Capital, and the Firm Investment Process

The firm's decision to invest in real assets is determined by the anticipated income possibilities on the one hand and on the other by the cost and availability of funds. What constitutes the cost of funds is a complex matter. Setting the base for the cost structure is the going level of the pure, nonrisk, interest rate. Rising from this are the rates that borrowers of increasing risk position have to pay. Finally, there is the cost of equity funds, influenced by the interest rate, the risk premium, and, very heavily, by anticipations.[11] The economic feasibility of investment projects is determined by matching the prospective returns against the costs of the capital mix, including the allowances for risks and uncertainties.

Expansion projects are undertaken when their probable profitability exceeds the cost of capital; thus, the aggregate amount of investment should increase if the cost of capital decreases. At any given level of anticipations, the volume of investment activity is an inverse function of the basic interest rate, assuming the level of the interest rate helps determine capital costs.[12] (The proportion of increased investment is determined by the elasticity of the investment function to changes in the interest rate.) The possible influence on the volume of investment spending is the rationale for monetary policy. The monetary authorities, by stimulating the money supply and moving the interest rate downward, can hope to stimulate investment and

---

[10]Two factors restrain this type of inflation: (1) As more bank credit is utilized, the interest rate rises. This rise in credit costs dampens the net profitability of many projects; (2) As the aggregate price level rises, the increase in the individual firm's dollar volume calls for an increased cash balance to carry the larger amount of sales safely. At this point, the firm's cash balance no longer seems excessive, and the firm is no longer willing to draw down its cash for other assets. When spending units in the aggregate find themselves with minimum cash "transaction" balances, no further increase in the velocity of the circulation of money is likely to take place.

[11]Estimating the cost of equity funds for new entrants is especially difficult because the supply of ownership funds may be discontinuous. The new entrepreneur may be able to raise a certain amount of capital at risk and then find no further amount available.

[12]Assuming a declining rate of return for successive increments of real capital investment, Keynesian theory constructs the schedule of the marginal efficiency of capital. This hypothetical function gives the amount of aggregate capital investment at various interest rates.

consumption spending. This should lead to a higher level of economic activity. Conversely, holding down monetary growth can lead to a rise in the interest rate, which can be used to dampen an inflationary rise in spending.

According to Keynesian economists, if the economy's reaction to monetary policy is not strong enough, the tools of fiscal policy may be used. The government may intervene directly in the income flow of the economy by spending more than it takes in revenue during a recession in business activity, or by running a surplus during an inflationary period (Weintraub, 1978). Antirecessionary policy can be initiated by cutting taxes, allowing tax revenues to drop, or by increasing government expenditures. Most economists generally agree that fiscal policy should at least be implemented if an economic downturn is sufficiently prolonged and severe.[13]

In contrast to advocates of fiscal policy and/or discretionary Central Bank policies, the monetarists led by Milton Friedman (1968) argue that a reasonably competitive market economy is inherently stable, under the proper financial condition. They argue for a stable forecasted quantity of the money supply as the necessary condition for economic stability. Equilibrating movements of the interest rates, saving, and investment will take place in the market.

20.5 Economic Growth and Firm Growth

The analysis of growth, both aggregate and that of the firm, has not been completely assimilated into the general corpus of macroeconomic theory, Price theory, although a fascinating subject in its own right, traditionally sets the task of solving the economic relations of output and price with a given set of tastes, usable resources, technology, population, and political, social, and economic institutions. Price theory is most concerned with equilibrium states and the movement from one such state to another. Growth, on the other hand, is a process of continuous disequilibrium.

Growth occurs when demand increases or production functions change. Under these two categories, we can classify shifts in population, tastes, income patterns, and innovations—new products, new techniques, or improved organization. In a free enterprise economy, adjustment to these changes is rewarded by increased returns. Producing to meet an increased demand brings profits, and adopting a successful innovation brings profits. Striving to obtain these profits, the entrepreneurs seek out and cajole more capital into potentially profitable activities. The result is capital formation and growth.

---

[13] See *Managing Our Money, Our Budget, and Our Debt*, by the President's Cabinet Committee on Price Stability for Economic Growth, U.S. Government Printing Office, October 1959, for a "moderate" approach to fiscal policy.

20.6   Firm Growth and Economic Growth

All instances of aggregate growth do not necessarily beget the growth of individual firms. For example, assume an industry conforming to the purely competitive model, with free entry, with firms already at the optimum size, and with perfect knowledge, including knowledge of the production function. Since all the firms know they are at the long-run optimum size, an increase in product demand does bring about growth of the industry by the entry of new firms but not by the growth in the size of individual firms.[14] On the other hand, the economic model will show firm growth if the equilibrium is upset by an innovation which changes the optimum scale of production. The first firm to introduce the change and increase its capacity would show profits. To obtain their share of the returns or to prevent losses other firms would imitate the innovation.

Since in most industries firms are of disparate size, it must be assumed that complex dynamic factors are at work and that imperfections and frictions hinder a perfectly smooth competitive adjustment. As a matter of fact, Professor Joseph Schumpeter emphasized the need for some competitive imperfections in encouraging economic growth (Schumpeter, 1989). If there were no imperfections or frictions in the competitive structure—no hurdles to entry, for example—a firm adopting an innovation would enjoy its profits for only a brief time. The entrepreneur who initiates an innovation must anticipate a reasonable period of economic advantage if the time, effort, and risk of working up new products, techniques, or expansion projects are to be justified.[15] However, although some competitive imperfection encourages research, development, invention, and innovation, too much tends to hinder widespread adoption of new improvements.

Once the pervasive frictions and imperfections of the real world enter the model, we have forces sheltering the individual firm which grows to meet an increase in demand.[16] Different firms in the same industry may not have the same optimum size. They differ in location, history, and the quality of the productive factors available to them; they may use different production techniques, and so they may differ in size. Moreover, in a dynamic society, a constant stream of improvements in communications skills, in organization and administration, and the increase in the engineering efficiency of large productive units tend continually to change the optimal scale of output. Thus, the pull of demand and the push of technological change may coincide to produce company growth.

---

[14] If we have a perfectly competitive model with imperfect knowledge, i.e., if the firms do not know they are at the lowest cost point for the long run, some may expand in response to increased profits. Those that do so may later suffer losses as new low-cost producers enter the industry.

[15] Thus in a highly competitive industry, research may have to be publicly sponsored. In agriculture, major technological improvements appear to be initiated and disseminated mainly by the various federal agencies, state research and extension services, and the agricultural research activities of state universities and colleges.

[16] It is also possible for an innovation to decrease the scale of operation. In this case, firms might shrink in size, and yet the productive potential of the economy would still grow.

Not all company growth is a response to increases in demand. Firm growth may be motivated by the power or status drives of management rather than by economic considerations. Expansion may be an attempt to acquire market dominance. These types of company growth do not necessarily increase general economic welfare. Not all the elements of a firm's environment favor expansion. Even if the forecasted profits seem substantial, management may reasonably fear the unknown, especially if the firm lacks experience in the new activity. Furthermore, in a dynamic environment, the possibility of future competition must be considered. Management may think twice if the present company is profitable and the new project is risky and entails a heavy commitment of fixed costs. However, the rate of future growth may have to be carefully controlled if increases in size are not to raise the cost structure disproportionately. A firm that expands too rapidly may find that its operating and financial structures have become inflexible. A company can be too big, unwieldy, and difficult to manage.

If a company expands by acquiring the already existing facilities or assets of another company through merger, combination, or purchase, the aggregate capacity of the economy has not increased. Corporate expansion can be counted as a net addition to national economic capital formation only when it brings into being new real assets or facilities.

From the viewpoint of macroeconomic analysis, corporation expansion which increases aggregate net investment coincides with either an increase in corporation retained earnings or an increase in the supply of financial assets held by the public. That is, either the book value of stocks increases or new bonds or stocks are floated; if the banking system financed the expansion, the money supply increases. (The organization of a new corporation using new assets also increases the financial holdings of the economy.) Company growth which is also economic growth either taps business savings, the community's pool of personal savings, or it creates an increase in monetary assets. On the other hand, a firm that expands by acquiring the assets of another company adds nothing to the net financial holdings of the economy, since all that is involved is an exchange of various financial assets.

References

Abramovitz, M. (1960). Economics of growth. In B. Haley (Ed.), *A survey of contemporary economics* (Vol. 2). Richard D. Irwin.

Boulding, K. E. (1958). *Principles of economic analysis*. Prentice-Hall, Chapters. 2, 3, 9.

Buchanan, N. S. (1950). *The economics of corporate Enterprise*. Holt, Rinehart and Winston, Chapters. 6, 11.

Friedman, M. (1968). The role of monetary policy. *American Economic Review, 58*(1968), 1–17.

Lucas, R. E., Jr. (1972). Expectations and the neutrality of money. *Journal of Economic Theory, 4*, 103–124.

Lucas, R. E., Jr. (1975). An equilibrium model of the business cycle. *Journal of Political Economy, 83*, 1113–1144.

Lucas, R. E., Jr. (1983). *Studies in business cycle theory*. The MIT Press.

Samuelson, P. A. (1961). *Economics* (5th ed.). McGraw-Hill, Chapters. 11, 12, 35, 36.

Schumpeter, J. A. (1989). *Business cycles*. McGraw-Hill Book Company. Abridged with an introduction by R. Fels.

Weintraub, S. (1978). *Keynes, Keynesians and monetarists*. University of Pennsylvania Press.

Zarnowitz, V. (1992). *Business cycles: Theory, history, indicators, and forecasting*. University of Chicago Press.

Zarnowitz, V. (2001). The old and the new in the U.S. economic expansion. The Conference Board. EPWP #01–01.

Zarnowitz, V., & Ozyildirim, A. (2001). On the measurement of business cycles and growth cycles. *Indian Economic Review, 36*, 34–54.

Chapter 21
International Business Finance

With the adoption of flexible exchange rates in 1973, international capital markets have become more completely integrated. This chapter discusses portfolio selection of international equities, and how international diversification lowers total risk of portfolios. Particular attention is paid to the diversification implications of Asian stocks, other emerging markets, and Latin American securities. The US equity selection model developed and estimated in Chap. 14 is used to rank global (ex-US) securities, and produces statistically significant information coefficients and excess returns. An investor owns foreign stocks because their inclusion into portfolios produces higher Sharpe ratios than using only domestic securities. Global and (domestic) US securities may produce portfolios of higher returns for a given level of risk.

21.1 Currency Exchange Rates

One of the additional risks affecting the returns on foreign investment is the possible movement of the exchange rate. The following data would be of interest to a US investor on July 9, 2021:

$$\$1.1869 \text{ U.S.} = 1\pounds \text{ (Pound)}.$$

What do these exchange rates mean? Let us take a trivial but interesting example. A buyer can purchase a Honda Odyssey minivan for \$25,000 in the United States. If the Honda minivan is priced at 19,000 pounds in London, should our US investor fly to London, buy the minivan, drive around England and Scotland, and ship it back to the United States for \$5000? The \$25,000 car in the United States should cost 19,000 X \$1.1869 in London, or \$22,551.10. The US investor should pay the \$25,000 domestic cost. The US investor should purchase his Japanese car in the United States, rather than in the UK. The cost of the Honda Odyssey of \$22,551.10 + \$5000 shipping equals \$27,551.10, exceeding the cost of the Honda Odyssey in the United States.

© The Author(s), under exclusive license to Springer Nature Switzerland AG 2022
J. B. Guerard Jr. et al., *Quantitative Corporate Finance*,
https://doi.org/10.1007/978-3-030-87269-4_21

Table 21.1 Cross rates, July 9, 2021
Currency cross rates table

Currency cross rates table

Base Currency	Dollar ($)	Euro (€)	Pound (£)	Yen (¥)
U.S. Dollar	--	0.8426	0.7211	110.18
E.U. Euro	1.1869	--	0.8465	130.76
U.K. Pound	1.3255	1.1813	--	152.80
Japan Yen	0.0091	0.0077	0.0066	--

Source: Bloomberg, L.P

A more traditional key cross currency rates matrix (see Table 21.1) is normally presented as the following on July 9, 2021:

The US dollar fell from 117.3 yen in 2003 to only 108.86 yen in 2005, and 110.18 yen on July 9, 2021. The United States continued to run a persistent balance of trade deficits that led to its continued currency depreciation. The US dollar depreciated versus the Euro, falling from 0.917 in 2003 to only 0.8254 in 2005.[1] However, despite the trade deficits, the US dollar ($USD) rose to 0.8426 Euro on July 9, 2021.

We introduced readers to call and put options on stocks in Chap. 16. We now introduce the reader to calls and puts on the Euro. On December 31, 2019, the Euro closed at $1.326. If one expects Boris Johnson to be correct and the Pound sterling (GBP) appreciates to, say, $1.35 by June 2020, then one could buy an option on the Pound, for the asking price of $1.90 (see Table 21.2). If the price appreciates to $1.35, in the next several weeks, then the price of the call could well exceed to $4.18, the current price of an at-the-money pound call. The gain on the option is 120% ($4.18 - $1.90)/$1.90 = 1.20. The gain on the holding of the (cash) pound is only 1.8% ((1.35–1.326)/1.326). Clearly, leverage enhances returns, as we saw with stocks in Chap. 16.

Economists have long held that in the flexible exchange rate regime in which we live, the purchasing power of currencies should be equated by changing currency

---

[1] A key cross currency rates table as of September 6, 2006, would be:

Currency	USD	EUR	JPY	GBP	CAD
USD	....	1.2785	.008574	1.8820	.9026
EUR	.7821	....	.00671	1.4720	.7060
JPY	116.63	149.11	....	219.50	105.27
CAD	1.1079	1.4165	.00950	2.0851	.....
AUD	1.3046	1.6680	.01186	2.4553	1.1776

Table 21.2 GBP June 2020 calls
Option Quotes, December 31, 2019
June 2020 (157d 6/5/20); CSize 62,500; BGAM0 1.33

Strike	Ticker	Bsze	Bid	Ask	Asze	Last	IVM	DM	Volm	Oint
129	BGAM0C	0	0.000	0.000	0	5.4400	0.000	0.000	0	1
130	BGAM0C	0	0.000	0.000	0	4.7100	0.000	0.000	0	3
131	BGAM0C	65	2.710	5.100	65	4.0300	7.746	0.635	0	0
132	BGAM0C	67	1.490	4.480	65	3.4000	6.757	0.587	0	35
132.5	BGAM0C	65	1.830	4.180	65	3.1100	7.623	0.549	0	7
133	BGAM0C	65	1.570	3.900	99	2.8400	7.588	0.520	0	1
134	BGAM0C	142	2.150	2.350	65	2.3500	7.543	0.460	1	2
135	**BGAM0C**	**142**	**1.730**	**1.900**	**85**	**1.9200**	**7.476**	**0.400**	**0**	**94**
135.5	BGAM0C	65	1.550	1.740	99	1.7300	7.515	0.372	0	1
136	BGAM0C	115	1.380	1.570	99	1.5500	7.515	0.344	1	7
137	BGAM0C	115	1.090	1.270	99	1.2400	7.534	0.292	0	2

Source: Bloomberg, L.P.
Contract size (CSize):
   100 shares per contract for equities
   one futures contract for 62,500 British pound
   one futures contract for 125,000 Euro
Risk-free rate (R)
Implied forward (IFwd)
On December 31, 2019:
IVM implied volatility based on mid-price, *DM* delta based on mid-price, *Volm* amount of options traded this session, *Oint* open interest: Total number of net option contracts outstanding

rates. That is, if a basket of commodities is cheaper in the United States than in Europe, then US exports should rise, US imports fall, and the US dollar should rise (appreciate) such that the purchasing powers of the US dollar and Euro are equal. The purchasing power parity theory (PPPT), put forth by Cassel (1916), holds that:

$$dE_x = \frac{P_{US_1}/P_{US_0}}{P_{\in_1}/P_{\in_0}} \tag{21.1}$$

where

dE_x = change in exchange rate
P_{US_1} = domestic (US) price level at time 1
P_{US_0} = domestic (US) price level at time 0
$P_{\in_1}$ = foreign (Euro) price level at time 1
$P_{\in_0}$ = foreign (Euro) price level at time 0

The PPPT holds that the rate of change in exchange rates equals the rate of change in price levels, which, of course, equals the relative ratios of inflation rates. See Yeager (1976) for a complete PPPT analysis.

A topic related to the purchasing power parity theory is the interest rate parity theory (IRPT). One may note the relationship between the nominal interest rate and the expected rate of inflation (see Chap. 9). Interest rates rise as the expected inflation rate rises. Irving Fisher put forth the relationship between interest rates and inflation in his *Theory of Interest* (1911). The Fisher Effect can be stated in terms of the domestic interest rates:

$$\frac{P_{US_0}}{P_{US_1}} = \frac{1 + r}{1 + Rn} \tag{21.2}$$

where r is the real rate of interest, equaling the marginal productivity of capital (the increase in output, GDP, due to an increase in the capital stock). The Fisher Effect can be stated in terms of the domestic and foreign price levels and interest rates:

$$\frac{E_{x_f}}{E_{x_s}} = \frac{1 + r_{f_0}}{1 + r_{d_0}} \tag{21.3}$$

E_{x_f} = current forward exchange rate; i.e., € = $1 US,
E_{x_s} = current spot exchange rate
r_{d_0} = current domestic interest rate
and r_{f_0} = current foreign interest rate

An alternative means of showing interest rate parity is in terms of the premium on the forward. Let E_{x_f} be the three-month forward exchange rate in dollars per pound, £, and E_{x_s} be the spot (current) exchange rate in dollars per pound. The premium on the forward, p, is:

$$p = \frac{E_{x_f} - E_{x_s}}{E_{x_x}}. \tag{21.4}$$

If r_f is the short-term (3 months) interest rate in London and r_d is the 3-month interest rate in New York, then:

$$1 + r_d = \frac{1}{E_{x_s}}(1 + r_f)E_{x_f} \tag{21.5}$$

Note that $E_{x_f}/E_{x_s} = H_p$. Interest rate parity holds that:

$$1 + r_d = (1 + p)(1 + r_f) \tag{21.6}$$
$$1 + r_d = 1 + p + r_f + pr_f$$

If p and r_f are small, then pr_f becomes almost zero and.

$$p = r_d - r_f \tag{21.7}$$

The forward premium on the pound can be expressed as a function of the excess of the domestic and foreign interest rates.

On August 26, 2003, the 3-month pound LIBOR (London interbank offered rate) was 3.5663%, whereas the US dollar LIBOR rate was 1.14%. The premium on the pound forward was:

$$p = \frac{1.5596 - 1.5733}{1.5733} = -.0087$$

where the spot price of the pound is $1.5733, and the current December futures price of the pound is $1.5596.[2] The pound, on August 26, had a forward discount (a negative premium) because UK interest rates exceeded US interest rates. Is this forward discount consistent with interest rate differentials? Let us assume that the relevant three-month rates are consistent for a fourth month, December. An investor could invest $100 in a 3-month US dollar security and have an ending value of $100.02. We calculated this terminal value by:

$$\$100 \times \left(1 + \frac{0.0114}{4}\right) = \$100.29.$$

An investor could invest $100 in a pound sterling (£) security and have a terminal wealth value of $100.29 by:

$$\$100 \times \left(1 + \frac{0.0357}{4}\right) \times \frac{\$1.5596}{\$1.5733} = \$100.21.$$

The US investor will do better to invest in the US security than to invest in the UK security. The Interest Rate Parity Theory (IRPT) holds that differences in interest rates determine the forward discount on currency. Most empirical studies, summarized in Cumby and Obstfield (1984) and Levich (2001), hold that deviations exceeding 3% of IRPT are relatively small in number for the short-term Eurocurrency market.

The spot price of the Euro was $1.0850 on August 26, 2003, and the closing futures price for the December Euro was $1.0847. Given the Euro LIBOR three-month rate was 2.1138%, and the US LIBOR rate was 1.01138%, one could calculate terminal wealth values for a US investor investing in a US dollar security:

---

[2]It is very difficult to beat the forward price forecast (Meese and Rogoff. (1983)). See Levich (2001), Levich and Potti (2015), and Kyriazi and Thomakos (2020) for recent evidence suggesting that statistically significant 10–15% forecast in error reductions can be achieved.

$$\$100 \; x \; \left(1 + \frac{.0114}{4}\right) = \$100.29$$

and a Euro dollar security:

$$\$100 \; \times \; \frac{\$1.0847}{\$1.0850} \; \times \; \left(1 + \frac{.02114}{4}\right) = \$100.50$$

The US investor has a slightly higher final wealth investing in the three-month Euro security.

Marston (1995) put forth several reasons why interest rate differentials might become distorted. Banking officials can impose controls on bank interest rates, and these controls can lead to distortions of IRPT. British government controls on resident capital outflows were eliminated in June 1979. Marston (1995) reported that the interest rate differentials between Euro sterling and British markets which averaged 1.50% during the flexible exchange rate period with controls fell to −0.03 for the July 1979–March 1991 period without controls. The average differential during a subset of the Bretton Woods period of fixed exchange rates (April 1961– April 1971) was 0.78%. The elimination of controls established a very close IRPT relationship between the Euro sterling and British markets. US controls, such as the Voluntary Foreign Credit Restraint Program, the Foreign Direct Investment Program, and the limitation placed on US CD rates, by the Federal Reserve with Regulation Q, led to a 1.33% interest rate differential between the Eurodollar and US (CD) markets during the 1966–December 1973 period. The postcontrol interest rate differential during the 1974–March 1991 period was only 0.48%.

21.2   International Diversification

Why would US investors want to own foreign or international stocks? First, one can achieve additional diversification by owning foreign stocks, particularly if the US stock returns are not correlated with foreign stock returns. Risk reduction was shown in Chap. 14 with US stocks, and a US investor can achieve additional diversification by owning foreign stocks. In our first edition, we examined the correlation matrix of country stock returns of the G7 (United States, Canada (CA), France (FR), Germany (GM), Italy (IT), Japan (JP), and the UK) countries during the 1970–2002 period, shown in Table 21.3. We use stock indexes of the Morgan Stanley Capital International (MSCI) database. One sees that the correlation of US and Canadian MSCI returns is 0.73, which is higher than the correlation between US and Japanese stock returns (0.31). We include the MSCI World Index (WI) in Table 21.4. A US investor can achieve tremendous diversification by buying Japanese stocks. The reader is reminded of the excess returns of Japanese stocks reported in Chap. 14.

Capital markets have become more integrated during the period of flexible exchange rates, and one sees the more complete integration in Table 21.4 where

Table 21.3 MSCI G7 stock market index return correlation matrix, 1970–2002
Pearson correlation coefficients, $N = 396$

	WI	CD	FR	GM	IT	JP	UK	U.S.	N	MEAN
WI	1.0000	0.7397	0.6612	0.6320	0.4706	0.6724	0.6948	0.8588	396	0.0052
CD	0.7397	1.0000	0.4810	0.4065	0.3384	0.3250	0.5243	0.7307	396	0.0044
FR	0.6612	0.4810	1.0000	0.6604	0.4844	0.3976	0.5667	0.4908	396	0.0056
GM	0.6320	0.4065	0.6604	1.0000	0.4471	0.3703	0.4671	0.4603	396	0.0050
IT	0.4706	0.3384	0.4844	0.4471	1.0000	0.3548	0.3693	0.2881	396	0.0027
JP	0.6724	0.3250	0.3976	0.3703	0.3548	1.0000	0.3777	0.3010	396	0.0070
UK	0.6948	0.5243	0.5667	0.4671	0.3693	0.3777	1.0000	0.5353	396	0.0052
U.S.	0.8588	0.7307	0.4908	0.4603	0.2881	0.3098	0.5353	1.0000	396	0.0053

Table 21.4 MSCI G7 stock market index return correlation matrix, 1995–2002
Pearson correlation coefficients, $N = 96$

	WI	CD	FR	GM	IT	JP	UK	U.S.	N	MEAN
WI	1.0000	0.8392	0.8250	0.8046	0.6111	0.6368	0.8400	0.9427	96	0.0026
CD	0.8392	1.0000	0.6839	0.6583	0.4674	0.5001	0.6516	0.8062	96	0.0050
FR	0.8250	0.6839	1.0000	0.8469	0.7063	0.4326	0.7649	0.6845	96	0.0043
GM	0.8047	0.6583	0.8470	1.0000	0.6342	0.3097	0.7093	0.7196	96	0.0002
IT	0.6111	0.4674	0.7063	0.6341	1.0000	0.2816	0.5180	0.5103	96	0.0034
JP	0.6368	0.5006	0.4326	0.3097	0.2816	1.0000	0.4676	0.4574	96	-0.0077
UK	0.8400	0.6516	0.7649	0.7093	0.5180	0.4676	1.0000	0.7462	96	0.0027
U.S.	0.9427	0.8062	0.6845	0.7196	0.5103	0.4574	0.7462	1.0000	96	0.0068

the MSCI country index correlations of the G7 countries are shown for the 1995–2002 period. The US–Canada stock return had a 0.81 correlation, and the US–Japan correlation rises to 0.46. One can (still) achieve diversification with the G7 countries.

If a US investor looked to invest within a particular region of the world, such as Europe (EU), North American (United States and Canada), Emerging Markets (EM), Asia, or Latin America (LA), where could one minimize the US stock returns correlations? The reader is referred to Table 21.5, which shows the MSCI region return correlation matrix for the 1995–2002 period. It should be obvious that the US and NA correlations are very high (0.99) and offer limited diversification. A US investor could invest in Europe (a correlation coefficient of 0.78), the Emerging

Table 21.5 MSCI regional stock market index correlation matrix, 1995–2002
Pearson correlation coefficients, $N = 96$

	WI	U.S.	EU	NA	EM	Asia	LA	N	MEAN
WI	1.0000	0.9427	0.9009	0.9465	0.7536	0.7357	0.7051	96	0.0026
U.S.	0.9427	1.0000	0.7777	0.9995	0.6861	0.5745	0.6473	96	0.0068
EU	0.9001	0.7777	1.0000	0.7828	0.6567	0.5557	0.6417	96	0.0036
NA	0.9465	0.9995	0.7828	1.0000	0.6967	0.5824	0.6553	96	0.0066
EM	0.7536	0.6861	0.6567	0.6967	1.0000	0.6931	0.8941	96	-0.0054
Asia	0.7357	0.5745	0.5557	0.5824	0.6931	1.0000	0.5555	96	-0.0075
LA	0.7051	0.6473	0.6417	0.6553	0.8941	0.5555	1.0000	96	-0.0032

Source: MSCI Data

Fig. 21.1 Risk reduction through domestic and international diversification

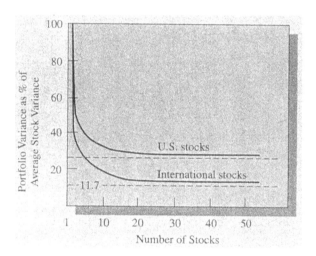

Markets (0.69), Latin American (0.65), or Asia (0.57). It would appear that Asia offered the highest diversification for a US investor, given its relatively low correlation coefficient with US returns. The risk reduction of a US investor owning foreign stocks is shown in Figs. 21.1 and 21.2, courtesy of Professor Solnik.

The relatively low correlation between US stocks and UK and Asian stocks creates greater diversification benefits for global investors (see Figs. 21.1 and 21.2).

In Table 21.6, we report USD-based stock market returns for global stock markets.

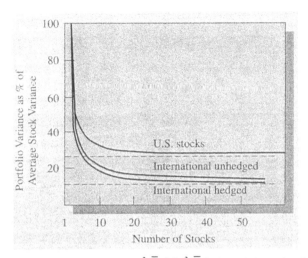

Note: The vertical axis of the graph is the ratio $\sigma^2(\overline{R}_N)/\sigma^2(\overline{R}_1)$ where N is the number of stocks in the portfolio.
Source: Bruno Solnik, "Why Not Diversify Internationally Rather Than Domestically" *Financial Analysis Journal*
30, no. 4 (July-August 1974), pp. 48-54.

Fig. 21.2 Risk reduction through domestic and international diversification hedged and unhedged results. Note: The vertical axis of the graph is the ratio $\sigma^2(\overline{R}_N)/\sigma^2(\overline{R}_1)$, where N is the number of stocks in the portfolio. Source: Bruno Solnik, "Why Not Diversify Internationally Rather Than Domestically" *Financial Analysis Journal* 30, no. 4 (July–August 1974), pp. 48–54

21.3   International Stock Selection and Portfolio Construction and Optimization

In Chap. 14, we estimated 8-, 9-, and 10-factor model of US equity holding period returns during the 2003–November 2018 period. We can estimate the corresponding nine-factor model for non-US securities (the World less the United States) and EM universes, producing statistically significant models for non-US asset managers during the 2003–November 2018 period. The components of the REG10 model are:

$$TR_{t+1} = a_0 + a_1 EP_t + a_2 BP_t + a_3 CP_t + a_4 SP_t + a_5 REP_t + a_6 RBP_t + a_7 RCP_t$$
$$+ a_8 RSP_t + a_9 CTEF_t + a_{10} PM_t + e_t$$

$$(21.8)$$

where $\alpha = (a_1, \ldots, a_{10})$ is the rolling average of normalized coefficients of $\alpha_{t-1}, \ldots, \alpha_{t-12}$.

PM = Price Momentum, P(t − 1)/T(t − 7).

Our portfolio analysis uses the regression-based composite security valuation models shown in Eq. (21.8). We create monthly optimized portfolios using international stocks which produce higher information ratios, IRs, Sharpe ratios, ShRs, and geometric means than US and Japanese stocks for the 2003–November 2018 time

Table 21.6 WEI, returns for 2019

1) Americas	Market Value	%Change YTD	%YTD Cur (USD)
DOW Jones	28,536.07	+22.33 %	+22.33 %
S&P500	3,230.49	+28.87	+28.87
NASDAQ	8,972.61	+35.23	+35.23
S&P/TSX Comp	17,063.43	+19.13	+25.12
S&P/ BMV IPC	43,548.32	+4.58	+8.69
2) EMA			
Euro Stoxx 50	3,745.15	+24.78	+22.08
FTSE 100	7,542.44	+12.10	+16.41
CACA 40	5,978.06	+26.37	+23.63
DAX	13,249.01	+25.48	+22.76
IBEX 35	9,549.20	+11.82	+9.40
FTSE MIB	23,506.37	+28.28	+25.50
OMX STH30	1,771.85	+25.78	+18.89
SWISS MKT	10,616.94	+25.95	+27.84
3) Asia/Pacific			
TOPIX	1,721.36	-15.21	+16.42
NIKKEI 225	23,656.62	+18.20	+19.44
JPX Nikkei 400	15,376.38	+16.02	+17.24
HANG SENG	28,189.75	+9.07	+9.61
CSI 300	4,096.58	+36.07	+34.41
Shanghai Comp	3,050.12	+22.30	+20.82
Shenzhen Comp	1,722.94	+35.89	+34.24
HS China Ent	11,168.06	+10.30	+10.86
Taiwan TAIEX	11,997.14	+23.33	+25.63
KOSPI	2,197.67	+7.67	+3.86

Source: Bloomberg, L.P

period. The portfolios are rebalanced monthly: 8% monthly buys and 8% monthly sells. The maximum weight is 4%; a minimum (threshold) position is 25 basis points, and we use the ITG transaction costs information. REG8 is more effective in Japan stocks, and the CTEF is more effective in US stocks. International stocks, emerging markets stocks, and Chinese A-Share stocks produce the highest specific returns with CTEF, whereas REG8, REG9, and REG10 are all statistically significant in XUS, EM, and China A-Shares stocks.

On July 9, 2021, as we went to press for the third edition, we updated the non-US portfolios, MSCI ACW ex USA, to run through December 2020, including the COVID-19 pandemic of 2020. We used the Axioma version 4 statistical risk model. The CTEF, REG8, REG9, and REG10 models continued to produce very substantial excess returns (see Table 21.8). The total excess returns exceeded transactions costs by over 315 basis points, annually, on all the non-US portfolios. The REG8 portfolios produced the largest stock selection, producing 220 basis points of stock-specific returns (stock selection) that was statistically significant, having a t-statistic of 3.21. The non-US REG8 portfolio produced 330 basis points of risk factor returns, led by value risk returns of 337 basis points annually, market yield sensitivity of 110 basis points, dividend yield factor returns of 29 basis points, and earnings yield returns of 24 basis points. A volatility factor exposure costs the portfolio 197 basis points, annually, in returns. The total effect, the active returns of the REG8 non-US portfolio, was 442 basis points, annually, 1996–2020. The CTEF non-US portfolio produced 422 basis points, annually, of actual returns, composed of 151 basis points of stock selection and risk factor returns of 361 basis points. Medium-term momentum contributed 157 basis points to the non-US CTEF portfolio, whereas value, dividend yield, and earnings yield risk returns were positive, but less than 100 basis points, respectively. The volatility exposure costs the CTEF portfolio 83 basis points, annually (Table 21.7).

The REG9 portfolios produced the largest stock selection, producing 44 basis points of stock-specific returns (stock selection) that was statistically not significant, having a t-statistic of 0.73. The non-US REG9 portfolio produced 362 basis points of risk factor returns, led by value risk returns of 286 basis points annually, market yield sensitivity of 86 basis points, dividend yield factor returns of 41 basis points, and earnings yield returns of 42 basis points. A volatility factor exposure costs the portfolio 121 basis points, annually, in returns. The total effect, the active returns of the REG9 non-US portfolio, was 315 basis points, annually, 1996–2020.

The REG10 portfolios produced the negative stock selection, producing -50 basis points of stock-specific returns (stock selection) that was statistically not significant, having a t-statistic of -0.78. The non-US REG10 portfolio produced 451 basis points of risk factor returns, led by value risk returns of basis points annually, market yield sensitivity of 91 basis points, dividend yield factor returns of 18 basis points, and earnings yield returns of 49 basis points. A volatility factor exposure costs the portfolio 107 basis points, annually, in returns. The total effect, the active returns of the REG10 non-US portfolio, was 382 basis points, annually, 1996–2020.

21.4 International Corporate Finance Decisions

If access to global capital markets allows multinational enterprises (MNE) to lower its equity costs and costs of capital relative to domestic firms, the MNE could maintain a higher desired debt ratio relative to domestic firms, even with large new capital issues. Eiteman, Stonehill, and Moffett (Eiteman et al., 2004) believe that the marginal cost of capital may be constant for a great range of the MNE capital budgets.[3] Moreover, despite the favorable effect of international diversification of MNE cash flow, bankruptcy risk of MNE is approximately equal to that of domestic firms. If a US firm issues foreign currency-denominated debt, such as a US firm issuing 8% debt in Germany, and the Euro appreciates in the coming year, then the US firm's dollar cost of obligations for repaying principal and interest must rise due

---

[3] E. Schwartz put forth a similar position for US firms in Chap. 10.

Table 21.7 Portfolio dashboard

Portfolios	Sharpe Ratio	Info Ratio	Risk Stock Specific Effect	Risk Stock Specific Effect T-Stat	Risk Factors Effect	Risk Factors Effect T-Stat	Risk Total Effect	Risk Transaction Effect	Total Effect
R3000_REG9_8TE	0.68	0.38	3.01	**2.04**	0.06	0.55	3.07	-1.20	1.88
R3000_REG8_6TE	0.60	0.18	1.86	**1.66**	-0.88	-0.24	0.98	-0.94	0.04
R3000_CTEF_8TE	0.75	0.52	1.61	1.36	3.30	2.24	4.92	-2.26	2.65
R3000_REG8_8TE	0.51	0.06	1.77	1.36	-1.53	-0.50	0.24	-1.28	-1.03
R3000_REG9_6TE	0.64	0.25	1.35	1.14	0.15	0.55	1.50	-1.12	0.38
R3000_CTEF_6TE	0.71	0.42	0.35	0.57	2.59	2.26	2.95	-1.55	1.39
R3000_CTEF_4TE	0.72	0.42	-0.02	0.20	2.03	2.73	2.02	-0.92	1.10
R3000_REG9_4TE	0.62	0.12	0.02	0.05	0.31	0.67	0.33	-0.55	-0.22
R3000_REG8_4TE	0.58	0.02	-0.06	-0.05	-0.11	0.17	-0.17	-0.45	-0.62
R3000_REG10_8TE	0.53	0.08	-0.42	-0.14	0.85	0.86	0.43	-1.96	-1.54
R3000_REG10_4TE	0.62	0.12	-0.17	-0.17	0.52	0.89	0.35	-0.81	-0.46
R3000_REG10_6TE	0.53	-0.01	-0.88	-0.62	0.53	0.68	-0.35	-1.33	-1.68

Portfolios	Sharpe Ratio	Info Ratio	Risk Stock Specific Effect	Stock Specific Effect T-Stat	Risk Factors Effect	Risk Factors Effect T-Stat	Risk Total Effect	Risk Transaction Effect	Total Effect
JAPAN_REG8_8TE	0.56	0.54	5.64	**4.00**	-0.18	0.44	5.46	-2.24	3.26
JAPAN_REG8_6TE	0.54	0.51	4.73	**3.74**	-0.23	0.20	4.50	-1.49	3.04
JAPAN_REG8_4TE	0.50	0.46	2.75	**2.88**	0.17	0.63	2.92	-0.79	2.17
JAPAN_REG9_8TE	0.42	0.28	2.14	**1.65**	0.74	1.34	2.87	-2.00	0.95
JAPAN_REG9_4TE	0.44	0.31	1.54	1.59	0.55	1.45	2.09	-0.82	1.32
JAPAN_REG9_6TE	0.41	0.25	1.83	1.50	0.44	1.04	2.27	-1.35	0.98
JAPAN_REG10_8TE	0.38	0.19	0.23	0.44	1.48	1.98	1.71	-1.80	-0.01
JAPAN_REG10_6TE	0.37	0.16	-0.13	0.14	1.40	2.01	1.26	-1.48	-0.16
JAPAN_CTEF_8TE	0.35	0.15	-0.93	-0.01	2.20	2.50	1.28	-6.73	-5.40
JAPAN_CTEF_6TE	0.12	0.09	-0.90	-0.18	2.32	3.13	1.42	-3.87	-2.38
JAPAN_CTEF_4TE	0.30	0.19	-0.82	-0.38	1.96	3.97	1.13	-1.35	-0.17
JAPAN_REG10_4TE	0.35	0.07	-0.97	-0.76	1.39	2.70	0.42	-1.06	-0.58

Portfolios	Sharpe Ratio	Info Ratio	Risk Specific Effect	Risk Stock Effect T-Stat	Factors Effect	Risk Effect T-Stat	Total Effect	Risk Transaction Effect	Total Effect
XUS_EP_4TE	0.53	0.62	4.88	**3.67**	1.55	2.79	6.43	-1.38	5.05
XUS_EP_8TE	0.53	0.51	7.28	**3.61**	2.06	2.74	9.34	-2.81	6.53
XUS_EP_6TE	0.54	0.55	6.16	**3.53**	1.61	2.68	7.77	-2.11	5.66
XUS_CTEF_4TE	0.59	0.79	3.98	**3.45**	3.62	4.58	7.60	-1.38	6.22
XUS_CTEF_6TE	0.63	0.79	4.43	**3.33**	5.00	4.18	9.43	-2.10	7.33
XUS_CTEF_8TE	0.65	0.77	4.37	**3.07**	6.33	3.97	10.71	-2.65	8.06
XUS_REG8_4TE	0.39	0.07	2.94	**2.21**	1.12	1.93	4.05	-1.28	2.77
XUS_REG8_6TE	0.35	0.03	3.57	**2.05**	0.99	1.74	4.56	-1.88	2.68
XUS_REG9_6TE	0.40	0.16	2.98	**1.77**	2.78	3.08	5.76	-1.94	3.83
XUS_REG9_4TE	0.42	0.16	2.22	1.63	2.46	3.45	4.67	-1.30	3.38
XUS_REG8_8TE	0.28	-0.04	3.39	1.60	0.68	1.60	4.07	-2.36	1.71
XUS_REG9_8TE	0.38	0.12	2.00	1.29	3.76	2.79	5.77	-2.31	3.46
XUS_REG10_4TE	0.45	0.26	1.45	1.23	3.30	4.17	4.75	-1.40	3.35
XUS_REG10_8TE	0.44	0.28	0.90	0.93	6.02	3.95	6.92	-2.52	4.40
XUS_REG10_6TE	0.44	0.26	1.00	0.89	4.94	4.25	5.94	-2.07	3.88

Table 21.7 (continued)

Portfolios	Sharpe Ratio	Info Ratio	Specific Effect	Effect T-Stat	Factors Effect	Effect T-Stat	Total Effect	Transaction Effect	Total Effect
EM_REG8_4TE	0.51	0.31	3.99	**2.38**	1.79	2.11	5.78	-2.05	3.73
EM_REG9_4TE	0.55	0.49	3.45	**2.20**	3.28	3.60	6.73	-2.11	4.63
EM_CTEF_8TE	0.59	0.61	3.54	**2.18**	5.76	3.64	9.30	-3.58	5.72
EM_CTEF_4TE	0.58	0.72	3.20	**2.16**	3.84	4.42	7.04	-2.10	4.94
EM_CTEF_6TE	0.60	0.65	3.67	**2.13**	4.67	4.02	8.34	-2.75	5.59
EM_REG9_8TE	0.51	0.30	3.40	**1.73**	5.09	3.31	8.50	-3.62	4.88
EM_REG8_6TE	0.44	0.06	3.30	1.64	2.32	2.17	5.63	-2.79	2.83
EM_REG10_4TE	0.54	0.45	2.25	1.58	4.06	4.24	6.31	-1.92	4.39
EM_REG9_6TE	0.48	0.22	2.62	1.54	4.37	3.49	6.99	-2.80	4.19
EM_REG8_8TE	0.43	0.10	3.37	1.36	2.59	2.27	5.96	-3.52	2.44
EM_REG10_6TE	0.52	0.33	1.33	1.07	5.63	4.17	6.96	-2.96	4.00
EM_REG10_8TE	0.49	0.24	0.30	0.59	6.34	3.68	6.64	-3.54	3.10

Portfolios	Sharpe Ratio	Info Ratio	Stock Specific Specific Effect	Specific Effect T-Stat	Risk Factors Factors Effect	Factors Effect T-Stat	Risk Total Effect	Risk Transaction Effect	Total Effect
CHINAA_CTEF_4TE	0.60	1.23	7.12	**6.75**	2.22	6.32	9.35	-1.74	7.68
CHINAA_CTEF_6TE	0.59	1.05	7.30	**5.16**	2.54	5.31	9.84	-2.39	7.52
CHINAA_CTEF_8TE	0.57	0.95	7.19	**4.87**	2.75	4.98	9.94	-2.61	7.40
CHINAA_REG10_4TE	0.51	0.87	4.62	**4.43**	2.31	5.60	6.93	-1.78	5.19
CHINAA_REG9_4TE	0.44	0.54	3.45	**3.41**	1.42	3.77	4.88	-1.66	3.24
CHINAA_REG10_6TE	0.52	0.72	4.77	**3.31**	3.06	5.51	7.84	-2.51	5.39
CHINAA_REG8_4TE	0.42	0.47	3.17	**3.29**	1.29	3.29	4.47	-1.65	2.83
CHINAA_REG10_8TE	0.50	0.63	4.38	**2.89**	3.18	5.05	7.56	-2.79	4.83
CHINAA_REG8_6TE	0.41	0.39	3.45	**2.67**	1.49	2.84	4.94	-2.41	2.58
CHINAA_REG8_8TE	0.41	0.37	3.65	**2.56**	1.64	2.75	5.30	-2.67	2.67
CHINAA_REG9_8TE	0.41	0.38	3.40	**2.40**	1.76	3.06	5.16	-2.63	2.57
CHINAA_REG9_6TE	0.42	0.42	3.18	**2.37**	1.82	3.52	5.01	-2.37	2.69

Source: FactSet Investment Research Systems, Inc.

to the appreciation of the Euro.[4] Many firms attempt to manage currency exposures through hedging. Hedging is taking a position that will change to offset the change in value of an existing position. Currency risk can be defined as the variance in expected cash flow due to unexpected changes in exchange rates. Hedging can reduce the risk of future cash flow, and reduce the likelihood that cash flow will fall below a necessary minimum balance. If a firm has an obligation payable in another currency, then it may want to hedge the obligation. Imagine that a US firm purchases an input, such as cocoa from a German firm, that is payable in 90 days. The firm could hedge the obligation, purchasing, for example, 100 million Euro of cocoa, could purchase 100 million Euros in the forward market for delivery in 90 days. If the Euro increases in value, the potential loss on the transactions (payable) is offset by the gain of the forward exchange purchase.

Exchange rate variability can affect the firm's financial decisions. In a perfectly integrated international capital market, the equilibrium forward rate is determined

---

[4]In the 1990s, German and Japanese firms used greater leverage than US and UK firms, primarily due to lower debt costs and equity costs (in the case of Japan). These relative advantages disappeared in the late 1990s and early 2000s, leading the German and Japanese firms to reduce their leverage. The leverage ratio was relatively constant for US and UK firms [Eun & Resnick, 2004; McCauley & Zimmer, 1994].

such that hedging will not produce a profit. Aliber (1978) found that the average deviation of the forward rate from the future spot rate approached zero as the time horizon expands. There is no need to hedge if long-term nominal foreign currency assets are not exposed in the long run. Commodity price parity (CPP) holds that physical assets are not exposed to exchange rate risk as long as the purchasing power parity theory (PPPT) holds. However, Adler and Dumas (Adler & Dumas, 1983) state that the prices of input goods and output may have different exposures such that PPPT and CPP may not be sufficient for sound financial management. In an efficient market, there may be minimal advantages to hedging. A financial manager for Hershey would be extremely naïve (or lucky) if he or she did not hedge their chocolate purchases.

References

Adler, M., & Dumas, B. (1983). International portfolio choice and corporation finance: A synthesis. *Journal of Finance, 38*, 925–984.

Adler, M., & Lehmann, B. (1983). Deviations from purchasing power parity in the long run. *Journal of Finance, 38*, 1471–1487.

Aliber, R. Z. (1978). *Exchange risk and corporate international finance.* Wiley.

Browne, F. X. (1983). Departures from interest rate parity: Further evidence. *Journal of Banking and Finance, 7*, 253–272.

Cornell, B. (1979). Relative price changes and deviations from purchasing power parity. *Journal of Banking and Finance, 3*, 263–279.

Cornell, B. (1977). Spot rates, forward rates and exchange market efficiency. *Journal of Financial Economics, 5*, 55–65.

Cornell, B., & Dietrich, J. K. (1978). The efficiency of the market for foreign exchange under floating exchange rates. *Review of Economics and Statistics, 60*, 111–120.

Cornell, B., & Reinganum, M. R. (1981). Forward and futures prices: Evidence from the foreign exchange markets. *Journal of Finance, 36*, 1035–1045.

Eiteman, D. K., Stonehill, A., Moffett, M. H., & A. I. (2004). *Multinational business finance* (10th ed.). Addison-Wesley.

Errunza, V. R., & Senbet, L. W. (1981). The effects of international operations on the market value of the firm: Theory and evidence. *Journal of Finance, 36*, 401–417.

Eun, C. S., & Resnick, B. G. (2004). *International financial management* (3rd ed.). Irwin.

Feiger, G., & Jacquillat, B. (1979). Currency option bonds, puts and calls on spot exchange and the hedging of contingent foreign earnings. *Journal of Finance, 34*, 1129–1139.

Geweke, J., & Feige, E. (1979). Some joint tests of the efficiency of the markets for forward foreign exchange. *Review of Economics and Statistics, 61*, 334–341.

Giddy, I. H. (1977a). Exchange risk: Whose view? *Financial Management, 6*, 23–33.

Giddy, I. H. (1977b). An integrated theory of exchange rate equilibrium. *Journal of Financial and Quantitative Analysis, 11*, 883–892.

Kyriazi, F. S., Thomakos, D. D., Guerard, J. B., & J.B. (2019). Adaptive learning forecasting with applications in forecasting agricultural prices. *International Journal of Forecasting, 35*, 1356–1369.

Levich, R. M. (2001). *International financial markets: Prices and policies.* McGraw-Hill/Irwin.

Levich, R., & Potti, V. (2015). "Predictability and good deals in currency markets", 2015. *International Journal of Forecasting, 31*, 454–472.

Logue, D. E., & Oldfield, G. S. (1977). Managing foreign assets when foreign exchange markets are efficient. *Financial Management, 16,* 16–22.

Logue, D. E., Sweeney, R. J., & Willett, T. D. (1978). Speculative behavior of foreign exchange rates during the current float. *Journal of Business Research, 6,* 159–174.

Maronna, R. A., Martin, R. D., Yohai, V. J., & Salibian-Barrera, M. (2019). *Robust statistics: Theory and methods (with R).* Wiley.

Maldonado, R., & Saunders, A. (1983). Foreign exchange futures and the law of one Price. *Financial Management, 12,* 19–23.

McFarland, J. W., Pettit, R. R., & Sung, S. K. (1982). The distribution of foreign exchange Price changes: Trading day effects and risk measurement. *Journal of Finance, 37,* 693–715.

McCauley, R., & Zimmer, S. (1994). Exchange rates and international differences in the cost of capital. In Y. Amihud & R. Levich (Eds.), *Exchange rates and corporate performance.* Irwin.

Marston, R. C. (1995). *International Financial Integration.* Cambridge University Press.

Meese, R. A., & Rogoff. (1983). Empirical exchange rate models of the seventies: Do they fit out of sample? *Journal of International Economics, 14,* 3–24.

Mehra, R. (1978). On the financing and investment decisions of multinational firms in the presence of exchange risk. *Journal of Financial and Quantitative Analysis, 13,* 227–244.

Mincer, J., & Zarnowitz, V. (1969). The evaluation of economic forecasts. In J. Mincer (Ed.), *Economic forecasts and expectations.* Columbia University Press.

Nikolopoulos, K. I., & Thomakos, D. D. (2019). *Forecasting with the theta method.* Wiley.

Reilly, D. R. (1980). Experiences with an automatic box-Jenkins modelling algorithm. In O. D. Anderson (Ed.), *Time series analysis.* North-Holland.

Senbet, L. W. (1979). International capital market equilibrium and the multinational firm financing and investment policies. *Journal of Financial and Quantitative Analysis, 14,* 455–480.

Solnik, B. (1983a). International arbitrage pricing theory. *Journal of Finance, 38,* 449–457.

Solnik, B. (1983b). The relation between stock prices and inflationary expectations: The international evidence. *Journal of Finance, 38,* 35–48.

Solnik, B. (1978). International parity conditions and exchange risk: A review. *Journal of Banking and Finance, 2,* 281–293.

Stehle, R. (1977). An empirical test of the alternative hypotheses of national and international pricing of risky assets. *Journal of Finance, 32,* 493–502.

Stulz, R. M. (1981). A model of international asset pricing. *Journal of Financial Economics, 9,* 383–406.

Tashman, L. (2000). Out-of-sample tests of forecasting accuracy: An analysis and review. *International Journal of Forecasting, 16,* 437–450.

Thomakos, D., & Guerard, J. B., Jr. (2004). Naïve, ARIMA, Transfer Function, and VAR models: A comparison of forecasting performance. *The International Journal of Forecasting, 20*(2004), 53–67.

Yeager, L. B. (1976). *International monetary relations.* Harper & Row.

Chapter 22
Management–Stockholder Relations: Is Optimal Behavior All That Is Necessary?

So long as a company is closely held, the control group and the stockholders are identical, and seldom is there a conflict of interest between them.[1] However, once a company goes public, it acquires a group of shareholders who depend on the management for the safety and profitability of their investment. In short, an agency principal relation is established where the management is the agent and the shareholder is the principal. This relation implies a commitment by management that the outside shareholders will be treated fairly in such matters as cash payouts, expansion policies, accounting probity, and the level of executive compensation, and that in general, the company affairs will be directed vigorously and conscientiously.

Discussions on the problem of the separation of corporate control and ownership date back to the 1930s.[2] Nevertheless, it may be well to note that the shareholder's abdication of direct concern in managing the firm may be a natural consequence of the organizational development of the large corporation. The average shareholder desires and expects a reasonably competent and just management. Beyond that, he is not much concerned for in most cases his major economic activities lie elsewhere.

The history of corporations is full of instances where an inside group abused its position of power to enrich itself at the expense of the shareholders. Boards of directors, promoters, and management have at times taken a position of caveat emptor ("let the buyer beware") toward the investors in corporations. The exact legal obligation of the board of directors to the shareholders has never been clearly defined. The NYSE requires that the majority of all directors be "independent," as must be all directors on the audit nominating, executive compensation, and corporate

---

[1] Nevertheless, divergences of interest can exist in family-held companies. For example, a family member holding shares chiefly for their yield and a family shareholder drawing an income mainly from his managerial activity may differ on what is best for the company. Such conflicts are generally settled intramurally, although on occasion they have gone to court.

[2] Adolph Berle and Gardiner C. Means, *The Modern Corporation and Private Property*, MacMillan, 1933.

© The Author(s), under exclusive license to Springer Nature Switzerland AG 2022
J. B. Guerard Jr. et al., *Quantitative Corporate Finance*,
https://doi.org/10.1007/978-3-030-87269-4_22

governance committees.[3] Insiders, such as top executives of the firm, cannot be considered independent. Nevertheless, by a process of financial education, of legal and governmental administrative decisions, and by a voluntary assumption by management themselves, management has come to be viewed as having a quasi-fiduciary responsibility to the shareholders. Successful management is somewhat of a trustee of the stockholders' interests.

The inherent structure of the corporation no doubt permits management possible laxity or indifference to the welfare of the shareholders. But the fact is that shareholders as a group have received substantial economic rewards over time. It is only logical to assume that the management in the majority of corporations possesses a considerable sense of duty and loyalty.[4]

If a firm succeeds in generating exceptional earnings, then how should the profits be divided among the inside managers and the shareholders? The argument is made that the average stockholder does not contribute much to the success of his company; if he buys in after the firm is well established (as most stockholders do), he has not shouldered much of the initial risk. Thus, the argument runs, the stockholders should be satisfied with a "normal return"; anything in excess of a "normal return" is ascribed to the superior talents or energy of the initial innovating or managerial group and is properly their reward. A normal return can be derived from its beta and the required rate of return from the CAPM. There is some merit to this view insofar as economic incentives should be provided for innovation and good management under the theory of agency. The argument fails to recognize, however, the overall function of risk capital in the economy. The tremendous gains which have on occasion accrued to the shareholders of a spectacularly successful company are in large part offset by the losses taken on many enterprises that fail. If the chances of obtaining potentially large rewards were not available, the supply of venture capital would dry up. The return for risk may seem to come after the risk is over, but to reduce this return ex-post is to reduce the supply of new risk-bearers on the next round.

22.1 General Agreement and Potential Conflicts in Management and Control

Before going on to some of the possible conflicts between shareholders and management, one might point out that the interests of these two groups are largely parallel rather than divergent. If the company is profitable, the management obtains increases in income and prestige. Company executives are rewarded with higher

---

[3]J.R. Emshwiller and J.S. Lublin, "Boardrooms, 'Independent' is Debatable," *The Wall Street Journal*, March 3, 2005, C1, C4.

[4]Mr. Benjamin Graham writes "The typical management is honest, competent, and fair-minded." *Intelligent Investor*, Harper, 1959, p. 222.

salaries, bonuses, stock option plans, and increased respect from their peer groups. Furthermore, although the control group's major income may derive from their salaries and other managerial perquisites and their stock holdings represent only a small fraction of the shares outstanding, these holdings may still be valuable enough to constitute a strong economic incentive to do well for the company. One cannot discount the subtle but powerful psychological motivations toward purposeful activity. If the executive groups have any morale at all, they identify with the company. If their actions have any meaningful social goal, they are directed toward achieving prosperity for the corporation. A management that is indifferent, or at worst is engaged in systematically looting the firm, has reached a state of group disorganization or disintegration. That the majority of managements are fundamentally honest in protecting the interests of the stockholders is an implicit assumption of a workable free enterprise system.[5]

22.2 Areas of Potential Conflict

Although there is an identity of major interests between shareholders and management, it is unrealistic not to recognize that these interests are likely to diverge at some important points. Usually, the conflicts are only potential, and the corporate policy (especially if the firm has good earnings) can easily accommodate the differing economic positions of the two groups.

22.2.1 Managerial and Board of Directors Compensation

A strongly entrenched management is able, within wide limits, to set its own compensation. Obviously, the higher the officer and executive salaries, the smaller the earnings available for the common shareholders. On the other hand, a level of managerial compensation that moves up with improved company earnings provides incentives for top-notch performance. If a firm wishes to attract exceptional administrative talent, it has to offer a competitive economic reward. One needs to only read the popular business press to see the wide range of opinions concerning Dick Grasso's compensation as Chairman of the New York Stock Exchange and its implications for management and the board. Nevertheless, at some point, the boundary is crossed between sufficient and too much. Stockholders could contribute to the health of their company by asking intelligent questions about executive

---

[5] Some people think that all corporation managements are dishonest. Then, of course, others think all government officials are grafters, all policemen are bribe takers, all union leaders gangsters, and all professors time servers. People who have these cynical and essentially narrow beliefs rarely suggest meaningful policies for a democratic society.

compensation at the annual shareholders meetings. Should the company refuse to moderate grossly unreasonable levels of officers' compensation, the stockholders' suit is a strong weapon.

About 70 years ago, Berle and Means wrote a best seller on the clash of interests involved in the separation of ownership and control in the modern corporation.[6] More recent economic and management *gurus* have attacked the problem under a new name. It is called agency theory (Jensen and Meckling 1976). The gist of agency theory deals with the construction of a compensation or reward system so that the interests of the agent (the management) largely coincide with those of the principal (the shareholders).

The massive abuse of agency/principal relations is at the heart of the current Enron, World Com, and Adelphia debacles.[7] The perceived evidence is that the inside owners and managers raided the till for the enhancement of their personal enjoyment and wealth at the expense of the outside shareholders. Beneath the froth and turmoil, the basic question remains: How is corporate governance to be structured and how is management to be paid so that its interests coincide with the long-run prosperity of the shareholders?

22.2.2 Executive Compensation

Economists since the time of Roberts (1959) and Lewellen (1968) have empirically studied the problem of executive (CEO) compensation. Several issues are involved. Compensation should include a cash salary, a cash bonus, stock option exercises, or better yet restricted grants of stock which must be held for a specified time before they can be sold. The determinants of executive compensation may often be placed within two categories: those reflecting the size of the firm, and those reflecting the performance of the firm. Roberts used size variables such as sales, assets, and profits as independent variables in the compensation equation estimation. There is, of course, severe multicollinearity among sales, assets, and profits, for many firms. One will use only the sales variable in this chapter. One can use 5-year least squares (5LS) growth rates in sales (Sales5LS), earnings per share (EPS5LS), and total returns to stockholders (TR5LS), as well as ROE, as performance measures. One can use the Compustat EXECCOMP database from the WRDS system, which derives its data from proxy statements and 7OK filings with the SEC. The Compustat executive compensation means are shown in Table 22.1 and indicate that cash bonuses and total direct compensation levels have risen during the 1992–2002 period, particularly relative to a cash salary.

One can pool the data during the 1992–2002 period and run regressions for three compensation variables the reader is referred to in Table 22.2 for pooled regression

---

[6]Berle and Means, op. cit.

[7]See Chap. 19.

Table 22.1   Mean executive compensation ($ thousands)

Year	N	Salary	Cash Bonus	Total Direct Compensation
1992	433	$622.7	$514.5	$2810.6
1993	1157	542.2	427.9	2159.4
1994	1548	514.2	436.7	1646.4
1995	1600	530.7	490.4	2009.9
1996	1651	548.4	594.7	2587.2
1997	1674	563.9	620.8	3570.4
1998	1731	579.9	604.8	4620.1
1999	1811	582.8	692.8	4187.5
2000	1791	607.1	734.4	6130.7
2001	1637	646.3	664.8	4549.7
2002	538	805.9	1083.7	5801.7

Table 22.2   Pooled regression determinants of executive compensation 1992–2002

Compensation Variable	Constant	Sales	ROE	Sales5LS	EPS5LS	TR5LS	F	$\bar{R}^2$
Salary	.054	.449	.155	-.169	0.021	0.019	554	.249
(t)	(4.84)	(51.23)	(3.35)	(-6.76)	(-1.39)	(-1.24)		
Bonus	.010	.305	.144	.011	.008	.202	216	.114
	(0.74)	(28.85)	(2.56)	(0.35)	(0.46)	(11.09)		
Total Direct	-.022	.237	.177	.061	0.008	.229	163	.089
Compensation	(-1.64)	(22.09)	(3.13)	(2.01)	(0.44)	(12.42)		

results. CEO salary is positively, and significantly, associated with sales level and the return on equity, ROE. CEO salary is not a function of sales and EPS growth, or the 5-year return to stockholders. CEO bonus is positively and significantly associated with sales, ROE, and 5-year return to stockholders. Total direct CEO compensation, including stock options exercised, is a function of sales, ROE, 5-year sales growth, and the 5-year return to stockholders.

The empirical evidence supporting size and performance measures is reassuring, particularly given the growth in bonuses and total direct compensation. A well-known executive compensation study was produced by Lewellen (1968), who analyzed salary and bonus and total compensation of 50 large industrial companies during the 1945–1963 period. Lewellen found that compensation, defined as salary plus bonus, was statistically associated with sales, net income, and the market value of the firm. There was no statistically significant correlation between the performance and size measures and total compensation. Using the EXECCOMP database

Table 22.3 Executive compensation correlations Lewellen (1968)

	Salary + Bonus (t)	Total Compensation (t)
Sales	.285 (1.71)	.213 (1.43)
NI	.253 (1.77)	.155 (1.03)
MKTVAL	.275 (1.88)	.203 (1.43)
Executive Compensation Database		
Sales	.349 (46.4)	.192 (24.4)
NI	.225 (28.7)	.112 (14.0)
MKTVAL	.324 (42.6)	.302 (39.4)

where   NI = Net Income,
       MKTVAL = Market Value of Firm.

during the 1992–2002 period, one finds statistically significant correlations between salary and bonus and sales, net income, and market value of the firm. Moreover, one finds highly statistically significant correlations between total compensation and size and performance measures. One finds very similar correlation coefficients in the Lewellen (1968) study and the analysis of the EXECCOMP database. Executive compensation is correlated with the size and performance measures of a large sample of firms during the 1992–2002 period (Table 22.3).

A residual plot reveals no evidence of heteroscedasticity, and the condition number of 2.28 reveals no evidence of multicollinearity in the total direct compensation pooled regression.

Recent CEO compensation research by Dow and Raposo (2005) recommends that shareholders should commit to a policy of high pay, with high-powered initial incentive packages, to provide an atmosphere for CEOs to improve strategy making and curb CEO tendencies regarding restructuring and other dramatic events. Dow and Raposo further suggest that shareholders never precommit to pay very high compensation packages required to implement dramatic change.

22.2.3   Board of Directors

Supposedly, the Board of Directors are elected by the shareholders; in fact, they are usually nominated by the management, and the subsequent election is perfunctory. In general, the members of the board are honorable citizens, competent in their various professions. The managers of TIAA (the largest pension fund in the country) assert that the majority of the board should be outside directors, as distinct from inside directors who hold managerial positions. Most importantly, the outside directors should comprise the majority of the audit and the managerial compensation committees. Given the proper incentives, the directors could perform their job as unbiased questioners and broad supervisors of managerial policies.

Shareholders should look at the director's cash pay. The best source in this information is the stockholder proxy information distributed prior to the Annual stockholder meeting. The members of the board at Enron were getting $350,000 a year, although most directors of publicly traded firms make $35,000–40,000, annually. Beware when the cash compensation of the directors seems too high! Their job has become too precious, and they are unlikely to question or scrutinize any managerial report.

Of course, the board should be rewarded. They have an important balancing function within the governing framework of the corporation. However, their current annual cash pay and the accompanying perks should be moderate. Their major compensation should be an award of substantial amounts of common stock, restricted, not saleable for 5 years or after retirement whichever comes later.

Compensation in the form of restricted shares would focus the directors on the interests of the long-run stockholders. The board should be concerned about helping shape policies for a viable firm that generates a satisfying return for its shareholders over time.

The compensation scheme for the top executives might differ somewhat from that proposed for the board of directors. Because, in contrast to the directors, the manager's main current income comes from their employment with the company, their cash pay should be higher. It should be set at a level enabling them to maintain a lifestyle proper to their status, perhaps at about the amount of the average pay for a major league ballplayer. Bonuses and other major additional rewards should be in the form of restricted or deferred saleable shares.

22.2.4 Stock Options

The short-term stock option, a most harmful and deleterious form of award, should be abolished. The stock option entitles the holder to buy shares in the near future at a fixed price which is set below the forecasted market. Presumably, the option ties the managers to the shareholder's interest of maximizing the value of the company's stock. But the stock option fixes the manager's attention on the short-term horizon. Unfortunately, the most lucrative way to cash in on an option is to manipulate the books, defer expenses, and anticipate revenues, so that the market is fooled and the price of the shares rises sharply just at the time you prepare to sell your option acquired stock.

A stock option plan giving the officers an opportunity to purchase company shares at a previously fixed price is a type of compensation whose presumed value is contingent on good economic performance by the company. Unless the management's activities cause an appreciation of the corporation's shares in the market, the executives do not realize any significant gain from their options. However, as noted previously, in many instances, short-term options have proven a disaster. They have encouraged short-sighted managers to indulge in manipulative activities that boomed the speculative market value of the shares at the expense of the longer-run interests of the company.

Moreover, the shareholders may fail to realize how much economic value they may be allowing to the management in the form of stock options. Because no asset leaves the company nor is any explicit liability created, the corporation's accounts do not register the impact of the option. Yet these options represent a potential dilution of the equity of the shareholders, and if the number of shares under option is a large enough fraction of the total shares outstanding, exercising the option could reduce the earnings per share. The existence of the options can act as a drag on the potential appreciation of the common stock.

Worse yet is the case where large options and saleable shares are granted to a top manager just before his retirement. Here the manager was happy if financial activities could be manipulated so that the stock reached the high point at the time of his leaving. It didn't matter if this price was not likely to be supported by the market over any length of time. As the French expression goes, *aprés moi l' deluge*. The disaster comes after I'm gone.

22.2.5 Bonuses

In addition to their regular salary, many corporation officers receive bonuses based on the company's earnings. Generally, the bonus increases if the firm's profits exceed a base level. A bonus plan has many merits since it lowers managerial costs in poor years, when the company can least afford them, and increases them in years when the company's economic position is better. By making part of the officers' incomes contingent on profitable operations, it furnishes a strong incentive for conscientious management. It is important to re-examine the bonus formula periodically to ascertain whether the base level of earnings has been left at an inappropriately low level without proper consideration to the subsequent growth of the company.

Pension plans are another means of executive remuneration that have come into favor. The promise of a pension may be more valuable to the recipient than an immediate increase in salary of the same amount, since pension payments are likely to be made when his income tax bracket is considerably lower.[8] A generous pension plan gives the firm an advantage in attracting and holding good managerial talent. But pensions may seem unduly high when, as sometimes happens, they are only slightly lower than the active officer's salary and generous pensions that were not voted in advance but are settled upon shortly before the executives' retirement may be somewhat suspect.

---

[8]Even a highly paid executive may find it difficult to save as much as he would like. A pension plan represents a method of forced or involuntary saving.

A number of sharp changes have taken place in the structure of the economy. Contrary to the Professor Joseph Schumpeter prognosis, management of modern industry may not have devolved into a set of staid, standard procedures. The old standard industries are gone, and the new industries staffed by technicians may be considerably more difficult to manage. (As the expression goes, managing technicians may be compared to herding cats.) We are in an era of innovation and rapid change. Catching the right wave leads to success and big stock returns, while missing the opportunity results in failure. If the risk/return ratio has widened, perhaps we might also expect a rise in the compensation paid for decision making.

22.2.6   Dividends, Buy Backs, and Retained Earnings

The theoretical value of common stock is the discounted value of its future cash dividends and net buybacks over purchase price plus a possible final sale or liquidation value. Since the present value of a future dividend or buyback cannot equal that of the immediate payments, the current level of buybacks and dividends is important in setting the intrinsic value of the stock, as we discussed in Chap. 8. Stockholders are, therefore, extremely interested in the management's dividend or buyback policies. Retained earnings have value to the stockholder only as they can be invested profitably at a rate of return within the company that at a minimum equals the market equity discount rate and thereby generates increased earnings and future dividends.[9]

Management may evaluate the importance of dividends or buybacks versus retained earnings differently than the shareholders. Usually management obtains more of its current income from salaries and bonuses than from the cash throw off on the shares it may hold. Salaries are likely to continue as long as the firm is operating. Retained earnings appear to be a source of capital carrying little obligation. Its use increases the safety and survival chances of the firm. To a conservatively oriented management, keeping the firm alive (and assuring continuing salaries) may appear more worthwhile than keeping the return on the shareholder's investment as high as possible.

A conflict between the stockholders and the company officers over the level of retained earnings can also arise if the management's time preference for future income differs from that of the shareholders. Or the management may be niggardly with cash outflow because it overestimates the relative profitability of investment in the company with the alternatives open to the shareholders in other areas of the economy.

---

[9] A complicating factor in valuing future returns may be the differing tax status of the individual shareholders. A stockholder with a high marginal present tax rate may prefer increased future dividends and growth over present returns. If the dividend policies are consistent in pattern, probably each firm eventually acquires stockholders of appropriate tax status.

22.2.7 Excessively Conservative Financial or Asset Structures

An excessively conservative financial structure has a disproportionate amount of equity financing and relatively little debt. Of course, the degree of conservatism, or risk, in the financial structure must be judged in the light of risks the firm faces. An overly conservative asset structure contains a high proportion of liquid, low-earning assets in relation to the needs of the business. Quite possibly a company could have both an ultraconservative asset structure and an excessively conservative financial structure. Both conditions are the results of a timorous or greedy management.

An extremely conservative financial structure may develop if all financing is accomplished through new common stock issues. More than likely it is the corollary of a niggardly cash buyback or dividend policy. An ultraconservative financial structure results if internally generated funds are used to finance the firm to the exclusion of other sources.

A financial structure containing a minimal debt component relieves the management of the fear of failure; however, it raises the cost of capital to the firm. Stockholders are deprived of the extra profits that would accrue through the use of judicious leverage. Indeed, some writers have suggested that a firm should carry some debt as a matter of principle since it is a constant reminder to the management that there is a cost of capital. The effort and planning required to service the debt should help to keep the managers from becoming complacent.

On the asset side of the ledger, excessive or redundant net working capital is another symptom of an unenterprising company administration. A large net working capital base helps assure survival of the firm, but survival without profits is hardly desirable. A firm that does not economize on its use of net current funds, other things being equal, has a low rate of return on its total assets.

Excessive net working capital can develop because of a low pay-out policy, with the retained funds invested in liquid assets. If a high rate of depreciation is covered by operations and the funds obtained do not go back into capital assets, the same condition results. A firm undergoing a decline in its scale of operations is more likely to have redundant net working capital than one that is vigorously expanding. Nevertheless, the managements of static firms are often reluctant to give up any of the funds they have under their control, although they may have no plans for their productive use.

Expansion In the shareholders' view, expansion is worthwhile when it increases their reinvested returns. The Gordon Model of stock valuation introduced in Chap. 8 recognizes the importance of reinvested profits and future growth. The use of retained earnings to finance expansion is justified when the company can make the going rate of return on equity on these funds. If the management, however, is not sufficiently concerned about the stockholders, it may pursue expansion for the economic or the psychological rewards that corporation officers receive simply from being associated with a larger firm. The shareholders may well feel that the basic objective of the corporation is sacrificed if the management follows a policy of "dry" or profitless expansion.

In many instances, "dry expansion" is a further reflection of the failure of some corporation administrations to recognize the implicit cost of retained earnings. The company may have a consistently low dividend payout rate and fail to find profitable outlets for its funds. Yet the management may be content because it feels that nothing needs be paid on the "surplus" portion of the equity. In other cases, the management may tend to overrate investment opportunities in its own firm, or it may do its shareholders a disservice by acquiring subsidiaries at costs which their earnings potential do not justify.

The management's interests and the welfare of the shareholders are not equally served if the company officers bring about company growth purely in order to enhance management prestige, to raise managerial perquisites, or to assuage their power drives.

22.2.8   Liquidating or Selling the Firm

The natural desire of the company officers to continue operations conflicts with the interests of the shareholders if the foreseeable returns for the firm are not sufficient to cover the value of the capital which can be withdrawn from the firm. The situation is, of course, similar if the company or its assets can be sold as a unit for an amount exceeding the sum of the going market price of its securities. Because a voluntary liquidation or sale eliminates executive incomes, it may well meet with management resistance.

A firm whose common shares sell on the market at a price measurably below their liquidating asset value invites purchase by individuals or syndicates who intend to turn a quick profit by liquidating the company. Such purchasers are often called "raiders"—a term strongly connoting disapproval, especially when used by conservative management. Nevertheless, the raiders may perform a useful economic function. By their very existence, they serve as a warning to those managements who fail to use their capital productively. However, the shareholders may not do as well under the raiders as they would with the original management if they were to proceed with a careful and orderly liquidation. Nevertheless, a raid and subsequent liquidation may still be an economic improvement over having shareholder funds slowly go to waste under the aegis of listless management.

A management that does not wish to give up entirely may turn a liquidated operating company into a closed-end investment corporation (i.e., a company which makes its income by holding other companies' stocks and bonds). For a company having substantial carry-forward tax loss credits, becoming a holding company may have some advantages.

22.2.9   Risky Acquisitions

In contrast, very speculative management may attempt to make a quick fortune by making many rapid acquisitions, obtaining new properties on low margins, and mortgaging or leveraging the original company to the hilt. Such speculative activities can bring great immediate gains for all, or it can result in the loss of the original equity values in what might have been at the start a small but solid company. The shareholders cannot legitimately complain if they knowingly accept these ventures. However, if the risk-taking proclivities of the management are covered by deceptive practices, the shareholders may have just reason to believe their interests had been breached. A management holding a plethora of exercisable stock options may be especially tempted to play the stock market for short-term speculative gain. Acquisitions must be rationale and that firms with low cash-adjusted leverage, CAL, those attractive as takeover targets, outperform high cash-adjusted leverage firms during the 1980–2003 period by over 11.2% annually.[10] Lower target leverage reduces the liability acquired in the takeover and large cash holdings increase acquirer liquidity. Cash-adjusted leverage is calculated by subtracting cash holdings from debt liabilities. The Cremers et al. (2005) CAL variable is calculated as (debt – cash)/total assets. The acquired firm's cash holdings can pay off the firm's debt, a concept very similar to the net current asset value (NCAV) concept of Graham and Dodd introduced in Chap. 14.

22.3   Turning Agents into Owners

The prescribed method for dealing with the Agency/Principal problem is to set up the compensation system so that the interests of the Agent are closely tied to that of the principal. Various devices have been developed so that at least part of the managerial compensation is based on the financial performance of the firm. At least a portion of the management's income is based on their being part or quasi-owners in the firm.

Illusory growth takes place when, in evaluating a stock, investors concentrate on an apparent periodic increase in earnings per share without checking on or correcting for the trend of the rate of earnings on total assets or the so-called core business. The true measure of growth is the earnings after the acquisition rises over the projected earnings of the company's independence. The phenomenon of illusory growth is developed by the use of acquisitions or mergers after the acquiring company has somehow been initially recognized as a growth company by the market.

The easiest way to explain illusory growth is by example. Suppose we have Athos Corporation, with 100,000 shares outstanding, earning $1.00 a share, currently

---

[10]Roll (1986) asks if acquisitions do not consistently increase the acquiring wealth, then hubris must explain why the firm does not withdraw its bid when it must have been developed with an incorrect valuation.

selling at $20.00 per share (a P/E ratio of 20 times earnings) because of a market belief in a past record of growth. Athos' total earnings is $100,000. Another firm, Zanox Inc., has 100,000 shares outstanding, an BPS (earnings per share) of $1.00, and is considered a moderately successful firm selling at $10.00 a share (a P/E ratio of 10 times earnings). Zanox also has total earnings of $100,000.

Athos now acquires Zanox by an exchange of Athos's shares for Zanox's on a value-for-value basis. One share of Athos ($20.00 per share) is exchanged for two $10.00 Zanox shares. (In the process, the Zanox shares are retired.) At the end of the transaction, Athos has 150,000 shares outstanding, but it has added $100,000 to its earnings, bringing the total to $200,000. Dividing through by the new number of shares, $200,000 divided by 150,000 total shares now gives us an EPS of $1.33. Lo and behold, Athos has raised its earnings per share by 33.0%. It may be now further confirmed in its reputation as a growth company, and its high P/E ratio justified.

Eventually, the illusory growth firm disappoints its investors, and the price of its shares drops sharply. As the original growth company merges with more and more average or slow growth firms, the growth rate of earnings on the total asset base must begin to decline. Externally, it becomes more and more difficult to find net new desirable acquisitions, and internally, the managerial problems of the conglomerate become unwieldy. But more basically, a simple mathematical relationship begins to undermine the whole process of illusory growth. When the original company was relatively small, the acquisition of another fair-sized company on the basis of a favorable relationship of P/E ratios raised the EPS of the acquiring company noticeably. When the so-called growth company reaches a large size, it requires larger and larger acquisitions to maintain the picture of rising earnings per share.

Again, a small example may illustrate the problem. Suppose Athos Conglomerate Inc. now has 2,000,000 shares, $2,000,000 of earnings (EPS of $1.00), and because of a growth reputation sells at $20.00 per share (P/E of 20X). Silon Corporation has 100,000 shares, earnings of $100,000 (EPS of $1.00), and a moderate price of $10.00 per share (P/E of 10X). Athos acquires Silon through an exchange of shares, value for value, and one share of Athos for two of Silon. After the acquisition, Athos has total earnings of $2,100,000 and total shares outstanding of 2,050,000. The earnings per share goes up from $1.00 to $1.024; but this is hardly a noticeable increase and nowhere near what its followers have expected from Athos in the past. In short, as the illusory growth company becomes larger and larger, it begins to outgrow its environment. It requires larger and larger acquisitions to maintain the appearance of earnings growth, and new gulps of these sources of financial nourishment might not be as available as they were earlier in the conglomerate's career.

Finally, as the investors become disillusioned, a more reasonable P/E ratio is given to the stock and the price of the shares falls sharply. The last round of buyers may experience severe losses on their holdings. Illusionary growth was behind many of the recent financial scandals.

22.4   The Diseconomies of Financial Scams

From a broad point of view, management perfections and frailties are but small glitches in the functioning of the financial markets. Yet they do entail a cost. The unnecessary volatility they add to the markets, the risk of loss because of fraudulent or near-fraudulent behavior, adds to the risk premium carried by the economy's cost of capital. The rise in the cost of capital reduces worthwhile real investment and slows the rate of economic improvement and growth. Moreover, the incidents of wild behavior or outright crookedness bring forth a strong social/psychological reaction of anger and resentment so that some reasonable and normal operations or innovations are mistrusted and hindered. This reaction adds friction and inefficiencies to the function of the financial markets.

The operations of the financial and capital markets are an essential element in the functioning of a free enterprise, price-directed economy. Yet in many ways, the financial markets are fragile institutions; in a large part, their operation is made possible only by a considerable degree of trust between the suppliers and users of funds. (In the current scholarly literature, this connection has been extensively examined under the heading of *agency–principal relations*.) Insofar as fake-out devices lead to a waste of resources, a reduction of trust, and the subsequent employment of more elaborate examining and policing devices, they impose an extra cost on the overall cost of capital.

22.5   Insider Trading

The chief officers of a corporation, the directors, and major stockholders are in a position to obtain advance information on important developments in the company. Such "inside" information, whether favorable or unfavorable, can easily be turned into quick trading profits on the stock market.[11] Clearly unethical would be an insider program to create speculative activity by placing tips and financial news in the market or by actually manipulating financial reports and the timing of financial announcements.

The Securities Exchange Act of 1934 substantially restrained the possible speculative activities of management and other insider groups. No longer can such groups operate secretly; any changes in the shareholdings of corporation officers, directors, and major stockholders must be reported, and these changes are published weekly by the Securities and Exchange Commission. Prison terms and/or substantial fines may await insiders guilty of deliberately leaking information or of engaging in other manipulative practices. Bettis et al. (2000) surveyed some 1900 member firms of the American Society of Corporate Secretaries in November 1996 with regard to

---

[11] Trading profits can be made on unfavorable developments, which would tend to depress the price of the stock, by "selling short."

corporate insider trading restrictions. Some 35% of firms (633) responded to the survey. Of the firms responding to the survey, 576, or 92%, had corporate policies regulating or restricting insider trading. The most common policy, involving approximately 75% of the sample, required prior approval of insider trades. Moreover, most companies had blackout periods. Blackout periods are periods during which trading by insiders is not allowed. The most common blackout period was 10 days, generally days +3 through +12 of quarterly earnings announcements. Bettis et al. (2000) find that blackout days reduce the bid-ask spread by two basis points, or 8.5%, and insider trading profitability is less in blackout periods than in trading windows during the 1992–1997 period. Insiders earned abnormal profits in allowed trading windows of 0.58%. Bettis et al. (2015) reported that (legal) insider trader (LIT) data could be incorporated into traditional stock selection modeling and used with multifactor risk models, see Chap. 15, such that LIT can enhance stockholder returns.

22.6   Conflict of Interests

No legal rule prohibits company officers or directors from dealing with the corporation on their own account, nor from having the corporation contract with other companies in which they have a personal interest. The interest of the corporation official must be fully disclosed, however, and the transaction must be open. A company official who has a concealed interest in another firm which is a supplier or customer of the corporation is in an especially vulnerable position. A corporation officer who has profited personally at the expense of the company is subject to suit by the corporation or by stockholders on behalf of the company. Major corporation officers should divest themselves of conflicting interests or at least abstain from participating in decisions when these interests are involved.

22.7   Stockholder Remedies

What can the stockholder do if his management seems incompetent or flouts his basic economic interests? A number of alternatives have been suggested, although none of them appear to be entirely efficacious.

22.7.1   Sell His Shares

Some authorities advise the unhappy stockholder to sell his shares. This is the so-called practical approach. It ignores the fact that although this course of action may work for one shareholder, if any large part of the shareholders attempts the same remedy, the price of the stock would drop sharply on the market, entailing a considerable loss for all the stockholders. The management would suffer only insofar as they held any shares themselves.

22.7.2   Institute a Stockholder Suit

A stockholder suit is valid only if the management has been guilty of breaching its quasi-fiduciary position. It will not hold for many of the situations described in this chapter. In addition, stockholder suits are difficult and expensive. The mores and culture of the investment community frown upon them; and indeed even shareholders who have benefited by the successful outcome of a "derivative" suit[12] often look askance upon the shareholders who instituted the action. It is true that the remedy of the stockholder suit has often been abused; some shareholders have used it to further their own special interests. Nevertheless, it is a salutary device even if its existence—the possibility of a suit—serves only to remind the management of the limitation of its powers.

Attend and Vote at Annual Stockholders' Meetings   The vast majority of shareholders do not attend the annual meetings. But, contrary to common belief, it is possible (assuming a responsive management) to have considerable influence at these meetings. Intelligent, incisive questions force the management to consider their policies carefully. Independent stockholder proposals can be submitted at the annual meetings. These proposals are usually voted down by an overwhelming majority, but if the proposals pick up support from 1 year to the next, they exercise considerable impact on a responsible management.[13] Many such independent proposals are eventually adopted by management itself.

22.7.3   Organize a Proxy Fight to Vote Out the Management

If the stockholder is sufficiently outraged by the management, he may attempt to vote in a new board of directors who will appoint a new management at the shareholders' annual meeting. The shareholder can succeed only if he can enlist a majority of the voting stock in his campaign. The management has all the advantages of the "ins" in such a struggle. It has better access to the mailing list of stockholders,

---

[12]The stockholders' action is called a "derivative" suit because they sue on behalf of the corporation.

[13]Unfortunately, the notion of the "divine right" of management is so imbedded in the investment community's thinking that the questioning, active, independent shareholder must often expect to arouse some hostility—mostly from his fellow shareholders.

to whom it can send proxy requests.[14] It can use company funds and personnel to help run its proxy solicitations and can count on a considerable number of proxies that are almost automatically returned to the management.

Indeed, the smaller "independent" stockholder can scarcely hope to win a proxy fight. Even if he finds himself allied with some powerful financial interest seeking to gain control of the management, the outcome is uncertain. And, indeed, in such a case, it may be better to stay with the old management.

"Respectable" entrenched managements look upon the organizers of proxy fights (even when they may represent major, responsible shareholders) with disfavor. The insurgents are called "raiders." Nevertheless, successful management seldom faces a proxy fight. Indeed, an occasional proxy fight probably does the investment community no harm, but rather encourages incumbent managements to seek improvement in their stockholder relations.

22.8 To Whom Is Management Responsible?

Social minded writers have long called for the arrival of a new type of management, one concerned with economic and social justice.[15] Management's primary concern would no longer rest with the maximizing of profits—i.e. with the economic interests of the shareholders. Because the "separation of ownership and control" is largely an accomplished fact, the corporation administrators have acquired tremendous powers independent of any likely stockholder interference. They should, therefore, devote themselves largely to deciding "fair prices," to developing pleasant communities, and to determining "just wages." The exponents of such developments seem not to fear that the corporate managements will abuse their powers. They feel that a properly trained manager would acquire an ethic of self-discipline or, better, of self-restraint, which could be relied upon to moderate any undue selfishness of the managerial groups.

To the more orthodox scholars of economics and finance, this sociological-derived thesis seems somewhat romantic. It is doubtful whether the widening area of managerial independence represents a complete divorce from the dominant interests of the risk-taking suppliers of ownership capital, which is to maximize the value to its owners.

It is questionable that even if it is granted that benevolent corporation administrations are free to make their own economic policy, whether it is desirable that they

---

[14] A proxy is a grant to a designated person to vote one's stock if one cannot attend the meeting. Proxies can be requested for voting on the membership for the board of directors or for voting on specific proposals. The SEC requires full disclosure of the issue on the solicitation of the proxies.

[15] See Adolph Berle and Gardiner C. Means, Op. Cit. (especially the last chapter). In this book, Berle merely hoped for the day when management would mediate between the stockholders, the public, and labor. In a later book, *Twentieth Century Capitalist Revolution*, by Harcourt, Brace, and World, 1954, he is deceptively pleased that that day had arrived.

do so. No matter how fine the ethics and social sensibility of the managers are, is it wise to encourage them to believe that they should make decisions that transcend the market? This argument should not be misinterpreted. We can agree that managers should be honest, ethical, and law-abiding and that they act within the framework of accepted social standards. As a matter of enlightened self-interest, management should not debase their product, produce unworkable or dangerously defective goods, put men to work under unsafe or unsanitary conditions, cheat on taxes, advertise falsely, or engage in collusive marketing practices. On the positive side, the management of large corporations can practice good community relations, contribute to worthwhile public or educational projects, engage in good labor practices, and encourage cultural and intellectual development, and the free exchange of ideas in the community. Nevertheless, these are not the main activities of corporation managements. We include Appendix A to discuss recent research in socially responsible investing, SRI. They should not consider themselves the dispensers and allocators of benefits to the rest of society, for nothing could be surer to raise the resentment of the rest of the citizenry. Independent men will tend to dislike being beholden to feudalism—even a benevolent feudalism.

22.9   Summary and Conclusions

The "new capitalism" implies a denigration of the economic function of management. Maximizing the long-run profits of the firm, in this thesis, becomes a crass money-grubbing activity. There is a considerable lack of appreciation of economic analysis in this view. Management performs important economic functions when it bargains hard but honestly. Management must use the criterion of economic efficiency if factors of production are not to be wasted or misemployed, if productivity is to be improved, if the investment of new capital is to be directed to areas where it is most needed, and if, to sum up, a system of free-market prices is to work at all. As we discussed in Chap. 1, management must work to maximize stockholder wealth. Clearly, with SRI, "one can do good while doing good."

Appendix A

Socially Responsible Investing

During the 1990s, many investors sought to invest in securities that were "socially responsible." There are, of course, many definitions of socially responsible investment (SRI). For example, in May 1990, the Domini 400 Social Index (DSI) was created to replicate the returns of the Standard & Poor's 500 (S&P500) Index. The Domini 400 Social Index was created by Kinder, Lydenberg, and Domini (KLD) to select companies with positive social and environmental records: the criteria involved community relations, diversity, employee relations, human rights, safety, environment, and (new) corporate governance. The DSI's social criteria led it to exclude stocks that: (1) derived more than 2% of gross revenues from military weapons; (2) firms with tobacco, gambling, and alcohol sales; and (3) firms that own or share in nuclear power plants. The returns of the DSI stocks have exceeded the returns of the S&P 500 stocks since 1990, although there is a beta bias (exceeding 1.0) and a growth bias to the DSI portfolio. The relative returns of the DSI and S&P 500 stocks, as of August 30, 2003, have been:

The relative returns on SRI portfolios can be volatile.[16] There can be additional social criteria. One may not want to invest in stocks with poor environmental records, or poor employment (unions, safety, pension concerns), or diversity (woman and/or minority CEO, or boards of directors, family or "gay" rights, hiring of disabled persons) records. There can be an almost infinite number of combinations of social criteria exclusions that can be used to create SRI portfolios. Do these social criteria affect portfolio performance and stock prices? The recent empirical evidence, as reported in studies winning the Moskowitz Prize for Research in Socially Responsible Investment, supports the notion that SRI portfolios do not perform statistically different from (traditional) equity portfolios, and stock prices of socially responsible stocks are not determined by different criteria than common stocks in general. The Moskowitz Prize winning studies are Guerard (1996), Waddock and Graver (1997), Russo and Fouts (1997), Repetto and Austin (1999), Dowell et al. (2000), and Bauer et al. (2001). There is no statistically significant cost of capital differences in the SRI stocks and non-SRI stocks.

The academic literature in the last decade has reported both the shadow cost and return enhancement views of ESG-screened investment portfolios. A number of literature reviews and meta-analyses find evidence on both sides of the SRI/ESG performance question. Friede et al. (2015) review more than 2200 studies and find that "the results show that the business case for ESG investing is empirically very well founded." Sjöström (2011) reviews 21 studies in a meta-analytic framework and finds, contra Friede et al., that "results point in all directions, and . . . there is no clear link between SRI and financial performance."

For stock portfolios, Evans and Dinusha (2010) find evidence of a positive relationship between ESG rating criteria and accounting measures of performance including return on assets and market-to-book ratios. Fulton et al. (2012) argue that a positive correlation exists between one aspect of ESG in particular, sustainability, and superior risk-adjusted returns. Park and Moon (2011) use KLD ratings and find that for S&P 500 constituents, top quantile firms outperform by as much as 6.24% over the 1991 to 2006 period controlling for standard factor model risks. From a global perspective and focusing on corporate social responsibility (CSR), Renneboog et al. (2008) find that some CSR measures are associated with shareholder value, but others are associated with destruction of shareholder value.

---

[16] As the authors updated this chapter and this text went to press, the relative Domini 400 Social Index returns were trailing the S&P 500 Index as August 31, 2006.

Statman (2005, 2007) analyzes SRI/ESG performance including comparing SRI indexes to the S&P 500 historically, finding little difference in performance. This is echoed in part by Geczy et al. (2005) who in a sophisticated framework create SRI/ESG mutual fund portfolios. They find that index-oriented portfolios under CAPM and zero alpha parameters seem to impose relatively small costs, as little as 1 or 2 basis points per month, compared to non-SRI portfolios. However, when informative alpha beliefs or beliefs in the importance of investment style are present, relative performance deteriorates. Barnett and Salomon (2004) argue that a nonlinear relationship exists between social and financial performance, finding that funds with intense screens outperform by weeding out "bad" firms, and funds that have weak screens outperform in risk-adjusted measures accounting for diversification, while funds in the middle underperform.

Corporate Governance has been studied by Shleifer and Vishny (1997) and Liang and Renneboog (2017). Liang and Renneboog reported that countries with higher social responsibility (sustainability) ratings were more likely to be civil law, rather than common law, countries, with the Scandinavian countries have the highest scores. Corporate social responsibility is positively associated with lower stockholder litigation risk and strong labor regulations. Liang and Renneboog (2017) stress that corporate social responsibility reflects the supply of socially responsible behavior by firms and the demand for corporate social responsibility practices by society.

Bialkowski and Starks (2017) reported that SRI funds attracted more flows, on average, than conventional mutual funds between 1999 and 2011. SRI funds had positive net flow in all but two quarters of the 13-year period; and SRI funds had statistically significant inflow following environmental disasters, such as the BP oil spill and the Fukushima Daiichi nuclear meltdown, and accounting scandals, Enron, Tyco, and Worldcom. SRI funds had higher exposure to (MSCI) ESG values that persisted across time. Riedl and Smeets (2017) surveyed some 3382 socially responsible investors and 35,000 randomly selected investors using administrative individual investor data in the Netherlands. Riedl and Smeets reported with an 8% response rate from conventional investors and a 12% response rate from SRI investors that SRI investors contribute more to charity than conventional investors, invest primarily in SRI equity funds, had longer holding periods than conventional investors, and expect lower (marginally) returns than conventional investors. Investors holding a university degree are more likely to be SRI investors. Our results are consistent with the recent findings on Diversity reported in Manconi et al. (2015) and Kim and Starks (2016). Manconi et al. (2015) reported that Diversity enhanced returns more than the Fama & French (2016) five-factor components and that the market-weighted diversity portfolios are enhanced relative to the equal-weighted diversity portfolios for the 2001–2014 time period. Kim and Starks (2016) tested for unique skills of female members of corporate boards.

How have the costs of SRI changed since the initial studies and enhancements to the database? In this context, Geczy and Guerard (2021) re-examined the KLD database from its 1991 inception through 2015. Over time, the KLD database was subject to enhancements resulting from acquisitions and other methodology changes. For example, in 2000 the human rights category was added (Galema et al. (2008)), in 2002 governance was added (Statman and Glushkov (2009)), and in 2010 KLD decided to rank companies only on issues relevant for their industry as opposed to all issues. ESG scores are naturally persistent in the short run, as reported and confirmed by Wimmer (2012) in reviewing SRI mutual funds. However, Wimmer (2012) also finds that persistence declines after approximately 2 years, making rebalancing crucial for SRI portfolios. Such instability of the dataset results in many versions of the final KLD scores constructed in academic literature, causing potential difficulties in comparing results across papers and time. It also highlights the potential dependency of results on the exact measurement methods used. As a result, recent literature has focused on ways to adjust and normalize the scores to account for the lack of stability in KLD dimensions over time.

From the early KLD studies (Sharfman (1996)) and continuing to the most recent ones (Statman & Glushkov, 2016), there has been an ongoing discussion about the challenges of creating a unique KLD-based score. The simplest way that sums all strengths and subtracts all weaknesses incurs its own set of biases and imbalances driven by data structure rather than companies' ESG attributes. Dorfleitner et al. (2014) study the relation between ESG score performance and stock performance in various markets worldwide, reiterating global evidence of positive association between firm ESG ratings and subsequent returns; however, the bias remains. The earlier literature attempted to address the implicit bias arising from weighting each issue equally. For example, in order to avoid treating each ESG strength and weakness as equally important, Waddock and Graves (1997) rely on the issues weighting scheme developed by Ruf et al. (1993). Because such weightings are highly subjective, they are no longer used in more recent studies. For example, Employee Relations strengths are evaluated on 10 individual variables, with a maximum score of 8, while Human Rights strengths are evaluated on 3 variables with a maximum score of 2. Hence, because of the uneven distribution, the raw score will be much more impacted by the Employee Relations strengths versus Human Rights strengths. The same issues affect the weights of strengths versus concerns. Depending on which area has a larger number of evaluated metrics, that area will get higher implied weightings in the overall raw calculations.

Another difficulty arises because of the changing nature of the KLD dataset over time. Specifically, as the number of strengths and weaknesses changes in each category, summing the raw strengths and weakness, such as was the earlier practice, creates score dynamics that are influenced by the dataset construction rather than only by the company's changing ESG policies. Kempf and Osthoff (2007) address this problem by normalizing the net scores within each of the six categories. In addition, they introduce a way to binary transform the weakness into the same direction as the strengths. Using decile spread portfolios, their paper documents statistically positive KLD score alphas for the 1992 to 2004 period. In addition, they introduce an important methodology to normalize the rankings by the sector to which the company belongs. This reduces the potential sector and resulting factor biases in the rankings. Finally, they create a version of the rank that excludes all companies that have at least one of the six controversial scores. Statman and Glushkov (2009) test a variation of the Kempf and Osthoff (2007) methodology by excluding companies that have a zero value in both the strength and the weakness fields in each category, claiming that those companies were not reviewed by KLD. They exclude the governance component because of short history. Over the period 1992–2007, they find statistically significant positive alphas for KLD long–short portfolios.

Manescu (2011) adds an additional refinement that normalizes strengths and weaknesses separately. Since in the Kempf and Osthoff (2007) and Statman and Glushkov (2009) methodologies a lack of a weakness is considered a strength, Manescu recognizes that it is important to normalize strengths and weaknesses separately. Unlike prior studies, Manescu does not exclude any categories and uses all seven categories. She finds that only community relations scores have statistically significant abnormal positive returns during the 1992 to 2008 period.

From Raw to Normalized Score Definitions

Raw Score

Inside each of seven subcategories (Governance, Community, Diversity, Employee Relations, Environment, Human Rights, and Product Safety), KLD provides binary ratings on multiple individual measures of strengths and concerns criteria. For each of the seven categories and for each company in each year, the Category Raw Net Score is the sum of category strengths minus sum of category weaknesses. The Total Raw Net Score is the sum of the strengths across all categories minus the sum of all the weaknesses across all categories. There is Total Net Score only if both strength and weakness exist. If strengths or weaknesses are missing entirely, the Net Score is NaN for that company in that year.

Normalized Scores

(a) *Category score:* For each subcategory, for each company in each year, we first normalize strengths (weaknesses) by dividing the sum of strengths (weaknesses) Booleans by the concurrent dimension of strengths (weaknesses), where the dimension is defined as the number of evaluated variables in each category during each year As the number of evaluated variables varies over time and differs across ESG categories, doing this adjustment normalizes the data and allows for cleaner review of company's ESG information rather than KLD's evaluation methodology.

For example, for the strengths of the Diversity category, Company A in year 1 is measured by two variables (dimension is 2); its scores are 1 and 1, respectively;

Table 22.4 Time period returns

Stocks	Year-to-Date (%)	One Year	Three Years	Five Years	Ten Years	Since Inception
DSI	16.46	14.05	-10.96	3.09	10.91	373.0
S&P500	15.93	12.07	-11.41	2.51	10.09	307.9

Source: www.kld.com/performance

hence, its normalized strength score is $1 + 1/2 = 1$, while its raw strength score is 2 Then, let's say that next year, Company B has three evaluated Diversity strengths criteria with scores of 1, 1, and 0. While the raw score would be the same (2), the normalized score in year 2 is 2/3. This normalization captures the fact that in year 2 the company could have earned a maximum rating of 3, while in year one, the maximum rating was 2. Manescu (2011) points out that because the dimensions of each substrength and subweakness are different, this is important.

Next, we deduct normalized weaknesses from the normalized strengths to get the Category Normalized Net Scores. We then subtract the corresponding industry average normalized net score from each company's normalized score, making scores industry-neutralized. Industries are defined by the ten Fama-French Sectors (from Kenneth French's website[17]). By subtracting the average industry score, we neutralize large industry biases present in ESG rankings, which remain even after the normalization. The energy sector has the lowest average scores, while consumer nondurables has the highest.

(b) **Total score:** Each subcategory-normalized and industry-neutralized score is further ranked from 1 to 100, and then total score is calculated as the average of all subcategory scores. If one or more subcategory score is missing, the total score equals the average of all other subcategory scores. Creating a percentile rank in order to combine across subcategories is important because combining even the normalized scores across seven subcategories is not balanced—as can be seen in Panel D of Table 22.4, the Min/Max distributions of each score are not even, which would result in varying weightings of each category. Ordinal rankings create an equal contribution to the final score from each subcategory. We suggest that this adjustment process should be the new default, a more robust starting point, for reviewing the ESG information content of the KLD data, making it more comparable over time and across categories, with equal representation of various ESG issues. If materiality or other reasons require for varying weightings across ESG issues, these methodologies can be described and tested explicitly and compared against the robust equally weighted normalized benchmark.

---

[17] http://mba.tuck.dartmouth.edu/pages/faculty/ken.french/data_library.html

Fama and French-Weighted Portfolios

For what we term "simply-weighted portfolios," we construct both equally-weighted and capitalization-weighted portfolios from the normalized KLD rankings by splitting up all the stocks in the KLD universe into high and low groups every year. As the normalized KLD scores range from 1 to 100, the high group holds all the companies with scores equal or greater than 50 and the low group contains the rest. Long–short portfolios are also formed by investing in the high versus low group, and the spread return is analyzed. In addition, we create a version of the overall portfolio that excludes controversial companies, traditionally labeled as "sin stocks." Portfolios are rebalanced annually. The net score is lagged by 3 months to accommodate KLD's update timing after year end. Although portfolios are constructed annually, we measure returns on a monthly frequency. The monthly spread portfolio returns and associated factor regressions are presented in Table 22.4.

The total score as well as several subcategory long–short portfolios have a positive return, although economically small. For example, the capitalization-weighted long–short portfolio that is based on the total score and excludes the controversial companies returns 2.9% per year, statistically significant at the 5% confidence level. The return remains positive in the more recent sample from 2004 to 2015, although lower than in the early history. The Human Rights capitalization-weighted long–short portfolio has the highest spread of 4.68% per year, also statistically significant. The Environment long–short portfolio has the lowest return of −0.42% that is not statistically significant. The factor loadings of these normalized and neutralized scores appear understandably different and more neutered from the traditionally reported raw scores because many of the database biases have been restored, taking out implied factor tilts. For example, the overall spread portfolio that excludes controversial companies has a positive loading on Value, Momentum, and Size Fama and French (1992, 1995, 2008, 2016) factors, with a positive intercept, although the R-squared is less than 4%. Importantly, these factor loadings do not appear stable across portfolio weighting schemes and time and hence should not be interpreted as definitive. Instead, these regressions validate the effectiveness of normalization and support a weak positive intercept. In summary, our simple portfolio return analysis strongly supports the no-cost argument associated with the ESG rankings. In addition, it weakly supports some positive association with return in the simple portfolio settings.

The tables show performance and factor exposures of KLD ratings ranked long/short portfolios during 1992–2015 and the subperiods. To conduct long/short periods, first, high (low) portfolios are formed as top 20%[1] (bottom 20%) of all companies according to the normalized and industry neutralized net scores. Long/short portfolio is equivalent to long high portfolio and short low portfolio. Portfolios are formed each year lagging 3 months of KLD ratings. Monthly portfolio returns are regressed on 4 Fama-French factors.

Table 22.5 KLD ratings ranked long/short portfolios different periods

March,1992 – December,2015								
Total Score	Return	Std.	Intercept	RMRF	HML	SMB	MOM	R^2
Cap. Weighted	0.83%	7.0%	1.6%	-0.03	-0.07	0.08*	-0.07*	3.6%
Eq. Weighted	1.08%	6.2%	0.5%	-0.02	0.13*	-0.03	0.07*	8.9%
Total ex Controversial	Return	Std.	Intercept	RMRF	HML	SMB	MOM	R^2
Cap. Weighted	2.90%*	7.1%	2.5%	-0.03	0.11*	0.03	0.05	3.4%
Eq. Weighted	1.67%	7.1%	0.7%	-0.02	0.25*	-0.06	0.09*	20.4%
Governance	Return	Std.	Intercept	RMRF	HML	SMB	MOM	R^2
Cap. Weighted	1.63%	6.4%	1.7%	-0.01	-0.01	0.1*	-0.02	1.9%
Eq. Weighted	0.04%	6.5%	0.3%	-0.04	-0.07	0.07*	0.01	3.1%
Community	Return	Std.	Intercept	RMRF	HML	SMB	MOM	R^2
Cap. Weighted	-0.22%	6.5%	0.0%	0.01	0.06	0.06	-0.09*	7.4%
Eq. Weighted	1.75%	5.0%	1.1%	0.05*	0.05	-0.01	0.02	0.9%
Diversity	Return	Std.	Intercept	RMRF	HML	SMB	MOM	R^2
Cap. Weighted	0.06%	7.5%	1.9%	-0.06	-0.21*	-0.31*	-0.01	22.2%
Eq. Weighted	0.03%	5.6%	0.4%	0.01	-0.04	-0.26*	0.03	25.1%
Employee Relations	Return	Std.	Intercept	RMRF	HML	SMB	MOM	R^2
Cap. Weighted	0.54%	6.2%	1.6%	-0.01	-0.12*	-0.1*	-0.07*	7.1%
Eq. Weighted	0.98%	4.9%	0.8%	-0.01	0.08*	-0.17*	0.05	23.5%
Environment	Return	Std.	Intercept	RMRF	HML	SMB	MOM	R^2
Cap. Weighted	-0.42%	6.8%	0.6%	-0.05	-0.03	0.03	-0.1*	4.1%
Eq. Weighted	0.20%	6.9%	0.2%	-0.09	0.22*	-0.03	0.01	19.4%
Human Rights	Return	Std.	Intercept	RMRF	HML	SMB	MOM	R^2
Cap. Weighted	4.68%	10.0%	2.7%	0.05	0.18*	0.12*	0.15*	8.3%
Eq. Weighted	1.84%	11.0%	1.1%	-0.03	0.05	0.11	0.1*	2.9%
Product Safety	Return	Std.	Intercept	RMRF	HML	SMB	MOM	R^2
Cap. Weighted	0.65%	7.7%	1.4%	-0.01	-0.29*	0.19*	-0.03	32.3%
Eq. Weighted	-0.37%	5.8%	-0.5%	-0.02	0.01	0.11*	0.00	2.8%
Controversial	Return	Std.	Intercept	RMRF	HML	SMB	MOM	R^2
Cap. Weighted	0.52%	5.8%	1.0%	-0.04	-0.04	0.05	-0.03	0.8%
Eq. Weighted	0.81%	1.9%	0.7%	0.03*	0.03*	-0.02*	0.01	5.0%

*Significant at 5% level

[1]If at 20%, there are companies with the rating lying outside the 20% range, we include all of them. For equally weighted portfolios, companies are equally weighted. For cap weighted portfolios, we adjust the weights of the companies that share the same rating of the 20% threshold to only account for the original portion of that rating in the portfolio before including companies that fall outside the 20% range

(continued)

Table 22.5 (continued)

March, 1992 - December, 2004							
Total Score	Return	Std.	Intercept	RMRF	HML	SMB	MOM	R^2
Cap. Weighted	2.32%	8.4%	3.7%	-0.04	-0.13	0.07	-0.05	4.0%
Eq. Weighted	2.09%	7.0%	1.5%	-0.08	0.02	-0.05	0.12*	13.9%
Total ex Controversial	Return	Std.	Intercept	RMRF	HML	SMB	MOM	R^2
Cap. Weighted	4.17%	8.3%	2.0%	-0.01	0.13	0.04	0.13*	10.0%
Eq. Weighted	3.32%	8.4%	1.4%	-0.06	0.17*	-0.08	0.16*	25.0%
Governance	Return	Std.	Intercept	RMRF	HML	SMB	MOM	R^2
Cap. Weighted	1.91%	7.2%	0.8%	0.09	0.12*	0.17*	-0.07*	9.8%
Eq. Weighted	-0.02%	6.4%	0.1%	0.01	-0.02	0.1	-0.04	4.3%
Community	Return	Std.	Intercept	RMRF	HML	SMB	MOM	R^7
Cap. Weighted	0.27%	7.2%	1.9%	-0.04	-0.06	0.00	-0.09*	2.8%
Eq. Weighted	1.02%	4.6%	1.0%	0.01	-0.09*	-0.05	0.06*	6.3%
Diversity	Return	Std.	Intercept	RMRF	HML	SMB	MOM	R^2
Cap. Weighted	0.00%	8.1%	2.6%	-0.06	-0.21*	-0.26*	-0.03	13.1%
Eq. Weighted	-0.78%	6.3%	-0.3%	-0.01	-0.06	-0.27*	0.06*	27.0%
Employee Relations	Return	Std.	Intercept	RMRF	HML	SMB	MOM	R^2
Cap. Weighted	1.00%	7.3%	2.9%	-0.02	-0.12*	-0.1*	-0.08*	5.9%
Eq. Weighted	0.70%	5.8%	0.8%	-0.06*	0.03	-0.18*	0.07*	31.4%
Environment	Return	Std.	Intercept	RMRF	HML	SMB	MOM	R^2
Cap. Weighted	-0.45%	8.1%	1.0%	-0.05	-0.04	0.05	-0.08*	1.3%
Eq. Weighted	0.64%	8.2%	1.0%	-0.18*	0.12*	-0.05	0.05	24.0%
Human Rights	Return	Std.	Intercept	RMRF	HML	SMB	MOM	R^2
Cap. Weighted	8.78%	12.1%	5.1%	0.00	0.08	0.03	0.29*	21.7%
Eq. Weighted	0.95%	12.4%	-0.7%	-0.11	0.02	0.11	0.19*	13.2%
Product Safety	Return	Std.	Intercept	RMRF	HML	SMB	MOM	R^2
Cap. Weighted	-0.51%	9.4%	0.2%	0.06	-0.29*	0.22*	-0.02	43.2%
Eq. Weighted	1.42%	5.9%	2.0%	-0.05	-0.15*	0.06	0.04	15.4%
Controversial	Return	Std.	Intercept	RMRF	HML	SMB	MOM	R^2
Cap. Weighted	0.65%	6.9%	1.3%	-0.04	-0.01	0.12*	-0.05	3.7%
Eq. Weighted	1.44%	2.5%	1.0%	0.01	0.04*	-0.02	0.02	4.8%

*Significant at 5% level

[1] If at 20%, there are companies with the rating lying outside the 20% range, we include all of them. For equally weighted portfolios, companies are equally weighted. For cap weighted portfolios, we adjust the weights of the companies that share the same rating of the 20% threshold to only account for the original portion of that rating in the portfolio before including companies that fall outside the 20% range

(continued)

Table 22.5 (continued)

December, 2004 – December, 2015

Total Score	Return	Std.	Intercept	RMRF	HML	SMB	MOM	R²
Cap. Weighted	-0.89%	5.0%	-0.3%	-0.05	0.06	-0.01	-0.09*	7.5%
Eq. Weighted	-0.09%	5.1%	0.3%	0.00	0.29*	0.21*	0.00	25.1%

Total ex Controversial	Return	Std.	Intercept	RMRF	HML	SMB	MOM	R²
Cap. Weighted	1.43%	5.6%	2.0%	-0.04	0.04	-0.08	-0.09*	6.3%
Eq. Weighted	-0.24%	5.2%	0.1%	0.00	0.31*	-0.2*	0.00	25.2%

Governance	Return	Std.	Intercept	RMRF	HML	SMB	MOM	R²
Cap. Weighted	1.30%	5.5%	1.5%	-0.06	-0.04	0.11	0.05	4.8%
Eq. Weighted	0.11%	6.7%	0.4%	-0.08	-0.01	0.10	0.09*	8.2%

Community	Return	Std.	Intercept	RMRF	HML	SMB	MOM	R²
Cap. Weighted	-0.79%	5.6%	-0.6%	0.00	0.23*	0.08	-0.07*	22.9%
Eq. Weighted	2.60%	5.5%	2.7%	0.01	0.33*	-0.09	0.01	20.2%

Diversity	Return	Std.	Intercept	RMRF	HML	SMB	MOM	R²
Cap. Weighted	0.14%	6.8%	1.3%	-0.04	-0.06	-0.53*	0.03	45.3%
Eq. Weighted	1.04%	4.4%	1.2%	0.04	-0.08	-0.29*	0.00	21.6%

Employee Relations	Return	Std.	Intercept	RMRF	HML	SMB	MOM	R²
Cap. Weighted	0.01%	4.7%	0.3%	-0.01	-0.12*	-0.09	-0.07*	6.7%
Eq. Weighted	1.31%	3.8%	1.2%	0.04	0.04	-0.19*	0.01	11.6%

Environment	Return	Std.	Intercept	RMRF	HML	SMB	MOM	R²
Cap. Weighted	-0.38%	4.9%	0.3%	-0.04	0.05	-0.12*	-0.11*	16.1%
Eq. Weighted	-0.30%	5.0%	0.2%	-0.01	0.26*	-0.19*	-0.03	25.0%

Human Rights	Return	Std.	Intercept	RMRF	HML	SMB	MOM	R²
Cap. Weighted	1.06%	7.5%	0.8%	0.04	0.03	0.19*	-0.07	8.8%
Eq. Weighted	2.63%	9.6%	2.2%	0.07	-0.10	-0.05	-0.04	-1.1%

Product Safety	Return	Std.	Intercept	RMRF	HML	SMB	MOM	R²
Cap. Weighted	2.00%	5.0%	2.6%	-0.1*	-0.07	0.09	-0.01	7.2%
Eq. Weighted	-2.43%	5.5%	-1.5%	-0.09	0.4*	-0.05	-0.02	33.1%

Controversial	Return	Std.	Intercept	RMRF	HML	SMB	MOM	R²
Cap. Weighted	0.37%	4.1%	0.5%	0.01	0.01	-0.18*	-0.01	8.8%
Eq. Weighted	0.08%	0.4%	0.1%	0.00	-0.01	-0.02*	0.00	32.2%

*Significant at 5% level

[1]If at 20%, there are companies with the rating lying outside the 20% range, we include all of them. For equally weighted portfolios, companies are equally weighted. For cap weighted portfolios, we adjust the weights of the companies that share the same rating of the 20% threshold to only account for the original portion of that rating in the portfolio before including companies that fall outside the 20% range

A Return to Optimized Portfolio Construction and Management

The majority of academic work that measures the costs of ESG investing focuses on the simply-weighted portfolios, yet in practice, ESG criteria are often applied alongside expected return and risk models as well as some form of optimized settings. By introducing a realistic multifactor expected return and risk model, along with optimization settings, we can gain deeper insights into the potential costs of ESG investing as the various elements of portfolio construction interact in a more practical setting.

In an optimized setting, we build optimal portfolios using a multifactor expected return model and the APT risk model and optimizer and the Axioma statistical risk model and optimizer as we discussed in Chap. 15. We introduce the underlying stock selection model in the coming section, and combined with a KLD-based social score forecasted return, we can build Mean-Variance Tracking Error at Risk (MVTaR) portfolios such that ESG/SRI values can be effectively brought into the portfolio construction analysis.

In an updated SRI/ESG analysis, Geczy et al. (2020) constructed expected returns using the earnings forecasting variable CTEF, the I/B/E/S consensus-based composite variable composed of forecasted earnings yield, earnings revisions, and earnings breadth, that we discussed in Chap. 14. CTEF is the public form of the McKinley Capital Management (MCM) variable representing forecasted earnings acceleration, a key variable in our stock selection model. CTEF passes the robustness tests of portfolio construction and transaction cost management tests of statistical significance.

One can create US portfolios with the CTEF model for the Russell 3000 stocks.[18] Guerard (1997a, b) found that portfolios of US stocks using KLD screens for sin, nuclear, military, and environmental criteria did not significantly underperform unscreened portfolios of US stocks using a statistically-based stock selection model. In this section, we replicate the "no cost to being socially responsible in investing" test. Guerard (1997a, b) treated all KLD criteria as being equally weighted. There is an 8% monthly turnover constraint, a 4% upper bound on stock weight, and a 35-basis point threshold (minimum) stock weight upon initiation, consistent with Guerard et al. (2014). The CTEF is simulated for the Russell 3000 stocks using the APT MVTaR portfolio optimization procedure, and its results are reported in Table 22.5. We replicate Guerard (1997a) by creating portfolios using a composite score composed of CTEF, 80%, and 20% the KLD raw concerns, for January 2000 to December 2014. The CTEF model produces statistically significant active returns for the 2000–2014 period. The majority of the 924 basis points of total active returns (t = 3.33) is composed of asset selection: 501 basis points (t = 2.84). The APT MVTaR CTEF Model total active returns and asset selection portfolios are

---

[18]Guerard et al. (2014) used all stocks on the CRSP database in their analysis of the USER model and corporate exports.

highly statistically significant. If one creates a composite score composed of CTEF, 80%, and 20% the KLD raw concerns (the KLD concern score = 0 if there is a concern in any of the subcategories), then Specific Returns (stock selection) asset selection falls for the Corporate Governance (ECGOV), Environmental (EENV), Community Rights (ECOM), Diversity (EDIV), and Product (EPRO) KLD criteria from the CTEF-specific return of 5.01%. We find that all KLD concerns have costs, except Human Rights and the Total KLD criteria. The introduction of the Human Rights (EHUM) concern, composed primarily of indigenous people relations and labor strength, and Total KLD criteria (ETOTAL), enhances stock selection. Corporate Governance, primarily corporate compensation, reduces active returns by 130 basis points; the Environment (ENV) concern costs 159 active return basis points; and the Product (PRO) concern costs 110 basis points. The CTEF variable produces highly statistically significant active returns and specific returns in the Axioma Attribution Model with and without the SRI Concerns (see Table 22.5). There continues to be no statistically significant costs to be an SRI investor in the United States associated with TOTAL KLD Variable concern variables for January 2000 to December 2014, if one tests for equally weighted KLD criteria in an MVTaR model.

CTEF MVTaR with Normalized KLD Criteria

We use the normalized KLD criteria inside the 80/20 CTEF/KLD composite expected return model as an input into the Markowitz MVTaR optimization system, for the 2000–2014 time period. The CTEF model combined with Total KLD variables and the KLD Human Rights criteria in the Russell 3000 analysis produces higher portfolio stock selection, as measured by Axioma Specific Returns, confirming the simply-weighted portfolio results. See Table 22.6. We produce a Zephyr report that uses the ITG transactions costs model, and we report several major results for Russell 3000 stocks. First, the composite models of CTEF and KLD social criteria substantially reduce tracking errors during the 2000–2014 time period and increase the portfolio Information Ratios. Second, Human Rights and KLD Total criteria enhance annualized Excess Returns considering ITG transactions costs. Third, the composite models of CTEF and KLD social criteria substantially reduce portfolio standard deviations during the 2000–2014 time period and increase the portfolio Sharpe ratios (see Table 22.6).

The Interaction of Environmental Scores and Expected Return Models

Geczy and Guerard (2021) updated their analysis of the KLD environmental criteria. GG reported one-, three-, and five-factor model time-series regressions in which, for each of the expected return models (USER, EWC, EValue, MQ, and CTEF,

Table 22.6 APT risk model and optimizer and ITG transactions costs

January 2000 - December 2014
Russell 3000 Universe

	Portfolio Annualized Return	Portfolio Annualized STD	Annualized Excess Returns	Information Ratio	Sharpe Ratio	Tracking Error
CTEF	14.77%	21.01%	9.64%	0.89	0.613	10.81
ECGOV	13.41	17.89	8.28	0.97	0.644	8.53
ECOM	13.03	17.03	7.91	0.97	0.655	8.14
EDIV	12.88	17.16	7.75	0.96	0.641	8.06
EENV	13.11	17.23	7.98	0.98	0.651	8.10
EHUM	15.69	18.41	10.57	1.17	0.750	9.01
EPRO	13.63	17.98	8.50	0.99	0.653	8.57
ETOTAL	15.32	18.24	10.19	1.19	0.736	8.59
Russell 3000	5.13%	15.68%			0.207	

discussed in Chap. 14), returns from portfolios constructed based on their ENV and model numerical values are calculated. High and low ENV and model groups are defined yearly by the 30/40/30 criteria (see Fama & French, 1992 and Carhart (1997)). Each year, the 30% of firms with the highest (low) normalized environmental scores are included in the high (low) ENV group. Independently firms are included in the high (low) USER groups yearly based on their raw USER scores. Firms that are in both high environmental score and high USER score groups are characterized as High ENV + High USER and so on. Firms are equal-weighted within groups.

Geczy and Guerard (2021) reported that High ENV have excess returns (alpha) holding USER constant and that firms that have high expected returns via the models along with high environmental scores produce the greatest pricing errors (alphas) in the various factor model estimations. For example, in the CAPM (RMRF) regression, the High ENV + High USER portfolio produces an annualized intercept (alpha) of about 3.6%, while the Low ENV + Low USER portfolio produces a historical alpha of 0.1%. The (High+High)–(Low+Low) spread produces a nearly zero-beta return (alpha) of about 3.5%. Moreover, when isolating the USER effect by going long and short along the USER dimension and examining the resulting ENV differentials, High ENV firms produce an 80 bps excess returns, while the Low ENV score portfolio produces an excess return of −0.40%, yielding an alpha of approximately 1.20%.

This results for the CAPM are robust across the additional factor models where interesting patterns in factor loadings emerge, indicating the interaction between ESG characteristics and traditional factor exposures. For example, while patterns in market betas remain intact in moving from the CAPM to the Carhart (1997) four-factor model, small size effects emerge and in particular value (HML) factor loadings are significantly larger for High ENV + High USER portfolios than for Low ENV + Low USER portfolios. Specifically, the Fama-French HML (value less growth) factor loading for the (High+High)–(Low+Low) spread is -0.35, the momentum spread loading estimate is 0.18, and the intercept (pricing error or alpha) is an annualized 3.5%. In other words, perhaps as expected, firms that have low environmental scores seem to have a more pronounced value exposure than those with high scores, corresponding to received wisdom that ESG stock portfolios are generally tilted toward growth and away from the asset-heavy traditional value sectors and firms. These results are borne out by the five-factor model extending the Carhart (1997) model with the Fama-French Quality spread. Interestingly, quality subsumes the momentum loading and to a lesser extent value. Geczy and Guerard (2021) found that equity portfolios load positively on the Quality factor portfolio, but Low ENV tends to load more strongly than High ENV, suggesting once again that positive environmental characteristics are negatively associated with realized measures of profitability, which is obliquely measured with respect to momentum and value, as is also known. Nonetheless, the alpha for the (High+High)–(Low+Low) spread is strong at an estimated 6.4% annualized value.

The evidence construed across model subcomponents reinforces the story for the aggregate USER model. For instance, for the naïvely equal-weighted EWC model as well as the value-bagged model (EVALUE), the MQ model incorporating only price momentum and CTEF, and for CTEF itself, the CAPM and Carhart models load essentially the same with nearly identical intercepts. However, in the five-factor model, we see an inversion of the quality loadings. In other words, in the unoptimized EWC model (recall that the input weightings in the USER model were optimized long ago, essentially out of sample, while the EWC model treats all inputs the same and the others break out subcomponents) as well as the others, it is quite clear that quality loadings are higher for High EWC, High EVALUE, High MQ, and High CTEF versus their low counterparts. The information ratio optimization inherent in the definition of USER seems to weigh components that invert the relationship. Nonetheless, the intercepts in all five-factor regressions remain economically and statistically significant for all models including for CTEF. Taken together, the results strongly suggest an interaction between ENV and various models for expected returns, which in turn indicates that when one is creating portfolios (ESG or not), one would do well to consider both sources of information and that ESG information in this important and currently very relevant case of environmental scores may be additive in creating portfolios.

Summary of SRI/ESG Portfolio Construction and Analysis

Analysis of socially responsible investment portfolios, since the introduction of the social investing discipline, has focused on the expected costs of the constraint, but also on the possibility of portfolio design, incorporating social responsibility factors, within which the expected return could be the same or greater than a portfolio without the SRI factors. We find that a number of important investment screens based on KLD social investment variables do not cost investors in our portfolio analyses. Stocks with good KLD Human Rights, Diversity, and Total KLD concern do not cost investors holding risk constant, and the KLD Total Social Criteria, Human Rights, and Diversity complement CTEF to enhance Active Returns and Specific Returns in the Russell 3000 and KLD-only universes for Human Rights and Total KLD criteria. Apparently, it may be possible to have one's cake and eat it too in several KLD-based universes.

KLD STATS[19]

KLD STATS (STATISTICAL TOOL FOR ANALYZING TRENDS IN SOCIAL AND ENVIRONMENTAL PERFORMANCE) is a data set with annual snapshots of the environmental, social, and governance performance of companies rated by KLD Research and Analytics, Inc. KLD STATS is now sold and serviced by RiskMetrics Group, RMG. KLD covered 650 stocks in annual spreadsheets from 1991 to 2000; 1100 stocks in 2001–2002; and 3100 stocks from 2003.

Strength and Concern (Positive and Negative Indicator) Ratings

RMG covers approximately 80 indicators in seven major qualitative issue areas including Community, Corporate Governance, Diversity, Employee Relations, Environment, Human Rights, and Product.

RMG also provides information for involvement in the following Controversial Business Issues: Alcohol, Gambling, Firearms, Military, Nuclear Power, and Tobacco. KLD STATS presents a binary summary of positive and negative ESG ratings. In each case, if RMG assigned a rating in a particular issue (either positive or negative), this is indicated with a 1 in the corresponding cell. If the company did not have a strength or concern in that issue, this is indicated with a 0.

KLD STATS data are organized by year. Each year, RiskMetrics takes a snapshot of its ratings and index membership to reflect the data at the calendar year-end. Each spreadsheet contains identifying information about the company, index membership, a listing of positive and negative ratings, involvement in controversial business issues, and total counts for each area.

Additionally, at the end of each spreadsheet is a summary count of all strengths and concerns the company received in a general category (either qualitative issue area or controversial business issue) in that year.

ESG Ratings Definitions
Qualitative Issue Areas
Diversity (DIV-)
Strengths

---

[19] See "Getting Started: An Introduction to KLD STATS," WRDS.

CEO (DIV-str-A). The company's chief executive officer is a woman or a member of a minority group.

Promotion (DIV-str-B). The company has made notable progress in the promotion of women and minorities, particularly to line positions with profit-and-loss responsibilities in the corporation.

Board of directors (DIV-str-C). Women, minorities, and/or the disabled hold four seats or more (with no double counting) on the board of directors, or one-third or more of the board seats if the board numbers less than 12.

Work/life benefits (DIV-str-D). The company has outstanding employee benefits or other programs addressing work/life concerns, e.g., childcare, elder care, or flextime. In 2005, KLD renamed this strength from Family Benefits Strength.

Women and minority contracting (DIV-str-E). The company does at least 5% of its subcontracting, or otherwise has a demonstrably strong record on purchasing or contracting, with women- and/or minority-owned businesses.

Employment of the disabled (DIV-str-F). The company has implemented innovative hiring programs; other innovative human resource programs for the disabled, or otherwise has a superior reputation as an employer of the disabled.

Gay and lesbian policies (DIV-str-G). The company has implemented notably progressive policies toward its gay and lesbian employees. In particular, it provides benefits to the domestic partners of its employees.

Concerns

Controversies (DIV-con-A). The company has either paid substantial fines or civil penalties as a result of affirmative action controversies, or has otherwise been involved in major controversies related to affirmative action issues.

Nonrepresentation (DIV-con-B). The company has no women on its board of directors or among its senior line managers.

Human Rights (HUM-)
Strengths

Positive record in South Africa (HUM-str-A). The company's social record in South Africa is noteworthy.

Indigenous peoples relations strength (HUM-str-D). The company has established relations with indigenous peoples near its proposed or current operations (either in or outside the United States) that respect the sovereignty, land, culture, human rights, and intellectual property of indigenous peoples.

Labor rights strength (HUM-str-G). The company has outstanding transparency on overseas sourcing disclosure and monitoring, or has particularly good union relations outside the United States, or has undertaken labor rights-related initiatives that KLD considers outstanding or innovative.

Other strength (HUM-str-X). The company has undertaken exceptional human rights initiatives, including outstanding transparency or disclosure on human rights issues, or has otherwise shown industry leadership on human rights issues not covered by other KLD human rights ratings.

Concerns

South Africa (HUM-con-A). The company faced controversies over its operations in South Africa.

Northern Ireland (HUM-con-B). The company has operations in Northern Ireland.

Burma concern (HUM-con-C). The company has operations or direct investment in, or sourcing from, Burma.

Mexico (HUM-con-D). The company's operations in Mexico have had major recent controversies, especially those related to the treatment of employees or degradation of the environment.

Labor rights concern (HUM-con-F). The company's operations have had major recent controversies primarily related to labor standards in its supply chain.

Indigenous peoples relations concern (HUM-con-G). The company has been involved in serious controversies with indigenous peoples (either in or outside the United States) that indicate the company has not respected the sovereignty, land, culture, human rights, and intellectual property of indigenous peoples.

References

Bassen, A., Kovacs, A. M., & A. M. (2008). Environmental, social and governance key performance indicators from a capital market perspective. *Zeitschrift für Wirtschafts- und Unternehmensethik, 9*(2), 182–192.

Bender, J., Sun, X., & Wang, T. (2016). *Thematic indexing, meet smart beta! Merging ESG into Factor Portfolios.* State Street Global Advisors white paper.

Bettis, J. C., Coles, J. L., & Lemmon, M. L. (2000). Corporate policies restricting trading by insiders. *Journal of Financial Economics, 57*, 191–220.

Bettis, C. J. G., Guerard, J., & McAuley, D. (2015). Is US insider trading still relevant? A quantitative portfolio approach. *Journal of Investment Management, 13*, 33–56.

Bialkowski, J., & Starks, L. T. (2017). *SRI Funds: Investor demand, exogenous shocks and ESG profiles.* Working paper.

Borgers, A., Derwall, J., Koedijk, K., & ter Horst, J. (2013). Stakeholder relations and stock returns: On errors in investors' expectations and learning. *Journal of Empirical Finance, 22*, 159–175.

Borgers, A., Derwell, J., Koedijk, K., & ter Horst, J. (2015). Do social factors influence investment behavior and performance? Evidence from mutual fund holdings. *Journal of Banking and Finance, 60*, 112–126.

Breedt, A., Cilibertu, S., Gualdi, S., & Seager, P. (2018). *Is ESG an equity factor or just an investment guide?* Capital Fund Management white paper.

Buchanan, N. S. (1950). *The economics of corporate Enterprise.* Halt. Ch. 15.

Carpenter, J., & Yermack, D. (1999). *Executive Compensation and shareholder value: Theory and evidence.* Kluwer Academic Publishers.

Chatterji, A. K., Levine, D. I., & Toffel, M. W. (2009). How well do social ratings actually measure corporate social responsibility? *Journal of Economics and Management Strategy, 18*, 125–169.

Cremers, K. J. M., Nair, V. B., & John, K. (2005, March). *Takeovers, governance and the cross-section of returns.* Yale School of Management working paper.

Cremers, M., Ferreira, M. A., Matos, P., & Starks, L. T. (2016). Indexing and active fund management: International evidence. *Journal of Financial Economics, 120*(2), 539–560.

Cremers, M., Fulkerson, J. A., Jon, A., & Riley, T. B. (2018). *Challenging the conventional wisdom on active management: A review of the past 20 years of Academic Literature on Actively Managed Mutual Funds.* Working paper.

Dimson, E., Karaka, O., & Li, X. (2015). Active ownership. *Review of Financial Studies, 28*(12), 3225–3268.

Dorfleitner, G., Utz, S., & Wimmer, M. (2014). *Patience pays off – Financial long-term benefits of sustainable management decisions.* Working paper.

Dow, J., & Raposo, C. R. (2005). CEO compensation, change, and corporate strategy *Journal of Finance, 60,* 2701–2728.

Dowell, G., Hart, S., & Yeung, B. (2000). Do corporate global environmental standards create or destroy value? *Management Science, 46,* 1059–1074.

Durand, R. B., Koh, S., & Limkriangkrai, M. (2013). Saints versus sinners. Does morality matter? *Journal of International Financial Markets, Institutions and Money, 24*(2013), 166–183.

Elias, E. (2017). Is it costly to introduce SRI into Islamic portfolios? *Islamic Economic Studies, 25,* 23–54.

Evans, J. R., & Peiris, D. (2010). *The relationship between environmental social governance factors and stock returns.* UNSW Australian School of Business Research Paper No. 2010ACTL02. Working paper.

Fabozzi, F. J., Ma, K. C., & Oliphant, B. J. (2008). Sin Stock Returns. *The Journal of Portfolio Management, 35,* 82–94.

Fama, E. F. (1991). Efficient capital markets II. *Journal of Finance, 46,* 1575–1617.

Fama, E. F., & French, K. R. (1992). Cross-sectional variation in expected stock returns. *Journal of Finance, 47,* 427–465.

Fama, E. F., & French, K. R. (2008). Dissecting anomalies. *Journal of Finance, 63,* 1653–1678.

Fama, E. F., & French, K. R. (2016). A five-factor asset pricing model. *Journal of Financial Economics, 118*(1), 1–22.

Friede, G., Busch, T., & Bassen, A. (2015). ESG and financial performance: Aggregated evidence from more than 2000 empirical studies. *Journal of Sustainable Finance and Investment, 5*(4), 210–233.

Fulton, M., Kahn, B. M., & Sharples, C. (2012). *Sustainable investing: Establishing long-term value and performance.* Working paper.

Galema, R., Plantinga, A., & Scholtens, B. (2008). The stocks at stake: Return and risk in socially responsible investment. *Journal of Banking and Finance, 32,* 2646–2654.

Geczy, C., & Guerard, J. B. Jr. (2021). *ESG and expected returns on equities: The case of environmental ratings,* Working Paper.

Geczy, C., Stambaugh, R. F., & Levin, D. (2005). *Investing in socially responsible mutual funds.* Working paper, 2003.

Geczy, C., Guerard, J., Jr., & Samonov, M. (2020). Warning: SRI need not kill your Sharpe and information ratios—Forecasting of earnings and efficient SRI and ESG portfolios. *Journal of Investing, 29.*

Geczy, C., Stambaugh, R., & Levin, D. (2021). Investing in socially responsible mutual funds. *The Review of Asset Pricing Studies, 11*(2), 309–351.

Gilbert, L. D. (1956). *Dividends and democracy.* American Research Council.

Graham, Benjamin, and Dodd, David, Security analysis, , 1951., Chs. 47, 48, 49, 50.

Graves, S. B., & Waddock, S. A. (1994). Institutional owners and corporate social performance. *Academy of Management Journal, 3,* 1034–1046.

Guerard, J. B., Jr. (1997a). Is there a cost to being socially responsible in investing? *Journal of Forecasting, 16,* 475–490.

Guerard, J. B., Jr. (1997b). Is there a cost to being socially responsible in investing? *Journal of Investing, 6,* 11–18.

Guerard, J. B., Jr., & Mark, A. (2003). The optimization of efficient portfolios: The case for an R and D quadratic term. In A. Chen (Ed.), *Research in finance* (Vol. 20).

Guerard, J. B., Jr., Rachev, R. T., & Shao, B. (2013). Efficient global portfolios: Big data and investment universes. *IBM Journal of Research and Development, 57*(5), Paper 11.

Guerard, J. B., Jr., Markowitz, H. M., & Xu, G. (2014). The role of effective corporate decisions in the creation of efficient portfolios. *IBM Journal of Research and Development, 58*, 6.1–6.11.

Guerard, J. B., Jr., Markowitz, H. M., & Xu, G. (2015). Earnings forecasting in a global stock selection model and efficient portfolio construction and management. *International Journal of Forecasting, 31*, 550–560.

Halbritter, G., & Drofleitner, G. (2015). The wages of social responsibility – Where are they? A critical review of ESG investing. *Review of Financial Economics, 26*, 25–35.

Harvey, C. R. (2017). Presidential address: The scientific outlook in financial economics. *Journal of Finance, 72*, 1399–1440.

Harvey, C. R., Lin, Y., & Zhu, H. (2016). . . .the cross-section of expected returns. *Review of Financial Studies, 2*(1), 5–69.

Hong, H., & Kacperczyk, M. (2009). The Price of sin: The effects of social norms in markets. *Journal of Financial Economics, 93*, 15–36.

Jensen, M. C., & Meckling, W. H. (1976). Theory of the firm: Managerial behavior, agency costs and ownership structure. *Journal of Financial Economics, 3*, 305–360.

Jin, I. (2018). Is ESG a systematic risk factor for US equity mutual funds? *Journal of Sustainable Finance and Investment, 8*(1), 72–93.

Kempf, A., & Osthoff, P. (2007). The effect of socially responsible investing on portfolio performance. *European Financial Management, 13*, 908–922.

Khan, M., Serafeim, G., & Yoon, A. (2015). *Corporate sustainability: First evidence on materiality*. Working paper.

Kim, D., & Starks, L. T. (2016). Gender diversity on corporate boards: Do women contribute unique skills? *American Economic Review, 106*(5), 267–271.

Kinder, P., & Domini, A. (1997). Social screening: Paradigms old and new. *Journal of Investing, 6*, 12–20.

Kinder, P., Lydenberg, S. D., & Domini, A. L. (1993). *Making money while being socially responsible*. Harper Business.

Klement, J. (2018). *Does ESG matter for asset allocation?* Fidante Partners white paper.

Kotsantonis, S., Pinney, C., & Serafeim, G. (2016). ESG integration in investment management: Myths and realities. *Journal of Applied Corporate Finance, 28*(2), 10–16.

Kurtz, L. (1997). No effect, or no net effect? Studies on socially responsible investing. *Journal of Investing, 6*(1997), 37–49.

Kurtz, L., & De Bartolomeo, D. (1996). Socially screened portfolios: An attribution analysis of relative performance. *Journal of Investing, 5*, 35–41.

Laermann, M. (2015). *The significance of ESG ratings for socially responsible investment decisions: An examination from a market perspective*. Working paper.

Levy, H. (2012). *The capital asset pricing model in the 21st century*. Cambridge University Press.

Lewellen, W. G. (1968). *Executive compensation in large industrial companies*. Columbia University Press.

Luck, C., & Pilotte, N. (1993). Domini social index performance. *Journal of Investing, 2*, 60–62.

Manconi, A, Rizzo, E. A., & Spalt, O. G. (2015). *Diversity investing*. Working paper.

Manescu, C. (2011). Stock returns in relation to environmental, social and governance performance: Mispricing or compensation for risk? *Sustainable Development, 19*, 95–118.

Margolis, J. D., Elfenbein, H. A., & Walsh, J. P. (2009). *Does it pay to be good. . .And does it matter? A Meta-analysis of the relationship between corporate social and financial performance*. Working paper.

Markowitz, H. M. (1952). Portfolio selection. *Journal of Finance, 7*, 77–91.

Markowitz, H. M. (1959). *Portfolio selection: Efficient diversification of investment* (Cowles foundation monograph no. 16). Wiley.

Markowitz, H. M. (2013). *Risk-return analysis*. McGraw-Hill.

Mason, E. S. (1958). *The apologetics of Managerialism*. Journal of Business.

Mason, E. S. (Ed.). (1960). *The Corporation in Modern Society*. Harvard University Press.

Mattingly, J. E., & Berman, S. L. (2006). Measurement of corporate social action: Discovering taxonomy in the Kinder Lydenburg Domini ratings data. *Business and Society, 45*, 20–46.

McCahery, J. A., Sautner, Z., & Starks, L. T. (2016). Behind the scenes: The corporate governance preferences of institutional investors. *Journal of Finance, 71*, 2905–2932.

McGuire, J. W., Chiu, J. S. Y., & Elbing, A. O. (1962). Executive income, sales and profits. *American Economic Review*, 753–761.

McWilliams, A., & Siegel, D. (1997). The role of money managers in assessing corporate social responsibility research. *Journal of Investing, 6*, 98–107.

McWilliams, A., & Siegel, D. (2000). Corporate social responsibility and financial performance: Correlation of misspecification. *Strategic Management Journal, 21*, 603–609.

McWilliams, A., & Siegel, D. (2001). Corporate social responsibility: A theory of the firm perspective. *Academy of Management Review, 26*(2001), 117–127.

McWilliams, A., & Siegel, D. (2006). Corporate social responsibility: Strategic implications. *Journal of Management Studies, 43*, 1–22.

Moskowitz, M. (1972). Choosing socially responsible stocks. *Business Review, 1972*, 72–78.

Moskowitz, M. (1997). Social investing: The moral foundation. *Journal of Investing, 6*(4), 9–11.

MSCI ESG Research Inc. (2015). MSCI ESG KLD STATS: 1991–2014 Data Sets.

Naaraayanan, L., Sachdeva, S., & Sharma, V. (2020). *Real effects of environmental activist investing*. European Corporate Governance Institute – Finance Working Paper No. 743/2021. https://ssrn.com/abstract=3483692

Rapapport, A. (1978). Executive incentives versus corporate growth. *Harvard Business Review*, 82–88.

Renneboog, L., ter Horst, J., & Zhang, C. (2008). Socially responsible investments: Institutional aspects, performance, and investor behavior. *Journal of Banking and Finance, 32*, 1723–1742.

Riedl, A., & Smeets, P. (2017). Why do Investors hold socially responsible mutual funds? *Journal of Finance, 62*, 2505–2549.

Roberts, D. R. (1959). *Executive Compensation*. Free Press.

Roll, R. (1986). The hubris hypothesis theory of corporate takeovers. *Journal of Business, 59*, 186–217.

Rosen, B. N., Sandler, D. M., & Shani, D. (1991). Social issues and socially responsible investment behavior: A preliminary empirical investigation. *The Journal of Consumer Affairs, 25*, 221–233.

Russo, M. V., & Fouts, P. A. (1997). A resource-based perspective on corporate environmental performance and profitability. *Academy of Management Journal, 40*(3), 534–559.

Sharfman, M. (1996). The construct validity of the Kinder, Lydenberg and Domini social performance ratings data. *Journal of Business Ethics, 15*, 287–296.

Sjöström, E. (2011). *The performance of socially responsible investment – A review of scholarly studies published 2008–2010*. Working paper.

Statman, M., & Glushkov, D. (2009). The wages of social responsibility. *Financial Analysts Journal, 65*(4), 33–46.

Verheyden, T., Eccles, R. G., & Feiner, A. (2016). ESG for all? The impact of ESG screening on return, risk and diversification. *Journal of Applied Corporate Finance, 28*(2), 47–55.

Waddock, S. A., & Graves, S. B. (1997). The corporate social performance-financial performance link. *Strategic Management Journal, 18*, 303–319.

Waddock, S., & Graves, S.. *Finding the link between stakeholder relations and quality of management*.

Wimmer, M. (2012). *ESG-persistence in socially responsible mutual funds*. Working paper.

Yang, R. (2018). *The market of environmental, social, and governance (ESG)*. Columbia Business School Research Paper No. 18–37.

Zyglidopoulos, S. C., Georgiadis, A. P., Carroll, C. E., & Siegel, D. (2012). Does media attention drive corporate social responsibility? *Journal of Business Research, 65*, 1622–1627.

Index

Printed in the USA
CPSIA information can be obtained
at www.ICGtesting.com
LVHW020736231023
761793LV00006B/596